JOURNAL OF CHROMATOGRAPHY LIBRARY – volume 41B

high-performance liquid chromatography of biopolymers and biooligomers

part B: separation of individual compound classes

JOURNAL OF CHROMATOGRAPHY LIBRARY - volume 41B

high-performance liquid chromatography of biopolymers and biooligomers

part B: separation of individual compound classes

O. Mikeš

Institute of Organic Chemistry and Biochemistry, Czechoslovak Academy of
Sciences, Flemingovo náměstí 2, 166 10 Prague 6, Czechoslovakia

ELSEVIER
Amsterdam — Oxford — New York — Tokyo   1988

ELSEVIER SCIENCE PUBLISHERS B.V.
Sara Burgerhartstraat 25
P.O. Box 211, 1000 AE Amsterdam, The Netherlands

Distributors for the United States and Canada:

ELSEVIER SCIENCE PUBLISHING COMPANY INC.
52, Vanderbilt Avenue
New York, NY 10017, U.S.A.

ISBN 0-444-43034-2 (Vol. 41B)
ISBN 0-444-41616-1 (Series)

Printed in The Netherlands

CONTENTS

X

To Irene
my wife, effective helper, and my best friend

PREFACE TO PART B

   The overall aim of the author of this monograph and the basic characteristics of the work of putting it together have been outlined in the Preface to
Part A. In Part B descriptions are included of the chromatography of individual
types of high- and medium-molecular-weight biopolymers and biooligomers, both
simple and complex. Their "Register" is also included, together with a full
title Bibliography, a chapter devoted to the utilization of high-performance
liquid chromatography (HPLC) in studies of the structures of these compounds
(especially biopolymer and biooligomer sequencing) and a final prognostic chapter.
   Putting together Part B started with the registration of studied compounds
and the setting up of the Bibliography. The elaboration of the complete Register
required the use of auxiliary files, which was time consuming. Even more lengthy
was the subsequent writing of chapters on the separation of proteins, enzymes,
peptides, nucleic acids and oligonucleotides, polysaccharides and compound biopolymers, and also the chapter on the applications of HPLC in studies of the
structure of biopolymers. Thus, when Part B was completed I found that the Register and Bibliography were slightly out of date and an Appendix and Addendum containing the most recent citations were necessary.
   A great advantage was that I worked in the building of two large sister institutes of the Czechoslovak Academy of Sciences (the Institute of Organic Chemistry and Biochemistry and the Institute of Molecular Genetics), where many specialized laboratories were "just round the corner". Many colleagues were willing
not only to lend various reprints that I had missed and made their card-indexes
available to me in order to complete references, but also to read chapters already written. Of course, it was also necessary to request similar services from
colleagues in other institutes, especially the Institute of Chemical Technology
in Prague and the College of Medicine, Charles University, Prague.
   Here I wish to thank above all the colleagues from my mother institute (Institute of Organic Chemistry and Biochemistry) for critical reading of some chapters: Dr. B. Meloun (chapters on proteins, enzymes and protein sequencing), Dr.
M. Lebl and Dr. M. Ryba (peptides), Dr. Z. Točík and Dr. L. Arnold (polynucleotides and oligonucleotides), Dr. M. Zaoral and Dr. J. Ježek (compound biopolymers,
especially immunoadjuvant glycopeptides and synthetic vaccines), Dr. L. Morávek
(synthetic vaccines), Dr. M. Pavlík (sequencing of proteins) and Dr. Z. Točík and
Dr. M. Ledvina (saccharides). Dr. Z. Hostomský and Dr. V. Pačes of the Institute
of Molecular Genetics lent some literature and read critically the chapters on
nucleic acids chromatography and sequencing. Dr. J. Staněk of the Prague Institute
of Chemical Technology lent missing index cards and read critically the chapter

XVIII

on polysaccharides and oligosaccharides. Dr. F. Šmíd of the Institute of Pathology, College of Medicine, Charles University, lent some missing literature and read critically the chapters on compound biopolymers (especially gangliosides) and studies of the structure of oligosaccharides and saccharide components of glycoconjugates, especially glycolipids. Assoc. Prof. Dr. K. Macek (Editor of the *Journal of Chromatography, Biomedical Applications*) of the Institute of Physiology, Czechoslovak Academy of Sciences, kindly lent some groundwork for writing of the final prognostic chapter and also read it critically. This last chapter was further reviewed by Dr. M. Ryba.

I further thank other helpers already mentioned in the Preface to Part A: Dr. Ivana Zemanová of the Research Institute for Pharmacy and Biochemistry, Prague, for reading the whole manuscript and preliminary language revisions; my wife, Dr. Irene Mikešová, for the transfer of all corrections and insertions from the manuscript to the final copy, for ordering and typing of all the literature and for some organizational work; my daughter, Dr. Eve Zažímalová-Mikešová, for carefully setting up the subject index; and last but not least, the staff of the publisher.

I am especially grateful to Dr. Milan Zaoral of the Institute of Organic Chemistry and Biochemistry, who took me into his research group and enabled me to finish Part B, for many prolific discussions, encouragement and interest in this work.

I hope that this book will be a useful and practical aid in the development of rapid chromatography of biopolymers, not only in basic research in the fields of organic chemistry, biochemistry and other life sciences, but also in practical applications in modern biotechnology, medical diagnostics, foodstuff research and control, agrochemical research and control and many other fields of research, control and production, including research on and production of special chromatographic packings.

Prague, Czechoslovakia   OTAKAR MIKEŠ

ACKNOWLEDGEMENTS

I am most grateful to the authors cited, to the publishers of the journals and books listed below and to copyright owners for their kind permission to use some of the figures and data from their papers.

Journals and periodicals

Advances in Hemoglobin Analysis; by courtesy of Alan R. Liss, Inc., New York, NY, U.S.A., Fig. 7.11.
Analytical Biochemistry; by courtesy of Academic Press, Inc., Orlando, FL, U.S.A., Figs. 7.3, 7.7, 9.8, 9.10, 10.12, 11.7, 11.10, 11.11, 11.13, 12.9, 13.2 and 13.4.
Analytical Chemistry; by courtesy of the American Chemical Society, Washington, DC, U.S.A., Fig. 9.2.
Biochemistry; by courtesy of the American Chemical Society, Washington, DC, U.S.A., Fig. 10.6.
Blood; by courtesy of Grune and Stratton, Inc., Orlando, FL, U.S.A., Fig. 7.12.
Journal of the American Oil Chemists' Society; by courtesy of the American Oil Chemists' Society, Champaign, IL, U.S.A., Fig. 11.3.
Journal of Biochemistry (Tokyo); by courtesy of authors and The Japanese Biochemical Society, Tokyo, Japan, Fig. 7.9.
Journal of Biological Chemistry, published by The American Society of Biological Chemists, Baltimore, MD, U.S.A.; by courtesy of authors and redaction, Bethesda, MD, U.S.A., Fig. 13.1.
Journal of Chromatography; by courtesy of Elsevier Science Publishers, Amsterdam, The Netherlands, Figs. 7.1, 7.2, 7.4, 7.5, 7.6, 7.8, 7.10, 8.1, 8.2, 8.3, 9.3, 9.4, 9.5, 10.1, 10.2, 10.3, 10.4, 10.5, 10.7, 10.8, 10.9, 11.1, 11.2, 11.5, 11.6, 11.9, 11.12, 11.14, 12.3, 12.4, 12.7, 12.8, 12.10 and 13.5.
Journal of Food Science; by courtesy of The Institute of Food Technologists, Chicago, IL, U.S.A., Fig. 11.4.
Journal of Lipid Research; by courtesy of the executive editor (L.I. Gidez, Albert Einstein College of Medicine, New York, NY, U.S.A.), Fig. 11.2.
Journal of Liquid Chromatography; by courtesy of Marcel Dekker, New York, NY, U.S.A., Figs. 9.7 and 9.9.
Methods in Enzymology; by courtesy of Academic Press, Inc., Orlando, FL, U.S.A., Fig. 10.11.

Nature (London); by courtesy of Macmillan Journals Limited, London, U.K.,
    Fig. 12.1.
Nucleic Acids Research; by courtesy of IRL Press Limited, Oxford, U.K., Figs.
    10.10 and 10.13.
Starch/Stärke; by courtesy of Verlag Chemie Verlagsgesellschaft, GmbH, Weinheim,
    F.R.G., Fig. 11.8.
Submolecular Biology and Cancer; by courtesy of Elsevier Science Publishers (Bio-
    medical Division), Amsterdam, The Netherlands, Fig. 7.13.
Trends in Biochemical Sciences; by courtesy of Elsevier Science Publishers, Cam-
    bridge, U.K., Fig. 9.1.

Books and dissertations

Chemical Synthesis and Sequencing of Peptides and Proteins (T.Y. Liu, A.N.
    Schechter, R.L. Heinrickson and P.G. Condlife, Editors), published in 1981
    by Elsevier Science Publishers, Amsterdam, The Netherlands, Fig. 13.3.
Developments in Analytical Methods in Pharmaceutical, Biomedical and Forensic
    Sciences (Proceedings of the International Conference, Verona, 1986) (F. Tag-
    liaro, Editor), published in 1987 by Plenum Publishing Corporation, New York,
    NY, U.S.A., Table 16.5.
Laboratory Handbook of Chromatographic Methods (O. Mikeš, Editor), published in
    1979 by Ellis Horwood, Ltd., Chichester, U.K., Table 9.1.
Synthetic Immunomodulators and Vaccines (Proceedings of the International Sympo-
    sium, Třeboň, 1985) (M. Zaoral, Z. Havlas, O. Mikeš and Ž. Procházka,
    Editors), published in 1986 by the Institute of Organic Chemistry and Bio-
    chemistry, Czechoslovak Academy of Sciences, Prague, Czechoslovakia, Fig.
    12.6.
Thesis for the Candidate of Sciences Degree (J. Ježek), Institute of Organic Chem-
    istry and Biochemistry, Czechoslovak Academy of Sciences, Prague, Czechoslova-
    kia, 1981, Fig. 12.5.

LIST OF MATHEMATICAL SYMBOLS FOR BASIC CHROMATOGRAPHY TERMS USED IN PART B

This table is a continuation of Table 2.1 published in Part A (pp. A18-A21). Arabic numerals are numbers of equations where these symbols are defined or first used; roman numerals indicate the number of the Chapter.

Symbol	Term	Equations; Chapter
K_i	Chromatographic distribution coefficient for separation of proteins by IEC	109-111; VII
$k'_{app\ i,j}$	Apparent capacity factors of the same peptide, eluted by two different gradient systems	114; IX
$k'_{i,app}$	Capacity factor for peptide retention in RPC	112; IX
k'_{ie}	Capacity factor for IEC separation of proteins (identical with k'_{iec} in eqn. 72; Chapter III)	110, 111; VII
t_{Ri}	Retention time of a peptide derived theoretically using retention constants D_j for amino acid residues	113; IX
z	Number of molecules of displacing agents required for the desorption of a biopolymer from the chromatographic support	108; VIII

Chapter 7

PROTEINS AND THEIR HIGHER MOLECULAR WEIGHT FRAGMENTS

7.1 INTRODUCTION

Barry L. Karger (1977), reviewing the Third International Symposium on Column Liquid Chromatography (September 20-30, 1977, Salzburg, Austria) had to state that, "Surprisingly, there were no lectures on the separation of proteins using bonded phases in liquid chromatography... High-performance separation of proteins certainly remains one of the major challenges". It is the aim of this chapter to survey how this challenge has been accepted during the past few years. High-performance (pressure) liquid chromatography (HPLC) of proteins requires special macroporous rigid or semi-rigid packing materials, because the mobile phase must be forced through microparticulate columns under pressure. The result is, that HPLC or medium-pressure liquid chromatography (MPLC) reduced the time necessary for the separation of proteins from hours (sometimes even days) to several tens of minutes (even to a few minutes). Such fast methods spread rapidly from research and development laboratories to numerous areas of pure and applied biochemistry, biology, medicine, foodstuffs technology, agricultural research, control laboratories, etc. They are used not only for analytical but also for semipreparative purposes, and in some instances even in production.

A concise history of the HPLC of proteins may be interesting for the reader. Haller (1965) was the first to prepare a rigid macroporous material suitable for these purposes, namely controlled porosity glass, even though he did not use pressure for column chromatography or the term HPLC (cf., Fig. 1.1). Haller et al. (1970) successfully fractionated human serum on columns made of 175 Å pore glass and separated immunoglobulin IgM and α_2-macroglobulin from albumin. In 1973, Čoupek et al. prepared ethylene glycol methacrylate - ethylene glycol bismethacrylate copolymers in spherical form, suitable for chromatography; later this material was produced under the commercial name Spheron or Separon HEMA and was used for protein separations. In 1974, Shechter first reported the high-pressure chromatography of proteins using silica gel columns, protected against irreversible sorption by Carbowax (cf., Fig. 1.2). Doležálek et al. (1975) used gel chromatography on Spheron for the fractionation of whey proteins.

The year 1976 was very rich in works dealing with rapid chromatography of proteins. Regnier and Noel (1976) and Chang et al. (1976a) prepared hydrophilic glycerolpropylsilane bonded phases (suggesting a carbohydrate surface) for the steric exclusion chromatography of biopolymers. Chang et al. (1976c) synthesized ion exchangers for the high-speed chromatography of proteins, and also Chang et al., 1976b) used both of these types of packings for the HPLC of enzymes and other proteins. To those who have studied the evolution of modern liquid column chromatography of proteins, it is clear that the contribution of Regnier's laboratory to the development of these separation methods was fundamental. Vondruška et al. (1976) described the gel permeation chromatography (GPC) parameters of Spheron gels, Vytášek et al. (1976) used unsubstituted Spheron for the HPLC of naturally occurring macromolecular species, and Mikeš et al. (1976) began a series of studies describing ion-exchange (IE) derivatives of Spheron, developed for the rapid separation of biopolymers (for a review, see Mikeš, 1979b).

In 1977, Engelhard and Mathes prepared N-acetylaminopropyl-bonded phases suitable for the size exclusion chromatography (SEC) of proteins; for the HPLC of proteins using chemically modified silica, see also other papers by Engelhard and Mathes (1979, 1981). Soon afterwards, Mönch and Dehnen (1978) described the HPLC of proteins on reversed-phase supports [octadecylsilane (ODS)-coated particles, Nucleosil 10 C_{18}]. Also Hancock et al. (1978) separated higher poly-peptides and lower molecular weight proteins on a C_{18} column using mobile phases with the addition of phosphoric acid. Important contributions to the HPLC of proteins were made by Japanese workers and producers. Hashimoto et al. (1978) reported new gels and described the HP-GPC of series of peptides and proteins, from the small diglycine (M_r = 132) up to the high-molecular-weight human fibrinogen (M_r = 341 000); TSK Gels G 3000, 4000 and 5000 PW, produced by Toyo Soda (Tokyo, Japan), were used for chromatography. Fukano et al. (1978) pre-sented similar information on TSK SW gels (Toyo Soda), developed for the chromatography of water-soluble biopolymers. However, no detailed chemical characterization of these gels was reported. In the same year Ohlson et al. (1978) published the first report on high-performance liquid affinity chromato-graphy (HPLAC) (of enzymes and antigens), Mikeš et al. (1978) on the exact characterization of the Spheron matrix and Štrop et al. (1978) on the rapid hydrophobic interaction chromatography (HIC) of proteins and peptides on un-substituted Spheron.

In the following year, Becker and Unger (1979) and Roumeliotis and Unger (1979) described the preparative separation of proteins in the mean molecular weight range 10 000-100 000 by SEC on LiChrosorb Diol. Rokushika et al. (1979)

and Wehr and Abbott (1979) further studied the application of the new Japanese TSK Gels SW to the high-speed aqueous SEC of biopolymers, including various proteins. O'Hare and Nice (1979) and Nice et al. (1979) described their experiments on the HP hydrophobic interaction chromatography of proteins; Rubinstein (1979), in addition to reversed-phase studies, also investigated the preparative normal-phase chromatography (NPC) of hydrophobic proteins; LiChrosorb RP-8 was used for RPC and LiChrosorb Diol for NPC (a linear gradient from 80 to 50% n-propanol in water was used for the elution of foetal calf proteins). The role of ion-pair (IP) reversed-phase HPLC in peptide and protein chemistry was described and discussed by Hearn and Hancock (1979).

Interesting contributions to the modern chromatography of proteins using the SEC, IE and HIC modes were made by Hjertén, who developed the original low-speed chromatography of proteins on agarose gels and their derivatives into a rapid, high performance method suitable for MPLC using highly concentrated cross-linked gels. A number of papers by Hjertén and his co-workers were cited in sections 4.2.4 and 4.4.3. Hjertén briefly surveyed these works in the eleventh Arne Tiselius Memorial Lecture (Hjertén, 1983) and in a review (Hjertén, 1984).

The present author feels that with the year 1979 this "historical" overview of pioneer work in the area of modern column chromatography of proteins should be terminated. From 1980 onwards the many papers published on this theme must be dealt with in individual specialized sections. However, no treatment of any branch of rapid column chromatography of proteins can be complete without the citation of important works carried out before 1980.

7.2 THEORY OF PROTEIN CHROMATOGRAPHY

We know that it is possible to predict the retention data of peptides based on their amino acid composition (e.g., Meek and Rossetti, 1981; Browne et al., 1982). Until now, no paper has appeared describing the prediction of chromatographic data for high-molecular-weight proteins based on their amino acid composition, in spite of the fact that O'Hare and Nice (1979) discerned a certain correlation between the total number of hydrophobic amino acid residues in some proteins and their retention order in RPC. This illustrates the difficulties with respect to the influence of higher protein structures on the chromatographic process. Amino acids alone and the primary structure of short-chain peptides interact both with solvents (including other dissolved substances) and bonded phases of chromatographic packings, and all these interactions influence the separation. When the length of a peptide chain reaches about 15-20 amino acid residues or more, then intramolecular interactions of amino acids of the peptide chain (forming the secondary and higher structures) become the third parameter to play a role in the chromatographic process. If the length of a peptide chain is much longer, then the tertiary structure typical of proteins is formed in addition to the secondary structure. In the case of proteins composed of subunits, a quaternary structure is the final spatial form. Some parts of a peptide chain are buried in the central "core" of a protein macromolecule (or hidden in the contact areas of quaternary structures) and therefore they cannot take part in interactions with either the solvent or the bonded phase of the chromatographic packings.

Only the peptide chains (mainly the amino acid side-chains) situated on the protein surface can influence the chromatographic retention. Most of them are hydrophilic in nature, especially those which are ionic, but there are also hydrophobic areas on the protein surface important for reversed-phase chromatography and responsible for hydrophobic interactions. Only the complete solution of the problem of predicting tertiary structures based on known primary structures can open up the pathway for predictive efforts in protein chromatography without using electrophoretic or chromatographic experimental information. However, even for proteins the tertiary structure of which has already been determined by X-ray crystallographic methods, and also in instances where the quaternary structure was determined, the known spatial structure has not yet been used as a basis for predicting their retention data.

The theory of protein separation has been attacked from other sides. For example Regnier and Gooding (cf., Appendix in review, 1980) and Sjödahl (1980) summarized the main approaches of general chromatographic theory to the HPLC of proteins, evaluating in a descriptive way the degree of separation achieved. Regnier (1983a), in his review, discussed the principles of "surface-mediated separation modes" [i.e., IEC, HIC (RPC) and BAC] in relation to the separation of macromolecular substances. It is highly probable that macromolecules will be adsorbed on the surface of the packing material at more than one site and that such an adsorption process will be cooperative: adsorption at one site would increase the probability of adsorption at other sites. For the desorption process multiple molecules of displacing agents will be required. The equilibrium between a biopolymer P and a displacing agent D may be represented by the equation

$$P_o + z \, D_b \rightleftharpoons P_b^n + z \, D_o \tag{108}$$

where P_o is the polymer in solution, D_o is the displacing agent in solution, D_b is the displacing agent bound to the surface, P_b^n is the protein bound to the support surface at n sites or residues and z is the number of molecules of the displacing agent required to desorb the polymer form the surface; z may

equal to n, but in most instances it will probably be larger. Another relationship was derived by Regnier (1983a) from eqn. 108, describing the capacity factor
$\bar{k}'$ (chromatographic retention) as being related to the adsorption process at
surfaces (cf., eqn. 93/III); it shows that there is an exponential relationship
between $\bar{k}'$ and z. This conclusion is in good agreement with the usual experience
that desorption curves for proteins become increasingly concave with increasing
molecular weight. Short columns have majority of resolving power in comparison
with longer columns in surface-mediated modes; columns separate more on the
basis of a selective desorption process. This is advantagenous for practical
purposes, because shorter columns can be easily packed, do not require very high
pressures for elution and are cheaper.

The SEC of proteins seems to be the simplest and most predictable and its
main features have been described in Section 3.2. Roumeliotis and Unger (1981)
worked on the assessment and optimization of system parameters in the size
exclusion separation of proteins on Diol-modified silica columns and found not
only that the retention is governed by the SEC effect, but also that some
secondary effects (ionic and Diol-ligand interactions) are superimposed, which
can be controlled and adjusted reproducibly by varying the composition of the
eluent. Some of these findings have already been discussed in Section 4.4.1.
Barford et al. (1982) studied the mechanism of protein retention in RP-HPLC
using alkylsilicas and elution with phosphate buffer (pH 2.1)-isopropanol,
buffer-ethanol and buffer-detergent mobile phases at various temperatures. The
percentage of organic component was varied over a wide range without causing
appreciable retention. However, they found a solution composition for which
very small changes (less than 1% for alcohols) caused dramatic increases in
protein retention. The mobile phase composition was found to be characteristic
for each protein at a given temperature. To explain all the phenomena found,
they calculated the energy of interaction from surface tension and contact angle
measurements and concluded that protein retention by alkylsilicas may be
explained by Van der Waals interactions and that the interactions may be
attractive or repulsive, depending on the surface tension of the mobile phase.

The theory of the IEC of proteins has been intensively studied. The influence
of pH changes can reverse the charge of the whole protein macromolecule and in
this way pH can have a considerable influence on the mobility of proteins in
electrophoresis and their retention on ion exchangers. Proteins are amphoteric
substances that contain both acidic groups (β- and γ-carboxyl groups of aspartic
and glutamic acid residues, respectively, and C-terminal carboxyl groups of
peptide chains) and basic groups (ε-amino groups of lysine residues, guanidyl
groups of arginine and N-terminal amino groups of peptide chains). In acidic
media the dissociation of carboxyl groups is supressed, amino groups are fully

B6

protonated and the protein becomes positively charged. In alkaline media, carboxyl groups are fully dissociated and the ionization of basic groups is supressed, so that the protein gains a net negative charge. There is a certain pH characteristic of a given protein in salt solution at which the acidic and basic groups are in equilibrium and the protein has zero external net charge and does not move in an electric field. This pH is called the isoelectric point (pI) of the protein and occupies an important position in the theory of ion-exchange chromatography and electromigration methods. Righetti and Caravaggio (1976) and Righetti et al. (1981) reported isoelectric points and molecular weights of many proteins, and Malamud and Drysdale (1978) selected 400 papers and prepared tables of pI values of proteins. These tables are of great help to biochemists. It was generally believed that there is minimal or no sorption of proteins on to ion exchangers at their pI values, that pH values below the pI of proteins are suitable for their sorption on cation exchangers and that pH values above the pI are optimal for sorption on anion exchangers. In order to reach a charge high enough for sorption and ion interactions in IEC, regions below pH = pI - 1 were usually recommended for cation-exchange chromatography and regions above pH = pI + 1 for separations on anion exchangers (e.g., Mikeš, 1979a). However, in spite of various successful applications of these rules, studies of the IEC of proteins using modern packings for HPLC or MPLC have shown that they are not generally valid and that they are not correct in some instances; strong sorption was often found also at the pI of proteins or at pH values close to the pI (e.g., Mikeš et al., 1983, 1984), as is illustrated in Fig. 7.1.

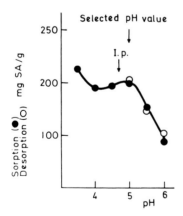

Fig. 7.1. Search for the optimal pH for the static sorption (using the batch method) of serum albumin on Spheron 300-Phosphate of nominal capacity 3.4 mequiv./g, equilibrated with 0.01 M sodium acetate at various pH values. For the desorption the same buffers were used, adjusted to 1 M with NaCl. I.p. = Isoelectric point of serum albumin. (Reprinted from Mikeš et al., 1983.)

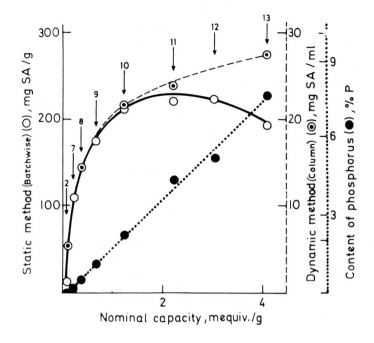

Fig. 7.2. Relationship between the nominal capacity of Spheron 1000-Phosphates and the phosphorus content, and between the static and dynamic capacities for the sorption of serum albumin and the nominal capacity for small ions (Na$^+$). The static capacity was determined by a batch method in the presence of excess of protein in a sorption buffer (0.01 M sodium acetate, pH 5), as long as the equilibrium had been established, followed by desorption with the same buffer enriched with 1 M NaCl. The dynamic capacity was determined with a flow through the column; for the sorption serum albumin dissolved in 0.01 M sodium acetate was pumped on to the column until saturation was attained, then the column was washed with the 0.01 M sorption buffer until the absorbance declined to the baseline. The bound protein was then released from the column by washing with the desorption buffer of high ionic strenth. (Reprinted from Mikeš et al., 1983.)

Great differences were found between the static and dynamic sorption of proteins on ion exchangers (Mikeš et al., 1979, 1980, 1983, 1984); the methods of determination were illustrated in Chapter 4 by Figs. 4.14 and 4.15, respectively. The mutual relationships of both capacities for serum albumin to the nominal capacity of a phosphate ion exchanger are illustrated in Fig. 7.2. Static sorption on all the ion exchangers usually follows a steep convex curve at low capacities, whereas the sorption of small ions (e.g., Cl$^-$ or Na$^+$, expressed in the form of nominal capacity) has a linear dependence on the concentration of functional groups. Ion exchangers with low nominal capacities can bind relatively large amounts of proteins. The dynamic sorption of proteins is usually several times lower, even if it is recalculated to the same units, e.g., weight units. This may be explained by the hypothesis of multiple sorption of proteins under static conditions, where the final equilibrium of the sorbed protein is measured towards the protein solution; the first layer of sorbed protein behaves like an ion exchanger and sorbs another layer. However, in the dynamic procedure the final equilibrium is evaluated towards the pure sorption buffer, so that the multiple sorption of proteins is very limited or disappears. The protein sorption values determined by the dynamic method more adequately describe the possibility of loading the ion exchanger in chromatography, whereas values ascertained by the static method give a better picture of the utilization of ion exchangers using the batch method.

Titration curves of proteins, in which their amphoteric nature is revealed, were used in order to establish the optimal pH for both electrophoretic and IEC separations of protein mixtures, because it was assumed that there is a strong correlation between these two phenomena. In early work on protein chemistry, titration curves of proteins were determined by acid-base titrimetry (for a survey, see Edsall, 1965) and later by series of free electrophoresis experiments (Alberty, 1953). Repeated electrophoretic experiments at multiple pH were tedious and simplifications were necessary. Rosengren et al. (1977), Bosisio et al. (1980) and Ek and Righetti (1980) developed a rapid and simple technique for obtaining titration curves of proteins by two-dimensional gel electrophoresis, in which isoelectric focusing (first dimension) was combined with classical gel electrophoresis (second dimension). For example, Righetti and Gianazza (1980), using carrier ampholytes, prepared a pH gradient in the electrophoretic field in one dimension of a slab gel (composed of a low percentage of polyacrylamide), then the sample was applied into a narrow trough cut in the same dimension throughout the gradient in the centre of the gel, and electrophoresis was performed in the perpendicular dimension. Because the series of sigmoidal curves obtained after staining were proportional to the theoretical protein titration curves, the authors used this technique for the prediction of optimal pH for charge-dependent separation methods. As the optimum pH for separating a value was selected at which the system of titration curves opened maximally. Fägerstam et al. (1982) showed that such an "electrophoretic titration" was useful in predicting the best conditions for purifying creatine kinase from chicken muscle extract. Richey (1983), using this technique, illustrated the correlation between electrophoretic mobility and chromatographic retention on the example of the separation of an enzyme carbonic anhydrase on Mono S, Mono P and Mono Q IE columns.

Haff et al. (1983) examined the correlation between this "electrophoretic titration" and the chromatographic behaviour of 15 proteins in a systematic manner. The whole method and the preparation of polyacrylamide and agarose-Sephadex gels was described in detail. The behaviour of the proteins tested was checked by cation- and anion-exchange chromatography as to whether it was in agreement with the "titration curves". Retention maps of proteins were prepared by plotting retention volumes, obtained by chromatography on ion exchangers, against pH. The results indicated that the retention of proteins in IEC (in terms of salt concentration required for elution) was generally dependent on the charge density of a protein (it was linearly proportional to charge density), which was in turn closely related to the electrophoretic mobility in low-percentage gels. The main reason for this correlation is that for globular proteins electrophoretic methods have the advantage of measuring only the external, exposed charged groups, which are presumably the groups responsible for the IEC characteristics of the protein. However, exceptions to the dependence of chromatographic parameters on titration curves were also found. Shape must also play a role in the determination of the IEC characteristics of a protein, as highly fibrous proteins would probably be bound in a manner more dependent on their total charge (similarly to oligonucleotides). The discrepancies were also found with globular proteins. The authors explained them as probably being due to asymmetry in shape or charge inhomogeneity within the protein. Electrophoresis measures a time-averaged protein net surface charge and is not expected to reveal charge inhomogeneity. Therefore, in such a case, where surface charge inhomogeneity of a protein is typical, the resolving power of HP-IEC exceeds that of electrophoresis and retention mapping may be more predictive than "electrophoretic titration" (it can show that two proteins with apparently identical electrophoretic mobilities could easily be resolved by IEC). A practical example of this phenomenon was given, viz., the IEC separation of two M_2H_2 lactic dehydrogenase isoenzymes.

A single protein in aqueous solution may be in equilibrium in many different ionic forms and because also the ion exchanger may alter the conformation of bound protein, the details of IEC interactions of proteins are not yet known. During salt-gradient elution, proteins may change their contact places with ion exchangers, and the salt concentration at which a protein is finally eluted should be determined by the site of the highest surface charge density, which cannot be predicted only on the basis of electrophoretic experiments. Haff et al. (1983) discussed this problem and, appreciating the prediction method developed for the IEC of proteins based on the "electrophoretic titration", they postulated: "The exact mechanism of ion exchange of proteins is still unknown, as many details, such as the kinetics of adsorption and desorption, are largely unknown. It is likely that proteins are bound to ion exchangers at many different sites on the ion exchanger, although any localized site with relatively higher charge density should have a higher probability of binding".

According to Regnier (1983a), recent research has cast doubt on the assumption that there is a strong correlation between the titration curves of proteins and their IEC retention, because the examination of IEC retention relative to net charge has failed to show a complete correlation with more than 70% of the proteins tested. Many proteins are retained on ion-exchange columns at their pI, where they have no net charge, and at a distance several pH units away from the pI some proteins show less retention than would be expected from their net charge. This phenomenon can be explained by charge asymmetry in polypeptides, which causes a heterogeneous distribution of charge at their surface, and by steric limitation of the contact of some charged groups with the surface of ion exchangers. According to Regnier (1983), "... maximum resolution of components will be achieved at the pH where the relative difference in the number of (their) ionic residues interacting with the surface is maximum. Unfortunately, there is no way to predict this pH a priori".

Kopaciewicz et al. (1983) obtained many experimental data to support the above conclusions. It was shown that other sorptive effects (in addition to electrostatic force), such as solvophobic effects and hydrogen bonding, which may be considered in an effort to explain the retention of proteins at their pI, are negligible with Pharmacia columns (Mono S, Mono Q). Strong ion exchangers were selected owing to their constant charge over a wide pH range. Retention mapping studies of a series of proteins on strong ion-exchange columns showed many deviations from the "net charge model", which resulted from the charge asymmetry, because only a fraction of the protein surface interacts with the stationary phase. Not only the net charge, but also the distribution of the net charge within a molecule is important. This opens the way for the IEC separation of proteins with the same pI, which is in contradiction to the "net charge concept"; the experimental verification of this possibility was discussed using published results (Brautigan et al., 1978) for cytochrome c derivatives.

A contribution to the explanation of the mechanism of IEC was the determination of a double-layer thickness of an ion-exchange support (SynChropak CM 300 cation exchanger) as a function of ionic strength (μ); a thickness of 1-14 Å was found between $\mu = 0.3$ and 0.002 using the measurement of the ion-exclusion effect. This is much less than the diameter of most proteins, so that only a small portion of the total surface of a protein may encounter the double layer. However, this space is large enough to allow the area of greater electrostatic potential (due to the charge asymmetry) on the surface of a protein to cause a proper molecular orientation at the support surface; this "steering effect" may be decisive during the adsorption-desorption process. The quality of displacing ions is also important: some of them change the retention of all proteins in nearly the same way, but the selectivity of a strong anion-exchange column can be altered by changing the mobile phase anion. With a strong cation exchanger this phenomenon is not so striking. Trisodium citrate enhanced the selectivity the most for strong anion-exchange columns, whereas sodium fluoride was superior for strong cation-exchange columns. When a macromolecule with its own double layer encounters the double layer on the support surface, a redistribution of ions in both double layers occurs. The authors assumed that ions are expelled during this process (because an increasing concentration of ions must be used for desorption), but the exact mechanism of this process is still unknown.

A non-mechanistic model has been developed (Kopaciewicz et al., 1983), which shows a positive correlation between the protein retention and the number of charges associated with the adsorption-desorption process. First an equation for the chromatographic distribution coefficient, K_i, was formulated:

$$K_i = \frac{C_s}{C_m} = \frac{[P_b]}{[P \cdot C_i]} = \frac{P_m}{[P \cdot C_i]} \cdot f \qquad (109)$$

where C_s is the concentration of a solute on the stationary phase (mol/m^2), C_m the mobile phase concentration (mol/l), P_b the bound protein, P_m the maximum load capacity, $P \cdot C_i$ the concentration (mol/l) of the protein in solution above the surface with the accompanying counter-ion concentration (C_i) and f the fraction of the surface loaded with protein. The above equation can be rearanged to give

$$K_i = K_b \cdot \frac{(D_{bi})^z}{a^z b^z D_o^z C_i^z} \cdot (1 - f) \qquad (110)$$

where K_b is the binding constant, D_{bi} the initial ligand concentration, a and b are constants needed to adjust for valence, activity coefficient and relative displacing power differences between ions, D_o is the displacing ion concentration of the mobile phase (mol/l) and z the number of charges that are associated with the adsorption-desorption process. If a very small amount of protein is chromatographed, the expression can be simplified to

$$K_i = K_y/(D_o C_i)^z \qquad (111)$$

in which K_b, D_{bi}^z, a^z and b^z in eqn. 110 have been incorporated into a single constant K_y. Eqn. 111 can be transformed to express the capacity factor, k'_{ie}, which is more usual in IEC, as

$$k'_{ie} = K_i A_s / V_m \qquad (112)$$

where A_s and V_m designate the available surface area (m^2/g) and the mobile phase volume, respectively, (cf., eqn. 70/III) and are constants for a given solute-support combination. If K_y, A_s and V_m are incorporated into a new constant K_z, the following equation is derived:

$$k'_{ie} = K_z / (D_o C_i)^z \qquad (113)$$

which relates the retention of a solute to the displacing agent concentration of the mobile phase and the number of charged groups involved in the adsorption-desorption process. The authors illustrated the applicability of this equation on examples of the determination of the number of charge interactions (z) occurring between β-lactoglobulin and the surface of a strong anion-exchange column. Calculations indicated that the number of charged sites involved in binding may be greater or less than the net charge of proteins. It seems that there is a positive correlation between z and protein retention. This work of Kopaciewicz et al. (1983) contributed considerably to the modern theory of the IEC of proteins.

The role of hydrodynamics and physical interactions in the adsorption and desorption of hydrosols of globular proteins was dealt with by Ruckenstein and Kalthod (1981).

7.3 MODES OF CHROMATOGRAPHY OF PROTEINS

7.3.1 Size exclusion chromatography

Rapid size exclusion chromatography (SEC) has often been used for analytical and preparative separations of proteins, owing to its simplicity. Barford et al. (1979) studied retention behaviour on commercial siliceous supports and found that sorption was observed on both surface-modified and unmodified supports. The increase in the retention time of ovalbumin and ribonuclease with increasing salt concentration suggests that hydrophobic interactions may take place. When dodecylsulphate was added to the same phosphate buffer and used with the same siliceous supports, the expected elution order with respect to the molecular weight (MW) of proteins was observed. Pfannkoch et al. (1980) characterized some commercial SEC columns for water-soluble polymers. The test procedures utilized molecular probes of various sizes. Several columns under certain conditions exhibited ion-exclusion, cation-exchange and hydrophobic partitioning. These non-permeation effects were minimal with the mobile phase ionic strengths between 0.1 and 0.6 M at pH 7.05. A contribution to the theory of SEC was derived (cf., Section 3.2). Column characteristics and performance data were reported that may serve as a guide to the optimization of columns for the strict SEC mode. The separation time was 40 min or less and protein recoveries were over 86%. Engelhardt et al. (1981) described studies with a bonded N-acetylamino-propylsilica stationary phase of nominal pore diameter 100 Å for the aqueous HPLC of polypeptides and small proteins. The data obtained on the recoveries and elution characteristics showed that the so-called "amide phase" can be successfully applied for fractionation in the MW range $2 \cdot 10^3$-$40 \cdot 10^3$ daltons. The presence of residual silanols leads to ionic interactions most noticably only with very basic ($pI > 9$) or very acidic ($pI < 5$) proteins.

Roumeliotis et al. (1981) gave a brief critical survey of the separation of proteins on diol-modified silica packings. The potential and limitations of SEC-HPLC in analytical and preparative separations of biopolymers were discussed. Gemershausen and Karkas (1981) described the preparative high-speed GPC of proteins on the inexpensive packing Toyopearl HW 55F (Toyo Soda, Tokyo, Japan). Using this material, the determination of molecular weight in the range 10^4-10^6 daltons was achieved in a few minutes. Large-scale enzyme purification (up to 1.6 g of starting material with a 105 x 3.2 cm I.D. column) was achieved with very good recoveries of enzymic activity. Data for the optimization of separation were presented. The same chromatographic support material, which is stable in the pH range 1-14, is available also as Fractogel TSK and its use in the aqueous gel filtration chromatography of enzymes and other proteins, oligosaccharides and nucleic acids was published by Gurkin and Patel (1982); packing procedures were described in detail. An optimal buffer for reliable SEC determinations of molecular weights of proteins, which minimizes the ionic and hydrophobic interactions of the solute with the stationary phase, was sought by Hefti (1982); Fig. 7.3 illustrates the separation achieved. The system used enabled the M_r of proteins between 10 000 and 70 000 to be determined reliably.

The non-ideal SEC of proteins and the effects of pH at low ionic strength were discussed by Kopaciewicz and Regnier (1982); their paper was discussed in Section 3.2. Gel permeation chromatography of asymmetric proteins was examined by Meredith and Nathans (1982); collagen, fibrinogen and lysozyme were studied with a variety of gel and HPLC media and compared. Hjertén and Eriksson (1984) described the HP molecular sieve chromatography of proteins on agarose columns; the relationship between concentration and porosity of the gel was studied. The exclusion limits of 5, 9, 12 and 20% agarose gels corresponded to proteins with molecular weights above 1 000 000, 600 000, 450 000 and 280 000, respectively. Examples of separations were given. Kato (1984) described Toyo Soda high-performance gel filtration columns.

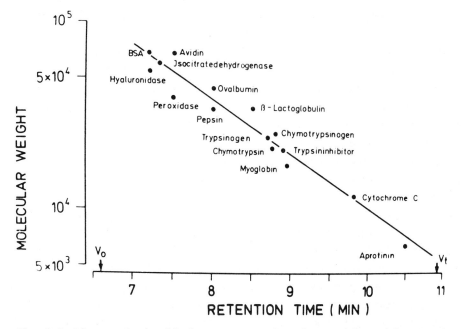

Fig. 7.3. Linear relationship between retention times and logarithm of molecular weight for standard proteins. An I-125 SEC column (Waters Assoc.) was used with a mobile phase consisting of 0.08 M sodium phosphate buffer (pH 7.0) containing 0.32 M sodium chloride and 20% (v/v) ethanol. Flow-rate, 1 ml/min; pressure, 400-600 p.s.i.; detection, 280 nm. The void volume (V_0) was determined with ferritin, the total volume (V_t) with ^3H_2O and with the tripeptide Tyr-Gly-Gly. The correlation graph was fitted using the least-squares method; correlation coefficient r = 0.97. (Reprinted from Hefti, 1982.)

7.3.2 Ion-exchange chromatography

Ion-exchange chromatography (IEC) is probably the most important chromato-graphic technique for the separation of proteins. Basic proteins (with high pI values) are best separated on cation exchangers, and for acidic proteins (with low pI) anion exchangers are usually the best choice. For neutral proteins both types of ion exchangers can be used. All proteins are sorbed on all ion ex-changers at low ionic strength (e.g., μ = 0.01) and eluted (desorbed) by gradual elevation of the ionic strenth (e.g., up to μ = 0.5 or more). Changes in pH can also be used for elution, but an increase in ionic strength is more important, and gradient elution using elevated concentration of a salt is often used at constant pH. For a good separation of a protein mixture long columns are not necessary; nearly the same separation can often be achieved using 5- or 30-cm columns (Regnier, 1982); of course, longer columns have a higher total capacity. In general, ion exchangers usually have a high loading capacity for proteins and even packings with a low nominal capacity sorb relatively large amounts of proteins (cf., Fig. 7.2); the high loading capacity of ion exchangers is advantageous when the preparation of proteins is planned.

Berford et al. (1979) studied the interactions of proteins with bonded-phase ion exchangers and found deviations from the expected behaviour consistent with the ion-exchange theory. The anion exchangers studied showed ambigous properties, some proteins (bovine serum albumin and ovalbumin) were retained on cation exchangers at a pH above their isoelectric points and their retention times increased rather then decreased with increasing eluent molarity. The authors concluded that a mixed mechanism was involved in the separation and elevation of the HETP in some instances. However, mixtures of reference proteins and protein isolates were resolved satisfactorily. The bonded phases represent one group of ion exchangers suitable for rapid protein chromatography. Another group are fully organic macroporous packing materials. Mikeš et al. (1979) introduced diethylaminoethyl derivatives of Spheron, which were followed by carboxylic derivatives (Mikeš et al., 1980). Both of these medium-basic anion exchangers and weakly acidic cation exchangers were successfully used for effective protein separations, as is illustrated by Fig. 7.4. These packing materials can be highly loaded; 60 mg of proteins in Fig. 7.4 does not represent the maximal load. Later, medium-acidic phosphate derivatives (Mikeš et al., 1983), strongly acidic sulphate derivatives (Mikeš et al., 1984) and strongly basic triethylaminoethyl derivatives (Mikeš et al., 1988) were prepared and used for the fractionation of various proteins. For example, in the isolation of cellulolytic enzymes (Hostomská and Mikeš, 1984), 560 mg of proteins from cultivation media of *Trichoderma viride* Reesei were applied on a 33 x 1.65 cm I.D. preparative column made of Spheron 1000-Phosphate and fractions of 9.7 ml were taken at 60-s intervals.

Schifreen et al. (1980) studied accuracy, precision and stability in the measurement of glycosylated haemoglobin A_{1c} by HP cation-exchange chromatography using Bio-Rex 70 resin (200-400 mesh); they found that temperature control and the addition of ethanol to the stepwise elution with phosphate increased the precision by stabilizing the peak shape and column activity. Vanecek and Regnier (1980) studied the variables in the HP anion-exchange chromatography of proteins using a 10-μm SynChropak AX-300 support. Up to 10 mg of protein per injection could be applied on a 250 x 4.1 mm I.D. column with good resolution. In a subsequent paper Vanecek and Regnier (1982) developed another macroporous HP anion-exchange support and showed that silicas of pore diameter 1000 and 4000 Å coated with heavy layers of polyethylenimines and cross-linked with 1,4-butanediol diglycyl ether (cf., Fig. 4.8) have an enhanced anion-exchange capacity and give good resolution of proteins. Saint-Blancard et al. (1981) described the

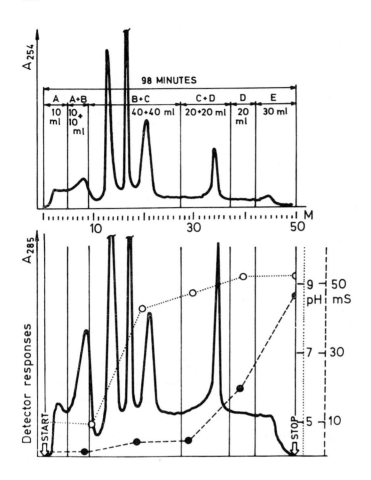

Fig. 7.4. Chromatographic analysis of 60 mg of a mixture of egg proteins on a column (20 x 0.8 cm I.D.) of CM-Spheron 300 (20-40 μm), nominal capacity 2.20 mequiv./g. Detection was performed with a tandem system of UV analysers (cf., Fig. 5.28). Buffers: (A) 0.05 M ammonia + acetic acid (pH 5.0); (B) 0.3 M Tris + acetic acid (pH 7.5); (C) 0.5 M Tris-HCl (pH 9.5); (D) buffer C, 1 M in NaCl. Gradients were linear. The column was finally washed with 2 M NaCl (E). Fractions of 4.2 ml were collected at intervals of 118 s. M = Step marks of the fraction collector. Right-hand scales: for the measurement of pH (o) and conductivity in mS (•). (Reprinted from Mikeš et al., 1980.)

use of Trisacryl ion exchangers and their application to the fractionation of human plasma proteins. Komiya et al. (1981) evaluated TSK-Gel IEX-500 columns (modification of TSK-Gel SW) for high-speed ion-exchange protein chromatography, and Kato et al. (1982) studied the effect of gradient conditions, flow-rate, column length and sample loading on retention and resolution using TSK-Gel IEX-545 DEAE Sil columns. Variables in the ion-exchange of proteins on silica-based cation exchangers were studied by Frolik et al. (1982). Cation exchange has proved to be a useful technique for basic proteins (Tsuda et al., 1982).

Alpert (1983) described the cation-exchange HPLC of proteins on polyaspartic acid-silica. This ion exchanger was mentioned in Chapter 4 (cf., Fig. 4.7); it has a high capacity for proteins (430 mg/g of haemoglobin) and the columns featured excellent performance. Protein standards were well resolved in tens of minutes. Ou et al. (1983) used this polyaspartic acid-silica cation exchanger for the HPLC of human haemoglobins. Very good results were obtained, complete separation of haemoglobin variants being achieved by gradient elution within 30 min. The high resolution and accuracy of the method combined with complete automation make this procedure useful for the diagnosis of haemoglobin disorders in both research and clinical laboratory applications. Gupta et al. (1983) and Toren et al. (1983) also showed that improvements in the IEC resolution of haemoglobin variants, important in the diagnosis of certain metabolic disorders, continue to be made.

Kato et al. (1983b) described TSK-Gel IEX-645 DEAE anion exchanger (prepared by introducing a diethylaminoethyl group into TSK-Gel 5000 PW) and its application to the separation of proteins and nucleic acids; it was found superior especially at high pH and for high-molecular-weight samples. Tandy et al. (1983) studied the HPLC purification of the hydrophobic ω-subunit of the chloroplast energy coupling complex. As this complex is insoluble in water, it was chromatographed on a SynChropak AX-300 column using unusual solvents for proteins: chloroform-methanol solution for application to the column and elution with a gradient of 3-20 mM ammonium acetate in chloroform-methanol-water. This procedure had the advantage over RPC of permitting the chromatography of greater loads of crude protein solution. Ion selectivity in the HP cation-exchange chromatography of proteins was studied by Gooding and Schmuck (1984). The weakly acidic cation exchanger SynChropak CM-300 was used and the effects of varying the ionic composition of aqueous mobile phases were observed. Cations generally followed the same order for retentive properties as for smaller molecules, but anions affected the retention of proteins in the reverse order to that for small molecules on an anion exchanger. The data obtained suggest that an inadequate separation of proteins on a given ion-exchange column might be made satisfactory simply by changing the salt composition used for gradient elution. Hjertén (1984) studied the application of high-concentration agarose gels in the HPLC separation of biopolymers and gave an example of a rapid separation of human serum on DEAE-agarose (12% cross-linked, 5-40 μm beads); a linear gradient of ionic strength 0.08 in Tris-acetic acid buffer (pH 8.8) was used for elution, which was finished in 50 min.

Burke et al. (1986) reported rapid protein profiling with a novel anion-exchange material, available from Bio-Rad Labs. (Richmond, CA, U.S.A.). Small (7 μm) non-porous beads of polymethacrylate resin were covered with a covalently coupled layer of polyethyleneimine (PEI). Ion exchangers resembling pellicular material were prepared in this way. They had relatively low capacities (1-300 µg of proteins for a 30 x 4.6 mm I.D. column) but the speed of the ionic strength gradient elution was very high (from tens of seconds to several minutes). The support was designated Microanalyser MA7P.

7.3.3 Reversed-phase chromatography

Separations with chemically bonded hydrophobic phases have been used for the RPC of proteins (Hearn and Hancock, 1979; O'Hare and Nice, 1979). This method is based on hydrophobic interactions between hydrocarbon chains and hydrophobic domains of chromatographed protein molecules. The mobile phase is polar. Gradual elution of individual components of a mixture can be achieved by decreasing the polarity of the mobile phase by the addition of alcohol or acetonitrile. An example of the separation of simple proteins is shown in Fig. 7.5. Lewis et al. (1980) developed supports for the RP-HPLC of large proteins (M_r up to 300 000); the supports were based on silica to which octyl, cyanopropyl or diphenyl groups were attached. These materials were used for the separation of α- and β-components of human collagen, chicken Type I collagen (M_r = 280 000), tyrosinase (M_r = 128 000), α_1 and α_2 chains of chicken collagen (α_1 chain, M_r = 95 000), bovine serum albumin (M_r = 68 000) and cytochrome c (M_r = 15 000). The authors concluded that normal commercial RP supports have pore sizes too small to be optimally utilized with proteins of M_r higher than 40 000-50 000. The developed 50-nm pore size C_8 suport had good elution characteristics both for small proteins and for proteins as large as Type I collagen. α_1 and α_2 chains of chicken Type I collagen were separated within 1 h. The diphenyl support offers a different type of interaction. As with the cyanopropyl support, the resolution probably depends on more than just hydrophobic interactions. With the diphenyl support aromatic ring stacking is probably involved. Proteins bind much tighter to this support than to the other RP-supports and the resolution is improved. To obtain an optimal resolution, lower flow-rates were necessary (30 ml/h or less with a 250 mm x 4.6 mm I.D. column) and a lower pH also gave better resolution.

Petridges et al. (1980) used the RP-HPLC of proteins for the separation of haemoglobin chain variants. Using an octadecylsilylsilica stationary phase and propane-pyridine formate as the solvent system, normal α- and β-chains of human haemoglobin were separated from several mutant chains. The results were discussed from the point of view of looking for an effective method for the separation of closely related proteins differing in single amino acids only. Pearson et al. (1982) studied various silicas of large pore diameter coated with n-alkylchloro-

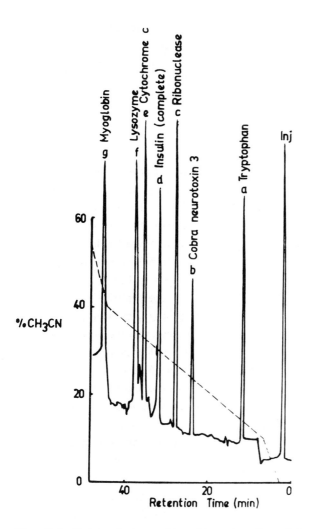

Fig. 7.5. Reversed-phase chromatography of a mixture of six simple proteins on octadecyl phase bound to Hypersil ODS in a 10 x 0.5 cm I.D. column. The hydrophobic amino acid tryptophan was applied as an internal standard. Primary solvent, 0.1 M NaH$_2$PO$_4$-H$_3$PO$_4$ (pH 2.1) with a total phosphate concentration of 0.2 M; secondary solvent, acetonitrile. The dotted line represents the gradient of acetonitrile. Inj means injection artifact. (According to O'Hare and Nice, 1979.)

silanes for RP protein separations, tested their efficiency and characterized the optimal silica. Vydac TP was found to be the best for both protein and peptide separations. Theoretical plate values determined for small unretained molecules were found not to be correlated with protein resolution. Wilson et al. (1982) compared the HPLC of peptides and proteins on 100- and 300-Å RP supports and found the latter to be superior for the chromatography of high-molecular-weight proteins ($M_r \gtrsim 15\ 000$). Gurley et al. (1983a,b) fractionated histones by HPLC on μBondapak C_{18} and cyanoalkylsilane (CN) RP columns; the latter support was superior for the separation of histones. Low trifluoroacetic acid concentrations (0.1%) in the water-acetonitrile eluent were used with CN columns and increased the sensitivity of histone detection by measurement of UV absorbance at 206 nm. Contaminating non-histone protein concentrations were greatly reduced by isolating nuclei prior to histone preparation. Tandy et al. (1983) described the HPLC purification of a very hydrophobic ω-subunit of the chloroplast energy coupling complex, which contains 80% hydrophobic residues. It is therefore insoluble in aqueous buffers and can be chromatographed in organic solvents. For rapid purification a procedure was developed using preliminary purification by Sephadex LH-20 chromatography (to remove lipids and other proteins which clogged the column), followed by chromatography on a Whatman M9 column of Partisil 10 ODS packing with stepwise elution with acetonitrile, methanol and trifluoroacetic acid. An alternative IEC procedure was also developed (cf., Section 7.3.2). Pollak and Campbell (1984) described Waters columns for rapid protein chromatography.

From the methodical point of view, the investigation of buffers suitable for the RPC of proteins is important. Rivier (1978) studied the use of triethylammonium phosphate (TEAP) as one of trialkylammonium phosphates (TAAP) compatible with RP-HPLC for the high resolution of peptides and simple proteins (such as insulin or cytochrome c), using μBondapak CN, alkylphenyl and C_{18} columns. This buffer is transparent to UV radiation of wavelength down to less than 200 nm and is very suitable for the separation of biological substances. By combining TEAP with suitably derivatized capped silica supports, good recoveries in the RP-HPLC of peptides and simple proteins were obtained (Rivier et al., 1979). Henderson et al. (1981) studied the purification of proteins (up to M_r = 80 000) and peptides by RP-HPLC using volatile solvents. μBondapak Phenylalkyl or C_{18} supports were used. Elution was achieved with simple linear gradients employing two solvents: (A) 0.05% trifluoroacetic acid (TFA) in water and (B) 0.05% TFA in acetonitrile. The solvent is completely volatile and compatible with UV detection at 206 nm and above. However, model studies with ovalbumin, chymotrypsinogen A, lysozyme and ribonuclease have shown that the yield of proteins is often less than 100%. Denatured parts of proteins cannot be eluted from the RP supports with the above-mentioned solvents. The authors proposed methods that allow refolding of denatured proteins to be catalysed, e.g., elution with 70% acetic acid saturated with guanidinium chloride; this method simultaneously cleans the RP support. Jones et al. (1980) tested those types of mobile phases for the RPC of proteins: (a) 0.5 M pyridine formate,

(b) 0.1% formic acid and (c) 0.1% phosphoric acid. The last eluent produced the lowest plate count values, which were affected very little by differences in flow-rate.

RPC may become a very efficient method for the separation of some biopolymers, but it can be used reasonably only for proteins that retain their solubility in relatively hydrophobic media and can be recovered in the native state after some additional operation.

Retention times of proteins and peptides in RPC can be modified by the addition to the mobile phase of compounds, that form ion pairs with the sample. Non-polar ions, such as tetraalkylammonium, increase retention times. Hydrophobic ion-pairing agents also include dodecylamines, dodecylsulphonates and *tert.*-butylammonium compounds. Hollaway et al. (1980) studied the hydrophilic ion-pair RP-HPLC of peptides and proteins: For low-polarity samples (where the retention times in the usual RP columns are inconveniently long) the retention times can be decreased by the formation of ion pairs with highly polar ions, such as acetate, formate, phosphate, citrate or perchlorate. For hydrophilic ion pairing of peptides and proteins, mobile phases that are transparent in the UV at 200 nm (the region of absorbance of amide or peptide bonds) should be employed. The authors gave examples of the hydrophilic ion-pair RPC of CNBr fragments and partial hydrolysates of various proteins. Dilute acetic, formic or phosphoric acid was used as the mobile phase. Both isocratic conditions and acetonitrile gradients were employed. Asakawa et al. (1981a) described the isocratic RP-HPLC of proteins with Nucleosil CN as the stationary phase and sodium heptanesulphonate as the IP reagent; insulin, RNase, myoglobin, lysozyme and cytochrome were separated. Asakawa et al. (1981b) also separated soy-bean trypsin inhibitor, trypsin, trypsinogen, α-chymotrypsin, α-chymotrypsinogen and elastase using an isocratic elution system with Nucleosil CN as the stationary phase and sodium ethanesulphonate, pentanesulphonate or octanesulphonate as the ion-pairing reagent.

Tweeten and Tweeten (1986) described the RPC of proteins using the newly developed wide-pore styrene-divinylbenzene packing PLRP-S (Polymer Labs., Church Stretton, U.K.).

7.3.4 Hydrophobic interaction chromatography

Totally hydrophobic supports with a high ligand density and elution in the presence of organic solvents are typical in RPC. This implies a certain risk in maintaining the native tertiary structure of proteins and enzymic activity. Protein or activity recoveries are sometimes low after RP-HPLC. Modern trends in the hydrophobic interaction chromatography (HIC) of proteins are towards hydrophilic supports bonded with hydrophobic ligands with a low ligand density. These types of supports are more compatible with the requirement of protein spatial structure stability. Hjertén et al. (1982) studied the MPLC separation of plasma proteins on columns packed with hydrophilic crosslinked agarose gel spheres of high concentration (9-15%), modified with octyl groups. Desorption was achieved by decreasing the ionic strength of the mobile phase in the absence of organic solvents to avoid denaturation. In spite of the medium pressure techniques used, good performance was obtained. Nishikawa et al. (1982) also studied this problem: commercially available RP C_8 phases have the tendency to bind a variety of proteins very avidly at neutral pH. Gradient elution at pH 2.1 allows better recoveries of proteins and sharper bands. However, many proteins are denatured or altered at this low pH. To avoid this problems, the authors developed hybrid bonded-phase silica gels containing both hydrophilic and hydrophobic ligands (cf., Section 4.4.3). Hybrid gels, with a low ratio of hydrophobic to hydrophilic groups, permitted for the first time the hydro-phobic interaction chromatography of several test proteins at pH 6.2. The problems of the HIC of proteins were discussed by the authors from various points of view.

Kato et al. (1983a) started from the hydrophilic TSK Gel G 3000 SW (10 μm particles), which proved suitable for the SEC of proteins, and modified it by introducing butyl and phenyl groups. Butyl-G 3000 SW and Phenyl-G 3000 SW were prepared with ligand contents of 0.12 and 0.14 mmol/g (corresponding to 38 and 44 μmol/ml), respectively. Using these supports it was possible to separate proteins with high efficiency under mild elution conditions, such as isocratic elution with 0-2 M ammonium sulphate in 0.1 M phosphate buffer (pH 6), and linear gradient elution with ammonium sulphate of concentration decreasing from 1.5 or 2 to 0 M in 0.1 M phosphate buffer (pH 6). The recovery of proteins or enzymic activity was almost quantitative. The dependence of capacity factors on ammonium sulphate concentration was measured for eight proteins using both butyl and phenyl derivatives. The capacity factors increased from almost zero to very large values with increasing ammonium sulphate concentration from 0 to 2 M. It is clear that proteins can be adsorbed and desorbed under very mild conditions and chromatographed without the risk of denaturation. Gooding et al. (1984) prepared a series of packings that had alkyl and aryl groups incorporated into a hydrophilic polymer matrix of SynChropak (SynChrom, Linden, IN, U.S.A.) and applied them to the HPLC of proteins in the HIC mode (see also Section 4.4.3). Using an inverse salt gradient of ammonium sulphate (or sodium chloride) a series of proteins were purified. A typical example is illustrated in Fig. 7.6.

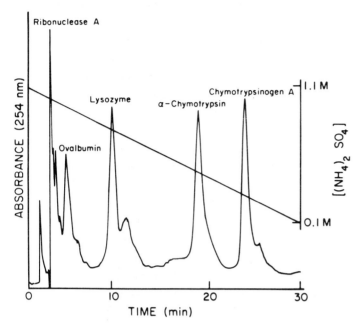

Fig. 7.6. Analysis of a protein mixture by HIC. Column, SynChropak Propyl (250 x 4.1 mm I.D.); flow-rate, 1 ml/min; pressure, 80 atm; 30-min gradient from 1.1 to 0.1 M ammonium sulphate in 0.02 M potassium phosphate (pH 7). (Reprinted from Gooding et al., 1984.)

It is interesting that the elution order of proteins is not identical with that on reversed-phase columns; bovine serum albumin and ovalbumin, for example, showed opposite elution orders. The authors explained this phenomenon as probably being due to the tertiary structure of proteins, which is maintained in HIC but disrupted in RP-HPLC (see also Regnier, 1983b).

Kato et al. (1985a) described a new resin-based support, TSK gel Phenyl-5 PW RP (Toyo Soda), designated for the RPC of proteins, which is a derivative of the hydrophilic TSK Gel G 5000 PW. Kato et al. (1985b) applied this support to the preparative HIC of proteins. Alpert (1986) continued the studies mentioned in Chapter 4 (cf., Fig. 4.7) and in Section 7.3.2. He described the high-performance HIC (HPHIC) of proteins on series of poly(alkyl aspartamide)silicas. Poly(propyl aspartamide)silica was found to be a good general-purpose material for the HPHIC of proteins, whereas the ethyl derivative was found to be useful for more hydrophobic proteins. A broad study of applications of these packings for various types of proteins was reported.

HIC has great potential in comparison with RPC in the HPLC of native proteins (especially of enzymes) based on hydrophobic interactions. Good separations and high recoveries are obtained using this method.

7.3.5 Other modes for rapid separation of proteins

One of the approaches to the HPLC of proteins was adsorption chromatography.
Mizutani (1980) studied the adsorption of proteins by porous glass and its
application to chromatography. The method may cause denaturation of proteins
after prolonged contact with porous glass, which may be prevented by surface
treatment of the glass. Mizutani (1981) described the siliconization of porous
glass, which prevents irreversible sorption and denaturation of proteins, but
diminishes the sorption properties to such an extent that hardly any protein
is retained at a low concentration of salts. Such a support and conditions may
be suitable for the SEC of proteins, but not for adsorption chromatography.
On the other hand, at high concentrations of salts, the coated glass adsorbs
significant ammounts of proteins by hydrophobic bonding and also lymphocytes
can be separated on siliconized glass beads. The author investigated the elution
profiles of a mixture of bovine albumin and haemoglobin, and of bovine globulin,
on a silicone-coated porous glass column in various buffers. Adsorption did
not seem to be the best choice for effective protein separations. Subsequently,
Mizutani and Narihara (1982) studied the conditions for the adsorption chromatography of proteins on siliconized porous glass using standard proteins and
rabbit serum. A cholate detergent solution was tested as the eluent, but an
acetonitrile gradient was found to be better.

Wunderwald et al. (1983) described a new principle of "sandwich affinity
chromatography", which is based on strong chelate binding of α_2-macroglobulin
(α_2M) to Zn^{2+}-biscarboxymethylamino-Sepharose (Zn-chelate Sepharose) and on
its ability to complex most active endoproteinases. This principle was used
for the removal of endoproteinases from biological fluids. In an effective
procedure α_2M (which is known to form complexes with a large number of active
endoproteinases, such as trypsin) should first be bound to Zn-chelate Sepharose,
followed by adsorption of the proteinases to α_2M-Zn chelate Sepharose using
elevated salt concentrations. The advantages of sandwich affinity chromatography
are (1) the simple loading procedure by adsorption, (2) the high capacity for
the gel material and (3) the possibility of re-using the Zn-chelate Sepharose
after eluting the reacted α_2M and reloading with new α_2M. This method can be
used for the specific sorption of chymotrypsin, trypsin, thermolysin, elastase,
bromelain, ficin and papain, but not for exoproteinases such as carboxypeptidases.

Fägerstam et al. (1983) studied the fast chromatofocusing of serum proteins
with special reference to α_1-antitrypsin and Gc-globulin. Mono P ion exchanger
was used as the chromatographic support in a pre-packed 20 x 0.5 cm I.D. column,

and Polybuffer (Pharmacia, Uppsala, Sweden) as the chromatofocusing eluent.
The time of chromatography was less than 1 h. Separations in a broad pH range
of 6.0-3.8 were checked by fused rocket immunoelectrophoresis and gradient
electrophoresis and was followed by chromatofocusing in narrow pH intervals of
about 0.5 pH units to study the microheterogeneity of both of the above-mentioned
proteins. The authors recommended chromatofocusing for the first dimension in
two-dimensional techniques for the resolution of complex protein mixtures.

Another principle of specific protein separation is ligand-exchange chromatography (LEC). According to Davankov et al. (1973) (see also Davankov, 1984),
LEC can be defined "as a chromatographic technique in which the formation of
the coordination bonds is the dominant mechanism for the separation of the
solute species". The term "ligand exchange" was first used by Helfferich (1961)
in order to describe the process in which one type of coordinated ligand is
displaced by another. The complex formation is a highly selective process
because atoms of the ligands can occupy only strictly sterically fixed positions.
Therefore, LEC was not only applied for the specific separation of amino acids
and peptides, but also experiments were carried out on its use for the separation of proteins in the "metal chelate affinity chromatography" or "labile
ligand affinity chromatography" mode. Porath et al. (1975) suggested polysaccharides (agarose or dextrans) as a suitable matrix, carrying transition metal
iminoacetate, bonded through a long spacer arm. Agarose, activated with
1,4-bis(2,3-epoxypropoxy)butane and coupled with iminodiacetate:

$$Agarose-OCH_2CHCH_2-O(CH_2)_4O-CH_2CHCH_2-N \cdots\cdots M^{2+} \cdots\cdots$$

was used for chelate binding of metal ions, M^{2+}, most often Cu^{2+} or Zn^{2+}, but
Cd^{2+}, Co^{2+}, Ni^{2+}, Hg^{2+}, Fe^{3+} and Al^{3+} have also been used. As the immobilized
ligand is only tridentate, the metal ion is capable of coordinating an addi-
tional mobile ligand from the external solution. For example, it can bind an
amino acid, a peptide or a protein with selective affinity to His, Cys and
probably also Trp, Tyr and Lys residues on the peptide or protein surface.

In addition to Sepharose and Sephadex matrices, strongly acidic sulphonated
polystyrene (Amberlite IR-120) has also been used as a carrier of metal ions,
e.g., Hg^{2+} and Al^{3+}, which can form active sites for the fractionation of some

proteins. The way is also opened for the application of macroporous silica-bonded ligand exchangers. In order to overcome difficulties with desorption, Nexø (1975) described a new principle in biospecific affinity chromatography used for the purification of cobalamin-binding proteins: cobalamin was attached to Sepharose matrix through the temperature-labile linkage to insolubilized 3,3'-diaminodipropylamine. Cobalamin-binding proteins were adsorbed selectively, but with only a weak bond at lower temperature ($4^{\circ}C$); on increasing the temperature ($37^{\circ}C$) they were eluted in a complex with cobalamin. In this way the so-called "labile ligand affinity chromatography" was introduced and applied successfully to the separation and purification of the whole group of this type of proteins.

The principle and applications of LEC were reviewed by Davankov (1984), who with his co-workers, developed this method very broadly, especially for the LEC of enantiomers. In his review proteins were also dealt with. Caude et al. (1984) extended the review of LEC to separations in which the ligand is bound to a normal-phase column, and Sugden (1984) to separations in which the ligand is bound to a reversed-phase system.

7.3.6 Conclusion

Three chromatographic modes are the most important in protein separations for both analytical and preparative purposes:

(a) Size exclusion chromatography allows (1) separations based on the hydrodynamic molecular dimensions of the wetted macromolecules, and (2) the rapid determination of molecular weights of proteins.

(b) Ion-exchange chromatography allows high loadings of chromatographic columns and separations based on ionic properties of the surfaces of protein macromolecules. Four variables can be changed: (1) type of exchanger (exchange of anions or cations); (2) pH of the mobile phase, which influences the charge on the protein surface; (3) ionic strength, which generally controls the degree of sorption of proteins to ion exchangers; and (4) the type of small ions used for the preparation of buffers, which specifically influences the desorption of particular proteins. A suitable choice of these variables permits a very sensitive approach to separation problems. Modern chromatofocusing opens up another possibility of ion-exchange fractionation of proteins, namely separation based on a linear gradient of isoelectric points.

(c) Hydrophobic interaction chromatography on hybrid (hydrophilic/hydrophobic) supports is based on a different principle to the above two approaches and covers the whole spectrum of mild separation modes.

We believe that by using a combination of these modes it will be possible
to fractionate with success even very complicated protein mixtures.

7.4 TECHNIQUES FOR THE SEPARATION OF PROTEINS

General HPLC techniques for biopolymers and biooligomers, dealt with in
Chapter 6, are valid also for the separation of proteins. In this section only
a few details specific to proteins will be discussed.

7.4.1 Determination of molecular weights of proteins

The term "molecular weight" should be replaced by "relative molecular mass"
and designated M_r, according to IUPAC recommendations. However, even the term
M_r is not correct in connection with SEC, because this method enables one to
determine the average hydrodynamic diameter of the fully hydrated macromolecule
and not directly the relative molecular mass. Because M_r is roughly proportional
to this diameter (the size of a wetted macromolecule), we can generally speak
about the determination of M_r. Also, the old term "molecular weight" is very
often used in the current literature. The principle and theory of the method
were explained in Section 3.2, so that here only a few papers commenting on
the method will be cited.

Jenik and Porter (1981) found and discussed limitations in the confidence
of molecular weight determinations by gel permeation chromatography. Two columns
were used: a Waters Assoc. I-125 protein column (30 cm x 7.8 mm I.D., nominal
exclusion limit M_r 2000-80 000) and a Varian TSK Gel 4000 SW column (30 cm x
7.5 mm I.D., nominal exclusion limits M_r 5000-1 000 000). Ionic strength and
the nature of buffers (ammonium acetate, potassium phosphate) influenced the
retention time when native proteins were analysed. Deviations from the usual
linear relationship between log M_r and k' were found. The selection of a
suitable buffer for the SEC of native proteins was mentioned in Section 7.3.1
(Cf., Fig. 7.3).

Many workers have measured the M_r of proteins with good results using
denaturing solvents. Deviations due to specific differences in tertiary structure of proteins are suppressed in this way. Blagrove and Frenkel (1977)
measured the M_r of eight proteins (from insulin B chain to oyster paramyosin)
using a glyceryl-CPG support and 8 M urea, 6 M guanidinium chloride or 0.1% SDS
solution as the solvent. Ui (1979) described the rapid determination of the M_r
of protein polypeptide chains using HPLC on TSK Gel G 3000 SW or G 4000 SW in
6 M guanidinium chloride and found good correlation, reproducibility, resolving
power and sensitivity. One simple run was completed within 50 min. Twelve

proteins with M_r from 2380 (bovine insulin A chain) to 160 000 (*E. coli* RNA polymerase β,β'-subunits) were chromatographed. Kato et al. (1980a) used 0.1% SDS solution in phosphate buffers (pH 7) of various concentrations and chromatographed proteins with M_r from 2900 (insulin) to 165 000 (thyroglobulin) on TSK Gel G 2000, 3000 and 4000 SW, and examined the range of linearity between log M_r and elution volume. A similar study with the application of 6 *M* guanidinium chloride was published by Kato et al. (1980b).

Takagi et al. (1981) examined the effect of salt concentration (sodium phosphate buffer, pH 7) on the elution properties of complexes formed between sodium dodecylsulphate and protein polypeptides. TSK Gel G 3000 SW was used. The retention time was markedly dependent on the buffer concentration. The resolution of protein polypeptides was satisfactory only at buffer concentrations between 0.05 and 0.15 *M*. This effect was discussed in detail in order to explain the phenomenon.

7.4.2 Importance of a low flow-rate

It is generally known that the velocity of the mobile phase influences substantially the quality of chromatographic separations. With proteins this effect is much more pronounced. Regnier and Gooding (1980) measured plate height versus velocity curves for a gel permeation column (250 x 10 mm I.D.) packed with SynChropak GPC 100, using 0.1 *M* phosphate buffer (pH 7.0) (see Fig. 3 in the paper cited). For a velocity v = 1.0 mm/s (corresponding to an analysis time of 8.5 min) they found that the plate height H = 0.07 mm for dipeptide glycyltyrosine (GT), 0.7 mm for chymotrypsinogen (CHYM) and 1.1 mm for bovine serum albumin (BSA). For v = 1 mm/s (analysis time 4.2 min) H = 0.11 mm for GT, 1.15 mm for CHYM and > 2.0 mm for BSA; for v = 3 mm/s (analysis time 2.8 min) H = 0.15 mm for GT and 1.65 mm for CHYM. It can be concluded that the price that must be paid for high-speed gel permeation analysis is a decrease in resolution. This decrease is substantially greater for large molecules (such as BSA) than for smaller molecules (GT) and is due to the smaller diffusion coefficient of macromolecules. Mobile phase velocities greater than 0.5-1 mm/s result in large decreases in resolution with large molecules.

Regnier (1983) discussed the measurement of molecular weight resolution, R_M, published by Pfannkoch et al. (1980) (cf., eqns. 59-61) versus mobile phase velocity. A TSK Gel 2000 SW column with 0.1 *M* phosphate buffer (pH 6.0) was used for the SEC of cytochrome *c* (CYT-*c*), CHYM, ovalbumin (OVA) and BSA with a mobile phase velocity in the range 0.1-0.6 mm/s (see Fig. 2 in the first and Fig. 5 in the second paper cited). For v = 0.1 mm/s all R_M values were in the range (in the order given below) 1.8-2.1, but at v = 0.6 mm/s the R_M values were 2.1 for CYT-c, 2.6 for CHYM, 2.9 for OVA and 3.6 for BSA. Also, R_M is proportional to the solute size, owing to lower diffusion and mass transfer. On the other hand, it is possible to conclude from the measurements that little is gained by using separation times of more than 2-4 h with HP-SEC columns of 10 µm particle size (Regnier, 1983).

The effect of eluent flow-rate on the efficiency of the RP- and IE-HPLC of proteins was studied by Jones et al. (1980). Both isocratic and gradient elution led to the same conclusions. The use of flow-rates much lower than those generally employed was found to be important for the efficient separation of large molecules. RP-C_8 packing with particle diameters d_p = 10 and 5 µm and the IE packing CM-Glycophase with d_p = 10 µm, were tested for the separation of amino acids (Phe and Asp), ribonuclease (RNase), CYT-c, BSA and collagen α (COLL). The lowest H (about 0.1 mm) was found for the separation of Phe, Asp and RNase under isocratic conditions on the C_8 column with d_p = 10 µm if 1 M pyridine-0.5 M acetic acid (pH 5.5) was used as the mobile phase at a velocity of v = 0.1 mm/s. Using 0.5 M formic acid-0.4 M pyridine containing n-propanol (20-26%), H = 0.45 mm was obtained for CYT-c on the C_8 column with d_p = 5 µm (10 nm pore size) with v = 0.25 mm/s. Under the same conditions on the C_8 column with d_p = 10 µm (50 nm pore size) H was 2.2 mm for CYT-c, 2.3 mm for BSA and 2.55 mm for COLL. The authors stated that reducing the flow-rate four-fold (from 60 to 15 ml/h) resulted in an approximately 50% reduction in the peak width at half-height.

CYT-c was eluted isocratically at various flow-rates from a CM-Glycophase ion exchanger (d_p = 10 µm, pore size 10 nm) using 0.36 M acetic acid-0.09 M pyridine (pH 4). The lowest H value (1.5 mm) was obtained at v = 0.25 mm/s. This is much slower than the flow-rate of 3 ml/min used in early experiments with HP-IEC (Chang et al., 1976). Also, in series of protein separations carried out by the present author with coarse Spheron ion exchangers of d_p = 20-40 µm, the usual flow-rate was high, 3-6 ml/cm^2/min, i.e., v = 0.5-1 mm/s, and good results were obtained from the point of view of practical protein isolation. However, all these rapid separations can be made more precise from the analytical point of view if the velocity of the mobile phase is decreased to $v <$ 0.5 mm/s.

Hancock and Harding (1984) reviewed the separation conditions for proteins.

7.4.3 Special techniques

Hancock et al. (1981b) described the use of RP-HPLC with radial compression for the analysis of peptide and protein mixtures. A C_{18} microparticulate support

was packed in a polyethylene cartridge and subjected to a radial compression
of ca. 2600 p.s.i. in order to minimize inhomogeneities in the column packing.
The effectivness of the technique was demonstrated by the efficient separation
of C-apolipoproteins from very low density human lipoproteins and by similar
examples with other lipoproteins. The mobile phase consisted of a 1% aqueous
solution of triethylammonium phosphate (pH 3.2) with acetonitrile or isopropanol
as the organic modifier. Radial compression greatly improves the efficiency of
the column, so that rapid preparative separations can be achieved. White et al.
(1981) applied HPLC to the immunological identification of compounds using the
rapid separation of high-molecular-weight complexes of antigen and antibody
from other proteins. Bovine serum albumin and anti-bovine serum albumin were
studied as a model system. Using molecular sieve HPLC, nanogram levels of the
antigen could be identified within 10 min after mixing the antigen and antiserum. The technique was also applied to the identification of a polysaccharide
and a hapten. It was found that HPLC-facilitated immunological analysis is a
versatile, sensitive, simple and rapid technique.

Kojima et al. (1982) described a new type of separation system, which combined isoelectric focusing with HPLC. A two-dimensional technique was applied
to the analysis and fractionation of serum proteins. Carrier-free isoelectric
focusing was used in the first dimension, which separated proteins according
to their electric charge (pI). A multichamber instrument was constructed for
this purpose. High-performance GPC in the second dimension was used to separate
the fractions obtained according to their molecular size. Human serum was
subjected to such a two-dimensional separation to demonstrate the application
of the technique. Williams et al. (1983) developed an HPLC application of the
Hummel and Dryer method for the determination of colchicine-tubulin binding
parameters. The method was used to determine the dissociation constant for
colchicine-tubulin interaction at 25°C. The technique, which does not require
radioisotopes and employs modern SEC columns, is rapid and sensitive. The analysis requires only 15 min.

7.5 DETECTION AND QUANTIFICATION OF SEPARATED PROTEINS

7.5.1 General methods

One of the oldest but a very useful method used in biochemistry for the
detection and quantification of non-coloured proteins is the measurement with
the Folin-Ciocalteu phenol reagent after alkaline copper treatment according
to Lowry et al. (1951). The colour intensity depends on the tyrosine content,
as in the original Folin reaction. The Lowry method can be used both for pro-

teins in solution and also for proteins that have been precipitated with acid. As little as 0.2 µg of protein can be measured using a micro-scale procedure. The method is empirical and for quantitative determinations it is necessary to apply a calibration with a protein, the concentration of which is determined by some other method. The method is as sensitive as Nessler's reagent (yet requires no digestion), much more specific than the measurement of UV absorbance at 280 nm, much less liable to disturbance by turbidity, several times more sensitive than the ninhydrin reaction (free amino acids give much more colour than proteins with ninhydrin, whereas the opposite is true with Folin reagent) and 100 times more sensitive than the biuret reaction. Because the amount of colour varies with different proteins, this method is less constant than biuret reaction and also the colour is not strictly proportional to concentration. In spite of these disadvantages, the method has been used in biochemistry for many years. The experimental procedure consists of two steps: (a) reaction with copper in alkali and (b) reduction of the phosphomolybdic-phosphotungstic reagent by the copper-treated protein. The colour formed has an absorption maximum near 750 nm and it is desirable to make readings at this wavelength for the range 5-25 µg of protein per millilitre of the final volume; for stronger solutions reading near 500 nm is recommended. Bennett (1967) modified the procedure in order to eliminate problems with some interferring substances.

Since 1984, Pierce Chemical Co. [Pierce, Rockford, IL., U.S.A.; Rodgau (Weiskirchen), F.R.G.] has offered the Lowry alternative "BCA Protein Assay Reagent", which has been developed to a high degree of perfection and is not expensive (U.S. $ 0.05 per assay). It consists of two reagents: (A) sodium carbonate, sodium hydrogen carbonate, BCA detection reagent and sodium tartrate in 0.1 M NaOH and (B) 4% $CuSO_4 \cdot 5H_2O$. In a standard procedure 2 ml of BCA working reagent (a 50:1 mixture of BCA assay reagents A + B) are added to 100 µl of sample containing 1-120 µg of protein. The mixture is incubated at $37^{o}C$ for 30 min. An intense purple colour is formed conditioned by copper(I). The absorbance at 562 nm is read against a blank. The BCA protein assay reagent is compatible with non-ionic detergents and many other substances that interfere with Folin reagent, and the working reagent is stable at least for 1 week at laboratory temperature (in contrast to Folin-Lowry reagent, which should be prepared fresh before use). This detection method has a wide linear working range of 1-2000 µg/ml and there is negligible protein-to-protein variation.

The biuret assay, which is also sometimes used to measure the protein content in solutions, is based on the reaction of Cu^{2+} with an NHCO (peptide) group. A coloured complex is formed with an absorption maximum at about 555 nm. This method is not very sensitive and is suitable for the determination of higher

concentrations of proteins (0.25-25 mg/ml). The procedure was described by,
e.g., Prusík (1976) and can be also carried out in a continuous flow-through
system.

Another rapid and sensitive method for the quantitation of microgram amounts
of protein utilizing the principle of protein-dye binding was published by
Bradford (1976). The Bradford method involves the binding of Coomassie Brilliant
Blue G-250 to protein, which causes a shift in the absorption maximum of the
dye from 465 to 595 nm, and it is the increase in absorption at 595 nm that
is monitored. The assay is very reproducible and the dye-binding process is
completed in 2 min with good colour stability for 1 h. Cations and carbohydrates
do not interfere, but large amounts of detergents, including sodium dodecyl
sulphate (often used in protein chemistry), do interfere. The influence of small
amounts of detergent may be eliminated by the use of proper control. The assay
procedure is very simple: Coomassie Brilliant Blue is dissolved in ethanol,
mixed with phosphoric acid and diluted, so that the final protein-reagent stock
solution contains 0.01% (w/v) of dye, 4.7% (w/v) of ethanol and 8.5% (w/v) of
phosphoric acid. Protein solution containing 10-100 µg of protein in a volume
of up to 0.1 ml is pipetted into 100 x 12 mm test-tubes. The volume in the
test-tube is adjusted to 0.1 ml with appropriate buffer. A 5-ml volume of protein reagent is added to the test-tube and the contents are well mixed. The
absorbance at 595 nm is measured after 2 min and before 1 h against a reagent
blank, prepared from 0.1 ml of the appropriate buffer and 5 ml of protein
reagent. The weight of protein is plotted against the corresponding absorbance,
resulting in a calibration graph that is used to determine the protein in
unknown samples. A microprotein assay based on the same principle was described.
There is a slight non-linearity in the response pattern, owing to spectral
overlap of two different colour forms of the dye. However, this presents no
real problem, as the degree of curvature is only slight. This dye-binding assay
is approximately four times more sensitive than the Lowry assay. Strongly
alkaline buffers interfere.

Barford et al. (1979) used the Bradford dye-binding test in a slightly
modified form for the evaluation of chromatographic peaks of proteins in their
studies of protein interactions with bonded-phase ion exchangers. If alcohols
or acetone were present in the column effluent, false positive tests were
obtained. Organic solvents can easily be removed by evaporation with nitrogen
prior to addition of Bradford reagent.

Colorimetric assays for proteins are used if some other foreign substances
that absorb in the UV region and interfere are present in the sample. If this
is not the case and if also the mobile phase does not contain strongly absorbing

components (such as volatile pyridine buffers), the measurement of UV absorption is the most often used method for the detection and quantification of proteins. The main advantages of this method are (a) the simple on-line arrangement and the possibility of using a flow-through UV absorbance cell connected with a inexpensive and generally available spectrophotometer, (b) the sensitivity and (c) the fact that UV monitoring obeys the Lambert-Beer law (cf., Section 5.1.8), which was not the case with some of the above-mentioned colorimetric assays. Proteins are most often measured at a wavelength of 275-280 nm, where most proteins have absorption maxima. Above 280 nm the absorbance is distinctly influenced by pH owing to the ionization of the phenolic group of tyrosine in strongly alkaline media. This measurement is based on UV absorption of aromatic amino acids, present in proteins, mainly of tryptophan (in 0.1 M HCl λ_{max} = 278 nm, ε = 5450; in 0.1 M NaOH λ_{max} = 280.5 nm, ε = 5250) and tyrosine (at pH 1.09 λ_{max} = 277.7 nm, ε = 1500; at pH 8.0 λ_{max} = 275, ε = 1500; at pH 12 λ_{max} = 293, ε = 2600). In some work, monitoring of the absorbance at 254 nm was used, because a cheaper UV spectrophotometer could be used (cf., Section 5.1.8). At this wavelength, however, nucleic acids and their fragments absorb strongly. Thacker et al. (1970) recorded simultaneously the absorbances at 253.7 and 280 nm and were able to determine the purity of protein eluates from the point of view of contamination with nucleic acid components (which often appear in biological materials) on the basis of the absorbance ratio $A_{280}/A_{253.7}$. Schlabach and Abott (1980) monitored the HPLC separation of serum proteins by detection at multiple wavelengths (254, 280 and 305 nm). Dual wavelegth profiles greatly aided the identification of several peaks. There are distinct differences in the absorbance ratios of some serum proteins, e.g., A_{280}/A_{254} for immunoglobulin G (IgG) is 2.47 and for serum albumin 1.74; the A_{305}/A_{280} ratio for the two proteins is 0.17 and 0.11, respectively.

Towards even shorter wavelengths, a minimum appears in the UV spectrum of proteins, followed by a steep increase in absorbance. If an eluent with a low absorbance is used, the sensitivity of detection may be increased several-fold when measurements are made below 235 nm in comparison with measurements at 280 nm. At 210-220 nm very sensitive detection of proteins is possible, but buffer components containing carbonyl groups (such as acetates) are unsuitable for use in this region. With even shorter wavelengths the sensitivity is increased more (192-194 nm is the specific region for the peptide bond), but the mobile phase composition has a considerable effect and only aqueous solutions of alkali metal fluorides can be used. The increase in sensitivity can be seen from a comparison of the molar absorptivities of various proteins, e.g., with human serum albumin $E_{1\,cm}^{1\%}$ = 6 for λ = 280 nm and 210 for λ = 210 nm.

TABLE 7.1

CHARACTERISTIC ABSORPTION MAXIMA OF SOME METALLOPROTEINS IN THE VISIBLE REGION
OF THE SPECTRUM (ACCORDING TO PRUSÍK, 1975)

Metalloprotein	λ_{max} (nm)
Caeruloplasmin	605-610
Cytochromes	550-560
Erythrocuprein	655
Haemocyanins	563-580
Haemoglobin (human)	415
Haemoproteins	412
Haemovanadin	425
Plastocyanin	597

Some proteins contain various molecular moieties with characteristic absorption bands in the longwave UV region or in the visible region, which can be
utilized for detection. The important condition is that such groups should
remain firmly bound to apoprotein in the course of the separation process.
Typical examples are nucleoproteins. In the visible region some types of metalloproteins may be detected by absorption spectrophotometry (Table 7.1). In such
instances it must be borne in mind that molar absorptivities and λ_{max} are
dependent on the oxidation state of the metal. The absorbance ratio $A_{visible}/A_{280}$
can be used as a criterion for the purity of metalloproteins.

A very sensitive method for the chromatographic separation of proteins is
the measurement of fluorescence (cf., Fig. 5.18 and Section 6.6.1). The UV
radiation at 340-350 nm after irradiation of the protein molecule by shorter
wavelengths (about 280-290 nm) is measured. The fluorescence is due to the
excitation of tryptophan (greater part) or tyrosine; the fluorescence of phenylalanine is less important. The sensitivity of detection of proteins by fluorescence is approximately 1000 times higher than that of absorbance at 280 nm.
Another method for fluorescence monitoring of proteins is to label the protein
with fluorescent reagents, such as fluorescein isothiocyanate, 5-dimethylamino-
naphthalene-1-sulphonyl chloride (dansyl chloride) or fluorescamine (cf., Section 6.6.2). Schultz and Wassarman (1977) published a (non-chromatographic)
application of [3H]dansyl chloride as a useful reagent for the quantitation and
molecular weight determination of nanogram amounts of protein under denaturing
conditions. In this method the labelling reactivity of dansyl chloride (carrying
3H) with primary amino groups of proteins is the main principle, and the detection is realized not by measurement of the fluorescence of the dansyl group,
but by the usual liquid scintillation counting of 3H.

Takaki (1980) described the application of a low-angle laser light-scattering photometer as a detector for the measurement of the molecular weight of proteins (RNase, serum albumin, amylase, chymotrypsinogen) by silica gel high-performance chromatography.

7.5.2 Special methods

In this section, specialized detection methods for narrow classes of proteins will be covered.

Schlabach et al. (1977) described a method suitable for the on-line post-column detection of sulphydryl compounds including proteins. The method is based on the reaction of a sulphydryl compound with dithionitrobenzoic acid (DTNB) in alkaline medium:

$$RSH \quad + \quad DTNB \xrightarrow{pH\ 8.0} SR \quad + \quad NTB\ (\lambda_{max}\ 412\ nm) \quad + \quad H^+$$

The nitrothiobenzoate (NTB) formed has an absorption maximum at 412 nm, and at this wavelength it can be measured. The linearity of the detector response with the amount of protein applied was illustrated for the example of yeast alcohol dehydrogenase monitored in a packed column reactor.

Lu et al. (1979) developed a rapid method for the determination of bilirubin in neonatal serum, based on the finding that a gel permeation column (SynChropak GPC 100) binds free bilirubin while allowing the passage and quantitation of protein-bound bilirubin. Subsequent injection of a desorbing agent (0.1 mmol/l bovine serum albumin) releases the adsorbed bilirubin from the column, permitting the quantitation of free bilirubin. The absorbance of the column effluent is measured at 453 nm. Bound and free serum bilirubin may be determined directly in less than 15 min using 10 µl of serum.

Suzuki (1980) described the direct connection of HPLC, using a gel permeation column of TSK Gel 3000 SW, with flame atomic-absorption spectrometry for the

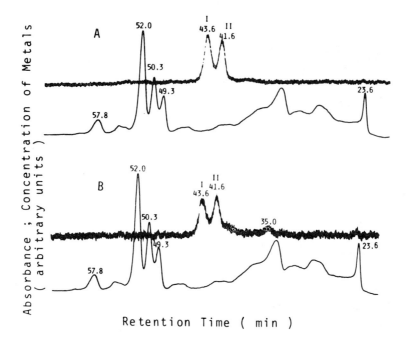

Fig. 7.7. Gel permeation-atomic absorption chromatograms of rat liver supernatant obtained by injection of cadmium chloride. Liver supernatant (0.2 ml) was applied to a TSK Gel 3000 SW column (600 x 21.5 mm I.D.) and eluted with 50 mM Tris-HCl buffer solution (pH 8.6 at 25°C). The absorbance at 280 nm (lower curves) and the concentration of cadmium (A) or zinc (B) (upper curves) were continuously monitored. I and II indicate metallothionein-I and -II, respectively. Numerals indicate retention times of peaks. (Reprinted from Suzuki, 1980.)

determination of metalloproteins (Fig. 7.7). Metallothioneins of the same molecular mass were separated into two peaks, indicating that the column has both gel chromatographic and ion-exchange properties. The method can also be developed for the detection of other types of metalloproteins.

A highly sensitive method for the determination of blood haemoglobin by high-performance gel permeation chromatography combined with atomic absorption spectrometry was published by Istrii (1981). The supernatant of haemolysed blood passed through a 600 x 7.5 mm I.D. column of TSK Gel 3000 SW X2-TSK Gel 2000 SW X2 and was eluted with 0.1 M phosphate buffer (pH 7.3)-0.1 M KCl followed by analysis of an aliquot of the haemoglobin fraction for Fe by atomic-absorption spectrometry. A linear calibration graph over the range 20-500 µg/ml of haemoglobin was obtained and the method showed good accuracy (average relative standard deviation 2.1%). The combination of HPLC and atomic-absorption spectrometry for the detection of metalloproteins was reviewed by Suzuki (1981).

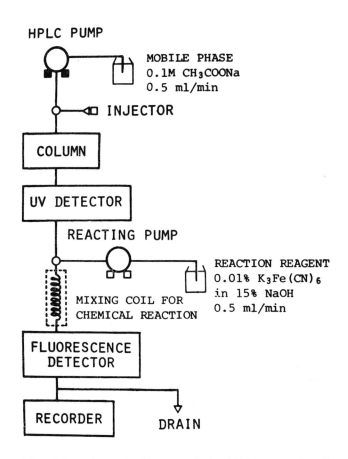

Fig. 7.8. Schematic diagram of the HPLC system for the detection and determination of thiamine-binding proteins. (Reprinted from Kimura and Itokawa, 1981.)

Miura et al. (1981) described a method for the detection of drug-binding proteins in chromatographic column effluents using a parallel-flow dialysis technique. The method was described in detail in the Section 5.1.8 (see also Fig. 5.26). Kimura and Itokawa (1981) described the separation and determination of thiamine-binding proteins in rats by HPLC (Fig. 7.8). A TSK Gel 3000 SW column was used for the SEC of proteins and UV detection (A_{280}) for the primary evaluation of the effluent. A fluorescence-forming reaction reagent was added with a proportioning pump in an on-line process and thiamine-binding proteins were converted into fluorophores in a mixing coil. The fluorescence was measured in a 12-μl flow cell with a spectrofluorimeter (excitation 375 nm, emmision maximum 450 nm). This method opened the way to specific determinations and studies of biologically important thiamine-binding proteins in various tissues,

B38

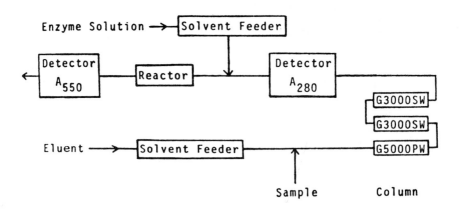

Fig. 7.9. Schematic flow diagram for the enzymic quantification of cholesterol
in human serum lipoproteins by HPLC. TSK Gels 5000 PW and 3000 SW were used as
GPC packings. (Reprinted from Okazaki et al., 1981a.)

as the method is sensitive and rapid. Drug protein binding determination by
chromatography was reviewed by Sebille and Thuaud (1984).

Okazaki et al. (1981a) developed a simple and rapid method for the quantitation of cholesterol in human serum lipoproteins (VLDL, LDL, HDL_2 and HDL_3) by
HPLC. The content of cholesterol in each lipoprotein fraction was determined
by means of a commercial enzymic reaction kit (Determiner TC 555; Kyowa Hako,
Tokyo, Japan) after separation by HPLC with a GPC column. A 10-20 μl volume
of serum was sufficient for the quantitation in less then 50 min, after passing
the mixed eluate and enzyme solution through the on-line reactor system of a
high-speed chemical derivatization liquid chromatograph. The principle of the
method is illustrated in Fig. 7.9. The first detector (A_{280}) measures the total
protein profile of the sample and the content of cholesterol in each lipoprotein
fraction can be calculated from the A_{550} peak area (Fig. 7.10). The detailed
biochemical scheme of enzymic reactions in this method catalysed by the Determiner TC 555 kit, was explained by Okazaki et al. (1981b) as follows:

$$\text{cholesterol ester} + H_2O \xrightarrow{\text{cholesterol esterase}} \text{free cholesterol} + \text{fatty acid}$$

$$\text{free cholesterol} \xrightarrow{\text{cholesterol oxidase}} \text{cholest-4-en-3-one} + H_2O_2$$

$$\left.\begin{array}{l}\text{4-aminoantipyrine} + \\ \text{N-ethyl-N-(3-methylphenyl)-N'-} \\ \text{acetylethylenediamine (EMAE)} + \\ H_2O_2\end{array}\right\} \xrightarrow{\text{peroxidase}} \text{quinone diimine dye} + 4H_2O$$
$$(\lambda_{max}\ 555\ nm)$$

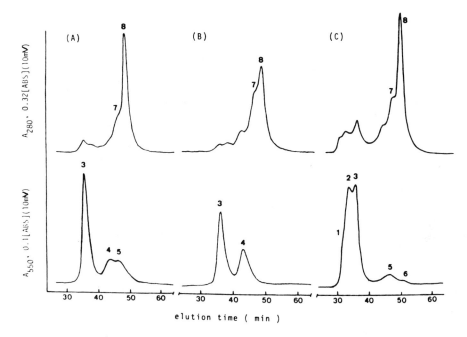

Fig. 7.10. Analyses of protein and cholesterol in human sera. Elution curves were monitored in terms of A_{280} and A_{550} by HPLC according to Fig. 7.9. Samples: (A) normal female subject; (B) liver cirrhosis subject; (C) hyperlipidaemic type subject. Loaded volume, 20 μl of whole serum. Peaks: 1 = VLDL; 2 = IDL; 3 = LDL; 4 = HDL$_2$; 5 = HDL$_3$; 6 = VHDL; 7 = γ-globulin; 8 = albumin. (Reprinted from Okazaki et al., 1981b.)

Optimum conditions for the cholesterol measurement were: temperature of the reactor, $40^{\circ}C$; dimensions of the reactor, 20 m x 0.25 mm I.D.; flow-rate of the main stream, 1.0 ml/min; and flow-rate of the enzyme solution [80 ml of 0.025 M potassium hydrogen phthalate buffer (pH 6) containing detergent, + lyophilized enzyme content of one vial], 0.35 ml/min.

Hara et al. (1982) developed a method for the selective detection of tri-glycerides in human serum lipoproteins (chylomicron, VLDL, LDL and HDL) after HPLC, based on similar enzymic reactions. The instrumentation principle was GPC using TSK gels, combined with reaction detection, catalysed by Determiner TG reaction kit (Kyowa Medex, Tokyo, Japan), containing lyophilized enzymes in vials and buffer solution containing detergent in a special bottle. Absorption and emission at 280 and 550 nm, respectively, were measured. The biochemical principles of the detection are as follows:

$$\text{triglycerides} + H_2O \xrightarrow{\text{lipoprotein lipase}} \text{free glycerol} + \text{fatty acid}$$

$$\text{free glycerol} + O_2 \xrightarrow{\text{glycerol oxidase}} \text{glyceraldehyde} + H_2O_2$$

$$\text{4-aminoantipyrine} + \text{EMAE} + H_2O_2 \xrightarrow{\text{peroxidase}} \text{quinone diimine dye} + H_2O$$

The establishment of selective detection methods for triglycerides in serum lipoproteins by HPLC should encourage rapid progress in the study of lipoprotein metabolism and the diagnosis of various diseases (such as hyperlipidaemia, acute liver hepatitis, liver cirrhosis and primary biliary cirrhosis).

Hagiwara et al. (1982) described a reagent specific for choline-containing phospholipids for the HPLC quantitation of serum proteins. TSK gels were used for the SEC separation of HD, LD and VLD lipoproteins from sera of healthy subjects and hyperlipidaemic patients. Cholesterol and choline-containing phospholipids were determined by a combination of enzymic methods. The influence of the concentration of phenol, 4-aminoantipyrine, Triton X-100 (necessary for enzymic action of phospholipase), NaCl and Tris-HCl buffer was examined. This procedure is suitable as a detection method for HPLC analysis.

7.6 APPLICATIONS

Examples of HPLC applications from various fields of protein chemistry and biochemistry will be reviewed in this section.

7.6.1 Structural proteins

One of the oldest examples of a practical application of the rapid chromatography of proteins was the paper by Persiani et al. (1976), describing the aqueous GPC of water-soluble polymers by HPLC using glyceryl CPG columns. Industrial protein glues were analysed and it was found that this method can be used to detect microbial infection of collagen, because the decrease in molecular weight is a function of bacterial degradation. Barford et al. (1977) monitored the biodegradation of keratin in activated sludge (tannery wastes) by rapid GPC using Glycophase G/CPG (glycerolpropylsilyl glass). These pioneer works opened up the way for the analysis of the biodegradation of other proteins in waste waters by HPLC in the SEC mode.

Macek et al. (1981) and Deyl et al. (1982) used Separon HEMA 1000 Glc (Spheron covered with a hydrophilic layer of glucose) for the HPLC of collagen types, chains and fragments. The separation of these fibrous proteins was reviewed by Deyl and Macek (1984). Collagen polypeptide chains [α_1 (types I, III and IV),

α_2 (I, II and IV), β (I) and γ (I) and multiple or combined chains] and some
other fragments of collagen were chromatographed on Separon HEMA Glc using the
SEC mode. The log (molecular mass) versus retention time plot was non-linear.
Inter-species differences could be found between chains of the same molecular
mass with mobile phases of low ionic strength, which opened up the possibility
of unusual separations. When the concentration of NaCl in the elution buffer
reached 0.2 M, these differences disappeared. It was evident that mixed modes
of HPLC influenced the separation of these fibrillar proteins on Separon HEMA
Glc using SEC conditions.

In contrast, Fallon et al. (1981) described the separation of the major
species of intestinal collagen by RP-HPLC using bonded cyanopropyl (CN) support
columns, pyridine buffers and stepwise propanol elution. Human collagen types
I, II and III and collagen types I and II from lathyric chick cartilage extract
were separated.

7.6.2 Milk proteins

Kearney and McGann (1978) applied controlled-pore glass chromatography to
the separation of reconstituted low-heat skim milk powder casein micelles. The
support was treated in advance with Carbowax 20M. Micelles were separated
according to size and analysed for their content of α_s-, β- and κ-caseins.
Diosady et al. (1980) applied HPLC to the rapid separation of whey proteins
using SEC and RPC modes with SynChropak GPC-100 and RP-8 column packings,
respectively. The methods were shown to be reproducible and allowed quantitative
analysis. Hartman and Persson (1980) described the GPC of proteins in cheese
using an LKB Ultropac TSK Gel 3000 SW column; thirteen different components
could easily be detected. Dimenna and Segall (1981) applied HPLC-GPC to the
separation of skim milk proteins. TSK Gel (Type SW) was sued and skim milk was
injected directly. The identified peaks were casein, IgG, bovine serum albumin,
β-lactoglobulin, α-lactalbumin and α_{s1}- and β-casein. HPLC of whey proteins was
reviewed by Humphrey (1984).

7.6.3 Blood proteins

7.6.3.1 Haemoglobins, globins

Haemoglobin variants have been intensively studied using HPLC methods. For
example, Congote (1981) described the RP-HPLC of globin chains and its application to the prenatal diagnosis of β-thalassaemia. The diagnosis is based on
the globin chain β/γ synthetic ratio. The cells from foetoscopy samples were
incubated 2 h with [³H]leucine and the globin chains from the cell lysates were

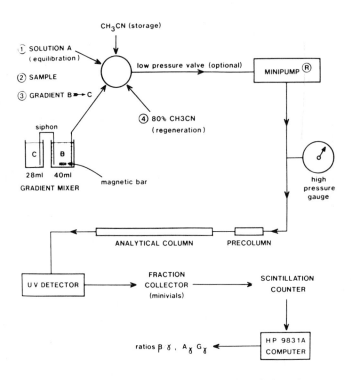

CH$_3$CN (storage)

① SOLUTION A
(equilibration)

② SAMPLE

③ GRADIENT B ➤ C

low pressure valve (optional)

MINIPUMP ®

④ 80% CH3CN
(regeneration)

siphon

C   B

28ml   40ml   magnetic bar

GRADIENT MIXER

high
pressure
gauge

ANALYTICAL COLUMN   PRECOLUMN

U V DETECTOR

FRACTION
COLLECTOR
(minivials)

SCINTILLATION
COUNTER

ratios β·δ, A$_\gamma$ G$_\gamma$

HP 9831A
COMPUTER

Fig. 7.11. Diagram of HPLC system utilized for the separation and quantitation of labelled human globin chains by RPC for the purposes of the prenatal diagnosis of β-thalassemia. (Reprinted from Congote, 1981.)

precipitated with acidified acetone. Very good chromatographic separation of β- from both α- and γ-chains was required. Columns of octadecylsilica were used with solvent mixtures of water and acetonitrile, acidified with phosphoric acid. The equipment is illustrated in Fig. 7.11. The addition of the hydrophobic ion-pairing reagent trifluoroacetic acid (TFA) or the chaotropic agent sodium perchlorate to the mobile phase increased the resolution. Decreasing gradients of TFA or NaClO$_4$ at 40°C allowed a good separation of α- and β-chains, a prerequisite for the successful prenatal diagnosis of β-thalassaemia. Huisman et al. (1981) modified the RPC method of Shelton et al. (1979) in order to achieve the complete separation and quantitation of the $^A_\gamma$T, $^A_\gamma$I and $^G_\gamma$-chains in human foetal haemoglobin (Hb). This rapid method requires 5-2000 μg of Hb F. The purity of Hb F is not essential (admixtures of up to 70% of adult Hb do not interfere). A Waters Assoc. μBondapak C$_{18}$ column, equilibrated with acetonitrile-methanol-phosphate solution (pH 2.84), was used (Fig. 7.12). The method has been applied to the Hb F of 64 Black SS (sickle cell anaemia) patients and seven other subjects with the Hb S-HPFH ($^G_\gamma$ $^A_\gamma$ type) condition. Another modification

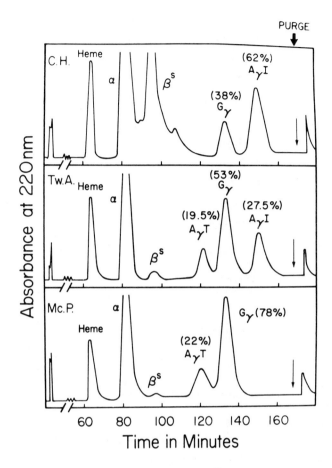

Fig. 7.12. Separation of the α, β^S and the three γ-chains ($^A_\gamma T$, $^G_\gamma$ and $^A_\gamma I$) by the modified HPLC procedure. All three subjects were suffering from sickle cell anaemia. (Reprinted from Huisman et al., 1981.)

of RPC was used by Shelton et al. (1982) for the detection of HB-Papio B, a silent mutation of the baboon β-chain. Solvents containing $NaClO_4$ solution + methanol + acetonitrile + 85% H_3PO_4 + nonylamine were used for the elution of differing globin chains from a Waters Assoc. μBondapak C_{18} column. No similar separation could be achieved using IEC.

Other groups of workers applied HP-IEC to the rapid separation of Hbs. For example, Hanash and Shapiro (1981) used SynChropak anion exchanger with UV monitoring at 410 nm for the separation of Hbs in haemolysates derived from newborns and from adult individuals with Hb disorders. The use of this method for the diagnosis of Hb disorders was discussed. Gardiner et al. (1982) used a SynChropak AX 300 anion-exchange column (SynChrom, Linden, U.S.A.) together

with a SynChrosorb AX pre-column and a SynChrosorb ASC guard column for the
rapid separation of Hbs A, S, C and F in newborn babies with different haemoglobinopathies. About 800-1000 µg of Hb in a red cell haemolysate were injected
on to the column. Various quantitation procedures were described and discussed.
This method was applied to blood samples from more than 300 newborn babies and
the results were discussed. In contrast, Toren et al. (1983) described the
cation-exchange HPLC determination of Hb A_{1c} using SynChropak CM-300 cation
exchanger. Ionic strength gradient elution was used; both weak and strong buffers
were phosphate solutions. The absorbance was measured with a Waters Assoc.
Model 440 dual-channel detector equipped with 405-nm filters and phosphors. The
response from either detector was treated with the analog-to-digital converter
of a DEC LAB 11/VO3 computer (Digital Equipment, Maynard, MA, U.S.A.) and the
data were reduced and plotted with the DEC GT-46 system described previously.
About five measurements were obtained per hour using small amounts of whole
blood. No centrifugation or washing of cells was required for sample preparation. The method was insensitive to small changes in pH and temperature. Different chromatographic profiles arising from haemoglobinopathies illustrated
the importance of this method for diagnostic screening.

The use of HPLC in the identification of human haemoglobin variants has been
reviewed, e.g., by Schroeder et al. (1981), including instrumentation and packings.

7.6.3.2 Lipoproteins and apolipoproteins

Schwandt et al. (1981) separated human C-apolipoproteins by R-HPLC in order
to prepare highly purified antigens, necessary for the preparation of specific
antisera for immunological quantification of apolipoproteins. Reversed-phase
C_{18} material (particle size 10 µm) was used. Solvent A was methanol and buffer
B was 0.01 M phosphate (pH 6). In 120 min a linear gradient from 40 to 28%
buffer separated nineteen C-apolipoprotein peaks of one sample peak prepared
in advance by GPC of 50 mg of tetramethylurea-soluble VLDL lipoproteins on a
Sephadex G-200 Fine column. The authors found this RPC method to be superior
to preparative isoelectric focusing of C-apolipoproteins owing to the close
isoelectric points, the low yield obtained with the latter method and the
difficulty in removing the ampholyte. Hancock et al. (1981a) separated apoprotein components of human VLD lipoproteins by IP-RP-HPLC. A Waters Assoc.
µBondapak-alkylphenyl column (10-µm particles) was used, the mobile phase consisting of a 1% solution of the polar ion-pairing reagent trimethylammonium
phosphate. A slow non-linear gradient of acetonitrile (37-42%) eluted the
apolipoproteins in order of known polarity: C_x, C-I, C-III$_2$, C-III$_1$, C-III$_0$ and
the most non-polar C-II. The recovery was 80-95%. In another study Hancock et

al. (1981c) separated apolipoproteins A-I and A-II (from human HD lipoproteins) by IP-RP-HPLC using, in addition to the above-mentioned column, a Zorbax C_8 column or a Radial-Pak C_{18} cartridge. The same ion-pairing reagent was used with linear gradients of acetonitrile.

Ohno et al. (1981) fractionated human serum lipoproteins using SEC on TSK Gel 5000 PW and 3000 SW columns. VLDL/chylomicron, LDL, HDL_2 and HDL_3 peaks were identified. Linear quantitative relationships were obtained when peak areas monitored at 280 nm were compared with the results of Lowry's method of protein determination (cf., Section 7.5.1). HPLC patterns of total lipoprotein fractions from a normal subject and a pathological case (hyperlipidaemia) were compared after monitoring at 280 nm and detection of cholesterol, phosphorus and tri-glyceride (cf., Section 7.5.2). The diagnostic significance of these findings was briefly discussed. Okazaki et al. (1984) also used rapid SEC for the separa-tion of apolipoproteins in serum high-density lipoproteins. TSK Gel 3000 SW column was used with sodium phosphate buffer (pH 7.0) containing 0.1% of sodium dodecylsulphate as eluent. Elution patterns monitored by measuring the absor-bance at 280 nm could give precise qualitative and quantitative information about apolipoproteins of MW between 10^4 and 10^5. HPLC patterns of HDL apolipo-proteins were compared between healthy humans and dogs and persons with various diseases (lecithin:cholesterol acyltransferase deficiency; acute hepatitis; hyper-α-lipoproteinaemia; liver cirrhosis). Identification or an attempt at the identification of individual peaks of the elution pattern was discussed.

Lipoprotein separation by high-performance GPC was reviewed by Okazaki and Hara (1984), and Edelstein and Scanu (1984) reviewed the application of molec-ular sieve HPLC to the separation of human plasma apolipoproteins.

7.6.3.3 Plasma and serum proteins

SEC was used by Tomono et al. (1979) for the efficient separation of fibrinogen, albumin, ovalbumin, cytochrome c and myoglobin from a model synthetic mixture using a TSK Gel 3000 SW column. In subsequent experiments α-, β- and γ-globulins were separated from albumin and aggregates. Phosphate or acetate buffers were used with the addition of 0.1 or 0.2 M Na_2SO_4. Later, Tomono et al. (1983) separated plasma proteins using IEC on Pharmacia Mono Q and Polyanion S anion exchangers and FPLC equipment. Pooled human plasma was separated into ten or more fractions within 10 min. From a 13-min experiment fifteen fractions were analysed for the presence of fourteen plasma proteins (albumin, transferrin, fibrinogen, ceruloplasmin, haptoglobin, α_1-acid glycoprotein, α_1-antitrypsin, α_1-lipoprotein, α_2-HS-glycoprotein, α_2-macroglobulin, β_1-lipoprotein, C_3-com-ponent, prealbumin and immunoglobulin IgG). In some instances the IEC was fol-lowed with SEC using a TSK Gel 3000 SW column. The Mono Q column was applied to the plasma analysis of a normal person and of subjects suffering from IgG-myeloma, lung cancer and, for comparison, lung cancer after membrane plasma-pheresis therapy.

Margolis and Rhoades (1981) described the preparation of high-purity factor VIII by controlled-pore glass chromatography. Co-Sarno et al. (1983) studied in detail the determination of polymers and the purification of albumin using SEC on TSK-250 (Bio-Rad Labs., Richmond, CA, U.S.A.) analytical column. Strahler et al. (1983) developed the separation of transferrin types in human plasma by anion-exchange HPLC with SynChropak AX 350 (SynChrom, Linden) as the packing material. Four molecular forms of transferrin (differing with respect to bound iron) were separated from each other and from other plasma proteins. Transferrin variants including B and D types could also be identified. Crowley and Walters (1983) studied the determination of immunoglobulins in blood serum by HP affinity chromatography. Protein A (from *Staphylococcus aureus*) was immobilized on 10-μm LiChrospher Si 4000 diol-bonded silica. Immunoglobulin-containing samples were injected into the column at pH 7 and eluted by stepwise changes to pH 3. In 4-min chromatographic experiments only immunoglobulin A and M were retained, and no immunoglobulin IgG. The protein A column had a long lifetime, being used more than 200 times over the experimental period of four months. This technique may be very useful as a rapid screening method for the immunoglobulins in blood serum.

7.6.4 Hormonal proteins and receptors

Dinner and Lorenz (1979) studied the determination of bovine insulin in the presence of the by-products most commonly encountered during its purification. Eight by-products were considered. A LiChrosorb RP-8 (10 μm) column and isocratic elution with acetonitrile-0.2 M ammonium sulphate solution (pH 3.5) were used for RP-HPLC separation. Bovine insulin, porcine insulin and porcine desamido insulin were well separated. The elution times of seven insulin-like proteins were tabulated. Stanton et al. (1983) described analytical and semi-preparative separations of several pituitary proteins by IE-HPLC. Gel HPLC studies on the elution behaviour of chemically deglycosylated human chorionic gonadotropin and its subunits were described by Shimohigashi et al. (1983).

Optimization of a melanotropin-receptor binding assay by RP-HPLC was studied by Lambert and Lerner (1983). Hutchens et al. (1983) described the rapid analysis of estrogen receptor heterogeneity by chromatofocusing with HPLC. SynChropak AX-300 and AX-500 columns were used for analytical and preparative focusing and receptor proteins were eluted from the anion exchanger with a mixture of Polybuffers, diluted with glycerol solution. Ten different ^{125}I-labelled binding proteins were identified in cytosols from the mammary gland and uterus. In contrast to isoelectric focusing, this technique is compatible with the inclusion of a commonly used receptor stabilizing agent, sodium molybdate, and its application allowed the identification of two acidic receptor species not previously reported. Lonsdorfer et al. (1983) used HPLC in the evaluation of the synthesis and binding of fluorescein-linked steroids to estrogen receptors. Preparative C_{18} RP-HPLC columns were applied to the purification of the final synthetic product, which was used in inhibition studies with other steroids. High- and low-molecular-weight forms of estrogen receptors from cytosols of human breast carcinomas were separated on TSK Gel 3000 SW and 4000 SW columns by SEC.

7.6.5 Some low-molecular-weight proteins

Macklin et al. (1981) purified [^{3}H]phenylalanine-labelled microsomal proteolipids from *Neurospora crassa* (hydrophobic proteins of MW ≈ 6500 daltons) by HPLC on silica gel. A LiChrosorb (5 μm) column was equilibrated with benzene-ethanol (95:5) and, after injection of the crude proteolipid extract, a 20-min gradient of 5-66% ethanol was applied. The fractions obtained were analysed for protein, radioactivity and lipid content. A 174-fold purification of chloramphenicol-sensitive proteolipid was obtained. Klee et al. (1981) described the determination of Ca^{2+}-binding proteins (calmodulin, troponin C, parvalbumin and calcineurin B) by RP-HPLC using a μBondapak phenyl column and elution with acetonitrile-buffer mixtures. Recoveries of 80-90% were obtained. Oldenwurtel et al. (1983) extended the above experiments to a two-step procedure consisting in the application of affinity chromatography on phenothiazine-Affigel-10 followed by RP-HPLC. This method was used also for the fractionation of calmodulin tryptic fragments, which were tested for their ability to bind to and activate cyclic nucleotide phosphodiesterase. RPC of calcium binding proteins and their fragments was reviewed by Manalan and Klee (1984).

Gurley et al. (1983c) fractionated histones by RP-HPLC, using a μBondapak C_{18} column and a linear elution gradient running from water to acetonitrile with the addition of trifluoroacetic acid. The recovery was greater than 90%. Gurley et al. (1983a) extended the method for the HPLC of chromatin histones to the application of cyanopropylsilane (CN) bonded phase and to Radial-Pak cartridges. The order of elution of histones was H1, H2B, (LHP)H2A, (MHP)H2A, H4, (LHP)H3 and (MHP)H3 (where LHP and MHP refer to less hydrophobic and more hydrophobic histone variants, respectively).

Rubinstein and co-workers (1978, 1979) developed the high-performance liquid partition chromatography of human leukocyte interferon. LiChrosorb RP-8 was used for RP partition chromatography and LiChrosorb Diol for NP partition chromatography. Production, purification to homogeneity with fluorescence detection (fluorescamine) and clinical characterization were described. Herring and Enns (1983) studies the rapid purification of leukocyte interferons by SEC on two Waters Assoc. I-125 columns in 0.05 M sodium phosphate buffer (pH 6) containing 0.2 M sodium chloride. SEC was followed by RP-HPLC on a Protesil 300 diphenyl column (Whatman) eluted with a gradient system at pH 2.4 consisting of (A) 0.05 M KH_2PO_4-methoxyethanol (19:1) and (B) 1-propanol-methoxyethanol (19:1). The presence of more than one form of leukocyte interferon was proved.

Welinder and Linde (1984) reviewed the high-performance ion-exchange chromatography of insulin and insulin derivatives, and Welinder (1984) covered with homogeneity of crystalline insulin determined by GPC and RP-HPLC. The separation of small proteins by RPLC was reviewed by Petrides (1984).

7.6.6 Some higher molecular weight proteins and protein complexes

Himmel and Squire (1981) studied the HPLC of sea worm chlorocruorin and other large complex proteins (viruses, ribosomes) in the SEC mode on TSK Gel 5000 PW preparative column. The pigmented protein chlorocruorin (MW $2.9 \cdot 10^6$) was found to serve as an excellent high-molecular-weight marker for SEC. Calibration constants for the tested columns were calculated for both molecular weight and molecular radii. The TSK Gel 5000 PW preparative column was found useful for the purification and characterization of biological macromolecules and supramolecular aggregates of molecular weight up to $1.4 \cdot 10^6$ (radius 105 Å).

Upreti and Holoubek (1981) separated proteins of 30-40S rat liver nuclear ribonucleoprotein particles using a SynChropak AX-300 anion-exchange column. The major core proteins of the particles had a strong affinity to each other and formed aggregates, which were eluted from the anion exchanger before the more acidic higher molecular weight minor particle proteins. The separation of the total proteins of the nuclear ribonucleoprotein particles by HPLC was similar to preparative electrophoresis in the results, but it was substantially quicker and adaptable to large-scale preparations. Dalrymple et al. (1983) also reported an ion-exchange HPLC separation of ribosomal proteins.

E. coli ribosome is a large ribonucleoprotein complex (M_r = $2.3 \cdot 10^6$) containing 53 proteins in two dissimilar subunits. Kerlavage and co-workers (1982, 1983) applied RP-HPLC to the separation of this complex protein mixture. A column (250 mm x 4.1 mm I.D.) of SynChropak RP-P C_{18} silica (6.5 μm) (SynChrom) was eluted with a gradient system of solvents: (A) 0.1% (w/v) trifluoroacetic acid (TFA) in water and (B) 0.1% (w/v) TFA in acetonitrile. The authors demonstrated the superiority of RP-HPLC over SEC for this purpose. The physical

properties and elution characteristics of all 53 *E. coli* ribosomal proteins were
tabulated. The recoveries of particular proteins ranged from 27 to 91% (average
70%); the load could be increased from several micrograms up to several milligrams with a minimal decrease in resolution. [^{3}H]Puromycin-labelled ribosomal
proteins were also analysed.

The separation of large proteins was reviewed by Lewis and Stern (1984).

7.6.7 Other miscellaneous proteins

According to Fox and Dose (1972), a number of proteinoids (so-called "thermal
proteins", containing tens to hundreds of covalently bound amino acid residues)
have been prepared by thermal synthesis from dry amino acids. Fox et al. (1979)
studied thermal copolymers that inhibit glyoxalase I. The poly(Glu,Trp) pro-
teinoid was chromatographed by RPC using a Sep-Pak C_{18} cartridge and also by
SEC. The pattern of molecular weight distribution is illustrated in Fig. 7.13.
Calam et al. (1983) investigated the allergens of cockfoot grass (*Dactylis
glomerata*) pollen, an important cause of allergic reactions in man. SEC-HPLC
on TSK Gel 3000 or 2000 SW in 0.1 *M* sodium phosphate (pH 7) and RP-HPLC on
Spherisorb S 5 ODS 2 (Phase Separations, Queensferry, U.K.) in a linear gradient
of solvent systems (A) 0.1 *M* ammonium sulphate (pH 2) and (B) 0.1 *M* ammonium
sulphate-acetonitrile (40:60) were used for the separation of cockfoot extract,
in addition to other non-chromatographic methods. The fractions were examined
by the radioallergosorbent RAST test to measure the IgE binding capacity in
vitro. Two active constituents were purified. Calam et al. (1982) also described
experiences obtained with other allergenic substances and the application of
IEC to the isolation of active principles from allergen extracts.

Kusunose et al. (1981) investigated the isolation of cytochrome P-450 from
solubilized microsomes of rabbit liver, kidney cortex and intestinal mucosa
using the IE-HPLC method developed by Kotake and Funae (1980). The occurence
of multiple forms (at least three different fractions) of cytochrome P-450 was
found in kidney cortex and intestinal mucosa microsomes. Power et al. (1983)
purified subunits of oligomeric membrane proteins (yeast cytochrome *c* oxidase)
by RP-HPLC. Additional sample handling techniques, which are not necessary for
soluble proteins, were required for the RPC of subunits of an oligomeric membrane protein (such as cytochrome *c* oxidase), and these techniques were
considered and discussed. Their use enables the complex mixture to be simplified
so that the hydrophobic proteins can be purified chromatographically in high
yield. Williams et al. (1983) studied staphylococcal enterotoxin B using HPLC
methods in the SEC, IEC modes and RPC. The active enterotoxin of M_r = 28 000

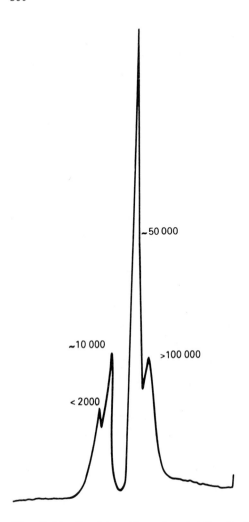

Fig. 7.13. Fractionation of poly(Glu,Trp) thermal copolymer in methanol using μBondagel E-125 and E-300 columns in series. Molecular weights are estimates by Dr. H. Jordi of Waters Assoc. (Reprinted from Fox et al., 1979.)

was isolated and purified from contaminating proteins. Low et al. (1983) isolated thymosin α_1 from thymosin fraction 5 (Sephadex G-25) of different species (calf, pig, sheep and mouse) by RP-HPLC. The isolated proteins were similar to bovine thymosin α_1. A μBondapak C_{18} column was used with 0.05% trifluoroacetic acid in acetonitrile as the solvent system. The results obtained suggest that thymosin α_1 may be synthesized in a precursor form in animal tissues, because no detectable protein of this type was found in fresh thymus tissue extracts. Lindblom et al. (1983) described the isolation of urinary proteins by HP-IEC (anion-exchange chromatography and chromatofocusing). The selection of the column and conditions was based on data obtained from electrophoretic titration curves (cf., Section 7.2). The anion exchangers used were Pharmacia Mono Bead Q (for IEC) and Mono Bead P (for chromatofocusing). The purity of selected peaks was evaluated by polyacrylamide gel electrophoresis in sodium dodecylsulphate solution.

The separation of multiple protein forms was reviewed by Aoshima (1984).

7.6.8 Large-scale and industrial applications

Tayot et al. (1978) described the industrial ion-exchange chromatography of proteins on DEAE-dextran derivatives of porous silica beads. Placental globulins, γ-globulins from human serum, the separation of haemoglobin from albumin and associated globulins and the preparation of albumin were dealt with. For non-pressurized industrial-scale gel filtration and ion-exchange chromatography with special reference to plasma protein fractionation, see also Curling (1978). Mirabel (1980) also described the preparative chromatography of proteins using French Spherosil and its ion-exchange derivatives. SEC and IEC of lactoproteins from whey in the production scale was illustrated and discussed.

Petrides et al. (1981) used RP-HPLC as an efficient procedure for the rapid purification of mouse epidermal growth factor (EGF) on the laboratory preparative scale. Up to 20 mg could be loaded on an analytical column (25 x 0.46 cm I.D.) of RP-8 (5 μm). O'Keefe and Sharry (1984) developed the IP-RP-HPLC of DEAE-murine epidermal growth factor into a macroscale procedure, in which an amount of 50 mg per run could be obtained (see also Section 6.9).

The development of preparative-scale HPLC for peptide and protein purification was reviewed, e.g., by Gabriel et al. (1981).

7.7 SEPARATION OF HIGHER MOLECULAR WEIGHT PROTEIN FRAGMENTS

Hurrell et al. (1980) have shown that RP-HPLC, which was successfully used for bovine proinsulin separation, can be applied also for the purification and analysis of bovine proinsulin C-peptide fragment, prepared by a solid-phase synthetic method. Congote et al. (1979) used a similar RP-HPLC system for the purification of cyanogen bromide (CNBr) fragments of human globulin chains, which was found to be good for the separation of the intact protein chain.

Black et al. (1980) studied the separation of CNBr-derived peptides of collagen by means of HPLC using a LiChrosorb C_{18} RP column and 10 mM potassium phosphate buffer (pH 8.5)-acetonitrile (9:1) as the fundamental mobile phase system; a 10-30% acetonitrile linear gradient was used for elution. Van der

Rest et al. (1980) also separated collagen CNBr-derived peptides by HPLC. They used µBondapak CN, µBondapak C_{18} and Vydac TP 201 C_{18} columns; 0.01 M hepta-fluorobutyric acid was used as a counter ion in the 1 h linear gradient of 12.8-44.8% (v/v) acetonitrile in water.

Pearson et al. (1981) investigated RPC supports for the resolution of large denatured protein fragments. CNBr-derived peptides of human globin chains were used as model fragments. Large pores ($\geq$300 Å) and spherical silica particles possesing C_{18} or C_8 hydrocarbon ligands were suitable for the separation of the denatured protein fragments. As short columns (5 cm) appeared to be as effective as columns five times longer, it was concluded that large peptides are adsorbed on the matrix rather than partition between the stationary and mobile phases.

Lozier et al. (1983) purified CNBr-derived fragments using a SynChropak RP-P packing and a gradient of 0-60% 1-propanol for elution. Ultrasphere ODS was used for the separation of proteinase subdigests of CNBr-derived fragments. The results indicate that the combination of the described methods can be used for sequence studies of carbohydrate-rich glycoproteins. Ortel et al. (1983) sepa-rated the limited tryptic fragments of human ceruloplasmin by HPLC in the SEC mode. A TSK Gel 3000 SW column with 8 M urea - 0.1 M Tris-HCl - 0.15 M sodium chloride (pH 8) was used for fractionation. A 67 000-dalton tryptic fragment from single-chain ceruloplasmin was isolated in the pure state for amino acid sequence analysis.

7.8 COMMENTS ON LITERATURE

Many valuable papers describing the HPLC of proteins and their fragments can be found in the proceedings of several symposia. The Proceedings of the First International Symposium on the HPLC of Proteins and Peptides, which took place in Washington, DC, in 1981, were edited by Hearn et al. (1983a). The Second International Symposium, held in Baltimore in 1982, had the extended title "HPLC of Proteins, Peptides and Polynucleotides" and the Proceedings were edited by Hearn et al. (1983b) in the form of a special issue of *Journal of Chromatography*. The Proceedings of the Third International Symposium (Monte Carlo, 1983) were edited by Unger et al. (1984) in the form of a special issue of the same journal. The Fourth International Symposium took place in 1984 in Baltimore and the Proceedings were edited by Regnier et al. (1985); those of the Fifth Internatio-nal Symposium (Toronto, 1985) were edited by Wehr et al. (1986), and those of the Sixth International Symposium (Baden-Baden, 1986) and of the Seventh Inter-national Symposium (Washington, 1987) by Hearn et al. (1987 and 1988, respec-tively), all in the same journal.

In 1984 the "International Symposium on HPLC in the Biological Sciences" was organized in Melbourne, Australia, and the Proceedings, containing also papers on the HPLC of proteins and peptides, were edited by Hearn (1984). Henschen et al. (1981) edited the Proceedings of the International Symposium on HPLC in Protein and Peptide Chemistry, held in Martinsried, F.R.G., in 1981. Also in the Proceedings of the Eighth International Symposium on Column Liquid Chromatography (New York, 1984), edited by Horvath and Heftman (1985) in the form of special issues of *Journal of Chromatography* there are many contributions that are important for the life sciences. Citations to other books and proceedings describing the HPLC of peptides and proteins can be found in the Chapter 9.

Many reviews on the rapid chromatographic separation of proteins have been published since the latter half of the last decade, and some will be cited here. Regnier et al. (1977) were among the first who dealt with this subject in detail. Hancock et al. (1977) reviewed the ion-pair HPLC of peptides and proteins, Mikeš (1979b) the ion-exchange MPLC of proteins and Regnier and Gooding (1980) and Alvarez et al. (1981) presented general reviews. Barford (1981) discussed the potential of HPLC for protein separations for the purposes of 33rd Annual Reciprocal Meat Science Conference, Clark and Kricka (1981) reviewed high-resolution techniques for proteins and peptides and their applications in clinical chemistry and Engelhardt (1981) for the separation of amino acids, proteins and their derivatives. Hanash et al. (1981) presented a specialized review describing the HPLC of haemoglobins, whereas Regnier and Gooding (1981), in addition to haemoglobin, described especially the separation of isoenzymes and some other proteins. Hearn (1982) broadly reviewed applications of HPLC to protein chemistry, and Hancock and Sparrow (1983) presented a specialized treatise on the separation of proteins by RP-HPLC. Richey (1983) described the FPLC technique and its importance for the separation of biopolymers and Hearn (1985) surveyed the ion-pair chromatography of peptides and proteins.

Cox and Dale (1981) reviewed computer simulation of transport experiments for interacting systems, in order to predict the behaviour of macromolecular mixtures in migration experiments, including, in addition to other non-chromatographic methods, also gel chromatography.

Hancock (1984) edited a two-volume Handbook, written with the collaboration of many contributors, devoted to the HPLC of amino acids, peptides and proteins; citations of reviews dealing with proteins were given in the preceding sections. Parvez et al. (1985) wrote a book on the gel permeation and ion-exchange chromatography of proteins and peptides, in which HPLC methods were also covered.

B54

REFERENCES

Alberty, R.A., in Neurath, H. and Bailey, K. (Editors), *The Proteins*, Academic Press, New York, 1953, p. 505.
Alpert, A.J., *J. Chromatogr.*, 266 (1983) 23-37.
Alpert, A.J., *J. Chromatogr.*, 359 (1986) 85-97.
Alvarez, V.L., Roitsch, C.A. and Henriksen, O., *Immunol. Methods*, 2 (1981) 83-103; *C.A.*, 95 (1981) 111 063d.
Aoshima, H., in Hancock, W.S. (Editor), *CRC Handbook of HPLC for the Separation of Amino Acids, Peptides, and Proteins*, Vol. II, CRC Press, Boca Raton, FL, 1984, pp. 343-348.
Asakawa, N., Tsuno, M., Hattori, T., Ueyama, M., Shinoda, A. and Miyake, Y., *Yakugaku Zasshi*, 101 (1981a) 279-282; *C.A.*, 95 (1981) 43 643r.
Asakawa, N., Tsuno, M., Hattori, T., Ueyama, M., Shinoda, A. and Miyake, Y., *Yakugaku Zasshi*, 101 (1981b) 708-712; *C.A.*, 95 (1981) 164 872y.
Barford, R.A., *Proceedings of the 33rd Annual Reciprocal Meat Science Conference, American Meat Science Association*, National Livestock Meat Board, Chicago, IL, 1981, pp. 102-106; *C.A.*, 95 (1981) 38 240c.
Barford, R.A., Kupec, J. and Fishman, M.L., *J. Water Pollut. Control Fed.*, (1977) 764-767.
Barford, R.A., Sliwinski, B.J. and Rothbart, H.L., *Chromatographia*, 12 (1979a) 285-288.
Barford, R.A., Sliwinski, B.J. and Rothbart, H.L., *J. Chromatogr.*, 185 (1979b) 393-402.
Barford, R.A., Sliwinski, B.J., Breyer, A.C. and Rothbart, H.L., *J. Chromatogr.*, 235 (1982) 281-288.
Becker, N. and Unger, K.K., *Chromatographia*, 12 (1979) 539-544.
Bennett, T.P., *Nature (London)*, 213 (1967) 1131-1132.
Black, C., Douglas, D.M. and Tanzer, M., *J. Chromatogr.*, 190 (1980) 393-400.
Blagrove, R.J. and Frenkel, M.J., *J. Chromatogr.*, 132 (1977) 399-404.
Bosisio, B.A., Loeherlein, Ch., Snyder, R.S. and Righetti, P.G., *J. Chromatogr.*, 189 (1980) 317-330.
Bradford, M.M., *Anal. Biochem.*, 72 (1976) 248-254.
Brautigan, D.L., Ferguson-Miller, S. and Margoliash, E., *J. Biol. Chem.*, 253 (1978) 130-139.
Browne, C.A., Bennett, H.P.J. and Solomon, S., *Anal. Biochem.*, 124 (1982) 201-208.
Burke, D.J., Duncan, J.K., Dunn, L.C., Cummings, L., Siebert, J. and Ott, G.S., *J. Chromatogr.*, 353 (1986) 425-437.
Calam, D.H., Davidson, J. and Ford, A.W., *Chromatographia*, 16 (1982) 216-218.
Calam, D.H., Davidson, J. and Ford, A.W., *J. Chromatogr.*, 266 (1983) 293-300.
Caudé, M.H., Jardi, A.P. and Rosset, R.H., in Hancock, W.S. (Editor), *CRC Handbook of HPLC for the Separation of Amino Acids, Peptides, and Proteins*, Vol. I, CRC Press, Boca Raton, FL, 1984, pp. 411-422.
Chang, S.H., Gooding, K.M. and Regnier, F.E., *J. Chromatogr.*, 120 (1976a) 321-333.
Chang, S.H., Gooding, K.M. and Regnier, F.E., *J. Chromatogr.*, 125 (1976b) 103-114.
Chang, S.H., Noel, R. and Regnier, F.E., *Anal. Chem.*, 48 (1976c) 1839-1845.
Clark, P.M.S. and Kricka, L.J., *Adv. Clin. Chem.*, 22 (1981) 247-296; *C.A.*, 96 (1982) 48 404u.
Congote, L.F., *Prog. Clin. Biol. Res. 60 (Adv. Hemoglobin. Anal.)* (1981) 39-52; *C.A.*, 95 (1981) 128 574u.
Congote, L.F., Bennett, H.P.J. and Solomon, S., *Biochem. Biophys. Res. Commun.*, 89 (1979) 851-858.
Co-Sarno, M.E., Tapang, M.A. and Luckhurst, D.G., *J. Chromatogr.*, 266 (1983) 105-113.

Čoupek, J., Křiváková, M. and Pokorný, S., *J. Polym. Sci., Part C*, 42 (1973) 185-190.

Cox, D.J. and Dale, R.S., in Frieden, C. and Nichol, L.W. (Editors), *Protein-Protein Interaction*, Wiley, New York, 1981, pp. 173-211; *C.A.*, 96 (1982) 177 143g.

Crowley, S.C. and Walters, R.R., *J. Chromatogr.*, 266 (1983) 157-162.

Curling, J., in Epton, R. (Editor): *Chromatography of Synthetic and Biological Polymers, Vol. 2, Hydrophobic, Ion-Exchange and Affinity Methods (Lect. Chem. Soc. Int. Symp., 1976)*, Ellis Horwood, Chichester, 1978, pp. 75-87.

Dalrymple, P.N., Gupta, S., Regnier, F. and Houston, L.L., *Biochim. Biophys. Acta*, 755 (1983) 157-162.

Davankov, V.A., in Hancock, W.S. (Editor), *CRC Handbook of HPLC for the Separation of Amino Acids, Peptides, and Proteins*, Vol. I, CRC Press, Boca Raton, FL, 1984, pp. 393-409.

Davankov, V.A., Rogozhin, S.V. and Semechkin, A.V., in Sladkov, A. (Editor), *Itogi Nauki i Tekhniki: Khimiya i Tekhnologiya Vysokomolekulyarnykh Soedi-nenii, (Chemistry and Technology of High Molecular Weight Compounds)*, 4 (1973) 5-44; *C.A.*, 80 (1974) 152 551v.

Deyl, Z. and Macek, K., in Kalász, H. (Editor), *New Approaches in Liquid Chromatography (Proceedings of 2nd Annual American-Eastern European Symposium on Advances in Liquid Chromatography, Szeged, 1982)*, Akademiai Kiadó, Budapest, and Elsevier, Amsterdam, 1984, pp. 13-21.

Deyl, Z., Macek, K., Adam, M. and Horáková, M., *J. Chromatogr.*, 230 (1982) 409-414.

Dimenna, G.P. and Segall, H.J., *J. Liq. Chromatogr.*, 4 (1981) 639-649.

Dinner, A. and Lorenz, L., *Anal. Chem.*, 51 (1979) 1872-1873.

Diosady, L.L., Bergen, I. and Harwalkar, V.R., *Milchwissenschaft*, 35 (1980) 671-674.

Doležálek, J., Drábiková, M. and Čoupek, J., *Prům. Potravin*, 26 (1975) 82-85; *C.A.*, 83 (1975) 77 075.

Edelstein, C. and Scanu, A.M., in Hancock, W.S. (Editor), *CRC Handbook of HPLC for the Separation of Amino Acids, Peptides, and Proteins*, Vol. II, CRC Press, Boca Raton, FL, 1984, pp. 405-412.

Edsall, J.T., in Cohn, E.J. and Edsall, J.T. (Editors), *Proteins, Amino Acids and Peptides*, Hafner, New York, 1965, p. 447.

Ek, K. and Righetti, P.G., *Electrophoresis*, 1 (1980) 137-140; *C.A.*, 94 (1981) 135 302m.

Engelhardt, H. and Mathes, D., *J. Chromatogr.*, 142 (1977) 311-320.

Engelhardt, H. and Mathes, D., *J. Chromatogr.*, 185 (1979) 305-319.

Engelhardt, H. and Mathes, D., *Chromatographia*, 14 (1981) 325-332.

Engelhardt, H., in Lottspeich, F., Henschen, A. and Hupe, K.-P. (Editors), *High Performance Chromatography in Protein and Peptide Chemistry*, Walter de Gruyter, Berlin, 1981, pp. 55-69.

Engelhardt, H., Ahr, G. and Hearn, M.T.W., *J. Liq. Chromatogr.*, 4 (1981) 1361-1379.

Fägerstam, L.G., Lizana, J., Axiö-Fredriksson, U.-B. and Wahlström, L., *J. Chromatogr.*, 266 (1983) 523-532.

Fägerstam, L., Söderberg, L., Wahlström, L., Fredriksson, U.-B., Plith, K. and Waldén, E., *Protides Biol. Fluids Proc. Colloq.*, 30 (1982) 621-628; *C.A.*, 98 (1983) 139 951c.

Fallon, A., Lewis, R.V. and Gibson, K.D., *Anal. Biochem.*, 110 (1981) 318-322.

Fox, S.W. and Dose, K., *Molecular Evolution and the Origin of Life*, Freeman, San Francisco, CA, 1972, p. 359.

Fox, S.W., Syren, R.M. and Windsor, Ch.R., *Submolecular Biology and Cancer (Ciba Foundation Series, No. 67)*, Excerpta Medica, Amsterdam, 1979, pp. 175-193.

Frolik, C.A., Dart, L.L. and Sporn, M.B., *Anal. Biochem.*, 125 (1982) 203-209.

Fukano, K., Komiya, Sasaki, H. and Hashimoto, T., *J. Chromatogr.*, 166 (1978) 47-54.

Gabriel, T.F., Michalewsky, J. and Meienhofer, J., in Eberle, A., Geiger, R.
and Wieland, T. (Editors), *Perspectives in Peptide Chemistry*, Karger, Basle,
1981, pp. 195-206; *C.A.*, 94 (1981) 116 960h.
Gardiner, M.B., Carver, J., Abraham, B.L., Wilson, J.B. and Huisman, T.H.J.,
Hemoglobin, 6 (1982) 1-13.
Gemershausen, J. and Karkas, J.D., *Biochem. Biophys. Res. Commun.*, 99 (1981)
1020-1027.
Gooding, K.M. and Schmuck, M.N., *J. Chromatogr.*, 296 (1984) 321-328.
Gooding, D.L., Schmuck, M.N. and Gooding, K.M., *J. Chromatogr.*, 296 (1984)
107-114.
Gupta, S.P. and Hanash, S.M., *Anal. Biochem.*, 134 (1983) 117-121.
Gurkin, M. and Patel, V., *Am. Lab. (Fairfield, Conn.)*, 14 (1982) 64-73.
Gurley, L.R., Prentice, D.A., Valdez, J.G. and Spall, W.D., *J. Chromatogr.*, 266
(1983a) 609-627.
Gurley, L.R., Prentice, D.A., Valdez, J.G. and Spall, W.D., *Anal. Biochem.*, 131
(1983b) 465-477.
Gurley, L.R., Valdez, J.G., Prentice, D.A. and Spall, W.D., *Anal. Biochem.*,
129 (1983c) 132-144.
Haff, L.A., Fägerstam, L.G. and Barry, A.T., *J. Chromatogr.*, 266 (1983) 409-425.
Hagiwara, N., Okazaki, M. and Hara, I., *Yukagaku*, 31 (1982) 262-267; *C.A.*, 97
(1982) 20 256u.
Haller, W., *Nature (London)*, 206 (1965) 693-696.
Haller, W., Tympner, K.-D. and Hannig, K., *Anal. Biochem.*, 35 (1970) 23-31.
Hanash, S.M. and Shapiro, D.N., *Hemoglobin*, 5 (1981) 165-175; *C.A.*, 94 (1981)
204 702f.
Hanash, S.M., Kavadella, M., Amanullah, A., Scheller, K. and Bunnell, K., *Prog.
Clin. Biol. Res. 60 (Adv. Hemoglobin Anal.)*, (1981) 53-67; *C.A.*, 95 (1981)
164 791w.
Hancock, W.S. (Editor), *CRC Handbook of HPLC for the Separation of Amino Acids,
Peptides, and Proteins*, CRC Press, Boca Raton, FL, 1984, Vol. I, 489 pp.;
Vol. II, 522 pp.
Hancock, W.S. and Harding, D.R.K., in Hancock, W.S. (Editor), *CRC Handbook of
HPLC for the Separation of Amino Acids, Peptides, and Proteins*, Vol. II, CRC
Press, Boca Raton, FL, 1984, pp. 303-312.
Hancock, W.S. and Sparrow, J.T., in Horváth, C. (Editor), *High Performance
Liquid Chromatography (Advances and Perspectives*, Vol. 3), Academic Press,
New York, 1983, 49-85.
Hancock, W.S., Bishop, C.A. and Hearn, M.T.W., *Chem. N.Z.*, 43 (1979) 17-24;
C.A., 91 (1979) 123 968h.
Hancock, W.S., Bishop, C.A., Gotto, A.M., Harding, D.R.K., Lamplugh, S.M. and
Sparrow, J.T., *Lipids*, 16 (1981a) 250-259.
Hancock, W.S., Bishop, C.A., Prestidge, R.L., Harding, D.R.K. and Hearn, M.T.W.,
Science (Washington, D.C.), 200 (1978) 1168-1170.
Hancock, W.S., Pownall, H.J., Gotto, A.M. and Sparrow, J.T., *J. Chromatogr.*,
216 (1981c) 285-293.
Hancock, W.S., Capra, J.D., Bradley, W.A. and Sparrow, J.T., *J. Chromatogr.*,
206 (1981b) 59-70.
Hara, I., Shiraishi, K. and Okazaki, M., *J. Chromatogr.*, 239 (1982) 549-557.
Hartman, A. and Persson, B., *Sci. Tools*, 27 (1980) 57-58.
Hashimoto, T., Sasaki, H., Aiura, M. and Kato, Y., *J. Chromatogr.*, 160 (1978)
301-305.
Hearn, M.T.W., *Adv. Chromatogr.*, 20 (1982) 1-82.
Hearn, M.T.W. (Editor), *J. Chromatogr.*, 336 (1984) 1-238.
Hearn, M.T.W., in Hearn, M.T.W. (Editor), *Ion-Pair Chromatography*, Marcel Dekker,
New York, 1985, pp. 207-257; *C.A.*, 102 (1985) 128 024c.
Hearn, M.T.W. and Hancock, W.S., *J. Chromatogr. Sci.*, 12 (1979) 243-271.
Hearn, M.T.W., Janson, J.-C., Regnier, F.E., Unger, K.K. and Wehr, C.T.
(Editors), *J. Chromatogr.*, 397 (1987) 1-448; 443 (1988) 1-402; 444 (1988) 1-396.
Hearn, M.T.W., Regnier, F.E. and Wehr, C.T., (Editors), *High Performance Liquid
Chromatography of Proteins and Peptides* (Proceedings of the 1st International
Symposium, Washington, DC, November 16-17th, 1981), Academic Press, New York,
1983a, pp. 288.

Hearn, M.T.W., Regnier, F.E. and Wehr, C.T. (Editors), *J. Chromatogr.*, 266 (1983b) 1-666.

Hefti, F., *Anal. Biochem.*, 121 (1982) 378-381.

Helfferich, F.G., *Nature (London)*, 189 (1961) 1001-1002.

Henderson, L.E., Sowder, R. and Oroszlan, S., in Liu, Schechter, Henrikson and Condliffe (Editors), *Chemical Synthesis and Sequencing of Peptides and Proteins*, Elsevier-North Holland, Amsterdam, 1981, pp. 251-260; *C.A.*, 96 (1982) 213 548d.

Herring, S.W. and Enns, R.K., *J. Chromatogr.*, 266 (1983) 249-256.

Himmel, M.E. and Squire, P.G., *J. Chromatogr.*, 210 (1981) 443-452.

Hjertán, S., *Protides Biol. Fluids Proc. Colloq.*, 30 (1983) 9-17.

Hjertén, S., *Trends Anal. Chem.*, 3 (1984) 87-90.

Hjertén, S. and Eriksson, K.-O., *Anal. Biochem.*, 137 (1984) 313-317.

Hjertén, S., Yao, K. and Patel, V., in Gribnau, T.C.J., Visser, J. and Nivard, R.J.F. (Editors), *Affinity Chromatography and Related Techniques*, Elsevier, Amsterdam, 1982, pp. 483-489.

Hollaway, W.L., Prestidge, R.L., Bhown, A.S., Mole, J.E. and Bennett, J.C., in Frigerio, A. and McCamish, M. (Editors), *Recent Developments in Chromatography and Electrophoresis*, Elsevier, Amsterdam, 1980, pp. 131-139.

Horváth, C. and Heftman, E. (Editors), *J. Chromatogr.*, 316 (1985) 1-628; 317 (1985) 1-628.

Hostomská, Z. and Mikeš, O., *Int. J. Peptide Protein Res.*, 23 (1984) 402-410.

Huisman, T.H.J., Altay, C., Webber, B., Reese, A.L., Gravely, M.E., Okonjo, K. and Wilson, J.B., *Blood*, 57 (1981) 75-82.

Humphrey, R.S., in Hancock, W.S. (Editor), *CRC Handbook of HPLC for the Separation of Amino Acids, Peptides, and Proteins*, Vol. II, CRC Press, Boca Raton, FL, 1984, pp. 471-478.

Hurrell, J.G.R., Fleming, R.J. and Hearn, M.T.W., *J. Liq. Chromatogr.*, 3 (1980) 473-494.

Hutchens, T.W., Wiehle, R.D., Shahabi, N.A. and Wittliff, J.L., *J. Chromatogr.*, 266 (1983) 115-128.

Ishii, M., *Kyorin Igakukkai Zasshi*, 12 (1981) 247-252; *C.A.*, 96 (1982) 48 592d.

Jenik, R.A. and Porter, J.W., *Anal. Biochem.*, 111 (1981) 184-188.

Jones, B.N., Lewis, R.V., Pääbo, S., Kojima, K., Kimura, S. and Stein, S., *J. Liq. Chromatogr.*, 3 (1980) 1373-1383.

Karger, B.L., *J. Chromatogr. Sci.*, 15 (1977) 515.

Kato, Y., Nakamura, K. and Hashimoto, T., *J. Chromatogr.*, 266 (1983b) 385-394.

Kato, Y., Kitamura, T. and Hashimoto, T., *J. Chromatogr.*, 266 (1983a) 49-54.

Kato, Y., Kitamura, T. and Hashimoto, T., *J. Chromatogr.*, 333 (1985a) 93-106.

Kato, Y., Kitamura, T. and Hashimoto, T., *J. Chromatogr.*, 333 (1985b) 202-210.

Kato, Y., Komiya, K. and Hashimoto, T., *J. Chromatogr.*, 246 (1982) 13-22.

Kato, Y., Komiya, K., Sasaki, H. and Hashimoto, T., *J. Chromatogr.*, 193 (1980a) 29-36.

Kato, Y., Komiya, K., Sasaki, H. and Hashimoto, T., *J. Chromatogr.*, 193 (1980b) 458-463.

Kato, Y., Nakamura, K. and Hashimoto, T., *J. Chromatogr.*, 266 (1983b) 385-394.

Kato, Y., in Hancock, W.S. (Editor), *CRC Handbook of HPLC for the Separation of Amino Acids, Peptides, and Proteins*, Vol. II, CRC Press, Boca Raton, FL, 1984, pp. 363-369.

Kearney, R.D. and McGawn, T.C.A., in Epton, R. (Editor), *Chromatography of Synthetic and Biological Polymers, Vol. 1*, Column Packings, GPC, GF and Gradient Elution *(Lect. Chem. Soc. Int. Symp., 1976)*, Ellis Horwood, Chichester, 1978, pp. 269-274; *C.A.*, 89 (1978) 127 803g.

Kerlavage, A.R., Kahan, L. and Cooperman, B.S., *Anal. Biochem.*, 123 (1982) 342-348.

Kerlavage, A.R., Weitzmann, C.J., Hasan, T. and Cooperman, B.S., *J. Chromatogr.*, 266 (1983) 225-237.

Kimura, M. and Itokawa, Y., *J. Chromatogr.*, 211 (1981) 290-294.

Klee, C.B., Oldewurtel, M.D., Williams, J.F. and Lee, J.W., *BioChem. Int.*, 2 (1981) 485-493.

Kojima, K., Manabe, T., Okuyama, T., Tomono, T., Suzuki, T. and Tokunaga, E.,
  J. Chromatogr., 239 (1982) 565-570.
Komiya, K., Kato, Y., Furukawa, K. Sasaki, H. and Watanabe, H., *Toyo Soda Kenkyu
  Hokoku*, 25 (1981) 115-132; *C.A.*, 96 (1982) 65 065b.
Kopaciewicz, W. and Regnier, F.E., *Anal. Biochem.*, 126 (1982) 8-16.
Kopaciewicz, W., Rounds, M.A., Fausnaugh, J. and Regnier, F.E., *J. Chromatogr.*,
  266 (1983) 3-21.
Kotake, A.N. and Funae, Y., *Proc. Natl. Acad. Sci. U.S.A.*, 77 (1980) 6473-6475.
Kusunose, E., Kaku, M., Nariyama, M., Kusunose, M., Ichihara, K., Funae, Y. and
  Kotake, A.N., *Biochem. Int.*, 3 (1981) 399-406.
Lambert, D.T. and Lerner, A.B., *J. Chromatogr.*, 266 (1983) 567-576.
Lewis, R.V. and Stern, A.S., in Hancock, W.S. (Editor), *CRC Handbook of HPLC
  for the Separation of Amino Acids, Peptides, and Proteins*, Vol. II, CRC
  Press, Boca Raton, FL, 1984, pp. 313-325.
Lewis, R.V., Fallon, A., Stein, S., Gibson, K.D. and Udenfriend, S., *Anal. Chem.*,
  104 (1980) 153-159.
Lindblom, H., Söderberg, L., Cooper, E.H. and Turner, R., *J. Chromatogr.*, 266
  (1983) 187-196.
Lonsdorfer, M., Clements, N.C., Jr. and Wittliff, J.L., *J. Chromatogr.*, 266
  (1983) 129-139.
Lottspeich, F., Henschen, A. and Hupe, K.P. (Editors), *High Performance Liquid
  Chromatography in Protein and Peptide Chemistry* (International Symposium on
  HPLC in Biochemistry: Proteins, Peptides, Amino Acids, Martinaried, F.R.G.,
  1981), Walter de Gruyter, Berlin, 1981, 388 pp.
Low, T.L.K., McClure, J.E., Naylor, P.H., Spangelo, B.L. and Goldstein, A.L.,
  J. Chromatogr., 266 (1983) 533-544.
Lowry, O.H., Rosebrough, N.J., Farr, A.L. and Randall, R.J., *J. Biol. Chem.*,
  193 (1951) 265-275.
Lozier, J., Takahashi, N. and Putnam, F.W., *J. Chromatogr.*, 266 (1983) 545-554.
Lu, K.-C., Gooding, K.M. and Regnier, F.E., *Clin. Chem.*, 25 (1979) 1608-1612.
Macek, K., Deyl, Z., Čoupek, J. and Sanitrák, J., *J. Chromatogr.*, 222 (1981)
  284-290.
Macklin, W.B., Pickart, L. and Woodward, D.O., *J. Chromatogr.*, 210 (1981) 174-
  179.
Malamud, D. and Drysdale, J.W., *Anal. Biochem.*, 86 (1978) 620-647.
Manalan, A.S. and Klee, C.B., in Hancock, W.S. (Editor), *CRC Handbook of HPLC
  for the Separation of Amino Acids, Peptides, and Proteins*, Vol. II, CRC
  Press, Boca Raton, FL, 1984, pp. 461-469.
Margolis, J. and Rhoanes, P.H., *Lancet*, ii (1981) 446-449; *C.A.*, 96 (1982)
  40 782v.
Meek, J.L. and Rossetti, Z.L., *J. Chromatogr.*, 211 (1981) 15-28.
Meredith, S.C. and Nathans, G.R., *Anal. Biochem.*, 121 (1982) 234-243.
Mikeš, O., in Mikeš, O. (Editor), *Laboratory Handbook of Chromatographic and
  Allied Methods*, Ellis Horwood, Chichester, 1979a, p. 240.
Mikeš, O., *Int. J. Peptide Protein Res.*, 14 (1979b) 393-401.
Mikeš, O., Hostomská, Z., Štrop, P., Smrž, M., Slováková, S., Vrátný, P., Rexová,
  L'., Kolář, M. and Čoupek, J., *J. Chromatogr.*, 440 (1988) 287-304.
Mikeš, O., Štrop, P. and Čoupek, J., *J. Chromatogr.*, 153 (1978) 23-36.
Mikeš, O., Štrop, P., Hostomská, Z., Smrž, M., Čoupek, J., Frydrychová, A. and
  Bareš, M., *J. Chromatogr.*, 261 (1983) 363-379.
Mikeš, O., Štrop, P., Hostomská, Z., Smrž, M., Slováková, S. and Čoupek, J.,
  J. Chromatogr., 301 (1984) 93-105.
Mikeš, O., Štrop, P., Smrž, M. and Čoupek, J., *J. Chromatogr.*, 192 (1980) 159-
  172.
Mikeš, O., Štrop, P., Zbrožek, J. and Čoupek, J., *J. Chromatogr.*, 119 (1976)
  339-354.
Mikeš, O., Štrop, P., Zbrožek, J. and Čoupek, J., *J. Chromatogr.*, 180 (1979)
  17-30.

Mirabel, B., *Actual. Chim.*, (1980) 39-44; *C.A.*, 93 (1980) 148 214p.

Miura, K., Nakamura, H. and Tamura, Z., *J. Chromatogr.*, 210 (1981) 461-467.

Mizutani, T., in Osawa, K., Tanaka, Y. and Kitami, S. (Editors), *Kyodai Ryushi No Gerupamieishion Kuromotogurafi: Seitai Ryushi No Ryukei Bunri*, 1980, pp. 243-260; *C.A.*, 95 (1981) 146 292e.

Mizutani, T., *J. Chromatogr.*, 207 (1981) 276-280.

Mizutani, T. and Narihara, T., *J. Chromatogr.*, 239 (1982) 755-760.

Mönch, W. and Dehnen, W., *J. Chromatogr.*, 147 (1978) 415-418.

Nexø, E., *Biochim. Biophys. Acta*, 379 (1975) 189-192.

Nice, E.C., Capp, M. and O'Hare, M.J., *J. Chromatogr.*, 185 (1979) 413-427.

Nishikawa, A.H., Roy, S.K. and Puchalski, R., in Gribnau, T.C.J., Visser, J. and Nivard, R.J.F. (Editors), *Affinity Chromatography and Related Techniques*, Elsevier, Amsterdam, 1982, pp. 471-482.

O'Hare, M.J. and Nice, E.C., *J. Chromatogr.*, 171 (1979) 209-226.

Ohlson, S., Hansson, L., Larsson, P.-O. and Mosbach, K., *FEBS Lett.*, 93 (1978) 5-9.

Ohno, Y., Okazaki, M. and Hara, I., *J. Biochem. (Tokyo)*, 89 (1981) 1675-1680.

Okazaki, M. and Hara, I., in Hancock, W.S. (Editor), *CRC Handbook of HPLC for the Separation of Amino Acids, Peptides, and Proteins*, Vol. II, CRC Press, Boca Raton, FL, 1984, pp. 393-403.

Okazaki, M., Kinoshita, M., Naito, C. and Hara, I., *J. Chromatogr.*, 336 (1984) 151-159.

Okazaki, M., Ohno, Y. and Hara, I., *J. Biochem. (Tokyo)*, 89 (1981a) 879-887; *C.A.*, 94 (1981) 135 262y.

Okazaki, M., Shiraishi, K., Ohno, Y. and Hara, I., *J. Chromatogr.*, 223 (1981b) 285-293.

O'Keefe, J.H., Sharry, L.F. and Jones, A.J., *J. Chromatogr.*, 336 (1984) 73-85.

Oldenwurtel, M.D., Krinks, M.H., Lee, J.W., Williams, J.F. and Klee, C.B., *Protides Biol. Fluids*, Proc. Colloq., 30 (1983) 714-716; *C.A.*, 98 (1983) 139 959m.

Ortel, T.L., Takahashi, N. and Putnam, F.W., *J. Chromatogr.*, 266 (1983) 257-263.

Ou, C.-N., Buffone, G.J., Regnier, G.L. and Alpert, A.J., *J. Chromatogr.*, 266 (1983) 197-205.

Parvez, H., Kato, Y. and Parvez, S., *Gel Permeation and Ion Exchange Chromatohraphy of Proteins and Peptides*, VNU Science Press, Utrecht, 1985, 224 pp.

Pearson, J.D., Lin, N.T. and Regnier, F.E., *Anal. Biochem.*, 124 (1982) 217-230.

Pearson, J.D., Mahoney, W.C., Hermodson, M.A. and Regnier, F., *J. Chromatogr.*, 207 (1981) 325-332.

Persiani, C., Cukor, P. and French, K., *J. Chromatogr. Sci.*, 14 (1976) 417-421.

Petrides, P.E., in Hancock, W.S. (Editor), *CRC Handbook of HPLC for the Separation of Amino Acids, Peptides, and Proteins*, Vol. II, CRC Press, Boca Raton, FL, 1984, pp. 327-342.

Petrides, P.E., Jones, R.T. and Böhlen, P., *Anal. Biochem.*, 105 (1980) 383-388.

Petrides, P.E., Levine, A.E. and Shooter, E.M., In Rich, D.H. and Gross, E. (Editors), *Peptides: Synthesis-Structure-Function (Proceedings of the Seventh American Peptide Symposium)*, Pierce, Rockford, IL, 1981, pp. 781-783; *C.A.*, 96 (1982) 193 556z.

Pfannkoch, E., Lu, K.C., Regnier, F.E. and Barth, H.G., *J. Chromatogr. Sci.*, 18 (1980) 430-441.

Pollak, J.K. and Campbell, M.T., in Hancock, W.S. (Editor), *CRC Handbook of HPLC for the Separation of Amino Acids, Peptides, and Proteins*, Vol. II, CRC Press, Boca Raton, FL, 1984, pp. 371-378.

Porath, H.J., Carlsson, J., Olsson, J. and Belfrage, G., *Nature (London)*, 258 (1975) 598-599.

Power, S.D., Lochrie, M.A. and Poyton, R.O., *J. Chromatogr.*, 266 (1983) 585-598.

Prusík, Z., in Deyl, Z., Macek, K. and Janák, J. (Editors), *Liquid Column Chromatography. A Survey of Modern Techniques and Applications (Journal of Chromatography Library*, Vol. 3), Elsevier, Amsterdam, 1975, pp. 773-806.

Regnier, F.E., *Anal. Biochem.*, 126 (1982) 1-7.
Regnier, F.E., *Science (Washington, D.C.)*, 222 (1983a) 245-252.
Regnier, F.E., *LC, Liq. Chromatogr. HPLC, Mag.*, 1 (1983b) 350.
Regnier, F.E. and Gooding, K.M., *Anal. Biochem.*, 103 (1980) 1-25.
Regnier, F.E. and Gooding, K.M., in Marton, J. and Kabra, P.M. (Editors), *Liquid Chromatography in Clinical Analysis*, Humana Press, Clifton, NY, 1981, pp. 323-353.
Regnier, F.E. and Noel, R., *J. Chromatogr. Sci.*, 14 (1976) 316-320.
Regnier, F.E., Gooding, K.M. and Chang, S.-H., *Contemp. Top. Anal. Clin. Chem.*, 1 (1977) 1-48; *C.A.*, 87 (1977) 196 576f.
Regnier, F.E., Hearn, M.T.W., Unger, K., Wehr, C.T. and Janson, J.C. (Editors), *J. Chromatogr.*, 326 (1985) 1-446; 327 (1985) 1-384 (special issues).
Richey, J., *Int. Lab.*, 13 (1983) 50-75.
Righetti, P.G. and Caravaggio, T., *J. Chromatogr.*, 127 (1976) 1-28.
Righetti, P.G. and Gianazza, E., in Radola, B.J. (Editor), *Electrophoresis '79*, Walter de Gruyter, Berlin, New York, 1980, pp. 23-38.
Righetti, P.G., Tudor, G. and Ek, K., *J. Chromatogr.*, 220 (1981) 115-194.
Rivier, J.E., *J. Liq. Chromatogr.*, 1 (1978) 343-366.
Rivier, J., Desmond, J., Spiess, J., Perrin, M., Vale, W., Eksteen, R. and Karger, B., in Gross, E. and Meienhofer, J. (Editors), *Peptides: Structure and Biological Function (Proceedings of the Sixth American Peptide Symposium)*, Pierce, Rockford, IL, 1979, pp. 125-128.
Rokushika, S., Ohkawa, T. and Hatano, H., *J. Chromatogr.*, 176 (1979) 456-461.
Rosengren, A., Bjellquist, B. and Gasparic, V., in Radola, B. and Graesslin, D. (Editors), *Electrofocusing and Isotachophoresis*, Walter de Gruyter, Berlin, 1977, pp. 165-171; *C.A.*, 87 (1977) 163 655c.
Roumeliotis, P. and Unger, K.K., *J. Chromatogr.*, 185 (1979) 445-452.
Roumeliotis, P. and Unger, K.K., *J. Chromatogr.*, 218 (1981) 535-546.
Roumeliotis, P., Unger, K.K., Kinkel, J., Brunner, G., Wieser, R. and Tschank, G., in Lottspeich, F., Henschen, A. and Hupe, K.-P. (Editors), *High Performance Chromatography in Protein and Peptide Chemistry*, Walter de Gruyter, Berlin, 1981, pp. 71-82.
Rubinstein, M., *Anal. Biochem.*, 98 (1979) 1-7.
Rubinstein, M., Rubinsten, S., Familletti, P.C., Gross, M.S., Miller, R.S., Waldman, A.A. and Pestka, S., *Science (Washington, DC)*, 202 (1978) 1289-1290.
Rubinstein, M., Rubinstein, S., Familletti, P.C., Miller, R.S., Waldman, A.A. and Pestka, S., *Proc. Natl. Acad. Sci. U.S.A.*, 76 (1979) 640-644.
Ruckenstein, E. and Kalthod, D.G., in Hallstroem, B., Lund, D.B. and Traegardth, Ch. (Editors), *Fundam. Appl. Surf. Phenom. Assoc. Fouling Clean. Food Process, (Proceedings of International Workshop, 1981)*, Lund University Division of Food Engineering, Alnarp, 1981, pp. 115-147; *C.A.*, 96 (1982) 177 218k.
Saint-Blancard, J., Fourcard, J., Limonne, F., Girot, P. and Boschetti, E., *Ann. Pharm. Fr.*, 39 (1981) 403-409; *C.A.*, 96 (1982) 74 550h.
Schifreen, R.S., Hickingbotham, J.M. and Bowers, G.N., Jr., *Clin. Chem.*, 26 (1980) 466-472.
Schlabach, T.D. and Abbott, S.R., *Clin. Chem.*, 26 (1980) 1504-1508.
Schlabach, T.D., Chang, S.H., Gooding, K.M. and Regnier, F., *J. Chromatogr.*, 134 (1977) 91-106.
Schroeder, W.A., Shelton, J.B. and Shelton, J.R., *Prog. Clin. Biol. Res. 60 (Adv. Hemoglobin Anal.)* (1981) 1-22; *C.A.*, 95 (1981) 146 235p.
Schultz, R.M. and Wassarman, P.M., *Anal. Biochem.*, 77 (1977) 25-32.
Schwandt, P., Richter, W.O. and Weisweiler, P., *J. Chromatogr.*, 225 (1981) 185-188.
Sebille, B. and Thuaud, N., in Hancock, W.S. (Editor), *CRC Handbook of HPLC for the Separation of Amino Acids, Peptides, and Proteins*, Vol. II, CRC Press, Boca Raton, FL, 1984, pp. 379-391.
Shechter, I., *Anal. Biochem.*, 58 (1974) 30-38.

Shelton, J.B., Shelton, J.R. and Schroeder, W.A., *Hemoglobin*, 3 (1979) 353-358;
 C.A., 92 (1980) 37 044t.
Shelton, J.B., Shelton, J.R. and Schroeder, W.A., *Hemoblogin*, 6 (1982) 451-464;
 C.A., 98 (1983) 68 206a.
Shimohigashi, Y., Lee, R. and Chen, H.-C., *J. Chromatogr.*, 266 (1983) 555-562.
Sjödahl, J., *Sci. Tools*, 27 (1980) 54-56.
Stanton, P.G., Simpson, R.J., Lambrou, F. and Hearn, M.T.W., *J. Chromatogr.*,
 266 (1983) 273-279.
Strahler, J.R., Rosenblum, B.B., Hanash, S. and Butkunas, R., *J. Chromatogr.*,
 266 (1983) 281-291.
Štrop, P., Mikeš, F. and Chytilová, Z., *J. Chromatogr.*, 156 (1978) 239-254.
Sugden, K., in Hancock, W.S. (Editor), *CRC Handbook of HPLC for the Separation
 of Amino Acids, Peptides, and Proteins*, Vol. I, CRC Press, Boca Raton, FL,
 1984, pp. 423-427.
Suzuki, K.T., *Anal. Biochem.*, 102 (1980) 31-34.
Suzuki, K.T., *Kagaku No Ryoiki Zokan*, 133 (1981) 79-92; *C.A.*, 95 (1981) 164 835p.
Takagi, T., Takeda, K. and Okuno, T., *J. Chromatogr.*, 208 (1981) 201-208.
Takaki, T., in Osawa, K., Tanaka, Y. and Kitami, S. (Editors), *Kyodai Ryushi
 No Gerupamieishion Kuromotogurafi: Seitai Ryunshi No Ryukei Bunri*, 1980,
 pp. 319-335; *C.A.*, 95 (1981) 146 293f.
Tandy, N.E., Dilley, R.A. and Regnier, F.E., *J. Chromatogr.*, 266 (1983) 599-607.
Tayot, J.L., Tardy, M., Gattel, P., Plan, R. and Roumiantzeff, M., in Epton, R.
 (Editor), *Chromatography of Synthetic and Biological Polymers. Vol. 2
 Hydrophobic, Ion-Exchange and Affinity Methods (Lect. Chem. Soc. Int. Symp.,
 1976)*, Ellis Horwood, Chichester, 1978, pp. 95-110; *C.A.*, 90 (1979) 18 449w.
Thacker, L.H., Scott, C.D. and Pitt, W.W., Jr., *J. Chromatogr.*, 51 (1970) 175-
 181.
Tomono, T., Ikeda, H. and Tokunaga, E., *J. Chromatogr.*, 266 (1983) 39-47.
Tomono, T., Yoshida, S. and Tokunaga, E., *J. Polym. Sci., Polym. Lett. Ed.*, 17
 (1979) 335-341.
Toren, E.C., Jr., Vacik, D.N. and Mockridge, P.B., *J. Chromatogr.*, 266 (1983)
 207-212.
Tsuda, T., Nomura, K. and Nakagava, G., *J. Chromatogr.*, 248 (1982) 491-492.
Tweeten, K.A. and Tweeten, T.N., *J. Chromatogr.*, 359 (1986) 111-119.
Ui, N., *Anal. Biochem.*, 97 (1979) 65-71.
Unger, K., Regnier, F.E., Hearn, M.T.W., Wehr, C.T. and Janson, J.-C. (Editots),
 J. Chromatogr., 296 (1984) 1-406; 297 (1984) 1-420.
Upreti, R.K. and Holoubek, V., *Anal. Chim. Acta*, 131 (1981) 239-245.
Van der Rest, M., Bennett, H.P.J., Solomon, S. and Glorieux, F.H., *Biochem. J.*,
 191 (1980) 253-256.
Vanecek, G. and Regnier, F.E., *Anal. Chem.*, 109 (1980) 345-353.
Vanecek, G. and Regnier, F.E., *Anal. Biochem.*, 121 (1982) 156-169.
Vondruška, M., Šudrich, M. and Mládek, M., *J. Chromatogr.*, 116 (1976) 457-461.
Vytášek, R., Čoupek, J., Macek, K., Adam, M. and Deyl, Z., *J. Chromatogr.*, 119
 (1976) 549-556.
Wehr, C.T. and Abbott, S.R., *J. Chromatogr.*, 185 (1979) 453-462.
Wehr, C.T., Hearn, M.T., Janson, J.-C., Regnier, F.E. and Unger, K.K. (Editors),
 J. Chromatogr., 359 (1986) 1-556.
Welinder, B.S., in Hancock, W.S. (Editor), *CRC Handbook of HPLC for the Separa-
 tion of Amino Acids, Peptides, and Proteins*, Vol. II, CRC Press, Boca Raton,
 FL, 1984, pp. 413-419.
Welinder, B.S. and Linde, S., in Hancock, W.S. (Editor), *CRC Handbook of HPLC
 for the Separation of Amino Acids, Peptides, and Proteins*, Vol. II, CRC
 Press, Boca Raton, FL, 1984, pp. 357-362.
White, R.R., Montgomery, E.H. and Williams, G.A., *J. Immunol. Methods*, 43 (1981)
 313-319; *C.A.*, 96 (1982) 50 214a.
Williams, R.F., Aivaliotis, M.J., Barnes, L.D. and Robinson, A.K., *J. Chroma-
 togr.*, 266 (1983) 141-150.

Williams, R.R., Wehr, C.T., Rogers, T.J. and Bennett, R.W., *J. Chromatogr.*, 266 (1983) 179-186.

Wilson, K.J., Van Wieringen, E., Klauser, S., Berchtold, W. and Hughes, G.J., *J. Chromatogr.*, 237 (1982) 407-416.

Wunderwald, P., Schrenk, W.J., Port, H. and Kresze, G.-B., *J. Appl. Biochem.*, 5 (1983) 31-42.

Chapter 8

ENZYMES

8.1 INTRODUCTION

Enzymes are proteins and therefore the methods and problems described and discussed in Chapter 7 are valid also for enzymes. However, there are some special features that must be emphasized when the separation of enzymes is considered. In general, enzymes are often very sensitive to denaturation, usually resulting in a total or partial loss of catalytic activity. The yield of an enzyme fractionation must be expressed not only in terms of the protein recovery, but also in terms of the recovery of enzymic activity. The increase in the ratio of the activity to the protein content is the usual measure of the degree of purification in the process of enzyme isolation. The activity/protein ratio is also an important constant characterizing the pure and homogeneous enzyme preparation.

The risk of denaturation of enzymes arises not only with elevated temperature. Fractionation at temperatures several degrees above zero is usually used not only to eliminate the possibility of thermal denaturation of enzymes, but also because many raw materials of biological origin contain certain amounts of various proteinases, which can partly decompose the enzymes to be isolated, especially at the usual ambient temperatures. Unknown enzyme preparations should certainly not be exposed to extreme conditions of pH, which very easily change their tertiary structure.

Organic solvents represent another threat to the activity of some enzymes especially at elevated temperatures. This must be considered when reversed-phase high-performance liquid chromatography (RP-HPLC) experiments are planned. The same applies to detergents: sodium dodecylsulphate solution is useful for the determination of molecular weights of enzymes using the size exclusion chromatography (SEC) mode, but the activity is destroyed. Some enzymes form special quaternary structures and the detergents dissociate this structure into individual subunits. Other surface-active compounds also endanger enzymes. Many enzymes cannot tolerate contact with ions of heavy metals, and this must be borne in mind when buffers are selected for the elution of enzymes from stainless-steel chromatographic columns: some buffers attack the stainless steel (e.g., acidic solutions of ammonium formate) and trace amounts of heavy metal ions can destroy the enzyme activity in some instances; glass or other non-corrosive columns and pumps are to be preferred.

Oxygen in the air (also dissolved in the mobile phase) can deactivate some sensitive enzymes. In particular, the enzymes of anaerobic microbes and many other enzymes of anaerobic metabolism may be sensitive to oxygen to such an extent that the addition of reducing agents to the mobile phase is not sufficient to protect the activity and the separation must be performed in a special anaerobic laboratory. Strong reducing agents will also destroy the activity of some enzymes.

Surface denaturation of enzymes in contact with unsuitable chromatographic sorbents that allow strong hydrophobic interactions has often been observed; hydrophilic adsorbents are preferably used. For this reason, the newly developed hybrid hydrophilic/hydrophobic packings for hydrophobic interaction chromatography (HIC) of enzymes are very important (cf., Sections 3.6 and 4.4.3).

The total loss of enzymic activity due to the denaturation of native enzymes is not the only risk when enzymes are separated. What may be neglected are slight changes in conformation, or limited proteolysis, which slightly modify enzymes so that they differ in their chromatographic retention from the original native form, but still retain their activity. Artificial enzyme polymorphism is created in this way and instead of native enzymes the active artifacts may be isolated. "Artificial isoenzymes" are formed that can hardly be distinguished from natural isoenzymes at the beginning of the study because multiple forms of enzymes very often occur in nature.

Special requirements for the low-speed chromatography of enzymes were discussed from this point of view, e.g., by Mikeš (1975).

Enzymes are biochemical catalysts and their activities have been measured by many workers in various ways. It was extremely difficult to compare results reported on the same enzyme. Hence the Fifth International Congress of Biochemistry adopted the recommendations of the Commissions on Enzymes of IUPAC (International Union of Pure and Applied Chemistry) and IUB (International Union of Biochemistry) (1965, 1972) for the definition of an enzyme unit. The recommended terms will be briefly presented here according to Guilbault (1976), as follows.

One unit (U) of any enzyme is that amount which will catalyse the transformation of 1 µmol of substrate per minute; where more than one bond of each substrate is attached, the transformation of 1 µequiv. of the group concerned per minute is considered, under defined conditions. Recommended parameters: temperature, 25^{0}C; pH, optimal; reaction rate, initial; and kinetics with respect to the substrate, zero order.

Concentration of an enzyme in solution: units per millilitre (U/ml) or per litre (U/l).

Specific activity of an enzyme: units of enzyme per milligram of protein (U/mg protein).

Molecular activity: units per micromole of enzyme, i.e., the number of molecules of the substrate transformed per minute per molecule of the enzyme.

The katal refers to the conversion of 1 mol of substrate per second = $6 \cdot 10^7$ µmol of substrate per minute = $6 \cdot 10^7$ units (one unit = 1 U = 16.67 nkatal).

8.2. PRINCIPLES OF SPECIFIC DETECTION METHODS FOR ENZYMES AND ISOENZYMES

One of the simplest approaches is to measure the absorbance in the long-wave UV region (Prusík, 1975), which is used for the sensitive detection of some enzymes with firmly bound prosthetic groups. However, when dealing with the fractionation of crude mixtures of proteins, specific enzymes must be quantitated by enzyme assay of fractions rather than by spectrophotometric monitoring of the effluent. The chromatographic column effluent can be analysed using an off-line or an on-line procedure. For the off-line procedures aliquots of the fractions are taken manually or by some automatic equipment (see, e.g., a short review in an essay by Mikeš, 1975). Raschbaum and Everse (1978) also described an apparatus for the automatic detection of enzyme activity in column chromatographic effluents, which consists of a simple stopped-flow instrument, equipped with a multi-shutoff valve to direct the reagent and eluent flows: assays of the eluate can be carried out at pre-set intervals. The chemical principles and analytical procedures for the quantitation of many important enzymes were described, e.g., in the handbook by Guilbault (1976).

From the point of view of the present state of evolution of the liquid chromatography of enzymes, off-line detection procedures seem to be useful only for pioneer works in attempts to isolate an unknown enzyme, or with some single-action preparative experiments in order to isolate a known required enzyme, and never for routinely repeated analytical or preparative procedures. When HPLC separations of proteins in 10 min became possible, time-consuming off-line enzymic assays of the components were the limiting step. Therefore, rapid on-line detection methods for enzymes and isoenzymes were sought. The instrumentation of detectors developed for this purpose was dealt with in detail in Section 5.2.3. In the following sections the biochemical principles of the specific detection methods will be explained.

This approach was especially important in clinical applications for the identification of isoenzymes. Meister (1950) and Neilands (1952) were the first to recognize multiple forms of the enzyme lactate dehydrogenase (LD). Vessel and Bearn (1957) opened a new dimension in clinical diagnosis by their finding that the serum profile of LD isoenzymes changes dramatically with myocardial infarction or acute myelogenous leukaemia. Cohen et al. (1964) described in detail serum LD patterns in cardiovascular and other diseases, with particular reference to acute myocardial infarction. The diagnostic importance of isoenzymes has been evaluated in many papers (e.g., Galen et al., 1975). Usually, isoenzymes were separated from each other using mainly electrophoretic, ion-exchange, immunochemical and various sorption techniques. Morin (1977) evaluated the methods current at that time for creatine kinase isoenzyme fractionation. Kudirka et al. (1975) were the first to apply modern rapid HPLC separation to creatine kinase isoenzymes, but the on-line detector was missing. However, the authors discussed its importance.

A new possibility for the development of serum isoenzyme profiling was opened with the introduction of HPLC methods into protein chemistry and biochemistry. The speed of the separation initiated a search for an equivalent rapid on-line detection method. The diagnostically important isoenzymes were separated from each other on HPLC columns, but were not separated from various additional serum proteins, so that detection methods were necessary that would be able to "see" the isoenzymes but not the other proteins covering their peaks. The biochemical principles of such techniques are described in the following sections.

8.2.1 Alkaline phosphatase

The specific detection of alkaline phosphatase (AP) represents the first simple example on which on-line post-column detection can be explained (Chang et al., 1976; Schlabach et al., 1977; Schlabach and Regnier, 1978). The analytical column effluent is mixed with the substrate p-nitrophenyl phosphate (NPP) in the equipment illustrated in Fig. 5.29, and the mixture is heated ($40^{\circ}C$) in the packed column reactor in which the product p-nitrophenol (NP) is formed according to the equation

$$O_2NC_6H_4OPO_3Na_2 \xrightarrow[H_2O]{AP} O_2NC_6H_4OH + HOPO_3Na_2$$
$$\quad\quad NPP \quad\quad\quad\quad\quad\quad\quad\quad\quad NP$$

The product (NP) can be readily detected in alkaline solution by measurement of the absorbance at 400 or 410 nm, and the presence of proteins (λ_{max} 280-285 nm) has no effect. An example of such an analysis is illustrated in Fig. 5.30. In addition to the selectivity, this detection principle is more than 20 times more sensitive than conventional absorbance monitoring of proteins at 280 nm.

If a wider reactor column (250 x 8 mm I.D.) was used and the temperature increased to 60°C, the detection limit was 25 ng/ml of alkaline phosphatase
(Regnier et al., 1977).

8.2.2 Trypsin and chymotrypsin

Another simple reaction suitable for on-line post-column detection is the
spectrophotometric determination of trypsin (TRY), studied for this purpose by
Schlabach et al. (1977). Bender et al. (1966) described a number of stoichiometric reagents that were developed for the titration of active site of enzymes and
made it possible to determine the concentration of an enzyme in solution with
an end-point reaction. A very effective active site titrant for TRY is nitro-
phenyl guanidinoboenzoate (NPGB) (Chase and Show, 1967), which reacts with TRY
according to the equation

The *p*-nitrophenol (NP) product can be monitored by spectrophotometry at 400 nm.
NPGB reacts only with active TRY and is insensitive to other enzymes and to
inactive TRY. Schlabach et al. (1977) demonstrated the linearity of the response
and the suitability of the method for quantitating TRY in a flow-through reactor.
The addition of NaCl at a 0.5 *M* concentration to the measured solution was
recommended in order to suppress the adsorption of TRY on the equipment surfaces.

Another post-column on-line detection method for TRY was published by Gooding et al. (1984). The exit from the analytical HI-HPLC column was connected, using a union tee, with the substrate pump and a packed column reactor containing SynChropak PCR (a microparticulate non-porous packing). The effluent from the reactor was monitored at 410 nm. The substrate for TRY pumped to the union tee was benzoyl-D,L-arginine p-nitroanilide (BANA), dissolved in 0.1 M potassium phosphate solution (pH 7). BANA is hydrolysed to p-nitroaniline (NA) by TRY (Hoverbach et al., 1960) according to the equation

BANA

$NA(A_{410})$

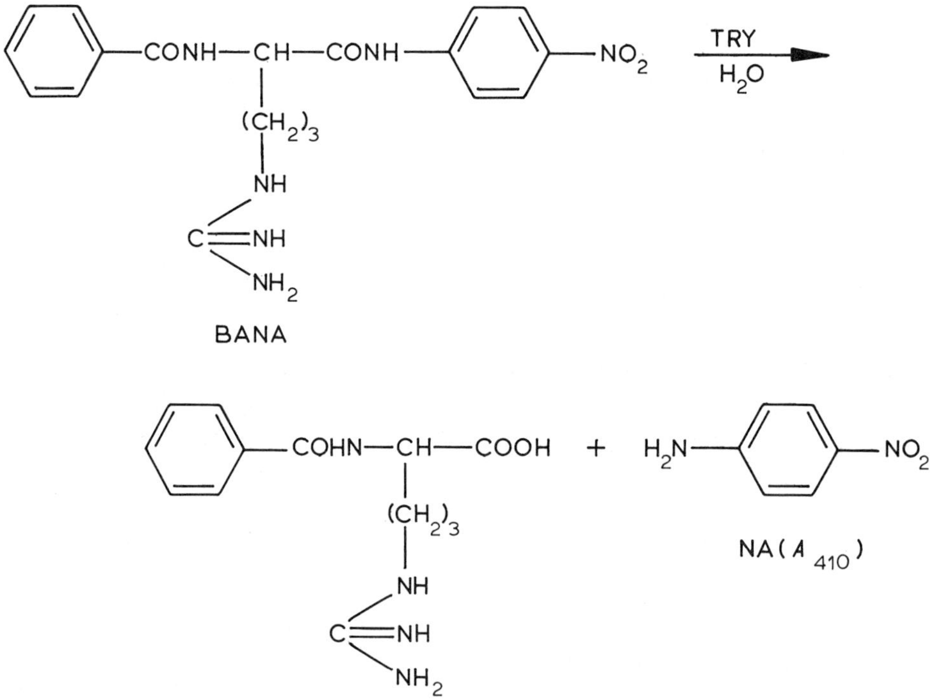

This is a normal tryptic catalysis and not a stoichiometric reaction as was the case in the first-mentioned equation.

A similar method was used by Gooding and Schmuck (1983) for the post-column on-line monitoring of chymotrypsin (CHY). Glutaryl-L-phenylalanine p-nitro-anilide (GPNA) was used as a specific substrate, which is hydrolysed by chymotrypsin to p-nitroaniline (NA) (Remy et al., 1981) according to the equation

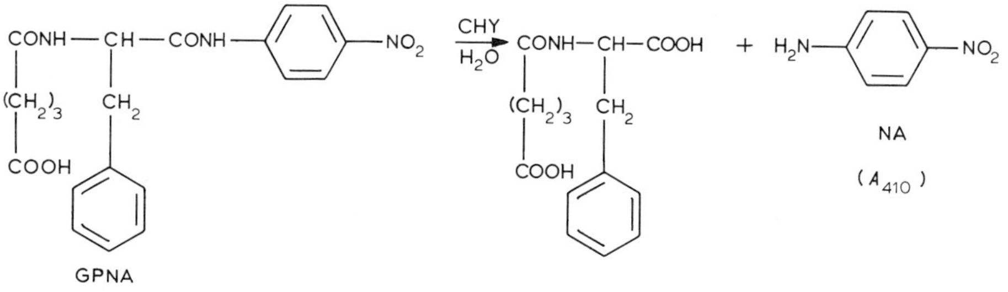

$$\text{CONH—CH—CONH—}\bigcirc\text{—NO}_2 \xrightarrow[\text{H}_2\text{O}]{\text{CHY}} \text{CONH—CH—COOH} + \text{H}_2\text{N—}\bigcirc\text{—NO}_2$$

with side chains $(CH_2)_3$, CH_2, $COOH$ and a phenyl group (GPNA on the left; product in the middle); NA on the right, (A_{410}).

The effluent was monitored by the measurement of the absorbance at 410 nm.

8.2.3 Lactate dehydrogenase isoenzymes

The most intensively studied isoenzyme family are lactate dehydrogenases, owing to their importance for diagnostic purposes. Lactate dehydrogenase (LD) is a tetramer of M_r = 150 000, containing two structurally distinct subunits (M from skeletal muscle and H from heart muscle). Various combinations of these subunits lead to five different isoenzymes of the same activity, LDH_1 - LDH_5 (i.e., H_4, H_3M, H_2M_2, HM_3 and M_4, respectively). A third subunit, C, was found in testicular tissue and spermatozoa. C usually occurs as tetramer C_4, but some hybridized C+H+M isoenzymes were also found in nature (Evrev, 1975).

The oxidation of lactate to pyruvate using the oxidized form of nicotinamide adenine dinucleotide, NAD^+, is the principal reaction used in all instances for the post-column detection of these isoenzymes:

$$CH_3CH(OH)COOH + NAD^+ \xrightarrow{LD} CH_3COCOOH + NADH + H^+$$
$$(A_{340}, E_{457})$$

The reduced cofactor, NADH, can be continuously monitored either by measurement of the absorbance at 340 nm or by fluorescence (excitation at 457 nm). The fluorescence method has the great advantage of higher sensitivity. More enzyme-catalysed reactions can be coupled to this redox reagent (cf., Section 8.2.4).

Kudirka et al. (1976) used for the detection of LD after anion-exchange HPLC separation the above reaction, realized with a Technicon AutoAnalyzer II single-channel colorimeter, where the absorbance at 340 nm was measured. Chang et al. (1976) used the principle illustrated in Fig. 5.29 (i.e., the supply of the substrate through a union tee into the line leading to a packed reaction column). In contrast, Schroeder et al. (1977) described the principle of single-stream

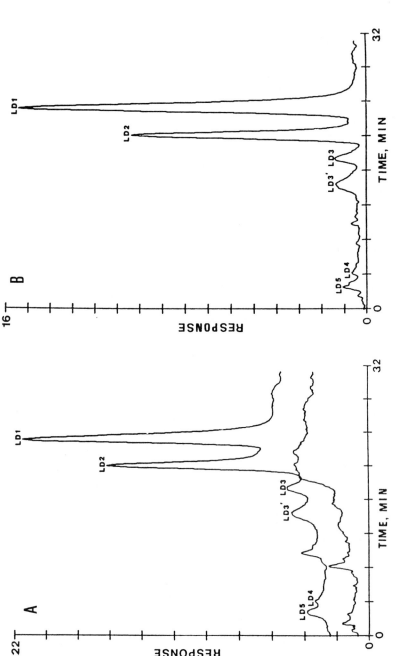

Fig. 8.1. (A) Chromatogram of serum LD isoenzymes from a patient with myocardial infarct. The upper trace was recorded at the downstream detector (detector 2). The lower trace shows the background absorbance, as it was observed at the upstream detector (detector 1) (cf., Fig. 5.31). The total serum LD activity was 502 U/l. (B) Profile of serum LD isoenzymes resulting from the correction for background absorbance (cf., A). The percentages of the total area are as follows: LD5, 2.5; LD4, 1.2; LD3', 6.5; LD3, 4.4; LD2, 28.4; LD1, 56.9%; the high values of LD1 and LD2 are significant for the diagnosis of infarction. (Reprinted from Fulton et al., 1979.)

or parallel-stream reaction detectors based on an open-tubular system (an empty
reaction coil). Both absorbance and fluorescence monitoring were used. The post-
column on-line detection of LD isoenzymes was further investigated and developed
by Schlabach and co-workers (1977, 1978, 1980a,b), Schlabach and Regnier (1978),
Denton et al. (1979) and Fulton et al. (1979).

Liver or muscle damage, myocardial damage and some other accidents and
pathological cases can be quickly detected by LD chromatographic profiling of
a small sample of serum. Fig. 8.1 shows the result of the application of a
computer-controlled dual-detector post-column reaction system to LD serum
profiling using background subtraction. Details of the various detector systems
were discussed in Section 5.2.3.

LD chromatographic profiling with on-line post-column detection methods was
reviewed by Regnier et al. (1977), Regnier and Gooding (1981) and Vacik and
Toren (1982).

8.2.4 Hexokinase isoenzymes and 3-phosphoglycerate kinase

The post-column on-line detection of hexokinase isoenzymes is an example of
coupled enzyme assays. The enzyme monitoring methods can be classified into two
groups: (a) direct detection, where the immediate enzymic product is monitored
(Sections 8.2.1 - 8.2.3 give several examples); (b) coupled enzyme assays,
where an additional enzyme or more additional enzymes are required to convert
the product of the primary enzyme reaction into a more easily detectable form.
Schlabach and Regnier (1978) described the first method and also the application
of the second method to the detection of hexokinase (HK) isoenzymes. They also
evaluated the efficiency of both free and immobilized coupling enzyme(s).

The measured hexokinase catalyses the phosphorylation of D-glucose to
D-glucose-6-phosphate (G-6-P) using adenosine-5-triphosphate (ATP), and
adenosine-5-diphosphate (ADP) is released. The G-6-P formed is then oxidized by
nicotinamide adenine dinucleotide (oxidized form NAD^+) to gluconolactone-6-
phosphate; this reaction is catalysed by a coupling enzyme, glucose-6-phosphate
dehydrogenase (G-6-PDH). The reduced coenzyme NADH can be determined by known
methods (cf., Section 8.2.3), i.e., by absorbance measurement at 340 nm or,
better, by fluorescence measurement at 457 nm. The sequence of reactions is as
follows:

$$\text{D-Glucose + ATP} \xrightarrow{\text{HK}} \text{D-Glucose-6-phosphate + ADP}$$

$$\text{D-Glucose-6-phosphate + NAD}^+ \xrightarrow{\text{G-6-PDH}} \text{NADH + D-Gluconolactone-6-phosphate + H}^+$$
$$(A_{340})$$

The coupling enzyme G-6-PDH must be added together with the primary substrate D-glucose and other reagents (ATP + NAD^+) to the solution pumped into the reactor. However, the coupling enzyme is only a catalyst and is not consumed in the reaction. It can be saved when used in the reactor in an immobilized form. This is the general principle of how to save the auxiliary enzymes, which are sometimes expensive, if they have to be added to the substrate solution in a relatively large amount and then flow from the reactor to waste. Hexokinase isoenzymes were detected in rat liver and testicular tissue extracts using immobilized G-6-PDH in a column reactor.

For the detection of HK, Lowe et al. (1981) used in their "universal assay medium" the oxidized form of nicotinamide adenine dinucleotide phosphate ($NADP^+$) instead of NAD^+ in the above coupling reaction, and the reduced form NADPH was obtained as the reaction product, which absorbed at 340 nm. For the detection of 3-phosphoglycerate kinase (PGK) activity, the following sequence of reactions was used, where the coupling enzyme glyceraldehyde-3-phosphate dehydrogenase (G-3-PDH) was applied:

$$3\text{-Phosphoglycerate} + ATP \xrightarrow{PGK} \text{Glycerate-1,3-biphosphate} + ADP$$

$$\text{Glycerate-1,3-biphosphate} + NADH + H^+ \xrightarrow{G\text{-}3\text{-}PDH} \text{Glyceraldehyde-3-phosphate} +$$
$$(A_{340})$$
$$+ NAD^+ + \text{Phosphate}$$

The "universal" assay medium contained both $NADP^+$ (in the oxidized) and NADH (in the reduced) form. It is evident that the presence of hexokinase in the column effluent produces an increase in absorbance at 340 nm, whereas 3-phosphoglycerate kinase activity decreases the absorbance at 340 nm. The problems connected with this approach were discussed.

8.2.5 Creatine kinase isoenzymes

Creatine kinase (CK) is a dimer of M_r = 86 000 and is composed of M (muscle) of B (brain) subunits. Sometimes this enzyme was designated creatine phosphokinase, CPK. Three isoenzymes, CK_1 (or CPK_1) to CK_3 (CPK_3), can be derived from subunit combinations, i.e., BB, MB and MM, respectively. CK in brain consists of BB or contains B subunit, whereas MM predominates in skeletal muscle and MB is typical of heart muscle. CK isoenzyme profiling is very important for the precise diagnosis of myocardial infarction, in which MB isoenzyme is elevated, in contrast to the situation in coronary insufficiency.

CK catalyses the phosphorylation of creatine using adenosine-5-triphosphate (ATP), and therefore the systematic name of this enzyme is ATP:creatine phosphotransferase. However, this reaction tends to go in the opposite direction:

$$\text{Phosphocreatine + ADP} \underset{}{\overset{CK}{\rightleftharpoons}} \text{Creatine + ATP;} \; \Delta G^0 = -3 \; \text{kcal/mol}$$

The liberated ATP can be determined in different ways. A method using a bio-luminiscence reaction with firefly luciferin (Bostick et al., 1980) was mentioned in Section 6.6.4. Other methods employing coupling enzymes will be described here. ATP can react with D-glucose using catalysis by hexokinase (HK) and the D-glucose-6-phosphate formed can be determined by oxidation with nicotinamide adenine dinucleotide (NAD^+) (Sections 8.2.3 and 8.2.4), or with nicotinamide adenine dinucleotide phosphate $[NAD(P)^+]$:

$$\text{ATP + D-glucose} \overset{HK}{\longrightarrow} \text{D-glucose-6-phosphate + ADP}$$

$$\text{D-Glucose-6-phosphate + NAD(P)}^+ \overset{G\text{-}6\text{-}PDH}{\longrightarrow} \text{D-gluconolactone-6-phosphate +}$$
$$+ \text{ NAD(P)H + H}^+$$

Schlabach et al. (1977) and Schlabach and Regnier (1978) described this procedure. The use of either NAD^+ or $NAD(P)^+$ depends on whether the G-6-PDH coupling enzyme is from yeast or from the microbe *Leuconostoc mesenteroides*. The latter prefers NAD^+, whereas the yeast enzyme is specific for $NAD(P)^+$. The authors also tested and discussed the application of immobilized coupling enzymes. Denton et al. (1978, 1979) described the on-line monitoring of CK isoenzymes by use of an immobilized enzyme microreactor. Other contributions to the development of CK post-column detection were published by Schlabach et al. (1978, 1980a). Interferences appearing in the fluorimetrically measured profiles of CK iso-enzymes in human serum, due to serum albumin, lipoprotein and prealbumin, were discussed by Schlabach et al. (1980b). Elkins (1977) described an alternative method for the determination of CK isoenzyme activity by the application of an anion-exchange column and a centrifugal analyser.

The HPLC on-line monitoring of CK isoenzymes was discussed and reviewed by Regnier et al. (1977), Regnier and Gooding (1981) and Vacik and Toren (1982).

8.2.6 Arylsulphatase isoenzymes

Bostick et al. (1978) described the anion-exchange separation of two arylsulphatase isoenzymes, A and B, in human urine using continuous detection based
on p-nitrocatechol sulphate hydrolysis. The product first passes through the
first (reference) cell and, after alkalinization with NaOH, it passes through
the second (sample) cell, in which the colour specific for the arylsulphatase
product is monitored.

This enzyme also hydrolyses p-nitrophenol sulphate, but the natural substrate
for arylsulphatase A is cerebroside sulphate, which accumulates in the body in
the case of a deficiency of this enzyme.

Bostick et al. (1978) also described the separation and analysis of arylsulphatase isoenzymes in other human body fluids. Sera from patients with
colorectal cancer were examined.

Arylsulphatase was also mentioned in a review by Regnier and Gooding (1981),
describing the rapid separation of proteins including isoenzymes for the purpose
of clinical analysis.

8.3 EXAMPLES OF ENZYME SEPARATIONS

8.3.1 Proteolytic enzymes

Buchholz et al. (1982) used SEC on LiChrosorb Diol to study changes in
enzymes in solution, e.g., for the rapid analysis of trypsin autolytic degradation. Fig. 8.2 illustrates that the decrease in the height of the first chromatographic peak follows the decrease in enzymic activity of the enzyme solution,
suggesting that the first peak in HPLC corresponds to the integral active enzyme.
Stricker et al. (1981) purified milligram amounts of commercially prepared bovine
trypsin (TRY) by RP-HPLC. Titani et al. (1982) described a simple and rapid
purification of commercial TRY and chymotrypsin (CHY) by RP-HPLC, using acetonitrile in dilute trifluoroacetic acid at pH 2. The enzymes were prepared in
amounts appropriate for the structural analysis of proteins. Each purified
enzyme showed the single expected substrate specificity. Štrop and Čechová (1981)
separated α- and β-trypsin by HIC on both analytical and preparative scales.
Gooding et al. (1984) described the analysis of proteins with new, mildly
hydrophobic HPLC packing materials (cf., Fig. 7.6). In another part of this
work, the activity of TRY was monitored using post-column on-line detection with
benzoyl-D,L-arginine p-nitroanilide (BANA) (cf., Section 8.2.2).

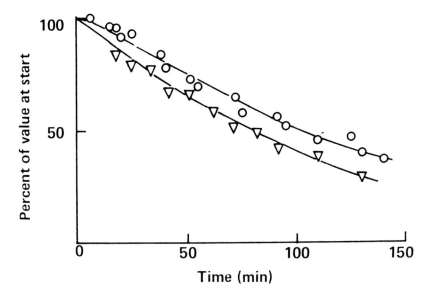

Fig. 8.2. Autolytic degradation of trypsin in solution as a function of time, followed by the decrease in the height of the first peak in the HPLC trace (∇) and by measurement of the enzymic activity using the substrate N-α-benzoyl-arginine p-nitroanilide (o). (Reprinted from Buchholz et al., 1982.)

Another group of workers used affinity chromatography for the rapid isolation of proteolytic enzymes. Kasche et al. (1981) immobilized soybean trypsin inhibitor as a biospecific adsorbent by the glutardialdehyde method on amino-silanized LiChrospher. α- and β-TRY were separated with a pH gradient and chymotrypsinogen and α-CHY were separated at constant pH. The kinetics of the separation were studied (cf., Section 3.9). Turková et al. (1981) used Separon H (Laboratory Instrument Works, Prague, Czechoslovakia) derivatized with epichlorohydrin as a support for large-scale HPLAC. For the preparation of a specific sorbent for carboxylic proteinases, the ligand ε-aminocaproyl-L-Phe-D-Phe-OCH$_3$ was synthesized and covalently attached to the support (see also Section 4.4.5). The proteinases from raw pepsin and *Aspergillus oryzae* were isolated using this sorption material. Specific sorbents for high-performance liquid affinity chromatography (HPLAC) and the large-scale isolation of proteinases were also dealt with by Turková (1982). Small et al. (1981) used packings containing immobilized triazine dyes (cf., Section 4.4.5) for the HPLAC of carboxypeptidase (CP) G-2, in addition to other enzymes and proteins; for the detection of CP the change in absorbance at 320 nm was followed, when methotrexate (4-amino-N^{10}-methylpteroylglutamate) was hydrolysed to 2,4-diamino-N^{10}-methylpteroate (McCullough et al., 1971). Marceau et al. (1983) described the rapid assay of human plasma carboxypeptidase N by HPLC separation of hippuryl

lysine and its products. Wunderwald et al. (1983) studied the removal of endoproteinases from biological fluids by "sandwich affinity chromatography" with
α_2-macroglobulin bound to zinc chelated Sepharose. TRY, CHY, thermolysin,
elastase, bromelain, ficin and papain were bound, but not exoproteinases such
as carboxypeptidases A and Y. The simple loading procedure, simple regeneration
and high capacity are advantages of the method (see also Section 7.4.3). Shimura
et al. (1984) described the HPLAC of plasmin and plasminogen on a hydrophilic
vinyl polymer gel (Toyopearl HW 65 S) with p-aminobenzamidine. The column packed
with this material retained both plasmin and plasminogen. Plasminogen was eluted
with 6-aminohexanoic acid (a haptenic compound for the lysine-binding site of
plasminogen). For the elution of plasmin, the coexistence of 6-aminohexanoic
acid and leupeptin (a competitive inhibitor of plasmin) was necessary. Monitoring
was effected by fluorimetric detection of the eluted protein and one-line assay
of plasmin activity using peptidylmethylcoumarylamide, a fluorogenic substrate
(cf., Fig. 5.18).
Gooding and Schmuck (1983) purified TRY and other basic proteins (CHY, lysozyme
and cytochrome c) by cation-exchange HPLC. A SynChropak CM 300 column was used
and sodium acetate proved best for ionic strength gradient elution. TRY activity
was monitored using BANA (see Section 8.2.2) and for the detection of CHY the
reagent GPNA was used (see Section 8.2.2). Cohen et al. (1984) demonstrated
multiple peak formation in the RP-HPLC of papain: the first peak was native and
the second was part of the enzyme, denatured in contact with the column.

8.3.2 Cellulolytic, pectolytic and amylolytic enzymes

Montenecourt et al. (1980) studied the biochemical nature of cellulases from
two physical hypercellulolytic mutants of *Trichoderma reesei* using HPLC on DEAE-
silica (Applied Science Labs., State College, PA, U.S.A.). Buchholz et al. (1982)
correlated the activity and the molecular size of cellulase components using
high-performance SEC on LiChrosorb Diol columns. For the determination of the
cellulase activities, glucanases were analysed by incubation with Avicel; samples
were taken at intervals, centrifuged and analysed for glucose and cellobiose by
HPLC on a LiChrosorb-NH$_2$ (Knauer) column with acetonitrile-water as the eluent
and a differential refractometer detector. β-Glucosidases were determined by
incubation with cellobiose and analysis as before. The main activities were
found in the main peaks. Hostomská and Mikeš (1983) described the analytical
MPLC of cellulolytic enzymes on Spheron ion-exchangers and also (Hostomská and
Mikeš, 1984) separated the cellulolytic system of *Trichoderma viride-reesei*
mutant by MPLC on the large preparative scale and isolated a new *exo*-cellobio-
hydrolase.

Mikeš et al. (1981) studied the MPLC separation of pectic enzymes (*endo*-D-galacturonase, *exo*-D-galacturonase, pectin lyase and pectin esterase) from Pectinex Ultra (originating from *Aspergillus niger* fermentation) and Rohament P technical pectolytic preparations. All available Spheron ion exchangers were tested for the separation of these pectic enzymes. The experience from these studies was used for a similar analysis of pectic enzymes (Rexová-Benková et al., 1982) in the Czechoslovak technical pectolytic preparation Leozym, which is a by-product in the manufacture of citric acid by fermentation of *Aspergillus niger*. Mikeš and Rexová (1988) reviewed techniques for the HPLC of pectic enzymes.

Fourmy et al. (1982) described a rapid quantitative method for the detection of macroamylase in human serum by HPLC (SEC). Isoamylases from saliva and pancreatic juice were also analysed. Serum from normal persons contained two amylase peaks, distant from the void volume. The second peak was markedly higher in serum from patients with acute pancreatitis and mumps. In contrast, the amylase activity in serum from patients with macroamylasaemia was eluted in the void volume.

8.3.3 Oxidoreductases

High-performance liquid affinity chromatography (HPLAC) has been used successfully for the separation of many enzymes. Horse liver alcohol dehydrogenase (LADH) and pig heart lactate dehydrogenase (LDH) were well separated from each other and from bovine serum albumin on an N^6-(6-aminohexyl)-AMP bonded silica column. Biospecific elution using dilute solutions of NAD^+-pyrazole and NAD^+-pyruvate were used, in addition to a high concentration of sodium chloride solution. When a gradient of NADH was applied, it was possible to separate H_4 and M_4 LDH isoenzymes (Ohlson et al., 1978). Lowe et al. (1981) used Cibacron Blue F3G-A bonded silica for the fully automated HPLAC resolution of dehydrogenases (such as LADH and LDH), in addition to hexokinase, 3-phosphoglycerate kinase and other enzymes. Simultaneous detection was achieved by monitoring the change in absorbance at 340 nm (cf., Section 8.2.4). Small et al. (1981) immobilized a number of triazine dyes to microparticulate silica and studied their application for the separation of LDH and other enzymes. Borchert et al. (1982) immobilized concanavalin A to porous silica and used this support for the analysis and purification of glucose oxidase and glucoproteins peroxidase. This HPLAC method was studied in detail.

IEX has also often been used for the rapid separation of enzymes. Matsumoro (1981) described the theory and use of serum LDH isoenzyme analysis on IEX 525

QAE; with a gradient of NaCl concentration the total separation of isoenzymes was achieved within 20 min. Van der Wal and Huber (1980) studied the performance of classical non-rigid ion-exchange packings of small particle size in the separation of cytochrome *c* and derivatives by HPLC, but better results were obtained with (meth)acrylic cation exchangers and hydroxyapatite. Vanecek and Regnier (1982) described a rapid separation of lipoxygenase I using a column of LiChrospher Si 4000 coated with a heavy layer of polyethylenimine and cross-linked with 1,4-butanediol diglycidyl ether. Lindblom (1983) developed a simple method for the isolation of glucose-6-phosphate dehydrogenase from a yeast enzyme concentrate. A column of Polyanion SI (8 μm) was used for analytical and Polyanion SI (17 μm) for preparative purposes.

Hearn et al. (1980) studied the SEC of sheep liver aldehyde dehydrogenase (and thyroglobulin) on μBondagel E-linear. Indications of the influence of a mixed-mode separation principle were found. Power et al. (1983) used RP-HPLC for the purification of nuclear coded subunits of a membrane oligomeric protein - yeast cytochrome *c* oxidase.

8.3.4 Enzymes of phosphate metabolism, other miscellaneous enzymes and enzymic reactions

Small et al. (1981) used dye-ligand chromatography for the fractionation of yeast hexokinase on Procion Green H-4G silica, and of L-tryptophanyl-tRNA synthetase of Cibacron Blue F3G-A silica and Procion Brown MX-5BR silica. Lowe et al. (1981) studied chromatography on Cibacron Blue F3G-A silica and detection methods for hexokinase, 3-phosphoglycerate kinase (cf., Section 8.2.4) and also of pancreatic ribonuclease A.

Mikeš et al. (1978) described the rapid IEC separation of technical enzymes [e.g., crude protease from *Aspergillus sojae* and glucose oxidase from *Aspergillus niger* (Fig. 8.3)]. Kato et al. (1980) purified crude β-galactosidase from bacterial cells and commercial urease by SEC on TSK Gel 3000 SWG. Roughly 15-fold purification was achieved in a single filtration.

Lim (1979) developed HPLC methods for the determination of enzymes of the haeme biosynthetic pathway (δ-aminolaevulinic acid synthetase and dehydratase, uroporphyrinogen I synthetase). The enzymes were not isolated, but determined by the separation of their low-molecular-weight products. Studebaker (1979) reviewed the analysis of enzymic reactions by HPLC.

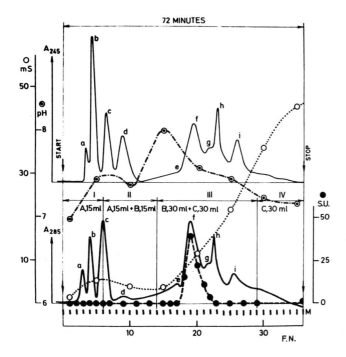

Fig. 8.3. Chromatography of technical glucose oxidase on a DEAE-Spheron column (20 cm x 0.8 cm). Load: 15 mg of preparation in 0.2 ml of buffer A. The ion exchanger was equilibrated with buffer A. I, Buffer A without gradient. Linear gradients II (A + B) and III (B + C). IV, buffer C without gradient. Flow-rate, 2 ml/min; 4-ml fractions; temperature, 14°C; counter pressure, 3-7 atm; chart speed, 2 mm/min. Buffers: A, 0.01 M acetic acid + NaOH, pH 6.8; B, 0.3 M acetic acid + NaOH, pH 5.5; C, buffer B, 1 M in NaCl, pH 5.3. Broken line, glucose oxidase activity of effluent in Sarret units (S.U.; 1 S.U. corresponds to the consumption of 600 μl of oxygen at 30°C). F.N. = Fraction number; M = automatic marking of fraction collection; mS = electric conductivity in mS. Corresponding peaks in both parts of the chromatogram are designated by a-i. The UV spectrum of the chromophore of compound d differed significantly from the others. (Reprinted from Mikeš et al., 1978.)

8.4 COMMENTS ON LITERATURE

Guilbault's (1976) Handbook is not devoted to HPLC methods, but it describes in detail assay procedures for 39 important enzymes, enzymic analyses of more than 55 substrates and the application of immobilized enzymes in enzymic analyses, and provides much additional practical information. Regnier et al. (1979) and Regnier and Gooding (1981) reviewed the HPLC of proteins, including the HPLC of enzymes; these reviews cover specialized applications in clinical analysis.

Ishiguro and Shinohara (1981) reviewed the use of IEC, SEC, affinity, adsorption and high-performance liquid chromatography in the separation of isoenzymes (especially of lactate dehydrogenase, creatine phosphate kinase, glutamate-oxalacetate transaminase, alkaline phosphatase and malic dehydrogenase). Mikeš (1981/1982 and 1982) dealt with the rapid chromatographic analysis of enzymes and other proteins for application to food chemistry. Vacik and Toren (1982) reviewed the separation and measurement of isoenzymes and other proteins by HPLC; this treatise covered predominantly biomedical applications.

Enzyme separation by RP-HPLC was reviewed by Strickler et al. (1984). Jakoby (1984) edited a special volume of *Methods in Enzymology* dealing with the purification of enzymes, where HPLC techniques for enzyme separations were described in detail; Wehr dealt with the care of HPLC columns, Unger explained SEC methods, Regnier IEC, Richey optimal pH conditions for IEC, Hearn RPC and Larsson HPLAC.

REFERENCES

Aoshima, H., *Anal. Biochem.*, 95 (1979) 371-376.
Bender, M.L., Begue-Canton, M.L., Blakely, R.L., Brubacher, L.J., Feder, J., Gunter, C.R., Kezdy, F.J., Killheffer, J.V., Marshall, T.H., Miller, C.G., Roeske, R.W. and Stoops, J.K., *J. Am. Chem. Soc.*, 88 (1966) 5890-5913.
Borchert, A., Larsson, P.-O. and Mosbach, K., *J. Chromatogr.*, 244 (1982) 49-56.
Bostick, W.D., Denton, M.S. and Dinsmore, S.R., *Clin. Chem.*, 26 (1980) 712-717.
Bostick, W.D., Dinsmore, S.R., Mrochek, J.E. and Waalkes, T.P., *Clin. Chem.*, 24 (1978) 1305-1316; *C.A.*, 89 (1978) 142 370g.
Buchholz, K., Gödelmann, B. and Molnar, I., *J. Chromatogr.*, 238 (1982) 193-202.
Chang, S.H., Gooding, K.M. and Regnier, F.E., *J. Chromatogr.*, 125 (1976) 103-114.
Chase, T. and Shaw, E., *Biochem. Biophys. Res. Commun.*, 29 (1967) 508-514.
Cohen, L., Djordjevich, J. and Ormiste, V., *J. Lab. Clin. Med.*, 64 (1964) 355-374.
Cohen, S.A., Benedek, K.P., Dong, S.T.Y. and Karger, B.L., *Anal. Chem.*, 56 (1984) 217-221.
Denton, M.S., Bostick, W.D. and Dinsmore, S.R., *Clin. Chem.*, 24 (1978) 1408-1413; *C.A.*, 89 (1978) 142 373k.
Denton, M.S., Bostick, W.D., Dinsmore, S.R. and Mrochek, J.E., *Biological/Biomedical Applications of Liquid Chromatography 2 (Chromatographic Science Series*, Vol. 12), Marcel Dekker, New York, 1979, pp. 165-191; *C.A.*, 92 (1980) 36 513h.
Elkins, B.N., *Clin. Chem.*, 23 (1977) 1510.
Evrev, T.I., in Markert, C. (Editor), *Isoenzymes*, Vol. 2, Academic Press, New York, 1975, p. 129.
Fourmy, D., Pradayrol, L., Bommelaer, G. and Ribet, A., *Gastroenterol. Clin. Biol.*, 6 (1982) 249-251; *C.A.*, 96 (1982) 176 658s.
Fulton, J.A., Schlabach, T.D., Kerl, J.E. and Toren, E.C., *J. Chromatogr.*, 175 (1979) 283-291.
Galen, R.S. and Raymond, G.S., *Clin. Chem.*, 21 (1975) 1848-1850; *C.A.*, 84 (1976) 15 416y.
Galen, R.S., Reiffel, J.A. and Gambino, S.R., *J. Am. Med. Assoc.*, 232 (1975) 145.
Gooding, D.L., Schmuck, M.N. and Gooding, K.M., *J. Chromatogr.*, 296 (1984) 107-114.
Gooding, K.M. and Schmuck, M.N., *J. Chromatogr.*, 266 (1983) 633-642.
Guilbault, G.G., *Handbook of Enzymatic Methods of Analysis (Clinical and Biochemical Analysis*, Vol. 4), Marcel Dekker, New York, Basle, 1976, 738pp.
Haverback, B.J., Dyce, B., Bundy, H. and Edmonson, H.A., *Am. J. Med.*, 29 (1960) 424.
Hearn, M.T.W., *Methods Enzymol.*, 104C (1984) 190-212.

Hearn, M.T.W., Grego, B., Bishop, C.A. and Hancock, W.S., *J. Liq. Chromatogr.*,
   3 (1980) 1540-1560.
Hostomská, Z. and Mikeš, O., *J. Chromatogr.*, 267 (1983) 355-366.
Hostomská, Z. and Mikeš, O., *Int. J. Peptide Protein Res.*, 23 (1984) 402-410.
International Union of Biochemistry (IUB), *Enzyme Nomenclature*, Elsevier,
   Amsterdam, 1965, 1972.
Ishiguro, I. and Shinohara, R., *Rinsho Kensa*, 25 (1981) 183-188; *C.A.*, 94 (1981)
   170 153k.
Jakoby, W.B. (Editor), *Enzyme Purification and Related Techniques* (*Methods
   Enzymol.*, Vol. 104C), Academic Press, New York, 1984, 528 pp.
Kasche, V., Buchholz, K. and Galunsky, B., *J. Chromatogr.*, 216 (1981) 169-174.
Kato, Y., Komiya, K., Sawada, Y., Sasaki, H. and Hashimoto, T., *J. Chromatogr.*,
   190 (1980) 305-310.
Kudirka, P.J., Bushy, M.G. and Toren, E.C., Jr., *Clin. Chem.*, 21 (1975) 450-452.
Kudirka, P.J., Schroeder, R.R., Hewitt, T.E. and Toren, E.C., Jr., *Clin. Chem.*,
   22 (1976) 471-474.
Larsson, P.-O., *Methods Enzymol.*, 104C (1984) 212-223.
Lim, C.K., *Proc. Anal. Div. Chem. Soc.*, 16 (1979) 305-307; *C.A.*, 92 (1980)
   71 551r.
Lindblom, H., *J. Chromatogr.*, 266 (1983) 265-281.
Lowe, Ch.R., Glad, M., Larsson, P.-O., Ohlson, S., Small, D.A.P., Atkinson, T.
   and Mosbach, K., *J. Chromatogr.*, 215 (1981) 303-316.
Marceau, F., Drumheller, A., Gendreau, M., Lussier, A. and St-Pierre, S., *J.
   Chromatogr.*, 266 (1983) 173-177.
Matsumoto, K., *Kensa to Gijutsu*, 9 (1981) 359-366; *C.A.*, 94 (1981) 187 510x.
McCullough, J.L., Chabner, B.A. and Bertino, J.R., *J. Biol. Chem.*, 246 (1971)
   7207-7213.
Meister, A., *J. Biol. Chem.*, 184 (1950) 117-129.
Mikeš, O., *Ernährung/Nutrition*, 5 (1981) 88-98 (in English, references ommited);
   C.A., 94 (1981) 152 760b. Czech translation with full references: *Chem.
   Listy*, 76 (1982) 59-79; *C.A.*, 96 (1982) 118 297u.
Mikeš, O. and Rexová, L'., *Methods Enzymol.*, 161B (1988) 385-399.
Mikeš, O., in Baltes, W., Czedik-Eysenberg, P.B. and Pfannhauser, W. (Editors),
   Recent Developments in Food Analysis (Proceedings of the 1st European Con-
   ference on Food Chemistry, EURO FOOD CHEM I, Vienna, Austria, February 1981),
   Verlag Chemie, Weinheim, 1982, pp. 306-321.
Mikeš, O., in Deyl, Z., Macek, K. and Janák, J. (Editors), *Liquid Column Chro-
   matography, A Survey of Modern Techniques and Applications* (*Journal of
   Chromatography Library*, Vol. 3), Elsevier, Amsterdam, 1975, pp. 807-830.
Mikeš, O., Sedláčková, J., Rexová-Benková, L. and Omelková, J., *J. Chromatogr.*,
   207 (1981) 99-114.
Mikeš, O., Štrop, P. and Sedláčková, J., *J. Chromatogr.*, 148 (1978) 237-245.
Montenecourt, B.S., Kelleher, T.J., Eveleigh, D. and Pettersson, L.G.,
   Biotechnol. Bioeng. Symp., 10 (1980); *C.A.*, 94 (1981) 26 631x.
Morin, L.G., *Clin. Chem.*, 23 (1977) 205-210.
Neilands, J.B., *Science (Washington, D.C.)*, 115 (1952) 143-144.
Ohlson, S., Hansson, L., Larsson, P.-O. and Mosbach, K., *FEBS Lett.*, 93 (1978)
   5-9.
Power, S.D., Lochrie, M.A. and Poyton, R.O., *J. Chromatogr.*, 266 (1983) 585-598.
Pressey, R., *HortScience*, 19 (1984) 572-573.
Prusík, Z., in Deyl, Z., Macek, K. and Janák, J. (Editors), *Liquid Column
   Chromatography, A Survey of Modern Techniques and Applications*, (*Journal of
   Chromatography Library*, Vol. 3), Elsevier, Amsterdam, 1975, p. 803.
Raschbaum, G.R. and Everse, J., *Anal. Biochem.*, 90 (1978) 146-154.
Regnier, F.E., *Methods Enzymol.*, 104C (1984) 170-189.
Regnier, F.E. and Gooding, K.M., in Laurence, J., Marton, J. and Kabra, P.M.
   (Editors), *Liquid Chromatography in Clinical Analysis*, Humana Press, Clifton,
   NJ, 1981, pp. 323-353.

Regnier, F.E., Gooding, K.M. and Chang, S.H., *Contemp. Top. Anal. Chem.*, 1 (1977) 1-48; *C.A.*, 87 (1977) 196 576f.

Remy, M.H., Guillochon, D. and Thomas, D., *J. Chromatogr.*, 215 (1981) 87-91.

Rexová-Benková, L., Omelková, J., Mikeš, O. and Sedláčková, J., *J. Chromatogr.*, 238 (1982) 183-192.

Richey, J.S., *Methods Enzymol.*, 104C (1984) 223-233.

Schlabach, T.D., Alpert, A.J. and Regnier, F.E., *Clin. Chem.*, 24 (1978) 1351-1360.

Schlabach, T.D. and Regnier, F.E., *J. Chromatogr.*, 158 (1978) 349-364.

Schlabach, T.D., Chang, S.H., Gooding, K.M. and Regnier, F.E., *J. Chromatogr.*, 134 (1977) 91-106.

Schlabach, T.D., Fulton, J.A., Mockridge, P.B. and Toren, E.C., Jr., *Anal. Chem.*, 52 (1980a) 729-733.

Schlabach, T.D., Fulton, J.A., Mockridge, P.B. and Toren, E.C., Jr., *Clin. Chem.*, 25 (1979) 1600-1608.

Schlabach, T.D., Fulton, J.A., Mockridge, P.B. and Toren, E.C., Jr., *Clin. Chem.*, 26 (1980b) 707-711; *C.A.*, 93 (1980) 21 320q.

Schroeder, R.R., Kudirka, P.J. and Toren, E.C., Jr., *J. Chromatogr.*, 134 (1977) 83-90.

Shimura, K., Kazama, M. and Kasai, K.-I., *J. Chromatogr.*, 292 (1984) 369-382.

Small, D.A.P., Atkinson, T. and Lowe, Ch.R., *J. Chromatogr.*, 216 (1981) 175-190.

Strickler, M.P. and Gemski, M.J., in Hancock, W.S. (Editor), *CRC Handbook of HPLC for the Separation of Amino Acids, Peptides, and Proteins*, Vol. II, CRC Press, Boca Raton, FL, 1984, pp. 349-355.

Strickler, M.P., Gemski, M.J. and Doctor, B.P., *J. Liq. Chromatogr.*, 4 (1981) 1765-1775.

Štrop, P. and Čechová, D., *J. Chromatogr.*, 207 (1981) 55-62.

Studebaker, J.F., *Biological/Biomedical Applications of Liquid Chromatography 1 (Chromatography Science Series*, Vol. 10), Marcel Dekker, New York, 1979, pp. 261-281; *C.A.*, 90 (1979) 147 426k.

Titani, K., Sasagawa, T., Resing, K. and Walsh, K.A., *Anal. Biochem.*, 123 (1982) 408-412.

Turková, J., in Gribnau, T.C.J., Visser, J. and Nivard, R.J.F. (Editors), *Affinity Chromatography and Related Techniques*, Elsevier, Amsterdam, 1982, pp. 513-528.

Turková, J., Bláha, K., Horáček, J., Vajčner, J., Frydrychová, A. and Čoupek, J., *J. Chromatogr.*, 215 (1981) 165-179.

Unger, K., *Methods Enzymol.*, 104C (1984) 154-169.

Vacik, D.N. and Toren, E.C., Jr., *J. Chromatogr.*, 228 (1982) 1-31.

Van den Wal, Sj. and Huber, J.F.K., *Anal. Biochem.*, 105 (1980) 219-229.

Vanecek, G. and Regnier, F.E., *Anal. Biochem.*, 121 (1982) 156-169.

Vesell, E.S. and Bearn, A.G., *Proc. Soc. Exp. Biol. Med.*, 94 (1957) 96-99; *C.A.*, 51 (1957) 10 606a.

Wehr, C.T., *Methods Enzymol.*, 104C (1984) 133-154.

Wunderwald, P., Schrenk, W.J., Port, H. and Kresze, G.-B., *J. Appl. Biochem.*, 5 (1983) 31-42.

Chapter 9

PEPTIDES

9.1 INTRODUCTION

   In 1951, Moore and Stein introduced the ion-exchange chromatography (IEC) of
amino acids on sulphonated polystyrene resins and in 1958 Spackman et al.
automated this process. These developments opened the way for relatively rapid
column separations of peptides, because the principle of amino acid separations
and the equipment developed could be applied also to the separation of peptides,
especially short ones. Because the ninhydrin reagent used in the apparatus involved produced only a weak colour yield with peptides (in comparison with amino
acids), in some instances a hydrolytic step was inserted before the detection
unit in amino acid analysers; however, direct ninhydrin colorimetry was also
used (cf., Section 5.1.10). Many papers have been published describing applications of this approach and various modifications to the process and the equipment, and some of them will be shortly mentioned here.
   In 1966, Benson et al. described the accelerated chromatographic analysis of
peptides on a spherical resin using a Beckman 120C analyser and applied it to
the separation of tryptic hydrolysates of haemoglobin; at that time one run
lasted 7.5 h. Bennett and Creaser (1970) modified the Beckman 120B analyser, used
volatile pyridine-acetate buffers for the automated analysis and obtained good
results with tryptic digest of oxidized ribonuclease. Oshima et al. (1978)
applied a JEOL amino acid analyser to the separation of series of homo-oligopeptides. Voelter et al. (1978) used Durrum DC-1A resin (18 µm particles) and
volatile buffers in a peptide analyser (developed in collaboration with Biotronic) for the preparative high-performance liquid chromatography (HPLC) of
synthetic thyrotropin-releasing hormone analogues. The analysis time varied
from 3 to 6 h. Johnson (1979) described an effective peptide fractionation of
chymotryptic digest of carboxymethylated ribonuclease A and peptic digest of
carboxymethylated actin with essentially quantitative yields using a special
Beckman W-3 resin in a Beckman 119CL analyser and a volatile pyridine-acetate
buffer system. Applications of amino acid analysers and similar equipment for
the chromatography of peptides on ion-exchange resins were briefly reviewed by
Mikeš and Šebesta (1979). For some time chromatography on ion-exchange resins was
the method of choice for the separation of complicated peptide mixtures.

In the early 1970s another much quicker method was applied for the column chromatography of peptides and was rapidly improved in efficiency, namely reversed-phase chromatography (RPC). First Ryeszotarski and Mauger (1973) applied pellicular Corasil C_{18} to the separation of actinomycins. Tsuji et al. (1974) and Tsuji and Robertson (1975) successfully used Waters Bondapak C_{18}/Corasil and microparticulate μBondapak C_{18} for the quantitative HPLC of a raw polypeptide antibiotic, bacitracin powder. At the 14th European Peptide Symposium Burgus and Rivier (1976) presented a paper describing the HPLC purification of synthetic peptides, such as vasopressin, angiotensin, LRF, neurotensin and somatostatin, prepared by the solid-phase method; 10 μm μBondapak C_{18} (Waters) was used as the packing in two 30 cm x 4 mm I.D. columns in series; acetonitrile or ethanol, buffered with ammonium acetate (pH 4), was used as the mobile phase. The chromatography was monitored by measurement of the absorbance at 210 nm. The separation was achieved in 12 min. This approach to peptide purification represents an example of the main features of RP-HPLC, which were later modified in various ways in order to obtain optimal separations. Also in 1976, in New Zealand, Hancock et al. published the first part of a very well known series of papers dealing with the HPLC of peptides and proteins, which contributed considerably to the development of this area. The HPLC of underivatized tri- to hexapeptides was described, using RPC on Waters Bondapak C_{18}-Corasil and Bondapak Phenyl Corasil columns with methanol-water as the eluent and monitoring the absorbance at 212 nm. The analysis time was 8-10 min.

Other papers followed quickly. Hansen et al. (1977) used Phenyl-Corasil, Poragel PN and PS for the RPC of tri- to decapeptides with acetonitrile-water as the mobile phase, and also studied GPC on Hydrogel IV. The influence of the residual silanol groups on the quality of the separation was also investigated. Mönch and Dehnen (1977) chromatographed di- to octapeptides on ODS supports and showed that a lower pH and the addition of potassium phosphate to the solvent resulted in narrower peaks and better resolution. Molnar and Horváth (1977) separated 29 components of a complicated mixture of amino acids and peptides in one run on LiChrosorb RP-18 using 0.5 M perchloric acid (pH 0.2) with acetonitrile as a gradient former. They gave a detailed explanation of the separation process and its theory, which will be dealt with in Section 9.2.

Krummen and Frei (1977a) separated five nonapeptides and by-products on the commercial 5 μm reversed-phase material Nucleosil C_8, C_{18} (Machery, Nagel & Co., Düren, F.R.G.), Spherisorb S5-ODS (Phase Separations, Queensferry, U.K.) and 10 μm RP-8 (E. Merck, Darmstadt, F.R.G.). Columns 7.5, 15 and 25 cm long and of 3 and 4 mm I.D. were used for isocratic elution using mixtures of water with organic solvents (acetonitrile, dioxane, methanol or n-propanol). The choice of organic solvent did not have a strong influence on the separation pattern. A detection limit of ca. 30 ng at 220 nm rendered this technique suitable for routine quantitative analysis, which was developed by the same authors. Krummen and Frei (1977b) described the quantitative analysis of nonapeptides in pharmaceutical dosage forms by HPLC. Oxytocin, lysopressin and other nonapeptides and their by-products were determined in liquid and solid pharmaceutical dosage forms using Nucleosil C_8 and C_{18} (5 and 10 µm) and RP-8 (10 µm) as supports, water buffered at pH 7 (phosphate buffers)-acetonitrile (4:1) as eluent and UV detection at 210 nm.

Karger and Giese (1978) discussed reversed-phase liquid chromatography and its applications in biochemistry in an excellent review. Kroeff and Pietrzyk (1978) studied the separation of short-chain peptide diastereoisomers on C_8 bonded phases and contributed to a better understanding of the process of peptide separation (cf., Section 9.2); similarly, Lundanes and Greibrokk (1978) examined four different reversed phases and many small underivatized peptides. Stein et al. (1978) discussed the HPLC ultramicroanalysis of opiate peptides, such as endorphins, and nonapeptides (vasopressin, oxytocin) using fluorescence detection. Stoklosa et al. (1978) studied the separation of trace amounts of impurities of nonapeptides by RP-HPLC with emphasis on the influence of water-acetonitrile ratio in the eluent. O'Hare and Nice (1979) tested the HPLC of 32 hormonal polypeptides and nine proteins on alkylsilane-bonded silica using gradient elution with acetonitrile in acid phosphate buffer. Because the results were correlated with Rekker fragmental constants, this paper will be dealt with in Section 9.2. Rivier and Burgus (1979) reviewed the application of RP-HPLC to peptides and discussed the separation of diastereoisomers.

In the development of the RPC of peptides, special attention has been devoted to the study of the influence of buffers and other ionic components in the mobile phase, which led to the broader application of ion-pair reversed-phase chromatography (IP-RPC). Hancock et al. (1978b), in the second part of their series, described the use of phosphoric acid in the analysis of underivatized peptides by RP-HPLC. In the study of the separation of di- to decapeptides they found that the addition of phosphoric acid to the mobile phase was very advantageous, altered the retention times dramatically and favourably, permitted good UV detection in the range 195-220 nm and gave a significant reduction in the concentration of organic solvents in the mobile phase; this reduced the risks of denaturation or precipitation. The conditions allowed excellent resolutions of peptides differing by as little as a single amino acid. The influence of the addition of phosphoric acid can be partly explained by protonation of the carboxyl group, which loses its negative charge, but mainly by protonation of the amino group, which in an acidic medium becomes positively charged and can create an ion pair with the hydrophilic phosphate anion, $R\overset{+}{N}H_3$ $^-OPO(OH)_2$, and

this ion pair behaves then as a more hydrophilic unit with a decreased retention
time. Hancock et al. (1978a) also studied the rapid HPLC analysis of peptides
using hydrophobic ion pairing of amino groups. It was found that a hydrophobic
anion (such as hexanesulphonate) can form an ion pair with a cationic group of
a peptide, $R\overset{+}{N}H_3$ $^-OSO_2(CH_2)_5CH_3$, and this ion pair decreases the polarity of the
sample and behaves as a more hydrophobic unit with an increased retention time.
If the ion pair was formed with an acid containing a long aliphatic hydrocarbon
chain [e.g., $CH_3(CH_2)_{11}SO_4Na$], an extremely hydrophobic ion pair was formed. In
this way a suitable selection of the corresponding pairing ions allowed the retention times of the separated peptides to be influenced very sensitively.
Hancock et al. (1978c) reviewed the above-mentioned principles and illustrated
them by examples of successful separations of peptides and short-chain proteins.

Rivier (1978) (see also Rivier et al., 1979a) described trialkylammonium
phosphate buffers in RP-HPLC and their advantages with regard to high resolution
and high recovery. In addition to μAlkylphenyl and μBondapak C_{18}, μCyanopropyl
columns have often been used for the purification of tritiated LRF, somatostatin,
insulin and cytochrome c. Various species of endorphins were also well separated
and the system was applied to the fractionation of tryptic digest of myelin.
Hancock et al. (1979a) also studied the use of cationic reagents (tetraalkylammonium, alkylammonium and ionorganic salts) in the RPC of di- to pentapeptides
and their effects on retention time. Some results could be explained on the
basis of either ion pairing or ion-exchange interactions of the reagent with the
peptide sample. It was interresting that hydrophobic cations with long or bulky
carbon chains, such as tetrabutylammonium or dodecylammonium ions, caused a
substantial decrease in retention times. In contrast, tetraethylammonium salts
gave a modest increase in retention time relative to ammonium salts. These results could be better understood if the influence of dynamic ion-exchange effects (i.e., sorption of hydrophobic ions on the support) for cationic reagents,
and their interplay with ion-pair partition, were considered. More detailed investigations of the effects of pH and ion-pair formation on the retention of
peptides on chemically bonded stationary phases were published by Hearn et al.
(1979b). Different anionic and cationic reagents were studies. At low pH hydrophobic anions resulted in an increase in retention, whereas hydrophobic cations
caused a decreased in retention. The results were in agreement with an ion-
pairing and ion-exchange explanation. Short alkyl chains participated predominantly in ion pair formation, whereas with long-chain reagents an ion-exchange mechanism dominated. The derived mathematical equations will be dealt with in
Section 9.2.

The applications of ion-pair HPLC to the rapid analysis and isolation of underivatized peptides and proteins were reviewed by Hancock et al. (1979b) and

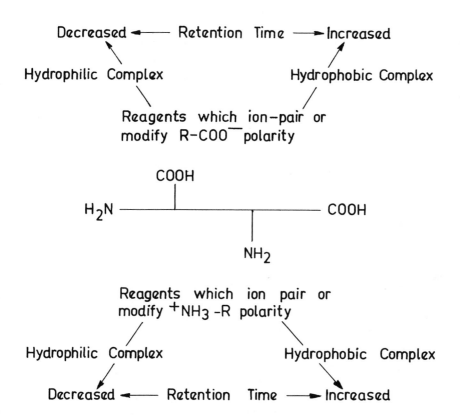

Fig. 9.1. Diagrammatic representation of polarity modes involved in ion-pair formation at free amino or carboxy groups and their effect on the retention times of peptides. With multi-component samples the selectivity and elution order can be manipulated by the use of either hydrophilic or hydrophobic ion-pairing reagents or mixtures involving both types of reagents. (According to Hearn and Hancock, 1979b.)

Hearn and Hancock (1979a,b). According to the last reference, the principle of ion-pair manipulation with the retention time of peptides can be explained by the scheme illustrated in Fig. 9.1.

In addition to chromatography on ion-exchange resins and classical bonded reversed-phase and ion-pair chromatography, some other approaches have also been tested for the efficient and rapid liquid column chromatographic separation of peptides. Kroeff and Pietrzyk (1978a) investigated the retention and separation of amino acids, their derivatives and peptides on porous hydrophobic co-polymers. Amberlite XAD-2, -4 and -7 were applied for RPC. Water-ethanol and water-acetonitrile mixtures were used for elution. An increased concentration of the organic solvent decreased the retention. At the isoelectric points the retention of peptides was low and in acidic or alkaline solution high, which is understandable from consideration of the scheme in Fig. 9.2. Eqn. 81, describing

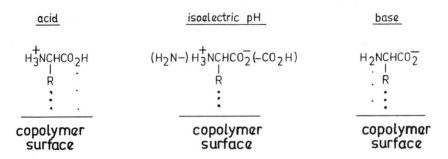

Fig. 9.2. Amino acid–Amberlite XAD interaction at two pH extremes and at the isoelectric point. The interaction between the (non-ionizable) R group and the polymer surface is constant throughout the entire pH range. Any variation of the capacity factor (k') is the result of change in the ionization of the COOH and/or NH$_2$ groups. In the region of the isoelectric point both groups are fully ionized (zwitterion) and are directed away from the non-polar copolymer surface. Bold dots indicate the constant interaction, light dots the variable interaction. (According to Kroeff and Pietrzyk, 1978a.)

the sorption of zwitterions on these materials, was dealt with in Chapter 3. Kikta and Grushka (1977) and Wing-Kin Fong and Grushka (1977, 1978) bonded tri-peptide stationary phases to silica, using the methods described in Chapter 4, and studied the separation of simple peptides on such supports, especially iso-meric dipeptides and PTH-amino acids. The retention order seemed to indicate that the retention mechanism is a combination of anion-exchange and hydrophobic interactions. Von Arx and Faupel (1978) described the changeover from strongly aqueous RP-HPLC systems to thin-layer chromatography and bioautography of re-versed-phase thin-layer chromatograms. Naider et al. (1979) studied the separa-tion of protected oligopeptides containing methionine by normal-phase HPLC. A μPorasil silica column was used with isopropanol-cyclohexane mixtures as the mobile phase. The retention behaviour of the separated peptides was correlated with the hydrophobicities of the component amino acid residues.

This historical survey of the beginnings of the HPLC of peptides finishes with the year 1979. In the author's opinion, starting from 1980 these methods were developed to a high degree from the points of view of both the main principles and the instrumentation and techniques involved.

9.2 THEORY OF CHROMATOGRAPHIC SEPARATION OF PEPTIDES

It is natural that the general theory of chromatography (cf., Chapter 2), valid for all soluble substances, has also been applied to the separation of peptides; as an example, the paper by Hansen et al. (1977) can be cited. How-ever, in this section only theoretical problems that are typical of or specific only for peptides will be discussed.

9.2.1 General features of peptides and their ionic forms

The structural properties of peptides lie somewhere between those of low-molecular-weight amino acids and high-molecular-weight proteins. The lower the molecular weight, the closer the oligopeptide behaviour is to that of amino acids, because a short peptide chain cannot create a complicated conformation in space. In contrast, very long polypeptides behave like small proteins, because the conformation of their chain shows characteristic features of secondary and tertiary structure. These general structural properties of peptides reflect themselves in their chromatographic behaviour.

Peptides are amphoteric substances and exist in dissociated form in aqueous solutions exclusively as amphoteric ions. These can change their charge depending on the pH of the medium. The relationships involved are summarized in Table 9.1. Both types of charge (a zwitterionic form) occur only in neutral medium. In acidic media the carboxyl group loses its charge owing to protonation and the peptide then behaves as a cation. In alkaline media the protonation of the amino group (to give an ammonium group) is eliminated and the zwitterion is converted into an anion. The degree of dissociation is determined by the dissociation constants, pK_1 and pK_2, and their dependence on pH is expressed by the Henderson-Hasselbalch equations. The pK value is the pH at which the respective group is 50% dissociated. At the isoelectric point, pI, peptide has net electroneutral properties.

The bonding of a neutral amino acid or a peptide to a cation exchanger is determined by its pK_1 value (because the ammonium group characterized by pK_2 is permanently fully protonated in the acidic medium of a cation exchanger and cannot contribute to fine differences in retention due to slight changes of pH). With increasing pK_1 value the affinity of peptides for cation exchangers increases; in other words, the elution of a peptide mixture from a cation exchanger starts with the peptides with the lowest pK_1 values. On an anion exchanger neutral amino acids or peptides are bound in the reversed order of their pK_2 values, i.e., a mixture of peptides is eluted starting with the species with the highest pK_2. (Carboxyl groups characterized by pK_1 constants are permanently fully dissociated in the alkaline medium of the anion exchanger.) With basic or acidic peptides, additional ionizable basic or acidic groups contribute to the equilibria with their own pK values, a basic peptide behaves as a cation over a wide pH range and an acidic peptide is predominantly in the form of an anion. Owing to hydrophobic and adsorption interactions of side-chains of amino acids or peptides with the network of the ion exchanger, this theoretical sequence of elution is sometimes altered. The higher is the temperature, the lower are these disturbing side adsorption interactions.

TABLE 9.1

ELECTROLYTIC DISSOCIATION AND ION-EXCHANGE SORPTION OF NEUTRAL AMINO ACIDS AND PEPTIDES

Reprinted from Mikeš (1979).

$H_2NCH(R_1)CONHCH(R_2)COOH$

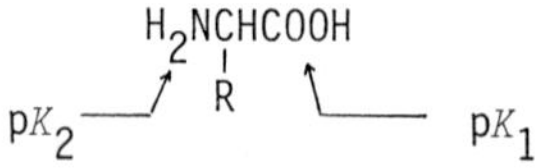

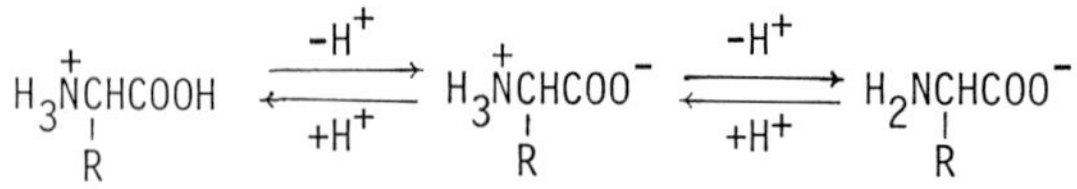

$$H_2NCHCOOH$$
$$\overset{pK_2}{\diagup}\underset{R}{|}\overset{\diagdown}{}\quad pK_1$$

$$\underset{R}{\overset{+}{H_3NCHCOOH}} \underset{+H^+}{\overset{-H^+}{\rightleftharpoons}} \underset{R}{\overset{+}{H_3NCHCOO^-}} \underset{+H^+}{\overset{-H^+}{\rightleftharpoons}} \underset{R}{H_2NCHCOO^-}$$

Medium

Acidic	Neutral	Basic
CATION $\overset{pK_1}{\longleftarrow}$	ZWITTERION $\overset{pK_2}{\longrightarrow}$	ANION

Sorbed on

Cation exchanger   Anion exchanger

K_1 applies for $COOH \rightleftharpoons COO^- + H^+$

K_2 applies for $\overset{+}{N}H_3 \rightleftharpoons NH_2 + H^+$

$$K_1 = \frac{[COO^-][H^+]}{[COOH]} \quad , \quad -\log K_1 = pK_1;$$

$$K_2 = \frac{[NH_2][H^+]}{[NH_3^+]} \quad , \quad -\log K_2 = pK_2$$

Henderson-Hasselbalch equations:

$pH = pK_1 + \log([COO^-]/[COOH])$

$pH = pK_2 + \log([NH_2]/[NH_3^+])$

Isoelectric point:

$$pI = \frac{pK_1 + pK_2}{2}$$

In principle, peptides are bound to the strongly polar ion-exchange support by electrostatic interactions (they exchange some bound small ions of low concentration; cf., Section 3.4), and can be liberated by elevation of the ionic strength of the mobile phase and by changes in pH. An increase in pH gradually elutes peptides from cation exchangers and a decrease in pH elutes peptides from anion exchangers.

9.2.2 Behaviour of peptides in contact with non-polar stationary phases

The mechanism of the retention of peptides on a reversed-phase support is very different from an ion-exchange mechanism, in which the interaction of the solute with the polar stationary phase is the main principle. Under conditions of reversed-phase chromatography, there is a strong driving force that primarily repulses lipophilic substances or lipophilic molecular moieties from polar solvents to hydrophobic stationary phases, and adsorption on the surface of the stationary phase is only a secondary force. The mechanism of this process, in which the orientation of water molecules plays an important role, was described in Sections 3.6 and 3.7, dealing with the principles of HIC and RPC, where the solvophobic theory was explained (cf., Horváth et al., 1976, 1977). Fig. 3.16 illustrated how the extent of the hydrocarboneous chain of an amino acid contributes to its retention on an RPC support. Side-chains of various amino acids differ in their hydrophobicity, which is the reason for the differences in the strengths of their binding to a non-polar support due to hydrophobic interactions. In peptides the sum of particular hydrophobic contributions of constituent amino acid residues is the decisive factor for their retention on non-polar stationary phases.

This opens the way for the "hydrophobic chromatography" of amino acids and peptides, which was studied by Molnar and Horváth (1977). Using aqueous solutions of acids and salts, peptides were eluted from non-polar chromatographic supports in order of their increasing hydrophobicity. The authors found this separation method to be superior in principle to that obtained on conventionally used ion-exchange columns, with the exception of hydrophilic peptides, which were poorly retarded. The following considerations from the discussion of their results are important. The balance of two antagonistic free energy changes determines the value of the capacity factor, k'(eqn. 6/II), of a given column: (a) the first is related to the decrease in the molecular surface area, which is exposed to the solvent, on binding of the solute to the stationary phase, and the effective surface tension of the eluent; (b) the other may be expressed by the sum of the free energy changes arising from the interaction of the solute with the solvent molecules. It can be predicted theoretically (and it was also confirmed experimentally; cf., Fig. 3.16) that log k' is linearly related to the carbon number of the members of a homologous series of substances separated under the same conditions. In general, the retention order for small molecules is determined by the contact area and is proportional to the size of the molecules.

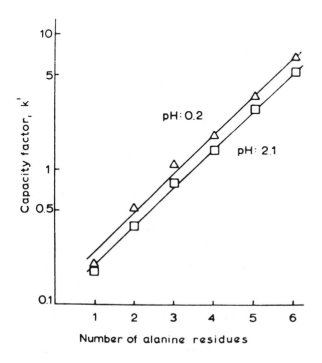

Fig. 9.3. Plots of the logarithm of the capacity factor, k', against the number of residues of alanine oligomers. Column, 5 μm LiChrosorb RP-18 (25 cm x 4.6 mm I.D.); eluent, 0.5 M HClO$_4$ (pH 0.2) or 0.1 M phosphate buffer (pH 2.1); temperature, 70°C; flow-rate, 2.0 ml/min; ΔP, 150 atm. (Reprinted from Molnár and Horváth, 1977.)

Molnár and Horváth (1977) verified this principle on the separation of small peptides. Oligomers of glycine containing up to six residues had no appreciable retention on an ODS support (phosphate buffers, pH 2.1, 70°C). This suggested that the peptide chain proper (the backbone) made no or only a very limited contribution to the retention of peptides. However, the oligomers of L-alanine were well separated and the plot of log k' versus the number of alanine residues gave a straight line (illustrated in Fig. 9.3). By comparison of this figure with Fig. 3.16, it is clear that the slopes of the straight lines are much smaller than those obtained with alkyl-α-amino acids under identical conditions. The log k' increment of a CH$_3$ group is roughly twice that of the -NHCH(CH$_3$)CO- residue.

Molnár and Horváth (1977) also employed gradient elution with an increasing concentration of acetonitrile and found this method to be useful for the separation of complicated mixtures, because the peak capacity (cf., Horváth and Lipsky, 1967, and Section 2.9 of Part A) was increased; the change in "eluent strength" was such that the adjusted retention of phenylalanine oligomers was a linear function of the number of residues. Hydrophobic amino acid residues increased retention and polar residues reduced retention, which was also influenced (in some instances) by the position of charged groups in the side-chains. The ionization of ionogenic groups reduced retention. The use of aqueous-organic solvents, which have a lower effective surface tension than water, gave a reduction in chromatographic retention.

Bij et al. (1981) showed that certain unprotected peptides do not exhibit the regular linear dependence of the logarithm of retention on the composition of binary aqueous-organic eluents on silane-bonded C_8 or C_{18} stationary phases; the plots had minima. A dual retention mechanism was proposed, in which, in addition to hydrophobic interaction, also interactions with silanol groups were incorporated.

Solvophobic considerations for the separation of unprotected peptides on chemically bonded hydrocarboneous stationary phases were dealt with in detail also by Hearn and Grego (1981b). A close linear dependence was found between the capacity factor (k') of peptides and surface tension (γ) [and hence on the volume fraction (ψ)] of acetonitrile- and methanol-water combinations. The effects of phosphate buffer, ionization and other pH-dependent effects were constant over the concentration range of acetonitrile examined. By increasing the ionic strength, the surface tension of an aqueous-organic mobile phase was also increased and the peptides were influenced by the "salting-out" effect. The authors also discussed a semi-empirical approach in order to rationalize the solute retention. They showed that "it should be possible to estimate the elution order for peptides on reversed phases under isocratic and linear gradient conditions using suitable topological indices, which take into account the effective hydrophobic contributions which each of the amino acid side-chain make to the retention process" (cf., Section 9.2.4). The influence of the water content of the mobile phase, ionic strength and pairing ions on the retention of peptides on ODS was examined.

Meek and Rossetti (1981) studied factors affecting the retention and resolution of peptides in HPLC. The characteristics of retention will be discussed in Sections 9.2.3 and 9.2.4. The resolution depends not only on the column, but also on the mobile phase composition and flow-rate, the gradient and the size and composition of the peptides themselves. The efficiency (HETP) of columns for peptides becomes worse with increasing solute molecular weight and with increasing flow-rate of the mobile phase. For high resolution the flow-rate should be maintained at not more than 1 ml/min (25 cm x 4 mm I.D. column), except for very small peptides (see also Section 7.4.2). A compromise must be made between analysis time and resolution. Replacement of $NaClO_4$ (a chaotropic agent) with NaH_2PO_4 decreased the retention of most peptides; the peak width was also decreased. The influence of the gradient was studied: the peak capacity (resolution) improved as the gradient rate decreased from very fast (5%/min, or 0-100% in 20 min) to very slow (0.5%/min, or 0-50% in 100 min). Little effect of the flow-rate on the resolution under gradient conditions was found.

It is interesting that in some instances Meek and Rossetti (1981) found the reverse elution order of the same substances with changes in gradient rate; the cause of this effect was not clear. The addition of a small amount (0.1%) of trifluoroacetic acid was useful, as it contributed to peak sharpness, but the low concentration was apparently not sufficient to cause ion pairing with basic residues; the advantage in its use was in the possibility that the peptide could be simply recovered from the eluate, as had been recommended by Bennett et al. (1980). Gradients of propanol gave a worse resolution than gradients of acetonitrile [propanol has been advocated by Rubinstein (1979) for larger polypeptides and small proteins].

Hearn and Grego (1981c) examined the effects of the organic solvent modifier in the separation of unprotected peptides by RP-LC. It is a general experience that the retention of small peptides on RP silicas decreases as the volume fraction of organic solvent, ψ_s, in the mobile phase increases (over the range $0 < \psi_s < 0.4$). Under these elution conditions the retention tends to be regular, i.e., peptides are eluted in order of increasing hydrophobicity. However, with more hydrophobic peptides, requiring a higher content of organic solvent in the mobile phase, this is not valid. Therefore, the authors studied in detail the elution behaviour of a variety of unprotected peptides using mobile phases of different composition. It was found that over a wide range of ψ_s, unprotected peptides did not show linear dependences of the logarithm of their capacity factors on the composition of binary aqueous-organic eluents; the curves showed minima at certain ψ_s values. Grego and Hearn (1981) continued their studies of the role of organic solvent modifier on the RP-HPLC of polypeptides (hormones) using μBondapak C_{18} and acetonitrile + TEA-phosphate buffers. The results were discussed in terms of the interplay of hydrophobic and silanophilic interactions which occured between ionized polypeptides and the stationary phase with variation in the mobile phase composition.

Su Seuw et al. (1981) analysed the group retention contribution for peptides separated by HPLC. The retention differences were evaluated from topological parameters that accommodated the influence of both amino acid side-chain and end-group contributions in the retention process. Divergences were discussed in terms of silanophilic acid and solvation interactions. In further studies, Hearn and Grego (1983a) extended the investigation of the role of organic modifiers in the RPC of peptides and examined the relationship between log k' and ψ_s over the range $0 < \psi_s < 0.8$. In spite of the experience of non-linearity of this relationship, the authors found that within the range of k' values of interest in the isocratic or gradient optimization of RPC of peptides, namely in the range $1 < k' < 10$ with water-rich eluents of low pH, the relationship between log k' and ψ_s was approximately linear, and the elution of peptides essentially followed the relative hydrophobicities of the polypeptides examined. It should be noted, however, that under isocratic conditions the utilizable composition of the mobile phase can sometimes be only very narrow: outside it the chromatographed compounds are eluted in the void volume, or are not eluted at all.

For a detailed study of the contribution of the lipophilic amino acid moieties to the retention of peptides in RPC, the hydrophobicity of amino acids had to be expressed in an exact way, and this is considered below.

9.2.3 Expression of hydrophobicity of amino acid residues in peptides

Hydrophobic interactions play a very important role in biochemistry and were discussed, e.g., by Kauzmann (1959) in the interpretation of protein denaturation. The role of the hydrophobic effects in biological processes and life sciences was well illustrated by Tanford (1973, 1978).

Amino acid residues are usually classified as hydrophobic and hydrophilic. In order to express this property quantitatively, Nozaki and Tanford (1971) used "group contributions", Δf_t, as a measure of the hydrophobicity of amino acid residues. The group contributions to ΔF_t (free energy of transfer of an amino acid from a pure organic solvent, such as ethanol, butanol, dioxane or acetone, to water at 25°C) were calculated for the amino acid side-chains and the backbone peptide unit, assuming additivity of the free energy of solvent interactions (the value for glycine was subtracted from that of the other amino acid). The calculated values were ordered on the hydrophobicity scale (cf., Table 9.2). Leo et al. (1971) measured the partition coefficients, P, of amino acids in octanol-water and used log P to express their hydrophobicity. Chothia (1974) studied hydrophobic bonding and the accessible surface area in proteins; he found linear relationships between the hydrophobicity of non-polar residues [calculated in cal/mol and derived according to Nozaki and Tanford (1971)] and their surface areas (measured in $\mathring{A}^2$) accessible to water molecules.

Segrest and Feldmann (1974) studied amino acid sequences and membrane penetration in membrane proteins and derived another hydrophobicity scale for all

TABLE 9.2

HYDROPHOBICITY SCALES FOR PROTEIN AMINO ACID RESIDUES, PEPTIDE END GROUPS AND SOME DERIVATIVES

Amino acid[a]	Symbol	Nozaki and Tanford[b] (1971) Δf_t	Leo et al.[c] (1971) Log P	Segrest and Feldmann[d] (1974) R_h	Bigelow and Channon[e] (1976) H_c	Rekker[f] (1977) Un-modified Σf	Modified Σf	Pliška and Fauchère[g] (1979) π	Meek[h] (1980b) pH 7.4 (min)	pH 2.1 (min)	Wilson et al.[i] (1981c) %B in A + B
Tryptophan	Trp W	3400[p]	-1.04	6.5	3400	2.31	(2.31)	1.87	14.9	18.1	7.9
(norleucine)		2600[q]									
Phenylalanine	Phe F	2500[p]	-1.43	5.5	2500	2.24	(2.24)	1.80	13.2	13.9	7.5
(Dihydroxyphenyl-alanine)		1800[r]									
Leucine	Leu L	1800[p]	-1.71	3.5	1800	1.99	(1.99)	1.66	8.8	10.0	6.6
Isoleucine	Ile I			5.0	2950	1.99	(1.99)	1.56	13.9	11.8	4.3
Tyrosine	Tyr Y	2300[p]	-2.26	4.5	2300	1.70	(1.70)	1.70	6.1	8.2	7.1
Valine	Val V	1500[r]	-2.10	3.0	1500	1.46	(1.46)	1.06	2.7	3.3	5.9
Cystine	Cys			0.0		1.11			-6.8	-2.2	6.3
Methionine	Met M	1300[r]		2.5	1300	1.08	(1.08)	1.39	4.8	7.1	2.5
Proline	Pro P			1.5	2600	1.01	(1.01)	0.56	6.1	8.0	2.2
Cystein	Cys C				1000	0.93		0.69			
Arginine	Arg R		-2.59		750	-0.82[s]	(0.10)	-1.23	0.8	-4.5	-1.1
Alanine	Ala A	500[r]	-2.94	1.0	500	0.53	(0.53)	0.38	0.5	-0.1	-0.3

Name	Code											
Lysine	K		-2.82			1500	0.52	(-0.10)	-0.93	0.1	-3.2	-3.6
Glycine	G		-3.03	0		0	0.00	(0.10)	0	0	-0.5	1.2
Aspartic acid	D					0	-0.02	(0.10)	-1.23	-8.2	-2.8	-1.4
Glutamic acid	E					0	-0.07	(0.20)	-1.20	-16.9	-7.5	0.0
Histidine	H	500[p]	-2.86		1.0	500	-0.23	(-0.10)	-1.30	-3.5	0.8	-1.3
Threonine	T	400[r]			0.5	400	-0.26	(0.10)	0.33	2.7	1.5	-2.2
Serine	S	-300[r]			-0.5	-300	-0.56	(0.10)	0.04	1.2	-3.7	-0.6
Asparagine	N				-1.5		-1.05	(0.10)	-0.27	0.8	-1.6	-0.2
Glutamine	Q				-1.0		-1.09	(0.20)	-0.09	-4.8	-2.5	-0.2
Amino group										2.4	-0.4	
Carboxyl group										-3.0	6.9	
Amide								(0.00)		7.8	5.0	
(Pyroglutamyl)										-1.1	-2.8	
(N-Acetyl)								(0.00)		5.6	3.9	
(Tyrosine sulphate)										10.9	6.5	
(Homoserine)								(0.10)				
[Aminoethyl cysteine (AEC)]								(-0.10)				
(O-Phospho)												
(N-Glyco)												
[Carboxymethyl cysteine (CMC)]								(0.10)				
[Trimethyllysine (TML)]								(-0.10)				
(Time for elution of unretained compounds, t_0)										2.0	2.0	

(Continued on p. B98)

TABLE 9.2 (continued)

Amino acid[a]	Symbol	Meek and Rossetti[j] (1981)		Su Suew et al. (1981)	Browne et al.[l] (1982)		Sasagawa et al.[m] (1982a)		Sasagawa et al.[n] (1982b)	
		$NaClO_4$ (min)	NaH_2PO_4 (min)	X_n	0.1% TFA, % CH_3CN	0.13% HFBA, % CH_3CN	pH 2, D_j	pH 8, D_j	Unweighted D_j	Weighted D_j
Tryptophan	Trp W	17.1	15.1	-0.28	16.3	17.8	1.79	2.48	35.8	2.34
Phenylalanine	Phe F	13.4	12.6	2.52	19.2	14.7	1.37	1.89	31.4	1.71
Leucine	Leu L	11.0	9.6	3.16	20.0	15.0	1.23	1.09	26.4	1.34
Isoleucine	Ile I	8.5	7.0	5.80	6.6	11.0	1.06	1.28	27.4	1.38
Tyrosine	Tyr Y	7.4	6.7	-0.79	5.9	3.8	0.77	0.67	21.0	1.33
Valine	Val V	5.9	4.6	1.08	3.5	2.1	0.73	0.85	7.4	0.38
Cystine	Cys	7.1	4.6							
Methionine	Met M	5.4	4.0	3.56	5.6	4.1	0.97	1.30	14.5	0.85
Proline	Pro P	4.4	3.1	-0.40	5.1	5.6	0.55	-0.32	7.9	0.48
Cysteine	Cys C			-1.25	-9.2	-14.3				
Arginine	Arg R	-0.4	-2.0	-1.37	-3.6	3.2	-0.21	0.32	0.0	0.26
Alanine	Ala A	1.1	1.0	-0.12	7.3	3.9	0.15	-0.03	2.4	0.13
Lysine	Lys K	-1.9	-3.0	-0.53	-3.7	-2.5	-0.50	0.04	-3.1	0.05
Glycine	Gly G	-0.2	0.2	-0.82	-1.2	-2.3	0.04	0.27	4.0	0.22
Aspartic acid	Asp D	-1.6	-0.5	1.92	-2.9	-2.8	-0.08	-1.04	-0.1	0.10
Glutamic acid	Glu E	0.7	1.1	0.44	-7.1	-7.5	0.17	-0.55	2.7	0.27
Histidine	His H	-0.7	-2.2	-2.67	-2.1	2.0	-0.12	0.66	8.8	0.34
Threonine	Thr T	-1.7	-0.6	-0.79	0.8	1.1	0.21	-0.07	7.4	0.12

Serine	Ser S	-3.2	-2.9	0.66	-4.1	-3.5	-0.16	0.60	1.1	0.18
Asparagine	Asn N	-4.2	-3.0	-1.76	-5.7	-2.8	-0.17	-0.25	-11.3	-0.45
Glutamine	Gln Q	-2.9	-2.0	-0.76	-0.3	1.8	-0.39	-0.22	3.2	0.36
Amino group		4.6	0.9	1.54 }	4.2	4.2				
Carboxyl group		2.2	1.6		2.4	2.4				
Amide		4.4	4.9		10.3	8.1			-13.2	-0.56
(Pyroglutamyl)		2.8	2.9							
(N-Acetyl)		6.6	3.8		10.2	7.0	0.69	0.71	12.4	0.81
(Tyrosine sulphate)		2.4	3.7							
(Homoserine)									12.3	0.23
(Aminoethyl-cysteine [AEC])									4.3	0.3
(O-Phospho)					-2.4	-4.1				
(N-Glyco)					-8.0	-6.5				
(Carboxymethyl-cysteine [CMC])							0.09	-0.86	32.5	1.57
(Trimethyllysine [TML])							-2.42	-1.69	-38.1	-1.38
(Time for elution of unretained compounds, t_0)		2.0	2.0							

[a] Non-protein amino acids are given in parentheses.
[b] The group contribution, Δf_t, is the free energy of transfer of the amino acid side-chain from pure organic solvent to water at 25°C (cal/mol).
[c] P is the partition coefficient of the amino acid in an octanol-water system. The values of log P are given according to Molnár and Horváth (1977).

(Footnotes continued on p. B100)

d R_h is the relative hydrophobicity of uncharged amino acid residues, derived as described in the text.

e H^d are amino acid hydrophobicity constants, cited according to Wilson et al. (1981c).

f Σf is the summation of the fragmental hydrophobic constants. Modified values (in parentheses) are given according to Sasagawa et al. (1982b) or Sasagawa and Teller (1984).

g π are the values of the hydrophobic parameters, given according to Wilson et al. (1981c).

h Retention coefficients are given in minutes for slightly basic (0.1 M NaClO$_4$, pH 7.4) or acidic (0.1 M NaClO$_4$, pH 2.1) conditions. Bio-Rad ODS columns were used. Time 0 min is for the above solutions; a gradient acetonitrile-0.1 M NaClO$_4$ in 80 min at room temperature was used for elution. The coefficients were calculated by repeated regression analysis.

i Amino acid hydrophobicity constants were calculated by multivariate regression analysis and are expressed in % of buffer B in a mixture A+B at which the peptides were eluted (see text).

j Retention coefficients of amino acid residues are given in minutes for two acidic media. The peptides were chromatographed on a Bio-Rad ODS (10 μm) column (24 cm × 4.0 mm I.D.) with a linear gradient of acetonitrile (0.75%/min) starting with either 0.1 M NaClO$_4$-0.1% H$_3$PO$_4$ or 0.1 M NaH$_2$PO$_4$-0.2% H$_3$PO$_4$, at 0 min after injection. HPLC-grade acetonitrile contained 0.1% H$_3$PO$_4$.

k X_{R_h} are group retention contributions for amino acid residues, derived from chromatography of peptides on a Waters Assoc. μBondapak C$_{18}$ column at 18°C. Linear gradients of acetonitrile (0.83%/min) commencing with 50 mM sodium dihydrogen phosphate-15 mM orthophosphoric acid (pH 2.65) at 0 min after injection were used for elution; the final elution condition was 50% acetonitrile-50% water-50 mM sodium dihydrogen phosphate-15 mM orthophosphoric acid. Sample size varied between 5 and 10 μg of peptide material injected in a volume of 5-10 μl; the flow-rate was 1.0 ml/min.

l The retention coefficients are given for chromatography on a Waters Assoc. μBondapak C$_{18}$ column (using a Waters Assoc. HPLC system with a Perkin-Elmer UV detector, 210 nm) which was eluted with linear gradients of aqueous acetonitrile (20%/h, flow-rate 1.5 ml/min) containing either 0.1% trifluoroacetic acid (TFA), or 0.13% heptafluorobutyric acid (HFBA). The retention coefficients are expressed in terms of acetonitrile concentration as indicated by the gradient former.

m D_j are retention constants according to eqn. 113. Retention times were measured on a Hamilton RP-1 column; the mobile phase was either 0.1% TFA (pH 2) or 5 mM ammonium hydrogen carbonate (pH 8). The modifier was acetonitrile, the flow-rate 2 ml/min and the slope of the gradient 2%/min.

n D_j are retention constants according to eqn. 113. Retention times were measured on a μBondapak C$_{18}$ column, the mobile phase was 0.1% TFA, the modifier was acetonitrile containing 0.07% TFA and the slope of the gradient was 1%/min. The weights used were 1/N_i.², where N_i is the number of amino acid in a peptide.

p Average of values for ethanol and dioxane.

q Average of values for ethanol, butanol and acetone.

r Values for ethanol only.

s Rekker (1977) gives zero for arginine. The value in the table is given according to Wilson et al. (1981c) and was calculated by minimization of the deviation for the arginine-containing peptides from the linear regression for those peptides lacking arginine.

the non-charged amino acid residues, which was also based on the free energy of transfer of amino acids from ethanol to water as measured by Nozaki and Tanford (1971); however, rather than using raw free energy values, a simplified linear scale was devised, which has zero and unity defined by glycine and alanine, respectively. Glycyl is defined as zero because it contains no side-chain, and alanyl as +1, because it has a single methyl side-chain. Increasing positive values on this scale represent increasing hydrophobicity (cf., Table 9.2). The increase is approximately linear for each additional methyl (methylene) group. The authors also proposed a hydrophobicity index, HI, for the characterization of protein sequence segments; it is obtained by adding the hydrophobicity values for residues of the sequence segment and dividing by the total number of its residues, in order to obtain a mean hydrophobicity per residue. The hydrophobicity index (HI) serves as a quantitative measure of the hydrophobicity of each sequence segment. The authors calculated these values by a computer program (available from them on request). Charged (polar) amino acids are not considered in the program. It is probable that a modification of such a simple approach to express the hydrophobicity using HI can also be used for individual peptides (perhaps using other fragmental hydrophobic constants in addition), if their relative hydrophobicity is to be quantified in relation to their size.

In 1976, Bigelow and Channon [cited by Wilson et al. (1981c)] compiled a more detailed survey of the constants presented by Nozaki and Tanford (1974).

In 1977 Rekker published a book on fragmental hydrophobic constants which included a hydrophobicity scale for the side-chains of various amino acids on the basis of hydrophobicity fragmental constants. Rekker's Σf values for amino acid residues have often been used for the calculation of the total peptide hydrophobicity by summation of particular amino acid contributions. Negative Σf values in Table 9.2 represent hydrophilic and positive values hydrophobic amino acids. The larger the positive value, the more hydrophobic is the amino acid. The hydrophobicity constants presented by Nozaki and Tanford (1971), Leo et al. (1971) and Rekker (1977) were compared and briefly discussed by Molnár and Horváth (1977).

In 1979, Pliška and Fauchére published values of hydrophobic parameters π for amino acid side-chains, derived from partition and chromatographic data available in the literature. The constant π_g of a substituent group R was calcualted by converting R_F values in a given system into the partition coefficient, P_c, for the same system, and transformation of P_c into partition coefficients in n-octanol-water. The group contribution π_g was calculated by multiple regression analysis. The π_R and π_g values were presented for 27 amino acids (both charged and non-charged) and for 15 groups. The π_R constants can be used for the calculation of the side-chain hydrophobicity.

Meek (1980b) described another derivation of "retention coefficients" that represented the contribution to the retention of each of the common amino acids and end-groups, based on the direct liquid chromatography of peptides. Peptide retention was measured in a chaotropic agent ($NaClO_4$; cf., Hatefi and Hanstein, 1969) to minimize conformational effects of the formation of the secondary structure. Retention coefficients of pertinent amino acid residues were computed by an automatic calculator programmed to change the retention coefficients for amino acid residues sequentially to obtain a maximum correlation.

Wilson et al. (1981c) investigated the behaviour of peptides on RP-HPLC supports; in a thorough study the behaviour of peptides ranging in length from 2 to 65 amino acid residues was examined in a A (0.125 M pyridine formate, pH 3.0) and B (1.0 M pyridine acetate, pH 5.5)-60% propan-1-ol system; the final concentration was based on pyridine. Peptide retention was independent of temperature between 25 and 55°C. The dependence of chromatographic retention of pH decreased with increasing peptide hydrophobicity. The chromatographic results obtained with C_8 and C_{18} supports were comparable. On the basis of the measured retentions of peptides, the authors calculated the hydrophobic constants for various amino acid side-chains, which are given in Table 9.2. The constants were compared with similar constants available from the literature. In 1981, Meek and Rossetti revised the data for several amino acids in the previous paper (Meek, 1980b) by examining 100 peptides, and listed the retention coefficients for two acidic media, (1) chaotropic $NaClO_4$ and (2) NaH_2PO_4 solution (which probably acts as an ion-pairing agent) (cf., Table 9.2).

Su Suew et al. (1981) also started from the solvophobic theory and discussed the group retention contributions for peptides separated by RP-HPLC. They concluded that with linear solvent strength gradient elution and with a restricted range of mobile phase compositions the capacity factor, k' (apparent), increased linearly if amino acids were added to a peptide in an ordered manner to form a homologous series. The capacity factor of a peptide, $k_{i,app}$, can be expressed in terms of summated group contributions

$$k_{i,app} = \overset{n}{\Sigma} C_n \chi_n + d \tag{112}$$

where C_n is a numerical factor indicating the incidence of a given fragment in the structure, χ_n represents the group retention contribution due to amino acid n and d is the intercept term, which under ideal circumstances (when only hydrophobic interactions mediate the retention process) should be zero. From data obtained for 57 peptides (including a variety of peptide hormones) eluted under the same conditions from a μBondapak C_{18} column, the authors tested the validity of the predicted retention parameters originating from a "forcing approach" using two methods of numerical analysis, and compared the predicted k'_{app} values with observed retention behaviour. The data for χ_n are listed in Table 9.2.

Alfredson et al. (1982) used the fragmental hydrophobic constants of amino acids to evaluate new microparticulate packings for aqueous SEC in order to describe the slight hydrophobicity of the packings in a quantitative way.

Browne et al. (1982) described the deviation and use of other predictive retention coefficients for the RPC of peptides, because they had found that the retention coefficients described by Meek (1980b) and Meek and Rossetti (1981) were not completely compatible with the data obtained using their own chromatographic system. Further, they found that the retention coefficients changed on going from their TFA to an HFBA system (cf., Table 9.2, footnotes), so that the derivation of new constants was necessary; 25 rat neurointermediary lobe peptides were chromatographed. The detailed chromatographic conditions were described in another paper (Bennett et al., 1981a). The position of each peptide, from which the new retention coefficients were derived, was expressed as acetonitrile concentration. The estimates took into account the total dead volume of the Waters Assoc. HPLC system used. The calculated retention times were derived by using the appropriate $y = mx + c$ relationship, where $m = 4.46$ and $c = -56.3$ for the TFA system and $m = 4.59$ and $c = -88.9$ for the HFBA system.

The retention coefficients for 20 amino acid residues were calculated by interactive linear regression analysis as described by Meek (1980b) and the set of "acid" Meek's retention coefficients were used as the starting set. The derived retention coefficients are listed in Table 9.2. The observed and calculated elution positions of 25 peptides were compared and very small deviations were found. The correlation analysis was performed on a Hewlett-Packard HP 9831 desk-top computer, a listing of the program for which is available from the authors on request; the discussion of the results will be commented on in Section 9.2.4.

Sasagawa et al. (1982a) derived retention constants, D_j, using data from the chromatography of a larger amount of peptides on polystyrene supports, where no undesirable silanol interactions could take place and influence the "hydrophobic chromatography". The constants were derived for both acidic (pH 2) and alkaline (pH 8) conditions. Sasagawa et al. (1982b) studied the chromatographic separation of 100 peptides on a μBondapak column in an acidic medium (0.1% TFA) with acetonitrile linear gradient elution and derived a new set of retention constants, D_j (Table 9.2), which were compared with Rekker's constants (both original and slightly modified for hydrophilic amino acid residues) and with Meek's constants.

The prediction of peptide retention using hydrophobicity constants is explained in Section 9.2.4.

9.2.4 Prediction of retention data of peptides in RPC on the basis of hydro-
  phobicity of amino acid residues and their retention coefficients

In general, the accurate quantitative prediction of some effect in nature is
usually considered to be striking proof that the theory of the process is valid.
This was the main reason why many scientists have endeavoured to predict the
retention of peptides theoretically, in addition to the practical importance of
such calculations.

Molnár and Horváth (1977) found a good correlation between the retention times
of various peptides and the sum of the side-chain fragmental hydrophobic constants
given by Rekker (1977) (cf., Table 9.2). They declared that "the retention order
of peptides may be estimated from data pertinent to the amino acid constituents"
and drew readers' attention to the paper by Bate-Smith and Westall (1950), where
the retention values of the individual amino acids were used to predict the retention of peptides in paper chromatography. It seemed likely that the RPC retention of a relatively large peptide could be estimated from the retention values
of the amino acid residues obtained under identical conditions. With increasing
chain length the conformation of the peptide must be taken into consideration.
Most workers describing the RPC of peptides agreed that separations of peptides
are dictated primarily by their hydrophobicity. For citations see the paper by
O'Hare and Nice (1979), who confirmed that the retention orders of smaller
peptides (<15 residues) on alkylsilane-bonded silica were generally correlated
with the sum of the Rekker fragmental constants of their strongly hydrophobic
residues (Trp + Phe + Leu + Ile + Tyr). Larger peptides showed numerous anomalies
and the precise behaviour of proteins was unpredictable with this system. In addition to the sum of the above-mentioned strongly hydrophobic amino acid residues, the authors expressed the peptide hydrophobicity also by adding the sum
of fragmental constants of less hydrophobic amino acids (Val + Cys-Cys + Met)
to the sum of strong hydrophobic amino acids. Analogous calculations based on
the hydrophobicity scale of Nozaki and Tanford (1971) were found to be less accurate in predicting retention orders. By evaluating the retention positions of
more than 30 polypeptides and comparing the data with their hydrophobicity, the
authors made a significant contribution to the efforts to determine whether
there are prominent correlations between chromatographic behaviour and the
structure of polypeptides.

The deviations found between the observed order of elution and the lipophilicity estimates may be explained by the fact that retention on ODS-silica
is a different process from octanol-water partition, from which the hydrophobicity
constants were derived. Also, the hydrophobicity data for the terminal groups of
peptides were not available.

In 1980, Meek published his first paper on the prediction of peptide retention times in HPLC on the basis of complete amino acid composition. The "retention coefficients" (cf., Table 9.2) characterizing amino acid residues were derived directly from RPC data using a chaotropic solution of $NaClO_4$ (in order to suppress the influence of the secondary structure). Meek found that the retention time of peptides containing up to 20 amino acid residues can be predicted solely on the basis of their amino acid composition, and can be calculated as the sum of the retention coefficients for each amino acid residue and end-groups, plus the t_0 value for the void volume. An approximately linear increase in retention time with increasing number of phenylalanine residues in phenylalanine oligomers was observed. The slope thus equaled the retention added per side-chain and peptide bond, but did not include the contribution of terminal amino or carboxyl groups. Extrapolation of the line to zero residues gave a positive value, representing the contribution to the retention of the end-groups. This contribution could not be derived from phenylalanine itself, because the pK values for a single amino acid differ considerably from the dissociation constants of peptides and the extent of ionization markedly affects retention. Aromatic or aliphatic side-chains showed a marked positive contribution, which changed relatively little with variation in pH. Basic and neutral residues had little effect, but acidic side-chains showed a marked negative contribution, which increased with increasing ionization. It should be possible to predict retention when using various chromatographic conditions. Meek found a linear relationship between the retention times of phenylalanine oligomers in two different chromatographic systems (a triphasic acetonitrile gradient containing NaH_2PO_4 on a Hypersil ODS column, and a linear acetonitrile gradient containing $NaClO_4$ on a Bio-Rad ODS column). A good correlation with the sum of pertinent constants was found at both pH 7.4 and pH 2.1 for 25 peptides.

For small linear peptides, the conformation and sequence of amino acids had relatively little effect on the retention of peptides with up to 20 residues (with the exceptions discussed in Section 9.2.6). The application of $NaClO_4$ was believed (a) to favour the transfer of non-polar groups to water by altering the water structure (Hatefi and Hanstein, 1969), (b) to increase their solubility and (c) to break down secondary and tertiary structures of polypeptides. Meek also discussed the selection of suitable chromatographic conditions.

Wilson et al. (1981) measured the elution points of 96 peptides from a LiChrosorb RP-8 or RP-18 column using the buffer system A (0.125 M pyridine-formate, pH 3.0)-B [0.1 M pyridine acetate (pH 5.5)-60% (v/v) 1-propanol]. They compared the values they obtained with values calculated using Rekker constants, and discussed the deviations. The actual measured elution points were expressed as a percentage of buffer B in the A+B buffer mixture. Calculated values were derived from linear regression analysis of all the peptides that gave a defined slope of the plot of peptide hydrophobicity against elution point. Comparison of the data was performed by computer analysis. The less strongly retained peptides eluted essentially independent of the gradient. Conversely, not only was the separation of the more strongly interacting peptides dependent on the steepness of the gradient, but also the points of elution decreased with decreasing steepness in a hyperbolic manner.

In many instances the elution points correlated with the sum of the hydrophobicities of their constituent amino acids. However, closer examination of the measured data indicates that factors other than side-chain hydrophobicity alone are responsible. A possible explanation of the differences might be the presence of secondary structure in larger peptides. Other differences, notably in the separation of diastereoisomers, have not been satisfactorily explained (see also Section 9.2.6).

As the hydrophobic constants of Rekker (1977) were used without corrections, discrepancies in the predicted retention times were expected. They could not be simply explained by the influence of the gradient or pH. Peptide length up to 18 residues had little or no effect on the retention deviation. Larger fragments tended to elute more rapidly than predicted from hydrophobic considerations alone. The authors derived hydrophobicity constants of their own, which showed higher correlation coefficients with the values found experimentally, but the differences remained independent of the source of the hydrophobicity constants. The authors compared graphically the predicted peptide retentions calculated using various hydrophobic constants and concluded that "It appears that predicting exact peptide retention, at least for this particular chromatographic system, is difficult". The elution time could be established for 30-50% of peptides, depending on their position in the gradient. Hearn and Grego (1981b) also compared the capacity factors, $k'_{appar.}$ and $k'_{pred.}$ (calculated in two different ways), for eight peptides and discussed the deviations. They explained the anomalies by ionic peptide-stationary phase silanol-interactions and, for larger polypeptides, by the involvement of peptide secondary structures due to folding, which modifies the number of exposed hydrophobic residues.

In their second paper, Meek and Rossetti (1981) tested the retentions of a large number of small peptides (20 residues or less) and found that they can be predicted by adding the contributions to retention of each amino acid and endgroup. The correlation between actual and predicted retention times for 100 peptides was very high ($r = 0.98$) and indicated that the conformation and sequence have minor effects on retention. The authors exploited the new corrected retention coefficients (cf., Table 9.2). The predicted retention time for a peptide = the sum of the retention coefficients for all the amino acids present

+ the coefficients for end-groups + t_0 (= 2 min). The good correlation was illustrated graphically and showed a linear dependence between the measured and
calculated retention times (Fig. 9.4).

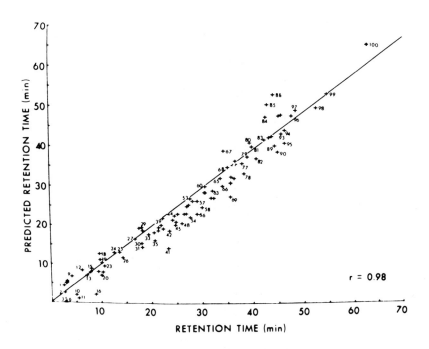

Fig. 9.4. Correlation of actual retention times of 100 peptides versus times
predicted by summation of retention coefficients for amino acid residues and
end-groups. For number identification of peptides, see the original paper.
(Reprinted from Meek and Rossetti, 1981.)

The retention of peptides cannot be simply correlated only with the number
of amino acid residues. In their discussion, Meek and Rossetti (1981) drew attention to the fact that the idea of the prediction of the chromatographic
behaviour of peptides is very old and was tested more than 35 years ago by
Knight (1951) and Pardee (1951) in their efforts to predict the R_F values of
peptides in "normal-phase" paper chromatography. The high correlation between
the actual and predicted retention times in RPC indicates that the composition
must be the major determining factor. However, the difference between the actual
and predicted times (the authors gave an average value of 1.9 min) indicates
that smaller roles are played also by other factors (conformation, size, sequence, charge distribution, polarity). They considered it "somewhat surprising
that the contribution of each residue appears to be nearly additive, i.e., that
each residue of small peptides interacts with the chromatographic matrix without

positive or negative effects on the interaction of the other residues with the matrix". The described results were obtained with peptides containing only L-amino acids. The replacement of D- and L-amino acids can markedly change the retention and will be discussed in Section 9.2.6. The efforts of Su Suew et al. (1981) to predict the retention behaviour of peptides in RPC and eqn. 112 were mentioned in Section 9.2.3. The results provided further evidence that the hydrophobic group retention contributions of amino acid residues (cf., χ_n in Table 9.2) in small peptides have an essentially additive effect on the peptide retention. Divergences in the retention behaviour on alkylsilicas were interpreted in terms of specific silanophilic and solvation interactions.

In 1982, Browne et al. described the isolation of peptides by HPLC using predicted elution positions. The peptides had not been isolated before and their amino acid sequences were postulated from cDNA sequences that were obtained for bovine and rat precursors of ACTH/β-LPH (Nakanishi et al., 1979; Drouin and Goodman, 1980); these two peptides were γ_3-MSH and an acidic linking or "hinge" peptide between γ_3-MSH and ACTH.

Browne et al. (1982) originally attempted to predict the elution position of peptide ACTH/β-LPH 77-95 using the data published by Meek (1980) for acidic conditions (pH 2.1) and obtained a negative value (-20.4). When the refined data of Meek and Rossetti (1981) were used, a small positive value (+19.8) was obtained. Although these values did not show the exact position of the desired peptide, they provided information that the peptide may be eluted very early, at the beginning of the chromatography. Indeed, such a peptide was found at about 11% acetonitrile. After repurification it had the composition of ACTH/β-LPH 77-94. The authors similarly isolated Lys-γ_3-MSH from rat neurointermediary lobes (Browne et al., 1981b) using a modification of the retention values of Meek. cDNA information was used for the predicted sequence, and the knowledge that the peptide contains tryptophan was used as an aid in its localization in the chromatogram. The authors had great luck, because the peptide differed from γ_3-MSH by having an extra lysine and a carbonate group. They concluded that this approach was of great potential, but it was necessary to deduce a set of retention coefficients that were more appropriate for their own chromatographic conditions. [The Meek values were derived using an $NaClO_4$-H_3PO_4-CH_3CN mobile phase and a Bio-Rad column. Brown et al. (1982) used a Waters Assoc. μBondapak C_{18} column and mobile phases with hydrophobic counter ions, which have a profound effect on the elution order of peptides].

The retention coefficients derived by the authors are applicable only to the chromatographic system that they used and, in particular, to their HPLC columns. Brown et al. (1982) recommended: "For readers to obtain an accurate set of coefficients, we suggest that they generate their own coefficients using their own

chromatographic conditions and reversed-phase column". However, they noted that
Waters Assoc. μBondapak C_{18} columns are efficiently "end-capped" and have constant batch-to-batch characteristics. This allows other variables (concentration
of counter ions, acetonitrile gradient, solvent flow-rate) to be kept comparably
constant in various experiments. The column is first "calibrated" with a suitable
set of peptides. If the details of the chromatographic conditions given by the
authors are fullfiled, their coefficients can be used for the prediction of the
elution positions of looked- for peptides. For natural peptides the coefficients
for the post-translational modification groups are important. The method can
also be used for predicting the fragmentation pattern obtained from the proteolytic digestion of peptides and proteins, and probably also for the prediction
of positions of deleted peptides, the by-products of solid-phase synthesis.

It is well known that free silanol groups of chromatographic support matrices
often interact with peptides, especially with the basic ones (Bij et al., 1981).
Therefore, Sasagawa et al. (1982a) examined polystyrene resin, which did not
exhibit such a secondary interaction. It was shown that retention constant D_j
derived from such a chromatography of peptides (cf., Table 9.2) could be used
for predictive purposes and were tested using 47 peptides. Sasagawa et al.
(1982b) studied the prediction of peptide retention times in RP-HPLC during
linear gradient elution. The behaviour of 100 peptides was studied on a C_{18}
column using aqueous trifluoroacetic acid as the mobile phase and acetonitrile
as modifier. It was found that retention time was not linearly dependent on the
sum of retention constants when the size of the peptide increased. The authors
showed that the retention is linearly related to the logarithm of the sum of
modified Rekker's constants for constituent amino acids. Assuming this relationship, the best fit constants for the chromatographic system used were calculated by non-linear multiple regression analysis. The new constants derived,
D_j, are listed in Table 9.2. Using these constants, it was possible to predict
retention times for a wide variety of peptides at any slope of the linear gradient. The retention time of a peptide, t_{Rj}, can be calculated according to the
equation

$$t_{Rj} = A \ln \left(1 + \sum_j D_j n_{ij}\right) + C \tag{113}$$

where A and C are constants and n_{ij} is the number of residues of amino acid j
in peptide i. If the relationship for the studied peptides was expressed in the
form of a plot of t_R versus the logarithmic term, then C was the intercept and
A the slope of the straight line. The authors illustrated these relationships
in a graphical form for various peptides. If unweighted fit retention constants

were used for D_j, $A = -30.3$, $C = 12.4$ and the correlation coefficient was 0.984 (the mean deviation of the retention time was 9.9%). If weighted fit retention constants were used for D_j, $A = -7.04$, $C = 13.6$ and the correlation coefficient was 0.981 (the mean deviation of the retention time was 9.2%).

In 1983, Grego et al. studied the predicted elution behaviour of tryptic peptides of human growth hormone, isolated by RPC under a variety of chromatographic conditions. Water-acetonitrile gradients containing phosphate, hydrogen carbonate, trifluoroacetate and heptanesulphonate buffers were used and the influence of these ionic modifiers on the peptide selectivity with different alkylsilicas was examined. The predicted values for various tryptic peptides were compared with the observed retentions in different systems. The real retention times were compared with the predicted values using the retention coefficients of Meek and Rossetti (1981), Su Suew et al. (1981) and Browne et al. (1982). The correlations between the observed and predicted values were $r^2 = 0.87$, 0.96 and 0.79, respectively. The degree of linearity between the predicted and the measured values was illustrated graphically. It should be noted, however, that the first two sets of constants were derived from a chromatographic system using μBondapak C_{18} columns with a phosphate-based water-acetonitrile linear gradient system, whereas the third set of constants were obtained with a TFA system.

The authors also studied the relationship between the observed retention based on one system and the predicted retention using retention coefficients based on different elution systems. These studies illustrated the difficulties with such a prediction. Comparison between elution systems of similar selectivities (e.g., phosphate-based with TFA-based mobile phases) resulted in modest to good correlations. If the overall selectivity of the chromatographic systems diverged (e.g., comparison of the phosphate with the pyridine-formate systems), the prediction became much less reliable. Comparison between buffer effects can be made reliable only when all the following criteria are satisfied: (1) the same stationary phase must be employed, (2) the same organic modifier must be used and (3) the linear gradient compositional limits must be identical. In addition to other reasons, the deviations of the predicted and measured values can also be explained by the fact that the retention coefficients derived both by Wilson et al. (1981) and by Browne et al. (1982) are expressed by the solvent concentration in the influent entering the column and not by the effluent, in which the zone of the eluted peptide is situated.

The authors also calculated coefficients that allowed in some instances the interconversion of retention data between different elution systems, using the equation

$$k'_{app,i} = M k'_{app,j} + C \tag{114}$$

where $k'_{\text{app},i}$ and $k'_{\text{app},j}$ are the apparent capacity factors of the same peptide
eluted by two different gradient elution systems with the same stationary phase,
M is the relative eluotropic strength difference due to their buffer components
and C is the system constant. This relathionship seems to have only limited
practical applicability, but it is interesting from the theoretical point of
view.

Lebl (1982) utilized the possibilities of predicting peptide retention data
for another purpose. He studied the correlation between the hydrophobicity of
substituents in the Phe moiety of oxytocin carba-analogues and RPC k' values in
order to check possible conformational changes of the hormone due to the substitution. This can be done if only one amino acid in the whole structure is
changed. Oxytocin analogues were synthesized and chromatographed using a 25 x
0.4 cm I.D. column of Separon SI-C$_{18}$ (Laboratory Instrument Works, Prague,
Czechoslovakia) and mixtures of methanol with water or with buffers (0.1% trifluoroacetic acid, pH 2; 0.01 M triethylammonium borate, pH 8.1; 0.05 M trimethylammonium carbonate, pH 5.8-9.2; 0.05 M sodium phosphate, pH 4.0-7.5; and
0.1 M ammonium acetate, pH 7.0) as mobile phases. Hydrophobicities (π values)
of various substituents attached to the aromatic ring of the Phe moiety were
correlated with the retention data. For calculation, π values published by
Hansch and Leo (1979) were used and the π value for the benzyloxycarbonylamino
group was derived on the assumption of additivity of values for methyl, ethyl,
benzyl, methyloxycarbonylamino and ethylcarbonylamino groups. Eleven analogues
were prepared and chromatographed and the results were calculated and compared.
As all the compounds studied obeyed a linear correlation between π values of
substituents on the aromatic nucleus and log k', the author assumed that the
conformations of all the analogues were probably very similar.

The prediction of peptide retention times in RP-HPLC was reviewed by Sasagawa
and Teller (1984); tables were presented listing observed and predicted retention values of 186 peptides.

*9.2.5 Capacity factor in RPC of peptides as a function of pH and the influence
of ion pairs*

Horváth et al. (1976, 1977) described an expression for the capacity factor
(k') of zwitterions in RPC, which was dealt with in Section 3.7 (eqn. 80/III).
Kroeff and Pietrzyk (1978a) investigated the separation of peptides on porous
hydrophobic copolymers (such as Amberlite XAD-2, -4 and -7) by HPLC. Water,
water-ethanol and water-acetonitrile were used for elution with or without pH
control. The authors derived an equation that described the influence of three
ionization steps on the relationship between k' and [H$^+$] (eqn. 81/III).

Hearn et al. (1979b) studied the effect of pH and ion-pair formation on the retention of peptides on bonded hydrocarboneous phases and discussed in detail the eqns. 80/III and 81/III. A local minimum of capacity factors occurred when the pH of the mobile phase corresponded to the pI of the peptide. It is possible to predict elution conditions based on pH for closely related peptides, but there is little practical advantage in using pH variation as the sole method of influencing selectivity. Larger selectivity differences can be easily achieved by the addition, at a suitable pH, of a low concentration (e.g., 5 mM) of suitable reagents that can either undergo ion-pair formation with peptides or modify the stationary phase to a dynamic ion exchanger. The authors studied the effect of ion-pair formation on k' and derived eqns. 84/III and 85/III, which were discussed in Section 3.8. Small polar counter ions [$H_2PO_4^-$, ClO_4^- or $(CH_3)_4N^+$] tended to have minimal interactions with the stationary phase and operated via a hydrophilic ion-pairing mechanism (retention of peptides was reduced). Reagents of intermediate polarity [heptanesulphonate, $(C_4H_9)_4N^+$] acted via a hydrophobic ion-pairing mechanism (retention of peptides was enhanced). Amphiphatic counter ions (dodecylsulphate, dodecylammonium) bonded to the stationary phase then acted as dynamic ion exchangers. Ion pairing and the modification of stationary phase represent a powerful method for varying the retention of peptides in RPC.

Bidlingmeyer et al. (1979) studied in detail the retention mechanism of reversed-phase ion-pair liquid chromatography (cf., Section 3.8). An ion-interacting model, describing Bidlingmeyer et al.'s "ion-interacting chromatography", was rewieved by Deming (1984).

9.2.6 Retention of peptide isomers and selectivity effects

There are two main types of peptide isomers: (a) sequential or positional isomers and (b) diastereoisomers.

9.2.6.1 Sequential isomers

The prediction of retention parameters of peptides in RPC, discussed in Section 9.2.4, was based only on the summation of hydrophobicity constants, and the retention was derived only from the amino acid composition of a peptide and not from its amino acid sequence. However, Meek (1980) stated that "amino acid composition cannot be the only factor determining the extent of retention because it is possible to separate positional isomers with the same composition (e.g., Gly-Trp and Trp-Gly; see also Molnár and Horváth, 1977) and stereoisomers". Meek and Rossetti (1981) drew attention to the work of Sanger (1952), who showed that "the relationship between R_F (in paper chromatography) and composition was not absolutely accurate as peptides containing the same amino acids in different order could frequently be separated". Wilson et al. (1981) stated that "other separations, notably of diastereoisomers and of positional isomers of the same peptide, have not been satisfactorily explained". However, they recalled that "differences in their molecular hydrophobicities, rather than the hydrophobic properties of the peptide constituents, have been implicated (Terabe et al., 1979a)".

Very interesting findings on the retention of sequential isomers of proline-containing dipeptides were published by Melander et al. (1982) in their study of the effect of molecular structure and conformational changes of these peptides in RPC. Peptides containing a proline residue (not at the N-terminus) may yield multiple peaks in HPLC. This phenomenon is caused by the slow kinetics of iso-merization, the speed of which is of the same order of magnitude as that of chromatographic separation.

Although most peptides exist exclusively in the *trans* form, certain peptides linked via the imino nitrogen of proline can be present in both *cis* and *trans* forms under the usual conditions (for references, see Melander et al., 1982) and the change in conformation, initiated by a slight change in conditions, is re-latively slow. Because the retention factors of *cis* and *trans* forms are usually not identical, peak splitting or excessive band spreading occurs.

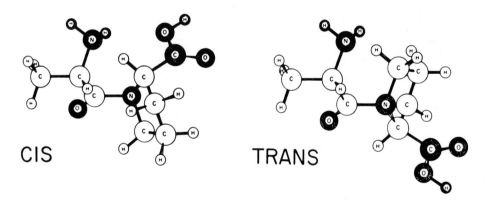

CIS   TRANS

Fig. 9.5. Perspective molecular structures of *cis*- and *trans*-L-alanyl-L-proline (prepared by the use of the PROPHET system) with the polar functions shaded. Whereas polar and non-polar residues in the *cis*-conformation can be placed on opposite sites of a hypothetical plane no such plane exists for the *trans*-con-formation. Consequently, retention of a *cis*-conformer in RPC is expected to be greater than that of the *trans*-conformer. (Reprinted from Melander et al. 1982.)

It is clear from Fig. 9.5 that the hydrophobic surface area, which affects the strength of binding of the eluite to the stationary phase, is different for the two conformers. In the *cis* conformer a plane through the two α-carbons separates non-polar residues from the polar NH_2 and COOH groups. No such plane can be found for the *trans* conformer. The solvophobic theory (cf., Sections 3.6 and 3.7) predicts that in RPC the retention of the *cis* conformer will be greater; the *trans* conformer has a smaller hydrophobic surface area.

Experiments by Melander et al. (1982) using an octylsilica column (plain aqueous buffer, pH 6, 25°C) verified the theoretical predictions. L-Pro-L-Ala, which cannot create conformers of the above-mentioned type, gave a normal sharp peak, whereas L-Ala-L-Pro yielded a broad "peak" with fairly sharp spikes on the leading and trailing edges. Similar results were obtained with L-Val-L-Pro and L-Pro-L-Val, and with L-Pro-L-Phe and L-Phe-L-Pro. In each instance the proline peptide (with C-terminal Pro) yielded "peaks" consisting of two spikes and an intermediate plateau region. On the other hand, prolyl peptides (with N-terminal Pro) showed only one sharp peak. The reason is that prolyl residues exist only in the *trans* form, whereas proline peptides can be present in both *cis* and *trans* forms and the *trans* conformer elutes faster. The first spike of the reaction "peak" of a proline peptide has the same retention as a single peak obtained when the corresponding prolyl peptide is chromatographed. When fractions of the column effluent corresponding to various portions of the "reactive peak" of proline peptides were rechromatographed, in each instance the original chromatogram was reproduced. The conditions (temperature, pH) for the acceleration of the isomerization process were also studied.

If these interesting findings are extended, multiple peaks would be expected in the chromatography of other molecules when the rate of conformational change is slow on the time scale of the separation by HPLC. This applies when the free rotation about a bond in the eluite molecule is hindered.

Hearn and Grego (1983b) studied selectivity effects of peptidic positional isomers and oligomers separated by RP-HPLC. The retention behaviour of peptide positional isomers was examined with aqueous acetonitrile mobile phases and the selectivity differences that arise between isomer pairs were discussed in terms of specific solvatation processes. The ln k' values of pairs of peptides (Val-Tyr, Tyr-Val; Leu-Tyr, Tyr-Leu; Ala-Tyr, Tyr-Ala), chromatographed on µBondapak C_{18}, were plotted against the volume fraction of the organic solvent in water-acetonitrile isocratic mobile phases. The minimum was found in the vicinity of 50% acetonitrile for all the peptides studied. The relative shifts of individual curves were explained using solvophobic considerations, and the observed logarithmic capacity factors were compared with the calculated values.

9.2.6.2 Diastereoisomers

   In this section the retention of diastereoisomers will be discussed. Lande
(1969) estimated and discussed the conformation of dipeptides. The amide bond of
both L,L- and L,D-derivatives was found to be *trans* and the preferred form contains α-hydrogens *trans* to each other and in the plane of the peptide bonds. In
this conformation, R^1-R^2 and amino-carboxyl distances are minimal in L,D-dia-
stereoisomers and maximal in L,L- or in D,D-diastereoisomers. These properties
of peptide conformation are the reason for the differences in the retentions of
diastereoisomers.

   Lundanes and Greibrokk (1978) studied systematically the separation of diastereoisomeric dipeptides. Of four types of columns tested, ODS columns were
found to be best and were able to separate L,L- and D,D-isomers from L,D- and
D,L-isomers. The L,L- and D,D-isomers generally had lower retentions than L,D-
and D,L-isomers. On the basis of model studies the authors concluded, that better
packing of the L,L- and D,D-isomers might partly shield the free amino groups
from interactions with the stationary phase. The resolution of diastereoisomers
was impressive: the capacity factors of L-Leu-L-Leu chromatographed with 0.01 *M*
ammonium acetate in 50% and 30% methanol were 1.4 and 3.0, respectively, whereas
the values for an L,D-D,L mixture increased from 2.7 to 10.6, respectively. For
dipeptides containing relatively large aliphatic or aromatic moieties, the steric
requirements were much more important than any other factor with regard to retention. The authors concluded, "Hence so far we cannot give a satisfactory explanation of the dramatic separation of diastereoisomers on ODS columns". Mixtures of L,L- and D,D-isomers and of L,D- and D,L-isomers could not be separated
on any of the columns examined.

   Kroeff and Pietrzyk (1978b) described an excellent HPLC study of the retention
and separation of short-chain peptide diastereoisomers on a C_8 bonded phase. They
suggested that the different conformations of peptide diastereoisomers had slight
differences in hydrophobicity and this influenced the retention order. First,
retentions of Ala-Ala diastereoisomers on the C_8 column were studied as a function
of pH. As with other types of peptides, the retention was low in the region of
isoelectric pH and high in acidic and basic eluents, regardless of the stereoisomeric form. At all pH values, L-Ala-L-Ala and its enantiomer D-Ala-D-Ala had
identical capacity factors. Similarly, L-Ala-D-Ala and the enantiomer D-Ala-L-Ala
had identical capacity factors, but the value was larger than for L-L and D-D
enantiomers. The difference in the *k'* values for the diastereoisomers was not
constant throughout the pH range; with decreasing pH the difference increased.

   The difference in the physical properties of peptide diastereoisomers was
attributed to differences in molecular conformation (cf., Lande, 1969, and Fig.
9.6). Physico-chemical investigations led several workers (for citations, see

Kroeff and Pietrzyk, 1978b) to conclude that amino acid side-chains of L-D or D-L enantiomers of dipeptides are *cis* to one another (with respect to peptide bond). In the L-L and its D-D enantiomer the side-chains are *trans* to one another. In L-D and D-L forms the methyl and R groups of the two subunits are on the same side of the peptide bond and in close proximity to one another. In this way a higher overall hydrophobic surface area is formed and the retention of the L-D and D-L enantiomers is increased in comparison with L-L and D-D enantiomers, where the hydrophobic groups are on the opposite sides.

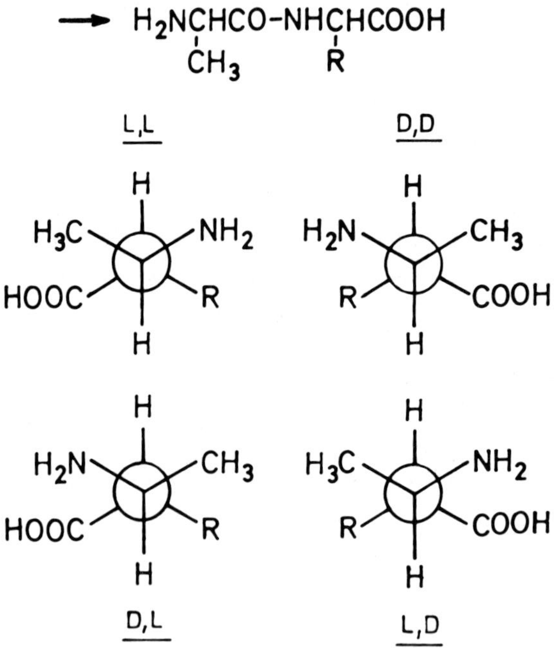

Fig. 9.6. The Newman projection (cf., Lande, 1969) of an alanyldipeptide given above, characterizing the diastereoisomeric configurations; a *trans*-conformation of the peptide bond is presuposed. The direction of view is indicated by the arrow. The alanyl residue is situated in front of the plane of the paper, the second residue behind it. The plane of the peptide bond is vertical and intersects both hydrogen atoms. (cf., Kroeff and Pietrzyk, 1978.)

The study started with Ala-Ala dipeptides and was expanded to many other diastereomeric dipeptides and tripeptides (such as Ala-Val, Ala-Leu, Ala-Phe, Leu-Tyr, Val-Val, Leu-Leu, Leu-Phe and Leu-Gly-Phe). The general trends observed were that L-L and D-D enantiomers had an identical k' value and this was less than the mutually identical k' values found for L-D and D-L enantiomers. This difference was large enough to allow the separation of the two pairs. The retention difference was enhanced using lower pH and/or reducing the amount of ethanol in the eluting mixture. The authors concluded that, for dipeptides having amino acid subunits of similar size and polarity, there is a significant increase in the peptide's hydrophobic surface area going from the L-L (or D-D) to the L-D (or D-L) form. The insertion of a symmetrical Gly subunit (such as in the tri-peptide Leu-Gly-Phe) has the effect of breaking up the large hydrophobic surface area formed by two non-polar subunits (Leu and Phe). The authors emphasized that, even though the chiral centres were separated from one another in the Leu-Gly-Phe tripeptide, the separation of diastereoisomers was still possible. Peptides with more central symmetrical glycyl subunits were not readily available, so that the influence of the greater distance between the chiral centres could not be ex-plained.

Kroeff and Pietrzyk (1978b) also studied a model with three chiral centres, Ala-Ala-Ala. In comparison with Gly-Gly-Gly, the retention was generally high, which indicated a high hydrophobicity and a great influence of alanine side-chains. The influence of pH was similar to that with other peptides, viz., mini-mum retention at the isoelectric point and higher values in acidic and basic media. The introduction of a D-subunit altered the retention of the L-Ala-L-Ala-L-Ala tripeptide, but it was shown that the position of the D-Ala subunit in the amino acid sequence also had an important influence. Starting from the construc-tion of a model and using the criteria mentioned above for dipeptides, the authors suggested that "For the LLL form the methyl groups of the Ala subunits are seen to alternate about the peptide backbone; for the LLD form, two adjacent methyl groups are on the same side and the third is on the other side; and finally, for the LDL form all three methyl groups are on the same side." On the basis of this idea, the following order of increase in the hydrophobic nature of the side-chain groups could be predicted: LLL < LLD < LDL. This should also be the order in which the retention increases. The same order was observed experimentally. As the retentions of diastereoisomeric Ala-Ala-Ala tripeptides are sufficiently different in an acidic medium, their mixture could be well separated. The authors also discussed briefly the theoretical and practical applications of the separa-tion of diastereoisomeric peptides.

Cahil et al. (1980) described the applications of *tert.*-butyloxycarbonyl-L-amino acid-N-hydroxysuccinimide esters (Boc-L-AA-OSu) in the chromatographic separation and determination of D,L-amino acids and diastereoisomeric dipeptides. A diastereoisomeric mixture of a dipeptide (D,L-D,L) was converted by means of Boc-L-AA-OSu into a mixture of LLL, LDD, LLD, and LDL tripeptides. The resulting diastereoisomeric peptides were separated on RPC columns. An example is illus-trated in Fig. 9.7.

The separation of diastereoisomers was reviewed by Blevins et al. (1984).

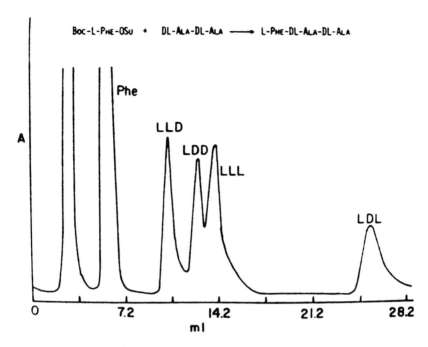

Fig. 9.7. A mixture of diastereoisomeric dipeptides L,D-Ala-L,D-Ala was converted by means of Boc-L-Phe-OSu (see text) into a mixture of diastereoisomeric tri-peptides L-Phe-D,L-Ala-D,L-Ala, which was chromatographically separated. Column: μBondapak C_{18}, 10 μm, 30 cm x 4 mm. Eluent: acetonitrile-water (5:95, v/v), I = 0.02, pH 2.5. Flow-rate: 1.87 ml/min. Detection at λ = 208 nm. (Reprinted from Cahil et al., 1980.)

9.3 COLUMN LIQUID CHROMATOGRAPHY MODES FOR SEPARATION OF PEPTIDES

There are three main modes that have proved to be general methods for the rapid separation of peptides: reversed-phase chromatography (RPC) (including ion-pair RPC), ion-exchange chromatography (IEC) and size exclusion chromatography (SEC), in addition to special methods. RPC and IP-RPC are the most important and most often used.

9.3.1 Reversed-phase chromatography

At the beginning of the development of this separation method, optimal chro-matographic supports were sought. Hansen et al. (1977) tested Bondapak Phenyl-Corasil, Poragel PN and Poragel PS under RP conditions with acetonitrile-water mixtures. It was found that residual silanol groups in Phenyl-Corasil and the functional groups in Poragels significantly influenced retention. None of the columns showed good efficiency (Phenyl-Corasil and Poragel PN were relatively better) because material of coarse particle size (37-75 μm) was tested. Lundanes and Greibrokk (1978) also tested various supports: Phenyl-Sil-X-I (13 μm), μBonda-pak NH_2 (10 μm), Nucleosil 5 CN (5 μm), Spherisorb S5W-ODS (5 μm), Spherisorb S5 Phenyl Bonded (5 μm) and ODS-Hypersil (5 μm). Of the four reversed phases examined, the ODS packings appeared to give the best selectivity and the best efficiency in the separation of most peptides, including diastereoisomers. However, a marked difference in retention was found on two ODS packings of different manufacture. Wehr et al. (1982) evaluated a series of silica-based chemically bonded stationary phases for the chromatography of hydrophilic or moderately hydrophobic peptides (ranging from 5 to 51 amino acid residues). ODS, hexyl, phenyl and cyanopropyl phases with or without secondary silanization (end-capping) were compared. The packing materials tested were: MicroPak MCH-10, MCH-N-CAP-10, MCH-5, CN-AQ-10 and CN-AQ-10EC, a 10 μm phenyl experimental column, a 10 μm hexyl experimental column and a 10 μm C_{18}-diol experimental column. The octadecyl phase of moderate carbon loading was found to provide optimal retention and selectivity for most separations. Wilson et al. (1982) studied the applicability of 100 and 300 Å reversed-phase supports for the separation of peptides and proteins in various buffer systems. LiChrosorb RP-18 (100 Å) and Aquapore RP 300 (300 Å) were tested. The larger pore size packings exhibited slightly more hydrophilic character and were found to be notably superior to the 100 Å supports, especially for the chromatography of higher molecular weight compounds (exceeding 15 000).

When using silica-based hydrocarbonaceous RPC supports, problems with the disturbing effects of surface silanols often occurred. Bij et al. (1981) studied the separation of certain unprotected peptides on such materials and found irregularities on C_8 and C_{18} stationary phases, which could be explained by the influence of the remaining uncovered ≡SiOH groups. The addition of a suitable amine to the buffer was recommended for silanol masking. The authors also presented a theory of silanol masking. On the other hand, the presence of small amounts of unmasked silanol groups may contribute to a better selectivity of such a packing, owing to a "dual retention" mechanism, which seemed to be a fairly common phenomenon. Olieman et al. (1981) studied the problem of strong tailing of peptide peaks on old, often used and deteriorated ODS columns. These phenomena were caused by the hydrolysis of silaxane bonds (by acids or traces of fluoride). In situ silylation of an octadecylsilylsilica stationary phase was described, involving washing of the exhausted column with a solution of chlorodimethyloctane-silane in toluene (0.1 g/ml). The good separation properties (for peptide hormones) of the RPC column were renewed after such a treatment.

The applicability to the RPC of peptides of packing materials that are not based on silica and that consequently cannot cause problems with silanol groups

was also studied. Iskandarani and Pietrzyk (1981) described LC separations on
the porous polystyrene-divinylbenzene copolymer PRP-1, which is chemically very
stable. Many short peptides including peptide diastereoisomers were successfully
separated. Pietrzyk et al. (1982) described the preparative LC separation of
amino acids and short peptides (including diastereoisomers) on Amberlite XAD-4
(polystyrene-divinylbenzene copolymer, 50 Å pores). The preparative separation
of a mixture of enkephalin peptides was also achieved using acetonitrile-water
(17:83) of pH 2.2 (H_3PO_4 buffer) with an ionic strength of 0.1 M as the mobile
phase, and a 250 mm x 8.0 mm I.D. column packed with 37-44 μm XAD-4; the flow-
rate was 5.6 ml/min.

Another group of papers dealt with the question of an optimal mobile phase
for the RPC of peptides. Cooke and Olsen (1980) discussed the role of mobile
phase in RPC with emphasis on ways of improving the peak shapes for problematic
samples. The use of special mobile phase additives were also briefly described
(cf., Section 3.7). Hearn and Grego (1981c, 1983a) and Grego and Hearn (1981)
studied organic solvent modifier effects on the separation of unprotected peptides by RPC. An octadecylsilica support was used and acetonitrile or methanol
was most often employed as the organic modifier. Unprotected peptides did not
show linear dependences of the logarithm of their capacity factors on the composition of binary aqueous-organic eluents; minima in bimodal plots were observed
(cf., Section 9.2.2). Henderson et al. (1981) described the application of
volatile solvents in the RP-HPLC purification of peptides and proteins, using
μBondapak Phenylalkyl or C_{18} supports. Elution was achieved with simple linear
gradients employing only two solvents: (A) 0.05% trifluoroacetic acid (TFA) in
water and (B) 0.05% TFA in acetonitrile. The advantage of this system is complete volatility and compatibility with UV detection at 206 nm and above. Wehr
et al. (1982) compared a variety of mobile phase conditions in the RP-HPLC of
peptides and found that aqueous buffers of low pH and ionic strength, containing
alkylamine or alkylammonium compounds, were most effective in eluting retained
species. High concentrations of an organic modifier (acetonitrile) led to an
increased retention of peptides, which suggests limitations on the gradient
elution of multi-component peptide mixtures.

Blevins et al. (1980b) studied factors affecting the separation of Arg-vasopressin (AVP) peptide diastereoisomers by HPLC, using several RPC-columns. The
effects of the percentage carbon loading on an ODS support, the carbon chain
length of the bonded phase, the concentration of the buffer and the organic
solvent were examined. The diastereoisomeric AVP analogues studied were 8-D-Arg,
4-D-Gln, 2-D-Tyr and 1-hemi-D-Cys, the columns tested were Spherisorb ODS,
μBondapak C_{18}, LiChrosorb RP-18 and LiChrosorb RP-8 and the organic modifiers
were methanol, acetonitrile and tetrahydrofuran and 0.025, 0.05 or 0.075 M TEA-

HOAc buffered the mobile phase to pH 4. The results allowed one to differentiate between the contributions of the mobile phase and that of the stationary phase compositon to the elution order (k') and selectivity (α) of the peptides studied. Mahoney and Hermondson (1980) investigated the separation of large denatured peptides by RP-HPLC using trifluoroacetic acid (TFA) as a peptide solvent. The cyanogen bromide (CNBr) fragments of haemoglobin α-, β- and γ-chains (13-91 residues long) were separated and recovered in 71-95% yields on a LiChrosorb C_8 column; 0.013 M TFA-water was used as the eluent, with a limiting organic solvent (1-propanol, acetonitrile, etc.). This procedure permitted the non-destructive detection of peptides at low wavelength (210 nm) on as little as 50 ng of sample. The peptides were recovered directly by evaporation or lyophilization. The 13-residue COOH-terminal CNBr peptide of human γ-chain was employed to test the relative effectiveness of various solvents for the elution of peptides from a C_8 column (5 μm LiChrosorb RP-8, 250 mm x 4.6 mm I.D.). The temperature was maintained at $30^{\circ}C$, the flow-rate was 0.7 ml/min and the rate of solvent change was 1.6%/min. Mahoney and Hermondson (1980) presented the following "eluotropic series":

Solvent	Elution time (min)	Solvent composition (%)
Methanol	34.6	57.6
Acetonitrile	28.1	46.8
Ethanol	27.6	46.0
Dioxane	26.2	43.6
Tetrahydrofuran	26.1	43.4
2-Propanol	22.5	37.5
1-Propanol	20.3	33.8

These data indicate that a methanol concentration of 57.6% was required to elute the γ-chain fragment, compared with 33.8% for 1-propanol. The authors also presented other compositional data for solvents allowing the successful elution of peptides of various hydrophobicity and length.

Browne et al. (1981a) described a procedure utilizing RP-HPLC for the extraction and purification of peptides from biological tissues. Corticotropin- and melanotropin-related peptides were isolated from the neurointermediary lobe of the rat pituitary and characterized. Peptides were isolated as follows. Tissue of 190 rat neurointermediary lobes was homogenized in a strongly acidic medium, the peptides were extracted using ODS-silica cartridges (C_{18} Sep-Paks; Waters Assoc.) and eluted with 3 ml of 80% acetonitrile containing 0.1% TFA. The eluate was diluted to 18 ml with 0.1% TFA and loaded directly on to a μBondapak C_{18} RP-HPLC column, where it was chromatographed (linear gradient of 20-40% acetonitrile, containing 0.1% TFA). Fractions containing ACTH and α-MSH were rechro-

matographed; they were diluted 1:1 with 0.13% heptafluorobutyric acid (HFBA)
and loaded back on to the same RP-HPLC column equilibrated with 12% acetonitrile containing 0.13% HFBA. Elution was achieved with a linear gradient from
20 to 45.6% acetonitrile, containing 0.13% HFBA. The ACTH and α-MSH peaks were
dried in vacuo. The main ACTH peaks after rechromatography were CLIP 1 and
CLIP 2 (corticotropin-like intermediary lobe peptides). Gazdak and Szepesi
(1981) studied the separation of large peptides by HPLC. Various solvent systems
[methanol-acetonitrile-buffered water (pH 2.2) with or without 0.1 M Na_2SO_4;
methanol-isopropanol; methanol-isopropanol-acetonitrile] were examined in different proportions for the separation of oxytocin, $ACTH_{1-32}$, aprotinine and
insulins on a Nucleosil 10 C_{18} column (250 mm x 4.6 mm I.D.). The use of isopropanol as an organic modifier in the eluent resulted in an improved separation. Meek's method of predicting retention times could be employed for isocratic elution with different elution systems.

The RP-HPLC preparative chromatography of peptides is one of the objectives
that has often been studied. For instance, Bishop et al. (1980) described the
application of this method for the purification of a synthetic underivatized
peptide. A polyethylene cartridge containing 75 μm octadecylsilica particles
(Waters Assoc. Prep-Pak 500-C_{18}) was used. A model tetrapeptide, L-Leu-Gly-Gly-
Gly, was prepared by standard solution-phase methods and purified to homogeneity
using this preparative HPLC approach. Amounts of the tetrapeptide up to 10 g
could be chromatographed in less than 30 min. The degassed mobile phase was
water-methanol-TFA (95:5:0.05) and the crude sample was loaded in amounts between
1 and 10 g in 10 ml of the eluting solvent. The flow-rate was 100 ml/min and
the back-pressure 100 p.s.i. Immediately after collection, each fraction was
neutralized to pH 7 with ammonia solution, concentrated in a rotary evaporator
and lyophilized. Bishop et al. (1981) used the same method for the preparative
separation of the synthetic peptides and derivatives Gly-Gly-OEt, Gly-Gly-Glu,
Gly-Gly-Lys, PyroGlu-His-Gly, Pro-Pro-Pro, Leu-enkephalin and Met-enkephalin.
A Pre-Pak 500-C_{18} cartridge (30 x 5.7 cm) was used for purification. Before
and after each chromatographic separation the cartridges were washed with 4 l
of methanol and 4 l of methanol-water (1:1) at a flow-rate of 50 ml/min, in order
to remove any adsorbed contaminants. The chromatographic method was the same as
mentioned in the preceding paper. In the case of $(Pro)_3$ peptide and Met-enkephalin the mobile phase was 0.05% TFA-water-25% methanol at a flow-rate of
100 ml/min. For the purification of Leu-enkephalin, 0.01% H_3PO_4-water-25% methanol was employed in addition to 0.05% TFA-water-25% methanol. Hearn et al.
(1981) also described the semi-preparative separation of peptides on RP-silica
packed into radially compressed flexible-walled columns.

Gabriel et al. (1981) reviewed the application of an RP-8 Lobar glass column
(30 cm x 2.5 cm I.D.) containing octyl-bearing silica for the purification of
β-endorphin on the 150-mg scale with the inexpensive apparatus illustrated in
the Fig. 6.6. The mobile phase, consisting of 0.01 M HCl modified with acetonitrile, was lyophilizable and transparent at 206 nm. Various mobile phases for
the preparative-scale RP-LC of peptides were listed.

9.3.1.1 Comments on literature

Bennett et al. (1982) reviewed the comprehensive approach to the isolation
and purification of peptide hormones using only RP-LC, and Lundanes and Greibrokk
(1984) the retention of dipeptides in RP-HPLC. The review of Ragnarsson et al.
(1984) was focused mainly on the separation of basic hydrophilic peptides by RPC.
Yamada and Imoto (1984) reviewed the application of hydrochloric acid as a mobile
phase modifier.

9.3.2 Ion-pair reversed-phase chromatography

Sometimes it is not simple to distinguish "pure" RPC from ion-pair (IP) RPC
at the beginning of chromatography, because the influence of the secondary equilibria on the separation process is not known, when some new atypical mobile
phases are used. The general principle of IP-RPC was explained in detail in
Section 3.8. The principle of the application of IP-RPC to the separation of
peptides was mentioned in Section 9.1 and illustrated by Fig. 9.1. In this section, only a few papers will be cited, describing the most important ion-pairing
reagents for peptide separations and the principles of some applications.

Bennett et al. (1979) described the purification of peptide hormones using
trifluoroacetic acid (TFA) as a hydrophobic anion-pairing reagent. They used
µBondapak C_{18} column and an aqueous acetonitrile solvent system containing
0.1% TFA. A linear gradient of 20-60% acetonitrile eluted the substances from
a mixture in the following order: Met-enkephalin, α-MSH, β-MSH, ACTH, calcitonin,
cytochrome c, bovine serum albumin, bovine prolactin and bovine growth hormone.
A great advantage of the solvent system used is its volatility. The application
of a Sep-Pak C_{18} cartridge for the batch extraction of peptides (tritiated corticotropin, somatostatin and their metabolites) from tissue homogenates was also
described. Water containing 5% (v/v) TFA, 1 M HCl and 1% (w/v) sodium chloride
was used for the homogenization of rat pituitaries, and reasons were given for
this composition. Following centrifugation of the homogenate the supernatant was
passed four times through the Sep-Pak C_{18} cartridge, which was washed with
0.1% aqueous TFA and the peptide fraction was eluted with 3 ml of aqueous 80%
acetonitrile containing 0.1% TFA. The pituitary extract was immediately subjected
to RP-HPLC.

Bennett et al. (1980) studied the relative effectiveness of completely volatile and UV transparent perfluorinated carboxylic acids, such as TFA, pentafluoro-propanoic acid (PFPA), heptafluorobutyric acid (HFBA) and undecafluorocaproic acid (UFCA), as hydrophobic counter ions in the RPC of peptides. In the given order of these acids, dissolved in the mobile phase (0.01 M solution), all the tested peptides showed increasing retention times. Because this effect was most apparent with the peptides having the greatest number of basic groups, this be-haviour could be applied to the purification of basic peptides using the rechro-matography with solvents containing different acids. μBondapak C_{18} was used in these experiments, with a linear gradient (20-58.4%) of acetonitrile, and the separation of the following peptides was tested: (1) [D-Ser 1, Lys 17,18]cortico-tropin(1-18)octadecapeptide amine, (2) Met-enkephalin, (3) luteinizing hormone-releasing hormone (LH-RH), (4) Leu-enkephalin, (5) synacthen, (6) α-melanocyte-stimulating hormone (α-MSH), (7) human adrenocorticotropin (h-ACTH), (8) somato-statin, (9) bovine insulin, (10) human β-endorphin, (11) human calcitonin and (12) bovine cytochrome c.

Grego and Hearn (1980) examined the effects of two ion-pairing agents, n-hexyl-sulphonate (I) and d-camphor-10-sulphonate (II) on the retention of peptides during RP-HPLC. A μBondapak C_{18} column was used with a methanol-water (5:95) mobile phase containing NaH_2PO_4 and either I or II. It was found that the capacity factor was dependent on and the selectivity factor (above 25 μmol/l) was inde-pendent of the concentration of the tested paired ions. The continuation of this work was described by Hearn et al. (1981), who studied the ion-pairing effects in the RP-HPLC of peptides in the presence of the same alkylsulphonates I and II. The influence of pH and the concentration of the two lipophilic pairing ions on the retention of a group of small peptides (unprotected and C-protected) on the hydrocarboneous microparticulate silica-bonded phases was described. With a low pH of aqueous methanol the capacity factors of unprotected and C-protected pep-tides showed similar dependences on the concentration of the pairing ions. At higher pH the influence of the pairing ions diminished (owing to competing protic equilibria).

Lin et al. (1980) applied aliphatic carboxylic acids as surfactants to separate basic, acidic and neutral dipeptides. Butanoic, pentanoic, hexanoic and octanoic acids were used at concentrations from 0.002 to 0.038% in 5% (v/v) butanol in water, with a μBondapak C_{18} column. The increase in the ln k' values of basic peptides varsus the chain length of the carboxylic acid followed sigmoidal curves, that of ln k' of neutral peptides was nearly linear, and for acidic peptides ln k' decreased. The addition of salts caused the decrease in capacity factors. Harding et al. (1981) described the use of diethyl ether-soluble perfluoroalkanoic acids (perfluoropropionic and perfluorobutyric acid) as volatile ion-pairing agents in the preparative RP-HPLC of underivatized peptides. Amounts of 1-5 g of the synthetic anorexigenic peptide Pyr-His-Gly (the crude product) was dissolved in 10 ml of the mobile phase (water containing 5 mM perfluorobutyric acid) and injected into a Waters Assoc. flexible-walled Per-Pak 500-C_{18} cartridge [mean particle size 75 μm, 30 x 5.7 cm I.D., washed in advance with 4 1 of isopropanol and 4 1 of isopropanol-water (1:1)], subjected to radial compression. The flow-rate was 100 ml/min and the back-pressure 100 p.s.i. The fractions were immediately neutralized to pH 7 with ammonia solution, concentrated on a rotary evaporator and lyophilized. The sample was dissolved in methanol and peptide was precipitated by the addition of diethyl ether. Ammonium salts of perfluorobutyric acid could be removed from the peptide by extraction with diethyl ether. The complete removal of the ion-pairing reagent is particularly important if the peptide is to be used in biological studies. The effect of perfluoroalkanoic acids on the retention of various peptides using water, 15% acetonitrile-water and 22.5% acetonitrile-water as mobile phases was also described.

Perfluoroalkanoic acids as mobile phase modifiers were reviewed by Olieman and Voskamp (1984).

The close functional connection between IP-RPC and other types of chromatography has also been described. Lin et al. (1980) analysed dipeptide mixtures by a combination of IP-RPC with gas chromatographic-mass spectrometric techniques. Acid and natural dipeptides, originating from the partial hydrolysis of poly-peptides (ribonuclease S peptide, Val-4-angiotensin) by dipeptidylaminopeptidase, were identified using this combined technique. Poll et al. (1982) described a simple method for monitoring preparative RP-HPLC by the use of the IP-RP thin-layer chromatographic analysis of peptides. This technique is useful especially for the evaluation of fractions from flexible-walled cartridges (subjected to radial compression) packed with C_{18} microparticulate silica and often used for the preparative separation of peptide mixtures, or for the extraction of peptides from biological fluids. The described system used Whatman K C_{18} F plates and aqueous solution of sodium chloride (3%) and sodium dodecylsulphate (SDS, 0.2%)-acetonitrile-methanol (50:10:10) (apparent pH = 2.50) as the mobile phase. The plates were dried in an oven (80°C), sprayed with ninhydrin solution [1% (w/v) in acetone-acetic acid (96:4)] and heated to develop the colour. The application of three solvent systems for the separation of groups of di- to octapeptides was described.

9.3.2.1 Comments on literature

Bishop (1984) discussed phosphates as mobile phase modifiers. The separation of basic hydrophilic peptides by RPC was reviewed by Ragnarsson et al. (1984), and Tomlinson (1984) dealt with the manipulation of the IP effect in HPLC. The IP chromatography of peptides was reviewed by Hearn (1985).

9.3.3 Ion-exchange chromatography

Ion-exchange chromatography (IEC) was the original oldest mode for the rapid separation of peptides (cf., Section 9.1). Later it was replaced by RPC, which showed higher efficiency, but in some instances, especially for the separation of basic, acidic and hydrophilic peptides and in special circumstances, IEC is still used.

Radhakrishnan et al. (1977) described the high-efficiency cation-exchange chromatography of synthetic polypeptide hormones in the nanomole range using a 25 cm x 0.46 cm I.D. column of the IEC derivative of spherical porous silica Partisil SCX (Whatman, Clifton, NJ, U.S.A.). Fluorescamine detection was used in an automated monitoring system. Good separation was observed but, in contrast to resinous ion exchangers, this packing material could be used only at pH below 7.6, because the silica beads were degraded by alkali.

Bradshaw et al. (1980) systematically studied the effect of copolymer cross-linking on the resolution of soluble tryptic peptides of human globulin (α- and β-chains), using columns containing ion-exchange derivatives of polystyrene resins in bead form (11 μm). The cation exchangers DC-X8-11 and DC-X12-11 and the anion exchangers DA-X12-11, DA-X8-11 and DA-X2-11 (Dionex, Sunnyvale, CA, U.S.A.) were compared. With both the cation- and anion-exchange resins polymers of lower cross-linkage provided better resolution; X-12 cross-linkage was found to be inferior. Microparticulate anion-exchange resins could be used in columns maintained at 55°C and the resolution yielded by a column 20 cm long was as good as or superior to that obtained with much longer columns of crushed resin. Various volatile pyridine-acetic acid buffers were used for both cation- and anion-exchange columns. Pyridine was redistilled after the addition of solid ninhydrin (1 g/l) to the distillation flask. Takahashi et al. (1981) examined an analytical and preparative method for the separation of peptides by HPLC on a macroreticular anion-exchange resin. Tryptic digests of bovine brain calmodulin, other bovine brain proteins and human urinary amyloid protein were used for separation. The ion exchanger was Diaion CDR-10 (5-7 μm) from Mitsubishi Kasei Industry (Tokyo, Japan). A convex gradient from distilled water to 0.25 *M* methanesulphonic acid containing 50% (v/v) acetonitrile and 25% (v/v) isopropanol (adjusted to pH 2.8 with aqueous ammonia) was used for elution at a flow-rate of 0.89 ml/min. The temperature was programmed from 20 to 70°C. The method can be applied both for peptide mapping and for preparation.

Dizdaroglu and Simic (1980) studied the separation of underivatized peptides by HPLC on weak anion-exchange bonded phases. MicroPak AX-10 (Varian, Walnut Creek, CA, U.S.A.) was used in a 30 cm x 0.4 cm I.D. column. Volatile mixtures of triethylammonium acetate (TEAA) buffer (which has low absorption at 200-220 nm)

and acetonitrile were applied as the eluent using both isocratic and gradient
conditions. Separations of about 50 dipeptides including sequence isomeric and
diastereoisomeric dipeptides were tested. The separation of oligomers into
classes according to chain length was also demonstrated. Dipeptides were generally not retained on MicroPak AX-10 and the addition of an organic solvent (acetonitrile) to the mobile phase increased the retention and facilitated the separation of peptides. Dizdaroglu et al. (1982) extended the above studies to the
separation of larger peptides (containing up to approximately 30 amino acid
residues) using a similar system. They studied peptides including somatostatin,
α-endorphin, ribonuclease-S-peptide, glucagon, various angiotensins and brady-
kinins, as well as tryptic digests of horse heart cytochrome c, calmodulin and
reduced and alkylated hen egg-white lysozyme. Multi-component mixtures of peptides
or closely related peptides were successfuly resolved by HPLC on a weak anion-
exchange bonded phase. The peaks were symmetrical and the peptides were obtained
with high recoveries. It was found that the composition, pH and column temperature
all contributed to the resolution. The authors found long column lifetimes (up
to 1 year with several daily injections) when using the experimental conditions
described.

Cachia et al. (1983) studied the separation of basic peptides by analytical
and preparative cation-exchange HPLC. Columns of 300 Å pore size SynChropak
CM 300 (250 mm x 4.1 mm I.D.) (SynChrom, Linden, IN, U.S.A.) were used for
analytical experiments, in which various pH values and eluent ionic strengths
were studied. Both isocratic conditions and gradients were used for elution.
No organic solvents were added to the mobile phase, so that the separation principle was true ion exchange. Buffers were based on KH_2PO_4 and KCl was added to
increase the ionic strength. A 50 mM KH_2PO_4-KCl gradient (pH 4.5) and isocratic
conditions at pH 4.5 and 6.5 were tested. Loads up to 10-20 mg (6.6-13.3 μmol)
could be applied on the preparative column (250 mm x 10 mm I.D.) before the
peaks of the crude peptide sample tested were seen to fuse. It was found that
the column capacity under gradient conditions was much greater at pH 6.5. Under
both isocratic and gradient conditions phosphate buffers containing KCl were more
effective in resolving peptides of similar charge than were phosphate buffers
alone.

The separation of peptides by ion-exchange HPLC was reviewed by Dizdaroglu
(1984).

B128

9.3.4 Miscellaneous modes

9.3.4.1 Size exclusion chromatography

Size exclusion chromatography (SEC) and gel-permeation chromatography (GPC) have only minor importance for the separation of peptides owing to the relatively low molecular weights of most peptides of interest and the low efficiency of the separation in comparison with RPC. In spite of this, several workers have endeavoured to exploit these methods, usually in connection with the separation of proteins. Hansen et al. (1977) chromatographed peptides on Hydrogel IV gel filtration packing material (37-75 μm) in a 3 ft x 1/8 in. I.D. column with aqueous eluents. Hydrogel IV (Waters Assoc., Milford, MA, U.S.A.) is based on a polar organic polymer compatible with aqueous solvents and does not collapse under the pressure applied in HPLC. The authors studied the retention of four low-molecular-weight peptides using elution with water and with 40% acetonitrile in water, Hydrogel IV is designed for the molecular weight range of ca. 500-40 000, but factors other than GPC played a significant role in retention with water as the eluent and shifted some peaks from their expected positions. A mixed-mode separation principle was proved. Acetonitrile (40% in water) was a stronger eluent than water itself in two instances, indicating some degree of RPC separation character. Gruber et al. (1979) evaluated a gel-permeation support, Syn-Chropak GPC-100 (100 Å) (SynChrom, Linden, IN, U.S.A.) packed into a 25 cm x 0.46 cm I.D. column for the SEC separation of proteins and peptides. Posterior pituitary extracts, tumour-associated foetal antigens and estrogen receptors were used for the separation and the experiments demonstrated the ability of this packing to separate biological samples. The column was able to give a linear separation of compounds of M_r between 5000 and 700 000. The elution time was 30-100 min. An automated fluorescamine method, radioactivity determination (^{35}S, ^{3}H) and radioimmunoassay were used for off-line detection. Shioya et al. (1982) examined the determination of the molecular weights of peptides by the determination of the HETP in SEC. TSK Gel 2000 SW (10 μm) with a 600 mm x 7.5 mm I.D. column was used for the separation and fluorescence detection with *o*-phthaldialdehyde for monitoring. The solvent system was 0.15 *M* phosphate buffer (pH 7.4), containing 1 *M* NaCl, 20% Methylcellosolve and 1% sodium dodecylsulphate (SDS). A straight line was obtained when log MW was plotted against the flow-rate which gave the optimal HETP. The typical amount of peptide required was 1 nmol and the analysis time was less than 2 h. The method was applied to peptides of MW 200-10 000 and the precision was ±20%.

9.3.4.2 Normal-phase chromatography

Naider et al. (1979) described the separation of protected hydrophobic oligopeptides by normal-phase chromatography (NP-HPLC). The resolution of structurally similar hydrophobic oligopeptides originating from the solution-phase synthesis was studied, and the relationship between residue location and side-chain lipophilicity and the retention behaviour was investigated. A 30 cm x 3.9 mm I.D. µPorasil silica column (Waters Assoc.) was used for the chromatography, 5-20 µg of the sample dissolved in 5-25 µl of dichloromethane being loaded. All samples were eluted isocratically with cyclohexane-isopropanol (96:4). *tert.*-Butyloxycarbonyl(BOC)-X-MetOCH$_3$ peptides were investigated. The retention time of individual dipeptides increased as a function of the X residue in the order Gly > Pro > Ala > Met > Val. The separation of isomeric Met$_{n=1}$ dipeptides Boc-Met$_n$-X-OCH$_3$ and Boc-X-Met$_n$-OCH$_3$ and isomeric Met$_2$ tripeptides and Met$_5$ hexapeptides was also investigated. A variable-wavelength UV detector operating at 220 nm was used for monitoring the effluent. Hara et al. (1981) studied the design of binary solvent systems for the separation of protected oligopeptides in silica gel liquid chromatography. They used differential refractometry in the study of the solvent systems, so that they were also able to test mobile phases based on strongly UV-absorbing solvents, such as benzene, dichloroethane, chloroform, ethyl acetate and acetone. Because this detection method is less sensitive, much higher loads had to be applied than in the above-mentioned procedure.

The separation of oligopeptides by NPC was reviewed by Naider et al. (1984); the mobility of Boc(γ-methyl)glutamate peptides on µPorasil was also studied and the results were tabulated.

9.3.4.3 *Ligand-exchange chromatography*

The principle of ligand-exchange chromatography (LEC) was explained in Section 7.3.5.

In 1978, Cooke et al. used a laboratory-made 5 µm silica-bonded diamine phase (the preparation was described) for the selective separation of aromatic dipeptides, in addition to other substances. The prepared support chelated Cd^{2+}, which was very efficient in optimizing the necessary interactions between the fixed and mobile ligands. This separation mode was related to "normal-phase" separations (cf., Section 3.10), as the retention decreased with increasing acetonitrile concentration. In another series of experiments, Cooke et al. (1978) applied as the mobile phase acetonitrile-water containing Zn^{2+} ions coordinated to a hydrophobic chelating agent, dodecyldiethylenetriamine, CH$_3$(CH$_2$)$_{11}$N(CH$_2$CH$_2$NH$_2$)$_2$, for the efficient separation of dipeptides. This reagent was bound to a silica RP packing (Waters Assoc. µBondapak C$_{18}$ or Merck LiChrosorb C$_8$) by hydrophobic interactions and the Zn^{2+} ion was then able to contribute to the LEC separation of dansylamino acids and dipeptides containing aromatic amino acids.

Hunter et al. (1980) studied the HPLC of peptides and diastereoisomers on ODS- and cyanopropylsilica gel column packing materials. In addition to other

procedures, they tested the resolution of D-Met[5] and L-Met[5] isomers of the pentapeptide L-Tyr-D-Ala-Gly-L-Phe-Met. A 25 cm x 0.46 cm I.D. column of 5 μm Spherisorb-CN was used for the separation at 87°C. The mobile phase was 0.035 mM cop-
per(II) acetate in acetonitrile-water (30:70) at a flow-rate of 1 ml/min and
UV detection at 210 nm was used. Good separation was achieved without any bonding or addition of special ligands. Probably the cyano-phase possesses a sufficient affinity to Cu^{2+}, and no additional special chelating agents are necessary. Diastereoisomers of pyroglutamylhistidyl-3,3-dimethylprolineamide were
also separated using LEC by simply adding 0.01 M copper(II) acetate to an aqueous
acetonitrile eluent (Sugden et al., 1981). A 250 mm x 4.6 mm I.D. column of
5 μm Spherisorb-CN was used with acetonitrile-0.01 M copper(II) acetate (30:70)
as the mobile phase at a flow-rate of 0.5 ml/min and 87°C, with UV detection at
210 nm. Guyon et al. (1979) described a very simple procedure for the LEC separation of small peptides originating from the enzymic degradation of Met-enkephalin. A pre-packed column of Partisil 5 (20 cm x 0.48 cm I.D.) was percolated
with copper-ammonia solutions and copper(II) silicate was formed on the silica
surface. The column retained its efficiency when peptides were analysed in dilute
ammonia solution in water-acetonitrile, because the organic modifier supressed
the solubility of silica in this alkaline medium. A good peptide separation was
obtained with water-acetonitrile solutions (13:87 or 70:30) containing ammonia
(0.125 or 0.2 M) for isocratic elution at room temperature.

   Ligand-exchange chromatography was reviewed by Davankov (1984). Caude et al.
(1984) reviewed separations in which the ligand is bound to a normal-phase
column, and Sugden (1984) surveyed separations in which the ligand is bound to
a reversed-phase system.

9.3.4.4 Other separation principles

   Naleway and Hoffman (1981) studied the use of acid metal salts of carboxylic
acids for the retention of hydrophilic peptides in RP-HPLC. Separations of phenylalanine and Phe-(Gly)$_n$ peptides (where n = 1-4) were studied. Spherisorb ODS
(Phase Separations, Queensferry, Clwyd, U.K.) was packed into a 30 cm x 0.42 cm
I.D. column. Alkali, alkaline earth and transition metal salts of carboxylic
acids dissolved in water were used as ion-pairing reagents, which were able to
interact with both the N (amino) and C (carboxyl) terminal groups. Using silver(I)
valerate or silver(I) octanoate as ion-pairing reagents in the mobile phase,
static retention of peptides was achieved. The separation of hydrophilic peptides,
unresolved by plain RPC, was possible using the system studied. A correlation
was found between the "softness" character of the cation of the chelate salt
used in the mobile phase and the retention in RP-HPLC.

   Mabuchi and Nakahashi (1984) reviewed the three-step HPLC separation of peptides. This approach consisted of (i) gel-permeation (size exclusion) chromato-

graphy in the presence of sodium dodecylsulphate (SDS), (ii) ion-pair reversed-
phase chromatography, using SDS as an ion-pairing agent and (iii) removal of
SDS (by high-performance cation-exchange chromatography or by sorption of SDS on
an ODS column or cartridge). The authors discussed various combinations and applications of this three-step HPLC technique, including the use of RP-HPLC
volatile solvent systems.

9.3.5 Comments on literature

Hancock and Harding (1984) published a general review of the separation conditions for the RPC of peptides, containing a detailed discussion of mobile
phases, including ion-pairing reagents and conditions for the preparative separations, and also the use of RP-TLC for the monitoring of preparative separations
of peptides.

9.4 DETECTION METHODS FOR MONITORING THE HPLC OF PEPTIDES AND QUANTIFICATION OF THE CHROMATOGRAPHY

9.4.1 Introduction

The commonest detection principle for peptides is the measurement of UV absorbance at 200-220 nm, because it is universal; the disadvantage is the sensitivity to impurities in buffers and solvents. Some types of pure solvents cannot
be used owing to their high UV absorbance. The wavelength of the maximum absorbance for the peptide bond is 187 nm. Usualy, longer wavelengths are used for
monitoring, because many available solvents and buffers are not transparent
enough below 195-200 nm. Peptides containing aromatic amino acids (and peptides
modified by aromatic protecting groups) can easily be detected with a simple UV
detector at 280 or 254 nm. The UV spectra of individual amino acids of interest
are illustrated in the review by Hancock and Harding (1984) (see also Wetlaufer,
1962). The monitoring of the chromatographic effluent using two wavelengths simultaneously (e.g., 280 and 210 nm) and calculation of the ratio of the two absorbances allows aromatic and aliphatic peptides to be distinguished. The sensitivity of UV detection at 195-205 nm is roughly 100-1000 ng.

The refractive index (RI) detector is a bulk property detector and its use is
not dependent on the spectral purity of the solvents and buffers, but, owing to
its low sensitivity, it is not often used in the chromatography of peptides. An
exception is preparative chromatography, because the RI detector is not so sensitive to high concentrations and is not easily overloaded in comparison with
the UV detector. Various wire-transport flame ionization detectors are seldom

applied in the chromatography of peptides, because the systems are not as simple
as photometric detection.

   A quantitative determination of peptides is often required, especially for
analyses of pharmaceutical dosage forms and for the determination of yields in
the isolation of natural products or in laboratory synthesis. For example,
Biemond et al. (1979) studied the determination of polypeptides by gradient
elution HPLC using a 30 cm µBondapak C_{18} (Waters Assoc.) column or two 30 cm
LiChrosorb RP-18 (E. Merck) columns in series. Tetramethylammonium phosphate
buffer (pH 3) in various water-methanol mixtures was used for elution, because
this buffer inactivated the residual silanol groups; the gradient was linear.
A variable-wavelength detector, set at 210 or 280 nm, was used. Arg- and Lys-
vasopressin, oxytocin, ACTH(1-24), ACTH, β-endorphin, glucagon and porcine and
bovine-insulin were analysed by repeated analyses and determined with a coefficient of variation below 2% (including sampling and weighing). The proportion-
ality of the peak area to the amount of oxytocin injected was measured in the
range 5-100 µg and the calibration graph was linear (correlation coefficient
0.9995). Similar results were obtained for other peptides. By repeated injections of the same solution the coefficients of variation of the chromatographic
procedure were found to be <1.2%.

   In the following sections the monitoring procedures that have been most often
used in the HPLC of peptides will be described.

9.4.2 Fluorimetric assays

   Fluorimetric detection is usually more sensitive than UV detection and can be
realized in three ways: (1) direct excitation of underivatized peptides, (2)
using pre-column derivatization, and (3) using post-column derivatization. The
general advantages and disadvantages of these procedures were briefly discussed
in Section 6.6.2.

9.4.2.1 Underivatized peptides

   Derivatization can complicate the analytical procedure and is not always
necessary, as there are natural compounds (indoles, phenols, naphthyls) that
yield relatively high fluorimetric intensities in the underivatized state. Krol
et al. (1977) used a spectrofluorimetric detector with a deuterium light source
(Model FS 970, Schoeffel, Westwood, NJ, U.S.A.) allowing excitation at wavelengths below 250 nm; KV 370 and 550 cut-off filters (Schott-Optical Glass,
Duryea, PA, U.S.A.) were used on the emission side. It was found that in this
case the observed detection limits were more sensitive by a factor of 10 than
those obtained with an excitation wavelength above 260 nm. Picogram-range direct

fluorimetric detection of indole decapeptide luteinizing hormone (LH), a naphthyl adrenergic blocking agent (propranolol) and phenol (estrogen) compounds,
separated by HPLC, was described. In addition to these underivatized compounds,
the dansyl derivative of Pro-Leu-Gly tripeptide was monitored.

Schlabach and Wehr (1983) described fluorescence techniques for the selective
detection of chromatographically separated peptides containing tyrosine or tryptophan, utilizing their strong native fluorescence. When this method was combined
in series with a lower wavelength absorbance detector (e.g., at 220 nm), the
fluorescence-to-absorbance ratios could be used to distinguish peptides containing a tyrosine residue from those having a tryptophan residue. A Varian Model
5060 ternary liquid chromatograph equipped with a UV 5 selectable wavelength
detector and a Fluorichrom fluorescence detector was used. For the native peptide
fluorescence, a deuterium source was used together with 220- and 330-nm interference filters for excitation and emission, respectively. The direct fluorescence
detection was more than six times more sensitive than 254-nm detection for peptides containing tryptophan.

9.4.2.2 Pre-column derivatization

Pre-column derivatization has the advantage that the modifying reaction need
not proceed very quickly (comparable to the speed of chromatography), but the
disadvantage is that the reaction product is a compound different from the original one and therefore the elution parameters differ. Completeness of the modification must also be ensured.

Gruber et al. (1976) described the fluorimetric assay of vasopressin and oxytocin as a general approach to the assay of peptides in tissues. They showed
that the sensitivity of detection of the fluorescamine (cf., Section 6.6.2)
derivatized peptides in HPLC is several orders of magnitude higher than UV detection and that the derivatization procedure is simple and efficient, so that
quantification of the results is possible. Live (1978) applied this technique
to a chemical assay of the purity of synthetic oxytocin and [Arg[8]]vasopressin
(AVP) and their analogues. A Whatman Partisil-10 ODS pre-packed column and a
Gibson Spectra/Glo fluorimeter were used with acetone-water solvents containing
0.03% ammonium formate and 0.01% thiodiglycol. The substitution procedure was
as follows. Fluorescamine solution was prepared by dissolving 20 mg of Fluram
(Hoffman-La Roche) in 100 ml of acetone. Samples for analysis were dissolved in
0.4 ml of 0.046 M Beckman standard phosphate buffer (pH 7) in a 7.5 x 1.0 cm
disposable borosilicate test-tube. Fluram solution (0.2 ml) was added and the
mixture was stirred briefly on a vortex mixer. After 10 min the volume was made
up to 2 ml with water containing 0.03% ammonium formate and 0.01% thiodiglycol
and the solution was again mixed. The sample was then injected. Seventeen pep-

tides (oxytocin, AVP and 15 analogues) were chromatographed and separated using
isocratic elution with 15% acetone-water and a linear gradient of 15-95% acetone-
water. The method was highly reproducible and very sensitive: a 5 ng sample of
derivatized oxytocin applied to the column yielded a peak with a signal-to-noise
ratio of about 30:1.

Krol et al. (1979) described the trace analysis of the melanocyte stimulating
hormone release-inhibiting factor (MIF) analogue (the pareptide) in blood plasma
by HPLC and short wavelength excitation fluorimetry. They endeavoured to monitor
the elution of the peptide L-Pro-N-methyl-D-Leu-Gly-NH$_2$, which had no natural
fluorescence or strong UV absorption, and the only readily derivatized functional
group was the secondary amine of the proline residue. For derivatization of the
fluorescent pareptide the application of 7-chloro-4-nitrobenzoyl-2-oxa-1,3-oxa-
diazole (NBD-Cl) (cf., Section 6.6.2) was investigated, which also reacted with
the prolyl secondary amino group. The advantages of this reagent (over the frequently used dansyl chloride; cf., Section 6.6.2) are a lower excess reagent
fluorescent background and higher solubility and stability of the derivative in
aqueous solution. The authors investigated the NBD derivatization yield as a
function of pH, time and reagent excess. A 30-cm microparticulate Spherisorb ODS
column with aqueous ammonium carbonate-methanol-acetonitrile (60:25:15, v/v) as
the eluent was used for the chromatography, and a Schoeffel FS 970 short wavelength excitation fluorimetric detector in connection with a Schott-Optical
Glass KV 370 cut-off filter for monitoring. A linear calibration graph for concentrations from 0.02 to over 0.10 μg/ml of pareptide in human plasma was obtained. The detection limit of pareptide in plasma samples was 5 ng/ml (17 pmol
per ml); in the absence of plasma, the corresponding on-column detection limit
was 0.5 pmol.

Hui et al. (1980) studied the separation of alkylaminonaphthylenesulphonyl
peptides by HPLC as a method suitable for measuring the breakdown of melanotropin
inhibiting factor (MIF). Five types of strongly fluorescent derivatives were
prepared, which permitted the detection of peptides at levels of 10^{-11}-10^{-9} mol.
Waters Assoc. μBondapak C$_{18}$ and μBondapak Phenyl columns were tested with ten
different solvent systems containing acetonitrile, methanol, acetone or isopropanol. For detection UV absorbance at 254 nm and fluorescence (excitation
360 nm, emission 487 nm) were used. HPLC with the aid of dansyl derivatization
was found to be reliable and fast, and was the preferred method for the study of
neuropeptide breakdown.

Gröningsson and Widahl-Näsman (1984) described the RP-HPLC determination of
felypressin in pharmaceutical formulations after pre-column derivatization with
fluorescamine. A Waters Assoc. μBondapak C$_{18}$ pre-packed column (300 x 3.9 mm
I.D.) was used with isocratic elution with the mobile-phase 56% (v/v) methanol

in phosphate buffer (pH 8.0) (ionic strength $\mu = 0.1$) at a flow-rate 1.3 ml/min. The effluent was measured using a Shimadzu RF-530 fluorescence spectromonitor with excitation and emission wavelengths of 390 and 470 nm, respectively. Felypressin was well separated from other sample components; the detection limit for felypressin was 0.3 ng and the relative standard deviation of the method was 1.2%.

9.4.2.3 Post-column derivatization

The main advantage of post-column derivatization is that the chromatographic retention data represent real chromatographic properties of the analysed substances. The disadvantage may be the slowness of the modifying reaction if on-line detection is planned.

Frei et al. (1976) studied the post-column fluorescence derivatization of pharmaceutically important nonapeptides containing a primary amino group that reacted with Fluram (cf., Section 6.6.2). The apparatus is illustrated in Fig. 5.16. The reaction conditions were examined; acetonitrile was found to be the best solvent for the mobile phase and pH 6-8 (phosphate buffer) gave the optimal net fluorescence. The optimal concentration of the Fluram reagent was 20 mg per 100 ml of acetonitrile at pH 7. The time needed to complete the reaction at room temperature was about 50 s. The detection limits were between 5 and 10 ng per injection, and the reproducibility was better than ±2% (relative standard deviation). Frei et al. (1977) further investigated the described system of nonapeptide-Fluram post-column fluorescence detection; the kinetics of the detection were studied, new aspects and technical improvements were published, and the reproducibility was increased to 1.5% (relative standard deviation). Radhakrishnan et al. (1977) examined high-efficiency cation-exchange chromatography of polypeptides in the nanomole range with an automated flurescamine column monitoring system. Approximately 10 nmol of synthetic peptide samples were applied to the column of Partisil SCX, volatile pyridine-acetate buffers were used for elution and 8% (800 pmol) of the sample were utilized for detection. Udenfriend and Stein (1977) studied fluorescence techniques for the ultramicroanalysis of peptides and, in addition to fluorescamine, also used 2-methoxy-2,4-diphenyl-3(2H)-furanone (MDPF) as a suitable reagent.

MDPF   +   RNH$_2$ $\longrightarrow$

An automated fluorimetric peptide microassay for carnosine in mouse olfactory
bulb was described by Wideman et al. (1978). Tissue extract was reacted with MDPF
and the resulting fluorophores were resolved on a Partisil 10/25 PXS (Whatman,
Clifton, NJ, U.S.A.) RP-HPLC column (25 cm x 0.46 cm I.D.). Quantitation was performed in a filter fluorimeter equipped with a flow cell. Carnosine was found to
be present at a level of 1.9 μmol/mg. Rubinstein et al. (1979) described the
fluorescence monitoring of tryptic digests of peptides and proteins (ovalbumin,
β-endorfin, prolylhydroxylase and a related protein) after RPC. Automatic fluorescamine detection analysed only a portion of the effluent.

Schlabach and Wehr (1983) used fluorescent derivatization techniques for the
selective detection of chromatographically separated lysine peptides. The ε-amino
group of lysine reacts far better with o-phthalaldehyde (OPA) than any other peptide group, including the primary amino terminus. This detection approach is very
important in chemical studies of partial hydrolysates of peptides and proteins,
where lysine peptides often play a key role, and the spectroscopic methods cannot distinguish lysine peptides from others. The isoindole ring of the OPA reaction products is highly fluorescent (cf., Section 6.6.2). A Fluorichrom fluorescence detector was used with the appropriate excitation and emission filters for
the OPA reaction (340 and 450 nm, respectively). MicroPak MCH-5 or SP-3 RPC
columns (15 x 0.4 cm I.D.) and a gradient of 0.05% v/v trifluoroacetic acid
(TFA) in water and 90% v/v acetonitrile (containing the same amount of TFA) were
used for the chromatography. The freshly prepared reagent, which was mixed with
the column effluent before the reaction coil, contained 0.25 mg/ml of OPA and
0.05% of mercaptoethanol in 0.1 M potassium borate (pH 10.4). The detection
limit for the lysine octapeptide xenopsin was about 15 ng or 15 pmol. For other
larger lysine peptides the detection limits were 5-10 pmol.

The post-column fluorimetric detection of peptides was reviewed by Lewis
(1984).

9.4.3 Other detection principles

9.4.3.1 Derivatization techniques

Chang et al. (1981) described the isolation of polypeptides at the picomole level using detection with dimethylaminoazobenzene isothiocyanate (DABITC), which couples through peptide amino group to form dimethylaminoazobenzenethio-carbamyl (DABTC) peptides.

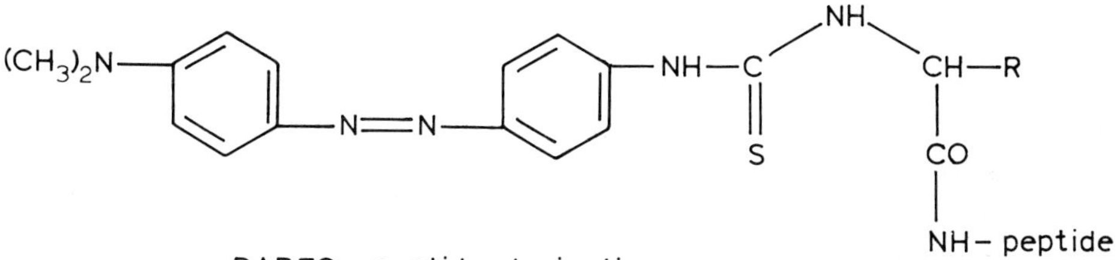

DABTC peptide derivative

The derivatives could be separated by RP-HPLC and detected in the visible region (436 nm). As little as 1 ng (2 pmol) of DABTC peptide can be identified against a stable baseline with a signal-to-noise ratio of 10. The isolated peptides can be further characterized. This method can be sucessfully used in sequence studies of proteins (for the separation of enzymic digests); N-terminal amino acids of peptides can be also analysed at the picomole level (cf., Section 13.1.5.2).

Meek (1983) dealt with the other derivatization reagents for the HPLC detection of peptides at the picomole level. The reason for this work was that the sensitivity of peptides to measurement of UV absorbance at 210 nm was too low to allow the determination of neuropeptides in tissue extracts. Derivatives have been sought that fulfil the following requirements: (1) small hydrophilic molecule, (2) easy electrochemical detection and (3) clean reaction with multiply substituted peptides giving single derivatives. Two reagents were found to be most promising: 3,6-dinitrophthalic anhydride (DNPT)

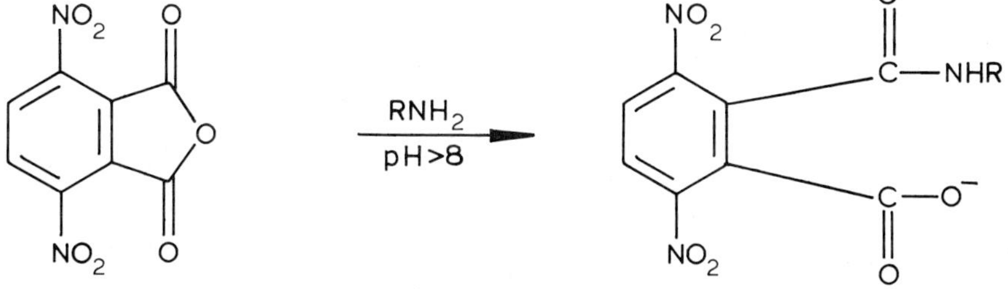

DNPT

and 2-carboxy-4,6-dinitrofluorobenzene (CDNFB)

The synthesis of these two compounds was described by Meek (1983). Both reagents react in alkaline media with amino groups as shown. DNPT derivatives can be detected electrochemically by reduction at -0.24 V and CDNFB derivatives can be measured by UV absorbance at 360 nm. A Bio-Rad ODS-25 column was used for the separation and an Altex-Hitachi variable-wavelength detector and a Bioanalytical System LC2A detector with dual glassy carbon electrodes for monitoring. CDNFB allows 50-fold and DNPT 500-fold sensitivity improvements in the detection of peptides.

9.4.3.2 *UV-absorbing counter ions*

Some strongly UV-absorbing ions, such as naphthalene-2-sulphonate can be sorbed on a suitable support (LiChrospher Si 1000, 10 μm) and applied as stationary phases in normal-phase chromatographic separations of amino acids and short peptides with chloroform - 1-pentanol mixtures as the mobile phase. UV detection at 254 nm is possible (Crommen et al., 1977; Crommen, 1979). Strongly UV-absorbing counter ions can also be used for the separation of short peptides by RPC. Denkert et al. (1981) described reversed-phase ion-pair chromatography with UV-absorbing ions in the mobile phase for the separation and detection of amino acids and short peptides. μBondapak Phenyl was used as the stationary phase and naphthalene-2-sulphonate ($4\cdot10^{-4}$ M) was added to the 0.05 M phosphoric acid (pH 2) mobile phase so that the peptides were separated in cationic form. The detection wavelength was 254 nm and the sample peaks were negative because the cationic peptides forming ion pairs with naphthalene-2-sulphonate were eluted before the system zone of the UV-absorbing counter ion material. This method allows detection of down to 7 pmol (about 1 ng) of non-UV-absorbing samples. The naphthalene-2-sulphonate system is very stable provided that careful thermal stabilization is maintained.

Improvements in detection by the use of UV-absorbing counter ions were reviewed by Crommen (1984).

9.4.4 Combined liquid chromatography with mass spectrometry or with ESR spectroscopy

An important method for the detection of peptides is the combination of HPLC with mass spectrometry (HPLC-MS). This topic was dealt with in the first volume of this book from the point of view of instrumentation and detection techniques (in Sections 5.1.8 and 6.5.5). The off-line combination of liquid chromatography with mass spectrometry was discussed, in addition to direct HPLC-MS coupling. Therefore, in this section only brief comments will be presented.

Games et al. (1980) described the potential of HPLC-MS in clinical studies. Morris et al. (1980) reported studies on mass spectrometric methods for the structure determination of neuropeptides. Peptides were separated by HPLC on μBondapak C_{18} and characterized by negative-ion chemical-ionization mass spectrometry. A general review of the possibilities of applying HPLC-MS in the life sciences, including peptides, was published by Games (1981). The analysis of N-acetyl-N,O,S-permethylated peptides combined with on-line HPLC-MS using a moving belt interface was described by Yu et al. (1981); this method can be used for sequence studies of peptides. Alcock et al. (1982) published a general review of HPLC-MS with transport interfaces.

The application of field desorption mass spectrometry to studies of small amounts of neuropeptides was discussed by Desiderio and co-workers in a series of papers (e.g., Desiderio and Cunningham, 1981; Desiderio et al., 1981; Desiderio and Yamada, 1982; Desiderio and Stout, 1983) which were dealt with in Section 5.1.8. Desiderio (1984) reviewed the field desorption mass spectrometry of peptides.

γ-Irradiated peptides in aqueous solution contain short-lived radicals, which can be converted into fairly stable nitroxide radicals [spin-trapped radicals or spin adducts using 2-methyl-2-nitrosopropane (MNP) as a spin trap], which can be separated by HPLC and identified by ESR spectroscopy. Moriya et al. (1980, 1982) studied this method on model systems of short peptides (Gly-Gly, Gly-L-Ala, L-Ala-Gly and L-Ala-L-Ala). Aqueous solutions of MNP (5 mg/l) were prepared by stirring for 1 h at $45^{\circ}C$ in the dark and the concentration of the peptide in MNP solution was 0.1 M. The sample solution was cooled with ice and irradiated with ^{60}Co γ-rays ($6 \cdot 10^{5}$ rad/h, total dose $3 \cdot 10^{5}$ rad), 2 M phosphate buffer (Na_2HPO_4-NaH_2PO_4, 1:9) was added immediately after irradiation and 1 ml of the stirred solution was chromatographed on a 60 cm x 0.95 cm I.D. column of IEX-210 SC cation exchanger (Toyo Soda, Tokyo, Japan). The eluate passed through UV and ESR detectors (JEOL PE-3X), operated at 100 kHz modulation frequency in the X band. For the detection of the radicals during chromatography the magnetic field was mixed. The magnetic field modulation was applied for the study of separated peptides at high amplitude (10 G).

Most of the short-lived radicals in aqueous solutions of Gly-Gly and Gly-Ala were spin trapped. Some adducts were derived from deaminated radicals (producted by the addition of e^- to the peptide carboxy groups), some were produced by hydrogen abstraction (by OH radicals) from the carbon adjacent to the carboxyl group and others were produced by hydrogen abstraction (by OH radicals) from the methyl group. Some ESR spectra changed considerably with variation in pH.

9.4.5 Peptide analysers, monitors and detectors

Nika and Hultin (1979) described an automatic analyser and monitor for rapid micro-scale peptide separations. A microbore ion-exchange column (150 cm x 2 mm I.D.) of DC-4A cation exchanger (Durrum Chemical, Palo Alto, CA, U.S.A.) in combination with volatile pyridine-acetate buffers was used. Ninhydrin solution was applied for detection. The equipment is illustrated in Fig. 9.8. According to the authors (unpublished work), the sensitivity of the peptide analyser could be substantially increased by using a flow cell with a path length up to 60 mm and by shortening the ninhydrin reaction coil (residence time 3 min) and increasing the temperature of the heating bath ($115^{\circ}C$). With these modifications, utilized by Hare (1977) for amino acids in the subnanomole range, peptide determinations at the 1 nmol level should be possible. The analysis time could be shortened to 150-200 min.

Schlabach and Abbott (1980) studied protein chromatographic profiles with detection at multiple wavelengths. Simple peptide peak absorbances were measured at 254, 280 and 305 nm and the ratios were calculated. Tyr-Gly had the highest and Phe-Ala-Gly the lowest 280:254 nm absorbance ratio; the highest 305:280 nm absorbance ratios were found for tryptophan dipeptides. These results are in conformity with the UV spectra of these amino acids (Wetlaufer, 1962).

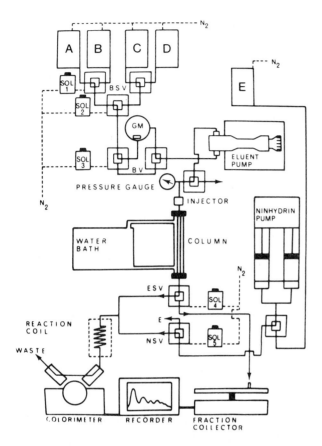

Fig. 9.8. Schematic flow diagram of the peptide analyser and monitor. A–D = Buffer reservoirs; E = ninhydrin reservoir; SOL 1,2 = solenoid valves penumatically actuating buffer selection three-way slider valves; SOL 3 = solenoid valve actuating bypass valve system BV; SOL 4,5 = solenoid valves actuating eluate and ninhydrin split valves ESV, NSV; GM = gradient mixer. Solutions sensitive to oxidation are protected by nitrogen. (Reprinted from Nika and Hultin, 1979.)

Nabeshima et al. (1982) described the HPLC determination of Met-enkephalin and some amines in brain tissue with electrochemical detection. A 25 cm x 4 mm I.D. column of 10 μm Yanapak ODS-T (Yanagimoto, Kyoto, Japan) and 0.1 M phosphate buffer (pH 3.3) containing 20% of acetonitrile were used for the chromatography of Met-enkephalin. A Model VMD-101 electrochemical detector (Yanagimoto) with the oxidation potential set at 1.0 V vs. Ag-AgCl were used for detection. The detection sensitivity was 8 nA for full-scale deflection. The retention time of Met-enkephalin was 15 min. Dual detection methods for the selective identification of prolyl residues and amide-blocked N-terminal groups in chromatographically separated peptides were described by Schlabach (1983).

9.4.6 Comments on literature

Detection methods suitable for monitoring the separation of amino acids, peptides and proteins were reviewed by Hancock and Harding (1984a) and radioimmuno-assay by Loeber (1984). Reversed-phase thin-layer chromatography as an off-line detection method for the HPLC of peptides (cf., Section 9.3) was reviewed by Lepri et al. (1984).

9.5 HPLC AND PEPTIDE SYNTHESIS

9.5.1 Introduction

There is no doubt that the preparation of synthetic peptides in a pure form is one of the most difficult tasks in synthetic organic chemistry, and its importance in both biological and conformational studies has been emphasized repeatedly (see, e.g., Rudinger, 1972; Meienhofer, 1973; Zaoral, 1986). HPLC methods can not only help in the usual characterization and purification of intermediates and final products, but they have also been used as a basis for the development of specialized racemization tests and for the development of the continuous-flow solid-phase automated synthesizer. A brief survey of these approaches will be given.

9.5.2 Chromatography of protected peptides

The preparation of purified protected peptides is an important prerequisite for the successful synthesis of complicated polypeptides. HPLC is the main preparation method used in this field, and several papers will be cited as examples. Gabriel et al. (1977) described a simple preparative system for the rapid and efficient purification of protected synthetic peptides on the gram scale. Peptides were modified with commonly used protecting groups, such as N-benzyloxy-carbonyl (Z), N-2-(*p*-biphenylyl)-2-propyloxycarbonyl, N-*tert.*-butyloxycarbonyl (Boc), O- and S-*tert.*-butyl and S-acetamidomethyl. A 43 cm x 3.8 cm I.D. heavy-walled glass pre-packed silica gel 60 (60 Å porosity) column was used for chromatography, with organic solvents, such as chloroform, isopropanol, ethanol, methanol, acetic acid or their mixtures for elution, both isocratic and gradient, or stepwise; the same solvents or dimethylformamide were applied for loading. Measurement of the absorbance at 254 nm was used for detection. A variety of synthetic peptides up to tetradecapeptides were chromatographed at pressures of 50-150 p.s.i. within 2-4 h. Simple and inexpensive equipment was described for this purpose (cf., Section 6.9).

Bakum et al. (1977) applied RP-HPLC for monitoring the synthesis of the gastro-intestinal hormone secretin (a heptacosapeptide) and analogues, by the repetitive excess mixed anhydride (REMA) method. Z and Boc were used for N^{α}-protection and Z, benzyl (Bzl) or NO_2 for side-chain protection. All intermediate protected peptides were purified on a micro-preparative scale (100-200 μg ⩾ 20 nmol). A 30 cm x 0.4 cm I.D. column of 10 μm LiChrosorb RP-18 (Merck, Darmstadt, F.R.G.) was used, with methanol-water (with the addition of 1% of acetic acid) in various proportions for gradient elution. If the larger peptides were scarcely soluble in aqueous methanol, dimethylformamide or acetic acid was used for loading. The α-β rearangement of aspartyl-glycyl Bzl-protected higher peptides was observed in non-acidic medium. This method also allowed the separation of protected dia-stereoisomeric [L-Ala^4]- and [D-Ala^4]-des-His^1-secretin.

The separation of diastereoisomers of protected di- and tripeptides using RP-HPLC was studied by Kuwata et al. (1980). Z-protected hydrophobic L,D-L dipeptide methyl esters and L-L,D-L tripeptide methyl esters were separated on a Develosil ODS (5 μm) column (150 mm x 4.6 mm I.D.) with aqueous methanol (55% or 65%) ap-plied for isocratic elution (at 30°C). All the diastereoisomers were well sep-arated. Ridge et al. (1982) described sulphur protection with the 3-nitro-2-pyridinesulphenyl group in solid-phase peptide synthesis. The HPLC purification of the intermediate product was applied in the synthesis of Lys^8-vasopressin.

The development of protected peptides chromatography was an important condi-tion for the possibility of monitoring peptide syntheses in which many amino acid residues in the peptide chain must be kept in a protected form.

9.5.3 Step-by-step monitoring and characterization of synthesized peptides

HPLC is very suitable for the gradual checking of the synthetic procedure in a peptide preparation and for the purification of intermediates. Some examples will be given in the following sections.

9.5.3.1 Monitoring of liquid-phase synthesis

Nachtman (1979) described the HPLC of intermediates in the oxytocin synthesis. RP-HPLC was found to be a powerful tool for the determination of both free and protected peptides formed during the synthesis. Carbobenzoxyproline and a variety of protected peptides (including S-benzyl-Cys derivatives) and unprotected pep-tides were chromatographed in the ester or amide forms, in addition to the re-sulting synthesized oxytocin. This approach allowed the optimization of in-process control of the complete synthesis. LiChrosorb RP-8 (5 μm) (E. Merck) was used as the stationary phase in a 15 cm x 3.2 mm I.D. column and 0.015 M phosphate buffer (pH 7)-acetonitrile in various proportions was used as the mobile phase for isocratic elution. The detection wavelength was 215 nm. Four peptides (three carbobenzoxy derivatives and one hexapeptide amine) were determined quantitatively and the detection limit was 17-50 ng. It was shown that both the protected and free peptides could be chromatographed simultaneously and this method could be used very well for controlling the peptide synthesis.

Kimura et al. (1981) studied the strategy for the synthesis of large peptides and applied it to the total synthesis of very pure human parathyroid hormone (hPTH). Some synthetic intermediates, products and by-products were chromatographed by RP-HPLC on a Nucleosil 5 C_{18} column (150 mm x 4 mm I.D.) at room temperature in dilute H_3PO_4-K_2HPO_4 buffers containing dilute sodium sulphate with a gradient of acetonitrile. Absorbance measurements at 210 nm were used for detection. Final success in the total synthesis of hPTH(1-84) demonstrated the usefulness of the described strategy.

9.5.3.2 Monitoring of Merrifield's solid-phase synthesis

Takahagi et al. (1978) described an HPLC method for monitoring the progress of the coupling reaction in solid-phase peptide syntheses by measuring the amount of remaining fragments or amino acids in the reaction solution. The reaction was completed in time at which the response curve indicated no further consumption of the fragment or amino acid uptake. The rapidity, accuracy and economy were great advantages of this method, because the time required for quantitative HPLC analysis in the Merrifield method was about 5 min without consumption of the synthesized peptide. The feasibility of this method was demonstrated through the synthesis of ACTH(1-24) by the oxidation-reduction condensation method [employing chain elongation via fragment condensation from N- to C-terminal amino acid; cf., Matsueda et al. (1970)] and in this instance the time of analysis was 15 min. For chromatographic monitoring of the model Merrifield synthesis sulphonated Hitachi Gel 3010 was employed using 0.2 M triethylamine-methanol (1:9) as the mobile phase at $60^{\circ}C$ and UV detection at 280 nm. For monitoring of the ACTH oxidation-reduction condensation aminated Hitachi Gel 3010 was used with the same mobile phase as above, but with UV detection at 240 nm. In both methods 1-nitronaphthalene was added to the sample as an internal standard. The applied methods allowed exact kinetic studies of the progress of the coupling reactions, which contributed to the highest possible reaction yield in each step.

Hearn et al. (1979a) studied the application of ion-pair RP-HPLC in the solid-phase peptide synthesis. Crude cleavage products of synthetic peptide preparations were analysed and purified, including angiotensins, Leu-enkephalin amide, a fragment of β-TSH and a linear antamanide. The rapid chromatographic recognition of incomplete coupling and deletion or partial deprotection were sought. Waters Assoc. μBondapak C_{18} and μAlkyl Phenyl columns (30 cm x 4 mm I.D.) were used for analytical purposes and Waters Assoc. Bondapak Phenyl-Porasil B

(37-50 µm) (60 cm x 7 mm I.D.) for preparations. Linear gradients of water-0.1% phosphoric acid to 40% methanol-water-0.1% phosphoric acid, or 10% acetonitrile-water-0.1% phosphoric acid to 75% acetonitrile-water-0.1% phosphoric acid were used for elution. For isocratic elution, 35% or 80% methanol-water-0.1% phosphoric acid or 55% methanol-100 mM potassium dihydrogen phosphate-0.1% phosphoric acid or 25%, 30% or 40% acetonitrile-water-0.1% phosphoric acid were applied. All chromatography was carried out at room temperature. The hydrophobicity parameters used (cf., Section 9.2.3) were examined and found to be useful for the prediction of the relative elution order (cf., Section 9.2.4) of these closely related peptides. Oroszlan (1979) published a model study for monitoring the Merrifield solid-phase peptide synthesis by HPLC, using as an example the preparation of the group specific antigen of cRNA animal virus. The course of the reaction could be analysed stepwise in this way: first, the amino groups remaining after incomplete coupling reactions were terminated with a UV-absorbing or fluorescent group (2,4-dinitrophenyl, 2,4,6-trinitrophenyl or dansyl). After the tagged peptides had been cleaved from the resin, they were separated by HPLC. The sensitivity of the detection of N-terminal amino acids was, e.g., 0.01 µg of DNP-Ala.

9.5.3.3 *Monitoring of protease-catalysed peptide synthesis and peptide semi-synthesis*

Kullman (1981) used HPLC for monitoring the protease-catalysed peptide synthesis in the course of experiments with the preparation of Leu-enkephalin. Stepwise elution from a pre-packed 40-63 µm LiChroprep Si 60 (E. Merck) silica gel column (24 cm x 1 cm I.D.) using as the solvent system dichloromethane-anhydrous ethanol-acetic acid in suitable proportions provided completely resolved peaks. Enzymatically prepared compounds and their chemically synthesized authentic analogues were co-chromatographed to permit the first assignment of eluted peaks. An example of such a chromatogram is illustrated in Fig. 9.9.

Čeřovský and Jošt (1984a) studied the papain-catalysed synthesis of phenyl-hydrazides of all the coded N^α-acylamino acids (with the exception of proline). The synthesis was checked analytically by means of RP-HPLC using a 15 x 0.32 cm I.D. column filled with Separon SIX C_{18} (Laboratory Instrument Works, Prague, Czechoslovakia), in addition to TLC. Preparative chromatography was carried out on a 25 x 1.27 cm I.D. column filled with packings of the same type and origin. The mobile phase was methanol-0.05% aqueous trifluoroacetic acid (1:1) in both instances and the flow-rates were 30 ml/h for analytical and 360 ml/h for preparative procedures. Absorbance measurement at 220 nm was used for detection. Čeřovský and Jošt (1984b) also described the enzymic resolution of γ-carboxy-D,L-glutamic acid by enantioselective reaction of benzyloxycarbonyl-γ-carboxy-D,L-glutamic acid with phenylhydrazine, catalysed by papain, and subsequent re-

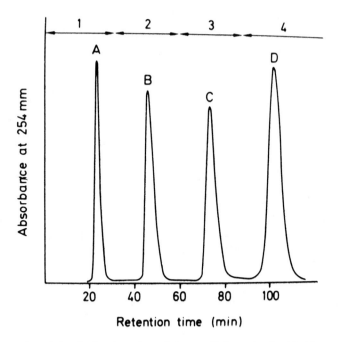

Fig. 9.9. Analysis of a mixture of four derivatized peptides under conditions of four-step elution. Column, LiChroprep Si 60 (24 x 1 cm I.D.); flow-rate, 2 ml/min. Mobile phases: dichloromethane-acetic acid in all steps; proportions: (1) 100:1:1; (2) 100:3:3; (3) 100:4:4; (4) 100:5:5. Chromatographed components: (A) Boc-Tyr(Bzl)-Phe-N_2H_2Ph; (B) Boc-Tyr(Bzl)-Gly-Gly-Phe-OEt; (C) Boc-Tyr(Bzl)-Gly-Gly-PheN_2H_2Ph; (D) Boc-Tyr(Bzl)-Gly-Gly-Phe-Leu-N_2H_2Ph. (Reprinted from Kullmann, 1981.)

moval of protecting groups from the optically active benzyloxycarbonyl-γ-carboxy-L-glutamic acid α-phenylhydrazide obtained. Analytical checking was carried out as described above. Čeřovský and Jošt (1985a) later used an analogous papain-catalysed enantioselective reaction for the enzymatically catalysed synthesis of dipeptides: benzyloxycarbonyl-γ-carboxy-D,L-glutamic acid reacted with phenyl-hydrazides of various amino acids and with leucine, protected with various carb-oxy-protecting groups. A 15 x 0.35 cm I.D. column filled with Separon SIX C_{18} and methanol-0.05% aqueous trifluoroacetic acid mobile phases at a flow-rate of 30 ml/h were used for analytical checking. For preparative purposes the column mentioned above was used with a mobile phase flow-rate of 420 ml/h.

Čeřovský and Jošt (1985b) studied the applicability of glutaminyl-tRNA cyclo-transferase in peptide syntheses. Peptides with N-terminal glutamine and their derivatives on treatment with papain containing glutaminyl-tRNA cyclotransferase were converted into the corresponding compounds with pyroglutamic acid. For all the HPLC separations Separon SIX C_{18} was used, with mobile phases composed of methanol and water containing 0.05% of trifluoroacetic acid. For analytical chro-

matography a 15 x 0.4 cm I.D. column and a mobile phase flow-rate of 42 ml/h and
for preparative purposes 25 x 0.8 cm and 25 x 1.27 cm I.D. columns and mobile
phase flow-rates of 120 and 240 ml/h were used, with UV detection at 224 nm.
Čeřovský and Jošt (1985c) also described the enzymically catalysed synthesis of
oxytocin fragments 1-6 and 7-9, utilizing papain, α-chymotrypsin, thermolysin
and elastase. The same analytical column as above was applied, with a mobile
phase flow-rate of 42 ml/h and detection at 222 nm.

Prestidge et al. (1981) published an approach to the semi-synthesis of acyl
carrier protein (ACP). Acetylated ACP_{1-6} was coupled via its activated pentachlorophenyl ester to native ACP_{7-77}, which had previously been acetylated and
converted into the S-5'-dithiobis(2-nitrobenzoate) (DTNB) derivative in order
to protect the peptide against oxidation during the coupling procedure. The
pentachlorophenyl active ester was purified on a series of two silanized Bondapak
Phenyl-Porasil B columns (60 cm x 7 mm I.D.) (Waters Assoc.) using as the mobile
phase water-methanol (40:60) containing 1% of acetic acid. The protected peptide
purified in this way was found to be homogeneous by analysis on a μBondapak Alkyl
Phenyl column with the same mobile phase. Removal of the DTNB moiety after the
coupling (with dithiothreitol) yielded active ACP in good yield. The semi-synthetic product was best analysed using peptide mapping by RP-HPLC of the thermolytic digest. The native and semi-synthetic products were found to be identical.
The results confirmed that the hexapeptide had been successfully rejoined in the
semi-synthesis.

9.5.4 Racemization studies and racemization tests

In some synthetic steps racemization of optically active derivatives occurred.
Because it is essential to obtain optically pure compounds in modern peptide
synthesis, various racemization studies were performed and racemization tests
were developed. HPLC was found to be very suitable for these purposes.

Goodman et al. (1977) described such experiments. They synthesized dipeptides
benzoyl-L-Phe-L-Ala benzyl ester and benzoyl-L-Phe-L-Ala methyl ester by both the
solid-phase and solution coupling methods. Each coupling reaction was analysed
by HPLC for the extent of racemization. Diastereoisomers from the coupling reactions in solution (benzoyl-Phe-Ala benzyl esters) were separated on a 60 cm
column of 10 μm Microporasil (Waters Assoc.) using a solvent mixture containing
0.6% of 95% ethanol in distilled chloroform at a flow-rate 2.0 ml/min. The reaction product was dissolved in 5% methanol-chloroform solution and 20-40 μl were
injected. Diastereoisomers from the solid-phase synthesis (benzoyl-Phe-Ala methyl
ester) were similarly analysed using 0.7% of 95% ethanol in chloroform at a flow-
rate 2.0 ml/min. The sample was dissolved in analytical-ragent grade chloroform

and 40-90 µl were injected. UV detection at 254 nm was used. Baseline separation
was achieved, which allowed the direct, rapid and reproducible determination of
dipeptide L-L and D-L diastereoisomers. Kent et al. (1978) described experiments
carried out in Merrifield's laboratory on tests for racemizaiton in a model peptide synthesis by a direct chromatographic separation of tetrapeptide Leu-Ala-
Gly-Val. Single D-amino acid diastereoisomers, L-Leu-D-Ala-Gly-L-Val and D-Leu-
L-Ala-Gly-L-Val, were separated from each other and from the all-L-amino acid
tetrapeptide. A standard amino acid analyser (based on ninhydrin detection) with
a cation-exchange column (58 cm x 0.9 cm I.D.) of Beckman AA-15-X8 sulphonated
polystyrene resin was used at $57^{O}C$ with 0.2 M sodium citrate buffer (pH 3.49) as
the eluent. The determination of the D-amino acid diastereoisomers was accurate
to better than 0.1% for a standard load of 4 µmol. The limit of detection was
<0.01% for a 12 µmol load. In the crude products from stepwise solid-phase syn-
theses almost no (<0.02%) D-amino acid-containing diastereoisomers were detected.
The model system described can be used to evaluate and adjust the experimental
conditions in both the solution and solid-phase methods of peptide synthesis.

Kuwata et al. (1980) described the separation of protected peptides and a
convenient test for racemization in peptide syntheses by RP-HPLC. The chromatographic conditions were mentioned in Section 9.5.2. Good separations of all
diastereoisomers stimulated the authors to examine the degree of racemization
in the reaction of Z-L-AA_1-L-AA_2-OH with H-L-AA_3-OMe, where AA represents amino
acids. Several coupling methods were examined and the contents of L-D-L isomer
were obtained. The advantages of racemization tests in peptide synthesis were
discussed. Kiso et al. (1981) endeavoured to develop a new method to test racemization in peptide synthesis and studied the extent of racemization during
peptide bond formation in the reaction

$$Z(OMe)-Gly-L-Ala \quad + \quad Phe-OBzl \quad \longrightarrow \quad \begin{cases} Z(OMe)-Gly-L-Ala-Phe-OBzl \\ \\ Z(OMe)-Gly-D-Ala-Phe-OBzl \end{cases}$$

Seven coupling methods were studied and the molar ratio of -L-Ala- to -D-Ala-
products was measured by RP-HPLC. A µBondapak C_{18} column (30 cm x 0.39 cm I.D.)
was used with methanol-water (55:45) as the eluent at a flow-rate of 1 ml/min
and detection at 280 nm. The difference in the retention volumes of the resulting protected diastereoisomers was large (44 ml for the L- and 51 ml for
the D-isomer). The lowest racemization (<0.5%) was found in three instances,
the highest (23.2%) with the dicyclohexylcarbodiimide coupling method.

9.5.5 Determination of purity and purification of synthetic peptides

The determination of the purity and homogeneity of physiologically active peptides is important from both the chemical and pharmacological points of view. Some problems may arise with large peptides.

Coy (1979) described the determination of the purity of large synthetic peptides using RP-HPLC. He studied the purity of β_h-endorphin and synthetic analogues, vasoactive intestinal polypeptide (VIP) and human pancreatic polypeptide (HPP). Products of their CNBr and tryptic cleavage were also studied. A LiChrosorb RP-18 column (25 cm x 0.4 cm I.D.) was used under conditions of isocratic elution (30% or 41% acetonitrile containing 0.1% triethylammonium acetate, pH 4) or linear gradient elution (10-40%, 25-30% or 30-50% acetonitrile-0.01 M ammonium acetate, pH 4, or 10-50% isopropanol-ammonium acetate, pH 4). Some mixtures of large peptides were well resolved and extrapolation from the behaviour of peptide analogues having very similar structures suggests that many types of impurities that can arise in rapid peptide syntheses should be readily separated by this technique. Viswanatha et al. (1979) described the synthesis of D,L-[2-^{13}C]leucine and its use in the preparation of (3-D,L-[2-^{13}C]leucine)oxytocin. The ^{13}C-labelled hormone derivative (8-[2-^{13}C]leucine)oxytocin was separated from its 8-positional diastereoisomer and side-products by partition chromatography on Sephadex G-25 using as the solvent system 1-butanol-3.5% aqueous acetic acid (1:1) containing 1.5% of pyridine. No suitable solvent system could be found for the partition chromatographic separation of diastereoisomers of analogous (3-D,L-[2-^{13}C]leucine)oxytocin. However, an excellent preparative separation was achieved using a Partisil 10 M9 ODS column (50 x 0.94 cm I.D.) with 0.05 M ammonium acetate (pH 4)-acetonitrile (81:19, v/v) as the eluent at a flow-rate of 6 ml/min. The separation of (8-D,L-[2-^{13}C]leucine)oxytocin diastereoisomers could also be accomplished by RP-HPLC.

Hruby et al. (1980) synthesized S-benzyl-D,L-[1-^{13}C]cysteine and described its incorporation into oxytocin and [Arg[8]]vasopressin and related compounds by total synthesis; the separation of Cys diastereoisomers by partition chromatography and HPLC was examined. The labelled S-benzylcysteine was converted into the *tert.*-butyloxycarbonyl derivative and incorporated into the 1- and 6-positions of a variety of oxytocin and [Arg[8]]vasopressin derivatives and analogues via total synthesis using the solid-phase method. The compounds were separated and purified by partition chromatography on Sephadex and their purity was checked by HPLC. Twelve compounds were studied. Two μBondapak C_{18} reversed-phase columns (30 cm x 0.39 cm I.D.) connected in series and acetonitrile-0.1 M ammonium acetate (pH 4) as the eluent were used. The effluent was monitored at 254 and 280 nm simultaneously. Analysis of the purity of commercial peptides by HPLC was also tested and discussed by Margolis et al. (1980). Octadecylsilyl columns were used for RPC with an eluent containing 0.1 M triethylamine phosphate (TEAP) aqueous buffer (pH 3.5) and acetonitrile. In isocratic elution experiments the proportion of acetonitrile was 17.5, 19 or 25% (v/v). Angiotensin I and II derivatives, somatostatin, substance P, LRH, eledoisin and Met- and Leu-enkephalins were chromatographed with detection at 254 nm.

Reynolds et al. (1981) used chromogenic p-[p-(dimethylamino)phenylazo] benzyl ester (Az ester) in the solution synthesis of Leu-enkephalin.

$$RO-CH_2 - \text{(benzene ring)} - N=N - \text{(benzene ring)} - N \begin{cases} CH_3 \\ CH_3 \end{cases}$$

Az ester; for AzOH , R = H

The application of the intensely coloured carboxyl-protecting Az moiety through-out the syntesis allowed the facile purification of protected intermediates on silica gel or ion-exchange columns and detection on TLC plates. The Az group readily withstood successive Boc deprotection treatments with 25% trifluoroacetic acid in dichloromethane, but it was cleanly removed by catalytic hydrogenation. A coupling reaction between AzOH and Boc-L-Leu was carried out using dicyclohexyl-carbodiimide with pyridine as the solvent. Each coupling reaction was monitored by TLC. Each of the synthetic intermediates was readily obtained in pure form by a combination of silica gel or ion-exchange chromatography and recrystallization. The final product after hydrogenation was purified by GPC on Bio-Gel P-2 with 2 M acetic acid and analysed by HPLC on a μBondapak Alkyl Phenyl column with 30% acetonitrile-water-0.1% phosphoric acid as the eluent.

9.5.6 Continuous-flow solid-phase synthesis using an HPLC system

Erickson and Prystowsky (1979) constructed an apparatus for the Merrifield synthesis of peptides using an HPLC pumping system and conventional polystyrene beads for the solid-phase synthetic method. During the coupling reaction activated Boc-amino acid was continuously pumped through the column, the detector (if present) and the injector in a closed loop to save the valuable reagent. Seven steps created a general synthetic procedure: (1) pre-swelling with dichloromethane; (2) deprotection with 0.1 M trifluoroacetic acid and 0.01 M methanesulphonic acid in dichloromethane; (3) washing with dichloromethane; (4) neutralization with

0.3 M N,N-diisopropylethylamine in dichloromethane; (5) washing with dichloromethane; (6) coupling with Boc-amino acid symmetric anhydride in dichloroethane;
and (7) washing with dichloromethane. A model peptide, Leu-Ala-Gly-Val, was
synthesized and analysed by ion-exchange chromatography for purity (98%) and
content of deletion peptides (individual values 0.05-0.82%). The results were
compared with those of the standard manual synthesis in glass vessels with the
addition of solvent and reagents in discrete portions. The advantages of the
continuous flow synthesis were discussed, mainly from the point of view of automation. A newer version of the solid-phase peptide synthesis under continuous-
flow conditions was published by Lukas et al. (1981). The model tetrapeptide
Leu-Ala-Gly-Val was formed (99.3%) in about 4 h on microporous X-1 styrene-divinylbenzene. The procedure consisted in 10 steps and one cycle lasted 83 min.
The products and by-products were analysed by RP-HPLC. A 17-residue peptide from
chicken ovalbumin was obtained in a similar purity and yield using the described
continuous-flow solid-phase synthesizer to those obtained when discontinuous
synthesis was applied.

Modern versions of continuous-flow solid-phase peptide synthesis using a
fluorenylmethoxycarbonyl (Fmoc)-polyamide method and a new rigid organic-inorganic
macroporous support (polydimethylacrylamide-Kieselgur-Macrosorb SPR, available
from Sterling Organics) were explained and described by Sheppard (1986); see also
Eberle et al. (1986).

9.5.7 Comments on literature

Sallay and Oroszlan (1984) published and discussed a model study for monitoring
the Merrifield solid-phase peptide synthesis through N-2,4-dinitrophenyl (DNP)
derivatization. Inouye et al. (1984) reviewed the RP-HPLC separation of synthetic
analogues of peptide hormones, emphasizing isocratic elution.

9.6 SURVEY OF EXAMPLES OF PEPTIDE SEPARATIONS

A detailed survey of HPLC and MPLC separations of peptides is tabulated in
Chapter 14 (Register), Table 14.2, paragraphs 7-12. Many citations of peptide
separations are listed in Chapter 15 (full title Bibliography). In this section,
papers will mainly be mentioned that, for various reasons, could not be included
in the Register or Bibliography. Glycopeptides are not commented on here, because they are treated in a Special chapter (Chapter 12), dealing with compound
biopolymers. In this section special attention will be paid to peptide hormones,
owing to their importance in pharmacology.

9.6.1 Neurohypophysial peptides, nonapeptides

In their pioneer work, Gruber et al. (1976) described a quantitative assay of vasopressin and oxytocin in pituitaries. The method was based on purification of the extract by filtration through a copper-Sephadex column, pre-column derivatization with fluorescamine and RPC on a Partisil ODS column using a linear gradient of acetone into an aqueous solution containing 0.03% ammonium formate and 0.01% thiodiglycol. This procedure was intended as a model system for the assay of natural peptides present in tissues in very small amounts. Blevins et al. (1980a) studied the experimental conditions and parameters involved in the RP-HPLC of seven neurohypophysial hormones: Arg-vasotocin, Lys-vasopressin, Arg-vasopressin, mesotocin, isotocin, oxytocin and glumitocin. Columns of µBondapak C_{18}, LiChrosorb RP-8 and LiChrosorb RP-2 were used with mobile phases consisting of tetrahydrofuran, acetonitrile and methanol in various proportions. An appropriate solvent composition and suitable columns were found that allowed the separation of all the studied peptides from each other. Major chromatographic parameters were discussed in relation to peptide structures. Franco-Bourland and Fernstrom (1981) examined the in vivo biosynthesis of L-[^{35}S]Cys-arginine-vasopressin, -oxytocin and -somatostatin. Rapid analytical determinations of small concentrations of the above peptides in pituitaries were achieved using RP-HPLC.

Burbach and Lebouille (1983) applied RPC to studies of the proteolytic conversion of Arg-vasopressin and oxytocin by brain synaptic membranes. Baláspiri et al. (1984) described the LC, TLC and HPLC of oxytocin, vasopressin and some of their analogues and fragments. Monoiodine- and diiodine-Lys-vasopressin (four analogues), oxytocin (six analogues), pareptide and oxytocin fragments were subjected to RPC on a µBondapak C_{18} column. Hrbas et al. (1985) examined mono- and diiodo derivatives of neurohypophysial hormones and their analogues and separated the components on both analytical and preparative scales by RP-HPLC.

9.6.1.1 Vasopressin

Live et al. (1977) described a rapid, efficient synthesis of oxytocin and [Arg^{8}]vasopressin. Benzyl, *p*-methoxybenzyl and *p*-methylbenzyl were compared as protecting groups for cysteine. RPC with fluorescamine detection was used for monitoring the separation.

Smith et al. (1980) studied a bidirectional synthesis of [Asp^{5}]arginine-vasopressin on poly-N-acryloylpyrrolidine resin. The purity of the product was examined by HPLC using fluorescamine derivatization and a pre-packed Partisil-10 ODS-2 reversed-phase column with gradient elution using 15-50% of acetone in water.

Möhring et al. (1982) compared radioimmunoassay, chemical assay (HPLC) and
bioassay for [Arg]vasopressin in synthetic standards and posterior pituitary.
Pre-column fluorescamine derivatization was used for monitoring and HPLC was
performed on a Partisil ODS reversed-phase column with a 15-85% acetone linear
gradient in 0.05% aqueous ammonium formate containing 0.01% of thiodiglycol.
Ridge et al. (1982) examined sulphur protection with the 3-nitro-2-pyridinesul-
phenyl group in the solid-phase synthesis of [Lys^8]vasopressin on benzhydrylamine
resin. Gel filtration and RP-HPLC were used for the isolation of material with
full biological activity. For HPLC a µBondapak column was used with a linear
gradient from 100% A [0.1% trifluoroacetic acid (TFA) in water] to 60% B [0.1%
TFA in acetonitrile-water (9:1)] with detection at 254 or 280 nm.

9.6.1.2 Oxytocin

Larsen et al. (1979) described the separation of peptide hormone diastereoisomers by RP-HPLC. Oxytocin was separated from each of the seven diastereoisomers
[hemi-D-Cys^1]-, [D-Tyr^2]-, [D-Gln^4]-, [D-Asp^5]-, [hemi-D-Cys^6]-, [D-Pro^7]- and
[D-Leu^8]oxytocin using two 30 cm x 0.39 cm I.D. columns of µBondapak C$_{18}$ connected
in series and 10% tetrahydrofuran-ammonium acetate buffer or 18% acetonitrile-
ammonium acetate buffer as the eluent. Krumen et al. (1979) studied the use of
HPLC in the quality control of oxytocin. A LiChrosorb RP-8 (or an octadecyl
silanized) column was applied with phosphate buffer-acetonitrile as the mobile
phase. The separation of [D-Tyr^2]oxytocin, [D-Gln^4]oxytocin, [D-Tyr^2,D-Gln^4]-
oxytocin and oxytocin was described. Bienert et al. (1981) described the synthesis
and application of S-*tert.*-butylsulphonium peptides; HPLC capacity factors of
substance P and oxytocin analogues with modified thioester bridges were described
in addition to applications of RPC in the analysis and purification of analogues.
Pask-Hughes et al. (1981) examined the assay of the combined pharmacological formulation of ergometrine and oxytocin by HPLC.

Procházka et al. (1982) studied two analogues of oxytocin with a modified
proline cyclic structure (in positions 7 and 8) in order to obtain metabolically
stable compounds. The analogues were purified by HPLC. Kanmera et al. (1983)
described the proteolytic processing of neurophysin/neurohypophysial hormone biosynthetic precursor using a synthetic precursor fragment. HPLC profiles of the
biotransformational converison of synthetic oxytocin dodecapeptide by neurosec-
retory granule lysate were examined. Lebl (1983) separated a group of diastereoisomers of oxytocin analogues using a 25 cm x 0.4 cm I.D. colum packed with
Separon SI C$_{18}$ (Laboratory Instrument Works, Prague, Czechoslovakia) and mixtures
of methanol or tetrahydrofuran with aqueous buffers as mobile phases. Pask-Hughes
et al. (1983) compared HPLC assays with the current pharmacopoeial assays for
the combined formulation of ergometrine and oxytocin. For the HPLC the Whatman

B154

cation exchanger Partisil 5 SCX was used. Good agreement between the two approaches was found. Hlaváček et al. (1984) studied oxytocin analogues modified
in position 8. The analogues were purified by ion-exchange, partition and preparative RPC. Procházka et al. (1984) described the synthesis and properties of
oxytocin analogues modified in the tripeptide side-chain. In addition to TLC and
paper electrophoresis, HPLC was used for the analysis and purification. A Separon
C_{18} (25 cm x 0.4 cm I.D.) column was applied with methanol-trifluoroacetic acid
or methanol-water as the mobile phase. In their studies in the oxytocin field
Hlaváček and Jošt (1985) described groups of analogues and fragments, which were
purified by ion-exchange chromatography, free-flow electrophoresis, size exclusion chromatography and RP-HPLC. Lebl et al. (1985a) dealt with oxytocin analogues
with inhibitory properties, containing in the 2-position a hydrophobic amino acid
of D-configuration (Tyr, Phe, p-methyl-Phe, p-ethyl-Phe and O-ethyl-Tyr). A
Partisil ODS column (50 x 0.9 cm I.D.) with methanol-0.05% aqueous TFA mixtures,
methanol-aqueous ammonium acetate buffer mixtures or methanol-water mixtures as
the mobile phase were applied for the purification of analogues.

Oxytocin intermediates were reviewed by Nachtmann and Gstrein (1984)

9.6.1.3 Oxytocin carba-analogues

Lebl (1980) described the HPLC of carba-analogues of oxytocin and a method
for the determination of sulphides and sulphoxides in peptides. Peptides derived
from deaminooxytocin, differing in the structure of the disulphide bridge region,
were studied on a 15 cm x 6 mm I.D. RP column of 6 μm Separon SI C_{18} (Laboratory
Instrument Works, Prague, Czechoslovakia). Methanol- or acetonitrile-buffer mixtures were used for elution. Buffer of pH 4.5 was prepared by addition of triethylamine to a 1% solution of trifluoroacetic acid (TFA); sodium phosphate buffer (pH 7.0) was of concentration 0.02 M. The optimal flow-rate was 2 ml/min and
detection was effected at 230 nm. Lebl et al. (1981) studied the chromatographic
and pharmacological properties of carba-analogues of neurohypophysial hormone
sulphoxides. A general method for the determination of sulphoxide groups in peptides was described. Chromatographic determination of the oxidation state of the
sulphur atom in peptides was examined and the RPC analysis and purification of
analogues was described. Procházka et al. (1981) dealt with the synthesis and
pharmacological properties of (2-p-fluorophenylalanine)deamino-1-carba-oxytocin
and its diastereoisomeric sulphoxides. The diastereoisomer mixture was obtained
after purification of the synthetic product by countercurrent distribution.
Chromatographic resolution was achieved by the use of a 6 μm Separon SI C_{18}
column (15 cm x 0.6 cm I.D.). Methanol-water or methanol-aqueous phosphate buffer (pH 4.4) was used as the mobile phase at a flow-rate of 1.2 ml/min and a
pressure of 25-30 MPa. Lebl et al. (1982b) studied the synthesis and properties

of analogues of neurohypophysial hormones containing *tert.*-leucine in position 8. Analogues of oxytocin, deamino-oxytocin and deamino-vasopressin were prepared by fragment condensation in solution. Stepwise synthesis in solution afforded analogues of 1-carba and 6-carba-deamino-oxytocin with *tert.*-leucine in position 2. Some of the analogues were purified by countercurrent distribution and gel filtration, followed by RP-HPLC, or by RP-HPLC only. Analytical chromatography was carried out on a 25 x 0.4 cm I.D. Separon SI C_{18} column and preparative runs on 50 x 0.9 cm I.D. Partisil ODS-2 column (Whatman, Clifton, NJ, U.S.A.) with methanol-water or methanol-phosphate buffer (pH 4.4) as the eluent. Lebl et al. (1982a) described the synthesis and properties of oxytocin analogues with high and selective natriuretic activity. Using a stepwise peptide synthesis in solution and cyclization in the last step, deamino-6-carba-oxytocin analogues with a modified aromatic amino acid in position 2 were prepared. RP-HPLC with methanol-water as the mobile phase was found to be the most effective purification method. Separon SI C_{18} (15 x 0.6 cm or 25 x 0.4 cm I.D.) and Partisil ODS (50 x 0.9 cm I.D.) columns were used. In some instances methanol-0.1% TFA in water were used as an efficient mobile phase.

Krojidlo et al. (1983) studied a method of cyclization of carba-analogues of oxytocin and used RP-HPLC on a Separon SI C_{18} column for both the analytical examination and purification of the products. Lebl et al. (1984a) described the synthesis and properties of [2-(3,5-^{3}H$_2$)-Tyr-4-Glu]deamino-1-carba-oxytocin; the product was purified by RP-HPLC on a Separon SI C_{18} column (25 x 0.46 cm I.D.) with methanol-0.05% aqueous TFA as the eluent. In other experiments Lebl et al. (1984b) synthesized two carba-6-analogues of oxytocin containing a deaminopenicillamine (dPen) residue in position 1 and described their properties. [dPen[1]]Carba-6-oxytocin and [dPen[1],Tyr(Me)[2]]carba-6-oxytocin were prepared and their biological activities were determined. Separon SI C_{18} (25 x 0.4 cm I.D.) and Partisil ODS (50 x 0.9 cm I.D.) columns were used for both analysis and purification, with methanol-water and methanol-aqueous TFA as eluents. The solid-phase synthesis and biological activities of oxytocin carba-analogues containing Thr in position 4 were described by Lebl et al. (1985). A Vydac C_{18} column (25 x 0.4 cm I.D.) was used for analytical purposes and a Waters Assoc. RCM column (15 x 0.8 cm I.D.) for the preparative work; 0.1% TCA was used as buffer and acetonitrile as organic modifier.

Carba-analogues of neurohypophysial hormones were reviewed by Lebl (1984).

9.6.1.4 Comments on literature

Smith et al. (1970) reviewed the determination of neurohypophysial hormones in body fluids (older methods were dealt with), Glasel (1984) reviewed neurophysins and Lebl (1986) the synthesis, purification and stability of the neurohypophysial hormone analogues.

9.6.2 Neuropeptides, endorphins, enkephalins

Rossetti (1981) described the separation of neuropeptides by HPLC. An ODS-10 column was used with linear gradients of acetonitrile, propanol or methanol. Six peptides were chromatographed, and acetonitrile gave the highest resolution. Phosphate buffer gave shorter retention times, but dilute volatile TFA could also be used. Bláha and Zaoral (1983) studied the biological importance of the carboxyl terminal part of endorphins (EP) and prepared five EP-(27-31)-penta-peptides, two EP-(1-5)-(27-31)-decapeptides and an EP analogue with GlnNH$_2$ in position 31. They repeatedly prepared β-human EP as a standard and reference substance. A solid-phase method in a laboratory-made automated peptide synthesizer was used for the preparation. The products were purified on carboxymethylcellu-lose, by partition chromatography on Sephadex G-50 and Bio-Gel P-2 and by RP-HPLC on LiChrosorb 10 RP-8 (250 mm x 4.6 mm I.D.) using methanol-water (80:20, v/v) + 0.1% TFA as the mobile phase at a flow-rate of 1 ml/min. The structure-activity relationships were discussed. Liat Tan (1983) described a simple and efficient ternary solvent system for the separation of luteinizing hormone and enkephalins by RP-HPLC. Gradient elution was used with methanol-water (25:75), containing linearly increasing amounts of acetonitrile-water (80:20) at a con-stant 0.1% TFA concentration. Five commercially available pre-packed RP columns were compared. Met-Enk (Enk = enkephalin), Leu-Enk, Leu-Enk-Arg-Lys, Leu-Enk-Arg-Arg, dynorphin$_{1-13}$, β-melanotropin, Lys-bradykinin, neurotensin, angiotensin and luteinizing hormone-releasing hormone (LH-RH) were chromatographed. The described gradient ternary solvent system was applied to the separation and purification of de novo biosynthesized enkephalins and LH-RH in thromboplastic shells of human placenta. In addition, the dependence of the chemical stability of organic modifiers in acidic media with time was compared.

Dennis et al. (1983) developed the characterization of β-endorphin (β-END) immunoreactive peptides in rat pituitary and brain by coupled gel-permeation HPLC and RP-HPLC in combination with radioimmunoassay (RIA). For gel-permeation HPLC four Waters Assoc. protein analysis columns (2 x I-125 + 2 x I-160) were connected in series. RP-HPLC was carried out on μBondapak C$_{18}$ columns (30 cm x 3.9 mm I.D.) (Waters Assoc.). Tritiated marker peptides were obtained by the in vitro incorporation of [^{3}H]leucine into neosynthesized proteins of rat neuro-intermediate pituitary, followed by separation and purification. The antiserum used recognized the mid-portion of β-END and cross-reacted on an equimolar basis with pro-opiomelanocortin, β-lipoprotein, β-END-31, β-END-1-27 and N-acetylated derivatives of β-END. After dilution 1/80 000 it had a sensitivity of 30-50 pg/ml. An extensive analysis of the molecular species comprising the total β-END immuno-reactivity in extracts of rat pituitary and brain regions was permitted by the application of the described coupled gel-permeation HPLC - RP-HPLC system. The gel-permeation HPLC allowed rapic screening for relative amounts of precursors and products involved in the β-END synthesis, and RP-HPLC revealed minor post-translational modifications (such as acetylation) that altered the biological activity of β-END. For a detailed description of the procedure and a thorough discussion, see the original paper. Bennett (1983) dealt with the isolation of pituitary peptides by RP-HPLC and an increase in the resolving power of RP columns by manipulating the pH and the nature of the ion-pairing reagent. Posterior pituitary glycopeptide and α-N-acetyl-β-endorphin(1-27) were chromatographed. A sequential approach to the purification of tissue peptides and RP-HPLC of pituitary extract at different pH valves were described.

9.6.2.1 *Comments on literature*

Boehlen (1979) reviewed the applications of HPLC in neuropeptide research, Meek (1980a) the separation of neuropeptides by HPLC (an algorithm of repetitive regression calculation of retention coefficients was also described) and Stein (1980) the ultramicro-scale isolation and analysis of peptides with emphasis on opiate peptides and fluorescence detection. Loeber and Verhoef (1981) reviewed the use of HPLC and RIA for specific and quantitative determinations of endorphins and related peptides. The separation of endorphins by RP-HPLC was reviewed by Gay and Lahti (1984).

9.6.3 *Angiotensins*

Margolis and Schaffer (1979) described the development of a standard reference material for angiotensin I and included the HPLC of angiotensin and analogues. Klickstein and Wintroub (1982) described the separation of angiotensins (ANG) and assay of ANG-generating enzymes by HPLC. $[Ile^5]$ANG II, $[Val^5]$ANG II, des$[Asp^1]$,$[Ile^5]$ANG II, $[Ile^5]$ANG I, des$[Asp^1]$,$[Ile^5]$ANG I, His-Leu and NH_2-terminal tetradecapeptide of equine ANG were chromatographed by IP-RPC on an ODS support. Dizdaroglu et al. (1982) studied the separation of angiotensins by HPLC on a weak anion-exchange bonded phase. A mixture of twelve angiotensins was chromatographed on a MicroPak AX column (Varian) using a volatile acetonitrile-buffer mobile phase. An excellent separation was obtained (cf., Fig. 9.10); the recoveries of all 12 peptides were over 90%. The retention was strongly dependent on temperature. A long column lifetime (up to one year with several daily injections) was reported.

Angiotensins were reviewed by Tonnaer (1984).

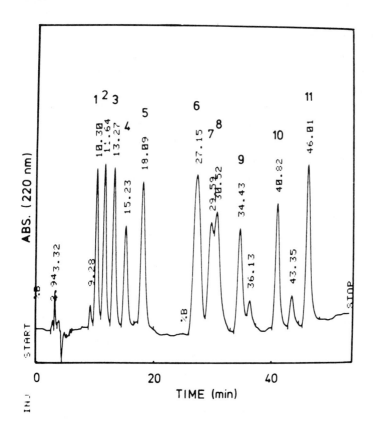

Fig. 9.10. Separation of angiotensins by HPLC. Column, MicroPak AX-10 (30 x 0.4 cm I.D.); eluent, (A) acetonitrile and (B) 0.01 M triethylammonium acetate (pH 6.0); gradient elution started from 25% B at a rate of 0.1% B per minute for 25 min, then 0.5% B per minute. Column temperature, 26°C; flow-rate, 1 ml/min; amount per peptide injected, ca. 1 µg (1 nmol based on angiotensin II). 0.1 a.u.f.s. at 220 nm. For peak identification, see the original paper. (Reprinted from Dizdaroglu et al., 1982.)

9.6.4 Pancreatic hormones, insulin

Hearn et al. (1979c) studied the analysis of insulin-related peptides by RP-HPLC. The solid-phase synthesis products bovine insulin C-peptide (41-53), porcine insulin C-peptide (41-53) and insulin B-chain fragment (22-27), in addition to natural bovine insulin, were chromatographed on a µBondapak-fatty acid column or two µBondapak C_{18} columns, using linear gradients from water-0.1% phosphoric acid to acetonitrile-water-phosphoric acid. In some instances a reducing agent (β-mercaptoethanol) was added, which allowed the direct analysis of insulin reduction products. Frank et al. (1983) described the HPLC preparation of single-site carrier-free pancreatic polypeptide hormone radiotracers. Iodinated

I^{125}-labelled proteins and polypeptides were isolated from unlabelled hormones and reaction products using HPLC methods. This approach was applied to a range of pancreatic polypeptides including insulins, proinsulins, C-peptides, glucagon, somatostatins and pancreatic polypeptide. A 25 cm x 0.46 cm I.D. column of Beckman-Altex ODS (C_{18}, 5 µm) was employed in a standard gradient chromatographic procedure (acetonitrile-aqueous ammonium sulphate), and a system developed in the Lilly Research Labs., consisting of LP-1/C_{18} silica gel (10-20 µm, 20% C) packed into 50 x 1.2 cm I.D. glass columns, was employed for both the analytical and preparative isocratic separation of iodinated products. Acetonitrile-aqueous ammonium acetate were used as the mobile phases. Miligram amounts of ^{127}I-label-led radiotracers were prepared, and this isocratic procedure can be readily and directly scaled up. These pure radiotracers have great advantages in biological assays.

9.6.4.1 Comments on literature

The high-performance ion-exchange chromatography of insulin and insulin de-rivatives was reviewed by Welinder and Linde (1984), and Welinder (1984) re-viewed the homogeneity of crystalline insulin determined by GPC and RP-HPLC.

9.6.5 Other peptide hormones, growth factors and inhibitors

9.6.5.1 Luteinizing hormone- and extrahypothalamic thyrotropin-releasing hormones

Rivier et al. (1979b) described the application of HPLC to the isolation of unprotected peptides. Pancreatic small somatostatin from pigeon was purified by RPC on a µBondapak C_{18} column using triethylamine-phosphoric acid (TEAP) buffer (pH 2.5) and TEAP-acetonitrile as the mobile phase. For desalting a µBondapak CN column and formic acid-triethylamine buffer (TEAF) (pH 3) with TEAF-acetonitrile were used. A 1000-fold purification was obtained. The isolation of [^{3}H]Pro^9-LH-releasing factor and somatostatin was also described. Flegel et al. (1982) studied the synthesis and gonadotropic activity of [D-Ile^6, Pro-NH-Et^9]LRF in order to find "superactive" analogues of luteinizing-releasing factor. A Separon SI C_{18} column (30 x 2.5 cm I.D.) was used with methanol-0.2% aqueous TFA (60:40) as the mobile phase at a flow-rate of 400 ml/h for the preparative purification of 250 mg of crude analogue, and 125 mg of HPLC-pure product (96% content) was obtained. In addition to the above analogue, [D-Ile^6, Arg(Tos)^8, Pro-NH-Et^9]LRF was also studied.

Hypothalamic releasing factors were reviewed by Coy (1984), extrahypothalamic thyrotropin releasing hormone (TRH) by Vuolteenaho and Leppäluoto (1984) and luteinizing hormone-releasing hormone (LH-RH) by Guyon-Gruaz et al. (1984).

9.6.5.2 Melanocyte stimulating hormones

Lebl et al. (1984c) described RP-HPLC studies of cyclic melanotropins.
α-Melanotropin (α-MSH) analogues (fourteen substances) were chromatographed on
two 25 x 0.46 cm I.D. RP columns (5 μm Altech and 16 μm Vydac) using several
mobile phases (mixtures of acetonitrile or methanol with aqueous TEAP or TFA).
The D-analogues were always eluted earlier than the L-analogues. The substitution of penicillamine for Cys in position 4 led to a conformational change in
the resulting peptide (the lipophilicity was reduced), and a similar effect was
observed if the size of the intramolecular disulphide was decreased or if the
carba-modification of the disulphide bridge was prepared. Lebl et al. (1984d)
described the modification of the disulphide bridge in cyclic melanotropins.
A series of peptides were prepared containing thiomethylene and variously modified bridges in place of the disulphide bridge in the $[Cys^4, Cys^{10}]-\alpha-MSH_{4-13}$
fragment. The influence of oxidation and reduction on the activity was also
studied. Altech C_{18} or Vydac C_{18} columns (25 x 0.46 cm I.D.) or a Waters Assoc.
C_{18} Radial-Pak column (10 x 0.8 cm I.D.) and a gradient of acetonitrile in
0.1% TFA were used for chromatographic analysis and purification.

9.6.5.3 Other peptide hormones

Amoscato et al. (1981) studied the analysis of contaminants in commercial
preparations of the hormone-like tetrapeptide tuftsin. A Whatman C_{18} ODS (10 μm)
column (25 cm x 4.6 mm I.D.) was applied with stepwise and gradient elution with
mobile phases consisting of aqueous phosphate solutions and acetonitrile as an
organic modifier. The heterogeneity of a commercial preparation of tuftsin was
demonstrated and a product with enhanced biological activity was prepared.
Bennett (1983) investigated the heterogeneity of the acidic joining peptide
(AJP) from the mouse pituitary. Extracts from pituitaries (prepared using silica
cartridges) were subjected to RPC using a μBondapak C_{18} column. Multiplied forms
of mouse AJP (M_1-M_5) were detected. Sullivan et al. (1983) described the use of
size exclusion and IE-HPLC for the isolation of biologically active growth factors. Chondrosarcoma growth factor (CHSA-GF), human milk growth factor (HMGF),
retinal-derived growth factor (RDGF) and mouse epidermal growth factor (EGF)
were studied. The columns used were TSK Gel 2000 (60 cm x 7.5 mm I.D.) (Altex)
of TSK 3000 (50 cm x 8 mm I.D.) (Varian) (for chromatography under denaturing
conditions) for SEC, SynChropak AX 300 (250 x 4.1 mm I.D.) (SynChrom, Linden,
IN, U.S.A.) for anion-exchange chromatography and SynChropak CM 300 (250 x 4.1 mm
I.D.) (SynChrom) for cation-exchange chromatography. It was shown that SEC and
IEC can be used without organic solvents or extreme pH values to purify a number
of different growth factors successfully with retention of biological activity.

Tyler and Rosenblatt (1983) studied the semi-preparative HPLC purification of a 28-amino acid synthetic bovine parathyroid hormone (bPTH) antagonist, [Tyr-34]-bPTH-(7-34)amide. Gel chromatography (Bio-Gel P-6, Bio-Rad Labs.) in a 100 x 2.0 cm I.D. column followed by semi-preparative RP-HPLC (30 x 7.8 mm I.D. µBonda-pak C_{18} column) using a lyophilizable solvent gradient system of acetonitrile in 0.1% TFA and, in parallel purification scheme, gel filtration followed by IEC (carboxymethylcellulose, 30 x 12 mm I.D. column) were utilized. The semi-preparative HPLC generated a highly purified product, whereas the products after IEC showed substantial heterogeneity. Murphy et al. (1984) described the measurement and chromatographic characterization of vasoactive intestinal peptide (VIP) from guinea-pig enteric nerves. Three chromatographic modes were applied: RPC (MicroPak MCH-10 column, acetonitrile-saline, or acetonitrile-dilute TFA, step-wise elution), cation-exchange chromatography (SynChropak CM 300) and GPC (Micro-Pak TSK Gel 2000 SW). Material of natural origin was found to be chromatographically identical with the synthetic product in all the experiments.

Human growth hormone (RPC separation) was reviewed by Li and Chung (1984).

9.6.6 Antibiotics and miscellaneous peptides

9.6.6.1 Antibiotics

Kalász and Horváth (1981) studied the preparative-scale separation of poly-myxins with an analytical HPLC system using displacement chromatography. Li-Chrosorb RP-8 (5 µ) (MCB, Cincinnati, OH, U.S.A.) was laboratory-packed into a 250 x 4.6 mm I.D. column and used for preparations (cf., Fig. 6.7). An aqueous solution of dodecylammonium chloride was used as displacer. Up to 100 mg of commercial polymyxin B sulphate could by separated and polymyxins B_1 and B_2 were obtained. RP-HPLC analysis of the fractions was performed with 5 µm LiChrosorb RP-18 (MCB), also laboratory packed into a 250 x 4.6 mm I.D. column, eluted with 0.005 M triethanolammonium phosphate buffer (pH 2.2) containing 0.025 M 1-butane-sulphonic acid and 25% (v/v) acetonitrile. The flow-rate was 1 ml/min, inlet pressure 2800 p.s.i. and temperature $22^{\circ}C$, detection was effected at 220 nm and the load was 60 µg. Fonina et al. (1982) described the synthesis, structure and membrane properties of gramicidin A dimer analogues. Three types of bis-gramici-dins were prepared (junction types head-to-head, head-to-tail and tail-to-tail); their homogeneity was monitored by RP-HPLC on a Separon SI C_{18} column with methanol-5 mM ammonium sulphate (9:1) as the mobile phase. Fonina et al. (1984) prepared thirteen new gramicidin A analogues, having head-to head, head-to-tail and tail-to-tail dispositions of the terminal groups, fixed by covalent methylene bridges of varying chain length. A 25 x 0.4 cm I.D. column of Separon SI C_{18} was used for their separation, with methanol-5 mM ammonium sulphate (pH 6.1) (9:1) as the mobile phase.

B162

TABLE 9.3

INTERNATIONAL SYMPOSIA ON PEPTIDES IN EUROPE AND NORTH AMERICA (SINCE 1975)

Symposium					Proceedings				
Year	No.	Continent	Place	Date	Editor(s)	Title	Publisher	Year	Pages
1975	4th	America	New York	June 1-6	Walter, R. and Meienhofer, J.	Peptides: Chemistry, Structure and Biology	Ann Arbor Sci. Publ., Ann Arbor, MI	1975	1053
1976	14th	Europe	Belgium, Wépion	April 11-17	Loffet, A.	Peptides 1976	Université de Bruxelles	1976	660
1977	5th	America	California, San Diego	June 20-24	Goodman, N. and Meienhofer, J.	Peptides	Halsted Press, Chichester; Wiley, New York	1977	612
1978	15th	Europe	Poland, Gdansk	Sept. 4-9	Siemion, I.Z. and Kupryszewski, G.	Peptides 1978	Wroclaw University Press	1979	697
1979	6th	America	Washington, Georgetown	June 17-22	Gross, E. and Meienhofer, J.	Peptides (Structure and Biological Function)	Pierce, Rockford, IL	1979	1079
1980	16th	Europe	Denmark, Helsingor	Aug.31-sept.6	Brunfeldt, K.	Peptides 1980	Scriptor, Copenhagen	1981	783
1981	7th	America	Wisconsin, Madison, Campere	June 14-19	Rich, D.H. and Gross, E	Peptides (Synthesis-Structure-Function)	Pierce, Rockford, IL	1981	853
1982	17th	Europe	Czechoslovakia, Prague	Aug.29-sept.3	Bláha, K. and Maloñ, B	Peptides 1982	Walter de Gruyter, Berlin	1983	846

1983	8th	America	Arizona Tuscon	May 22-27	Hruby, V.J. and Rich, D.H.	Peptides, Structure and Function	Pierce Rockford, IL	1983	927
1984	18th	Europe	Sweden Djurönäset	June 10-15	Ragnarsson, U.	Peptides 1984	Almquist and Wiksell, Stockholm	1984	633
1985	9th	America	Canada Toronto	June 23-28	Deber, Ch.M., Hruby, V.J. and Kopple, K.D.	Peptides, Structure and Function	Pierce, Rockford, IL	1985	976
1986	19th	Europe	Greece Porto Caras	Aug.31-Sept. 5	Theodoropoulos, D.	Peptides 1986	Walter de Gruyter, Berlin	1987	684

Anhalt (1979) reviewed clinical antibiotic assays by HPLC. Konaka and Shoji
(1981) reviewed the HPLC of peptide antibiotics (tridecapeptins, cerexins, poly-
myxins, octapeptins and brevistins) in Japanese and Sakakibara and Kimura (1984)
reviewed the separation of peptide antibiotics in English. A book on the chromatography of antibiotics, including HPLC, was written by Wagman and Weinstein
(1984).

9.6.6.2 *Miscellaneous peptides*

Fox et al. (1976) described a practical HPLC determination of L-Asp-L-Phe
methyl ester (aspartame sweetener) in various food products and formulations.
The method is based on IEC using a strong cation-exchange column (1 m x 2.1 mm
I.D.) packed with SCX based on Zipax (DuPont, Wilmington, DE, U.S.A.) and detection at 254 nm. The solvent system was 0.1 M citric acid-0.5 M sodium perchlorate,
adjusted to pH 4.7 with NaOH. Manberg et al. (1982) developed a radioimmunoassay
(RIA) method for Pro-Leu-GlyNH$_2$ (PLG) and presented evidence that PLG is not
present in rat brain. The C-terminal tripeptide tail of oxytocin has been reported to possess MSH-release-inhibiting activity. The authors developed a combined method involving HPLC followed by RIA, based on the addition of a synthetic
standard to tissue extract, which indicated that endogenous PLG was not present
in the rat hypothalamus and other brain or head tissues, and therefore cannot
be the physiological regulator of MSH secretion. Carnegie (1984) described and
reviewed the identification of kangaroo and horse meat in processed meats by the
IE-HPLC identification of histidine dipeptides.

9.6.7 *Comments on literature on physiologically active peptides*

In this section reviews and monographs devoted to biologically active peptides
will be dealt with. Krummen (1980) surveyed the use of HPLC in the analysis and
separation of pharmaceutically important peptides, and Richmond (1980) the separation of clinically important peptides and proteins. Yoshida and Imai (1981)
reviewed (in Japanese) the HPLC of biologically active peptides, and Greibrokk
(in English) the purification of peptide hormones by RP-HPLC. Inouye et al.
(1984) reviewed the separation of synthetic analogues of peptide hormones,
Verhoeff et al. (1984) ACTH- and MSH-like neuropeptides, Gröningsson and
Abrahamsson (1984) somatostatin, Van Nispen and Janssen (1984) synthetic β-lipotropin fragments and Starratt and Stevens (1984) proctolin.

Pradayrol et al. (1984) dealt with the separation of intestinal peptides,
Svoboda and Van Wyk (1984) the RPC of somatomedins and Corran and Zanelli (1984)
the RPC of calciotropic hormones.

Jošt et al. (1987) edited a handbook on neurohypophysial hormone analogues.

9.7 LITERATURE ON PEPTIDE SEPARATIONS

Many important citations of specialized reviews and monographs on the HPLC
of peptides can be found at the end of several sections of this chapter in the
form of "Comments on Literature". The list of references also containing reviews
on peptides is presented in the Chapter 15 (full title Bibliography). Because
many authors presented surveys on peptides together with proteins, important
citations can also be found in Section 7.8. In this section, information will be
given on general summarizing works concerning HPLC separations of peptides.

Many contributions describing the HPLC of peptides have been presented at
international symposia on peptides in Europe and North America and are summarized
in Table 9.3, covering the period since 1975, when HPLC became an important tool
in peptide chemistry. The same applies to the Japan symposia on peptide chemistry
(e.g., Shioiri, 1982). Another important series are the international symposia
on the chromatography of proteins, peptides (and polynucleotides), which were
dealt with in Section 7.8; many interesting papers on peptide separations can be
found in the corresponding Proceedings.

Porthault (1979) wrote an "Introduction to HPLC - a Promising Method for
Peptide Analysis and Preparation". Smith and McWilliams (1980a,b) published a
two-part treatise on "HPLC of Peptides". Stein (1981) discussed the ultramicroanalysis of peptides and proteins by HPLC and fluorescence detection, Waterfield
and Scrace (1981) wrote a treatise "Peptide Separation by HPLC Using Size
Exclusion and Reversed Phase Columns" and Wilson et al. (1981a) "The Use of HPLC
in Peptide/Protein Chemistry". Pradayrol (1982) discussed (in French) the
"Significance of HPLC in Polypeptide Chemistry", Hearn (1983) reviewed the separation of polypeptides by HPLC and Hearn et al. (1983) the HPLC of peptides and
proteins. Richey (1983) thoroughly described and discussed fast protein liquid
chromatography as a comprehensive separation technique for biopolymers, including
peptides. McMartin (1984) reviewed separation techniques in the study of peptide
metabolism.

Hancock (1984) edited a comprehensive two-volume monograph on the HPLC of
amino acids, peptides and proteins, a standard book in this field.

REFERENCES

Alcock, N.J., Eckers, Ch., Games, D.E., Games, M.P.L., Lant, M.S., McDowall,
   M.A., Rossiter, M., Smith, R.W., Westwood, S.A. and Hee-Yen Wong, *J.
   Chromatogr.*, 251 (1982) 165-174.
Alfredson, T.V., Wehr, C.T., Tallman, L. and Klink, F., *J. Liq. Chromatogr.*, 5
   (1982) 489-524.
Amoscato, A.A., Babcock, G.F. and Nishioka, K., *J. Chromatogr.*, 205 (1981) 179-
   184.

Anhalt, J.P., *Biological/Biomedical Applications of Liquid Chromatography 2* (*Chromatography Science Series*, Vol. 12), Marcel Dekker, New York, 1979, 1-16; *C.A.*, 92 (1980) 33 480j.

Bakkum, J.T.M., Beyerman, H.C., Hoogerhout, P., Olieman, C. and Voskamp, D., *Recl. Trav. Chim. Pays-Bas*, 96 (1977) 301-306; *C.A.*, 88 (1978) 105 770w.

Baláspiri, L., Tóth, M.V., Fekete, T., Janáky, T., László, F.A., Tóth, G. and Sirokmán, F., in Kalász, H. (Editor), *New Approaches in Liquid Chromatography* (*Proceedings of 2nd Annual American-Eastern Symposium on Advances in Liquid Chromatography, Szeged, Hungary, June 16-18, 1982*), Akadémiai Kiadó, Budapest, and Elsevier, Amsterdam, New York, 1984, pp. 217-230.

Bate-Smith, E.C. and Westall, R.G., *Biochim. Biophys. Acta*, 4 (1950) 427-440.

Bennett, D.J. and Creaser, E.H., *Anal. Biochem.*, 37 (1970) 191-194.

Bennett, H.P.J., *J. Chromatogr.*, 266 (1983a) 501-510.

Bennett, H.P.J. in Hruby, V.J. and Rich, D.H. (Editors), *Peptides. Structure and Function (Proceedings of 8th Amarican Peptide Symposium)*, Pierce, Rockford, IL, 1983b, pp. 257-260.

Bennett, H.P.J., Browne, C.A., Brubaker, P.L. and Solomon, S. in Hawk, G.L., Champlin, P.B., Hutton, R.F. and Mol, Ch. (Editors), *Biological/Biomedical Applications of Liquid Chromatography III (Chromatography Science Series,* Vol. 18), Marcel Dekker, New York, 1982, pp. 197-210; *C.A.*, 96 (1982) 136 024a.

Bennett, H.P.J., Browne, C.A., Goltzman, D. and Solomon, S., in Gross, E. and Meienhofer, J. (Editors), *Peptides: Structure and Biological Function (Proceedings of 6th American Peptide Symposium)*, Pierce, Rockford, IL, 1979, pp. 121-124; *C.A.*, 93 (1980) 234 214v.

Bennett, H.P.J., Browne, C.A. and Solomon, S., *J. Liq. Chromatogr.*, 3 (1980) 1353-1365.

Bennett, H.P.J., Browne, C.A. and Solomon, S., *Biochemistry*, 20 (1981) 4530-4538.

Benson, J.V., Jr., Jones, R.T., Cormick, J. and Patterson, J.A., *Anal. Biochem.*, 16 (1966) 91-106.

Bidlingmeyer, B.A., Deming, S.N., Price, W.P., Jr., Sachok, B. and Petrusek, M., *J. Chromatogr.*, 186 (1979) 419-434.

Biemond, M.E.F., Sipman, W.A. and Olivie, J., *J. Liq. Chromatogr.*, 2 (1979) 1407-1435.

Bienert, M., Lebl, M., Mehlis, B. and Niedrich, H., in Brunfeldt, K. (Editor), *Proceedings of 16th European Peptide Symposium*, Scriptor, Copenhagen, 1981, pp. 127-132.

Bigelow, C.C. and Channon, M., *Handbook of Biochemistry and Molecular Biology*, Vol. I, 1976, p. 209 [cited by Wilson et al. (1981c)].

Bij, K.E., Horváth, C., Melander, W.R. and Nahum, A., *J. Chromatogr.*, 203 (1981) 65-84.

Bishop, C.A., in Hancock, W.S. (Editor), *CRC Handbook of HPLC for the Separation of Amino Acids, Peptides and Proteins*, Vol. I, CRC Press, Boca Raton, FL, 1984, pp. 153-159.

Bishop, C.A., Harding, D.R.K., Meyer, L.J., Hancock, W.S. and Hearn, M.T.W., *J. Chromatogr.*, 192 (1980) 222-227.

Bishop, C.A., Meyer, L.J., Harding, D.R., Hancock, W.S. and Hearn, M.T.W., *J. Liq. Chromatogr.*, 4 (1981) 661-680.

Bláha, I. and Zaoral, M., in Bláha, K. and Maloň, P. (Editors), *Peptides 1982 (Proceedings of 17th European Peptide Symposium, Prague, Aug. 29-Sept. 3 1982)*, Walter de Gruyter, Berlin, 1983, pp. 301-304.

Blevins, D.D., Burke, M.F. and Hruby, V.J., *Anal. Chem.*, 52 (1980a) 420-424.

Blevins, D.D., Burke, M.F., Hruby, V.J. and Larsen, B.R., *J. Liq. Chromatogr.*, 3 (1980b) 1299-1318.

Blevins, D.D., Burke, M.F. and Hruby, V.J., in Hancock, W.S. (Editor), *CRC Handbook of HPLC for the Separation of Amino Acids, Peptides and Proteins*, Vol. II, CRC Press, Boca Raton, FL, 1984, pp. 137-143.

Boehlen, P., *Psychopharmacol. Bull.*, 15 (1979) 46-50; *C.A.*, 92 (1980) 18 175v.

Bradshaw, R.A., Bates, O.J. and Benson, J.R., *J. Chromatogr.*, 187 (1980) 27-33.
Browne, C.A., Bennett, H.P.J. and Solomon, S., *Biochemistry*, 20 (1981a) 4538-4546.
Browne, C.A., Bennett, H.P.J. and Solomon, S., *Biochem. Biophys. Res. Commun.*, 100 (1981b) 336-343.
Browne, C.A., Bennett, H.P.J. and Solomon, S., *Anal. Biochem.*, 124 (1982) 201-208.
Burbach, J.P.H. and Lebouille, J.L.M., *J. Biol. Chem.*, 258 (1983) 1487-1494.
Burgus, R. and Rivier, J., in Loffet, A. (Editor), *Peptides 1976 (Proceedings of 14th European Peptide Symposium, 1976)*, Université de Bruxelles, Brussels, 1976, pp. 85-94.
Cachia, P.S., Van Eyk, J., Chong, P.C.S., Taneja, A. and Hodges, R.S., *J. Chromatogr.*, 266 (1983) 651-659.
Cahill, Jr., W.R., Kroeff, E.P. and Pietrzyk, D.J., *J. Liq. Chromatogr.*, 3 (1980) 1319-1334.
Carnegie, P.R., in Hancock, W.S. (Editor), *CRC Handbook of HPLC for the Separation of Amino Acids, Peptides and Proteins*, Vol. II, CRC Press, Boca Raton, FL, 1984, pp. 45-48.
Caude, M.H., Jardy, A.P. and Rosset, R.H., in Hancock, W.S. (Editor), *CRC Handbook of HPLC for the Separation of Amino Acids, Peptides and Proteins*, Vol. I, CRC Press, Boca Raton, FL, 1984, pp. 411-422.
Čeřovský, V. and Jošt, K., *Collect. Czech. Chem. Commun.*, 49 (1984a) 2557-2561.
Čeřovský, V. and Jošt, K., *Collect. Czech. Chem. Commun.*, 49 (1984b) 2562-2565.
Čeřovský, V. and Jošt, K., *Collect. Czech. Chem. Commun.*, 50 (1985a) 879-884.
Čeřovský, V. and Jošt, K., *Collect. Czech. Chem. Commun.*, 50 (1985b) 2310-2318.
Čeřovský, V. and Jošt, K., *Collect. Czech. Chem. Commun.*, 50 (1985c) 2775-2781.
Chang, J.-Y., *Biochem. J.*, 199 (1981) 537-545.
Chothia, C., *Nature (London)*, 248 (1974) 338-339.
Cooke, N.H.C. and Olsen, K., *J. Chromatogr. Sci.*, 18 (1980) 512-524.
Cooke, N.H.C., Viavattene, R.L., Eksteen, R., Wong, W.S., Davies, G. and Karger, B.L., *J. Chromatogr.*, 149 (1978) 391-415.
Corran, P.H. and Zanelli, J.M., in Hancock, W.S. (Editor), *CRC Handbook of HPLC for the Separation of Amino Acids, Peptides and Proteins*, Vol. II, CRC Press, Boca Raton, FL, 1974, pp. 445-459.
Coy, D.H., in Hawk, G.L., Champlin, P.B., Hutton, R.F., Jordi, H.C. and Mol, C. (Editors), *Biological/Biomedical Applications of Liquid Chromatography II (Chromatography Science Series)*, Vol. 12, Marcel Dekker, New York, 1979, pp. 283-292); *C.A.*, 92 (1980) 37 022j.
Coy, D.H., in Hancock, W.S. (Editor), *CRC Handbook of HPLC for the Separation of Amino Acids, Peptides and Proteins*, CRC Press, Boca Raton, FL, 1984, pp. 197-203.
Crommen, J., *Acta Pharm. Suec.*, 16 (1979) 111-124; *C.A.*, 91 (1979) 78 936h.
Crommen, J., in Hancock, W.S. (Editor), *CRC Handbook of HPLC for the Separation of Amino Acids, Peptides and Proteins*, Vol. I, CRC Press, Boca Raton, FL, 1984, pp. 175-186.
Crommen, J., Frensson, B. and Schill, G., *J. Chromatogr.*, 142 (1977) 283-297.
Davankov, V.A., in Hancock, W.S. (Editor), *CRC Handbook of HPLC for the Separation of Amino Acids, Peptides and Proteins*, Vol. I, CRC Press, Boca Raton, FL, 1984, pp. 393-409.
Deming, S.N., in Hancock, W.S. (Editor), *CRC Handbook of HPLC for the Separation of Amino Acids, Peptides and Proteins*, Vol. I, CRC Press, Boca Raton, FL, 1984, pp. 141-152.
Denkert, M., Hackzell, L., Schill, G. and Sjögren, E., *J. Chromatogr.*, 218 (1981) 31-43.
Dennis, M., Lazure, C., Seidah, N.G. and Chrétien, M., *J. Chromatogr.*, 266 (1983) 163-172.
Desiderio, D.M., in Hancock, W.S. (Editor), *CRC Handbook of HPLC for the Separation of Amino Acids, Peptides and Proteins*, Vol. I, CRC Press, Boca Raton, FL, 1984, pp. 197-211.

Desiderio, D.M. and Cunningham, M.D., *J. Liq. Chromatogr.*, 4 (1981) 721-733.
Desiderio, D.M. and Stout, Ch.B., *Adv. Chromatogr.*, 22 (1983) 1-36.
Desiderio, D.M. and Yamada, S., *J. Chromatogr.*, 239 (1982) 87-95.
Desiderio, D.M., Yamada, S., Tanzer, F.S., Horton, J. and Trimble, J., *J. Chromatogr.*, 217 (1981) 437-452.
Dizdaroglu, M., in Hancock, W.S. (Editor), *CRC Handbook of HPLC for the Separation of Amino Acids, Peptides and Proteins*, Vol. II, CRC Press, Boca Raton, FL, 1984, pp. 23-43.
Dizdaroglu, M., Krutzsch, H.C. and Simic, M.G., *Anal. Biochem.*, 123 (1982a) 190-193.
Dizdaroglu, M., Krutzsch, H.C. and Simic, M.G., *J. Chromatogr.*, 237 (1982b) 417-428.
Dizdaroglu, M. and Simic, M.G., *J. Chromatogr.*, 195 (1980) 119-126.
Drouin, J. and Goodman, H.M., *Nature (London)*, 288 (1980) 610-612.
Eberle, A.N., Atherton, E., Dryland, A. and Sheppard, R.C., *J. Chem. Soc., Perkin Trans. 1*, (1986) 361-367.
Erickson, B.W. and Prystowsky, M.B., in Hawk, G.L., Champlin, P.B., Hutton, R.F., Jordi, H.C. and Mol, Ch. (Editors), *Biological/Biomedical Applications of Liquid Chromatography II (Chromatography Science Series*, Vol. 12, Marcel Dekker, New York, 1979, pp. 293-305; *C.A.*, 92 (1980) 164 280s.
Flegel, M., Pospíšek, J., Pícha, J. and Píchová, D., in Bláha, K. and Maloň, P. (Editors), *Peptides 1982 (Proceedings of 17th European Peptide Symposium, Prague, Aug. 29-Sept. 3, 1982)*, Walter de Gruyter, Berlin, 1983, pp. 551-554.
Fonina, L., Demina, A., Sychev, S., Irkhin, A., Ivanov, V. and Hlaváček, J., in Voelter, W., Wünsch, E., Ovchinnikov, J. and Ivanov, V. (Editors), *Chemistry of Peptides and Proteins*, Vol. 1, Walter de Gruyter, Berlin, 1982, pp. 259-267.
Fonina, L.A., Demina, A.M., Sychev, S.V., Ivanov, V.T. and Hlaváček, J., *Bioorg. Khim.*, 10 (1984) 1073-1079; *C.A.*, 102 (1985) 25 014v.
Fox, L., Anthony, G.D. and Lau, E.P.K., *J. Assoc. Off. Anal. Chem.*, 59 (1976) 1048-1050; *C.A.*, 86 (1977) 15 283u.
Franco-Bourland, R.E. and Fernstrom, J.D., *Endocrinology*, 109 (1981) 1097-1106.
Frank, B.H., Beckage, M.J. and Willey, K.A., *J. Chromatogr.*, 266 (1983) 239-248.
Frei, R.W., Michel, L. and Santi, W., *J. Chromatogr.*, 126 (1976) 665-677.
Frei, R.W., Michel, L. and Santi, W., *J. Chromatogr.*, 142 (1977) 261-270.
Gabriel, T.F., Jimenez, M.H., Felix, A.M., Michalewsky, J. and Meienhofer, J., *Int. J. Peptide Protein Res.*, 9 (1977) 129-136.
Gabriel, T.F., Michalewsky, J. and Meienhofer, J., in Eberle, A., Geiger, R. and Wieland, T. (Editors), *Perspectives in Peptides Chemistry*, Karger, Basle, 1981, pp. 195-206; *C.A.*, 94 (1981) 116 960h.
Games, D.E., *Biomed. Mass Spectrom.*, 8 (1981) 454-462.
Games, D.E., Eckers, Ch., Gower, J.L., Hirter, P., Knight, M.E., Lewis, E., Rao, K.R.N. and Weerasinghe, N.C., *Clin. Res. Cent. Symp. (Harrow, Engl.)*, 1980, No. 1 *(Curr. Dev. Clin. Appl. HPLC, GCMS)*, 97-118; *C.A.*, 94 (1981) 152 731t.
Gay, D.D. and Lahti, R.A., in Hancock, W.S. (Editor), *CRC Handbook of HPLC for the Separation of Amino Acids, Peptides and Proteins*, Vol. II, CRC Press, Boca Raton, FL, 1984, pp. 423-428.
Gazdak, M. and Szepesi, G., *J. Chromatogr.*, 218 (1981) 603-612.
Glasel, J.A., in Hancock, W.S. (Editor), *CRC Handbook of HPLC for the Separation of Amino Acids, Peptides and Proteins*, Vol. II, CRC Press, Boca Raton, FL, 1984, pp. 429-433.
Goodman, M., Keogh, P. and Anderson, H., *Bioorg. Chem.*, 6 (1977) 239-247.
Grego, B. and Hearn, M.T.W., *Proc. Otago Univ. Med. Sch.*, 58 (1980) 41-43; *C.A.*, 93 (1980) 182 155e.
Grego, B. and Hearn, M.T.W., *Chromatographia*, 14 (1981) 589-592.

Grego, B., Lambrou, F. and Hearn, M.T.W., *J. Chromatogr.*, 266 (1983) 89-103.
Greibrokk, T., *Z. Naturforsch.*, in press (Special Issue).
Gröningsson, K. and Abrahamsson, M., in Hancock, W.S. (Editor), *CRC Handbook of HPLC for the Separation of Amino Acids, Peptides and Proteins*, Vol. II, CRC Press, Boca Raton, FL, 1984, pp. 221-228.
Gröningsson, K. and Widahl-Näsman, M., *J. Chromatogr.*, 291 (1984) 185-194.
Gruber, K.A., Stein, S., Brink, L., Radhakrishnan, A. and Udenfriend, S., *Proc. Natl. Acad. Sci. U.S.A.*, 73 (1976) 1314-1318.
Gruber, K.A., Whitaker, J.M. and Morris, M., *Anal. Biochem.*, 97 (1979) 176-183.
Guyon, F., Foucault, A. and Caude, M., *J. Chromatogr.*, 186 (1979) 677-682.
Guyon-Gruaz, A., Raulais, D. and Rivaille, P., in Hancock, W.S. (Editor), *CRC Handbook of HPLC for the Separation of Amino Acids, Peptides and Proteins*, Vol. II, CRC Press, Boca Raton, FL, 1984, pp. 213-220.
Hancock, W.S. (Editor), *CRC Handbook of HPLC for the Separation of Amino Acids, Peptides and Proteins*, CRC Press, Boca Raton, FL, 1984, Vol. I, 489 pp.; Vol. II, 522 pp.
Hancock, W.S., Bishop, C.A. and Hearn, M.T.W., *FEBS Lett.*, 72 (1986) 139-142.
Hancock, W.S., Bishop, C.A., Mayer, L.J., Harding, D.R.K. and Hearn, M.T.W., *J. Chromatogr.*, 161 (1978a) 291-298.
Hancock, W.S., Bishop, C.A., Prestidge, R.L., Harding, D.R.K. and Hearn, M.T.W., *J. Chromatogr.*, 153 (1978b) 391-398.
Hancock, W.S., Bishop, C.A., Prestidge, R.L., Harding, D.R.K. and Hearn, M.T.W., *Science (Washington, D.C.)*, 200 (1978c) 1168-1170.
Hancock, W.S., Bishop, C.A., Battersby, J.E., Harding, D.R.K. and Hearn, M.T.W., *J. Chromatogr.*, 168 (1979a) 377-384.
Hancock, W.S., Bishop, C.A. and Hearn, M.T.W., *Chem. N.Z.*, 43 (1979b) 17-24; *C.A.*, 91 (1979) 123 968h.
Hancock, W.S. and Harding, D.R.K., in Hancock, W.S. (Editor), *CRC Handbook of HPLC for the Separation of Amino Acids, Peptides and Proteins*, Vol. I, CRC Press, Boca Raton, FL, 1984a, pp. 189-192.
Hancock, W.S. and Harding, D.R.K., in Hancock, W.S. (Editor), *CRC Handbook of HPLC for the Separation of Amino Acids, Peptides and Proteins*, Vol. II, CRC Press, Boca Raton, FL, 1984b, pp. 3-22.
Hansch, C. and Leo, A., *Substituent Constants for Correlation Analysis in Chemistry and Biology*, Wiley, New York, 1979, p.49.
Hansen, J.J., Greibrokk, T., Currie, B.L., Johansson, K.N.-G. and Folkers, K., *J. Chromatogr.*, 135 (1977) 155-164.
Hara, S., Ohsawa, A. and Dobashi, A., *J. Liq. Chromatogr.*, 4 (1981) 409-423.
Harding, D.R.K., Bishop, C.A., Tarttelin, M.F. and Hancock, W.S., *Int. J. Peptide Protein Res.*, 18 (1981) 214-220.
Hare, P.E., *Methods Enzymol.*, 47 (1977) 3-19.
Hatefi, Y. and Hanstein, W.C., *Proc. Natl. Acad. Sci. U.S.A.*, 62 (1969) 1129-1136.
Hearn, M.T.W., in Horváth, C. (Editor), *High Performance Liquid Chromatography (Advances and Perspectives*, Vol. 3), Academic Press, New York, 1983, pp. 87-155; *C.A.*, 99 (1983) 35 386b.
Hearn, M.T.W., in Hearn, M.T.W. (Editor), *Ion-Pair Chromatography. Theory, Biological and Pharmaceutical Applications*, Marcel Dekker, New York, 1985, pp. 207-257; *C.A.*, 102 (1985) 128 024c.
Hearn, M.T.W., Bishop, C.A., Hancock, W.S., Harding, D.R.K. and Reynolds, G.D., *J. Liq. Chromatogr.*, 2 (1979a) 1-21.
Hearn, M.T.W., Grego, B. and Bishop, C.A., *J. Liq. Chromatogr.*, 4 (1981a) 1725-1744.
Hearn, M.T.W. and Grego, B., *J. Chromatogr.*, 203 (1981b) 349-363.
Hearn, M.T.W. and Grego, B., *J. Chromatogr.*, 218 (1981c) 497-507.
Hearn, M.T.W. and Grego, B., *J. Chromatogr.*, 255 (1983a) 125-136.
Hearn, M.T.W. and Grego, B., *J. Chromatogr.*, 266 (1983b) 75-87.
Hearn, M.T.W. and Hancock, W.S., *Biological/Biomedical Applications of Liquid Chromatography 2 (Chromatography Science Series*, Vol. 12), Marcel Dekker, New York, 1979a, pp. 243-271.

Hearn, M.T.W., Grego, B. and Hancock, W.S., *J. Chromatogr.*, 185 (1979b) 429-444.
Hearn, M.T.W. and Hancock, W.S., *Trends Biochem. Sci.*, 4 (1979b) N58-N62; *C.A.*, 90 (1979) 164 146m.
Hearn, M.T.W., Hancock, W.S., Hurrell, J.G., Fleming, R.J. and Kemp, B., *J. Liq. Chromatogr.*, 2 (1979c) 919-933.
Hearn, M.T.W., Regnier, F.E. and Wehr, C.T., *Int. Lab.*, 13 (1983) 16-37.
Hearn, M.T.W., Su Suew, J. and Grego, B., *J. Liq. Chromatogr.*, 4 (1981b) 1547-1567.
Henderson, L.E., Sowder, R. and Oroszlan, S., in Liu, Schechter, Heinrikson and Condlife (Editors), *Chemical Synthesis and Sequencing of Peptides and Proteins (Developments in Biochemistry, 1981)* North-Holland, Amsterdam, 1981, pp. 251-260; *C.A.*, 96 (1982) 213 548d.
Hlaváček, J. and Jošt, K., *Chemistry and Biotechnology of Biologically Active Natural Products, Vol. 3 (Proceedings of FEBS 3rd International Conference, Sofia, 1985)*, Bulgarian Academy of Science, Sofia, 1985, pp. 244-261.
Hlaváček, J., Pospíšek, J., Slaninová, J., Barth, T. and Jošt, K., in Ragnarsson, U. (Editor), *Peptides 1984 (Proceedings of 18th European Symposium, Djurönäset, June 10-15*, Almqvist and Wiksell, Stockholm, 1984, pp. 415-418.
Horváth, C. and Lipsky, S.R., *Anal. Chem.*, 39 (1967) 1893.
Horváth, C., Melander, W. and Molnár, I., *J. Chromatogr.*, 125 (1976) 129-156.
Horváth, C., Melander, W. and Molnár, I., *Anal. Chem.*, 49 (1977) 142-154.
Hrbas, P., Barth, T., Lebl, M., Papsuevich, A.O. and Flegl, M., *Beitr. Wirkstofforsch.*, 24 (1985) 182-188.
Hruby, V.J., Viswanatha, V. and Young C.S. Yang, *J. Labelled Compd. Radiopharm.*, 17 (1980) 801-812; *C.A.*, 94 (1981) 175 527v.
Hui, K.-S., Salschutz, M., Davis, B.A. and Lajtha, A., *J. Chromatogr.*, 192 (1980) 341-350.
Hunter, C., Sugden, K. and Lloyd-Jones, J.G., *J. Liq. Chromatogr.*, 3 (1980) 1335-1352.
Inouye, K., Watanabe, K. and Konaka, R., in Hancock, W.S. (Editor), *CRC Handbook of HPLC for the Separation of Amino Acids, Peptides and Proteins*, Vol. II, CRC Press, Boca Raton, FL, 1984, pp. 111-129.
Iskandarani, Z. and Pietrzyk, D.J., *Anal. Chem.*, 53 (1981) 489-495.
Johnson, P., *J. Chromatogr. Sci.*, 17 (1979) 406-409.
Jošt, K., Lebl, M. and Brtník, F. (Editors), *Handbook of Neurohypophysial Hormone Analogs*, CRC Press, Boca Raton, FL, 1987, in press.
Kalász, H. and Horváth, C., *J. Chromatogr.*, 215 (1981) 295-302.
Kanmera, T., Feinstein, G. and Chaiken, I.M., in Hruby, V.J. and Rich, D.H. (Editors), *Proceedings of 8th American Peptide Symposium*, Pierce, Rockford, 1983, pp. 261-264.
Karger, B.L. and Giese, R.W., *Anal. Chem.*, 50 (1978) 1048A-1073A.
Kauzmann, W., *Adv. Protein Chem.*, 14 (1959) 1-63.
Kent, S.B.H., Mitchell, A.R., Barany, G. and Merrifield, R.B., *Anal. Chem.*, 50 (1978) 155-159.
Kikta, Jr., E.J. and Grushka, E., *J. Chromatogr.*, 135 (1977) 367-376.
Kimura, T., Takai, M., Masui, Y., Morikawa, T. and Sakakibara, S., *Biopolymers*, 20 (1981) 1823-1832; *C.A.*, 95 (1981) 204 420h.
Kiso, Y., Satomi, M., Miyazaki, T., Hiraiwa, H. and Akita, T., in Okawa, K. (Editor), *Peptide Chemistry 1980*, Vol. 18, Protein Research Foundation, Osaka, 1981, pp. 71-74; *C.A.*, 95 (1981) 115 988g.
Klickstein, L.B. and Wintroub, B.U., *Anal. Biochem.*, 120 (1982) 146-150.
Knight, C.A., *J. Biol. Chem.*, 190 (1951) 73-756.
Konaka, R. and Shoji, J., *Kagaku No Ryoiki, Zokan*, 133 (1981) 151-166; *C.A.*, 95 (1981) 156 629j.
Kroeff, E.P. and Pietrzyk, D.J., *Anal. Chem.*, 50 (1978a) 502-511.
Kroeff, E.P. and Peitrzyk, D.J., *Anal. Chem.*, 50 (1978b) 1353-1358.
Krojidlo, M., Flegel, M. and Lebl, M., in Bláha, K. and Maloň, P. (Editors), *Proceedings of 17th European Peptide Symposium*, Walter de Guyter, Berlin, 1983, pp. 199-202.

Krol, G.J., Bankovsky, J.M., Mannan, C.A., Pickering, R.E. and Kho, B.T., *J. Chromatogr.*, 163 (1979) 383-389.

Krol, G.J., Mannan, C.A., Pickering, R.E., Amato, D.V., Kho, B.T. and Sonnenschein, A., *Anal. Chem.*, 49 (1977) 1836-1839.

Krummen, K., *J. Liq. Chromatogr.*, 3 (1980) 1243-1254.

Krummen, K. and Frei, R.W., *J. Chromatogr.*, 132 (1977a) 27-36.

Krummen, K. and Frei, R.W., *J. Chromatogr.*, 132 (1977b) 429-436.

Krummen, K., Maxl, F. and Nachtmann, F., *Pharm. Technol. Int.*, 2 (1979) 37-43, 62; *C.A.*, 94 (1981) 7832x.

Kullmann, W., *J. Liq. Chromatogr.*, 4 (1981) 1121-1134.

Kuwata, S., Yamada, T., Miyazawa, T., Dejima, K. and Watanabe, K., in Okawa, K. (Editor), *Peptide Chemistry 1980*, Vol. 18, Protein Research Foundation, Osaka, 1981, pp. 65-70; *C.A.*, 95 (1981) 115 987f.

Lande, S., *Biopolymers*, 7 (1969) 879-886.

Larsen, B., Fox, B.L., Burke, M.F. and Hruby, V.J., *Int. J. Pept. Protein Res.*, 13 (1979) 12-21.

Lebl, M., *Collect. Czech. Chem. Commun.*, 45 (1980) 2927-2937; *C.A.*, 94 (1981) 152 773h.

Lebl, M., *J. Chromatogr.*, 242 (1982) 342-345.

Lebl, M., *J. Chromatogr.*, 264 (1983) 459-462.

Lebl, M., in Hancock, W.S. (Editor), *CRC Handbook of HPLC for the Separation of Amino Acids, Peptides and Proteins*, Vol. II, CRC Press, Boca Raton, FL, 1984, pp. 169-178.

Lebl, M., in Jošt, K., Lebl, M. and Brtník, F. (Editors), *Handbook of Neurohypophysial Hormone Analogs*, CRC Press, Boca Raton, FL, 1987, Ch. 5, in press.

Lebl, M., Barth, T., Crankshaw, D.J., Cerný, B., Daniel, E.E., Crover, A.K. and Jošt, K., *Collect. Czech. Chem. Commun.*, 49 (1984a) 1921-1926.

Lebl, M., Barth, T. and Jošt, K. in Brunfeldt, K. (Editor), *Peptides 1980 (Proceedings of 16th European Peptide Symposium)*, Scriptor, Copenhagen, 1981, pp. 719-724.

Lebl, M., Barth, T., Servítová, L., Slaninová, J. and Hrbas, P., *Collect. Czech. Chem. Commun.*, 49 (1984b) 2022-2023.

Lebl, M., Barth, T., Servítová, L., Slaninová, J. and Jošt, K., *Collect. Czech. Chem. Commun.*, 50 (1985a) 132-145.

Lebl, M., Cody, W.L. and Hruby, V.J., *J. Liq. Chromatogr.*, 7 (1984c) 1195-1210.

Lebl, M., Cody, W.L., Wilkes, B.C., Hruby, V.J., Castrucci, A.M. de L. and Hadley, M.A., *Collect. Czech. Chem. Commun.*, 49 (1984d) 2680-2688.

Lebl, M., Hrbas, P., Škopková, J., Slaninová, J., Machová, A., Barth, T. and Jošt, K., *Collect. Czech. Chem. Commun.*, 47 (1982a) 2450-2560.

Lebl, M., Hruby, V.J., Slaninová, J. and Barth, T., *Collect. Czech. Chem. Commun.*, 50 (1985b) 418-427.

Lebl, M., Pospíšek, J., Hlaváček, J., Barth, T., Maloň, P., Servítová, L., Hauzer, K. and Jošt, K., *Collect. Czech. Chem. Commun.*, 47 (1982b) 689-701.

Leo, A., Hansch, C. and Elkins, D., *Chem. Rev.*, 71 (1971) 525-616.

Lepri, L., Desideri, P.G. and Heimler, D., in Hancock, W.S. (Editor), *CRC Handbook of HPLC for the Separation of Amino Acids, Peptides and Proteins*, Vol. I, CRC Press, Boca Raton, FL, 1984, pp. 213-226.

Lewis, R.V., in Hancock, W.S. (Editor), *CRC Handbook of HPLC for the Separation of Amino Acids, Peptides and Proteins*, Vol. I, CRC Press, Boca Raton, FL, 1984, pp. 193-196.

Liat Tan, *J. Chromatogr.*, 266 (1983) 67-74.

Li, C.H. and Chung, D., in Hancock, W.S. (Editor), *CRC Handbook of HPLC for the Separation of Amino Acids, Peptides and Proteins*, Vol. II, CRC Press, Boca Raton, FL, 1984, pp. 435-437.

Lin Shen-Nan, Smith, L.A. and Caprioli, R.M., *J. Chromatogr.*, 197 (1980) 31-41.

Live, D.H., in Goodman, M. and Meienhofer, J. (Editors), *Peptides (Proceedings of 5th American Peptide Symposium)*, Halsted Press, Chichester, and Wiley, New York, 1977, pp. 44-47; *C.A.*, 88 (1978) 185 503z.

Live, D.H., Agosta, W.C. and Cowburn, D., *J. Org. Chem.*, 42 (1977) 3556-3561.

Loeber, J.G., in Hancock, W.S. (Editor), *CRC Handbook of HPLC for the Separation of Amino Acids, Peptides and Proteins*, Vol. I, CRC Press, Boca Raton, FL, 1984, pp. 227-232.

Loeber, J.G. and Verhoef, J., *Methods Enzymol.*, 73 (Immunochemical Techniques, Part B) (1981) 261-275; *C.A.*, 95 (1981) 217 173e.

Lukas, T.J., Prystowsky, M.B. and Erickson, B.W., *Proc. Natl. Acad. Sci. U.S.A.*, 78 (1981) 2791-2795.

Lundanes, E. and Greibrokk, T., *J. Chromatogr.*, 149 (1978) 241-254.

Lundanes, E. and Greibrokk, T., in Hancock, W.S. (Editor), *CRC Handbook of HPLC for the Separation of Amino Acids, Peptides and Proteins*, Vol. II, CRC Press, Boca Raton, FL, 1984, pp. 49-52.

Mabuchi, H. and Nakahashi, H., in Hancock, W.S. (Editor), *CRC Handbook of HPLC for the Separation of Amino Acids, Peptides and Proteins*, Vol. II, CRC Press, Boca Raton, FL, 1984, pp. 67-73.

Mahoney, W.C. and Hermodson, M.A., *J. Biol. Chem.*, 255 (1980) 11199-11203.

Manberg, P.J., Youngblood, W.W. and Kizer, J.S., *Brain Res.*, 241 (1982) 279-284; *C.A.*, 97 (1982) 49 855z.

Margolis, S.A. and Longenbach, P.J., in Barker, J.L. and Smith, Jr., T.G. (Editors), *The Role of Peptides in Neuronal Function*, Marcel Dekker, New York, 1980, pp. 49-67; *C.A.*, 94 (1981) 170 366g.

Margolis, S.A. and Schaffer, R., *Development of a Standard Reference Material for Angiotensin I (National Bureau of Standards, Report NBSIR 79-1847)*, National Intitutes of Health, Bethesda, MD, 1979, pp. 1-31.

Matsueda, R., Mukaiyama, T. and Suzuki, M., *Tetrahedron Lett.*, (1970) 1901-1904.

McMartin, C., in Hancock, W.S. (Editor), *CRC Handbook of HPLC for the Separation of Amino Acids, Peptides and Proteins*, Vol. II, CRC Press, Boca Raton, FL, 1984, pp. 147-152.

Meek, J.L., in Costa, E. and Trabucchi, M. (Editors), *Neural Peptides and Neuronal Communication (Advances in Biochemistry and Psychopharmacology, Vol. 22)*, Roven Press, New York, 1980a, pp. 145-151; *C.A.*, 93 (1980) 145 637t.

Meek, J.L., *Proc. Natl. Acad. Sci. U.S.A.*, 77 (1980b) 1632-1636.

Meek, J.L., *J. Chromatogr.*, 266 (1983) 401-408.

Meek, J.L. and Rossetti, Z.L., *J. Chromatogr.*, 211 (1981) 15-28.

Meienhofer, J., in Li, C.H. (Editor), *Hormonal Proteins and Peptides*, Vol. 2, Academic Press, New York, 1973, pp. 45-267.

Melander, W.R., Jacobson, J. and Horváth, C., *J. Chromatogr.*, 234 (1982) 269-276.

Mikeš, O., in Mikeš, O. (Editor), *Laboratory Handbook of Chromatographic and Allied Methods*, Ellis Horwood, Chichester, 1979, pp. 237-238.

Mikeš, O. and Šebesta, K., in Mikeš, O. (Editor), *Laboratory Handbook of Chromatographic and Allied Methods*, Ellis Horwood, Chichester, 1979, pp. 306-310.

Möhring, J., Böhlen, P., Schoun, J., Mellet, M., Suess, U., Schmidt, M. and Pliska, V., *Acta Endocrinol. (Copenhagen)*, 99 (1982) 371-378; *C.A.*, 96 (1982) 136 057p.

Molnár, I. and Horváth, C., *J. Chromatogr.*, 142 (1977) 623-640.

Mönch, W. and Dehnen, W., *J. Chromatogr.*, 140 (1977) 260-262.

Moore, S. and Stein, W.H., *J. Biol. Chem.*, 192 (1951) 663-681.

Moriya, F., Makino, K., Suzuki, N., Rokushika, S. and Hatano, H., *J. Phys. Chem.*, 84 (1980) 3614-3619; *C.A.*, 94 (1981) 121 927g.

Moriya, F., Makino, K., Suzuki, N., Rokushika, S. and Hatano, H., *J. Am. Chem. Soc.*, 104 (1982) 830-836.

Morris, H.R., Dell, A., Etienne, T. and Taylor, G.W., *Dev. Biochem.*, 10 *(Front. Protein Chem.)* (1980) 193-209; *C.A.*, 95 (1981) 20 660e.

Murphy, R., Furness, J.B. and Costa, M., *J. Chromatogr.*, 336 (1984) 41-40.

Nabeshima, T., Hiramatsu, M., Noma, S., Ukai, M., Amano, M. and Kameyama, T., *Res. Commun. Chem. Pathol. Pharmacol.*, 35 (1982) 421-442; *C.A.*, 96 (1982) 193 559c.

Nachtmann, F., *J. Chromatogr.*, 176 (1979) 391-397.

Nachtmann, F. and Gstrein, K., in Hancock, W.S. (Editor), *CRC Handbook of HPLC for the Separation of Amino Acids, Peptides and Proteins*, Vol. II, CRC Press, Boca Raton, FL, 1984, pp. 131-135.

Naider, F., Sipzner, R., Steinfeld, A.S. and Becker, J.M., *J. Chromatogr.*, 176 (1979) 264-269.

Naider, F., Steinfeld, A.S. and Becker, J.M., in Hancock, W.S. (Editor), *CRC Handbook of HPLC for the Separation of Amino Acids, Peptides and Proteins*, Vol. II, CRC Press, Boca Raton, FL, 1984, pp. 89-97.

Nakanishi, S., Inoue, A., Kita, T., Nakamura, M., Chang, A.C., Cohen, S.N. and Numa, S., *Nature (London)*, 278 (1979) 423-427.

Naleway, J.J. and Hoffman, N.E., *Anal. Lett.*, 14 (1981) 1711-1724.

Nika, H. and Hultin, T., *Anal. Biochem.*, 98 (1979) 178-183.

Nozaki, Y. and Tanford, C., *J. Biol. Chem.*, 246 (1971) 2211-2217.

O'Hare, M.J. and Nice, E.C., *J. Chromatogr.*, 171 (1979) 209-226.

Olieman, C., Sedlick, E. and Voskamp, D., *J. Chromatogr.*, 207 (1981) 421-424.

Olieman, C. and Voskamp, D., in Hancock, W.S. (Editor), *CRC Handbook of HPLC for the Separation of Amino Acids, Peptides and Proteins*, Vol. I, CRC Press, Boca Raton, FL, 1984, pp. 161-166.

Oroszlan, S., *Chromatogr. Sci.*, 10 (1979) *(Biol./Med. Appl. Liq. Chromatogr)* 199-224; *C.A.*, 90 (1979) 147 917w.

Oshima, G., Shimabukuro, H. and Nagasawa, K., *J. Chromatogr.*, 152 (1978) 579-584.

Pardee, A.B., *J. Biol. Chem.*, 190 (1951) 757-762.

Pask-Hughes, R.A., Corran, P.H. and Calam, D.H., *J. Chromatogr.*, 214 (1981) 307-315.

Pask-Hughes, R.A., Hartley, R.E. and Gaines Das R.E., *J. Biol. Stand.*, 11 (1983) 13-17.

Pietrzyk, D.J., Cahil, W.J. and Stodola, J.D., *J. Liq. Chromatogr.*, 5 (1982) 442-461.

Pliška, V. and Fauchère, J.-L., in Gross, E. and Meienhofer, J. (Editors), *Peptides, Structure and Biological Function (Proceedings of 6th American Peptide Symposium)*, Pierce, Rockford, IL, 1979, pp. 249-252; *C.A.*, 93 (1980) 233 189d.

Poll, D.J., Knighton, D.R., Harding, D.R.K. and Hancock, W.S., *J. Chromatogr.*, 236 (1982) 244-248.

Porthault, M., in Rosselin, G., Fromageot, P. and Bonfils, S. (Editors), *Hormone Receptors in Digestion and Nutrition (Proceedings of 2nd International Symposium on Horm. Recept. Dig. Tract Physiol.)*, Elsevier/North-Holland Biomedical Press, Amsterdam, 1979, pp. 69-77; *C.A.*, 92 (1980) 90 154e.

Pradayrol, L., *Gastroenterol. Clin. Biol.*, 6 (1982) 13-15 (in French); *C.A.*, 96 (1982) 118 298p.

Pradayrol, L., Fourmy, D. and Ribet, A., in Hancock, W.S. (Editor), *CRC Handbook of HPLC for the Separation of Amino Acids, Peptides and Proteins*, Vol. II, CRC Press, Boca Raton, FL, 1984, pp. 261-264.

Prestidge, R.L., Harding, D.R.K., Moore, C.H. and Hancock, W.S., *Bioorg. Chem.*, 10 (1981) 277-282.

Procházka, Z., Lebl, M., Barth, T., Hlaváček, J. and Jošt, K., in Bláha, K. and Maloň, P. (Editors), *Peptides 1982 (Proc. 17th Europ. Peptide Symp.)*, Walter de Gruyter, Berlin, 1982, pp. 441-444.

Procházka, Z., Lebl, M., Barth, T., Hlaváček, J., Trka, A., Buděšínský, M. and Jošt, K., *Collect. Czech. Chem. Commun.*, 49 (1984) 642-652.

Procházka, Z., Lebl, M., Servítová, L., Barth, T. and Jošt, K., *Collect. Czech. Chem. Commun.*, 46 (1981) 947-956.

Radhakrishnan, A.N., Stein, S., Licht, A., Gruber, K.A. and Udenfriend, S., *J. Chromatogr.*, 132 (1977) 552-555.

Ragnarsson, U., Fransson, B. and Zetterqvist, Ö., in Hancock, W.S. (Editor), *CRC Handbook of HPLC for the Separation of Amino Acids, Peptides and Proteins*, Vol. II, CRC Press, Boca Raton, FL, 1984, pp. 75-88.

Rekker, R.F., *The Hydrophobic Fragmental Constants*, Elsevier, Amsterdam, 1977, p. 301.

Reynolds, G.D., Harding, D.R.K. and Hancock, W.S., *Int. J. Peptide Protein Res.*, 17 (1981) 231-234.

Richey, J., *Int. Lab.*, 13, Jan./Feb. (1983) 50-75.

Richmond, W., *Clin. Res. Cent. Symp. (Harrow, Engl.)*, 1980, 1 *(Curr. Dev. Clin. Appl. HPLC, GCMS)*, 85-96; *C.A.*, 94 (1981) 152 730s.

Ridge, R.J., Matsueda, G.R., Haber, E. and Matsueda, R., *Int. J. Peptide Protein Res.*, 19 (1982) 490-498.

Rivier, J.E., *J. Liq. Chromatogr.*, 1 (1978) 343-366.

Rivier, J. and Burgus, R., *Chromatogr. Sci.*, 10 (1979) *(Biol./Biomed. Appl. Liq. Chromatogr.)* 147-161; *C.A.*, 90 (1979) 109 681j.

Rivier, J., Desmond, J., Spiess, J., Perrin, M., Vale, W., Eksteen, R. and Karger, B., in Gross, E. and Meienhofer, J. (Editors), *Peptides: Structure and Biological Fundtion (Proceedings of the 6th American Peptide Symposium)*, Pierce, Rockford, IL, 1979a, pp. 125-128; *C.A.*, 93 (1980) 234 151x.

Rivier, J., Spiess, J., Perrin, M. and Vale, W., *Biological/Biomedical Applications of Liquid Chromatography 2 (Chromatography Science Series*, Vol. 12), Marcel Dekker, New York, 1979b, pp. 223-241; *C.A.*, 92 (1980) 37 020g.

Rossetti, Z.L., *Riv. Farmacol. Ter.*, 12 (1981) 35-39; *C.A.*, 95 (1981) 146 324s.

Rubinstein, M., *Anal. Biochem.*, 98 (1979) 1-7.

Rubinstein, M., Chen-Kiang, S., Stein, S. and Udenfriend, S., *Anal. Biochem.*, 95 (1979) 117-121.

Rudinger, J., in Meienhofer, J. (Editor), *Chemistry and Biology of Peptides*, Ann Arbor Sci. Publ., Ann Arbor, MI, 1972, pp. 729-735.

Ryeszotarski, W.J. and Mauger, A.B., *J. Chromatogr.*, 86 (1973) 246-249.

Sakakibara, S. and Kimura, Y., in Hancock, W.S. (Editor), *CRC Handbook of HPLC for the Separation of Amino Acids, Peptides and Proteins*, Vol. II, CRC Press, Boca Raton, FL, 1984, pp. 153-168.

Sallay, S.I. and Oroszlan, S., in Hancock, W.S. (Editor), *CRC Handbook of HPLC for the Separation of Amino Acids, Peptides and Proteins*, Vol. II, CRC Press, Boca Raton, FL, 1984, pp. 99-110.

Sanger, F., *Adv. Protein Chem.*, 7 (1952) 1-67.

Sasagawa, T., Ericksson, L.H., Teller, D.C., Titani, K. and Walsh, K.A., in *2nd International Symposium on the HPLC of Proteins, Peptides and Polynucleotides*, 1982a, p. 12 (cited by Sasagawa and Teller, 1984).

Sasagawa, T., Okuyama, T. and Teller, D.C., *J. Chromatogr.*, 240 (1982b) 329-340.

Sasagawa, T. and Teller, D.C., in Hancock, W.S. (Editor), *CRC Handbook of HPLC for the Separation of Amino Acids, Peptides and Proteins*, Vol. II, CRC Press, Boca Raton, FL, 1984, pp. 53-65.

Segrest, J.P. and Feldmann, R.J., *J. Mol. Biol.*, 87 (1974) 853-858.

Schlabach, T.D., *J. Chromatogr.*, 266 (1983) 427-437.

Schlabach, T.D. and Abbott, S.R., *Clin. Chem.*, 26 (1980) 1504-1508; *C.A.*, 93 (1980) 200 302g.

Schlabach, T.D. and Wehr, C.T., in Hearn, M.T.W., Regnier, F.E. and Wehr, C.T. (Editors), *Proceedings of the 1st International Symposium on the HPLC of Proteins and Peptides*, Academic Press, New York, 1983, pp. 221-232.

Sheppard, R.C., *Sci. Tools*, 33 (1986) 9-16.

Shioiri, T. (Editor), *Peptide Chemistry 1981 (Proceedings of the 19th Symposium on Peptide Chemistry, Nagoya, Japan, 1981)*, Peptide Institute, Protein Research Foundation, Minoh, Osaka, 1982, 210 pp.

Shioya, A., Yoshida, H. and Nakajima, T., *J. Chromatogr.*, 240 (1982) 341-348.

Smith, C.W., Skala, G. and Walter, R., *Int. J. Peptide Protein Res.*, 16 (1980) 365-371.

Smith, J.A. and McWilliams, R.A., *Am. Lab. (Fairfield, Conn.)*, 12, No. 5 (1980a) 23-29; No. 6 (1980a) 25-30; *C.A.*, 93 (1980) 95 657b.

Smith, J.A. and McWilliams, R.A., *Int. Lab.*, Oct. (1980b) 29-32.

Smith, M.W., in Heller, H. (Editor), *Int. Encycl. Pharmacol. Ther.*, 41 (Vol. 1) (1970) 173-228; *C.A.*, 77 (1972) 85 121c.

Spackman, D.H., Stein, W.H. and Moore, S., *Anal. Chem.*, 30 (1958) 1190-1206.
Starratt, A.N. and Stevens, M.E., in Hancock, W.S. (Editor), *CRC Handbook of HPLC for the Separation of Amino Acids, Peptides and Proteins*, Vol. II, CRC Press, Boca Raton, FL, 1984, pp. 255-260.
Stein, S., *Miles Int. Symp. Ser.*, 12 (1980) 77-85; *C.A.*, 93 (1980) 234 144x.
Stein, S., *Peptides (N.Y.)*, 4 (1981) 185-216; *C.A.*, 95 (1981) 183 184s.
Stein, S., Rubinstein, M. and Udenfriend, S., *Psychopharmacol. Bull.*, 14 (1978) 29-30; *C.A.*, 90 (1979) 35 676k.
Stoklosa, J.T., Ayi, B.K., Shearer, C.M. and DeAngelis, N.J., *Anal. Lett.*, B11 (1978) 889-899; *C.A.*, 90 (1979) 87 873s.
Sugden, K., in Hancock, W.S. (Editor), *CRC Handbook of HPLC for the Separation of Amino Acids, Peptides and Proteins*, Vol. I, CRC Press, Boca Raton, FL, 1984, pp. 423-427.
Sugden, K., Hunter, C. and Lloyd-Jones, J.G., *J. Chromatogr.*, 204 (1981) 195-200.
Sullivan, R.C., Shing, Y.W., D'Amore, P.A. and Klagsbrun, M., *J. Chromatogr.*, 266 (1983) 301-311.
Su Suew, J., Grego, B., Niven, B. and Hearn, M.T.W., *J. Liq. Chromatogr.*, 4 (1981) 1745-1764.
Svoboda, M.E. and Van Wyk, J.J., in Hancock, W.S. (Editor), *CRC Handbook of HPLC for the Separation of Amino Acids, Peptides and Proteins*, Vol. II, CRC Press, Boca Raton, FL, 1984, pp. 439-444.
Takahagi, H., Matsueda, R. and Maruyama, H., *Sankyo Kenkyusho Nempo (Annu. Rep. Sankyo Res. Lab.)*, 30 (1978) 57-64; *C.A.*, 91 (1978) 57 485v.
Takahashi, N., Isobe, T., Kasai, H., Seta, K. and Okuyama, T., *Anal. Biochem.*, 115 (1981) 181-187.
Tanford, C., *The Hydrophobic Effect*, Wiley-Interscience, New York, 1973, 240 pp.
Tanford, C., *Science (Washington, D.C.)*, 200 (1978) 1012-1018.
Terabe, S., Konaka, R. and Inouye, K., *J. Chromatogr.*, 172 (1979a) 163-177.
Tomlinson, E., in Hancock, W.S. (Editor), *CRC Handbook of HPLC for the Separation of Amino Acids, Peptides and Proteins*, Vol. I, CRC Press, Boca Raton, FL, 1984, pp. 129-139.
Tonnaer, J.A.D.M., in Hancock, W.S. (Editor), *CRC Handbook of HPLC for the Separation of Amino Acids, Peptides and Proteins*, Vol. II, CRC Press, Boca Raton, FL, 1984, pp. 179-185.
Tsuji, K. and Robertson, J.H., *J. Chromatogr.*, 112 (1975) 663-672.
Tsuji, K., Robertson, J.H. and Bach, J.A., *J. Chromatogr.*, 99 (1974) 597-608.
Tyler, G.A. and Rosenblatt, M., *J. Chromatogr.*, 226 (1983) 313-318.
Udenfriend, S. and Stein, S., in Goodman, M. and Meienhofer, J. (Editors), *Peptides (Proceedings of the 5th American Peptide Symposium)*, Wiley, New York, 1977, pp. 14-26.
Van Nispen, J.W. and Janssen, P.S.L., in Hancock, W.S. (Editor), *CRC Handbook of HPLC for the Separation of Amino Acids, Peptides and Proteins*, Vol. II, CRC Press, Boca Raton, FL, 1984, pp. 229-253.
Verhoef, J., Codd, E.E., Burbach, J.P.H. and Witter, A., in Hancock, W.S. (Editor), *CRC Handbook of HPLC for the Separation of Amino Acids, Peptides and Proteins*, Vol. II, CRC Press, Boca Raton, FL, 1984, pp. 187-195.
Viswanatha, V., Larsen, B. and Hruby, V.J., *Tetrahedron*, 35 (1979) 1575-1580; *C.A.*, 92 (1980) 181 619b.
Voelter, W., Bauer, H., Fuchs, S. and Pietrzik, E., *J. Chromatogr.*, 153 (1978) 433-442.
Von Arx, E. and Faupel, M., *J. Chromatogr.*, 154 (1978) 68-72.
Vuolteenaho, O. and Leppäluoto, J., in Hancock, W.S. (Editor), *CRC Handbook of HPLC for the Separation of Amino Acids, Peptides and Proteins*, Vol. II, CRC Press, Boca Raton, FL, 1984, pp. 205-212.
Wagman, G.H. and Weinstein, M.J., *Chromatography of Antibiotics (Journal of Chromatography Library*, Vol. 26), Elsevier, Amsterdam, New York, 1984, 510 pp.
Waterfield, M.D. and Scrace, G.T., in Hawk, G.L., Champlin, P.B., Hutton, R.F. and Mol, Ch. (Editors), *Biological/Biomedical Applications of Liquid Chromatography III (Chromatography Science Series*, Vol. 18), Marcel Dekker, New York, 1981, pp. 135-158; *C.A.*, 96 (1982) 118 335y.

Wehr, C.T., Correia, L. and Abbott, S.R., *J. Chromatogr. Sci.*, 20 (1982) 114-119.

Welinder, B.S., in Hancock, W.S. (Editor), *CRC Handbook of HPLC for the Separation of Amino Acids, Peptides and Proteins*, Vol. II, CRC Press, Boca Raton, FL, 1984, pp. 413-419.

Welinder, B.S. and Linde, S., in Hancock, W.S. (Editor), *CRC Handbook of HPLC for the Separation of Amino Acids, Peptides and Proteins*, Vol. II, CRC Press, Boca Raton, FL, 1984, pp. 357-362.

Wetlaufer, D.B., *Adv. Protein Chem.*, 17 (1962) 303-390.

Wideman, J., Brink, L. and Stein, S., *Anal. Biochem.*, 86 (1978) 670-678.

Wilson, K.J., Berchtold, M., Honegger, A. and Hughes, G.J., in Lottspeich, F., Henschen, A. and Hupe, K.-P. (Editors), *High Performance Chromatography in Protein and Peptide Chemistry*, Walter de Gruyter, Berlin, 1981a, pp. 159-174; *C.A.*, 97 (1982) 88 030w.

Wilson, K.J., Honegger, A., Stoetzel, R.P. and Hughes, G.J., *Biochem. J.*, 199 (1981b) 31-41.

Wilson, K.J., Van Wieringen, E., Klauser, S., Berchtold, M.W. and Hughes, G.J., *J. Chromatogr.*, 237 (1982) 407-416.

Wing-Kin Fong, G. and Grushka, E., *J. Chromatogr.*, 142 (1977) 299-309.

Wing-Kin Fong, G. and Grushka, E., *Anal. Chem.*, 50 (1978) 1154-1161.

Yamada, H. and Imoto, T., in Hancock, W.S. (Editor), *CRC Handbook of HPLC for the Separation of Amino Acids, Peptides and Proteins*, Vol. I, CRC Press, Boca Raton, FL, 1984, pp. 167-174.

Yoshida, H. and Imai, H., *Tanpakushitsu Kakusan Koso*, 26 (1981) 1135-1141; *C.A.*, 95 (1981) 93 123s.

Yu, T.J., Schwartz, H., Giese, R.W., Karger, B.L. and Vorous, P., *J. Chromatogr.*, 218 (1981) 519-533.

Zaoral, M., in Zaoral, M., Havlas, Z., Mikeš, O. and Procházka, Ž. (Editors), *Synhtetic Immunomodulars and Vaccines (Proceedings of the International Symposium, Třeboň, Sept. 1985)*, Institute of Organic Chemistry and Biochemistry, Czechoslovak Academy of Sciences, Prague, 1986, pp. 1-2 and 317-322.

Chapter 10

NUCLEIC ACIDS, POLYNUCLEOTIDES AND OLIGONUCLEOTIDES

10.1 INTRODUCTION

The application of liquid chromatography (LC) to the separation of nucleic acid constituents in the last three to four decades was the key factor in the elucidation of the structure of nucleic acids. In the last 15 years this method has been predominantly used in the form of high-performance liquid chromatography (HPLC). Our knowledge of the structure of nucleic acids, their fragments and also their numerous specific interactions studied in modern branches of molecular biology, such as genetics and genetic engineering, has been greatly extended and deepened, especially in the last decade. This was due to the development of modern techniques in new specialized areas, such as nucleic acid sequencing, gene cloning, restriction enzymology and separation of restriction fragments. In addition to electrophoresis, HPLC methods have also contributed to the development of these modern techniques.

Polynucleotide structures are represented by linear polymeric molecules. Monomeric nucleotides are linked by phosphodiester bonds between the 3'- and 5'-positions of the ribose components of adjacent residues. If only one type of base is present in synthetic homopolymers, then the chromatographic behaviour is essentially a function of chain length. However, natural polynucleotide chains are heteropolymers, because they usually contain four types of pyrimidine or purine bases in different positions. The most common pyrimidine bases are uracil, thymine and cytosine and the most common purine bases are adenine, guanine and hypoxanthine. Some minor bases are also found in RNAs. The separation of heteropolymers is based not only on size, but also on the relative purine/pyrimidine contents and on the nature of the terminating groups; many competing interactions (such as hydrogen bonding and hydrophobic interactions) influence the separation.

The pyrimidine and purine bases have aromatic character and therefore show strong UV absorbance, which can be used for the simple and effective detection of nucleic acids or their fragments. In addition, these aromatic rings are also the source of hydrophobic interactions of oligo- and polynucleotide chains with the chromatographic support, which influence reversed-phase chromatographic (RPC) separations. The simultaneous presence of nucleobases and the secondary phosphoric acid groups determines the acid-base interactions which are important for ion-exchange chromatography (IEC) separations and others.

Nucleic acids (polynucleotides) possess a considerable degree of secondary and tertiary structure that can restrict the number of structural components available for interaction with chromatographic supports, e.g., the number of ionogenic phosphate groups that can interact with the anion exchanger may be diminished. The solution conditions (the presence and concentration of some important ions, or temperature) can affect the spatial structure of nucleic acids and hence influence the number of groups available for interactions. For example, the sorption of a nucleic acid on an anion exchanger need not be proportional to the molecular weight of the biopolymer (i.e., to the analytical number of phosphate groups).

For the separation of low-molecular-weight components, i.e., nucleobases, nucleosides and nucleotides, partition and ion-exchange chromatography were most often used at the beginning of this type of research. Later, reversed-phase chromatography proved to be a very effective method. All these methods have also been applied to the separation of low-molecular-weight oligonucleotides and the separation modes were clear and readily understandable. However, for complicated polynucleotides special, more effective chromatographic separation methods were sought and the detailed separation mechanism was found not to be a simple process.

About two decades ago, reversed-phase columns were introduced for the separation of tRNAs (Kelmers et al., 1965). Small particles of inert material (such as diatomaceous earth or resin beads) were coated with basic quaternary aliphatic ammonium derivatives of high molecular weight, insoluble in water. The latter had the function of an active extractant (cf., Sections 4.4.4 and 4.4.5). Both ionic and hydrophobic interactions between the solutes and coating material were involved in the separation process. In designing RPC systems for tRNA separation, Kelmers et al. (1965) assumed that tRNAs could be considered as long-chain polyphosphates, and thus an anion-exchange type of column would be needed for the sorption and chromatographic separation (see the review by Kelmers et al., 1971).

Later, inorganic supports were replaced with polychlorotrifluoroethylene resin (RPC-5) and the separation was improved (Pearson et al., 1971); a shorter analysis time was then sufficient and scale-up for preparative separations was easier. This method found numerous applications in the separation of nucleic acids, which will be discussed in the following sections. The RPC-5 column was found to be very useful for obtaining gram amounts of purified tRNAs. This column also proved to be an excellent tool for analysing minute differences in the tRNA structures, e.g., tRNAs containing a common sequence with a difference in only one nucleotide were resolved using these columns (Kelmers et al., 1971).

Reversed-phase chromatography was originally discovered as a means of achieving separations of individual tRNA species. In spite of the fact that it was developed at the macromolecular level, it has been successfully applied to the separation of smaller molecules. This RPC has also been used to separate oligonucleotides and monomers (Egan, 1973; Roe et al., 1973; Singhal, 1973, 1974b).

Singhal (1974a) found that smaller nucleic acid components are separated on these reversed-phase columns in a manner essentially identical with that observed in the usual anion-exchange chromatography on a polystyrene matrix. The difference between this RPC and conventional IEC lay in the difference in the composition and physical character of the inert support holding the anion-exchange groups, which were quaternary ammonium derivatives in all instances. Since aryl residues of polystyrene ion exchangers are more hydrophobic than the alkyl chains of RPC columns and because the conventional ion-exchange material is produced as uniform small beads, improved resolutions have been observed for oligonucleotides and monomers on conventional ion exchangers (Singhal, 1974b). This was why Singhal et al. (1976) extended the use of such a polystyrene anion-exchange material from nucleic acid components to the separation of tRNAs; evidence was obtained that it was superior in several respects to the widely used RP systems. However, residual amines found in certain brands of polystyrene anion exchangers caused base-catalysed deesterification of the aminoacyl-tRNA bond and the yields were reduced (Kopper and Singhal, 1979).

Affinity chromatography has also been used to isolate a given tRNA from nineteen other species in one step, and this process was followed by the resolution of isoacceptors by a combination of RP- and IE-HPLC (Vakharia and Singhal, 1979). Singhal (1983) reviewed recent advantages in tRNA separation methods and compared polystyrene anion exchangers and the RPC-5 material for the HPLC of tRNAs.

RPC-5 column chromatography was established as a powerful technique for the preparation of nucleic acids and for their purification (Wells et al., 1980). However, because great problems were encountered with the lack of a commercially available suitable matrix, Thompson et al. (1983) developed a new packing material, designated RPC-5 ANALOG, which is commercially available and serves for the rapid column separation of nucleic acids and their fragments, without requiring high pressures. They reviewed the purification of nucleic acids by RPC-5 ANALOG chromatography (peristaltic and gravity-flow applications). This material has primarily the function of anion exchanger.

A practical introduction to these modern HPLC techniques was presented by Wehr (1980); his book contains a short description of the structure and properties of nucleic acid constituents, important from the point of view of chromatography. The HPLC of DNA was reviewed by Wells et al. (1980), Hillen and

B180

Wells (1983),Thompson et al. (1983) and Wells (1984); see also the NACS Applications Manual. The HPLC of tRNAs was dealt with by Singhal (1983). An overview
of nucleic acid research (important from the point of view of HPLC applications
in this field) was written by Brown (1984b) and an overview of the structures,
properties and chromatographic behaviour of nucleic acid constituents was
compiled by Scoble (1984).

10.2 NUCLEIC ACIDS AND HIGHER MOLECULAR WEIGHT POLYNUCLEOTIDE FRAGMENTS

10.2.1 Size exclusion chromatography and adsorption chromatography

Separation according to the effective size of a macromolecule (which is
roughly proportional to molecular weight) has often been used for the fractionation of biopolymers. Wehr and Abbott (1979) studied the high-speed steric exclusion chromatography of proteins and nucleic acids in the molecular weight range
13 500-340 000. MicroPak TSK Gel SW columns of different pore sizes (2000 SW
and 3000 SW) were compared and the effects of molecular shape and denaturation
elution were investigated. The dimensions of the columns were 30 cm x 7.5 mm I.D.
Aqueous solutions of dilute buffers were used for elution: 0.067 M KH_2PO_4 - 0.1 M
KCl - $6 \cdot 10^{-4}$ M sodium azide (pH 6.8). The flow-rate was 1.0 ml/min and the
temperature $30^{\circ}C$. The columns were calibrated with both standard proteins and
standard dextran polymers. Tyrosine-specific tRNA and N-formylmethionine-specific
tRNA from *Escherichia coli* and rabbit liver tRNA were chromatographed. The
influence of hydrodynamic volume on permeation was evident. Relatively small
RNAs (M_r = 25 000-30 000) were eluted from a TSK Gel 2000 SW column in the position of globular proteins with M_r = 60 000. This reflects a relatively linear
tRNA structure. Tyrosine and N-formylmethionine species were partially resolved
on the TSK Gel 3000 SW column. Comparison of the elution volumes with those of
dextran polymers indicated that the tRNAs were eluted as would linear dextrans
with M_r = 23 000 and 30 000. Chromatography of a tRNA extract from rabbit liver
revealed a major peak of M_r = 24 000 (determined from the dextran calibration
graph), another component of M_r = 30 000 and a smaller amount of material of
M_r > 100 000 (possibly ribosomal RNA). The resolved fractions were further
chromatographed on an ion exchanger (cf., Section 10.2.2).

Mueller et al. (1979) studied the size fractionation of double-stranded DNA
fragments [150-22 000 base pairs (bp) in size] by liquid-liquid chromatography.
The procedure made use of the fact that the partitioning of DNA in a polyethylene
glycol-dextran system is dependent on size and can be altered by alkali metal
cations. Cellulose or Celite particles (100-36 and 36-25 μm) used as a support
were wetted with a poly(ethylene glycol)-dextran-rich stationary phase. Separa-

tion was controlled through gradient elution using buffer mobile phases. The fractionation of digests produced by DpnII and HindII restriction endonucleases, as well as λ DNA digest produced by HindIII and EcoRI restriction endonucleases, were shown as examples.

Gurkin and Patel (1982) reviewed the aqueous gel-filtration chromatography of enzymes, proteins, oligosaccharides and nucleic acids. The properties of Fractogel TSK was described. The purification of mRNA (M_r = 300 000) from *Bombix mori* silkworm was mentioned, which was accomplished by Izumi and Tomino and published commercially in "Fractogel TSJ News No. 15". First poly(A)-RNA was isolated by affinity chromatography from the total body RNA and purified on a Fractogel HW 65 F column (90 cm x 2.5 cm I.D.) using an aqueous mobile phase of pH 6 consisting of 0.02 M sodium citrate - 5 mM EDTA - 0.5% SDS - 6 M urea. The flow-rate was 24 ml/h (a peristaltic pump was sufficient) and the temperature was 25°C. UV detection at 254 nm was used, and mRNA activity was measured. The results of the separation were comparable to those for sucrose gradient centrifugation of similar constituents. For the gel-filtration HPLC of DNA fragments, see also the paper by Himmel et al. (1982).

Kato et al. (1983b) systematically studied operational variables in the high-performance gel filtration of double-stranded DNA fragments and ribosomal and transfer RNAs. The separation ranges and resolutions on columns of TSK Gel 2000 SW, 3000 SW, 4000 SW and 5000 PW were investigated, together with the effects of ionic strength and flow-rate on retention. It was found that the columns could separate double-stranded DNA fragments of molecular weight up to ca $1 \cdot 10^6$ and rRNA up to $5 \cdot 10^6$ daltons. Elution was delayed as the eluent ionic strength was increased. An eluent ionic strength of 0.3-0.5 seemed appropriate in most instances. Similarly to other biopolymers, the resolution was greatly increased as the flow-rate decreased. Systems consisting of two TSK Gel 2000 SW, two 3000 SW, two 4000 SW and two 5000 PW 60 cm x 0.75 cm I.D. columns were employed. The usual eluent was 0.1 M phosphate buffer (pH 7.0) containing 0.1 M sodium chloride and 1 mM EDTA. In the study of the influence of ionic strength, 0.1 M Tris-HCl buffer (pH 7.5) containing 0.025-1.6 M sodium chloride and 1 mM EDTA was used. The flow-rate was 1 or 1/3 ml/min, except in the study of its effect (when it was varied from 1/15 to 1 ml/min). The injection volume was 0.1 ml, the sample concentration was 0.01-0.1% (w/v) and UV detection at 260 nm was used. Double-stranded DNA fragments were prepared by cleaving plasmid DNA pBR 322 with restriction endonuclease HaeIII or BstNI; the first digest contained 22 fragments of 7-587 bp and the second contained 6 fragments of 13-1857 bp. Molecular weights of the fragments were determined by multiplying their base pair numbers (bp) by 650. Total *Escherichia coli* RNA contained four

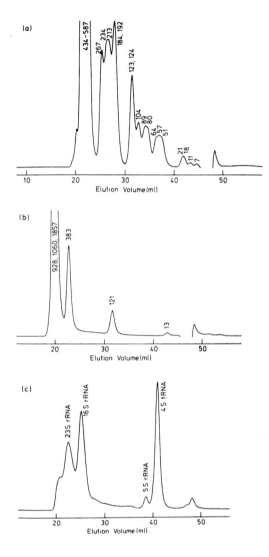

Fig. 10.1. Size exclusion chromatograms of (a) HaeIII-cleaved pBR322 DNA (b) BstNI-cleaved pBR322 DNA and (c) total *E. coli* RNA, obtained on a TSK Gel 4000 SW two-column system with 0.1 *M* phosphate buffer (pH 7) containing 0.1 *M* sodium chloride and 1 m*M* EDTA, at a flow-rate of 1 ml/min. Numbers on the peaks in (a) and (b) indicate the number of base pairs. Compare with Fig. 10.4. (Reprinted from Kato et al., 1983b.)

components: 4S tRNA (M_r = 25 000), 5S rRNA (39 000), 16S rRNA (560 000) and 23S rRNA (1 100 000). Some examples of the separations obtained are illustrated in Fig. 10.1.

The experiments on the size exclusion separation of proteins and nucleic acids on porous glass supports led to the observation of strong, partly irreversible sorption effects, which could be limited by the surface treatment of glass (cf., Mizutani, 1980a). Narihara et al. (1982) examined the fractionation of tRNA on siliconized porous glass (cf., Mizutani, 1980b) coated with trialkyl-methylammonium chloride (Adogen 464). Bovine liver tRNA was chromatographed on the above support using a sodium chloride gradient and transfer ribonucleic acids were eluted in the order $tRNA^{Met}$, $tRNA^{Val}$ and $tRNA^{Ser}$, and the recovery of tRNA from the glass column averaged 90%. Partially purified $tRNA^{Ser}$ (on benzoylated DEAE-cellulose) was isolated on the siliconized Adogen-coated porous glass. Mizutani and Narihara (1986) studied the adsorption chromatography of bovine liver tRNA, *E. coli* rRNA and calf thymus DNA on siliconized porous glass in 5 M NaCl-0.01 M Tris-HCl (pH 7.6) by lowering the NaCl concentration on the columns: tRNAs were eluted at 5-3 M NaCl in the order Pro, Val, Ile, Thr, Ser and Phe tRNA, unless the peaks were perfectly separated. RNA species in crude rRNA were also separated.

10.2.2 Ion-exchange chromatography

Nucleic acids and their higher molecular weight fragments are long linear polyanions and therefore suitable anion exchangers seemed to be the best material for their sorption and chromatographic separation. Singhal (1974a) reviewed the separation and analysis of nucleic acids and their constituents by ion-exclusion and ion-exchange chromatography. The major part of his chapter was devoted to nucleic acid constituents (nucleobases, nucleosides and nucleotides), and several examples of oligonucleotide separations on classic resinous or poly-saccharide ion exchangers were also discussed.

The separation of transfer ribonucleic acids on strongly basic polystyrene anion exchangers was systematically examined by Singhal et al. (1976). The separation was compared with the widely used reversed-phase chromatography (Kelmers et al. 1965). Transfer RNAs containing hydrophobic groups were adsorbed more strongly on a polystyrene anion exchanger and this type of support had twice the number of theoretical plates as RPC-5 material (the HETP of the polystyrene anion-exchange column for tRNA was 0.46 mm). Better resolution was found with the polystyrene anion exchanger as single peaks of $tRNA_2^{Glu}$ and $tRNA_1^{Phe}$ obtained from an RPC column gave multiple peaks on the polystyrene anion exchanger. Strong-base polystyrene anion-exchange chromatography was capable of separating tRNAs with minor structural differences. The recovery of tRNA from the polystyrene anion-exchange column was reported to be equal to or even better

than those obtained with RPC-5 by both UV absorption and radioactivity methods
(98% vs. 88% for the RPC column).

The anion exchanger used was preferentially Aminex A-28 (7-11 μm beads) (Bio-
Rad Labs., Richmond, CA, U.S.A.), but Aminex A-25 (17 μm) and A-27 were also
satisfactory (A-25 caused less back-pressure). The authors suggested that
probably any other brand of polystyrene strong anion exchanger of small and
uniform size which could tolerate extremes of temperature, pressure, pH and
organic solvents could be used for such work. Usually 200 μl (36 A_{260} units of
tRNA) were applied to the column (27 cm x 0.6 cm I.D.) and eluted with a linear
gradient of 0.5-0.75 M NaCl, 100 ml of each solution, at 22°C and a flow-rate
of 0.4 ml/min. The buffer used was 10 mM NaOAc - 10 mM MgCl$_2$ - 1 mM EDTA - 1 mM
Na$_2$S$_2$O$_7$ (pH 4.5). The resulting pressure was 245 p.s.i.

The influence of temperature was studied using [^{32}P]tRNAs (*E. coli*) with
carrier tRNAs and temperatures of 25 and 50°C. Although eluted under identical
conditions, the resolution was better at the lower temperature (tRNAs were more
sorbed to polystyrene at 25°C). Certain tRNA species could be desorbed from the
polystyrene column only by strong eluents, e.g., 1.5 M NaCl containing 5%
ethanol. Such tRNAs may contain strong hydrophobic groups [e.g., isopentenyladenine or "wye" (formerly "Y") base] in their exposed structures. All six
leucine tRNAs of *Escherichia coli* were resolved using Aminex chromatography,
and fine heterogeneity was found in the two major peaks. The polystyrene anion-
exchange column was also examined for the separation of "free" tRNA and the
aminoacyl tRNA. For example, 3.6 A_{260} units of tRNA$_1^{Phe}$ in 80 μl and 25 000 cpm
of [^{3}H]Phe-tRNA$_1^{Phe}$ (0.2 A_{260} units in 200 μl) were applied to the column and
first chromatographed using the above conditions, then a second linear gradient
was applied of 0.75-1.25 M NaCl (plus 10% ethanol), 50 ml of each, at a flow-
rate of 0.4 ml/min; the pressure was 250 p.s.i. and the temperature was 21°C.
The recovery of Phe-tRNA was 92%. "Charging" of tRNA$_1^{Phe}$ with phenylalanine
increased its adsorption on the anion exchanger. Such strongly adsorbed species
require dilute ethanol for elution. Differences between tRNAs from early and
late stages of the proliferating lymphocytes were demonstrated using Aminex
chromatography, and also differences in tyrosine isoacceptor tRNAs from normal
and tumour cells. Singhal et al. (1976) found the separation of tRNAs on the
Aminex A-28 columns to be superior to RPC in many ways.

Kopper et al. (1979) reviewed the problem of the stability and separation of
aminoacyl-transfer RNAs on polystyrene anion exchangers and RPC-5 chromatographic
columns. They used lysyl-tRNA for comparative studies and, in addition to an
RPC-5 column, they chromatographed tRNA on Aminex A-25 (X8, 18 μm), A-28 (X8, 9 μm),
Hamilton HA-X8 (9 μm) and Durrum DA-X12-11 (9 μm). The different yields of the
separated components were explained by the chemical deacylation of aminoacyl-

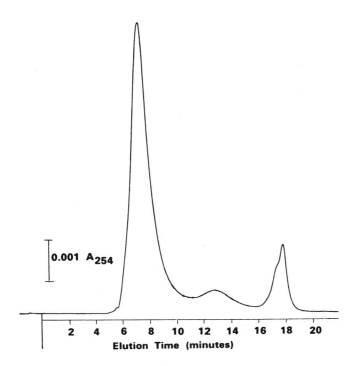

Fig. 10.2. Separation of rabbit liver tRNA extract on the column of MicroPak MAX-500. The extract was prepared by the extraction with phenol in acetate buffer (pH 4.5) followed by chromatography on DEAE-cellulose. Conditions on MicroPak MAX-500 column: solvent A, 0.1 M Tris, (pH 6.8); solvent B, 0.1 M Tris (pH 6.8) + 1 M NaCl. Gradient, 0% B, 5 min, then 0-45% B in 15 min; flow-rate, 0.5 ml/min; temperature, 30°C. (Reprinted from Wehr and Abbott, 1979.)

tRNAs catalysed by the residual amines present in certain brands of polystyrene anion exchangers. The catalytic process was influenced by the total structure of the chromatographic supports. The kinetics of deacylation in the mixtures of ion exchangers with lysyl-tRNA solution were also measured. Enhanced de-esterification was noted after prolonged exposure to Aminex A-25 and Hamilton HA-8. In contrast, RPC-5, Aminex A-28 and Durrum DA-12 did not disrupt the aminoacyl bond more than the control reaction.

Wehr and Abbott (1979) tested anion-exchange chromatography in the purification of N-formylmethionine tRNA and tyrosine tRNA, which eluted as a single peak in SEC (cf., Section 10.2.1) and on a MicroPak MAX-500 column resolved into at least three UV-absorbing species. The conditions were the same as in Fig. 10.2, except for the flow-rate (1.0 ml/min). Fig. 10.2 illustrates the separation of a rabbit liver tRNA extract on MicroPak MAX-500.

Kato et al. (1983a) described a new ion exchanger for the separation of proteins and nucleic acids. TSK Gel IEX-645 (Toyo Soda, Tokyo, Japan) was

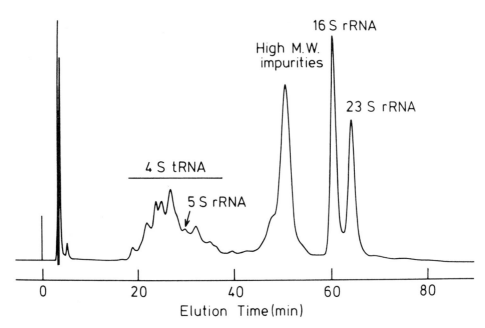

Fig. 10.3. Chromatogram of total *E. coli* RNA obtained by IE-HPLC on IEX-645 DEAE with a 300-min linear gradient from 0.1 *M* Tris-HCl buffer (pH 7.6) containing 0.3 *M* sodium chloride to 0.1 *M* Tris-HCl buffer (pH 7.6) containing 1 *M* sodium chloride. Other conditions are given in the text. (Reprinted from Kato et al., 1983a.)

developed by introducing diethylaminoethyl groups into the hydrophilic polymer TSK Gel 5000 PW (used in the rapid gel-permeation chromatography of macromolecules). Being a fully organic support, it is stable in alkaline media and can be used over a wide pH range. This ion exchanger is homoionic with a functional group of $pK_a \approx 11.5$ and shows little buffering capacity at pH < 10, which has advantages in column chromatography and makes the separation more lucid. The nominal capacity of the ion exchanger is 0.12 mequiv./ml. The exclusion limit for proteins is much higher than 10^6 and for polyethylene glycol it is ca. 10^6.

In addition to protein chromatography, Kato et al. (1983a) also described the separation of nucleic acids and their fragments. A column of 15 x 0.6 cm I.D. was used with linear gradient elution at a flow-rate of 1 ml/min at 25°C, with UV detection at 260 nm. Total *E. coli* RNA contained 4S tRNA (M_r = 23 000-27 000), 5S rRNA (39 000), 16S rRNA (560 000) and 23S rRNA (1 100 000). The separation is illustrated in Fig. 10.3. Peaks were collected and assigned by SE-HPLC according to Kato et al. (1983b) (cf., Section 10.2.1). The same material could not be separated well on a silica-based IEX-645 DEA SIL column. HaeIII-cleaved pBR322 DNA contained 22 double-stranded restriction fragments of 7, 11, 18, 21,

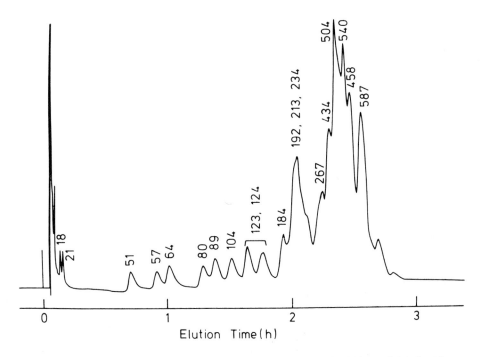

Fig. 10.4. Chromatogram of HaeIII-cleaved pBR322 DNA (0.027 mg in 0.1 ml) obtained by IE-HPLC on IEX-645 DEAE with a 300-min linear gradient from 0.1 M Tris-HCl buffer (pH = 7.6) containing 0.35 M sodium chloride to 0.1 M Tris-HCl buffer (pH 7.6) containing 0.5 M sodium chloride. Other conditions are given in the text. Numbers on the peaks indicate the number of base pairs. Compare with Fig. 10.1. (Reprinted from Kato et al., 1983a.)

51, 57, 64, 80, 89, 104, 123, 124, 184, 192, 213, 234, 267, 434, 458, 504, 540 and 587 base pairs; for their separation see Fig. 10.4. Peaks were assigned by polyacrylamide slab gel electrophoresis. The components were mostly eluted in order of increasing base pair number. The double-stranded DNA fragments were mainly separated according to the chain length.

Singhal (1983) surveyed recent advances in the HPLC of tRNAs. Polystyrene and RPC-5 anion exchangers were compared with respect to their ability to resolve tRNAs without a loss of the aminoacyl-tRNA bond. The polystyrene anion exchanger Aminex A-28 was considered to be the best, with the effluent being monitored at two wavelengths (254 and 280 nm). The polystyrene anion exchanger had a higher resolving power. The separation of six $[^{14}C]$leucyl-tRNA isoacceptors showed different orders of elution: 1, 2, 2', 3, 4 and 5 on the RPC-5 column and 1, 3, 2', 2, 4, 5 on the Aminex A-28 column. Singhal (1983) presented a general scheme for resolving queuosine-containing tRNA of mammalian origin, involving polystyrene anion exchange and RPC-5 HPLC, in addition to lectin affinity chromatography. New approaches to tRNA separation were discussed.

The RPC 1-5 methods are based mainly on IEC. However, because special supports have been developed for this purpose, they are dealt with in Section 10.2.4.

10.2.3 *Reversed-phase and hydrophobic interaction chromatography*

By the end of the 1960s, the "true" RP-HPLC of nucleic acids was limited to the separation of nucleic acid constituents and short oligonucleotides (cf., Section 10.3). In 1979 Hjertén et al. described a new support, naphthoyl-Sepharose, for tRNA fractionation, and in 1982 Hjertén et al. studied the high- or intermediate-performance hydrophobic interaction chromatography (HIC) of biopolymers (proteins) and also tested spherical silica derivatized with short alkyl chains (e.g., hexyl groups) for the HIC of tRNAs. Desorption was achieved by decreasing the ionic strength of the buffer in the absence of organic solvents (to avoid denaturation). Transfer RNAs, which are less hydrophobic than proteins, can be fractionated under these conditions, provided that short alkyl chains are used as ligands. In hexylsilica (Spherisorb, 5 μm) (Deeside Industrial Estate, Queensferry, Clwyd, U.K.), the silica particles were suspended in acetone and packed into a stainless-steel tube (300 mm x 4 mm I.D.) at a pressure of 300 atm in chloroform and equilibrated with 0.02 M sodium phosphate (pH 6.8) containing 0.2 M Na_2SO_4 (buffer R_1). About 10 μl of a tRNA ($E.$ $coli$) solution ($A_{260}^{1 cm}$ = 509) were applied. Buffer R_2 was 0.02 M sodium phosphate (pH 6.8). Two linear gradients, I and II, were used for elution: I, from 0% to 60% R_2 in 40 min; II, from 60% to 100% R_2 in 80 min. The flow-rate was 0.5 ml/min. A complicated but reproducible pattern was obtained with UV absorbance measurement at 260 nm. In another experiment with a different preparation of tRNA ($A_{280}^{1 cm}$ = 30) only one gradient was used starting with 0.02 M sodium phosphate buffer (pH 6.8) containing 0.8 M ammonium sulphate (buffer T_1), from 100% T_1 to 100% R_2 in 60 min; a different, complicated pattern was obtained.

Nguyen et al. (1982) reported preliminary experiments with partial resolution of an artificial mixture of yeast tRNA from globin mRNA by ion-paired reversed-phase HPLC using commercial available C_{18} columns and methanol in the presence of tetrabutylammonium counter ion for elution. Two Waters Assoc. μBondapak C_{18} columns (each 30 cm x 0.4 cm I.D.) were connected in series. A 40 ml methanol to water linear gradient was generated immediately after injection; each solvent contained 5 mM PIC A reagent (tetrabutylammonium phosphate, Waters Assoc.). Samples containing 5-25 μl (0.1-0.5 μg) of radiolabelled RNA were applied, the flow-rate was 1 ml/min. The operating pressure 55-69 bar and 45 fractions were taken. Aliquots of 200 μl were taken from each fraction for analysis by poly-acrylamide gel electrophoresis with autoradiographic detection. 4S yeast transfer RNA (80 nucleotides) was well separated from 9S rabbit globin mRNA

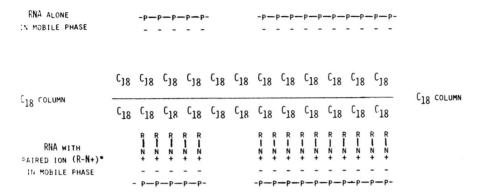

Fig. 10.5. Idealized view on the interaction of RNA with the C_{18} column matrix in the absence and presence of tetrabutyl counter ion. Above the C_{18} column two "RNA molecules" are shown; one has five phosphates and the second ten phosphates, each with a negative charge. Below the column these two molecules are shown paired with the counter ions containing the tetrabutyl group (R), which can interact with the C_{18} residues of the column. In this idealized view, the non-polar character contributed by the paired ion is proportional to the number of phosphates, and therefore the molecule with five phosphates would be eluted from the column before the longer molecule. (Reprinted from Nguyen et al., 1982.)

(ca. 650 nucleotides). In another experiment, RNA was isolated from a culture of *E. coli* growing in the logarithmic phase and ^{32}pCp labelled sample of this natural mixture was chromatographed as previously. Here also 45 fractions were taken and 200 µl aliquots were analysed. Clearly, molecules of increasing molecular weight were eluted with increasing time.

In their discussion Nguyen et al. (1982) explained the separation principle in detail. There is only a low probability that RNA molecules would interact with non-polar octadecyl residues of the RPC column. In order to promote this interaction, tetrabutylammonium counter ion was added to the mobile phase. The reason for the separation of shorter and longer molecules is explained in the legend to Fig. 10.5. The results obtained in the described experiments indicate that RNA molecules larger than tRNA can be fractionated by a simple technique employing commercially available "true" RPC columns. The separation was comparable to that obtained by sucrose gradient centrifugation, but it is more advantageous: the chromatographic process (40 min) is faster than centrifugation (several hours) and both solvents and ion pairs are volatile, so that subsequent ethanol precipitation is not necessary. This method opened the way for further methodological improvements in the field of rapid separations of nucleic acids.

Simonian and Capp (1983) described the purification of an enriched trans-latable poly(A)-messenger ribonucleic acid by RP-HPLC. The isolation was due to the presence of poly(A) residue on the 3'-terminus of the mRNA, so that RNA that lacks poly(A) can be well separated by RP-HPLC. Over 90% of the eukaryotic mRNAs (with the exception of histone mRNAs) that enter cytoplasm contain this 3'-polyadenylate segment of ca. 200 residues (cf., Darnell, 1982), and enrichment of their preparations can be achieved by affinity chromatography on poly(U)-Sepharose or oligo(dT)-cellulose; for references, see the original paper. RP-HPLC of the above-mentioned preparation and other RNA samples was performed on a 150 cm x 4.6 mm I.D. column packed with SynChropak RP-P C_8 (6.5 μm, 300 Å pore size) (SynChrom, Linden, IN, U.S.A.). The RNA samples dissolved in water were treated at 65°C for 4 min immediately prior to injection. On injection a linear gradient of acetonitrile in 0.01 M sodium sulphate (pH 6.0, 37°C, flow-rate 1 ml/min) was started. UV absorbance measurement at 260 nm was used for detection. Purified RNA fractions were tested for biological activity by in vitro translation methods in a rabbit reticulocyte lysate system. Using the described RP-HPLC method, highly purified poly(A)-mRNA was isolated from an enriched poly(A)-RNA preparation within 30 min in quantitative yield, and purified mRNA retained its biological activity.

Other specialized methods designated as RPC but based on a combination of various separation principles are described in Section 10.2.4.

The RPC of nucleic acid fragments was reviewed by Scoble and Brown (1983).

10.2.4 SPECIAL METHODS

10.2.4.1 RPC 1-5 chromatography of nucleic acids and their fragments

There is no doubt that the so-called reversed-phase chromatography (RPC 1-5) is one of the most important techniques for the separation not only of tRNAs but also for large and small DNA fragments. The method originated from a paper by Kelmers et al. (1965) and was first described in detail by Kelmers et al. (1971b) (cf., Kelmers et al., 1981b). Although termed "reversed-phase chromatography", the separations are actually based on a mixed reversed-phase and ion-exchange mechanism. The first RPC columns were based on inert diatomaceous earth support material of relatively high surface area, which was acid washed and then treated with dimethylchlorosilane to yield a hydrophobic surface with a minimum surface activity. This support was covered with higher molecular weight quaternary ammonium salts, and according to the chemical quality of this salt (the so-called extractant) and preparation techniques, the following RPC systems were distinguished:

RPC-1	Dimethyldicocoammonium chloride*	(Isoamylacetate solvent)
RPC-2	Tricaprylmethylammonium chloride	(Tetrafluorotetrachloropropane)
RPC-3	Trioctylpropylammonium bromide	(Dimethyldichlorosilane treated)
RPC-4	Dimethyldicocoammonium chloride*	(Dimethyldichlorosilane treated)

The RPC-4 system was found to be preferable to the others and has often been used for the separation of tRNAs. The sample was applied to the column in the chloride form in dilute sodium chloride solution, and chloride ions bound to quaternary ammonium extractant exchanged for the phosphate anion sites of the tRNA. This was the sorption process. At higher sodium chloride concentrations, mass action favoured chloride binding with the quaternary ammonium compounds and the tRNAs were thus released to the effluent. Gradient elution gave the best separation results. The elution of tRNA from RPC columns was controlled by the phosphate groups available for interaction with quaternary ammonium exchange sites. This concept supposed that the more tightly structured tRNAs would be eluted first and the more flexible, loosely structured tRNAs would be eluted later at higher NaCl concentrations; the chromatographic results were in agreement with this concept. The presence of magnesium ions (e.g. at 10 mM concentration), which is known to increase the tertiary structure of tRNAs, makes the tRNAs easier to elute. The addition or removal of Mg^{2+} ions can be used to shift the chromatographic position of neighbouring tRNAs. Also, the pH of the eluent affects the elution sequence of the tRNAs; changes in the number of protonated bases (depending on pH) modulate the ionic interactions between tRNA and the functional ion-exchange groups of the RPC column. The preparation of RPC-1 to RPC-4 columns and many important technical details and examples of separations were throuroughly described in a review by Kelmers et al. (1971b). However, RPC-4 columns were soon replaced by RPC-5 columns.

In 1971, Pearson et al. replaced diatomaceous earth by a new support material, Plaskon CTFE (polychlorotrifluoroethylene resin, particle size 10 μm), manufactured by Allied Chemical (Morristown, NJ, U.S.A.). The stationary phase, Adogen 464 (Ashland Chemical, Columbus, OH, U.S.A.), was trialkylmethylammonium chloride with a predominant alkyl chain length of C_8-C_{10}. Improved separation of tRNA isoacceptors was obtained with shorter analysis times and easier scale-up for preparative separations. The preparation of a 240 cm x 2 cm I.D. RPC-5 column was described by Wehr (1980): 300 g of Plaskon powder were combined with 12 ml of Adogen 464 dissolved in 600 ml of chloroform. After shaking for 2 h the slurry was evaporated nearly to dryness on a glass tray and in a closed

---

* The term "coco" referred to straight-chain saturated alkyl groups derived from coconut oil; they ranged from C_8 to C_{18}, but C_{12} and C_{14} were the major components. The base (Adogen 462) was produced by Ashland Chemical (Columbus, OH, U.S.A.).

container tumbled mechanically for 2 h to ensure an even distribution of the
Adogen on the Plaskon support. The material was then passed through a 100-mesh
sieve to reduce clumping. The RPC-5 packing was subsequently slurried with
aqueous buffer and poured into the column, which was compacted by pumping and
topped off as necessary.

In their study of the separation of tRNAs on polystyrene anion exchangers,
Singhal et al. (1976) described examples of RPC-5 separations, based on packings
consisting of beads of polychlorotrifluoroethylene (Kel-F) coated with a
quaternary ammonium derivative [methyltrialkyl(C_8-C_{10})ammonium chloride] prepared by the standard procedures of Kelmers et al. (1971a,b) (see also Kelmers
et al., 1981a, and Fig. 4.12). In addition to Kel-F, polytetrafluoroethylene
(Teflon) can also be used as a solid support for the coating.

Singhal et al. (1976) chromatographed *E. coli* tRNAs on such an RPC column
(25 cm x 0.6 cm I.D.). In 200 µl, 36 A_{260} units were applied and eluted with a
linear gradient of 0.5-0.75 *M* NaCl, 100 ml of each solution, at $22^{\circ}C$ and a flow-
rate of 0.4 ml/min. The buffer was 10 m*M* NaOAc, 10 m*M* $MgCl_2$, 1 m*M* EDTA and 1 m*M*
$Na_2S_2O_7$ (pH 4.5). The resultant pressure was 105 p.s.i. In another experiment,
two A_{260} units of [^{14}C]leucyl-tRNA (*E. coli*) were eluted from a 30 cm x 0.6 cm
I.D. RPC column using a linear gradient of 0.5-0.8 *M* NaCl (100 ml each) at 0.5
ml/min, 100 p.s.i. and $37^{\circ}C$. The buffer was the same as above. Also, co-chromatography of tRNA synthesized in early (^3H) and late (^{14}C) stages of proliferating human lymphocytes was achieved on an RPC column (30 cm x 0.6 cm I.D.)
with a linear gradient of 0.5-0.75 *M* NaCl (100 ml each) at 0.4 ml/min, 150 p.s.i.
and $37^{\circ}C$; the buffer composition was the same as above (see Fig. 6 in the
original paper). The co-chromatography of mouse liver and mouse plasma-cell
tumour [^3H]tyrosine-tRNAs was also carried out on a 30 cm x 0.6 cm I.D. RPC
column with a linear gradient of 0.45-0.55 *M* NaCl (100 ml each) at $25^{\circ}C$ and with
the buffer composition described above (see Fig. 7 in the original paper). In
all these examples, the RPC was accompanied by and compared with Aminex A-28
(polystyrene anion exchanger) chromatography. The authors found that the RPC
columns cannot operate at low temperatures or high pH or with organic solvents
(alkylamine is desorbed). RPC columns have a greater capacity (5 A_{260} units of
tRNA can be applied to 1 ml of Aminex resin, compared with 32 A_{260} units/ml to
an RPC column).

Dion and Cedergren (1978) also studied the HPLC of tRNAs. A rapid separation
of tRNA isoacceptor species on RPC-5 resin was described. The work was an
extension of the RPC methods developed by Pearson et al. (1971) and Kelmers and
Heatherly (1971). Dion and Cedergren (1978) reduced the time of analysis to
20 min. They used a precision-bore stainless-steel column (35 cm x 4 mm I.D.)
and modern HPLC equipment. The RPC-5 resin consisted of inert polychlorotri-

fluoroethylene beads of 10 μm diameter, on which a pellicular layer of trioctyl-
ammonium chloride (Adogen 464) had been adsorbed; the support was prepared
according to method B of Pearson et al. (1971). The column was packed by either
a dry or a wet method and washed with 0.01 M sodium acetate buffer (pH 4.5)
containing 0.4 M NaCl and 0.01 M MgCl$_2$. Prior to use, the resin was equilibrated
with the same acetate buffer containing NaCl at the molarity used to start the
salt gradient. The aminoacyl-tRNA was dissolved in 200 μl of the starting buffer
and after injection it was washed with the starting buffer at 4 ml/min (the
pressure was 1000-3000 p.s.i., depending on the age of the column). After
washing, a linear salt gradient (generally between 60 and 120 ml total) in the
acetate-magnesium buffer was started.

Three examples of separations can be given: (1) A 5 A_{260} unit sample of
tRNA (*E. coli*) was aminoacylated with 10 nmol of [^{14}C]serine of 50 mCi/mmol.
The sample (total radioactivity 12 000 cpm) was dissolved in 200 μl of 0.01 M
sodium acetate buffer at pH 4.5 containing 0.01 M MgCl$_2$ and 0.4 M NaCl. The salt
gradient was composed of 150 ml of 0.4-1.0 M NaCl in 0.01 M sodium acetate and
0.01 M MgCl$_2$ at pH 4.5, and the flow-rate was 2 ml/min (pressure 900 p.s.i.).
The radioactivity in the effluent was counted in the flow-cell counter. Four
peaks appeared (see Fig. 2 of the original paper). (2) A 4.0 A_{260} unit sample
of tRNA from *Klebsiella aerogenes* was aminoacylated with 10 nmol of [^{14}C]ty-
rosine of 420 mCi/mmol and dissolved in 200 μl of the starting buffer (total
radioactivity 32 000 cmp). The salt gradient was 120 ml of 0.4-1.2 M NaCl in
0.01 M sodium acetate and 0.01 M MgCl$_2$ (pH 4.5). The flow-rate was 4 ml/min
and the pressure 1400 p.s.i. The effluent was counted in the flow-cell counter.
Three well separated peaks appeared (see Fig. 4 of the original paper). (3)
A 2 A_{260} unit sample of tRNA from aerobically or photosynthetically grown
Rhodospirillum rubrum was aminoacylated with 1.25 nmol of [^{14}C]phenylalanine
of 450 mCi/mmole and dissolved in the starting buffer containing 0.4 M NaCl.
The salt gradient was 120 ml of 0.4-1.2 M NaCl in 0.01 M sodium acetate and
0.01 M MgCl$_2$ at pH 4.5. The flow-rate was 4 ml/min and the pressure 1400 p.s.i.
Photosynthetic and aerobic tRNA were well separated (see Fig. 5 of the original
paper).

Kopper and Singhal (1979) studied the stability and separation of aminoacyl-
tRNAs and, in addition to polystyrene anion exchangers, they also used RPC-5
columns. For the separation of rat liver lysyl-tRNA a 50 cm x 0.64 cm I.D. RPC-5
column was used. About 0.5 A_{260} unit of [^{3}H]lysyl-tRNA was applied and eluted
at 25°C with a 400 ml linear gradient of 0.50-0.60 M NaCl in 10 mM sodium
acetate - 10 mM magnesium acetate - 1 mM EDTA - 3 mM 2-mercaptoethanol (pH4.5) at
a flow-rate of 0.66 ml/min and a pressure of 200 p.s.i. Fractions of 2 ml were

collected and the radioactivity was determined. Two well separated peaks were
obtained (see Fig. 1b in the original paper). If pH 7.4 was used for the chromatography instead of pH 4.5, enhanced deacylation of the lysyl-tRNA ester bond
occurred in the RPC-5 column. Other examples of the RPC-5 chromatography of
tRNAs were given by Wehr (1980). The disadvantage of this method is the short
column lifetime, e.g., only ten chromatographic runs caused by the loss of
coated stationary phase, especially at low ionic strength (< 0.05 M) or on
addition of an organic modifier to the mobile phase. The use of an RPC-5 guard
column (or saturating the mobile phase with Adogen 464) and also operating at
higher pressures (up to 3000 p.s.i.) results in a longer column lifetime, in
addition to acceleration of the analysis (Dion and Cedergren, 1978).

RPC-5 chromatography has also been used for the separation of oligonucleotides
(cf., Section 10.3) and DNA restriction fragments. RPC-5 column chromatography
of DNA fragments was described in detail in the review by Wells et al. (1980).
This type of separation can be used for the high-resolution fractionation of
single-stranded DNA or RNA oligonucleotides, double-stranded DNA restriction
fragments and the complementary strands of DNA restriction fragments. Milligram
amounts of nucleic acids can be fractionated. In some instances this method can
completely resolve duplex restriction fragments of the same size.

For the separation of duplex restriction fragments, Wells et al. (1980)
described inexpensive analytical equipment (for the fractionation of 10-100 µg
of DNA) and also preparative equipment. RPC-5 was purchased from Miles Laboratory (Elkhart, IN, U.S.A.), but this material is no longer commercially available. RPC-5 could be easily prepared from Plascon 2300 (cf., Pearson et al.,
1971). Before packing, the RPC resin was suspended in 5 volumes of low-concentra-
tion elution mixture containing 0.06% sodium azide. The solution should have a
relatively high ionic strength to prevent stripping of the Adogen 464 from the
Plascon 2300. The suspension was stored in this solution and, just prior to
packing, it was shaken or stirred for several hours and degassed. It was then
packed into a column using a modified slurry method. The eluting solutions for
the separation of restriction fragments were based on the salt concentration
gradient (KCl, NaCl, sodium acetate, potassium acetate). All the salt solutions
contained 10 mM Tris buffer of the appropriate pH, 2 mM sodium thiosulphate and
10 mM EDTA (the presence of EDTA protects the DNA from possible nucleolytic
degradation and reduces the back-pressure). The restriction enzyme digest was
extracted with phenol to remove protein and with diethyl ether to remove phenol.
The DNA solution was dialysed into 10 mM Tris-HCl (pH 6.8), 2 mM sodium thiosulphate and 0.01 M EDTA. The whole chromatographic and auxiliary equipment was
washed before used with 0.06% sodium azide to prevent bacterial contamination.

The fraction collector was kept in the cold (4°C). The application of 1-10 mg of DNA per 30 ml of resin was recommended. The gradient was applied immediately at a flow-rate of 40-100 ml/h·cm^2 (5 p.s.i./cm bed). A 60-90% recovery of the loaded DNA was generally obtained. The column could be re-used many times and regenerated (without stripping the Adogen 464) by washing with 0.2 M NaOH plus 0.2 M salt. After many uses it showed a decreased capacity and recoating (Pearson et al., 1971) was necessary.

For the separation of complementary strands of DNA restriction fragments, a strongly alkaline medium (e.g., pH 12.2) should be used (Eshaghpour and Crothers, 1978). The fractionation of purified fragments under alkaline denaturating conditions yields two elution peaks and evidence can be provided that the two peaks are indeed complementary strands. Wells et al. (1980) gave an example of such a separation. The 210 bp (base pairs) HindII restriction fragment from φX174 replicative form DNA was isolated by RPC-5 chromatography in a separate experiment and dialysed against 0.25 M KCl-5 mM Tris (pH 7.4) - 0.5 mM EDTA. The pH of the solution was adjusted to 12.5 (using 1 M NaOH) and 7.5 µg of the fragment in 0.2 ml was loaded on to a 200 mm x 1.5 mm I.D. RPC-5 column equilibrated with 0.25 M KCl-12 mM NaOH (pH 12.5). After washing with 2 ml of the equilibration buffer the column was eluted with a 20-ml linear gradient of 0.90-0.95 M KCl in 12 mM NaOH (pH 12.5) at a flow-rate of 0.23 ml/min and a pressure of 150 p.s.i. Continuous monitoring at 265 nm was used for detection. Fractions of 0.2 ml were collected. The strands were perfectly separated in the form of two peaks eluted between 0.92 and 0.93 M KCl (see Fig. 6 of the cited paper). All solutions were degased before use and alkaline solutions were protected by a nitrogen atmosphere to maintain the high pH.

Wells et al. (1980) discussed the prospects of RPC-5 column chromatography and its potential uses.

Hillen et al. (1981) described the preparation of milligram amounts of 21 homogeneous DNA restriction fragments ranging from 12 to 880 bp, using recombinant methods to construct plasmids with multiple inserts of a desired fragment, together with an improved procedure for the selective purification of DNA restriction fragments from gram amounts of plasmid DNA, with final purification by RPC-5 column chromatography. The final yields from 1 kg of wet-packed cells varied from 0.6 mg for the 12-bp fragment to 87.5 mg for the 301-bp fragment. The recombinant plasmid containing 301 bp and dealt with below as an example contained *lac* operator-promoter region of *Escherichia coli* (cf., Hardies et al., 1979).

After cell lysis (using lysozyme) and phenol extraction, the nucleic acid was treated with RNase (to destroy the ribonucleic acids) and the purified plasmid DNA was cleaved with EcoRI. The digest was selectively precipitated with poly-

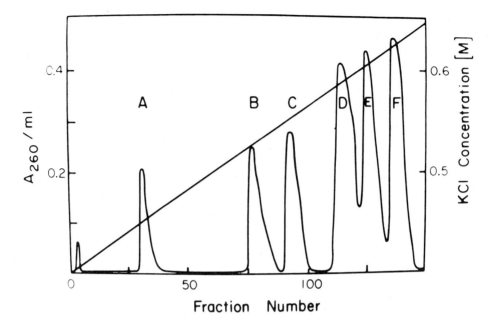

Fig. 10.6. RPC-5 elution profile of the HaeIII/AluI double digest of the 301-bp fragment. DNA digest (12 mg) consisting of 17-, 22-, 27-, 64-, 76- and 95-bp fragments was loaded on to a 25 cm x 1 cm I.D. RPC-5 column (thermostated at 45°C) and developed with a 2-1 gradient from 0.4 to 0.75 M KCl containing 10 mM Tris-HCl (pH 8) and 0.1 mM EDTA. Fractions of 9 ml were collected at a flow-rate 0.65 ml/min; 6% polyacrylamide gel analysis of the UV-absorbing fractions revealed that peaks A-F contained the pure fragments in order of increasing size. The small peak (fraction 3) did not contain DNA. (Reprinted from Hillen et al., 1981.)

ethylene glycol and subjected to RPC-5 chromatography to isolate the pure 301-bp insert (to remove the residual vector DNA). An 85 cm x 2.5 cm I.D. RPC-5 column thermostated at 45°C was loaded with 120 mg of UV-absorbing material and developed with a linear 3-1 gradient from 0.5 to 0.8 M KCl containing 10 mM Tris-HCl (pH 8) and 0.1 M EDTA (see Fig. 7 in the original paper). Fractions were collected at a flow-rate of 1.7 ml/min and some of them were analysed by 5% polyacrylamide gel analysis. The 301-bp fragment appeared in fractions 143-218 and the total yield (after thickening by evaporation and dialysis) was 87.5 mg. This pure 301-bp insert DNA was recleaved independently with other restriction enzymes (HpaII, HinfI or HaeIII/AluI) and the fragments obtained were also separated by RPC-5 chromatography. The results received with the last recleavage are illustrated in Fig. 10.6.

RPC-5 column chromatography proved to be a very important technique for the separation of nucleic acids and their fragments. Many observations led to the conclusion that the separation ability was not limited only to fractionation according to the size of the molecule. It was shown that single-stranded DNA was more retarded than double-stranded (duplex) DNA and that DNA fragments with unpaired ("sticky") ends were eluted with substantially higher elution volumes (at higher salt concentrations) than fragments of the same size with paired ("blunt") ends. Also, AT-rich DNA fragments were more retarded. It is clear that the separation mechanism is more complicated and comprises several chromato-graphic principles, and in this way RPC-5 chromatography can compete well with other separation techniques, such as gel electrophoresis. However, difficulties occurred with the commercial availability of the original matrix. In addition, the quaternary amine layers on the particulate support could not withstand many repeated chromatographic runs, because it was gradually released. Owing to these and other difficulties, a new matrix was sought in order to develop this useful technique further.

Thompson et al. (1983) presented a new modification of the RPC-5 technique under the name RPC-5 ANALOG column chromatography. The RPC-5 ANALOG material is composed of a thin film of trialkylmethylammonium chloride covering a solid, non-porous, non-compressible, inert microparticulate resin. It is readily avail-able commercially (from Bethesda Research Labs., Gaithersburg, MD, U.S.A.), the products are homogenous from batch to batch and the bleeding is substantially reduced. Defined particle size distributions of RPC-5 ANALOG are commercially available under the trade-name NACS (Nucleic Acid Chromatography System). In order to prepare RPC-5 ANALOG for use, 1 g of the matrix is suspended in 10 ml of 2.0 M NaCl, buffered with 10 mM Tris-HCl (pH 7.2) + 1 mM Na$_2$EDTA (this is the so-called TE buffer) and stirred gently for 1 h. The matrix is allowed to settle for 10 min and the supernatant is removed by aspiration; this procedure is repeated several times until the supernatant is clear. In order to ensure complete hydration, the suspension is kept overnight at 4°C with gentle stirring. After settling, the supernatant is removed by aspiration and the sediment is resuspended (10 ml/g) in fresh degassed 2 M NaCl in TE buffer and stored at 4°C until used. Details of the use of RPC-5 ANALOG can be found in the NACS Applica-tion Manual.

Similarly to the preceding RPC-5 columns, the separation with RPC-5 ANALOG is based primarily on an ion-exchange mechanism. Nucleic acids are bound at lower ionic strength and eluted by a salt gradient. Usually, lower molecular weight nucleic acids are eluted before higher molecular weight species, so that, as a rule, the separation follows the length of the nucleic acids (number of bases). However, the retention is influenced by the available phosphate groups only. Therefore, changes in the secondary and tertiary structure of nucleic acids influence the separation.

The analytical capacity of the matrix is 0.2 mg of nucleic acid per gram. The sample should be carefully prepared and contaminating molecules (proteins, organic solvents, blocking groups, divalent cations) should be removed. Thompson et al. (1983) presented examples of various applications (Table I of the original paper); the column length was from 5 to 88 cm and the I.D. from 0.2 to 0.9 cm. The load buffer was from 0.1 to 0.5 M NaCl in TE buffer (pH 4.5-12). The solvents for gradient elution contained from 0.1-0.7 M NaCl to 0.5-1.0 M NaCl in 12 mM NaOH (pH 12) or in TE buffer of various pH (4.5-7.2). The gradient volumes were roughly proportional to the column dimensions, from 50 to 1000 ml. The recovery of nucleic acids from RPC-5 ANALOG was routinely greater than 95%. Two specific applications were described in detail in the cited review: (a) the purification of supercoiled DNA and (b) the purification of synthetic oligodeoxyribonucleotides. In addition, analytical examples using RPC-5 plastic mini-columns (volume 1 ml, load 0.1 g of RPC-5 ANALOG) were summarized in Table III of the original paper. Mini-columns are widely used in biochemistry, but after use they must be discarded in order to ensure the expected performance.

10.2.4.2 *Affinity and boronate chromatography of nucleic acids and their fragments*

Singhal et al. (1980) presented a new principle of reversed-phase boronate (RPB) chromatography for the separation of O-methylribose nucleosides and aminoacyl-tRNAs. They synthesized phenyl boronates with hydrophobic side-chains about 1 nm long and coated inert 10-μm solid beads of polychlorotrifluoroethylene with this material. The synthesis was described in Section 4.4.6 and illustrated by Fig. 4.13. The polychlorotrifluoroethylene (Plaskon CTFE, 2300 powder) beads used for the preparation of RPB were originally obtained from Allied Chemical (Morristown, NJ, U.S.A.). The material is now unavailable from this source, but another source of similar material (Kel-F) was described by Usher (1979).

The principle of action of RPB can be explained according to Fig. 10.7. Boronate forms an anionic complex with free *cis*-2',3'-hydroxyl groups of unsubstituted ribonucleotides and the 2',3'-hydroxyl groups of the terminal adenosine of non-acylated tRNAs. The complexes are not formed with O-alkyl or O-acyl derivatives, which permits the separation of mammalian and bacterial aminoacyl-tRNAs from uncharged tRNAs in a single chromatographic step. In fact, both uncharged and aminoacyl-tRNAs are retained by the RPB material by the hydrophobic nature of the matrix. To weaken the interactions between aminoacyl-tRNAs and the matrix, relatively high concentrations of chloride and magnesium ions are necessary. However, these ionic conditions are not strong enough to weaken the boronate complex formation between the matrix and the *cis*-diol groups of

hydrophobic chain

reacts with cis—diols

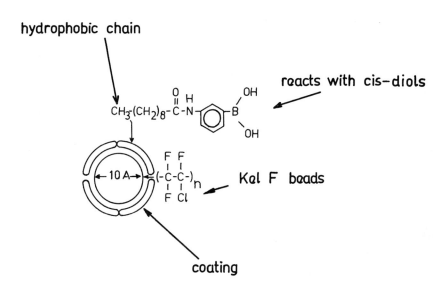

Kel F beads

coating

Fig. 10.7. Structure of reversed-phase boronate matrix and its interaction with the 3'-end of the uncharged tRNAs. The m-(N-decanoyl)phenylboronic acid is bound to polychlorotrifluoroethylene beads by hydrophobic interactions. (According to Singhal, 1983.)

uncharged tRNA. Strongly adsorbed uncharged tRNA requires significantly higher concentrations of chloride and a low eluent pH for desorption.

For the preparation of the column, the matrix (as a white powder) was packed into a jacketed glass column with the help of a spatula and a polyethylene rod 5 mm in diameter. The column was washed with 50 mM sodium acetate buffer (pH 4.5) containing 1 M NaCl to remove any poorly adsorbed affinity groups (the effluent was then clear at 260 and 280 nm). As the arm attached to the matrix is sensitive to light and tends to deteriorate at 20°C, the column was protected against light and maintained at 4°C at all times. The application of RPB will be illustrated by an example.

For the separation of aminoacyl-tRNA from uncharged tRNAs (Singhal et al., 1980), about 3 mg of rat liver tRNAs containing [^{3}H]lysyl-tRNA (ca. 4 nmol) in 0.2 ml of buffer A (50 mM sodium cacodylate-10 mM MgCl$_2$-0.5 M NaCl, pH 6.8) was applied to an RPB column (23 cm x 0.63 cm I.D.) and the UV-absorbing material was eluted with buffer A. The peak of lysyl-tRNAs appeared in an elution volume of 4-5 ml. The column was then eluted in a stepwise manner with buffer B (50 mM sodium acetate-10 mM MgCl$_2$-1.0 M NaCl, pH 4.5). The double peak of uncharged tRNAs appeared in of elution volumes 6-9, 9-13 ml (see Fig. 4 of the original paper). The temperature was 4°C and the flow-rate 0.3 ml/min. If a linear gradient of 0.2 to 1.0 M NaCl (at pH 4.5) was used instead of the stepwise elution, the uncharged tRNAs could be fractionated on this RPB column.

Examples of the purification of several *Escherichia coli* tRNAs by RPB chromatography were summarized in Table 2 of the original paper. Uncharged tRNAs form boronate complexes on RPB columns about 2 pH units away from the reported pK of the benzeneboronic acid. This lowering of the pK value was explained by the influence of the microenvironment on the hydrophobic matrix. Up to 30 satisfactory chromatographic runs were achieved on re-using the same RPB column. Eluents containing ethanol could not be used with the RPB matrix, as the affinity compound was partially solubilized. The RPB matrix appeared to deteriorate in the alkaline buffer (part of the impregnated boronate material was eluted). To overcome this problem, both the column and the unused matrix were stored routinely in acidic buffer (50 mM sodium acetate-10 mM MgCl$_2$-1.0 M NaCl, pH 4.5). The recoveries from the RPB columns were satisfactory (92-97%). Acetylated N-(m-dihydroxyborylphenyl)carbamylmethylcellulose (CMB-cellulose) was also prepared and tested for similar purposes.

The HPLC of tRNAs and the separation of tRNAs from mammalian sources under conditions of stability with respect to the amino acid-tRNA ester bond were reviewed by Singhal (1983). Both chemical (boronate) and biological (plant lectins) affinity groups and their interactions with tRNAs were described. The author also discussed some disadvantages of the described type of RPB matrix (time-consuming preparation, problems with the commercial availability of the inert support, light sensitivity and partial deesterification of some aminoacyl-tRNAs even at neutral pH). Therefore, the author's group is redesigning the affinity matrix in order to be able to form a complex with *cis*-diols (of the uncharged tRNA) at much lower pH (about 4-5) where the amino acid-tRNA bond is most stable.

Hjertén and Yang (1984) studied the HPLC separation of various biopolymers on dihydroxyboryl-agarose and found that small, rigid agarose beads derivatized with m-aminophenylboronic acid can be also effective for the boronate HPLC of aminoacylated and non-aminoacylated tRNAs.

For the coupling of m-aminophenylboronic acid to agarose (see also Sundberg and Porath, 1974), a suspension of 15 μm agarose beads (15% cross-linked, prepared by the authors from Agarose EEO = -0.17; IBF Villeneuve, La Garenne, France) was centrifuged (400 g for 10 min) and 3 g of the sediment were suspended in 2.5 ml of 1 M sodium borohydride, then 1 ml of 1,4-butanediol diglycidyl ether was added dropwise and the suspension was stirred at room temperature for 10 h. After the epoxy-activated gel had been washed with water, 400 mg of m-aminophenylboronic acid (prepared from hemisulphate according to Singhal et al., 1980; cf., Section 4.4.6), dissolved in 3 ml of 1 M sodium carbonate solution (pH 10), were added. The gel suspension was stirred at room temperature for 48 h, then deactivated with 0.1 ml of 2-mercaptoethanol overnight at 4°C. The gel was washed (by centrifugation at 400 g for 5-10 min) first with 1 M NaCl and then with water. The product obtained was referred to as dihydroxyboryl-agarose II. Another coupling procedure was also described based on the method of Bethel et al. (1981) and utilizing activation with 1,1'-carbonyldiimidazole. This product was referred to as dihydroxyboryl-agarose I.

For the separation of aminoacylated and non-aminoacylated tRNAs an 8.8 cm x 0.6 cm I.D. column was used, packed with dihydroxyboryl-agarose II. After equilibration with 0.05 M HEPES (pH 8.5), containing 0.05 M MgCl$_2$ and 20% 2-propanol, 2 μl of [^{14}C]leucine-tRNA were injected. The same buffer was passed through the column for 25 min at a flow-rate of 0.2 ml/min and a pressure of 2 atm, followed (in a stepwise manner) by 0.1 M sodium acetate (pH 4.5) containing 0.03 M MgCl$_2$ and 20% 2-propanol. Fractions of 0.4 ml were immediately frozen. After they had thawed, the radioactivity was determined in the acid-precipitated material. The radioactive peak of aminoacylated tRNA appeared in 10 min, washed with the first buffer of pH 8.5; the non-aminoacylated tRNA was eluted in 60 min (total time) with the second buffer of pH 4.5. The column of dihydroxyboryl-agarose II was used in this experiment, as the boronate group in this matrix was attached to agarose via a longer spacer than in the species I and was therefore more accessible to macromolecules and gave a higher capacity for non-acylated tRNAs. Propanol was used as the buffer component in order to suppress possible hydrophobic interactions. Aminoacylated tRNAs were exposed to pH 8.5 for only a short time during elution, which meant that negligible hydrolysis of the amino acid-tRNA ester bond occurred.

10.3 POLYNUCLEOTIDES AND OLIGONUCLEOTIDES

10.3.1 Instrumentation and detection

A number of commercially produced HPLC systems (cf., Chapter 5) are now available, and most of them can be used for the separation of oligonucleotides and polynucleotides. A concise review on instrumentation was given by Wehr (1980) in a booklet on the HPLC of nucleic acid constituents. Brief comments on HPLC equipment for the purification of synthetic oligonucleotides can be found in the review by McLaughlin and Piel (1984a). They discussed the question of the necessity to use pre-packed HPLC columns, recommended self-packing and described this procedure in detail in Appendix II (McLaughlin and Piel, 1984b) in the book edited by Gait (1984).

Owing to the high molar absorptivities exhibited by oligonucleotides in the range 250-280 nm, UV detectors (cf., Section 5.1.8) are almost exclusively used. The sensitivity is usually high enough, but preparative flow-through sample cells may be utilized that reduce the detector sensitivity by a factor of 10.

It is possible to carry out the preparative runs without the necessity for a special preparative flow-through cell, provided that the column effluent is monitored at a wavelength at which the molar absorptivity is not too high.

The common purine and pyrimidine bases usually show absorption spectra in the range 180-300 nm with two major maxima in the vicinity of 190-200 nm and 250-280 nm (cf., Voet et al., 1963). The longer wavelength is often used for the detection and the molar absorptivity is of the order of 10^4. The physico-chemical origin and meaning of these spectra were explained by Walton and Blackwell (1973) and tables of molar absorptivities of the maxima and minima of the UV spectra of nucleic acid components were compiled by Wehr (1980). As the UV spectra of bases are affected by protonation, the absorption spectra of nucleic acid constituents are sensitive to the pH of the solvent. Therefore, precise control of the pH of the mobile phase is necessary if absorption rela-tionships at multiple wavelengths are to be used in peak identification; naturally, wavelength stability of the detector is assumed. Because the phos-phate groups in nucleotides are removed sufficiently from the bases by the sugar residue, the ionization of phosphates has relatively little influence on the nucleobase chromophores.

UV detection in the region 240-280 nm is almost universally used for moni-toring nucleotides for both qualitative and quantitative purposes. The most common tool is the inexpensive UV detector with a fixed wavelength (usually 254 nm), in which the light is generated by the emission of a low-pressure mercury lamp; it is sensitive enough to detect nucleic acid constituents down to the picomole range. Other types of universal detectors (e.g., refractive index detectors) are hardly used in the separation of oligonucleotides.

For radioactivity monitoring, continuous flow cells based on particles of solid scintillators or on mixing with "scintillation cocktails" can be used. Flow monitoring of aqueous solutions containing weak β-emitters was decribed by Schram (1970). However, there are problems with such continuous measurements: the efficiency is low, or the system is complicated and expensive. According to Wehr (1980), "these detectors should in the future provide a fertile ground for improvement of sensitivity and reduction of cost". In many instances discontin-uous measurement of the radioactivity in aliquots of collected fractions is preferred.

Some special detection principles have been published. Schiebel and Schulten (1981) described the identification of protected deoxyribonucleotides by field desorption mass spectrometry in fractions obtained by HPLC. The production of abundant cationized molecules and the formation of structurally significant fragments by thermal- or field-induced processes allowed the direct and ambig-uous identification of the synthesized products. In addition, both organic and

inorganic impurities could be detected. Heine et al. (1982) described the
determination of nucleotides by LC with a phosphorus-sensitive inductively
coupled plasma detector.

Detection systems for HPLC in nucleic acid research were reviewed by Assenza
(1984).

10.3.2 Size exclusion chromatography

Size exclusion chromatography (SEC) has only minor importance for the separation of low-molecular-weight oligonucleotides. Sproat and Gait (1984), in their
review on the solid-phase synthesis of oligodeoxyribonucleotides, recommended
it as a possible method for the removal of formamide and desalting the samples
of ion-exchange-purified material: Bio-Gel P-2 was used for this purpose in a
column volume at least 10 times that of the volume of the solution to be desalted.
The sample was passed through the column and eluted with ethanol-water (2 : 8,
v/v). The desired material appeared at the void volume; these fractions were
then evaporated to dryness in vacuo.

Molko et al. described the HPLC of short oligonucleotides using an I-125
protein analysis column (Waters Assoc.). The influence of pH and molar concentration of the mobile phase was studied. Volatile buffers (e.g., triethylammonium acetate, TEAA) gave good and rapid purification of deprotected synthetic
oligonucleotides. For example, the deprotected pentadecanucleotide d(TpCpApAp-
CpCpApApGpApGpGpApCpA) was applied to the I-125 column for preparative separation
using 0.1 M TEAA (pH 6.4) at a flow-rate of 1.04 ml/min and monitored at 290 nm.
Peak 1, containing the main product, was eluted between 7 and 8 min. Reversed-
phase chromatography on a LiChrosorb RP-8 column (300 x 4.7 mm I.D.) using a
gradient of 10-40% methanol in 0.1 M TEAA (flow-rate 2 ml/min, detection at
280 nm) separated the main product (eluting between 22 and 24 min) from some
minor impurities, eluted later. A total of 1.40 A_{260} units of the pentadecanucleotide were recovered.

10.3.3 Ion-exchange chromatography

Ion-exchange chromatography (IEC) has been used for the separation of nucleic
acid constituents for nearly 40 years. The way for its application was opened up
by the pioneer works of Cohn (e.g., 1949a,b, 1950), who introduced polystyrene-
divinylbenzene ion exchangers to this field of biochemistry and organic chemistry; these classic endeavours were reviewed, e.g., by Mikeš (1954). Since that
time, the IEC of nucleic acid constituents has evolved steadily. The method is
based on differences in the dissociation constants of nucleic acid constituents,

which are decisive for ionic interactions with the functional groups of ion exchangers, together with hydrophobic interactions with the aromatic matrices. In addition to some data listed by Mikeš (1954), estimated pK values of purines, pyrimidines, nucleosides and nucleotides were compiled by Wehr (1980). Information on sites for the protonation and ionization and pK values of nucleotides can be found in the book by Bloomfield et al. (1974). In spite of the fact that from the historical point of view both cation and anion exchangers have been used for the separation of bases, nucleosides and nucleotides (cf., e.g., Khym, 1974; Cohn, 1975), anion exchange is the dominant technique for the separation of oligonucleosides owing to the numerous highly ionized phosphate groups.

In the early days of oligonucleotide research, column chromatography on DEAE-cellulose (especially under denaturing conditions in the presence of urea) was the method of choice for their separation (cf., Tomlinson and Tener, 1962, 1963). However, this method was very time consuming. Later, rigid pellicular ion exchangers provided improvements in both speed and resolution, but the capacity was low. The appearance of microparticulate silica-based ion exchangers provided a high speed of chromatography in addition to sufficient loading capacity. The only limitation was the instability of the bonded phases in alkaline media. A chronological survey of examples illustrating this development will be given below.

Gabriel and Michalewsky (1973) shortened the separation of oligonucleotides from tens of hours (and even days) to less than 90 min by replacing DEAE-cellulose with a more modern packing material. They used a Zipax weak anion-exchange (WAX) column (500 mm x 1.2 mm I.D.) (Du Pont, Wilmington, DE, U.S.A.) in the sulphate form and the sample was eluted with a linear gradient of 0.001 to 1 M sodium or potassium sulphate (40 ml each) in 0.001 M ammonium acetate (pH 4.4). The flow-rate was 1 ml/min and the temperature 50°C. Samples of oligonucleotides from the 4-mer to 14-mer were prepared by removing the acyl blocking groups from chemically synthesized oligomers with ammonia solution, which was evaporated. The sample size was from 0.01 to 1 absorbance unit, monitoring at 254 nm. Mono- and dinucleotides were eluted at or very near the void volume of the column and longer oligomers were retained in proportion to their charge and composition. Köster and Keiser (1974) improved the chromatographic separation of synthetic oligonucleotides on DEAE-cellulose by the addition of suitable alcohol or alcohol mixtures (such as methanol-isopropyl alcohol) to the mobile phase. Asteriadis et al. (1976) separated oligonucleotides, nucleotides and nucleosides on 100 cm x 0.4 cm I.D. columns of polystyrene anion exchangers (Dowex 1-X2, -400 mesh) with solvent systems containing ethanol at a flow-rate of 12-20 ml/h. The addition of 20% ethanol to the particular salt

gradient caused a 50% decrease in the retention volume; the addition of 40%
ethanol caused a further decrease in retention volume. Demushkin and Plyashevich
(1976) studied the application of high-speed ion-exchange chromatography in
nucleotide analysis; strongly basic and strongly acidic ion exchangers of the
Aminex type were used for the separation of nucleic acid hydrolysates. Aukaty
et al. (1978) tested the application of a new anion exchanger, an aminosilica,
for the HPLC analysis of synthetic oligodeoxynucleotides. This anion exchanger
was prepared from trimethylchlorosilane silylated silica gel, coated with
trimethylcetylammonium bromide (Cetavlon). The details of the preparation of
the ion exchanger were published. Both synthetic oligonucleotides and fragments
of native nucleic acids were rapidly separated.

Aukaty et al. (1977) used the commercial anion-exchange MicroPak NH_2-10
(Varian, Zug, Switzerland) column (25 cm x 0.2 cm I.D.), previously washed with
0.75 M KH_2PO_4 at pH 3.3 and equilibrated with 0.1 M phosphate buffer. Chromatographic analysis of deprotected synthetic oligodeoxyribonucleotides from different stages of oligonucleotide synthesis was investigated. By eluting with
0.1 M potassium phosphate (pH 3.3) the analysis of nucleotides could be established in 15 min. With a linear phosphate buffer (0.10-0.75 M) at neutral pH,
separation of oligonucleotides according to their length was preferred. In
acidic medium (pH 3.3-4.3), separation according to composition was possible.
At pH 7 (linear gradient of 0.10 to 0.75 M KH_2PO_4, slope 0.32, flow-rate
60 ml/h, pressure 3500 p.s.i.), mono- to hexathymidylic acids were perfectly
separated. At pH 3.3 (linear gradient of 0.1 to 0.75 M KH_2PO_4, slope 0.26,
flow-rate 60 ml/h, pressure 3500 p.s.i.) nine oligonucleotides were well
separated. Gait and Shepard (1977) used the strong anion exchanger Partisil
10 SAX to separate an oligonucleotide mixture, whereas Yanagawa (1978) applied
medium basic DEAE-2000 SW (Toyo Soda, Tokyo, Japan) in a 60 cm x 7.5 mm I.D.
column; using isocratic elution with 0.5 M acetate buffer (pH 7.0) containing
5 M ethylene glycol as mobile phase and flow-rate 1 ml/min, 1-10-mers of
oligodeoxyadenylic or oligothymidylic acid were perfectly separated.

Dizdaroglu et al. (1979) studied the separation of sequence isomers of
pyrimidine tetradeoxynucleoside triphosphates by IE-HPLC. Strongly basic
Partisil-10 SAX (Chrompack) of 10 μm particle size was used in these experiments,
based on linear gradient elution using triethylammonium acetate buffers at pH
3.1-3.4 (these buffers can be removed from the effluents by freeze-drying). The
authors reported all the observed elution orders of sequence isomers. It was
shown that the deoxycytidine component speeded up the elution more strongly the
nearer it was to the 3'-terminus of a given nucleotide, i.e., d(TpC) was eluted
before d(CpT). The elution order of (dC,dT_2) isomers was d(TpTpC), d(TpCpT),
d(CpTpT). An example of the separation of tetradeoxynucleoside triphosphates is

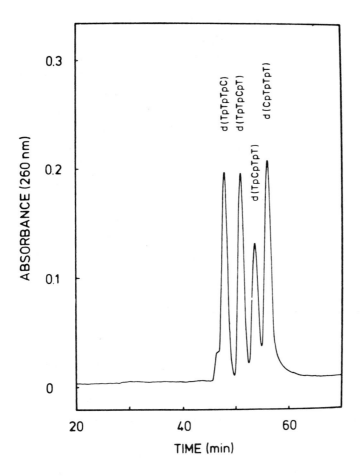

Fig. 10.8. Separation of sequence isomers (dC,dT$_3$) on a Partisil-10 SAX (10 μm) column (50 cm x 0.3 cm I.D.). Linear gradient of 0.15 M (pH 3.1) to 0.5 M (pH 3.4) triethylammonium acetate, duration of elution 70 min, flow-rate 1.0 ml/min, temperature 60°C, pressure 60 bar. (Reprinted from Dizdaroglu et al., 1979.)

given in Fig. 10.8. Later, this work was extended by Dizdaroglu et al. (1980) to the HPLC separation and sequencing of sequence isomers of pyrimidine penta-deoxynucleoside tetraphosphates. A 30 cm x 0.4 cm I.D. column of MicroPak AX-10 (Varian) (see below) was used for separation. Mixtures of p(dC,dT$_4$)p, p(dC$_2$, dT$_3$)p and p(dC$_3$,dT$_2$)p were isolated from the partial hydrolysate of herring sperm DNA and terminal phosphate groups were removed with alkaline phosphatase. The column was eluted with 0.5 M triethylammonium acetate (pH 3.4) at a flow-rate of 1.5 ml/min and a temperature of 70°C. Dizdaroglu and Hermes (1979) also examined the separation of small DNA and RNA oligonucleotides by IE-HPLC. Here

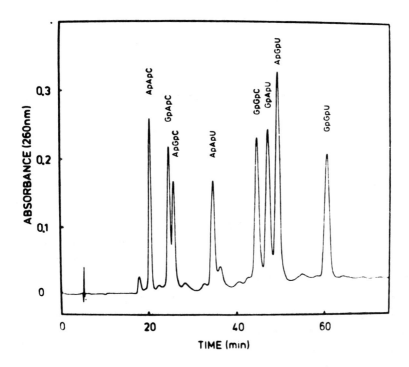

Fig. 10.9. Separation of trinucleoside triphosphates isolated from an alkaline phosphatase-pancreatic RNase digest of yeast RNA on a Partisil-10 SAX (10 μm) column (15 cm x 2 cm I.D.). Linear gradient from 0.03 M (pH 3.1) to 0.4 M (pH 3.4) triethylammonium acetate was used for elution, duration 80 min, flow-rate 1 ml/min, temperature 60°C, pressure 60 bar. (Reprinted from Dizdaroglu and Hermes, 1979.)

Partisil-10 SAX (Chrompack, Middelburg, The Netherlands) and triethylammonium acetate buffers were applied. These buffers exhibited negligible absorption at 260 nm, buffered well at the pH employed, and could easily be prepared from triethylamine and acetic acid. Laboratory packing of the column was described in detail. Fifteen of the sixteen possible deoxynucleoside monophosphates and all sixteen dinucleoside monophosphates were separated. All pairs of sequence isomers were well resolved. The separation ability of this system is illustrated here by an example of the chromatography of small nucleotides from enzymic digest of RNA (Fig. 10.9). Also, Gough et al. (1979) described the application of strongly basic AS-Pellionex SAX to separate an oligonucleotide mixture. In contrast, Edelson et al. (1979) studied the ion-exchange separation of nucleic acid constituents using a new silica-based bonded-phase weak anion-exchange MicroPak AX-10 column (Varian). The separation of ApA, ApApA and nicotinamide adenine dinucleotide was described, in addition to the chromatography of many nucleoside phosphates. Other examples of applications of MicroPak AX-10 and of other ion exchangers of the MicroPak series (MicroPak MAX and SAX) were given by Wehr (1980).

The application of weakly basic diethylamino-modified silica was described by Jost et al. (1979). The ion exchanger has the structure

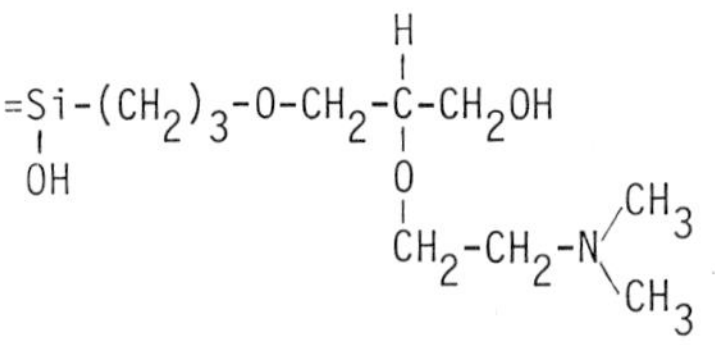

$$
\equiv Si-(CH_2)_3-O-CH_2-\underset{\underset{O}{|}}{\overset{\overset{H}{|}}{C}}-CH_2OH
$$
$$
OH \qquad \qquad \underset{CH_2-CH_2-N\begin{smallmatrix}CH_3\\CH_3\end{smallmatrix}}{}
$$

and was designated DMA-silica. A LiChrosorb Si 100 support was used for its preparation. The surface concentration of the functional groups was 2.0 $\mu mol/m^2$ and the nominal capacity was 0.78 mequiv./g. The separation of homologous oligonucleotide series was examined. For example, UpU up to $(Up)_8U$ were well separated on a 300 mm x 4 mm I.D. column of DMA-silica (5 μm) isocratically eluted with 0.115 M phosphate buffer (pH 7.0) at a flow-rate of 1.0 ml/min, with detection at 254 nm (see Fig. 9 of the original paper). Šatava et al. (1979) studied the separation of oligonucleotides on ion-exchange derivatives of Spheron. A specific diphenylamine-formic acid hydrolysate of *Bacillus subtilis* DNA (a mixture of pyrimidine oligodeoxyribonucleotides) was chromatographed using an ionic strength gradient on seven ion exchangers (five DEAE-Spheron 300 species of various capacities, two types of BD-Spheron 300 and DEAE-Sephadex A-25). The best results (see Fig. 1 of the original paper) were obtained with DEAE-Spheron of 1.8 mequiv./g (Lachema, Brno, Czechoslovakia). A 20-mg amount of the mixture was applied on a 23 cm x 0.8 cm I.D. column and unwanted by-products were eluted with 150 ml of buffer A (0.01 M sodium acetate, pH 5.3). Then a linear gradient of 600 ml of buffer A and 600 ml of buffer B (buffer A, 0.4 M in NaCl) was applied (this was the start of the chromatogram). Eleven peaks of oligonucleotides were eluted within this gradient and no other peak appeared when a higher ionic strength was applied. Vanderberghe et al. (1979) described the HPLC separation on pellicular weakly basic anion exchanger, AL Pellionex WAX (Reeve Angel, Clifton, NJ, U.S.A.), of particle diameter 50 μm. Oligonucleotides obtained by RNAse hydrolysis of *Escherichia coli* 5S rRNA were eluted from a 500 mm x 2.1 mm I.D. column with a linear gradient from 0 to 1 M NaCl in 0.001 M Tris-HCl-7 M urea (pH 7). The gradient was programmed over 100 min, but the largest oligonucleotides handled in these experiments were eluted in 30 min (at 0.3 M NaCl). Solutions buffered above pH 7 should be avoided owing to instability of the ion exchanger.

Horn et al. (1980) used the ion exchanger Permaphase AAX for the purification
of the synthetic product in their efforts to synthesize the gene for the gastric
inhibitory polypeptide. Vandenberghe et al. (1980) described the HPLC analysis
of oligo- and monoribonucleotide mixtures with special reference to ribosomal RNA
constituents. *E. coli* 5S rRNA was hydrolysed by pancreatic or T$_1$RNAse and the
oligonucleotides obtained were chromatographed using an AL Pellionex WAX column
(500 mm x 2.1 mm I.D.), which was eluted with linear gradients formed by mixing
0.01 *M* Tris-HCl-7 *M* urea (pH 7) with 0.01 *M* Tris-HCl-1 *M* NaCl-7 *M* urea, pH 7. Oligonucleotide mixtures were fractionated according to the chain length at neutral
pH in the presence of urea, and mononucleotides were separated according to the
base composition using acidic buffers. Edge et al. (1981) separated oligonucleotide mixtures using the strongly basic anion exchanger Partisil 10 SAX.
Hansson (1983) described the purification of NADP by high-performance anion-
exchange chromatography on Polyanion SI (Pharmacia, Uppsala, Sweden). For
optimization experiments on the analytical scale an 8-µm exchanger was used,
whereas for larger scale preparations a 17-µm exchanger was applied. Polyanion
SI (17 µm) was used in an HR 10/10 column (10 cm x 1 cm I.D.) and eluted with a 0-100%
gradient of buffer A (0.01 *M* phosphate, pH 6) and buffer B (0.50 *M* phosphate, pH
6). The flow-rate was 4 ml/min and detection at 254 nm was used. A 10-mg load
of commercial-grade NADP was applied in 500 µl of the starting buffer.

Lawson et al. (1983) studied the separation of synthetic oligonucleotides
on columns of microparticulate silica coated with cross-linked polyethyleneimine
(PEI). Columns were prepared in situ by adsorbing a layer of PEI on SupelcoSil
LC-Si porous silica (d_p 5 µm) (Supelco, Bellefonte, PA, U.S.A.) followed by
cross-linking with multi-functional oxirane diglycidylethylene glycol (the
procedure was described in detail). A small column (150 mm x 4.2 mm I.D.) had
a capacity from less than 1 µg for analytical applications (injection of 0.08-
0.5 A_{260} units) up to several milligrams for preparative runs (injection of
5-100 A_{260} units). This weak anion-exchange column was useful for the separation
of oligonucleotides of up to 20 residues in either homo- or heteropolymers.
Linear salt gradients were used for elution and sulphates proved best. The weak
buffer was 0.05 *M* potassium phosphate (pH 6) in 30% aqueous HPLC-grade methanol
and the strong buffer was 1 *M* ammonium sulphate-0.05 *M* potassium phosphate (pH 6)
in 30% aqueous methanol. The authors noted that commercial reagent-grade ammonium
sulphate often contained appreciable ammounts of UV-absorbing impurities. Therefore, they used food-grade material, or they prepared the salt by titration of
reagent-grade sulphuric acid with ammonia solution. Many successful separations
were presented and this technique was found to be extremely useful for the
analysis and purification of chemically synthesized oligonucleotides. The time

of the separation was 40-60 min (the usual flow-rate was 1.5 ml/min) and the
column life time was long, being at least 450-600 h.

Pearson and Regnier (1983) extended the above studies. The methodology for
both the static (batchwise) and in situ surface modification of silica using
PEI was oulined, with emphasis on the contributions of stationary phase thickness
and silica type. The types of silica tested were Hypersil, Nucleosil 100-5,
Spherisorb S5W, LiChrosorb and Vydac TP. Ethylene glycol diglycidyl ether (EDGE)
or 1,4-butanediol diglycidyl ether (BUDGE) was used for the cross-linking. The
best resolution was obtained with 3 μm Hypersil with PEI-EDGE coating. For the
mobile phases, the weak buffer was 0.05 M potassium phosphate-30% methanol,
adjusted to pH 5.9, and the strong buffer was 1 M ammonium sulphate in the weak
buffer. The flow-rate was 0.6 ml/min at ambient temperature. Relatively short
columns (5 cm x 0.41 cm I.D.) were capable of separating oligonucleotide
homologues of up to 35 bases.

Newton et al. (1983) described the IE-HPLC of oligodeoxyribonucleotides using
formamide. The buffer system containing denaturing formamide was generally
superior for IE-HPLC, and it was particularly valuable where the oligonucleotide
of interest was highly self-complementary and/or rich in deoxyguanosine residues.
An example of such a highly self-complementary 21-residue oligonucleotide is

```
     5'-CAATTG C T C CAGGAGCTCCTG-3' 5'-CAATTGCTCCAGGAGCTCCTG-3'
            | | | | | |                | | | | | | | | | | | |            | | | | | |
3'-GTCCTCGAGGACCTCGTTAAC-5' 3'-GTCCTCGAGGAC C T C GTTAAC-5'
```

Partisil 10 SAX anion exchanger (Whatman, Maidstone, U.K., or Jones Chromatography, Llanbradach, Glamorgan, U.K.) was applied using linear 75% or 100%
gradient systems: 0 to 0.4 M phosphate buffer (pH 6.5) in 5% or 30% aqueous
ethanol; 0 to 0.3 M phosphate buffer (pH 6) in 30% aqueous formamide; 0 to 0.4 M
phosphate (pH 6.5) in 60% formamide. These systems were combined as necessary
at a flow-rate of 2 ml/min. The temperature was ambient or 50°C. The authors
advocated the use of 30% formamide for routine purifications of most sequences
prepared by the automated solid-phase method, and 60% formamide for highly self-
complementary sequences, for oligonucleotides with a high content of dG or for
those which have more than two consecutive dG residues. Scanlon et al. (1984)
purified synthetic oligodeoxyribonucleotides from 11 to 37 nucleotides in length
(prepared by both the phosphotriester and phosphite procedures) by IE-HPLC using
Whatman Partisil 10 SAX columns (25 cm x 0.46 cm I.D.) and phosphate buffer
gradients. The effects of different buffer systems and elution times were tested
and the results were obtained with buffers containing formamide, particularly
with longer sequences (more than 30 residues) and in the case of 2'-deoxy-
guanosine-rich sequences. Examples of buffers and gradients are as follows:

(i) 0.001-0.2 M KH_2PO_4 (pH 6.5), 5% ethanol, over 40 min; (ii) 0.001-0.2 M KH_2PO_4 (pH 6.5), 30% acetonitrile, over 40 min; (iii) 0.001-0.3 M KH_2PO_4 (pH 6.3), 30% formamide, over 60 min; (iv) 0.001-0.3 M KH_2PO_4 (pH 6.3), 60% formamide, over 60 min. However, when using formamide buffers, the life of the Partisil 10 SAX column was generally decreased (to less than 2 months, in comparison with 4-6 months when an ethanol- or acetonitrile-based solvent system was used). The authors considered this method to be particularly useful for the purification of synthetic oligonucleotides prepared by automated solid-phase procedures.

Sproat and Gait (1984) reviewed the solid-phase synthesis of oligodeoxyribonucleotides by the phosphotriester method and described a purification procedure on a Partisil 10 SAX (or equivalent ion exchanger) column using a linear gradient from 1 mM to 0.3 M KH_2PO_4 (pH 6.3) in formamide-water (6:4, v/v). Purine-rich oligonucleotides are retained more strongly then pyrimidine-rich ones of the same length. Formamide prevents aggregation (caused by G-rich sequences). With self-complementary sequences (containing a restriction site linker) or if the sequence can form a stable hairpin structure, thermostating of the column at $50°C$ is recommended. Oligonucleotides of up to 40 bases can be purified by this method if an extended gradient (to 0.4 M KH_2PO_4) is used. Examples of the purification of particular synthetic oligonucleotides using SAX stationary phases and aminopropylsilyl bonded phases (APS) were also described in the review by McLaughlin and Piel (1984), on the chromatographic purification of synthetic oligonucleotides.

10.3.4 Reversed-phase chromatography

Reversed-phase chromatography (RPC) proved very useful for the rapid separation of bases and nucleosides (cf., e.g., Hartwick and Brown, 1976; Krstulovic et al., 1977; Rustum, 1978). At the very beginning of separation attempts this was not so for nucleotides. Anderson and Murphy (1976) studied the isocratic separation of some purine nucleotide, nucleoside and base metabolites from biological extracts by HPLC, and Wakazaka et al. (1979) described the rapid RPC separation of DNA constituents (bases, nucleosides and nucleotides) under the same chromatographic conditions; however, the selectivity was not high enough to resolve a multi-component mixture at several levels of phosphorylation owing to their frequent coelution. RP-HPLC was most successful when only a limited number of phosphorylated components were to be resolved. Affinity for the stationary phase was determined by the base and addition of methyl groups (usual in modified bases) increased the retention; in contrast, the addition of phosphate groups increased the affinity for the mobile phase, so that the RPC order of elution of tri-, di- and monophosphate nucleotides was the opposite of that in anion-exchange separations. However, the addition of suitable solvophobic ions (cf., Hoffman and Liao, 1977) such as tetrabutyl-ammonium chloride to the mobile phase enhanced the solute retention on ODS reversed-phase supports to such an extent that the retention resembled that of an anion-exchange separation (i.e. the elution order was mono-, di- and tri-phosphates). Yanagawa (1978) compared the separation of oligonucleotides (from the monomer to decamer of deoxyadenylic acid) for analytical purposes by HPLC using three chromatographic modes: (1) SEC, (2) RPC and (3) IEC. The best results were obtained with the last method.

Examples and a discussion of the utilization of RPC for the separation of nucleic acids constituents were presented by Wehr (1980). RP-HPLC is now routinely used to check the purity of synthetic oligonucleotides after IEC purification (Sproat and Gait, 1984). In the following part of this section, the development of the RPC of oligonucleotides will be discussed in greater detail.

Fritz et al. (1978) studied the application of HPLC in polynucleotide synthesis. RP-HPLC microparticulate ODS material was used as a support, and protected oligonucleotides, standard intermediates in the stepwise synthesis of polydeoxyribonucleotides, were analysed. The method was developed to the preparative scale and led to a marked reduction in the time required for the oligonucleotide synthesis. The 30 cm x 0.4 cm I.D. column used for analyti-cal purposes, packed with μBondapak C_{18} (10 μm), had approximately $1.5 \cdot 10^4$ theoretical plates for synthetic oligonucleotides. The preparative Waters Assoc. Bondapak C_{18}/Porasil B column had dimensions 183 cm x 0.7 cm I.D. Aqueous buffers consisting of 0.1 M ammonium acetate, 0.1 M trialkylammonium acetate (pH 7) and 0.1 M ammonium acetate containing varying percentages of acetonitrile were used as mobile phases. The flow-rate for the μBondapak C_{18} column was 2 ml/min. As examples, protected bz-(benzoyl), ib-(isobutyryl) and an-(p-anisoyl) derivatives of di- to tetranucleotides with a free 5'-phosphate group were eluted with 5-22% acetonitrile in 0.1 M aqueous ammonium acetate with retention times of 3.0-16.5 min. Various higher protected oligonucleotides carrying 5'-mono-methoxytrityl groups were eluted with 32-40% acetonitrile in 2.8-16.2 min. Unprotected oligonucleotides with 5'-hydroxy end-groups were eluted with 8-12% acetonitrile in 2.5-11.8 min. Extensive studies on the influence of various purine and pyrimidine bases, protecting groups, phosphate groups and chain length on the retention of oligonucleotides were reported.

Jones et al. (1978), in a paper on the use of a lipophilic *tert*.-butyldi-phenylsilyl protecting group in the synthesis and rapid separation of polynu-cleotides showed that the silyl group facilitated the isolation of the required product in synthetic reactions because of the selective and strong retention of the condensation product during RP-HPLC. Several examples of synthetic procedures including RP-HPLC analysis and purification of the products have been described and illustrate the scope and usefulness of the HPLC technique in polynucleotide synthesis.

Melander et al. (1979) studied the changes in the conformation and retention of oligo(ethylene glycol) derivatives with temperature and eluent composition as a model to study the chromatographic behaviour in RPC of biological substances such as peptides and oligonucleotides. McFarland and Borer (1979) described the separation of complex mixtures of oligoribonucleotides on an ODS silica column with gradient elution with acetonitrile-water-ammonium acetate. Larger oligomers (at least 20-mers) were well separated with a practical detection limit at 254 nm of ca. 1 nmol (absolute limit was ca. 1 pmol) of base. It was possible to purify ca. 1 mg of an oligonucleotide in a 10-30 min elution. The reproducibility was 1% absolute and 0.1% relative retention time. A 30 cm x 0.39 cm I.D. column of μBondapak C_{18} (Waters Assoc.) preceded by a 2 cm x 2 mm I.D. Corasil C_{18} guard column (Waters Assoc.) was used with gradient elution from 1% to 12% acetonitrile in 10% aqueous ammonium acetate (pH 5.9), at a flow-rate of 3 ml/min and a pressure of 1500-3000 p.s.i. Examples of the separation of homologous series are given in Fig. 10.10.

Knight et al. (1980) reported radioimmune, radiobinding and HPLC analysis of (2-5)A and related oligonucleotides from intact interferon-treated or EMC-virus-infected mouse L-cells. HPLC analysis was carried out on a μBondapak C_{18} column (Waters Assoc.) with 50 mM ammonium phosphate (pH 7.0) at a flow-rate of 1 ml/min, using two subsequent linear gradients made by means of methanol-water (50:50) differing in slope.

Brown et al. (1981) reviewed analysis of (2'-5')-linked oligo(A) and related oligonucleotides by HPLC. They originally used amine columns (isopropylamine bonded phases), which gave very good separations, but were unstable in phosphate buffers. Therefore, stable C_{18} columns, such as Bondapak C_{18}, were investigated. An example of the separation is illustrated in Fig. 10.11. Caley et al. (1982), in a paper on the HPLC of (2'-5')-linked oligonucleotides, discussed the instability of weak anion-exchange columns in phosphate buffers. They used RPC on C_{18} columns with satisfactory results. Ammonium phosphate (pH 7) with a linear gradient of methanol-water (1:1) gave a good separation of individual oligomers. For the separation of 5'-mono- to triphosphorylated oligomers or non-phosphorylated components, ammonium phosphate (pH 6.0) and potassium phosphate (pH 6.5) were used. The C_{18} columns proved very stable in phosphate buffers.

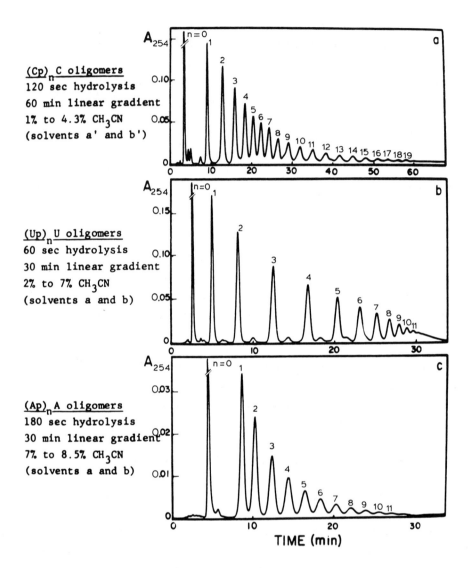

Fig. 10.10. Reversed-phase chromatography elution profiles of a homologous series of oligoribonucleotides, prepared by alkaline hydrolysis of the appropriate homopolymer; conditions: 3.3 mM polyX, 1.0 M KOH, 100 μl total volume, 60°C. Solvents a, a', b, b' are dilute solutions used for the preparation of the gradient and described in the original paper. All elutions were carried out at a flow-rate of 2 ml/min. The column system is described in the text. (Reprinted from McFarland and Borer, 1979.)

Crowther et al. (1981) compared the NPC and RPC of oligonucleotides (cf., Section 10.3.5). The RPC of deprotected oligonucleotides using Partisil 10 ODS-3 and Partisil 5 ODS-3 columns (Whatman, Clifton, NJ, U.S.A.) was described. The laboratory packing of columns was also described. For elution, gradients

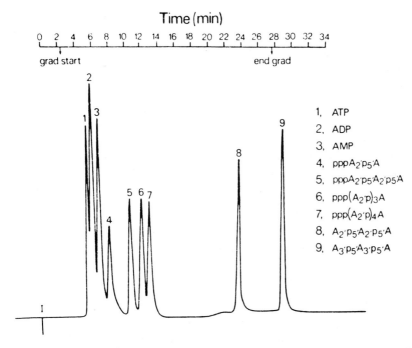

Fig. 10.11. HPLC analysis of (2'-5')-oligo(A) (2-5A). Reticulocyte 2-5A containing the trimer, tetramer and pentamer 5'-triphosphates and smaller ammounts of the corresponding di- and monophosphates were analysed together with AMP, ADP, ATP, A2'p5'A2'p5'A and A3'p5'A3'p5A on a μBondapak C_{18} column in 50 mM ammonium phosphate (pH 7). A linear gradient from 0 to 50% of methanol-water (1:1) was applied (flow-rate 1 ml/min) for 25 min. The injection point (I), the start (allowing for the void volume) and the end of the gradient are shown. The 5'-mono-, di- and triphosphates of the individual oligomers of 2-5A are not resolved in this system. (Reprinted from Brown et al., 1981.)

(6-100% in 20 min) of 0.1 M KH_2PO_4 (pH 5) containing 1.2% methanol and 0.1 M KH_2PO_4 (pH 5) containing 50% methanol (and occasional isocratic delays) were used at a flow-rate of 1 ml/min. In other experiments the application of ion-pair RPC on Partisil 10 C_8 and Partisil 5 C_8 allowed the simultaneous separation of ribo- and deoxyribooligonucleotides and their mononucleotides. Tetra-butylammonium iodide was used as the hetaeron ($3 \cdot 10^{-3}$ M) in aqueous mobile phase containing 7.5% methanol. The buffering salt was 0.05 M KH_2PO_4 (pH 4.9). Isocratic elution was used.

Knox and Jurand (1981) studied the mechanism of zwitterion-pair chromatography of nucleotides. Sokolowski et al. (1981) examined the IP-RPC of oligo-deoxynucleotides; DNA fragments ranging from one to four nucleotidyl units were chromatographed using tetraalkylammonium counter ions of different kinds. The regulation of the retention of oligonucleotides was possible by changing the concentration of counter ions. A LiChrosorb RP-18 (5 μm) column (100 mm x 4.5 mm

I.D.) was used as the stationary phase. The mobile phase consisted of equal
volumes of methanol and phosphate buffer (pH 7.0) (ionic strength μ = 0.032),
which was 0.0016 M in tetrapentylammonium ion. The linear flow-rate was
0.84 mm/s and the pressure 140-150 bar. Jost et al. (1982) studied the IP-RPC
of polyvalent ions using oligonucleotides as model substances. The objective
of the study was to shed more light on the retention mechanism of IP-RPC by
employing defined polyvalent ions. A column of LiChrosorb RP-18 (5 μm) was
applied and eluted with mobile phase containing tetra-n-alkylammonium compounds
as ion-pairing reagents [tetramethylammonium chloride (TMACl), tetraethylammonium chloride (TEACl) and tetrabutylammonium hydrogen sulphate (TBAHSO$_4$),
all from E. Merck, Darmstadt, F.R.G.]. The theoretical part of this paper was
reviewed in Section 3.8.

In their review, McLaughlin and Piel (1984) gave some practical examples
of the RPC purification of synthetic oligonucleotides using ODS Hypersil columns
(250 mm x 4.6 mm I.D. or 250 mm x 9.4 mm I.D.) with ammonium acetate buffer-
acetonitrile gradients or phosphate buffer-methanol gradients. Stec et al.
(1985a) studied the solid-phase synthesis, separation and stereochemical aspects
of P-chiral methane- and 4,4'-dimethoxytriphenylmethanephosphonate analogues
of oligodeoxyribonucleotides. A 30 cm x 7.8 mm I.D. column of μBondapak C$_{18}$
eluted with a linear gradient of acetonitrile vs. 0.1 M triethylammonium acetate
(TEAE) (pH 7) at a flow-rate of 4 ml/min was used for the analysis and collection of synthetic products and of enzymic digests. Acetonitrile was removed
from the coeluted products under a stream of nitrogen, and water and buffer
were removed using a vacuum centrifuge.

Stec et al. (1985b) described the RP-HPLC separation of diastereoisomeric
phosphorothienate analogues of oligodeoxyribonucleotides and other backbone-
modified congeners of DNA. An automated solid-phase phosphoroamidite-coupling
method was used for the synthesis and the diastereoisomers were conveniently
separated by RP-HPLC on 10 μm μBondapak C$_{18}$ (30 cm x 7.8 mm I.D.) with a linear
gradient of acetonitrile in 0.1 M TEAE buffer (pH 7.0) at a flow-rate of
4 ml/min, starting with acetonitrile-TEAA (20:80) for the 5'-dimethoxytrityl
component (gradient 1%/min) and acetonitrile-TEAA (5:95) for the 5'-hydroxy
component (gradient 0.5%/min). The base composition of backbone-modified
analogues of oligodeoxyribonucleotides was determined by the modified version
of the formic acid-catalysed degradation method (Fritz et al., 1982), which
involved heating at 120°C for 12 h and RP-HPLC on μBondapak C$_{18}$ with 2% acetonitrile in TEAA at a flow-rate of 4 ml/min, detection of the released cytosine,
guanine, thymine and adenine bases at 280 nm and normalization of peak areas
by comparison with a standard sample of GGAATTCC (Broido et al., 1984). Stec

et al. (1985b) discussed the limitations of the application of the described method for the separation of stereoisomers of structurally diverse backbone-modified oligodeoxyribonucleotide analogues.

Allinquant et al. (1985) studied the IP-RPC of nucleotides and oligonucleotides using a 5 μm Radial-Pak C_{18} column (100 mm x 8 mm I.D.) with two mobile phase systems: (A) 0.02 M monopotassium phosphate buffer (pH 6.5), containing 5 mM Pic A reagent (i.e. tetrabutylammonium phosphate used for ion pairing; Waters Assoc., Milford, MA, U.S.A.) and (B) a 1:1 (v/v) mixture of mobile phase A and acetonitrile. Elution was first performed with A (isocratic conditions) and after 3 min a linear gradient was started, leading in 45 min to final conditions of B. The constant flow-rate was 2.0 ml/min. Becker et al. (1985) described the use of a C_4 column for the RP-HPLC purification of synthetic oligonucleotides. For the separation of synthetic DNA fragments a large-pore C_4 column was found to be superior to small pore C_{18} columns. The C_4 functionalized column separated well not only DMT-bearing DNA from unsubstituted DNA, but also similar tritiated DNAs. Vydac C_4, 300 Å pore size, 5 μm particle size (Separation Group, Hesperia, CA, U.S.A.), was packed into an analytical (250 mm x 4.6 mm I.D.) or preparative (250 mm x 10 mm I.D.) column. A linear gradient of acetonitrile (e.g., 15-21%, 16-22% or 20-30%) against 0.1 M TEAA (pH 7.0) was used for elution. For a synthetic 51-mer, isocratic conditions (19% acetonitrile + 0.1 M TEAA) were used for the purification.

10.3.5 Normal-phase chromatography and special methods

Crowther et al. (1981) improved the HPLC separation of protected oligonucleotides using a polar bonded CN-phase with gradient elution with methylene chloride-methanol. High recoveries were obtained and problems associated with the application of highly polar silica columns were overcome. Some protecting groups (e.g., dimethoxytrityl) made the oligomers strongly lipophilic and therefore normal-phase chromatography (NPC) using only organic solvents gave more effective separations, owing to the limited solubility of many protected oligomers in the mobile phases used in RPC, which contain water. Originally silica was used for the NPC of such compounds but, owing to the low recoveries (e.g., De Rooij et al., 1979) and undesirable catalytic activity of silica, a more suitable chromatographic support was sought. A bonded cyanoamine phase (Whatman PAC) gave an excellent separation of the fully protected oligomer without apparent rearrangement reactivity. Gradient elution from methanol-dichloromethane (5:95) (solvent A) to 100% methanol-dichloromethane (50:50) (solvent B) in 20 min was routinely used. For semi-preparative separations a

Whatman Magnum-9 PAC column was applied, using isocratic elution with methanol-dichloromethane (7.5:92.5) at a flow-rate of 6.0 ml/min.

Another principle for the separation of oligonucleotides is based on the reaction of *cis*-diol groups of the non-phosphorylated terminal sugar moieties of oligoribonucleotides with an immobilized phenylboronate acid group: oligo-deoxyribonucleotides cannot react in this way. Phenylboronate chromatographic support is commercially available under the name Affi-Gel (Bio-Rad Labs.); cf., Sections 3.9 and 10.2.4. The preparation of the gel was described in detail by Uziel et al. (1976) and Gehrke et al. (1978), and the procedure for packing the phenylboronate column by Wehr (1980).

RPC-5 column chromatography (Wells et al., 1980), which was found to be very effective for the separation of nucleic acids and polynucleotides (cf., Section 10.2.4), suffers from a disadvantage with very short oligonucleotides, which require a low ionic strength for elution, that the support is not stable enough in media of low ionic strength. In spite of this, the method has been used for the separation of oligonucleotides [e.g., Hardies et al. (1976), Larson et al. (1979) and Patient et al. (1979) used this method for the separation of DNA restriction fragments].

Mixed-mode chromatography was also examined for the separation of oligonu-cleotides by a different approach (cf., Bishop et al., 1983; Bishop and McLaughlin, 1983; and comments by McLaughlin and Piel, 1984). Special supports were prepared either by hydrophobic modification of an anion exchanger or by anion-exchange modification of an ODS support. In some instances better resolu-tion than with classical packing materials was obtained using such special materials.

10.4 OLIGONUCLEOTIDES OF NATURAL ORIGIN

Natural oligonucleotides can be prepared from various extracts of biological origin, or they are products of digests of nucleic acids or polynucleotides. Extracts of biological materials are usually very complex and require a prelim-inary clean-up step to remove interfering components, in order to be capable of application to a chromatographic column. For this purpose anion-exchange chromatography is often used, both on resins and cellulose or Sephadex deriva-tives. Charged nucleotides and oligonucleotides (and, of course, also various natural zwitterions) are retained and uncharged or basic compounds are eluted. Nucleotides and oligonucleotides are then released by elution at higher ionic strength, and the combination of an ionic strength gradient with a pH gradient allows partial fractionation. Volatile buffers are used with advantage for such an elution. Sample clean-up of nucleic acid constituents was discussed by Wehr (1980). In the following part of this section, examples will be given of the high-speed fractionation of oligonucleotides from natural sources.

Wakizaka et al. (1978) described a rapid separation of dinucleotides and trinucleotides in DNase I digest of salmon sperm DNA according to base composition using HPLC with a weak anion-exchange column. The digest was preliminarily fractionated using conventional DEAE-Sephadex A-25 chromatography (according to Junowicz and Spencer, 1970) and the appropriate fractions of di- and trinucleotides were subjected to further separation on a 25 cm x 4 mm I.D. column of the weak anion exchanger LiChrosorb 10 NH$_2$ (d_p 10 μm) (E. Merck), using linear gradients of 1-61 mM sodium phosphate (pH 2.30 for di- and pH 2.25 for trinucleotides) at a flow-rate of 0.6 ml/min, at 60°C and 50 atm pressure. Thirteen dinucleotide peaks (including positional isomers) and thirty trinucleotide peaks were resolved. Differences in the separation patterns from those obtained by DEAE-Sephadex chromatography were discussed. Williams et al. (1979) studied the natural occurrence of 2'-5'-adenine dinucleotide inhibitors (2-5A) in interferon-treated EMC virus-infected L cells, which are produced in amounts consistent with the idea that they play a role in the inhibition of virus growth. For their characterization by HPLC analysis a LiChrosorb 10 NH$_2$ column was used and eluted with a linear gradient of 40-200 mM ammonium phosphate (pH 7.2). An excellent separation was obtained, but the columns of weak anion exchanger had a relatively short lifetime in phosphate buffers (only 30-50 runs were possible).

Jones (1981) described the determination of pyridine dinucleotides (NAD$^+$ and NADP$^+$) in an isolated liver cell extract. The dinucleotides were separated on a μBondapak C$_{18}$ column using the following programme: 5 min with 50% buffer A, 10 min gradient to 25% buffer A, hold for 4 min with 25% buffer A; buffer A was 100 mM potassium phosphate (pH 6.0) and buffer B was buffer A containing 5% (v/v) methanol. The flow-rate was 1 ml/min, the temperature was 18-22°C and the recycling time between runs was 1 min to 50% buffer A and 10 min before the next injection.

McLaughlin et al. (1981) presented a method for the rapid analysis of modified tRNA^{Phe} from yeast by HPLC involving chromatography of oligonucleotides after RNAse T$_1$ digestion on aminopropylsilica and assignment of the fragments based on nucleoside analysis by chromatography on C$_{18}$-silica. Oligonucleotide separations were performed on a 250 mm x 4.6 mm I.D. Zorbax-NH$_2$ column (DuPont, Bad Neuheim, F.R.G.). Nucleoside separations were performed on a 5 μm ODS-Hypersil column (250 mm x 4.6 mm I.D.) (Shandon Southern, Runcorn, U.K.). The laboratory filling of the last column was described in detail. The elution profile of a digest on aminopropylsilica is shown in Fig. 10.12.

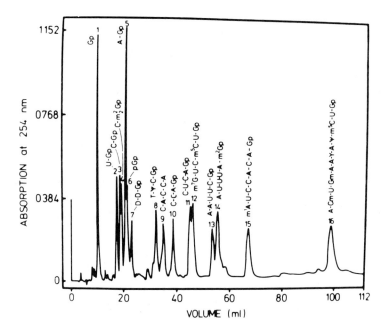

Fig. 10.12. Separation of RNase T$_1$ digest of yeast tRNA^{Phe} on a Zorbax-NH$_2$ (7 μm) column. Buffers: A = 0.05 M KH$_2$PO$_4$; B = 10% (v/v) methanol in 0.9 M KH$_2$PO$_4$. Gradients: 0-60% buffer B during 36 min, 60-100% buffer B during 5 min and finally 15 min with 100% buffer B. Flow-rate, 2 ml/min; detector 1.28 a.u.f.s.; temperature, 35°C. (Reprinted from McLaughlin et al., 1981.)

Block and Pingoud (1981) described the identification and analysis of nucleotides bound to elongation factor Tu from *Escherichia coli*. They found that aminoalkylsilica columns had an excellent resolving power for different nucleotides using NH$_4$H$_2$PO$_4$-salt gradients (0.1-0.7 M at pH 4.5) for elution. A glass column (30 cm x 0.3 cm I.D.) pre-packed with 10 μm amino-alkyl-silica reversed-phase particles (Riedel de Haen) was used at room temperature with a flow-rate of 1.5-2.0 ml/min and a back-pressure up to 100 bar.

Block et al. (1982) studied the use of HPLC for the analysis and determination of signal nucleotides derived from guanosine and adenosine. Acid-soluble nucleotides act as low-molecular-weight mediators in the regulation and coordination of intracellular reactions (cf., review by Koch and Richter, 1979). A reversed-phase amino column (25 cm x 0.46 cm I.D.) of 10 μm LiChrosorb RP-NH$_2$ (Merck), eluted with a linear gradient of NH$_4$H$_2$PO$_4$ of pH 4.5 (0.2-0.5 M, 20 min) at a flow-rate of 2 ml/min led to a good separation of nucleotides of varying degrees of phosphorylation. Blokhin and Poteshnykh (1983) described the separation and analysis of ribonucleotide mixtures of natural origin by anion-exchange HPLC. A Whatman Partisil 10 SAX column (250 mm x 4.6 mm I.D.) with an 80 mm x

4.6 mm I.D. guard column (pellicular anion exchanger, Whatman) were eluted with a gradient from A (3 mM KH_2PO_4-0.3 mM KCl, pH 3.3) to B (0.3 M KH_2PO_4-0.3 M KCl, pH 4.9) (5%/min) at a flow-rate 2 ml/min and 35°C. Problems associated with the extraction of nucleotides from biological material were discussed.

A special group of separation methods applied to natural oligonucleotides is represented by template chromatography, which is a variation of affinity chromatography (cf., Sections 3.9 and 4.4.5). Schott and Greber (1971) studied the column chromatographic separation of nucleoside-nucleotide and nucleoside-dinucleotide phosphate mixtures on nucleoside gels, prepared by the immobilization of nucleoside residues to polymer gels. In the presence of water nucleosides interacted more strongly with this support, allowing the quantitative separation of nucleosides from mono- and dinucleotides, and in the presence of buffers the separation of mononucleotides from nucleosides and dinucleotides was possible. Schott (1980) described a preparative isolation of oligonucleotides from chemically degraded DNA. This approach represented a simple alternative to complicated chemical synthesis aimed at the preparation of the desired oligonucleotides from readily available natural sources. Herring sperm DNA or DNA which occurs as a by product of biotechnological processes served as a raw material. The author developed various more or less specific methods for the preparative partial hydrolysis of DNA leading to various mixtures of oligonucleotides. Pre-separation was effected on DEAE-cellulose, followed by QAE-Sephadex chromatography in the presence of urea or preparative paper chromatography, and template chromatography. For the separation of higher molecular weight material (prepared by partial hydrolysis of oxidized DNA) a PVAL-p(dT)$_n$-DEAE-cellulose [PVAL = poly(vinyl alcohol)] column (20 cm x 2 cm I.D.) was prepared, equilibrated with 1 M NaCl-0.01 M Na_2HPO_4 (pH 6.5) and cooled to -4°C. The mixture was passed through the cooled column and washed with the same cold buffer until nearly all absorbing material was removed, then the temperature was increased to 30°C and elution with the same buffer was continued (40 ml/h). The material corresponding to the double peak was desalted, lyophilized, enzymatically dephosphorylated and rechromatographed using conventional supports.

Schott and Schrade (1981) examined the preparative isolation of purine oligonucleotides from chemically degraded DNA. Here also the readily available herring sperm DNA was used as a natural source, which was treated with hydrazine and subsequently with alkali. The purine nucleotide segments remained intact and were set free from depyrimidinated DNA. From the resulting partial hydrolysate defined purine oligonucleotides were isolated using conventional low-pressure chromatographic methods. RP-HPLC was applied only for the final separation of d(G-G-A-G-A-G) from a mixture of sequence isomers $d(A_2,G_4)$. Schott (1982)

also described the preparative isolation of oligothymidine phosphates from
partial hydrolysates of chemically degraded DNA using template chromatography.
Herring sperm DNA (500 g) was treated with formic acid-diphenylamine for 12 h
at $37^{\circ}C$ and the neutral partial hydrolysate was ultrafiltered and chromatographed on DEAE-cellulose. Long-chain pyrimidine oligonucleotide mixture (80 g)
was fractionated on QAE-Sephadex. The final fraction contained 1.5 g of oligothymidine phosphates, which were further separated by template chromatography
and rechromatography on QAE-Sephadex. Chromatographically pure single substances
(1-4 mg) of the $(dT)_{3-12}$ series were obtained.

Schott and Watzlawick (1982) studied the preparative isolation of oligoguanosine phosphates from partial hydrolysates of DNA using template chromatography on two differently substituted $PVAL-p(dC)_{n}$-DEAE-celluloses, the synthesis
of which was described. Herring sperm DNA (100 g) was treated with hydrazine
hydrate for 4 h at $60^{\circ}C$, and depyrimidinated DNA (37 g) after alkaline hydrolysis
(7 M KOH, 1 h, $100^{\circ}C$) yielded an oligoguanosine phosphate mixture (14 g), which
was fractionated on DEAE-cellulose, QAE-Sephadex A-25, $PVAL-p(dC)_{n}$-DEAE-cellulose
and part (after enzymic dephosphorylation) on QAE-Sephadex A-25 and by means of
HPLC on a LiChrosorb RP-18 (5 µm) column (250 mm x 4.6 mm I.D.). A linear
gradient (A to 60% A-40% B in 20 min) was applied in this instance, followed by
isocratic elution (60% A-40% B); buffer A was 0.01 M KH_2PO_4 (pH 5-6) and B was
methanol-water (60:40).

In most instances the described combined preparative separation methods
including template chromatography steps led to homogeneous oligonucleotides,
the purity of which was tested by homochromatography and "fingerprinting" (I,
electrophoresis; II, homochromatography).

10.5 HPLC IN THE SYNTHESIS OF OLIGONUCLEOTIDES

At present HPLC methods constitute, similarly to peptide chemistry, an integral part of oligonucleotide synthesis procedures. They are used mainly (a) for
purification of synthetic products and (b) for checking their degree of homogeneity. All LC modes can be employed for these purposes. Ultrapure oligonucleotides can be prepared using a combination of HPLC methods.

In this section a short survey of applications of HPLC techniques in the
synthesis of oligonucleotide will be given. One of the techniques used for many
years in oligonucleotide chemistry is ion-exchange chromatography. It was
elaborated from lengthy procedures using anion-exchange derivatives of cellulose
and polydextran, or from the application of anion-exchange resins, to pelliculars,
or modern microparticulate anion exchangers on bonded phases. For instance,
Leutzinger et al. (1978) studied the application of HPLC in the synthesis of

oligonucleotides using a weakly basic anion-exchange column (100 cm x 0.31 cm I.D. or 100 cm x 0.21 cm I.D.) of Pellionex AL WAX (Whatman) for the separation of synthetic protected products. Stepwise elution or linear gradients of ammonium acetate (1 mM to 4.0 M) or ammonium sulphate (1 mM to 2.0 M) in ethanol (40% or 60%), buffered at pH 5.0-6.2, were used for elution. The published examples demonstrated the possibility of both analytical and preparative applications of HPLC to the synthesis of protected 2'-deoxyribohexanucleotide. Miyoshi et al. (1980c) described the synthesis of polythymidylic acids by the block-coupling phosphotriester method and the final products were isolated by HPLC on Permaphase AAX [buffer A, 0.05 M KH_2PO_4 (pH 4.5); B, 0.05 M KH_2PO_4-1 M KCl (pH 4.5)] using linear gradient from A to 3% B/min at a flow-rate of 2 ml/min.

Hagemeier et al. (1982) examined HPLC in the separation of dinucleotides using an Altex Partisil PAC (10 μm) column (250 mm x 3.2 mm I.D.). In addition to experiments under isocratic conditions, a good separation was obtained with a linear gradient from water up to 0.8 M ammonium formate (pH 4.1), at a flow-rate of 2 ml/min and an inlet pressure of 2000 p.s.i.; ten components of a synthetic mixture were resolved. Efimov et al. (1982) developed an effective method for the synthesis of oligonucleotides via phosphotriester intermediates for work in solution. A number of DNA fragments up to 32 long were synthesized. The proposed method was adapted to solid-phase synthesis and a pentadecamer was assembled in 4-5 h using dinucleotides as coupling units. The deprotected oligonucleotides were purified by gel electrophoresis and ion-exchange chromatography on a Pellionex SAX column (250 mm x 4 mm I.D.) with a gradient from 0 to 1 M KCl in 0.02 M potassium phosphate (pH 4.5) - 40% ethanol over 30 min at 50°C at a flow-rate of 2 ml/min. The homogeneity of the desalted main peak was checked by RPC on a Zorbax C_8 column (250 mm x 4 mm I.D.) using elution at a flow-rate of 1 ml/min with a gradient of 5-35% methanol in 0.1 M ammonium acetate at 45°C.

Coomber (1985) studied the purification of synthetic oligodeoxyribonucleotides with a high guanosine content by anion-exchange HPLC. The usual difficulties associated with the IEC of these compounds when standard conditions were used were overcome provided that the base-protecting groups (isobutyryl [ib] and benzoyl [bz]) were retained on the otherwise deprotected oligomer. A Partisil 10 SAX column (25 cm x 0.5 cm I.D.) was applied for separation using gradient elution with 0.001 M potassium dihydrogen phosphate in formamide-water (4:6) to 0.4 M potassium dihydrogen phosphate in formamide-water (4:6) in 1 h at 50°C. The method was exemplified on the synthesis and purification of d(GGTCAGGGTGGAGAT).

Reversed-phase chromatography has also been found to be very effective for applications in oligonucleotide synthesis. The fundamental papers by Fritz et al. (1978) on the use of HPLC in the stepwise synthesis of deoxyribopolynucleotides and by Jones et al. (1978) have already been discussed in Section 10.3.4.

Balagaje et al. (1979), in their paper on the total synthesis of tyrosine
suppressor transfer DNA gene, emphasized that a noteworthy feature of the
described syntheses was the use of RP-HPLC for the rapid and efficient separation of synthetic reaction mixtures. A μBondapak C_{18} column (30 cm x 0.4 cm I.D.)
was used for analytical runs, and a μBondapak C_{18}-Porasil B column (183 cm x
0.7 cm I.D.) for preparation (the capacity of the latter column was 200-300 mg
of nucleotidic material) at a flow-rate of 9.9 ml/min. An auxiliary column
(61 cm x 0.2 cm I.D.) packed with the same support was employed to define the
necessary preparative conditions, such as the concentration (A) of acetonitrile
in 0.1 M triethylammonium acetate (TEAA) which eluted impurities, the concentration (B) which eluted oligonucleotides but kept the tritylated material on the
column, and the concentration (C) that eluted the mixture of tritylated compounds.
In general, two linear gradients (A to B and B to C) were used to fractionate
the reaction mixture. Finally, a steep gradient to 20% more acetonitrile was
used and the column was then eluted for 10 min at this concentration.

Caruthers et al. (1980), in a paper on the synthesis of nucleotides, also
described the use of HPLC. A Prep 500 C_{18} (Waters Assoc.) RP column was used
for the separation of dinucleotide 5'-phosphates. The mixture resulting from
a synthetic step was dissolved in 0.1 M TEAA containing 16% acetonitrile and
resolved isocratically using the same solvent. The products were pure, but the
yield was less than 50%. Broka et al. (1980) described the simplified synthesis
of trinucleotide blocks without purification at the dimer stage. The fully
protected trinucleotides were isolated in 50-60% yields by short column chromatography on silica gel to remove charged phosphate compounds, followed by column
chromatography on μBondapak C_{18} RP silica gel. A Prep LC System 500 on an RP
silica gel cartridge (Waters Assoc.) was used for the purification. The eluent
was acetone-water (7:3, v/v) containing 0.1 M diethyleneammonium acetate (pH 7).
The method was very convenient for both solution synthesis and solid-phase
synthesis. Matteucci and Caruthers (1981) examined the synthesis of deoxyoligo-
nucleotides on a silica-polymerized support. The expected products and all
polymer-bound intermediates were freed from the support and base-labile protecting groups by treatment with concentrated ammonia solution and analysed by
RP-HPLC. A μBondapak C_{18} column was used with 0.1 M TEAA (pH 7) and acetonitrile
(e.g., 26%) for the elution of dimethoxytrityl derivatives of oligodeoxynucleotides.

Crowther et al. (1981) studied the HPLC separation of deprotected oligonucleotides by IP-RPC using an ODS stationary phase and protected oligonucleotides
by NPC using a polar CN-bonded phase; these methods were discussed in Sections
10.3.4 and 1.3.5, respectively.

Chow et al. (1981) applied RP-HPLC on an RP C_{18} column (Altex) for the separation of both protected and deprotected products. First a gradient of 20-30% acetonitrile in 0.1 M TEAA (30 min) was used, and the required peak was detritylated and chromatographed again using a gradient of 10-15% acetonitrile in 0.1 M TEAA (flow-rate 1 ml/min). Tanaka and Letsinger (1982) described a simple procedure ("syringe method") for a stepwise chemical synthesis of oligonucleotides, based on phosphite chemistry. Oligonucleotide chains can be constructed rapidly with a minimum of equipment (a syringe and reagent bottles). The method was illustrated by the synthesis of d-TGCAGGTT. Pertinent supporting data on the effect of variations in the detritylation, condensation, oxidation, capping and cleavage steps in the synthetic approach and in isolation procedures were also presented. The final purification was effected by means of HPLC: a Whatman Partisil PX 10/25 ODS-2 column with a gradient of 10-15% acetonitrile in 0.1 M TEAA or a PXS 10/25 SAX column with a gradient from 0.125 M triethylamine-0.075 M citric acid to 0.25 M triethylamine-0.15 M citric acid was used. Ohtsuka et al. (1982) studied the synthesis of λ cro binding heptadecanucleotide by means of a new condensing reagent on a polymer support, and for the final purification of the protected product used RP-HPLC on a C_{18} silica gel column, eluting with a gradient of acetonitrile (5-40%) in 0.1 M TEAA. After removal of the dimethoxytrityl groups, the product was isolated by chromatography on C_{18} silica using a gradient of 5-25% acetonitrile and then analysed by HPLC on μBondapak C_{18}.

Ike et al. (1983) examined the possibility of the simultaneous synthesis of mixed oligonucleotides by the phosphotriester solid-phase method. For example, for the analysis of a mixture modified with 4',4'-dimethoxytrityl (DMT) groups, a 12.5-15% acetonitrile gradient in 40 min was used. After removal of all protecting groups, the mixture was again analysed by HPLC on a μBondapak C_{18} column. A linear gradient of 4-7% or 5-15% acetonitrile for 20 min was used. Dörper and Winnacker (1983) presented the results of a double HPLC separation of synthetic oligonucleotide (21-mer) prior to the removal of the DMT group and after deprotection. A μBondapak C_{18} column and isocratic elution with 25% acetonitrile in 0.1 M TEAA (pH 7.0) were used. Kalashnikov et al. (1983) systematically studied the application of RPC in the oligonucleotide triester solution synthesis and the purification of protected and fully deblocked preparations; silanized silica, silanized porous glass and GLC Polychrom-1 (an U.S.S.R. product) were tested. Hsiung et al. (1983) described further improvements to the phosphotriester synthesis, consisting also of a simplified isolation procedure that eliminated the tedious hydrogen carbonate extraction after each condensation reaction, replacing it with diethyl ether-hexane (1:1, v/v) or diethyl ether precipitation. The final products were purer than those obtained

TABLE 10.1

EXAMPLES OF HPLC CONDITIONS USED FOR PURIFICATION OR ANALYSIS OF SYNTHETIC OLIGO- AND POLYNUCLEOTIDES

Method	Support	Buffer or elution liquid	Gradient	Flow-rate and temperature	Notes	Ref.
IEC	Permaphase AAX (DuPont) (50 cm x 0.4 cm I.D.)	A = water; B = 1 M KCl (pH 4.5)	Linear, 3%/min	3 ml/min; 60°C		Crea and Horn (1980)
	Anion exchanger AX-10A (Brownlee)	A = 0.2 M TEAA (pH 2.9); B = 0.5 M TEAA (pH 3.5)	Linear, 0-100% B in 60 min	4 ml/min		Alvaro-Urbina et al. (1981)
	LiChrosorb AN (25 cm x 4.6 cm I.D.)					
	Partisil PXS 10/25 SAX (25 cm x 4.6 cm I.D.)					
	Permaphase AAX (Waters Assoc.)	A = 0.01 M KH_2PO_4; B = 0.05 M KH_2PO_4 and 0.7 M KCl (pH 4.45)	Linear	60 or 65°C		Balgobin et al. (1981)
	LiChrosorb	A = water; B = KH_2PO_4 solution	Linear, 0 to 0.07 M			Rosenthal et al. (1983)
RPC	Partisil PXS ODS-3	A = 0.1 M ETAA; B = acetonitrile	Linear from 11% B, 0.1%/min			Letzinger et al. (1982)
	Ultracil RP C_{18} (Altex)	7.5%, 8.5% or 11% aceto-nitrile in 0.1 M TEAA (pH 7)	Isocratic		Deblocked derivatives	Elmblad et al. (1982)
	Spherisorb RP C_{18} (CDC HPLC)		20-30% acetonitrile gradient in 0.1 M TEAA		DMT derivatives	

Column	Mobile phase	Gradient	Flow rate	Application	Reference
Micropak MCH-10 (Varian) (30 cm x 4 mm I.D.)	A = water; B = acetonitrile; C = 3 M TEAF* (pH 5)	A:B:C from 9:0:1 to 1:8:1 in 25 min			Patel et al. (1982)
µBondapak C_{18} (Waters Assoc.)	Acetonitrile in 0.1 M TEAA (pH 7.0)	20-30%, 5-20%		DMT derivatives, detritylated derivatives	Kohli et al. (1982)
µBondapak C_{18}	Various concentrations of acetonitrile in water	Isocratic			Fisher and Caruthers (1983)
Zorbax C_8 (DuPont)	A = 0.2 M ammonium acetate (pH 7.2); B = methanol	0 to 100% B	1 ml/min		Dobrynin et al. (1983)
Nucleosil 30 C_{18} (25 cm x 0.6 cm I.D.)	A = 0.1 M TEAA or ammonium acetate (pH 8); B = acetonitrile, dioxane	Linear	2-4 ml/min	Protected oligonucleotides	Efimov et al. (1983a)
Zorbax C_8 (250 mm x 4 mm I.D.)	A = 0.1 M ammonium acetate; B = methanol	5-20% B	1 ml/min	Unprotected derivatives	
LiChrosorb 10 RP 18	A = 0.025 or 0.1 M TEAA (pH 8); B = acetonitrile or dioxane	Linear, 45-75% acetonitrile in 0.025 M TEAA or in 0.1 M TEAA containing 10-20% dioxane; linear; 60-80% dioxane in 0.1 M TEAA		Protected	Efimov et al. (1983b)
Nucleosil 30 C_{18} (Machery Nagel Co.) (25 cm x 1 cm I.D.)					
Zorbax C_8 (250 mm x 4 mm I.D.)	A = 0.1 M TEAA; B = methanol	Linear		Deblocked	
µBondapak C_{18}	0.1 M TEAA (pH 6.8)-acetonitrile (70:30)	Isocratic	2.2 ml/min		Adams et al. (1983)

(Continued on p. B228)

TABLE 10.1 (continued)

Method	Support	Buffer or elution liquid	Gradient	Flow-rate and temperature	Notes	Ref.
RPC	RP C$_8$ resin (Alltech)	12% methanol in 0.05 M aqueous TEAA	Isocratic			Froehler and Matterucci (1983)
	μBondapak C$_{18}$ (30 cm × 7.8 mm I.D.)	A = 0.1 M TEAA; B = acetonitrile	Linear, 20-50% in 30 min;	2 ml/min	Purification, protected	Gaffney et al. (1984)
	Radial-Pak Bondapak C$_{18}$ cartridge (analytical)		Linear, 20-50% in 30 min;	2 ml/min	Purification, detritylated	
			Linear, 5 or 6-20% in 5 min	4 ml/min	Analytical	
	RP C$_{18}$ (Beckman)	A = 0.1 M TEAA; B = acetonitrile	10-50% B in 20 min	1.5 ml/min		Köster et al. (1984)
	Partisil PXS ODS 3 (Whatman)	A = 0.1 M TEAA; B = acetonitrile	In various proportions, from 10% (for 8-mer), from 11% (for 14-mer); 0.1%/min			Letsinger et al. (1984)
	IBM-C$_{18}$					

*TEAF = Triethylammonium formate.

by solid-phase chemistry, as each intermediate block was purified by chromatography. The synthetic oligonucleotide triester was deblocked and purified on
polyethylenimine (PEI)-impregnated cellulose TLC. The PEI-purified oligonucleotides were further eluted and purified by HPLC on a μBondapak C_{18} column
according to Brown et al. (1980).

Other examples of chromatographic conditions for HPLC separations in chemical
syntheses of oligonucleotides are listed in Table 10.1.

Modern separation methods have also been successfully applied in enzymic
syntheses. Shum et al. (1978) described simplified methods for the large-scale
enzymic synthesis of oligoribonucleotides using polynucleotide phosphorylase.
The main features of the method are the use of RPC-5 chromatography (cf.,
Section 10.3.5) for checking the reaction conditions and reaction course. An
example of such a chromatogram is given in Fig. 10.13. El Rassi and Horváth
(1983) examined the high-performance displacement chromatography of nucleic
acid fragments in a tandem enzyme reactor - liquid chromatographic system. The
separation of the reaction mixture in the ribonuclease T_1-catalysed synthesis
of GpU from cyclic GMP in the presence of a large excess of uridine was carried
out by displacement chromatography on an ODS column; 0.25 M n-butanol in the
carrier buffer (50 mM phosphate, pH 3) was used as a displacer. The product of
a small packed-bed reactor was introduced directly into the chromatographic unit,
consisting of two columns, 150 mm x 4.6 mm I.D. (Supelcosil LC-18) and 250 mm x
4.6 mm I.D. (Zorbax C_{18}), connected in series. The bulk uridine withdrawn from
the first column was further separated (also by a displacement development) in
the second column. The technical arrangement and experimental conditions were

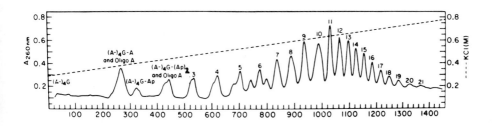

Fig. 10.13. RPC-5 chromatography of a product of large-scale reaction of A_4G
primer with ADP, [ADP]/[A_4G] = 28, in 0.2 M Tris-HCl (pH 9.1)-5 mM MgCl$_2$-800 mM
NaCl, with 0.2 mg/ml primer-dependent PNPase, incubated at 37°C for 3 h. Peaks
numbered N are (A-)$_4$G-(Ap)$_N$. The elution profile shown was obtained with a
10 cm x 2.6 cm I.D. column, gradient volume = 30 l, from 0.28 to 1.0 M KCl-0.01 M
imidazole (pH 7); flow-rate, 77 ml/h; fraction size, 15 ml. (Reprinted from Shum
and Crothers, 1978.)

described in detail. Nearly 100 mg of GpU with a purity of 99.7% could be obtained in a chromatographic experiment taking 2.4 h.

It is logical and might be expected that the combination of various chromatographic modes (at least two, but even more) would lead to much sharper separations of oligonucleotides. This approach is illustrated in the following part of this section. For example, Teoule et al. (1980), in experiments with oxytocin gene synthesis, used (1) a 10 μm Partisil SAX column (300 mm x 4.7 mm I.D.) with a KCl-KH$_2$PO$_4$ (pH 6.7) gradient at 1 ml/min for IEC, (2) a Waters Assoc. Protein Analysis Column eluted with 0.1 M TEAA (pH 7) for SEC and (3) a Merck 10 μm LiChrosorb RP-8 column (300 mm x 4.7 mm I.D.) with a gradient of methanol in 0.1 M TEAA (pH 7) for RPC. Gait et al. (1980a) applied Partisil 10 SAX with buffer A (1 mM KH$_2$PO$_4$-5% or 30% ethanol)-buffer B (0.2 M KH$_2$PO$_4$-5% or 30% ethanol) (pH 6) for gradient IEC, and μBondapak C$_{18}$ with (A) 0.1 M ammonium acetate and (B) acetonitrile-0.1 M ammonium acetate (8:2), using a gradient of 8-12% B (45 min) or 8-15% B (30 min) for subsequent RPC. Similarly, Miyoshi and Itakura (1980) chromatographed the products on the ion exchanger Permaphase AAX (DuPont) with buffers (A) 0.05 M KH$_2$PO$_4$ and (B) 0.05 M KH$_2$PO$_4$-1.0 M KCl (pH 4.5), using a linear gradient of 3% B/min at 55°C and a flow-rate of 2 ml/min. For checking the purity by RPC μBondapak C$_{18}$ was used.

Chakhmakhcheva et al. (1980), in work on the chemical synthesis of the structural gene Leu-enkephalin, applied the following combination of chromatographic methods for the separation of unprotected oligonucleotides: (1) IEC on DEAE-cellulose in the presence of 7 M urea (pH 7.5), (2) IEC on AS Pellionex SAX (250 mm x 4 mm I.D.) at pH 4.5 and (3) RPC on Zorbax C$_8$ (DuPont) using acetonitrile gradients in 0.1 M ammonium acetate (to 5% in 2 min and 5-12.5% in 10 min). Gait et al. (1980b), in an improved solid-phase synthesis of oligodeoxyribonucleotides through phosphotriester intermediates, carried out the HPLC separation by means of Partisil 10 SAX columns (PXS, analytical; M 9, preparative) eluted with a gradient of potassium phosphate (pH 6.5)-5% ethanol from 1 mM to 0.2 M at ambient temperature; a silica pre-column can prolong the otherwise short lifetime of the main column (30-50 injections). The purity of the samples was determined by means of RPC on a μBondapak C$_{18}$ column using 0.1 M ammonium acetate-acetonitrile solvent mixtures. Markham et al. (1980) studied the solid-phase phosphotriester synthesis of large oligodeoxyribonucleotides (up to heneicosanucleotide) on a polyamide support and the deblocked product was purified or analysed by HPLC: for IEC a Partisil 10 SAX column (PXS) was used and eluted with a gradient of B (0.4 M potassium phosphate, pH 6.8) to A (1 mM potassium phosphate, pH 6.8), both in 5% ethanol. For RPC of the 21-mer, a μBondapak C$_{18}$ column was eluted with a gradient to 10% B (6 min) and 10-25% B

(40 min); A = 0.1 M ammonium acetate, B = a mixture of A with acetonitrile (1:1). An approach to the purification of the synthetic octadecanucleotide without recourse to HPLC was also described.

Miyoshi et al. (1980b) studied the synthesis of polynucleotides on a poly-acrylamide support and for the separation of unprotected products they used Permaphase AAX chromatography at 55°C and RPC on a μBondapak C_{18} column (30 cm x 0.3 cm I.D.) at 55°C; linear gradients of acetonitrile in ethylendiammonium acetate (EDAA) were applied; A = 0.01 M EDAA (pH 7); B = 15% acetonitrile in A. The flow-rate was 2 ml/min and the time required was 15 min. Similar separation procedures for the isolation and analysis of products were used in other syn-thetic experiments on polystyrene resin supports (Miyoshi et al., 1980a). Duckworth et al. (1981) described a rapid and efficient mechanized synthesis of heptadecadeoxyribonucleotides by an improved solid-phase phosphotriester route. In addition to the synthetic procedure, improvements were also made in the purification of deprotected oligonucleotides by HPLC, which consisted of combinations of IEC and RPC. Partisil 10 SAX was applied for IEC using elution with a gradient system of buffer A (1 mM potassium phosphate, pH 6.2-30% ethanol) and B (0.3 M potassium phosphate, pH 6.2-30% ethanol) at 50-60°C; the flow-rate was 2 ml/min for analytical and 7 ml/min for preparative purposes. Various gradient intervals were used for particular oligonucleotides. A column of μBondapak C_{18} was used for RPC; buffer A was 0.1 M ammonium acetate and B was 0.1 M ammonium acetate-acetonitrile (2:8); various gradient steepnesses were applied, e.g., 4 min 10% B, 40 min 10-13% B, at 35°C. The samples were warmed before injection.

10.5.1 Notes on literature

The solid-phase synthesis of oligonucleotides by the phosphotriester method was described (including comments on IE-HPLC and RP-HPLC separations) in a practical review by Sproat and Gait (1984). The chromatographic purification of synthetic oligonucleotides was reviewed and illustrated by some detailed practical examples by McLaughlin and Piel (1984a).

10.6 COMMENTS ON LITERATURE

Several books describing separation principles and techniques in the chem-istry and biochemistry of nucleic acids and their fragments have been published. Wehr (1980) wrote a small practical booklet on the separation of nucleic acid constituents by HPLC. Flavel (1983) edited a book dealing with techniques in nucleic acid biochemistry, and Brown (1984a) a monograph on HPLC in nucleic acid

research and applications; in addition to methodology and applications, the
later book also contains an overview of nucleic acid research and properties
of nucleic acid constituents important from the point of view of chromatography.
Gait (1984) edited a short but very useful book on oligonucleotide synthesis
(a practical approach) describing also separation procedures. Schott (1984)
wrote a book on the template affinity chromatography of nucleic acids and proteins, describing the immobilization of oligonucleotides, polynucleotides and
nucleic acids, the chromatography of oligonucleotides, the isolation of DNA and
its fragments, ribonucleic acids, the enzymic synthesis and degradation of
polynucleotides (using polymer-bound primers, templates and substrates) and
studies of peptide-nucleotide interactions.

   Numerous reviews have been published on both general and special topics in
this field of chemistry and biochemistry and a short survey of some of them
will be given here. Among the first were reviews by Brown (1978) and Brown et
al. (1980), describing the development and current state of the art in the HPLC
of free nucleotides, nucleosides and bases in biological fluids. Wells et al.
(1980) described all methodical details of RPC-5 column chromatography for the
isolation of DNA fragments, Ikehara and Ohtsuka (1981) (in Japanese) dealt with
the chemical synthesis of genes including the separation and purification of
oligonucleotides, Hillen and Wells (1983) reviewed the HPLC of DNA and Scoble
and Brown (1983) the reversed-phase chromatography of nucleic acid fragments.
Thompson et al. (1983) presented many practical details of the purification of
nucleic acids by RPC-5 ANALOG chromatography (peristaltic and gravity-flow
applications).

   Crowther (1984) reviewed the application of HPLC for the separation of
oligonucleotides, Simpson (1984) for the separation of nucleic acids and Wells
(1984) the HPLC of DNA. Wu et al. (1984) described the purification and
sequence analysis of synthetic oligonucleotides. Zakaria (1984) dealt with chromatographic methods for nucleic acid constituents in connection with disease
processes.

   Floyd et al. (1985) have written an extensive review (101 references) of the
HPLC of nucleic acids, describing trends in the separation of synthetic single-
stranded oligonucleotides and double-stranded DNA fragments. Kwiatkowski et al.
(1985) dealt with some aspects of oligonucleotide chemistry, including HPLC
separations (38 references), and Perrone and Brown (1985) the ion-pair chromatography of nucleic acid derivatives.

REFERENCES

Adams, S.P., Kavka, K.S., Wykes, E.J., Holder, S.B. and Galluppi, G.R., *J. Am. Chem. Soc.*, 105 (1983) 661-663.

Allinquant, B., Musenger, C. and Schuller, E., *J. Chromatogr.*, 326 (1985) 281-291.

Alvaro-Urbina, G., Sathe, G.M., Liu, W.-Ch., Gillen, M.F., Duck, P.D., Bender, R. and Ogilvie, K.K., *Science (Washington, D.C.)*, 214 (1981) 270-274.

Anderson, F.S. and Murphy, R.C., *J. Chromatogr.*, 121 (1976) 251-262.

Assenza, S.P., in P.R. Brown (Editor), *HPLC in Nucleic Acid Research*, Marcel Dekker, New York, 1984, pp. 139-160; *C.A.*, 102 (1985) 92 346t.

Asteriadis, G.T., Armbruster, M.A. and Gilham, P.T., *Anal. Biochem.*, 70 (1976) 64-74.

Aukaty, M.F., Bubenschikova, S.N., Kagrammanova, V.K. and Baratova, L.A., *J. Chromatogr.*, 137 (1977) 351-356.

Aukaty, M.F., Bubenshchikova, S.N., Kagramanova, V.K. and Baratova, L.A., *Vestn. Mosk. Univ., Ser. Khim.*, 19 (1978) 350-352; *C.A.*, 89 (1978) 159 647a.

Balgobin, N., Josephson, S. and Chattopadhyaya, J.B., *Acta Chem. Scand., Ser. B*, 35 (1981) 201-212.

Becker, C.R., Efcavitch, J.W., Heiner, C.R. and Kaiser, N.F., *J. Chromatogr.*, 326 (1985) 293-299.

Belagaje, R., Brown, E.L., Fritz, H.J., Lees, R.G. and Khorana, H.G., *J. Biol. Chem.*, 254 (1979) 5765-5780; *C.A.*, 91 (1979) 170 105j.

Bethell, G.S., Ayers, J.S., Hearn, M.T.W. and Hancock, W.S., *J. Chromatogr.*, 219 (1981) 353-359.

Bischoff, R., Graeser, E. and McLaughlin, L.W., *J. Chromatogr.*, 257 (1983) 305-315.

Bischoff, R. and McLaughlin, L.W., *J. Chromatogr.*, 270 (1983) 117-126.

Block, W., Lüstorff, J. and Pingoud, A., *Fresenius' Z. Anal. Chem.*, 311 (1982) 422.

Block, W. and Pingoud, A., *Anal. Biochem.*, 114 (1981) 112-117; *C.A.*, 95 (1981) 57 414d.

Blokhin, D.Yu. and Poteshnykh, A.V., *Bioorg. Khim.*, 9 (1983) 673-677.

Bloomfield, V.A., Crothers, D.M. and Tinoco, I., Jr., *Physical Chemistry of Nucleic Acids*, Harper and Row, New York, 1974, 515 pp.; *C.A.*, 81 (1974) B59 729v.

Broha, Ch., Hozumi, T., Arentzen, R. and Itakura, K., *Nucleic Acids Res.*, 8 (1980) 5461-5471.

Broido, M.S., Zon, G. and James, T.L., *Biochem. Biophys. Res. Commun.*, 119 (1984) 663-670; *C.A.*, 100 (1984) 170 150z.

Brown, E.L., Belagaje, R., Ryan, M.J. and Khorana, H.G., *Methods Enzymol.*, 68 (Recomb. DNA) (1980) 109-151; *C.A.*, 93 (1980) 90 402e.

Brown, P., *Chem. Aust.*, 45 (1978) 257-264; *C.A.*, 90 (1979) 50 699p.

Brown, P.R. (Editor), *HPLC in Nucleic Acid Research. Methods and Applications (Chromatographic Science Series, Vol. 28)*, Marcel Dekker, New York, 1984a, 424 pp.; *C.A.*, 102 (1985) B109 357w.

Brown, P.R. in Brown, P.R. (Editor), *HPLC in Nucleic Acid Research. Methods and Applications (Chromatographic Science Series, Vol. 28)*, Marcel Dekker, New York, 1984b, pp. 3-15; *C.A.*, 102 (1985) 91 469y.

Brown, P.R., Krstulovic, A.M. and Hartwick, R.A., *Adv. Chromatogr.*, 18 (1980) 101-138.

Brown, R.E., Cayley, P.J. and Kerr, I.M., *Methods Enzymol.*, 79 (1981) 208-216; *C.A.*, 96 (1982) 158 497j.

Caruthers, M.H., Beaucage, S.L., Efoavitch, J.W., Fisher, E.F., Matteucci, M.D. and Stabinsky, Y., *Nucleic Acids Res., Symp. Ser.*, 7 (1980) 215-224.

Cayley, P.J., Brown, R.E. and Kerr, I.M., *J. Liq. Chromatogr.*, 5 (1982) 2027-2039; *C.A.*, 98 (1983) 49 798d.

Chakhmakhcheva, O.G., Efimov, V.A. and Ovchinnikov, Yu.A., *Nucleic Acids Res., Symp. Ser.*, 7 (1980) 345-364.

Chow, F., Kempe, T. and Palm, G., *Nucleic Acids Res.*, 9 (1981) 2807-2817.
Cohn, W.E., in Heftmann, E. (Editor), *Chromatography*, Van Nostrand Reinhold, New York, 1975, pp. 714-743.
Cohn, W.E., *J. Am. Chem. Soc.*, 71 (1949a) 2275-2276.
Cohn, W.E., *Science (Washington, D.C.)*, 109 (1949b) 377-378.
Cohn, W.E., *J. Am. Chem. Soc.*, 72 (1950) 1471-1478.
Coomber, B.A., *Biochimie*, 67 (1985) 798-800.
Crea, R. and Horn, T., *Nucleic Acids Res.*, 8 (1980) 2331-2348.
Crowther, J.B., in P.R. Brown (Editor), *HPLC in Nucleic Acid Research. Methods and Applications (Chromatographic Science Series)*, Vol. 28), Marcel Dekker, New York, 1984, pp. 195-203; *C.A.*, 102 (1985) 109 068d.
Crowther, J.B., Jones, R. and Hartwick, R.A., *J. Chromatogr.*, 217 (1981) 479-490.
Darnell, J.E., *Nature (London)*, 297 (1982) 365-371; *C.A.*, 97 (1982) 104 821n.
Demushkin, V.P. and Plyashkevich, Yu.G., *Bioorg. Khim.*, 2 (1976) 1652-1659; *C.A.*, 86 (1977) 52303w.
De Rooij, J.F.M., Arentzen, R., Den Hartog, J.A.J., Van der Marel, G. and Van Boom, J.H., *J. Chromatogr.*, 171 (1979) 453-459.
Dion, R. and Cedergren, R.J., *J. Chromatogr.*, 152 (1978) 131-136.
Dizdaroglu, M. and Hermes, W., *J. Chromatogr.*, 171 (1979) 321-330.
Dizdaroglu, M., Harmes, W., Von Sontag, C. and Schott, H., *J. Chromatogr.*, 169 (1979) 429-435.
Dizdaroglu, M., Simic, M.G. and Schott, H., *J. Chromatogr.*, 188 (1980) 273-279.
Dobrynin, V.N., Filippov, S.A., Bystrov, N.S., Severtsova, J.V. and Kolosov, M.N., *Bioorg. Khim.*, 9 (1983) 706-710; *C.A.*, 99 (1983) 158 773h.
Dörper, T. and Winnacker, E.-L., *Nucleic Acids Res.*, 11 (1983) 2575-2584.
Duckworth, M.L., Gait, M.J., Goelet, P., Hong, G.F., Singh, M. and Titmas, R.C., *Nucleic Acids Res.*, 9 (1981) 1691-1706.
Edelson, E.H., Lawless, J.G., Wehr, C.T. and Abbott, S.R., *J. Chromatogr.*, 174 (1979) 409-419.
Edge, M.D., Greene, A.R., Heathcliffe, G.R., Meacock, P.A., Schuch, W., Scanlon, D.B., Atkinson, T.C., Newton, C.R. and Markham, A.F., *Nature (London)*, 292 (1981) 756-762.
Efimov, V.A., Buryakova, A.A., Reverdatto, S.V. and Chakhmakhcheva, O.G., *Bioorg. Khim.*, 9 (1983a) 1367-1381; *C.A.*, 100 (1984) 175 199g.
Efimov, V.A., Buryakova, A.A., Reverdatto, S.V., Chakhmakhcheva, O.G. and Ovchinnikov, Yu.A., *Nucleic Acids Res.*, 11 (1983b) 8369-8387; *C.A.*, 100 (1984) 192 215b.
Efimov, V.A., Reverdatto, S.V. and Chakhmakhcheva, O.G., *Nucleic Acids Res.*, 10 (1982) 6675-6694.
Egan, B.Z., *Biochim. Biophys. Acta*, 299 (1973) 245-252.
Elmblad, A., Josephson, S. and Palm, G., *Nucleic Acids Res.*, 10 (1982) 3291-3301.
El Rassi, Z., and Horváth, C., *J. Chromatogr.*, 266 (1983) 319-340.
Eshaghpour, H. and Crothers, D.M., *Nucleic Acids Res.*, 5 (1978) 13-21.
Fisher, E.F. and Caruthers, M.H., *Nucleic Acids Res.*, 11 (1983) 1589-1599.
Flavell, R.A. (Editor), *Techniques in Nucleic Acid Biochemistry*, Elsevier, New York, 1983.
Floyd, T.R., Crowther, J.B. and Hartwick, R.A., *LC, Liq. Chromatogr. HPLC Mag.*, 3 (1985) 508-512; 516-520; *C.A.*, 103 (1985) 215 653w.
Fritz, H.J., Belagaje, R., Brown, E.L., Fritz, R.H., Jones, R.A., Lees, R.G. and Khorana, H.G., *Biochemistry*, 17 (1978) 1257-1267; *C.A.*, 88 (1978) 19 413w.
Fritz, H.J., Eick, D. and Werr, W., in Gassen, H.G. and Lang, A. (Editors), *Chemical and Enzymatic Synthesis of Gene Fragments*, Verlag Chemie, Deerfield Beach, FL, 1982, p. 199.
Froehler, B.C. and Matteucci, M.D., *Tetrahedron Lett.*, 24 (1983) 3171-3174.
Gabriel, T.F. and Michalewsky, J.E., *J. Chromatogr.*, 80 (1973) 263-265.
Gaffney, B.L., Marhy, L.A. and Jones, R.A., *Tetrahedron*, 40 (1984) 3-13.
Gait, M.J. (Editor), *Oligonucleotide Synthesis, A Practical Approach*, IRL Press, Oxford, Washington, 1984, 217 pp.
Gait, M.J., Popov, S.G., Singh, M. and Titmas, R.C., *Nucleic Acids Res., Symp. Ser.*, 7 (1980a) 243-257.

Gait, M.J. and Sheppard, R.C., *Nucleic Acids Res.*, 4 (1977) 1135-1158.
Gait, M.J., Singh, M., Sheppard, R.C., Edge, M.D., Greene, A.R., Heathcliffe, G.R., Atkinson, T.C., Newton, C.R. and Markham, A.F., *Nucleic Acid Res.*, 8 (1980b) 1081-1096.
Gehrke, Ch.W., Kuo, K.C., Davis, G.E., Suits, R.D., Waalkes, T.P. and Borek, E., *J. Chromatogr.*, 150 (1978) 455-476.
Gough, G.R., Singleton, C.K., Weith, H.L. and Gilham, P.T., *Nucleic Acids Res.*, 6 (1979) 1557-1570.
Gurkin, M. and Patel, V., *Am. Lab. (Fairfield, Conn.)*, 14 (1982) 64-73; *C.A.*, 96 (1982) 138 998v.
Hagemeier, E., Bornemann, S., Boss, K.S. and Schlimme, E., *J. Chromatogr.*, 237 (1982) 174-177.
Hansson, K.-A., *J. Chromatogr.*, 266 (1983) 395-399.
Hardies, S.C., Hillen, W., Goodman, T.C. and Wells, R.D., *J. Biol. Chem.*, 254 (1979) 10 128-10 134.
Hardies, S.C. and Wells, R.D., *Proc. Natl. Acad. Sci. U.S.A.*, 73 (1976) 3117-3121.
Hartwick, R.A. and Brown, P.R., *J. Chromatogr.*, 126 (1976) 679-691.
Heine, D.R., Denton, M.B. and Schlabach, T.D., *Anal. Chem.*, 54 (1982) 81-84; *C.A.*, 96 (1982) 16 806k.
Hillen, W., Klein, R.D. and Wells, R.D., *Biochemistry*, 20 (1981) 3748-3756; *C.A.*, 95 (1981) 38561h.
Hillen, W. and Wells, R.D., in Flavell, R.A. (Editor), *Techniques in Nucleic Acid Biochemistry*, Vol. B5, Elsevier, New York, 1983, pp. 1-17.
Himmel, M.E., Perna, P.J. and McDonell, M.W., *J. Chromatogr.*, 240 (1982) 155-163.
Hjertén, S., Hellman, U., Svensson, I. and Rosengren, J., *J. Biochem. Biophys. Methods*, 1 (1979) 263-273.
Hjertén, S. and Yang, D., *J. Chromatogr.*, 316 (1984) 301-309.
Hjertén, S., Yao, K. and Patel, V., in Gribnau, T.C.J., Visser, J. and Nivard, R.J.F. (Editors), *Affinity Chromatography and Related Techniques*, Elsevier, Amsterdam, 1982, pp. 483-489; *C.A.*, 96 (1982) 118 344a.
Hoffman, N.E. and Liao, J.C., *Anal. Chem.*, 49 (1977) 2231-2234.
Horn, T., Vasser, M.P., Struble, M.E. and Crea, R., *Nucleic Acids Res.*, *Symp. Ser.*, 7 (1980) 225-232.
Hsiung, H., Inouye, S., West, J., Sturm, B. and Inouye, M., *Nucleic Acids Res.*, 11 (1983) 3227-3239.
Ike, Y., Ikuta, S., Sato, M., Huang, T. and Itakura, K., *Nucleic Acids Res.*, 11 (1983) 477-488.
Ikehara, M. and Ohtuska, E., *Tanpakushitsu Kakusan Koso*, 26 (1981) 531-541; *C.A.*, 95 (1981) 62 506g.
Jones, D.P., *J. Chromatogr.*, 225 (1981) 446-449.
Jones, R.A., Fritz, H.J. and Khorana, H.G., *Biochemistry*, 17 (1978) 1268-1278.
Jost, W., Unger, K.K., Lipecky, R. and Gassen, H.G., *J. Chromatogr.*, 185 (1979) 403-412.
Jost, W., Unger, K. and Schill, G., *Anal. Biochem.*, 119 (1982) 214-223; *C.A.*, 96 (1982) 118 346c.
Junowicz, E. and Spencer, J.H., *Biochemistry*, 9 (1970) 3640-3648.
Kalashnikov, V.V., Samukov, V.V., Shubina, T.N. and Yamshnikov, V.F., *Bioorg. Khim.*, 9 (1983) 666-672; *C.A.*, 99 (1983) 158 772g.
Kato, Y., Nakamura, K. and Hashimoto, T., *J. Chromatogr.*, 266 (1983a) 385-394.
Kato, Y., Sasaki, M., Hashimoto, T., Murotsu, T., Fukushige, S. and Matsubara, K., *J. Chromatogr.*, 266 (1983b) 341-349.
Kelmers, A.D. and Heatherly, D.E., *Anal. Biochem.*, 44 (1971) 486-495.
Kelmers, A.D., Heatherly, D.E. and Egan, B.Z., *Methods Enzymol.*, 20C (1971a) 35-38.
Kelmers, A.D., Heatherly, D.E. and Egan, B.Z., in Moldave, K. (Editor), *RNA and Protein synthesis*, Academic Press, New York, 1981a, pp. 29-32; *C.A.*, 96 (1982) 118 292q.

Kelmers, A.D., Novelli, G.D. and Stulberg, H.P., *J. Biol. Chem.*, 240 (1965) 3979-3983.

Kelmers, A.D., Weeren, H.O., Weiss, J.F., Pearson, R.L., Stulberg, M.P. and Novelli, G.D., *Methods Enzymol.*, 20C (1971b) 9-34.

Kelmers, A.D., Weeren, H.O., Weiss, J.F., Pearson, R.L., Stulberg, M.P. and Novelli, G.D., in Moldave, K. (Editor), *RNA and Protein Synthesis*, Academic Press, New York, 1981b, pp. 3-28; *C.A.*, 96 (1982) 139 078g.

Khym, J.X., *Analytical Ion-Exchange Procedures in Chemistry and Biology*, Prentice-Hall, New York, 1974, pp. 168-205.

Knight, M., Cayley, P.J., Silverman, R.H., Wreschner, D.H., Gilbert, C.S., Brown, R.E. and Kerr, I.M., *Nature (London)*, 288 (1980) 189-192; *C.A.*, 94 (1981) 135 505e.

Knox, J.H. and Jurand, J., *J. Chromatogr.*, 218 (1981) 341-354; 355-363.

Koch, G. and Richter, D. (Editors), *Regulation of Macromolecular Synthesis by Low Molecular Weight Mediators* (Proceedings of a Workshop, Hamburg-Blankensee, May 29-31, 1979), Academic Press, New York, 1979, 370 pp.; *C.A.*, 92 (1980) B 142 117s.

Kohli, V., Balland, A., Wintrerith, M., Sanerwald, L., Staub, A. and Lecocq, J.P., *Nucleic Acids Res.*, 10 (1982) 7439-7448.

Kopper, R.A. and Singhal, R.P., *Int. J. Biol. Macromol.*, 1 (1979) 65-72; *C.A.*, 92 (1980) 1 802z.

Köster, H., Biernat, J., McManus, J., Wolter, A., Stumpe, A., Narang, Ch.K. and Sinha, N.D., *Tetrahedron*, 40 (1984) 103-112.

Köster, H. and Kaiser, W., *Justus Liebigs Ann. Chem.*, (1974) 336-341.

Krstulovic, A.M., Brown, P.R. and Rosie, D.M., *Anal. Chem.*, 49 (1977) 2237-2241.

Kwiatkowski, M., Heikkilae, J., Welch, C.J. and Chattopadhyaya, J., *Stud. Org. Chem. (Amsterdam)*, 20 (Nat. Prod. Chem.) (1985) 259-274; *C.A.*, 103 (1985) 105 225e.

Larson, J.E., Hardies, S.C., Patient, R.K. and Wells, R.D., *J. Biol. Chem.*, 254 (1979) 5535-5541; *C.A.*, 91 (1979) 71 156v.

Lawson, T.G., Regnier, F.E. and Weith, H.L., *Anal. Biochem.*, 133 (1983) 85-93.

Letsinger, R.L., Groody, E.P., Lander, N. and Tanaka, T., *Tetrahedron*, 40 (1984) 137-143.

Letsinger, R.L., Groody, E.P. and Tanaka, T., *J. Am. Chem. Soc.*, 104 (1982) 6805-6806.

Leutzinger, E.E., Miller, P.S. and Ts'o, P.O.P., *Nucleic Acid Chemistry, Methods and Techniques, Part 2*, Wiley, New York, 1978, pp. 1037-1043; *C.A.*, 93 (1980) 8430f.

Markham, A.F., Edge, M.D., Atkinson, T.C., Greene, A.R., Heathcliffe, G.R., Newton, C.R. and Scanlon, D., *Nucleic Acids Res.*, 8 (1980) 5193-5205.

Matteucci, M.D. and Caruthers, M.H., *J. Am. Chem. Soc.*, 103 (1981) 3185-3191.

McFarland, G.D. and Borer, P.N., *Nucleic Acids Res.*, 7 (1979) 1067-1080; *C.A.*, 92 (1980) 37 034g.

McLaughlin, L.W., Cramer, F. and Sprinzl, M., *Anal. Biochem.*, 112 (1981) 60-69.

McLaughlin, L.W. and Piel, N., in Gait, M.J. (Editor), *Oligonucleotide Synthesis, A Practical Approach*, IRL Press, Oxford, Washington, 1984a, pp. 117-133.

McLaughlin, L.W. and Piel, N., in Gait, M.J. (Editor), *Oligonucleotide Synthesis, A Practical Approach*, IRL Press, Oxford, Washington, DC, 1984b, pp. 207-210.

Melander, W.R., Nahum, A. and Horváth, C., *J. Chromatogr.*, 185 (1979) 129-152.

Mikeš, O., in Šmíd, J. (Editor), *Měniče Iontů, Jejich Vlastnosti a Použití (Ion Exchangers, Their Properties and Applications)*, SNTL, Prague, 1954, pp. 387-443.

Miyoshi, K., Arentren, R., Huang, T. and Itakura, K., *Nucleic Acids Res.*, 8 (1980a) 5507-5518.

Miyoshi, K. and Itakura, K., *Nucleic Acids Res.*, Symp. Ser. No. 7 (1980) 281-291.

Miyoshi, K., Huang, T. and Itakura, K., *Nucleic Acids Res.*, 8 (1980b) 5491-5505.

Miyoshi, K., Miyake, T., Hozumi, T. and Itakura, K., *Nucleic Acids Res.*, 8 (1980c) 5473-5490.
Mizutani, T., in Osawa, K., Tanaka, Y. and Kitami, S. (Editors), *Kyodai Ryushi no Gerupamieishion Kuromatogurafi: Seitai Ryushi no Ryukei Bunri*, Pharmacy College, Nagoya City University, 1980a, pp. 243-260; *C.A.*, 95 (1981) 146 292e.
Mizutani, T., *J. Chromatogr.*, 196 (1980b) 485-488.
Mizutani, T. and Narihara, T., *Nucleic Acids Res.*, Special Publ., in press.
Molko, D., Derbyshire, R., Guy, A., Roget, A., Teoule, R. and Boucherle, A., *J. Chromatogr.*, 206 (1981) 493-500.
Mueller, W., Schuetz, H.J., Guerrier-Takada, C., Cole, P.E. and Potts, R., *Nucleic Acids Res.*, 7 (1979) 2483-2499.
NACS Applications Manual, Bethesda Research Labs., Gaithersburg, MD.
Narihata, T., Fujita, Y. and Mizutani, T., *J. Chromatogr.*, 236 (1982) 513-518.
Newton, C.R., Greene, A.R., Heathcliffe, G.R., Atkinson, T.C., Holland, D., Markham, A.F. and Edge, M.D., *Anal. Biochem.*, 129 (1983) 22-30.
Nguyen Phi Nga, Bradley, J.L. and McGuire, P.M., *J. Chromatogr.*, 236 (1982) 508-512.
Ohtsuka, E., Tozuka, Z., Iwai, S. and Ikehara, M., *Nucleic Acids Res.*, 10 (1982) 6235-6241.
Patel, T.P., Millican, T.A., Bose, C.C., Titmas, R.C., Mock, G.A. and Eaton, M.A., *Nucleic Acids Res.*, 10 (1982) 5605-5620.
Patient, R.K., Hardies, S.C., Larson, J.E., Inman, R.B., Maquat, L.E. and Wells, R.D., *J. Biol. Chem.*, 254 (1979) 5548-5554; *C.A.*, 91 (1979) 71 157w.
Pearson, J.D. and Regnier, F.E., *J. Chromatogr.*, 255 (1983) 137-149.
Pearson, R.L., Weiss, J.F. and Kelmers, A.D., *Biochim. Biophys. Acta*, 228 (1971) 770-774.
Perrone, P.A. and Brown, P.R., in Hearn, M.T.W. (Editor), *Ion Pair Chromatography*, Marcel Dekker, New York, 1985, pp. 259-282; *C.A.*, 102 (1985) 128 025d.
Roe, B., Marcu, K. and Dudock, B., *Biochim. Biophys. Acta*, 319 (1973) 25-36.
Rosenthal, A., Cech, D., Veiko, V.P., Orezkaja, T.S., Kuprijanova, E.A. and Shabarova, Z.A., *Tetrahedron Lett.*, 24 (1983) 1691-1694.
Rustum, Y.M., *Anal. Biochem.*, 90 (1978) 289-299.
Šatava, J., Mikeš, O. and Štrop, P., *J. Chromatogr.*, 180 (1979) 31-37.
Scanlon, D., Haralambidis, J., Southwell, Ch., Turton, J. and Tregear, G., *J. Chromatogr.*, 333 (1984) 189-198.
Schiebel, H.M. and Schulten, H.-R., *Z. Naturforsch.*, *Teil B*, 36 (1981) 967-973; *C.A.*, 95 (1981) 187 583u.
Schott, H., *Nucleic Acids Res.*, *Symp. Ser.*, 7, (1980) 203-214.
Schott, H., *J. Chromatogr.*, 237 (1982) 429-438 (in German; English summary); *C.A.*, 96 (1982) 196 016j.
Schott, H., *Affinity Chromatography. Template Chromatography of Nucleic Acids and Peptides* (*Chromatographic Science Series*, Vol. 27), Marcel Dekker, New York, 1984, 256 pp.
Schott, H. and Greber, G., *Macromol. Chem.*, 149 (1971) 253-260; *C.A.*, 76 (1972) 46 437v.
Schott, H. and Schrade, H., *Nucleic Acids Res.*, *Symp. Ser.*, 9 (1981) 187-190; *C.A.*, 94 (1981) 116 994x.
Schott, H. and Watzlawick, H., *J. Chromatogr.*, 243 (1982) 57-70.
Schram, E., in Bransome, E.D., Jr. (Editor), *The Current Status of Liquid Scintillation Counting*, Grune and Stratton, New York, 1970, pp. 95-109.
Scoble, H.A., in Brown, P.R. (Editor), *HPLC in Nucleic Acid Research. Methods and Applications* (*Chromatographic Science Series*, Vol. 28), Marcel Dekker, New York, 1984, pp. 17-27; *C.A.*, 102 (1985) 109 309h.
Scoble, H.A. and Brown, P.R., in Horváth, C. (Editor), *High Performance Liquid Chromatography* (*Advances and Perspectives*, Vol. 3), Academic Press, New York, 1983, pp. 1-47; *C.A.*, 99 (1983) 35 385a.
Shum, Wai-King B. and Crothers, D.M., *Nucleic Acids Res.*, 5 (1978) 2297-2311; *C.A.*, 89 (1978) 125 612v.

Simonian, M.H. and Capp, M.W., *J. Chromatogr.*, 266 (1983) 351-358.
Simpson, R.C., in Brown, P.R. (Editor), *HPLC in Nucleic Acid Research. Methods and Applications* (*Chromatographic Science Series*, Vol. 28), Marcel Dekker, New York, 1984, pp. 181-193; *C.A.*, 102 (1985) 109 067c.
Singhal, R.P., *Biochim. Biophys. Acta*, 319 (1973) 11-24.
Singhal, R.P., *Sep. Purif. Methods*, 3 (1974a) 339-398; *C.A.*, 82 (1975) 166 788f.
Singhal, R.P., *Eur. J. Biochem.*, 43 (1974b) 245-252.
Singhal, R.P., *J. Chromatogr.*, 266 (1983) 359-383.
Singhal, R.P., Bajaj, R.K., Buess, Ch.M., Smoll, D.B. and Vakharia, V.V., *Anal. Biochem.*, 109 (1980) 1-11.
Singhal, R.P., Griffin, G.D. and Novelli, G.D., *Biochemistry*, 15 (1976) 5083-5087.
Sokolowski, A., Balgobin, N., Josephson, S., Chattopadhyaya, J.B. and Schill, G., *Chem. Scr.*, 18 (1981) 189-191; *C.A.*, 96 (1982) 65 071a.
Sproat, B.S. and Gait, M., in Gait, M.J. (Editor), *Oligonucleotide Synthesis. A Practical Approach*, IRL Press, Oxford, Washington, DC, 1984, pp. 83-115.
Stec, W.J., Zon, G., Egan, W., Byrd, R.A., Phillips, L.R. and Gallo, K.A., *J. Org. Chem.*, 50 (1985a) 3908-3916.
Stec, W.J., Zon, R. and Uznański, B., *J. Chromatogr.*, 326 (1985b) 263-280.
Sundberg, L. and Porath, J., *J. Chromatogr.*, 90 (1974) 87-98.
Tanaka, T. and Letsinger, R.L., *Nucleic Acids Res.*, 10 (1982) 3249-3260.
Téoule, R., Derbyshire, R., Guy, A., Molko, D. and Roget, A., *Nucleic Acids Res.*, *Symp. Ser.*, 7 (1980) 23-37.
Thompson, J.A., Blakesley, R.W., Doran, K., Hough, C.J. and Wells, R.D., *Methods Enzymol.*, 100B (1983) 368-399.
Tomlinson, R.V. and Tener, G.M., *J. Am. Chem. Soc.*, 84 (1962) 2644-2645.
Tomlinson, R.V. and Tener, G.M., *Biochemistry*, 2 (1963) 697-702.
Usher, D.A., *Nucleic Acids Res.*, 6 (1979) 2289-2306.
Uziel, M., Smith, L.H. and Taylor, S.H., *Clin. Chem.*, 22 (1976) 1451-1455; *C.A.*, 85 (1976) 155 922f.
Vakharia, V.N. and Singhal, R.P., *J. Appl. Biochem.*, 1 (1979) 210-212; *C.A.*, 92 (1980) 37 059b.
Vanderberghe, A., Van Broeckhoven, C. and De Wachter, R., *Arch. Int. Physiol. Biochim.*, 87 (1979) 848-849; *C.A.*, 92 (1980) 106 680m.
Vanderberghe, A., Nelles, L. and De Wachter, R., *Anal. Biochem.*, 107 (1980) 369-376.
Voet, D., Cratzer, W.B., Cox, R.A. and Doty, P., *Biopolymers*, 1 (1963) 193-208.
Wakizaka, A., Kurosaka, K. and Okuhara, E., *IRCS Med. Sci.: Libr. Compend.*, 6 (1978) 485; *C.A.*, 90 (1979) 50 797u.
Wakizaka, A., Kurosaka, K. and Okuhara, E., *J. Chromatogr.*, 162 (1979) 319-326.
Walton, A.G. and Blackwell, J., *Biopolymers*, Academic Press, New York, 1973, pp. 236-242.
Wehr, C.T., *Nucleic Acid Constituents by High Performance Liquid Chromatography*, Varian, Palo Alto, CA, 1980, 108 pp.
Wehr, C.T. and Abbott, S.R., *J. Chromatogr.*, 185 (1979) 453-462.
Weels, R.D., *J. Chromatogr.*, 336 (1984) 3-14.
Wells, R.D., Hardies, S.C., Horn, G.T., Klein, B., Larson, J.A., Neuendorf, S.K., Panayotatos, N., Patient, R.K. and Delsing, E., *Methods Enzymol.*, 65 (1980) 327-347.
Williams, B.R.G., Golgher, R.R., Brown, R.E., Gilbert, C.S. and Kerr, I.M., *Nature (London)*, 282 (1979) 582-586.
Wu, R., Wu, N.-H., Hanna, Z., Georges, F. and Narang, S., in Gait, M.J. (Editor), *Oligonucleotide Synthesis, A Practical Approach*, IRL Press, Oxford, Washington, DC, 1984, pp. 135-151.
Yanagawa, M., *Nucleic Acids Res.*, *Spec. Publ.*, 5 (1978) 461-464; *C.A.*, 90 (1979) 83 055d.
Zakaria, M., in Brown, P.R. (Editor), *HPLC in Nucleic Acid Research. Methods and Applications* (*Chromatographic Science Series*, Vol. 28), Marcel Dekker, New York, 1984, pp. 365-368; *C.A.*, 102 (1985) 110 711h.

Chapter 11

POLYSACCHARIDES AND OLIGOSACCHARIDES

11.1 INTRODUCTION

Liquid column chromatography is the main technique for the separation of both low- and high-molecular-weight saccharides. The classic column chromatographic methods for the separation of these strongly hydrophilic compounds have been reviewed in detail several times, e.g., in book chapters entitled "Carbohydrates" and "Polysaccharides" by Čapek and Staněk (1975a,b). In this chapter modern high- and medium-pressure techniques will be dealt with. As column chromatography was applied in nearly every study of the separation of saccharides, this chapter cannot include all published papers, and the Register with the Bibliography (Chapters 14 and 15) should be consulted for further information. However, the main separation techniques in this field will be described.

In spite of the fact that saccharides are rich in hydroxy groups and are therefore strongly hydrophilic, a certain level of weak hydrophobic interactions allows these substances to be separated by reversed-phase chromatography (RPC). Differences in the solubilities of individual saccharides in pairs of various non-miscible liquid phases or in a swollen solid phase and a mobile liquid phase create the basis for separation by partition chromatography. Differences in the size and shape of hydrated or swollen poly- and oligosaccharides open the way for separations using size exclusion chromatography (SEC). If ionogenic groups are present (sugar acids and amino sugars), the application of ion-exchange chromatography (IEC) is possible, and this is so also if neutral saccharides are chromatographed in borate buffers in which acidic borate complexes are formed. Other possibilities arise in the chromatography of chemically modified saccharides in which the modifying moiety can interact with the chromatographic support. Affinity chromatography represents another separation possibility. All these principles have been used for the fractionation of saccharides.

This chapter is devoted only to the separation of simple saccharides. Complex compounds that contain other moieties in addition to glycides will be dealt with in Chapter 12.

11.2 POLYSACCHARIDES

The most usual method for the chromatographic isolation, purification and characterization of polysaccharides was gel permeation chromatography (GPC) on polydextran or agarose gels or on Biogels [cf., reviews by Churms (1970) and Čapek and Staněk (1975b)]. Its modern version, size exclusion chromatography (SEC), will be illustrated by examples in this section. Only exceptionally have other chromatographic modes been applied for the separation of polysaccharides, e.g. reversed-phase (RP) or ion-exchange chromatography. As the SEC method is very simple both in principle and in experimental technique and because it has been sufficiently described in the Volume A of this book, no general commentary is required here and the subsequent parts of this section can be classified according to the substances separated.

The most usual method for monitoring polysaccharides in the column effluents is refractive index (RI) measurement. However, the use of dyed polysaccharides (e.g., with Procion Blue M_3G or Procion Brilliant Red M_2B) leads to a simple monitoring method in GPC, provided that the low degree of substitution does not affect the separation properties. Dudman and Bishop (1968) used these dyes in electrophoretic experiments and Anderson et al. (1969) in GPC. Procion Red M_2B was preferred and available from ICI Ltd., Dyestuffs Division, Manchester, U.K. The following procedure can be used (cf., Čapek and Staněk, 1975a). A solution of 2.5 mg of dye in 0.25 ml of water was added to a solution of 2.5 mg of poly-saccharide in 0.25 ml of water; after 5 min, 4 M sodium chloride (0.5 ml) was added to give a final concentration of 2 M. The dyed sample (1-3 mg) in 2 M sodium chloride solution (0.5 ml) was applied with care on the top of a GPC column (35 cm x 1.5 cm I.D.) and eluted with 1 M sodium chloride solution at a flow-rate of 0.5-1.0 ml/min. The column effluent was passed directly into the cell of a spectrophotometer.

11.2.1 Starch, pullulan, inulin

Meuser et al. (1979) reported the chromatographic separation of maize starch polymers modified during mechanical treatment. Five samples of controlled pore glass with defined porosity (85-2941 Å) and LiChrospher Si 100 silica gel (100 Å) were applied as chromatographic supports and dimethyl sulphoxide was used for dispersion. The fractions obtained were characterized using enzymic degradation methods (by treatment with amyloglucosidase, pullulanase or β-amylase). In order to check the separation system, a mixture of dextran T 5000, two amyloses [with degree of polymerization (DP) 1130 and 200], glucose and dimethyl sulphoxide was separated using differential refractive index (DRI) detection.

Dreher et al. (1979) compared open-column and high-performance (HP) GPC in the separation and molecular weight determination of polysaccharides. Ultrogel (LKB, Bromma, Sweden) columns and µBondagel column (Waters Assoc., Milford, MA, U.S.A.) were compared using pullulan (Calbiochem, Los Angeles, CA, U.S.A.) in addition to other testing materials (dextrans). For HPLC, two stainless-steel columns (300 mm x 3.9 mm I.D.) were used, one packed with µBondagel E-linear (fractionation range 200-2 000 000 daltons), the second packed with µBondagel E-300 (3000-100 000 daltons). The temperature was ambient (25°C) and the mobile phases were either methanol-water (2:3, v/v) (for some of the dextrans) or 0.1 M sodium acetate-acetic acid buffer solution, pH 5.5 (containing 0.02% of sodium azide) (for other dextrans and pullulan). The flow-rate was maintained at 1 ml/min by pressures of 1200 and 900 p.s.i., respectively. RI detection was used. The MW of pullulan determined with a µBondagel E-linear column was 1 660 000 and with Ultrogel AcA-22 1 623 000 daltons. The main advantage of HPLC is its speed: complete separation was achieved in 5 min or less.

Wada et al. (1985) described the GP-HPLC of water-soluble samples on poly-(vinyl alcohol) columns. Pullulans P-5 (MW 5300), -10 (12 000), -20 (20 800), -50 (46 700), -100 (95 400), -200 (194 000), -400 (338 000), and P-800 (758 000) (obtained from Showa Denko, Tokyo, Japan), peptides, proteins and poly(ethylene glycols) were chromatographed on newly developed pre-packed 50 cm x 7.6 mm I.D. columns (Asahipak GS series: GS-310, -320, -500 and -520 of particle size 9 µm) (Asahi, Tokyo, Japan). Pullulans and poly(ethylene glycols) were eluted according to the GPC mode, whereas peptides and proteins were slightly adsorbed on the columns. The calibration graphs for the determination of molecular weight are illustrated in Fig. 11.1.

Mino et al. (1985) reported the separation of acetylated inulin by RP-HPLC. Oligomers of this fructan (purchased from Merck, Darmstadt, F.R.G., or isolated from the underground parts of the plants *Dahlia pinnata* Cav. or *Helianthus tuberosus* L.) were successfully separated by RP-HPLC after pre-acetylation. A Zorbax ODS column (25 cm x 4.6 mm I.D.) and acetonitrile-water (73:27) as the mobile phase at a flow-rate of 1 ml/min were used for chromatography with a sample size of 10 µl. A sample of non-acetylated inulin which gave a simple peak in GPC on a TSK Gel 3000 PW column was resolved (after pre-acetylation) into many peaks (differing in molecular weight) with a degree of polymerization up to 22, on a Zorbax column. Similar results were obtained with a commercial sample of inulin and plant samples that differed in the relative amounts of oligomers, depending on the vegetation period. An increase in the acetonitrile content in the mobile phase was effective in the separation of larger polymers [acetonitrile-water (4:1, v/v) as eluent could clarify the distribution of DP in the range of

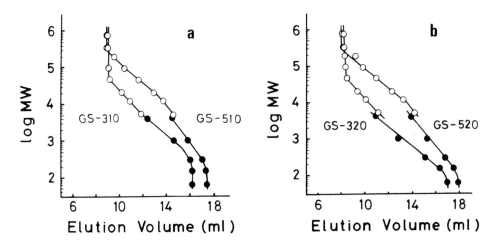

Fig. 11.1. Calibration graphs for (a) GS-310 and GS-510 poly(vinyl alcohol) Asahipak columns and (b) GS-320 and GS-520 columns. Samples: o, pullulans; ●, poly(ethylene glycols). Conditions: eluent, water; flow-rate, 1 ml/min; sample concentration, 0.1%; sample volume, 50 µl; detection, RI; temperature, ambient. (Reprinted from Wada et al., 1985.)

about 20-35]. This work indicated that pre-acetylation with RP-HPLC is a powerful technique for a wide range of investigations relating to inulin.

11.2.2 Cellulosics

GPC on soft polysaccharide supports (cross-linked polydextran columns), which was found to be very effective for separation of water-soluble polymers, was extended to the HPLC of biopolymers. The first experiments with the chromatography of cellulose derivatives were published by Buytenhuys and Van Der Maeden (1978), who tested GPC on unmodified silica using aqueous solvents. Totally porous unmodified spherical 10 µm silica particles (LiChrospher or LiChrosorb; Merck) packed into 30 cm x 4.6 mm I.D. stainless-steel columns using the slurry method were applied to the GPC of water-soluble polymers. The molecular weights of carboxymethylcellulose and other soluble cellulose ethers and esters were determined by means of two columns (LiChrospher Si 100 and Si 500) in 0.5 M sodium acetate solution (pH 6) at a flow-rate of 0.5 ml/min using a sample of 100 µl of a 0.5% solution. DRI detection was used and the time of analysis was 20 min. Strubert and Hovermann (1978) chromatographed cellulose acetate (in addition to various organic polymers) by GP-HPLC in a time not exceeding 10-20 min.

Barth and Regnier (1980) studied the HPLC of water-soluble cellulosics on a specially developed packing material for GP-HPLC, hydrophilic SynChropak (a derivative of LiChrospher macroporous silica, covered with Glycophase) (SynChrom, Linden, IN, U.S.A.). Particles of 10 μm with nominal pore sizes of 100, 500, 1000 and 4000 Å were pre-packed into 25 cm x 4.1 mm I.D. stainless-steel columns. The columns were arranged in series with the smaller pore-sized support placed first. The columns were calibrated using sets of dextran standards (Pharmacia, Uppsala, Sweden). Twelve samples of carboxymethylcellulose (average degree of substitution 0.7), four samples of hydroxyethylcellulose (average degree of substitution 2.5) and one sample of carboxymethylhydroxyethylcellulose (degree of substitution 0.37, molar substitution 1.89) were studied. The effects of mobile phase ionic strength and viscosity were examined. The recommended mobile phase had a high ionic strength, corresponding to 0.7 M concentration. Acetate buffer of pH 7 was fortified by the addition of sodium sulphate. The relative viscosities of the injected solutions must be below 1.5. Injection volumes greater that 50 μl result in a reduction in column efficiency. These conditions are at the limits of the sensitivity of refractometers and the authors therefore began to investigate the application of light-scattering detectors.

Marx-Figini and Soubelet (1982) described the SEC (GPC) of cellulose nitrate using modified silica gel as a stationary phase. Fractions of hydrolytically degraded cotton cellulose and technologically treated cellulose, both nitrated before fractionation, were analysed. The solvent was peroxide-free tetrahydrofuran (Merck), freshly distilled in a nitrogen atmosphere the day before, and degassed immediately before. Pre-packed series of columns filled with 10 μm LiChrospher CH-8 types 4000-4000-1000 (Merck) were used for chromatography. UV detection at 254 nm was used and the elution velocity was 0.3 ml/min. Calibration was carried out using fractions with molecular weights that had been derived from the respective intrinsic viscosities in acetone and compared with data for nearly monodisperse polystyrene (Pressure Chemical Co.).

The results showed that silica gel passivated with C-8 chains represents a very appropriate material for the SEC of cellulose nitrate. Unfortunately, the exclusion limit of the tested columns occurred at DP = 1900 (i.e., at MW 550 000) owing to the high viscosity of the samples. This is relatively low compared with the highest possible degree of polymerization of cellulose (DP = 14 000). Silica gel with a larger pore size should be used, but it was not available. UV detection was found to be useful for the given purpose.

11.2.3 Pectins

Pectins are very important polysaccharides used in the food industry as gelling agents (e.g., for jams and jellies) and as a suspension stabilizer (for milk products and frozen desserts). They are also used in the pharmaceutical industry and in cosmetics. For the purposes of application it is very useful to characterize these polymers in terms of molecular weight distribution (MWD), which strongly influences their properties. Formerly, such analyses were performed using Sepharose or Sephadex gels (see, e.g., Bock et al., 1977). Strubert and Hovermann (1978) published the first orientation experiments for the determination of the MWD of pectin by HPLC with an analysis time of 10-20 min.

The HPLC of pectins in the GPC mode was examined in detail by Barth (1980). He used commercial hydrophilic coated LiChrospher silica particles (available as SynChropak from SynChrom). The nominal pore sizes of these supports were 100, 500 and 1000 Å. The column dimensions were 25 cm x 4.1 mm I.D. and the columns were ordered in sets (cf., Section 1.2.2). The recommended mobile phase was sodium acetate (pH 3.7) with an ionic strength corresponding to 0.22 M concentration; it was fortified with sodium sulphate to an ionic strength corresponding to 0.7 M. High-methoxy pectins [with degree of methylation (DM) 67% and 63%), low-methoxy pectins (DM 34% and 30%) and amidated pectins (DM 27% and 26%) were analysed. The respective molecular weights were 137 000 and 140 000; 58 000 and 91 000; and 96 000 and 113 000. The sample size was 20 μl, the flow-rate 0.5 ml/min and RI detection or (later) low-angle laser light-scattering detection was used. The relative viscosity of the sample had to be below 1.5 and the maximum sample concentration was 1 mg/ml. The influence of ionic strength, pH and viscosity was examined. The time of analysis did not exceed 10 min. No evidence of a non-SEC mechanism was observed. The continuation of these studies was published by Barth and Regnier (in press), where further experimental details can be found.

Fishman et al. (1982) examined soluble pectins by HPLC using the SEC mode and a μBondagel E-linear support. Pectins with 0%, 37%, 50% and 73% methoxy groups were chromatographed in water and in phosphate buffer (pH 7.3) of 0.08 and 0.16 M DRI and UV (206 nm) detection was used. Differences in average partition coefficients, $\overline{K}_{AV}$, between experiments in water and buffers were found (being lower in water) and in 0.08 M phosphate increased in the order of DM 73% > 50% > 37% > 0%.

11.2.4 Plant gums

SEC methods developed by Barth (1980) and Barth and Regnier (1980), described in Sections 11.2.2 and 11.2.3 were used by Barth and Smith (1981) for the MWD characterization of guar gum. A boiled, filtered aqueous sample of guar gum (concentration below 0.5 mg/l) was applied to SynChropak columns (pore sizes 100 and 4000 Å). The MW of the sample was greater than that of a dextran standard

$(2 \cdot 10^{6})$. In the presence of 6 M urea a very slight increase in elution volume
was observed because the aggregation of molecules was suppressed, probably owing
to the elimination of hydrogen bonding. Sonication (30 min at 80 W) resulted in
a decrease in the molecular size.

Vandevelde and Fenyo (1985) described the macromolecular distribution of
Acacia senegal gum (gum arabic) by SEC. Sephacryl S-400 and S-500 were used as
supports. The MWD characterization and physico-chemical data (including CD)
showed that the main component (ca. 70%) of gum arabic is a material containing
homogeneous polysaccharide chains with a very low nitrogen content. The remaining
material is a combination of polysaccharide and nitrogen moieties, probably an
arabinogalactan-protein complex.

11.2.5 Microbial polysaccharides

Leuconostoc mesenteroides and some other microorganisms produce dextran polysaccharides when grown in sugar medium. These polymers can be fractionated to
poly- or oligosaccharidic chains, and fragments of defined degree of polymeriza-
tions are produced commercially. They are used in pharmaceutical and other
industries and some fractions are also used as standards for GPC. Higher molecular weight coloured Blue Dextran (Pharmacia, Uppsala, Sweden) is often applied
as a visible marker of void volume in GPC columns. Polydextran fragments with
defined DP (MW) are currently used also for the calibration of SEC columns in
HPLC experiments.

High-speed aqueous GPC of various polymers including dextrans using TSK gel
(type PW) supports was described by Hashimoto et al. (1978). Problems associated
with the column packings used in the characterization of dextran by GPC were
also studied by Barker et al. (1979); Spherosil-Porasil, DuPont SEC, Spheron
1000 and Hydrogel were compared in experiments on the MW characterization of
dextran fractions by SEC. Dreher et al. (1979) compared open-column chromatography and HPLC in the SEC mode in the separation and MW determination of polysaccharides. Starch and T-series dextrans (T-10, MW 10 400; T-20, 21 600; T-40,
44 400; T-70, 68 500; T-150, 154 000; T-500, 450 000) and Blue dextran (MW >
2 000 000) (all available from Pharmacia) were studied in addition to Dextrans
D 4133 and D 5251 (Sigma, St. Louis, MO, U.S.A.) and Dextran B and *D*-glucose
(BDH, Poole, U.K.). All samples used in HPLC were prepared in the form of 2%
(w/v) solutions in the mobile phase, and 4-5 µl were injected in each run. All
samples were passed through filters of pore size 0.4 µm (Nucleopore, Pleasanton,
CA, U.S.A.). The Chromatographic conditions were the same as for pullulan and
were described in detail in Section 1.2.1. Using µBondagel E-linear, the following molecular weights were determined: Dextran B (in methanol-water, 2:3),

126 000; Dextran D 2551 (in 0.1 M acetate), 430 400; Dextran 4133 (in 0.1 M acetate), 50 040. The results obtained showed that the separation and determination of the molecular weight of dextran on μBondagel and Ultrogel columns (LKB, Bromma, Sweden) gave similar values. The HPLC system had the advantage of speed (5 min or less) and sample size (down to 10 μg). The authors used the above system also for studies of marine algae polysaccharides (not described in the paper cited).

Barker et al. (1981) studied suitability of TSK Gel Toyopearl packing for the GPC analysis of dextran. The authors also mentioned experiments with the successfull application of TSK Gel type PW (AMTO, Amsterdam, The Netherlands) for dextran analyses (ca. 13 000 plates/m for glucose were achieved; the analysis time was 40 min). Toyopearl HW 55 S and 65 S (particle size 20-40 μm) were slurry packed into 1 m x 4 mm I.D. glass columns and the eluent rate was controlled to generate a pressure drop of 7 kg/cm^2. Dextran B 161 D 40 was chromatographed in 0.02% potassium hydrogen phthalate solution on such a column. Individual packing materials did not gave satisfactory results; however, using a mixture of HW 55 S and HW 65 S (1.0:1.5), glucose and dextran B 161 D 40 (MW 70 000) were nearly completely separated. The disadvantage was a long analysis time (ca. 2.5 h).

Himmel and Squire (1981) applied HPLC in the SEC mode to large proteins, viruses and polysaccharides on a TSK Gel 5000 PW preparative column. Dextran samples were purchased from Sigma Biochemicals. The fractions of dextran mentioned (produced by *Leuconostoc mesenteroides*) were obtained by alcohol fractionation and had a viscosity-average MW of 500 000, 70 000 and 9700. A column of 600 mm x 21.5 mm I.D. eluted with phosphate buffers was used for the separation and an RI detector for monitoring. The authors calculated the molecular radii, R, of dextrans from elution data (for MW = 9700, R = 36 Å; MW = 70 000, R = 61 Å; MW = 500 000, R = 100 Å) and compared them with the radii from published light-scattering data (34, 80 and 160 Å, respectively). The differences were discussed. Kim et al. (1982) described the characterization of dextrans by SEC using a DRI/LALLSP (differential refractive index low-angle laser light-scattering photometer) system. Columns of 4 ft x 3/8 in. I.D. were dry-packed with controlled porosity glass (CPG) of 200-400 and 120-200 mesh. The mobile phase was 0.05 M K_2HPO_4 (the pH was adjusted to 7.0 with NaOH). The flow-rate was 1 ml/min, the sample size 0.4 ml of 0.5-1.0% (w/w) solution and the temperature was ambient. Pharmacia dextran standards were chromatographed. Detection will be dealt with in Section 11.4.5.

Mencer and Grubisic-Gallot (1982) examined GPC on chemically modified silica using different solvents. The packing material was μBondagel (Waters Assoc.), which consists of a monomolecular layer of polyether chemically bonded on the

surface and pores of silica and which can be used with different organic solvents
and with aqueous mobile phases (Vivilechia et al., 1977). Four μBondagel columns,
E-1000, E-500, E-300 and E-125 (Waters designation), were employed. In addition
to polystyrene or poly(2-vinylpyridine) in tetrahydrofuran or dimethylformamide,
polydextrans in water were also examined, as eluents. Dextrans remained completely adsorbed on the μBondagel if no moderator was added. After the addition
of sodium dodecyl sulphate at a concentration of at least 0.4%, chromatography
was possible. Addition of further moderator had no effect. These findings
indicated some undesirable interactions of dextrans with the surface of the
packing in aqueous solution.

The determination of the MWD of a microbial polysaccharide by HPLC on
μBondagel was also published by Cheng (1980). The polysaccharide was produced by
Xanthomonas manihotis. Dextrans of MW $2 \cdot 10^4$-$200 \cdot 10^4$ were used as standards. The
molecular weights found were higher than those of standards ($1.49 \cdot 10^6$ or
$1.52 \cdot 10^6$) and had to be determined by extrapolation. No significant relationship
between molecular weight and viscosity was found.

11.3 SEPARATION MODES OF OLIGOSACCHARIDES

For the separation of neutral oligosaccharides, various types of columns can
be applied. Strongly acidic cation-exchange resins can effect separations in the
SEC or NPC modes, strong and medium anion exchangers can be applied in the IEC
mode, whereas medium and weak amino bonded phases and cyanoalkyl bonded phases
are often used for partition chromatography in the "normal-phase" mode. RPC
columns were also successfully applied for the separation of neutral oligosaccharides. In addition to these methods, any ionogenic group present in the
oligosaccharide opens up the possibility of application of ion-exchange columns;
thus oligomers with bound amino sugars are well separated on cation exchangers,
whereas for oligouronic acids and other acidic saccharide oligomers anion exchangers are the supports of choice.

High-performance liquid chromatography as a method for the investigation of
starch hydrolysates was dealt with by Richter and Woelk (1977). Liquid chromatography of sugars on silica-based stationary phases was discussed and explained
in detail, including sample preparation, by Verhaar and Kuster (1981).

11.3.1 Size exclusion chromatography

Monosaccharides and low-molecular-weight oligosaccharides have relatively
small molecular dimensions and therefore macroporous chromatographic supports
currently used for the SEC of higher molecular weight polymers are only seldom

used for the GPC of oligosaccharides. Natowicz and Baenziger (1980) described a
rapid method for the chromatography of oligosaccharides on Bio-Gel P-4 (-400
mesh) (Bio-Rad Labs., Richmond, CA, U.S.A.). Either a single column (100 cm x
0.6 mm I.D.) or two columns of this size permitted the analyses to be carried
out in 1.5 or 3.0 h, respectively, using a flow-rate of 64 ml/cm^2·h, a temperature of 55°C and a pressure of 55-100 p.s.i. Pre-warmed, degassed water was used
for elution. Lower molecular weight (DP 2-12) [^{3}H]dextran oligomers and various
glycoprotein-derived oligosaccharides were successfully analysed. Differences
of 1 glucose equivalent could be precisely distinguished for oligomers of at
least 15 glucose equivalents. Gurkin and Patel (1982) emphasized the fact that
Fractogel TSK neither contained saccharides, nor was there any interaction
between saccharides and the gel matrix, and therefore they recommended this
support for saccharide separations. Examples were given of SEC on two HW-40S
columns in series; oligosaccharides from the hydrolysate of β-cyclodextrin were
separated on two 60 cm x 4.4 cm I.D. columns, using water as the eluent at a
flow-rate of 1.4 ml/min, a temperature of 25°C, a sample load of 10 ml of a 10%
solution, optical rotation detection and fractions of 5.2 ml per tube. In another
example, a partial hydrolysate of dextran T 40 was analysed on two 60 cm x 2.2 cm
I.D. columns at a flow-rate of 1 ml/min and a sample injection of 0.5 ml of a 5%
solution.

In many instances microporous ion-exchange resins have been used for the GPC
of mono- and oligosaccharides, because the cross-linked network of the swollen
ion-exchange resins is compatible with the molecular size of monosaccharides and
lower molecular weight oligosaccharides, and these hydrophilic substances can be
fractionated according to the ease of their penetration. Oshima et al. (1979)
described the separation of acidic and neutral saccharides by HPLC on a cation-
exchange resin in the H$^+$ form with acidic eluents. The supports were strongly
acidic sulphonated styrene-divinylbenzene copolymers: TSK Gel LS-212, d_p 12 μm,
21X (pre-packed by Toyo Soda, Tokyo, Japan into 600 mm x 7.6 mm I.D. columns),
or Shodex Ionpak C-811, 12 μm, 8X (pre-packed by Showa Denko, Tokyo, into
500 mm x 8 mm I.D. columns). A laboratory-made column of cation-exchange resin
prepared by sulphonation of porous polystyrene gel, 35X (Hitachi Gel 3011-S,
12 μm), was also tested. Eluent was 0.1% or 1% orthophosphoric acid. Chromatography was carried out at room temperature at a flow-rate of 0.7-0.9 ml/min and
a pressure ca. 45 kg/cm^2. A 100-μl volume of sample solution (0.1% or less) was
applied to the column. Neutral saccharides were well separated from each other
in 0.1% acid based on differences in molecular size. Use of 1% acid for elution
resulted in a slight increase in the retention of uronic acids, which indicated
a certain degree of electrostatic repulsion of acidic solutes.

Other workers used cation-exchange resins loaded with metal counter ions for the SEC of mono- and oligosaccharides. Ladisch and Tsao (1978) reported a very thorough study of the theory and practice of rapid liquid chromatography at moderate pressures using water as eluent. The strongly acidic cation exchanger Aminex 50W-X4 (Ca^{2+}), 20-30 μm, packed into 45- or 60-cm columns of various diameters (2-8 mm) was used for the general investigation of flow-rate vs. pressure conditions and for the separation of malto- and cellodextrins. Very good separations were obtained in 12-15 min. The surprising phenomenon of a more than 50-fold increase in pressure due to an increase in column diameter from 4 to 8 mm was explained. The optimum performance of the 60-cm column occurred at 6 mm I.D. (pressure 130 p.s.i.g, plate height 0.11). Hokse (1980) used the above method of Ladisch and Tsao for the HPLC analysis of cyclodextrins. The system consisted of two 25 cm x 0.62 cm I.D. columns of Aminex 50W-X4 (Ca^{2+}), 20-30 μm, thermostated at $90^{\circ}C$ (the temperature of the DRI detector was only $37^{\circ}C$). Degassed water, filtered through a 0.2-μm filter and warmed to $90^{\circ}C$, was used for elution. The separation achieved is illustrated in Fig. 11.2.

Scobell and Brobst (1981) applied the silver form of strongly acidic cation-exchange resins of X4, X5, X6 and X8 cross-linking for the rapid high-resolution separation of sugars. The silver form retained oligosaccharides to a greater extent than the calcium form of the same resin, resulting in the separation of a larger number of components of mixtures. By varying the amount of bound silver, the column efficiency could be optimized. In addition to Aminex A-7 (7-11 μm), A-5 (10-15 μm), 50W-X4 (10-15 μm) and experimental 6% cross-linked 10-15 μm resin particles (Bio-Rad Labs.), Aminex HC-40 resin (10-15 μm) (Hamilton, Reno, NV, U.S.A.) was also examined. The chromatographic conditions were as follows: column, 30 cm x 7.8 mm I.D., thermostated at $85^{\circ}C$; solvent, deionized, degassed water at $90^{\circ}C$; DRI detector thermostated at $45^{\circ}C$; sample concentration, 2-5% (dry solid basis); and sample volume, 5-15 μl. Aminex HPX-42 had an optimum 70% silver loading level and separated corn-derived oligosaccharides up to DP 15. Brobst and Scobell (1982) reviewed modern chromatographic methods for the analysis of carbohydrate mixtures. In addition to HPLC, gas-liquid chromatography (GLC) was also dealt with and in the former method the improved separation on cation-exchange resins in the Ag^{+} form was predominantly discussed.

Bonn (1985) investigated the HPLC elution behaviour of oligosaccharides, monosaccharides and sugar degradation products on series-connected ion-exchange resin columns using water as the mobile phase. The non-additivity of the number of theoretical plates and the resolution on ion-exchange stationary phases was discussed. Ca^{2+}-, H^{+}- and Ag^{+}-loaded columns were studied. Djordjevic et al. (1986) described the preparative chromatography of acidic oligosaccharides using

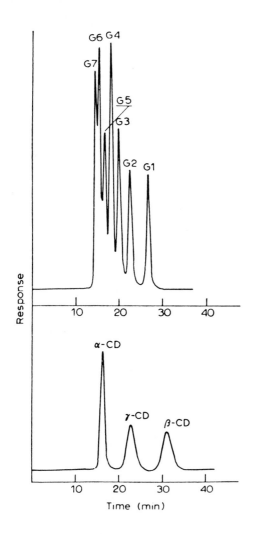

Fig. 11.2. Size exclusion separation of linear glucooligosaccharides (G2-G7) and cyclodextrins (CD) on the strongly acidic cation-exchange resin Aminex 50W-X4 (Ca2+ form), 20-30 μm. Eluent, water; temperature, 90oC; detection, DRI at 37oC; sample size, 50 μl; flow-rate, 0.3 ml/min; recorder chart speed, 0.2 cm/min. (Reprinted from Hokse, 1980.)

a 0.2 *M* volatile acidic buffer composed of formic acid and trimethylamine. Uronic acid-containing oligosaccharides were separated and a Bio-Gel P-2 support was used.

11.3.2 Partition chromatography in the normal-phase mode

When studying publications on oligosaccharide chromatography, the principles of separation are sometimes not clear at the first sight. For example ion exchangers are used not only for the IEC or ion-exclusion chromatography of ionogenic derivatives, but also for SEC, ligand-exchange or NPC; the alkylamine bonded phases can sometimes act both in an ion-exchange mechanism and in the normal-phase mode, and in other modified silicas the decision between the NPC and RPC modes is not clear in some particular instances if only isocratic elution was used. Often mixed modes participate in various separations of oligosaccharides, and in some instances authors do not specify the exact type of chromatographic separation they achieved (or sometimes they classify it incorrectly); usually their main goal is the best resolution of sample substances and the principle involved is not of prime importance.

The types of packings used for the NPC of oligosaccharides are summarized in Table 11.1. The examples of separations will be ordered in this sequence in this section, with the exception of the description of the chromatography on in situ (dynamically or physically) modified silica gel, which will be dealt with in Section 11.3.5.

TABLE 11.1

TYPES OF CHROMATOGRAPHIC SUPPORTS FOR THE NPC OF OLIGOSACCHARIDES

Basic support	Functional groups, modification	Eluent
Ion-exchange resins	$-SO_3^- Me^+$; $-NR_3^+ A^-$	Aqueous ethanol
Porous silica gel	Alkylamine derivative	Acetonitrile-water
Porous silica gel	Alkylnitrile derivative	Acetonitrile-water
Porous silica gel	Dynamic modification using amines	Acetonitrile-water
Porous silica gel	Unmodified	Water-organic solvents

11.3.2.1 NPC on ion exchangers

Partition chromatography on ion-exchange resins is based on an uneven distribution of the oligosaccharides between the resin and the external solution (Samuelson, 1972). In chromatography in aqueous ethanol, the relative amount of water is greater in the resin phase (due to ionogenic groups) than in the external solution, so that the polar solutes tend to be retained by the resin. Both strong cation- and anion-exchange resins can be applied for NPC, but the former (sulphonic derivatives) are preferred in various counter-ionic forms [Li^+,

Na^+, K^+, Ca^{2+}, Ba^{2+} and also in the form of organic counter ions, $CH_3NH_3^+$, $(CH_3)_2NH_2^+$, $(CH_3)_3NH^+$ or piperidinium]. In the case of anion exchangers HSO_3^- or Cl^- counter ions are used (Čapek and Staněk, 1975a).

The influence of the type of counter ion on the separation of saccharides on strong cation exchangers was examined by Lawrence (1974): the trimethylammonium form of the column was best for monosaccharides and the lithium form for di-saccharides. Ethanol diluted with water was used as a solvent. Oshima et al. (1980) improved the separation of anomers of saccharides by HPLC on a macro-reticular strongly basic anion-exchange resin in the sulphate form. The anion exchanger CDR-10, 5-7 μm (Mitsubishi, Tokyo, Japan) was used in a 250 mm x 4.0 mm I.D. column. The eluent was 80%, 90% or 95% ethanol and the temperature was 20 or $50^{\circ}C$. The sample concentration was 0.1-0.4% (w/v) and the sample volume was 20 μl. Mono- and disaccharides were analysed with RI detection.

11.3.2.2 NPC on alkylamine bonded phases

Majors (1977) in a review explained the NPC function of amino derivatives of bonded phases as follows: "The amino group displayed a preferential interaction with the hydroxyl groups of the sugar and gave good chromatographic resolution. An increase in the water concentration of the acetonitrile-water mobile phase decreased retention. Thus the bonded phase functioned in the "normal" mode since the more polar solvent, water, reduced the competitive interaction between the sugar and the polar phase". He added, "Amino bonded phases may replace ion exchangers as the standard carbohydrate separation column because of increased speed and resolution". Indeed, amino bonded phases have often been used for the separation of sugars.

Gum and Brown (1977) examined two alternative HPLC separation methods for reduced and normal cellooligosaccharides. A 30-cm μBondapak carbohydrate column (Waters Assoc.) was eluted with 25% (w/w) water in acetonitrile (the flow-rate increased regularly from 2.0 to 4.6 ml/min over a 45-min period). The second method investigated used a 25-cm long PXS-1025 PAC column (Whatman) and 29% (w/w) water in acetonitrile at a constant flow-rate of 1.5 ml/min. The first method separated reduced oligosaccharides up to G_5H and the second up to G_6H (cf., Fig. 11.11). Both methods allowed a good separation of the residual substrate and products of enzymic hydrolysis of cellohexaitol and/or cello-heptaitol. Black and Bagley (1978) described the determination of oligosaccha-rides in a water-ethanol extract of soybeans by HPLC using β-cyclodextrin as an inexpensive internal standard. The method is suitable for quantitative anal-ysis and is illustrated by Fig. 11.3, where the chromatographic conditions are given. Bergh et al. (1981) studied the HPLC of sialic acid oligosaccharides.

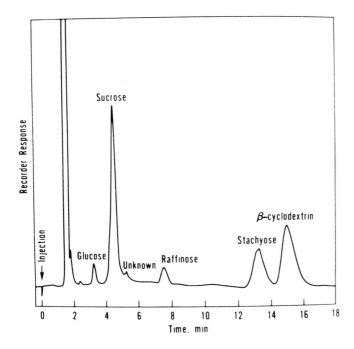

Fig. 11.3. Chromatogram of a defatted soya flour extract including the internal standard β-cyclodextrin. Column, μBondapak-carbohydrate (30 cm x 4 mm I.D.) (silica with amine functionality); eluent, acetonitrile-water (70:30, v/v); flow-rate, 2 ml/min; sample, 10 μl; detection, DRI. The column was ready for the next injection in 17 min. (Reprinted from Black and Bagley, 1978.)

A 250 mm x 4 mm I.D. LiChrosorb-NH$_2$ (5 μm) column (Merck) was eluted with acetonitrile-15 mM potassium phosphate (pH 5.2) (18:7, v/v) at a flow-rate of 2 ml/min. For other experimental conditions, see Section 11.5.7. An extension of this method was published by Bergh et al. (1982).

Orth and Engelhardt (1982) discussed in detail the separation of sugars on chemically bonded silica gel. The separation on amino phases depends on the partition of the solutes between a stagnant aqueous liquid phase and a moving acetonitrile-water mixture. The authors concluded that the bonded amino phase groups cause only demixing of the aqueous acetonitrile eluent, and do not other-wise participate in the separation process. The possible loading increases with increasing nitrogen content of the bonded phase. With amino groups a loading of 0.15 g/g was achieved and with triamino phases the liquid loading increased to 0.25 g/g. The authors systematically studied separations on monoamino-, diamino- and triamino-bonded phases eluted with acetonitrile-water mixtures of various proportions.

Escott and Taylor (1985) studied the determination of sugars by HPLC using a novel bonded phase column and selective post-column spectrophotometric detection. A (propylamino)ethanol column packing material was examined, which permitted the rapid separation of complex mixtures of mono- and disaccharides and components of non-ionic surfactants. Supelco (Supelco Reporter, 1986) offer a Supelcosil LC-NH$_2$ column (γ-aminopropylsilyl polar bonded phase) for NPC partitioning. Using acetonitrile-water (60:40) as the mobile phase at 35°C, a 25 cm x 4.6 mm I.D. column (d_p 5 μm) resolved oligosaccharides well up to DP 10; the sample contained 2 mg of oligomers from Karo light syrup. RI detection was used.

11.3.2.3 NPC on alkyl cyanide bonded phases

Alkyl cyanide phases are not used as often as alkylamine bonded phases. Havel et al. (1977) reported the total sugar separation of textured soya. The procedure consisted in the application of a 25 cm x 6.35 mm I.D. stainless-steel column packed with silica gel (d_p 10 μm) to which a cyanoamine polar phase had been bonded (Partisil 10 PAC; Whatman, Clifton, NJ, U.S.A.). The mobile phase was acetonitrile-2.5 mM aqueous potassium hydrogen phosphate (70:30, v/v), previously adjusted to pH 5 using dilute phosphoric acid. The flow-rate was 70 ml/h and the column pressure 300 p.s.i. The sample size was 10 μl and an RI detector was used. An example of the separation is illustrated in Fig. 11.4.

Clarke and Brannan (1978) described the separation of sugars in molasses. A Whatman PXS10/25 PAC (polar cyanoamino bonded phase) column for carbohydrate analysis was eluted with acetonitrile-water (83:17, v/v) at a flow-rate of 1.5 ml/min. The sample size was 10 μl. The column was washed with methanol-water (1:1) after use each day and stored in this mixture overnight. For longer periods the column was washed with methanol and chloroform and stored in hexane. Lee and Tieckelmann (1981) applied HPLC to enzymic assays of chondroitin sulphate isomers in normal human urine. A Whatman Partisil-10 PAC bonded cyanoamine column (25 cm x 4.6 mm I.D.) (d_p 10 μm) was eluted with acetonitrile-methanol-0.5 M ammonium formate (pH 4.5) (60:20:20, v/v) at a flow-rate of 2.0 ml/min. The pressure was 700 p.s.i. and UV detection at 254 nm was used.

11.3.3 Ion-exchange chromatography

"True" ion-exchange chromatography clearly occurs with acidic or basic sugar derivatives, which can be chromatographed on anion or cation exchangers. However, the opposite effect, ion exclusion, may also take place in the separation of saccharides: sugar acids are excluded from the particles of strongly acidic cation-exchange resins. In general, it is sometimes not easy to define all the

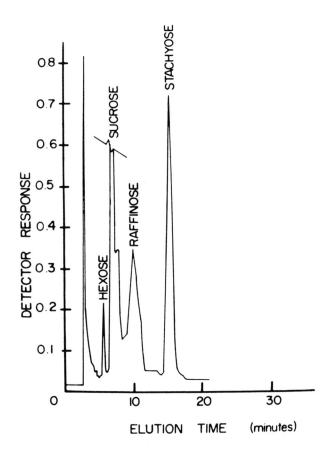

Fig. 11.4. Separation of the total sugars in a ground sample of textured soya. The stationary phase in HPLC was Partisil-10 PAC (cyano polar phase) and the mobile phase was acetonitrile-water (70:30, v/v). Other experimental details are given in the text. (Reprinted from Havel et al., 1977.)

mechanisms contributing to retention on ion exchangers. In addition to ion exchange partition effects, adsorption or a size-exclusion mechanism may also take part in the process. Induced ionogenic groups may also appear: neutral glycosides can be separated on strongly basic anion exchangers due to the ion-exchange process, involving the loss of a proton from one of the hydroxy groups according to the equation

$$R-OH \rightleftarrows R-O^- + H^+$$

The chromatography of borate complexes of sugars on anion exchangers involves a true ion-exchange mechanism. It is based on the interactions of vicinal hydroxy groups of sugars with boric acid, according to the reaction

$$\left[\begin{matrix}-OH \\ -OH\end{matrix}\right. \quad + \quad \begin{matrix}HO \\ HO\end{matrix}\right\rangle B-OH + OH^- \quad \rightarrow \quad \left[\begin{matrix}-O \\ -O\end{matrix}\right\rangle B-O^- + 3 \ H_2O$$

and the complex sugar anion (which is more acidic than boric acid) is chromatographed via the usual anion-exchange chromatography. We can see that the chromatographic separation of sugars on ion exchangers has considerable possibilities.

The classical process involving the separation of neutral sugars in a borate
buffer medium on anion exchangers was reported by Khym and Zill (1951). For this
purpose, strongly basic anion exchangers have been preferred (Kesler, 1967;
Walborg and Lantz, 1968; Jandera and Churáček, 1974; Čapek and Staněk, 1975a).
The strong anion exchangers used included Dowex 1-X8 (200-400 mesh) or Dowex
1-X4 (for di- and trisaccharides), Bio-Rad AG1-X8 (30-40 μm), Technicon 3/28/VI
and Aminex A-14 (X4), A-15 and A-25 (X8).

In contrast, Chytilová et al. (1978) developed a method employing medium
basic DEAE-Spheron (a macroporous glycol methacrylate copolymer, particle size
20-40 μm) in borate form for the chromatography of sugars. The ion exchanger had
to have a relatively high nominal capacity (2.2 mequiv./g). Using a glass column
(50 cm x 6 mm I.D.), isocratic elution with a relatively low concentration of
buffer (0.1 M sodium borate, pH 8.5) at 50°C and a flow-rate of 50 ml/h, a very
good separation of a mixture of trehalose, rhamnose, mannose, arabinose,
galactose and xylose was achieved. Orcinol-sulphuric acid detection (420 nm)
was used. This separation was compared with that obtained on strongly basic
Aminex A-15 (under comparable conditions using a more concentrated 0.355 M
buffer), and the DEAE-Spheron chromatogram was found to be better. This method
was extended by Vrátný et al. (1979) to the efficient chromatography of mixtures
of mono- and oligosaccharides. An example of the separation of twelve saccharides
is illustrated in Fig. 11.5. The retention data for 26 mono- and oligosaccharides
were measured.

Hostomská-Chytilová et al. (1982) further modified separations on Spheron
anion exchangers and, in order to permit the analytical control of the enzymic
hydrolysis of cellulose, they developed a method for the simultaneous analysis
of a mixture of glucose, cellobiose and cellodextrins by isocratic elution.
Strongly basic TEAE-Spheron (a quaternary ammonium base prepared by ethylation
of DEAE-Spheron) was found to be very suitable for the separation of cellodextrins G_2-G_5 in borate buffers. However, this anion exchanger strongly retained
glucose. In order to overcome the difficulties in isocratic elution with large
differences in the retention volumes of monosaccharides and oligosaccharides on

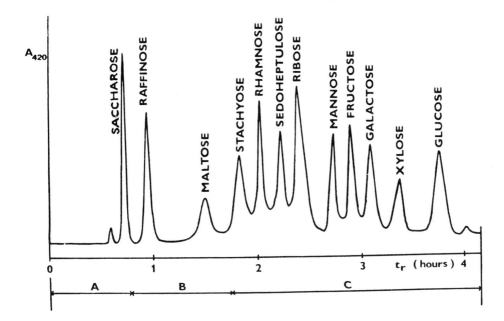

Fig. 11.5. Separation of a standard mixture of saccharides on a 487 mm x 6 mm I.D. column of DEAE-Spheron 300 (20-40 μm) in borate buffers (stepwise elution). Temperature, 60°C; flow-rate, 50 ml/h. Buffers: A, 0.03 *M* borate (pH 7.5); B, 0.1 *M* borate (pH 8.85); C, 0.25 *M* borate (pH 8.88). Weights of individual saccharides were from 16 to 80 μg. Orcinol-sulphuric acid detection was used. (Reprinted from Vrátný et al., 1979.)

strongly basic TEAE-Spheron, two columns were joined in a tandem system of medium basic (DEAE) and strongly basic (TEAE) anion exchanger. The function of this system is illustrated in Fig. 11.6.

Mopper et al. (1980) described borate complex IEC with fluorimetric detection for the determination of saccharides. A 250 mm x 6 mm I.D. column was packed with a strongly basic anion-exchange resin, (DA X-4, 20 μm; Durrum, Palo Alto, CA, U.S.A.). The mobile phase consisted of 0.7 *M* boric acid adjusted to pH 8.6 with 8 *M* sodium hydroxide solution. Ethylenediamine (EDA) was added to the mobile phase at a concentration of 5 ml/l, so that the final concentration of EDA was 7.5 mmol/l. The flow-rate was 0.7 ml/min. The detection principle will be dealt with in the Section 11.4.2. This chromatographic method was applied to environmental and natural product samples involving little or no sample pre-treatment.

Chromatograms obtained by the direct injection of 20 μl of filtered apple juice and direct injection of the neutralized hydrolysate of the calcified tissue of *Halimeda incrassata* (chlorophycease) were described.

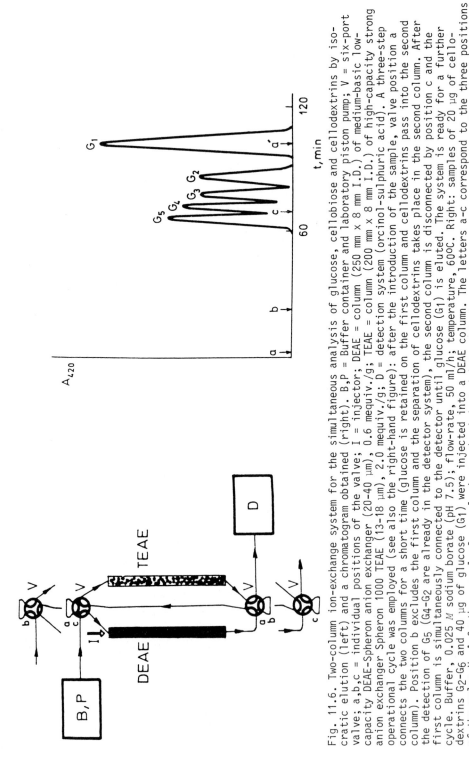

Fig. 11.6. Two-column ion-exchange system for the simultaneous analysis of glucose, cellobiose and cellodextrins by iso-cratic elution (left) and a chromatogram obtained (right). B,P = Buffer container and laboratory piston pump; V = six-port valve; a,b,c = individual positions of the valve; I = injector; DEAE = column (250 mm × 8 mm I.D.) of medium-basic low-capacity DEAE-Spheron anion exchanger (20–40 μm) 0.6 mequiv./g; TEAE = column (200 mm × 8 mm I.D.) of high-capacity strong anion exchanger Spheron 1000 TEAE (13–18 μm), 2.0 mequiv./g; D = detection system (orcinol–sulphuric acid). A three-step operational cycle was employed (see also the right-hand figure): after the introduction of the sample, valve position a connects the two columns for a short time (glucose is retained on the first column and cellodextrins pass into the second column). Position b excludes the first column and the separation of cellodextrins takes place in the second column. After the detection of G5 (G4–G2 are already in the detector system), the second column is disconnected by position c and the first column is simultaneously connected to the detector until glucose (G1) is eluted. The system is ready for a further cycle. Buffer, 0.025 M sodium borate (pH 7.5); flow-rate, 50 ml/h; temperature, 60°C. Right: samples of 20 μg of cello-dextrins G2–G6 and 40 μg of glucose (G1) were injected into a DEAE column. The letters a–c correspond to the three positions of the valve V. A further sample for analysis can be injected into this isocratic system at position a' (arrow). (Reprinted from Hostomská-Chytilová et al., 1982.)

A different type of application of IEC in saccharide chemistry is the separation of sugar compounds containing ionogenic groups. The separation of hexosamines, hexosaminitols and hexosamine-containing di- and trisaccharides on a cation exchanger (in an amino acid analyser) was described by Donald (1977); Locarte No. 12 resin (X8 cross-linked sulphonated polystyrene) was used and eluted at $50^{O}C$ with sodium citrate-borate buffers (see also Section 11.3.6). Delaney et al. (1980b) examined the quantitation of ^{3}H-labelled sulphated disaccharides of heparin by HPLC. A pre-packed Partisil-10 SAX anion exchange column (25 cm x 4.5 mm I.D.) (Whatman) was eluted with KH_2PO_4 buffers. The direct HPLC analysis of heparin deamination products was described (cf., Section 11.5.2). Baenziger and Natowicz (1981) developed a rapid separation of anionic ^{3}H-labelled oligosaccharide species by HPLC, utilizing a 30 cm x 4 mm I.D. column of MicroPak AX-10 anion exchanger (Varian), which was eluted with a mobile phase gradient (25-500 mM KH_2PO_4, pH 4.0). The flow-rate was 1 ml/min and 300-µl fractions were collected in mini-scintillation vials, to which 3.0 ml of 3a70 cocktail (Research Products) was added. Oligosaccharides bearing zero to five sialic acid residues were resolved in less than 45 min.

11.3.4 Reversed-phase chromatography

In spite of the fact that oligosaccharides are typical hydrophilic compounds, RPC based on hydrophobic interactions can also be successfully applied for their separation, especially if the oligosaccharides have been acetylated prior to chromatography; however, this is not an essential condition. Wells and Lester (1979) showed excellent examples. Peracetylated oligosaccharides were separated on two water-jacketed 100 cm x 0.32 cm I.D. columns with a Vydac reversed-phase octadecyl support, d_p 30-40 µm (Separation Group, Hesperia, CA, U.S.A.), maintained at $65^{O}C$. Equivalent results could also be obtained with Bondapak C_{18} Corasil, 37-50 µm (Waters Assoc.). An exponentially programmed gradient of water-acetonitrile from 9:1 to 3:7 was used for elution. The flow-rate was either 2 ml/min (starting pressure 1500 p.s.i. and total gradient time 80 min) or 1 ml/min (1000 p.s.i., 160 min). After the gradient run was completed, the column was prepared for the next sample by washing with 20 ml of acetonitrile followed by 20 ml of water-acetonitrile (9:1) at a flow-rate of 2 ml/min.

The peracetylation procedure was as follows. Dry carbohydrate samples (1-25 mg) were acetylated by treatment with 1 ml of pyridine-acetic anhydride (1:1) at $100^{O}C$ for 90 min. The reagents were removed with a nitrogen stream and the product was dried several times by the addition of toluene. The acetylated products were dissolved in 0.1-0.2 ml of acetonitrile. A total of 0.04-1.0 mg

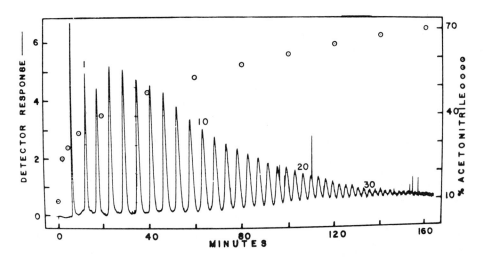

Fig. 11.7. Reversed-phase chromatography of a peracetylated sample of partially hydrolysed amylose on a Vydac RPC ODS support, eluted with an exponential gradient of acetonitrile-water (see text) at a flow-rate of 1 ml/min. The degree of polymerization is indicated by numbers above the peaks. About 3.5 μequiv. of hexose were injected in 60 μl. A moving-wire detector was employed. (Reprinted from Wells and Lester, 1979.)

of carbohydrate was injected in 0.01-0.05 ml. Deacetylation of acetylated oligosaccharides and the reduction of end-groups with $NaBH_4$ was also described.

Partial hydrolysates of purified potato amylose were analysed. The results of experiments with isocratic elution (in 30% or 50% aqueous acetonitrile) explained why an exponential gradient was required for a successful separation. For monitoring of all the non-volatile carbon compounds, a moving wire detector (Model LCM2; Pye Unicam, Cambridge, U.K.) was used. An example of this effective separation is illustrated in Fig. 11.7. Acetylated oligosaccharides from Karo corn syrup were also chromatographed and resolved up to DP 16 using a similar method.

Fonknechten et al. (1980) and Heyraud and Rinaudo (1980) analysed underivatized sugars on ODS columns with water as eluent. Almost complete separation from DP 1 to DP 10 could be obtained in 30 min. Lower temperatures ($3.5^{o}C$) contributed to a better separation (Heyraud and Rinaudo, 1980), and double peaks for the anomeric forms were obtained, as this system lacked a mutarotation catalyst. Vrátný et al. (1983) used water as eluent in the accelerated RPC chromatography of carbohydrate oligomers with RI detection. Two columns were used: (1) a standard 250 mm x 6 mm I.D. stainless-steel column packed with Separon Six ODS silica, d_p 10 μm, and (2) a special glass column (high-pressure CGC cartridge, 150 mm x 3.2 mm I.D.) packed with Separon Six ODS silica, d_p 5 μm, optimized for the separation of oligosaccharides. Both columns and packings were supplied by Laboratory Instruments Works, Prague, Czechoslovakia. The separation of cellodextrins (up to DP 6) on both columns at 70 and 80°C was completed in 10 min. Maltodextrins (up to DP 8) were resolved on a glass column at 24°C in 5 min. An increase in temperature accelerated the rate of interconversion between α- and β-anomers, thus eliminating the unwanted double peaks. The thermodynamic aspects of the chromatography were also studied.

Cheetham and Teng (1984) described some applications of RP-HPLC to oligosaccharide separations using aqueous solvents with various additives (anionic, cationic and non-ionic surfactants, tetramethylurea and organic solvents) that accelerated the separation. In contrast, the addition of neutral inorganic salts increased the retention, but allowed the good resolution of some compounds that were poorly resolved in water alone. Two columns were applied: (1) a Dextropak 10 cm x 0.8 cm I.D. cartridge containing spherical silica of 10 μm particle size, 125 Å pore size, with a 10% loading of ODS phase, and (2) a LiChrosorb RP-8 25 cm x 0.4 cm I.D. stainless-steel column (Merck) containing porous ODS bonded silica (LiChrosorb Si-60). Coating of the column with the additives was carried out by pumping an aqueous detergent solution (0.1%, w/v) through the column at a flow-rate of 1 ml/min. The standard compound used to determine the extent of column modification was isomaltotetraose, as a ca. 10 mg/ml solution in water. The surfactants tested were cetyltrimethylammonium bromide, sodium dodecyl sulphate and Triton X-100. The mechanism of the detergent effect and the effects of organic-aqueous solvents, tetramethylurea and added salts were studied in relation to solvophobic theory. Elution profiles of maltooligosaccharides (up to DP 11 or 13) and isomaltooligosaccharides (up to DP 8) on modified or unmodified Dextropak or C_8 columns were illustrated. The addition of Triton suppressed the separation of anomers (manifested by the doubling of peaks) and shortened the elution times of single peaks, so that both qualitative and quantitative analysis were possible.

McGinnis et al. (1986) studied the use of RP columns for the separation of unsubstituted carbohydrates. Maltodextrins, isomaltodextrins and cellodextrins were chromatographed using water eluent; for the separation of cyclodextrins methanol-water (1:9) was used. An analytical column (25 cm x 4.6 mm I.D.) was packed with LiChrosorb C_{18} (RP-18), LiChrosorb C_8 (RP-8), Whatman C_{18} (ODS-1 or ODS-2) and PRP-I (Hamilton divinylbenzene-styrene copolymer). A preparative column (50 cm x 9.4 mm I.D.) was packed with ODS-2 RP packing material (Whatman 6526-41). RI detection was used. The retention time was dependent on the molecular weight of the oligosaccharide and on the type of anomeric configuration of the linkage.

Also Rajakylä (1986) described the use of RPC in carbohydrate analysis. For the HPLC of carbohydrates different types of RP columns with RI detection and water as the eluent were applied. Maltooligosaccharides up to DP 9 and mixtures of various other oligosaccharides and their anhydrides were chromatographed. ODS packing materials from several suppliers were systematically compared. It was concluded that RPC is a very good alternative to either ion-exchange or amino-bonded columns for the analysis of sugars. A decrease in column temperature resulted in increased retention times and better resolution. A decrease in alkyl chain length or the use of silica with larger pore diameters decreased the resolution and retention times. Supelco (Supelco Reporter, 1986) offer Supelcosil LC-18 columns and describe separations on a 25 cm x 4.6 mm I.D. column, d_p 5 μm, protected by an LC-18 guard column (2 cm x 4.6 mm I.D.) with water as the mobile phase at a flow-rate of 1 ml/min, temperature 35°C and RI detection. A sample of 20 μl containing 2 mg oligomers from Karo light syrup was resolved in 14 min up to DP 10; anomers were also resolved in the form of double peaks.

11.3.5 Physically (in situ) modified silica columns

A very simple method was described by Aitzetmüller et al. (1978, 1979) for converting an unmodified silica column into an effective column for sugar analysis. A polyfunctional amine (HPLC Amine Modifier I; NATEC, Hamburg, F.R.G.) of concentration 0.01-0.02% was used for the impregnation of silica (such as LiChrosorb Si 60, μPorasil, Nucleosil or Hypersil) and results similar to those with chemically bonded phases were obtained if a slightly higher water content in the eluent (acetonitrile-water mixtures) was applied. This method was used with good results for separations such as glucose-fructose, glucose-maltose-maltotriose, propylene glycol-glycerol-sorbitol and even α-, β- and γ-cyclodextrins. Linear sugars were eluted according to the number of subunits present; cyclodextrins were eluted much earlier. On a 250 mm x 4 mm I.D. Hibar LiChrosorb Si 60 (d_p 5 μm) column (Merck) impregnated in situ and eluted with acetonitrile-water (63.2:36.8) + 0.02% of HPLC Amine Modifier I, a complete separation of eight saccharides (fructose, glucose, sucrose, maltose, lactose, raffinose, α-cyclodextrin and stachyose) was obtained in 35 min with RI detection. Deproteinized soyabean extracts and soya protein products were also successfully analysed for their saccharide content. According to Aitzetmüller (1980), the equilibration of the column can be accelerated using a 0.1% concentration of the Amine Modifier overnight and then changing to a 0.01% solution for washing the column.

Wheals and White (1979) and White et al. (1980) investigated the applicability of several amines modifiers, such as commercial 1,3-diaminopropane, 1,4-diaminobutane, 1,3-diaminobenzene, 1,5-diaminopentane, 1,6-diaminohexane, 1,8-diaminooctane, tetraethylenepentamine, pentaethylenehexamine, 3-aminopropionitrile (as fumarate), triethylenetetramine and di(4-aminodiphenyl)methane. 1,4-Diaminobutane was found to be the best amine modifier for oligosaccharides. D-Glucooligosaccharides (obtained from partially hydrolysed starch) were resolved up to DP 20 on a 200 mm x 8 mm I.D. column packed with physically modified ca. 5 µm irregular silica (H.S. Chromatography Packings), using isocratic elution with 50% aqueous acetonitrile containing 0.01% of 1,4-diaminobutane. A pre-column (50 mm x 5 mm I.D.) containing 4 Chroprep Si 60 (15-25 µm) (Merck) was fitted between the pump and the septum injector. Because isocratic elution was applied, RI detection could be used.

Aitzetmüller (1980) mentioned the following advantages of the physical modification: cheap support with easy modification; prolonged use; an eluent with a higher water content leads to a better solubility of oligomers; and a relatively large isocratic range. The disadvantages are that the presence of an amine in the effluent makes preparative application difficult, acid-containing samples remove amine from the column and cause baseline disturbances and large amount of salts may interfere. The physically modified amine columns were discussed in a review by Verhaar and Kuster (1981).

Hendrix et al. (1981) described the separation of carbohydrates and polyols on a radially and hydraulically compressed HPLC silica column modified with tetraethylenepentamine (TEPA). Radial-Pak silica cartridges (10 cm x 8 mm I.D.) (Waters Assoc.) were employed and protected by a Waters guard column filled with AX/Corasil. For column pre-treatment the Radial-Pak silica cartridge was initially conditioned by pumping 50 ml of acetonitrile-water (70:30) containing 0.1% (v/v) of TEPA (pH 9.2) through the column. The final elution solvent was then introduced [acetonitrile-water (81:19), pH 8.9, containing 0.02% of TEPA] and the column stabilized by recirculation of this solvent overnight. The injected sample contained 250 µg each of seven standard components dissolved in a volume of 50 µl. The above final eluent was used at a flow-rate 2 ml/min at 2.34 MPa (340 p.s.i.) and 26°C. The chromatographic conditions were studied and retention data for 22 carbohydrates or polyols were tabulated. The authors can provide data for about 100 chromatographed substances (sugars and polyols) in addition to those published in the table.

11.3.6 Separation of saccharides using amino acid analysers and sugar analysers

Structural determinations of oligosaccharide moieties in glycoproteins, glycosylaminoglycans, glycolipids and some oligosaccharides require the separation and identification of oligosaccharide fragments containing amino sugars. Donald (1977) described the separation of hexosamines, hexoaminitols and hexosamine-containing di- and trisaccharides on an amino acid analyser. Hexosamine-containing oligosaccharides were reduced and de-N-acetylated, and the resulting amino alditols were separated on an amino acid analyser. Borate-citrate buffers were used for elution and ninhydrin for detection. Ninhydrin reacts with amino and imino groups of many organic substances and coloured compounds are formed, which can be monitored by absorbance measurements at 570 and 440 nm, respectively (for a description of an amino acid analyser see e.g., Zmrhal et al., 1975). The principle used for the separation of amino sugars consists in the fact that the formation of a borate complex should lower the affinity of the amino sugar for a cation-exchange resin in a manner dependent on the acidity of such a complex.

The reduction and de-N-acetylation procedure was as follows. To a sugar solution (20 μg - 1 mg in 90 μl of water) in 5 cm x 0.4 cm I.D. Pyrex tubes, 10 μl of 10% (w/v) $NaBH_4$ were added. After 3 h, 20 μl of 6 M sodium hydroxide solution were added. The samples were frozen in an acetone-solid carbon dioxide mixture and sealed under vacuum. The tubes were then heated for 16 h at $100^{\circ}C$, opened and the contents diluted to 1 ml with pH 2.2 buffer (Moore et al., 1958).

The amino acid analyser (Evans Electroselenium, Halstead, U.K.) was modified by the addition of an automatic sample loader and a buffer change valve (Locarte, London, U.K.). All separations were performed on a 40 cm x 1 cm I.D. column of Locarte No. 12 resin (X8 cross-linked sulphonated polystyrene) at $50^{\circ}C$. The flow-rates of buffer and ninhydrin were 45 and 22.5 ml/h, respectively. The absorbance at 570 nm was measured in a 20-mm flow-through cuvette. The elution buffers were as follows: (I) 0.1 M in Na^+, containing trisodium citrate dihydrate (49.0 g), boric acid (1.55 g), 33% (w/v) Brij-35 solution (15 ml), water to 5 l and hydrochloric acid to pH 7.5; (II) 0.2 M in Na^+, prepared from trisodium citrate dihydrate (91.88 g), sodium tetraborate decahydrate (11.9 g), 33% (w/v) Brij-35 solution (15 ml), water to 5 l and hydrochloric acid to pH 8.0.

Aliquots containing 5-50 μg of the reduced and de-N-acetylated sugars (dissolved as described above in pH 2.2 buffer prepared according to Moore et al., 1958) were transferred to the autoloader of the amino acid analyser, washed with the pH 2.2 buffer and run with the following standard programme: buffer I (320 min), buffer II (200 min), 0.5 M sodium hydroxide solution (40 min), buffer I (120 min). The analyser was then prepared for the next run.

Donald (1977) described the separation of mixtures of up to fourteen reduced and de-N-acetylated sugars using various chromatographic conditions.

Kennedy (1974) reviewed the analytical ion-exchange chromatography of monosaccharides and smaller oligosaccharides, based on the formation of borate complexes. The automation of this process was analogous to that of amino acid analysis, where ninhydrin detection was used. Kennedy and Fox (1977) described the fully automatic ion-exchange and gel permeation chromatography of neutral monosaccharides and oligosaccharides with a Jeolco JLC-6AH analyser. The amino acid analyser was modified for carbohydrate analysis using sulphuric acid - orcinol detection in two ways: (1) for ion-exchange chromatography in borate buffers an LC-R-3 quaternary ammonium ion-exchange resin column was eluted in the descending mode and the separation lasted 6.5 h; (2) for gel permeation chromatography a Bio-Gel-P2 column was eluted in the ascending mode and the separation lasted 22 h. Chytilová et al. (1978) tested medium basic DEAE-Spheron 300 as a chromatographic support in a Model 71000 A modular sugar analyser (Developmental Workshops of the Czechoslovak Academy of Sciences, Prague, Czechoslovakia) equipped with borate elution and sulphuric acid - orcinol detection, and found DEAE-Spheron to be very suitable for this purpose (see Section 11.3.3 for examples of the extension of this method to oligosaccharide analyses).

Torii et al. (1979) developed a preparative use of the analytical column of a sugar autoanalyser for the resolution of glucooligosaccharides of the same molecular weight. In analytical experiments, solutions containing 10-50 µg of individual sugars were loaded on a small column (150 mm x 8 mm I.D.) of Jeol LCR-3 anion-exchange resin (Japan Electronic Optical Laboratory, Tokyo, Japan) and with detection by measurement of absorption at 440 nm. For preparative resolution, up to 100 mg of sugar mixtures were applied on a column of the same size and monitored at 490 nm. Elution was carried out stepwise at 55 or 65°C with borate buffers: (I) 0.13 M boric acid at pH 7.5; (II) 0.25 M boric acid at pH 9.0; (III) 0.35 M boric acid at pH 9.6. The following separations were carried out: (a) a trisaccharide fraction obtained by acetolysis of dextran B 1355; (b) a trisaccharide fraction obtained by acid hydrolysis of elsinan (a glucan of *Elsinoe leucospila*); and (c) a pentasaccharide mixture obtained from a dextranase digest of dextran B 1397.

Kennedy et al. (1980) described a detailed automated qualitative and quantitative analysis of starch components and related mono- and oligosaccharides. An analytical protocol for an extensive range of neutral mono- and oligosaccharides (DP 2-12) with on-line computer calculation of the data was developed. This approach is generally applicable to the analysis of polysaccharides, glycoproteins, etc., in solids, solutions and biological fluids. Applications in a wide area of starch chemistry were dealt with.

Experiments with a Jeol JLC-6AH carbohydrate analyser and the way in which it might be adapted for additional analytical applications were described. Jeol JLC-6AH amino acid analyser was converted for use as an ion-exchange chromatographic analyser for carbohydrates by replacement of the ninhydrin reagent piston pump with a piston pump suitable for pumping concentrated sulphuric acid. The 570- and 440-nm absorbance filters were replaced with 510- and 425-nm filters, respectively. A schematic diagram and a description of the sugar analyser are shown in Fig. 11.8. A particular feature not incorporated in standard amino acid analysers is the use of a second column to duplicate the action of the first.

Short jacketed 17 cm x 0.8 cm I.D. columns were packed with LC-R-3 quaternary ammonium anion-exchange resin (Jeol) up to a bed height of 12 cm and the columns were regenerated and equilibrated (at a flow-rate of 30 ml/h) with buffer I (0.13 M sodium tetraborate, pH 7.50). Sample solutions (1-1000 µg of each component) were loaded (after centrifugation) into the storage loops (capacity 800 µl) of the analyser, from which they were automatically loaded to one of the ion-exchange columns and eluted at 55°C at a flow-rate 30 ml/h using a buffer programme: buffer I 110 min, buffer II (0.25 M sodium tetraborate, pH 9.08) 90 min, buffer III (0.35 M sodium tetraborate, pH 9.60) 190 min. Buffer IV (0.50 M potassium tetraborate, pH 9.60), pumped for 90 min, was used for regeneration and buffer I (120 min) for re-equilibration. Buffer I eluted 2-deoxyribose, sucrose, cellobiose, maltose, lactose and rhamnose (elution times were tabulated), buffer II ribose and mannose and buffer III fucose, arabinose, fructose, galactose, xylose and glucose. The elution programme was made repetitive by using an appropriate punched tape in the elution programmer. A pair of ion-exchange columns were used in the programme, one for analysis while the other was regenerated.

The assay reagent, recrystallized orcinol dissolved in concentrated sulphuric acid 1.5 g/l), pumped at 24 ml/h, was mixed with the sample taken from the column eluate (the sampling rate was 12 ml/h). The residence time of the mixture in the heating coil (95°C) was 15 min. The unknown components in the sample were identified by comparison of their elution positions with those for standards run under the same conditions. They were quantitated by measurement of the areas under the peaks and comparing them with those for the appropriate standards.

In addition to the ion-exchange mode, gel permeation chromatography was also automated. The ion-exchange resin column was replaced with a jacketed 1.5 m x 2 cm I.D. column slurry-packed with deaerated Bio-Gel P-2 (-400 mesh) (Bio-Rad Labs.). This column was eluted at 65°C in the ascending mode with deaerated water at a flow-rate of 16 ml/h. The samples were loaded automatically. The orcinol method was used for detection.

The paper by Kennedy et al. (1980) provides very detailed descriptions of both modern apparatus and numerous applications in starch chemistry and similar areas.

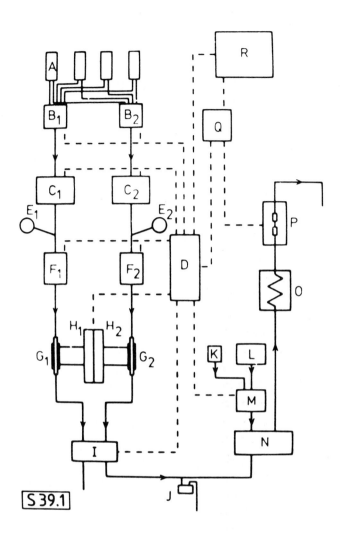

Fig. 11.8. Diagram of the modified Jeolco JLC-6AH automatic carbohydrate analyser.
The series of borate buffers (A) are selected by two eight-way valves (B$_1$, B$_2$)
and pumped by two double-action, high-pressure pumps (C$_1$, C$_2$). Buffer and sample
selection and flow switching are performed by a tape programmer (D) and the output from the pump passes the pressure gauges (E$_1$,E$_2$). The samples, previously
injected into sample storage loops (F$_1$, F$_2$), are pumped on to columns (G$_1$, G$_2$),
which are maintened at 55°C by independent heating circulating pumps (H$_1$, H$_2$).
The eluent from these columns passes to a two-way selector (I), which enables
either of the two columns to be connected to the detection system. An all-PTFE
pressure pump (N) draws the sample stream from I, excess running to waste (J),
and mixes it with orcinol-sulphuric acid reagent which is drawn from storage (K)
via a valve (M), which can also admit water (L). The reagent sample mixture is
then pumped through a reaction coil (O) at 95°C. The colour so formed is measured
by two flow cells (P) at wavelengths of 425 and 510 nm. The data are then printed
out on a point-plot chart recorder (Q). The data may be calculated and quantitated by means of the Nova 1220 computer (R), which is on-line to the analyser.
Control communications are illustrated by dashed lines. (Reprinted from Kennedy
et al., 1980.)

11.4 DETECTION PRINCIPLES IN HPLC OF OLIGOSACCHARIDES

There are two fundamental types of detection of chromatographed compounds:
(1) non-destructive detection is usually based on the examination of the effluent
by physical methods that do not alter the detected substances and (2) destructive
detection is accompanied by some chemical reaction (or reactions) during which
the substance of interest is altered or destroyed. In the field of saccharide
chromatography, the first class includes methods such as the measurement of
refractive index (RI), the absorbance of transmitted light or optical activity
(rotation). The second class of detection methods can be subdivided into pre-
column and post-column chemical modification types. Naturally, flame ionization
detection belongs to the destructive methods.

From another point of view, the detection methods can be classified as off-line
or on-line methods. For the HPLC of saccharides on-line methods are the most important because they can be automated and are compatible with the rapidity of the
HPLC process. Automated detection principles in the LC of saccharides were reviewed by Čapek and Staněk (1975a). In addition to brief comments on non-destructive methods, the authors described chemical modification methods in great detail.
Their treatise was accompanied by flow diagrams (for the Technicon AutoAnalyzer
system) of individual procedures and included the following methods, a concise
survey of which may be useful for the reader:

(1) The orcinol-sulphuric acid method in an automated form (Bathgate, 1970) was
probably the most frequently used method for the detection of sugars after chromatography in borate buffers. The column effluent was mixed with 1% aqueous
orcinol and then with 72% sulphuric acid. The reaction mixture was heated at
$95^\circ C$ for 60 min, cooled and its absorbance monitored at 420 nm.

(2) The anthrone-sulphuric acid method as modified by Otter et al. (1970)
consisted in mixing 0.1% anthrone in 85% sulphuric acid with the effluent from
partition chromatography [water-ethanol - n-butanol (42:25:33)] in the ratio 2:1.
The mixture was heated at $80^\circ C$ for 20 min and the absorbance was measured at
640 nm.

(3) The phenol-sulphuric acid method was also used for the detection of sugars
in borate buffer column effluents (Ohms et al., 1967; Jolley et al., 1970), which
were continuously treated with 5% aqueous phenol-concentrated sulphuric acid
(0.6:3.05) and monitored by absorption measurements at 480, 486 and 490 nm.

(4) Potassium hexacyanoferrate(III) assay (Samuelson and Swenson, 1963) was
used for detection in aqueous ethanol effluents, to which 0.3 M sulphuric acid
was admixed and the mixture was heated in a water-bath at $80^\circ C$ for 5 min in
order to hydrolyse oligosaccharides. An aliquot of the (incompletely) hydrolysed

solution was continuously mixed with 0.09% $K_3Fe(CN)_6$ in 2 M sodium hydroxide solution containing 0.5% of potassium cyanide, then heated at 80°C for 5 min and (after cooling) the absorbance was measured at 440 nm.

(5) The cysteine-sulphuric acid method was used for the determination of the total hexose concentration in dextrans (Baker et al., 1969). Part of the column effluent was mixed with water and a 0.07% (v/v) solution of cysteine hydrochloride in 86% (v/v) sulphuric acid was added. After heating at 95°C for 3 min and cooling, the absorbance was measured at 420 nm.

(6) In parallel with the previous determination, reducing end-group assay was also carried out by Barker et al., (1969). Part of the column effluent, diluted with water, was mixed with an aqueous solution containing 0.53% sodium carbonate, 0.065% potassium carbonate and 0.05% potassium hexacyanoferrate(III). After heating at 95°C for 15 min, a solution of ammonium iron(III) sulphate (0.75%) and sodium lauryl sulphate (0.5%) in 0.25 M sulphuric acid was added and the absorbance was measured at 660 nm.

(7) Periodate oxidation was used in the application of IEC to mixtures of alditols and aldoses (Samuelson and Strömberg, 1966). The aqueous ethanol column effluent was mixed with periodate reagent, consisting of (a) 0.015 M periodic acid neutralized with ammonia and buffered to pH 7.5 with phosphate buffer or (b) 0.015 M sodium periodate in 0.12 M hydrochloric acid; the solution had a pH of 1.0. To the mixture arsenite reagent was added (0.5 M sodium arsenite neutralized with hydrochloric acid to pH 7) and 2,4-pentanedione reagent (2 M ammonium acetate + 0.05 M acetic acid + 0.02 M 2,4-pentanedione). The solution was then heated at 80°C for 5 min and the absorbance was measured at 420 nm. The assay was based on the colour reaction of formaldehyde, resulting from periodate oxidation, with 2,4-pentanedione. At pH 7.5 (reagent a) formaldehyde was formed in a high yield from both alditols and aldoses, whereas at pH 1 (reagent b) most sugars gave rise to negligible amounts of formaldehyde, while the alditols reacted without difficulty.

(8) Chromic acid assay was used for the analysis of sugar acids after IEC (Johnson and Samuelson, 1966; Samuelson and Thede, 1967). The column effluent was mixed with the assay solution containing 5 volumes of concentrated sulphuric acid + 2 volumes of 2.45 g/l of potassium dichromate solution. The mixture was heated at 95°C for 7-8 min and the absorbance was measured at 633 nm. Chromic acid oxidation gave a response with all eluted hydroxy acids.

(9) Carbazole assay (Johnson and Samuelson, 1966; Samuelson and Thede, 1967) was used in parallel experiments with the previous chromic acid assay. The column effluent was mixed with sulphuric acid and a 0.15% (w/w) solution of carbazole in 70% (w/w) ethanol. The mixture was heated at 95°C for 3 min and the absorbance was measured at 531 nm.

(10) Simultaneous periodate, chromic acid and carbazole assays were used by Carlsson and Samuelson (1970) in order to distinguish various types of sugar acids.

(11) An acetic acid-aniline-orthophosphoric acid assay was developed by Walborg and Kondo (1970) for the quantitation of neutral mono- and oligosaccharides in buffers containing glycerol or 2,3-butanediol. The column effluent (containing, e.g., boric acid and 2,3-butanediol buffers) was mixed at $60^{o}C$ with the reagent (200 ml of glacial acetic acid was added to 6 ml of redistilled aniline, then 100 ml of 85% orthophosphoric acid was added to the mixture), the mixture was heated at $120^{o}C$ for 17.5 min using a back-pressure to prevent boiling and the absorbance was measured at 310, 365 or 390 nm in order to distinguish between classes of saccharides. The precission of this quantitative method was ±5%.

The above automated procedures were important pioneer methods in LC analyses of saccharides. However, not all of them were used later in HPLC separations and, if used, their application was an exception. In spite of this, they are presented here in a concise form, because some of them may be modified for special determinations using modern chromatographic techniques. In the following part of this section detection methods compiled mainly from papers on the HPLC of oligosaccharides will be dealt with.

The detection methods used in the LC of sugars on modern silica-based stationary phases were reviewed by Verhaar and Kuster (1981).

11.4.1 Refractive index detection and photometric detection

The refractometric method is universal and the photometric method is an almost universal detection method; they have often been applied for monitoring the HPLC of oligosaccharides. Especially refractometry, which has been widely used in the sugar industry to measure the concentrations of sugar solutions (Charles and Meads, 1962), has also been generally accepted in the HPLC of sugars. Numerous applications of a differential, automatically recording refractometer have been described (e.g., Brown, 1970). In most of the work cited in Chapter 11, RI monitoring of the effluent was mentioned as a technique used for evaluation, and therefore only general features of this method will be noted here.

The advantage of RI determination is that it is non-destructive, which is important in preparative separations. Disadvantages are the relatively low sensitivity[*] to separated substances (the minimum amount is 20-3 µg of sugars) and

---

[*] The sensitivity of interference refractometers is about two orders of magnitude higher (e.g., Optilab 5902 Interference Refractometer, available from Tecalor AB, Höganäs, Sweden).

the fact that it reflects also variations in the buffer concentration (some
"ghost peaks" may appear); usually this method is not compatible with gradient
elution, and also pulsation must be eliminated. Moreover, the relatively high
sensitivity to temperature requires that the detector must be thermostated very
preciselly (better than ±0.01^{o}C), especially if the column is heated. Variations
of ±0.01^{o}C cause changes in the mobile phase composition (due to the temperature-
conditioned differences in the solvent-sorbent interactions), and these variations give stronger detector responses than sugars present in small amounts;
this is valid especially at higher acetonitrile concentrations. However, in
spite of these disadvantages, RI detection is often used in the chromatography
of carbohydrates, because it is simple, does not require any modification of
the solutes and RI detectors are readily available. The response to sugars
usually increases with the increasing water content in acetonitrile mixtures.

Photometry is another important method that can be applied in a non-destructive mode. The direct detection of underivatized saccharides using UV light is
possible. Generally, the UV absorbance method is only moderately influenced by
temperature or pressure variations. Changes in solvent concentration do not
influence the detection so that the UV absorbance method can be applied to
gradient elution, which reduces the analysis time. Sugars can be detected in the
wavelength range 185-195 nm. This is a disadvantage, because most solvents used
in LC are not transparent enough in this range. Thus, e.g., acetonitrile must be
of analytical-reagent grade or specially purified for chromatography, as impurities absorb strongly. Aromatic solvents cannot be used at all. Also, some
buffer components absorbing below 200 nm and normally often used in LC cannot
be applied in this instance. In addition, traces of impurities in samples may
cause great difficulties, because they often have high molar absorptivities.
However, UV detectors have been applied for direct measurements in the HPLC of
sugars.

Majors (1977) applied a Varian variable-wavelength detector (Varichrom) for
direct UV detection (at 192 nm) in the NPC of a mixture of mono-, di- and trisaccharides (a gradient from 10% to 40% of water in acetonitrile was used for
elution). Fishman et al. (1982) applied direct UV detection (at 206 nm) for
monitoring high-molecular-weight pectins after GPC (in water or dilute phosphate
buffers) in addition to RI detection; UV detection revealed new information
regarding the chemical heterogeneity of pectins, their aggregation properties
and the chromatographic conditions necessary to prevent peak distortion. Supelco
(Supelco Reporter, 1986) recommend sensitive direct UV detection (at 190 nm) in
connection with the application of Supelcosil LC columns to saccharides.

Derivatized sugars can be detected at higher UV wavelengths and generally the detectability (sensitivity) and selectivity (of pre-column-derivatized samples) are improved considerably; e.g., Vrátný et al. (1985) examined p-aminobenzoic hydrazide and various other reagents; for other citations see the review by Verhaar and Kuster (1981). Detection by photometry in visible light was also described in HPLC sugar separations and an increase in sensitivity in these destructive post-column derivatization methods could be obtained. Hough at al. (1972) studied the automatic analysis of neutral monosaccharides in glycoproteins and polysaccharides using IEC in borate buffers at pH 7. Carbohydrates were analysed by following the reaction with cysteine hydrochloride solution (0.07%, w/v) in sulphuric acid [86%; cf., point (5) in Section 11.4] and the absorbance was measured at three wavelengths (395, 405 and 425 nm) in a 15-mm flow cell; the advantage of this system was discussed. Noël et al. (1979) and D'Amboise et al. (1980) described a systematic liquid chromatographic separation of poly-, oligo- and monosaccharides and their high-sensitivity determination. Wood extracts, hydrolysed xylan and starch were analysed and the separation of glucose, fructose, mannose, xylose and galactose was examined. The method using a reduced form of tetrazolium blue for colour detection (at 530 nm) gave fast and reproducible results with a detection limit of less than 10 µg for monosaccharides.

11.4.2 Fluorimetric or chemiluminescence detection and fluorescence labelling

Many papers have been published on this subject owing to the very high sensitivity of fluorimetric analyses. These methods are based on both pre-column and post-column derivatization of sugars. Post-column derivatization is currently realized in the column effluent in on-line system. As relatively narrow peaks are obtained in HPLC, the continuous flow reaction system in wide-bore glass columns used in earlier LC post-column derivatization experiments was replaced with a reaction system in capillary tubes, which brings about much less peak broadening. Of course, a higher pressure is required to force the liquid through at sufficient speed, because the narrow tubes have a high flow resistance.

Katz and Pitt (1972) described a versatile and sensitive post-column monitoring system for liquid chromatography based on cerate oxidation. The detector depends on the measurement of Ce^{III} (by fluorescence) produced from the reaction of Ce^{IV} with eluted reducing agents. This detector is more sensitive then the previous oxidation detectors and can be varied to discriminate among compounds according to the ease of oxidation. Mrochek et al. (1975) used it in LC analysis for neutral carbohydrates in serum glycoproteins. IEC in borate buffers was used for the separation and the effluent was detected by a cerate oxidimetric detection system, monitoring the fluorescence of Ce^{III}; sensitivity to 1 nmol was demonstrated.

Mopper (1978) developed a very sensitive detection of (mono)saccharides based on oxidation of copper bicinchoninate, which is used in a Biotronic LC 7000 sugar analyser.

Mopper et al. (1980) examined borate-complex IEC with fluorimetric detection of saccharides. The fluorescent products were formed by the reaction of reducing and non-reducing saccharides with ethylenediamine (EDA). What was interesting about this method was the fact that the reagent was added to the mobile phase containing borate buffer prior to chromatography (cf., Section 11.3.3) and, in spite of this, this method was not based on some form of pre-column modification, as the fluorescent product was formed only after warming of the effluent in an on-line post-column reactor. EDA was purchased in a highly purified form from Breda Scientific (Breda, The Netherlands); it is sold under the trade name Nanochrome II and consists of a solution of triply distilled aldehyde-free EDA in borate buffer (cf., Section 11.3.3).

A commercially available sugar analyser, the Fluorimetric Sugar Module (Breda Scientific) was used for the separation. The post-column reactor contained 30 m of 0.5 mm I.D. PTFE tubing, which provided a reaction time of about 9 min at given flow-rate (0.7 ml/min). For chromatographic runs the reactor temperature was set at 145°C; boiling in the reaction coil was suppressed with a restrictor (about 5 bar back-pressure) placed after the cuvette. The fluorescence of the reaction product was monitored continuously with a filter fluorimeter (Gilson Medical) containing a 45-μl flow-through cuvette. Broad band-pass filters centred at about 360 and 455 nm were used for excitation and emission, respectively. The detection limit was less than 1 nmol for most saccharides.

Kato and Kinoshita (1980) described the fluorimetric detection and determination of carbohydrates by HPLC using ethanolamine. The reaction of a mixture of ethanolamine and boric acid with reducing sugars gave intense fluorescence. Its application to post-column derivatization provided a rapid, simple and sensitive detection of carbohydrates under mild reaction conditions. A flow diagram of the relatively simple HPLC equipment was illustrated in Fig. 5.17 in Part A of this book. The chromatographic column (30 cm x 4 mm I.D.) was pre-packed with Aminex A-27 or LiChrosorb-NH$_2$ (Shimadzu Seisakusho, Kyoto, Japan). A μBondapak CH column (30 cm x 3.9 mm I.D.) was purchased from Waters Assoc. The LC-3A HPLC pumps (delivering the mobile phase and the fluorescent reagent) were obtained from Shimadzu Seisakusho. The mobile phase for the Aminex column (delivered at 2 ml/min) was 0.5 M potassium borate buffer (pH 8.7); for the LiChrosorb and μBondapak columns, acetonitrile-water (75:25) at 1.5 ml/min was used.

The fluorescent reagent was prepared as follows. To a mixture of 20 g of boric acid and 20 g of ethanolamine, redistilled water was added to give 1000 ml of solution, which was stable for 1 week at room temperature. It was delivered into the chromatograph at 0.5 ml/min.

Glucose showed excitation and fluorescence maxima at 357 and 436 nm, respectively. An aqueous solution containing 2% (w/v) ethanolamine and 2% (w/v) boric acid was sufficient to give a maximum detector response. The described method was applied to the separation and determination of carbohydrates in enzymic hydrolysates of amylose and human serum.

Honda et al. (1980) studied the fluorimetric analysis of carbohydrates using aliphatic amines and its application to the automated analysis of carbohydrates. Hitachi 2633 anion-exchange resin was used with stepwise elution with borate buffers, or Hitachi 2634 resin with a single-buffer elution (analysis time 80 min). The application of ethylenediamine, malonamide and 2-cyanoacetamide, which gave strong fluorescence in a weakly alkaline media, was used for labelling reducing carbohydrates eluted from the columns. Samples of serum and urine were analysed for medical diagnostic purposes.

Gandelman and Birks (1982) developed a novel and interesting photooxygenation-chemiluminescence (POCL) HPLC detector for the determination of aliphatic alcohols, aldehydes, ethers and saccharides, which was capable of quantitating many oxygen-containing compounds that could not be determined by either absorption or direct fluorescence. The POCL detector first oxygenated the analytes (e.g., sugars) in a sensitized photochemical reaction using a post-column reactor to produce H_2O_2, which was then quantitated using a cobalt(II)-luminol chemiluminescence reaction. As the photochemical reaction was sensitized by sodium anthraquinone-2,6-disulphonate (AQDS), the analytes themselves did not need to absorb light. The probable mechanism for AQDS-sensitized photooxygenation was discussed in detail. The H_2O_2 formed could be quantitated by the reaction with luminol (see the review by White and Roswell, 1970):

| Luminol | Excited compound | Stable anion |

Luminol = 5-amino-2,3-dihydro-1,4-phthalazinedione.

The transition metal [Co(II)] increases the intensity of chemiluminescence and this catalytic mechanism is believed to involve the formation of a Co(II)-peroxy complex, which then oxidizes the luminol molecule to produce an electronically excited aminophthalate anion, which is the emitting species. AQDS and luminol (Aldrich, Milwaukee, WI, U.S.A.) and cobalt nitrate hexahydrate (Fisher, Denver, CO, U.S.A.) were used without further purification.

The construction of an inexpensive detector and the optimization of the procedure were described in detail. The mobile phase [70% acetonitrile-4.85 · 10^{-5} M AQDS - 1 · 10^{-4} M Co(II)] was pumped at 0.5 ml/min through a 250 mm x 4.6 mm I.D. C_{18} column and the effluent entered the photochemical reactor, consisting of 1 m of 0.38 mm I.D. transparent PTFE tubing, coiled around a 26 cm x 2 cm O.D. Pyrex cylinder, which slid over the fluorescence lamp tube (Sylvania Model E8 TS IBLB); the reactor was covered with aluminium foil. The H_2O_2 solution from the reactor entered the mixing cell (laboratory made) into which the luminol solution (2 · 10^{-4} M in 0.1 M borate buffer at pH 10.9) was forced by a syringe pump (0.34 ml/min). The chemiluminescent solution immediately entered the detection cell (the coil of PTFE tubing wrapped around a 1-cm diameter metal tube, positioned near the first focus of an ellipsoidal mirror) and then the solution passed to waste. The second focus of the mirror was placed on the photocathode of the photomultiplier, the output current of which was measured with a picoammeter (Model 480; Keithley, Cleveland, OH, U.S.A.) and displayed on a stripchart recorder.

The detection limits for fructose and sucrose were 3.1 and 14.7 µg, respectively. The reproducibility of the whole equipment, estimated from ten repeated experiments and expressed by the relative standard deviation, was 7.8%.

Hase et al. (1979) described a method of linking the reducing sugar end-unit of an oligosaccharide and fluorescent 2-aminopyridine, during which a pyridyl-amino (PA) derivative is formed:

PA derivative

Hase et al. (1981a,b) used this pre-column reaction as a highly sensitive method for the analysis of sugar moieties of glycoproteins, and for the determination of molecular weights of neutral oligosaccharides by fluorescence labelling. The separation and identification of the PA derivatives were carried out either (1) by HPLC with a 250 mm x 4 mm I.D. C_{18} reversed-phase column (TSK Gel LS 410, 5 μm; Toyo Soda, Tokyo, Japan) eluted with 0.1 M phosphate buffer (pH 8.3) or 0.1 M ammonium acetate (pH 4.0) at a flow-rate 1.6 ml/min, or (2) by HPLC with a 600 mm x 7.5 mm TSK Gel 2000 PW column (Toyo Soda), eluted with 0.01 M ammonium chloride (pH 7.6). The separations were carried out at 20°C. The chromatograph was a Hitachi Model 638 or a Gaschro-Kogyo Model 570 B equipped with a fluorescence spectrophotometer (Hitachi Model 650-10 M). An excitation wavelength of 320 nm and an emission wavelength of 400 nm were used for phosphate buffer solution, and an excitation wavelength of 310 nm and emission wavelength of 380 nm for ammonium chloride solution. As little as 0.1 pmol of PA derivative could be detected. As little as 10 μg of Taka-amylase A were easily detected by this system.

Dutot et al. (1981) examined the HPLC of oligosaccharides in the form of dansylhydrazones. A method was described with fluorimetric detection following reaction with dansylhydrazine. For partition HPLC LiChrosorb Si 60 and various mobile phases were applied. The amount of hydrazones formed was proportional to the amount of dansylhydrazine. The detection limit for galactose was 4 pmol.

Coles et al. (1985) reported the pre-column fluorescent labelling of carbohydrates and their analysis by LC, including a comparison of derivatives using mannosidosis oligosaccharides. The fluorescent labels introduced into oligosaccharides by reductive amination were compared. Both labels (2-aminopyridine and 7-amino-1-naphthol) improved the chromatographic efficiency and detection sensitivity, but reductive amination with pyridinylamine (PA) derivatives was found to be incomplete. In contrast, naphthoamine (NA) derivatives:

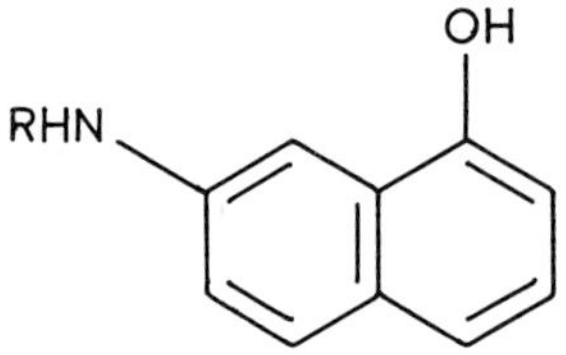

NA derivative

were found to be completely reduced and possessed enhanced fluorescence.

The fluorescent NA derivatives were prepared as follows. 7-Amino-1-naphthol (2 mg) was dissolved in dimethyl sulphoxide (50 μl), the solution was added to the dried oligosaccharides and the mixture was heated for 30 min at 105°C. A solution (50 μl) made by adding sodium cyanoborohydride (5 mg) to methanol (400 μl) and glacial acetic acid (40 μl) was added and the mixture was heated for 2.5 h at 105°C, dried and subjected to partition according to Folch et al. (1957). The upper layer was removed and dried, and re-extraction of the bottom layer ensured complete recovery of all derivatized components.

Liquid chromatographic separations were performed with equipment from Waters Assoc.; the column was amino Spherisorb (250 mm x 4.6 mm I.D.). For elution of the rechromatographed samples (cf., Section 11.5.1) a 20-70% gradient of 0.15 M aqueous ammonia in acetonitrile was used at a flow-rate of 2 ml/min. Fluorescence was detected with a Model FS 970 spectrofluorimeter (Kratos, Ramsey, NJ, U.S.A.). The excitation wavelength was either 240 or 232 nm with a 320-nm cut-off filter on the detector side. The reagents, 7-amino-1-naphthol and sodium cyanoborohydride, were obtained from Aldrich. The described methods of fluorescence labelling seem to be very promising, because they not only enhance the sensitivity of LC detection but also improve the chromatographic resolution of the products.

11.4.3 Optical activity detection

For monitoring sugars in chromatographic effluents, polarimetric detection has also been used. However, this principle has not been applied very often because there are some problems with its use. The quantitative evaluation of sugars is possible only with well separated solutes (owing to difficulties with the broad specific rotation spectrum of overlapping compounds) for which the specific rotation is known, and where the anomeric ratio remains constant and is known, in view of the fact that anomer forms give different responses. The advantage of the method is in its applicability to gradient elution (using optically inactive eluents) and also some additional information on the solutes can be derived from their polarization. Usually, the sensitivity of this method is not high.

Yeung et al. (1980) developed a detector for HPLC detection of trace organics based on optical activity. A micro-polarimeter based on Ar^+ laser optics was interfaced to the HPLC system, and with the use of selected gland prisms, cell-window material and air-based Faraday rotators, it was possible to obtain extinction ratios four orders of magnitude better than in a standard polarimetric instrument. A detection limit of 0.5 µg in the separation of sugars was obtained in a 200-µl detection volume. An absorption detector (254 nm) was compared with an optical activity detector in the separation of untreated human urine components. Böhme (1980) applied a micro-polarimeter to the detection of sugars in HPLC.

Kuo and Yeung (1981) described the determination of carbohydrates in urine by HPLC with optical activity detection. Their detector was subsequently improved. A laser was operated at 484 nm to match better the spectral response of the photomultiplier tube, the internal volume of the flow cell was reduced to 80 µl and other improvements were described. All these changes led to a detectability of 100 ng of fructose (signal-to-noise ratio 3) when two standard 10-µm C_{18} columns were used in series to increase the chromatographic efficiency. This allowed the simultaneous detection of six naturally occurring carbohydrates in 100-µl samples of human urine.

Gurkin and Patel (1982) described an experiment on the GPC separation of products of the partial hydrolysis of β-cyclodextrin (using a Fractogel HW-40 S column), optical rotation analysis (at a wavelength of 411.3 nm, path length 5 cm) being used for detection.

11.4.4 Mass spectrometric detection

Mass spectrometric methods have been used for special purposes of carbohydrate detection and simultaneous characterization. However, these methods require expensive equipment. Morris et al. (1980) reviewed mass spectrometric methods (after HPLC separation) for the determination of the structures of neuropeptides, proteins and glycoproteins, and presented some results of studies on the determination of the carbohydrate structure of prothrombin A by electron-impact (EI) mass spectrometry.

Games and Lewis (1980) described the combined liquid chromatography-mass spectrometry (LC-MS) of glycosides, glucuronides, sugars and nucleotides in an on-line system. Using ammonia chemical ionization (CI), good total ion current (TIC) traces were obtained together with mass spectra, which permitted the characterization of mono- and disaccharides, in addition to other compounds studied. The methods and equipment for LC-MS were explained in Sections 5.1.8 and 6.6.5. The TIC trace (ammonia CI) obtained during LC-MS effluent monitoring of a mixture of xylose, fructose, glucose and sucrose, chromatographed on a 200 mm x 5 mm I.D. Hypersil amide bonded column and eluted at 1.5 ml/min with acetonitrile-water (85:15) containing 0.1% of acetic acid, was presented, in addition to TIC traces of other substances. The ammonia CI spectrum of glucose obtained during LC-MS was also illustrated. The ammonia CI spectral data of other sugars and components were given in tabular form.

Games et al. (1981) studied combined LC-MS with a moving-belt interface. In addition to the application of this method to many natural substances, some sugars were also examined (xylose, fructose, glucose and sucrose). Dichloromethane negative CI spectral data were tabulated. Games (1981) reviewed this method very thoroughly, including a diagram of the moving-belt LC-MS equipment and the results obtained in studies of various natural compounds; saccharides were also discussed. Rajakylä (1986) described the use of RPC in carbohydrate analysis and showed that the combination of a chromatographic system and a modern thermospray quadrupole mass spectrometer worked satisfactorily with a starch hydrolysate sample. TIC chromatogram patterns of positive and negative ions of glucose oligomers up to $M_r > 600$ were presented, in addition to LC-MS negative ion spectra of glucose ($M_r = 180$) and maltose ($M_r = 342$).

Alcock et al. (1982) reviewed HPLC-MS with a transport interface. Applications of this method to studies of many (especially natural) substances, including sugars, were listed.

11.4.5 Other detection principles

Some special detection principles were developed for sugar analyses or could be applied in this area, e.g., nearly universal mass detection (Charlesworth, 1978). In the evaporation analyser (mass detector) designed by Applied Chromatography Systems the effluent stream, containing the solute, was nebulized and carried by an air stream through a heated column. The eluent evaporated and a fine mist of dry solute particles passed through a light beam and caused light scattering, which was detected by a photomultiplier. Various non-volatile synthetic and natural substances could be detected. Only volatile eluents were acceptable for chromatography without the addition of salts.

Verhaar and Kuster (1981), in a review, mentioned this analyser, the detection limit of sugars being given as about 500 ng.

The principle of the moving-wire detector was also applied to the separation of carbohydrates. Hobbs and Lawrence (1972a,b) examined the use of a flame ionization detector in the partition chromatography of sugars on a strong cation exchanger in the Li^+ form using 85% aqueous ethanol as the eluent. A modified Pye moving-wire detector was connected to the column and provided a direct trace of the column effluent on a potentiometric recorder with linear responses for carbohydrates. The moving-wire detector (Model LCM2, Pye Unicam) was also used by Wells and Lester (1979) to monitor gradient elution in the HPLC of per-acetylated oligosaccharides (cf., Fig. 11.7); it detected all non-volatile carbon compounds.

Vrátný et al. (1980) described a double-reaction detection principle for the LC of non-reducing oligosaccharides. In the first reaction oligosaccharides were hydrolysed by passing the effluent from the chromatographic column through a small reaction column (packed with a strongly acidic cation exchanger in the H^+ form). The reducing saccharides formed were then detected using the second reaction with p-hydroxybenzoic acid hydrazide (Lever, 1972) and photometry of the products at 410 nm; this was the final step. The construction of the detec-

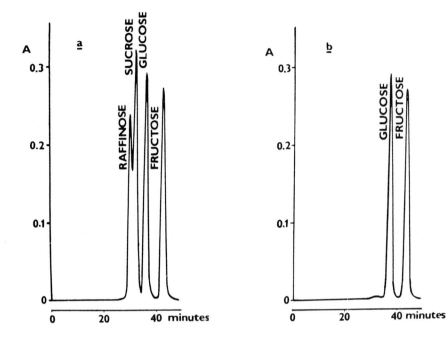

Fig. 11.9. Separation of a standard mixture of pure sugars in 24-μg amounts, using the double-reaction detector illustrated in Fig. 5.5 in Part A of this book. (a) Hydrolytic reaction column used; (b) reaction column disconnected. (Reprinted from Vrátný et al., 1980.)

tor was explained in Fig. 5.25 in Part A of this book. Here only an example of the resulting chromatograms is illustrated, in Fig. 11.9. Kim et al. (1982) characterized dextrans by SEC using a differential refractive index/low-angle laser light-scattering photometer (DRI/LALLSP) for monitoring the column effluent. As detectors, Waters R-403 DRI and Chromatix KMS-6 instruments with angle 6-7° and field stop 0.15 were used. A method for this purpose has been developed.

Honda (1984) reviewed detection systems for carbohydrates.

11.5 SURVEY OF EXAMPLES OF APPLICATIONS OF OLIGOSACCHARIDE CHROMATOGRAPHY

In this section only a few application examples of HPLC separations of oligosaccharides can be given. The reader is referred to the Register and Bibliography (Chapters 14 and 15) for further information.

11.5.1 Neutral homooligosaccharides

Torii and Sakakibara (1974) described the separation and quantitation of α-linked glucose oligosaccharides. Glucose and isomaltose up to isomaltoheptaose,

and also glucose, kojibiose, nigerose, nigeritol and isomaltitol, were well
separated on Jeol LCR-3 anion-exchange resin using stepwise elution with borate
buffers at elevated temperatures. White et al. (1980) used irregular silica
columns dynamically modified with various di- and polyamines (cf., Section
11.3.5) for analyses of underivatized D-glucooligosaccharides (DP 2-20 from
partially hydrolysed starch), which were separated excellently in less than
40 min using a simple isocratic system. Mellis and Baenziger (1981) separated
reduced neutral oligosaccharides by NPC on MicroPak AX-5 and other Varian columns
and on a μBondapak/carbohydrate Analysis Column (Waters Assoc.). All the columns
(30 cm x 4 mm I.D.) were eluted by gradient elution starting from acetonitrile-
water (65:35), the acetonitrile content decreasing at 0.5% min. The method was
applied to both the analysis and the preparative isolation of glycoprotein-
derived oligosaccharides obtained by enzymic release with endoglycosidase or
chemical release by hydrazinolysis. Introduction of ^{3}H by reduction with
[^{3}H]NaBH$_4$ permits the detection of sub-nanomole amounts of oligosaccharides.
Examples of the separations achieved are illustrated in Fig. 11.10.

Kainuma et al. (1981) described the quantitative HPLC analysis of maltooligosaccharides using a Jascopak SN-01 column (25 cm x 4.6 mm I.D.) (Nihon
Bunko, Tokyo, Japan) or a μBondapak carbohydrate column (30 cm x 4.0 cm I.D.)
(Waters Assoc.). Acetonitrile-water (65:35, v/v) was used for isocratic elution
at room temperature and DRI detector (Shodex SE-11) for monitoring. Mixtures
from G_1 up to G_{15} or G_7, respectively, were separated. Okada et al. (1985)
described the development of a fractionation method for maltooligosaccharides
with gel filtration chromatography on an industrial scale. An enzymatically
digest suspension of potato starch was fractionated on a 30 x 20 cm I.D. column
of Toyopearl HW 40 C gel and a 125 x 15 cm I.D. column of Toyopearl HW 40 S.
The recovery of 70% pure maltopentaose was 40%, with a productivity of 700 g of
dry matter per day on 100 l of gel.

Gum and Brown (1977) examined the HPLC separation of reduced and normal cellooligosaccharides on two columns (cf., Section 11.3.2). The best results were
obtained with a Whatman PXS-1035 PAC column eluted with 29% (w/w) water in
acetonitrile at a constant flow-rate of 1.5 ml/min. The results obtained are
illustrated in Fig. 10.11. Residual substrate and the products of the enzymic
hydrolysis of cellohexaitol and/or cellopentaitol could be separated. The possibility of a straightforward quantification was an advantage of this method
in comparsion with others. Ladisch et al. (1978) applied IEC on AG 50W-X4 resin
in the Ca^{2+} form to the chromatography of cellodextrins and achieved the separation of celloheptaose to glucose in 30 min using water as the eluent. Hostomská-
Chytilová et al. (1982) examined the application of anion-exchange derivatives

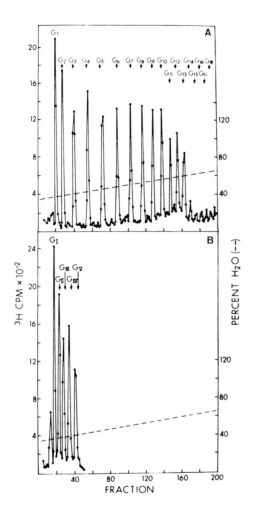

Fig. 11.10. Separation of [3H] dextran (A) and [3H] chitin (B) oligomers on
MicroPak AX-5. (A) G_1 is [3H]glucitol and G_2-G_{18} are reduced oligomers of glucose
containing 2-18 glucose residues per mole, respectively. (B) G_I is [N-3H]acetyl-
glucosaminitol and G_{II}-G_V are reduced oligomers of N-acetylglucosamine containing
two to five residues of N-acetylglucosamine per mole of oligomer, respectively.
Material eluting prior to G_1 and G_I consists of trace contaminants, usually
removed in the preparation of larger oligosaccharides by descending paper chromatography. (Reprinted from Mellis and Baenziger, 1981.)

of Spheron (cf., Section 11.3.3) for the separation of glucose and cellobiose
in kinetic studies of the enzymic hydrolysis of cellulose and for the separation
of lower cellodextrins resulting from enzymic hydrolysis of celloheptaose. The
results of these experiments are illustrated in Fig. 11.12.

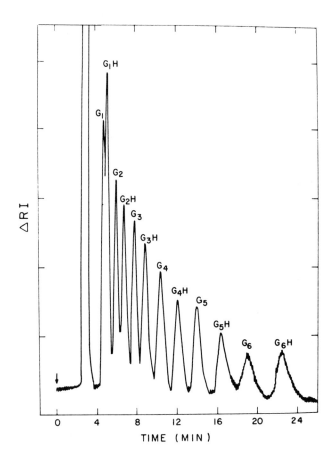

Fig. 11.11. Separation of reduced and normal cellooligosaccharides using a
Whatman column and isocratic elution with acetonitrile-water. The amount of each
oligosaccharide was 25 µg. (Reprinted from Gum and Brown, 1977.)

Zsadon et al. (1979) separated cyclodextrins by HPLC on a µBondapak carbohydrate column (30 cm x 4 mm I.D.) that contained amine functional groups.
Acetonitrile-water mixtures (25-35 vol.-% of water) were used for elution at
25°C. α-, β- and γ-cyclodextrins were completely separated from glucose; RI detection was used. Hokse (1980) analysed cyclodextrins using the strongly acidic
cation-exchange resin Aminex 50W-X4 in the Ca^{2+} form and water as eluent (cf.,
Section 11.3.1); the results are illustrated in Fig. 11.2.

Coles et al. (1985) examined the separation of mannosidosis oligosaccharides
after reductive amination using fluorescent labelling (cf., Section 11.4.2);
$(Man)_2$- to $(Man)_6$-1-acetamido-1,2-dideoxy-D-glucit-1-yl derivatives were studied.
Permethylated derivatives were also subjected to chemical ionization mass spec-

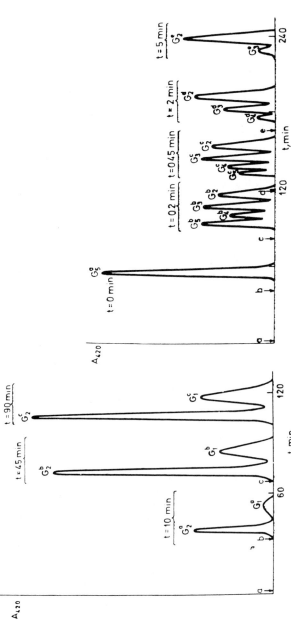

Fig. 11.12. Left: Chromatography of a hydrolysate of ground Whatman No. 1 chromatographic paper with cellulolytic enzymes of *Trichoderma viride-reesei* carried out on a 250 mm x 8 mm I.D. column of medium basic DEAE-Spheron 300 (0.6 mequiv./g, 20-40 μm) at 60°C and a flow-rate of 50 ml/h. The column was equilibrated and isocratic elution was carried out with 0.025 M borate buffer (pH 7.5). Samples of 100 μl, hydrolysed for 10, 45 and 90 min, were injected. The total time of analysis was about 60 min, 27 min of which the eluent spent in the detector, and further samples were injected every 30 min. Right: formation of lower cellodextrins by the action of cellulolytic enzymes from *T. viride-reesei* on cellopentaose. The 200 mm x 8 mm I.D. column of strongly basic Spheron 1000 TEAE (13-18 μm, 2 mequiv./g) was equilibrated with 0.025 M borate buffer (pH 7.5), which was also used for isocratic elution at 60°C and a flow-rate of 50 ml/h. The arrow and the letters a, b, c, d and e indicate the points of injection of hydrolysates. (a) Cellopentaose was not incubated with cellulases; (b) time of incubation 0.2 min; (c) 0.45 min; (d) 2 min; (e) 5 min. The total time of chromatography was 78 min. The effluent persisted for 27 min in the detector. New samples were injected before the preceding analysis was terminated, at approximately 45-min intervals. G1 = Glucose; G2-G5 = cellobiose to cellopentaose. The orcinol-sulphuric acid method was used for detection with monitoring at 420 nm. The flow-rate of orcinol-sulphuric acid through the detector was 1.93 ml/min plus 0.42 ml/min of the effluent. (Reprinted from Hostomská-Chytilová et al., 1982.)

trometry. At the beginning of the work human urinary oligosaccharides obtained
from a mannosidosis patient were chromatographed on an amino-Spherisorb column
(250 mm x 4.6 mm I.D.) using acetonitrile-water (11:9) at 1 ml/min as the eluent
and RI detection. Five major peaks were obtained, containing $(Man)_2$-GlcNAc to
$(Man)_6$-GlcNAc, and subjected to fluorescent modification with both the PA and NA
reaction methods, as mentioned in Section 11.4.2. The heterogeneity of the
original major peaks was proved in this way and the main rechromatographed fractions were subjected to permethylation and characterization by mass spectrometry.

11.5.2 *Acidic and hexosamine-containing oligosaccharides*

The simplest method for the isolation and separation of anionic oligosaccharides is IEC in conventional aqueous buffers. For example, Baenziger and
Natowicz (1981) used a MicroPak AX-10 anion-exchange column with a mobile phase
characterized by an increasing potassium phosphate concentration, and separated
well oligosaccharides bearing 0-4 sialic acid residues (cf., Section 1.3.3).
Oligosaccharides containing mannose-6-phosphate moieties in monoester or diester
linkages could be analysed in this system. This method was also applied to
preparative separations: the resolution of 20 mg of oligosaccharide could be
accomplished in a single run with quantitative yields. Vrátný et al. (1983) used
a partially quaternized anion-exchange derivative of DEAE-Spheron 40 (DEAE-
Separon HEMA 40, 14 µm) in a 25 cm x 0.6 cm column, isocratically eluted with
0.1 *M* sodium sulphate at 60°C, for the chromatography of oligogalacturonic acids
(DP 1-8) (see also Mikeš et al., 1988).
    In contrast, Bergh et al. (1982) studied the specificity of ovine submaxillary
gland sialyltransferases and applied NPC on LiChrosorb-NH$_2$ in acetonitrile-15 m*M*
potassium phosphate for the HPLC identification of sialooligosaccharide products
(for conditions, see Section 11.3.2). Portions of radioactive oligosaccharide
products (3-6 nCi of ^{3}H) were injected simultaneously with unlabelled reference
oligosaccharides (25-30 nmol), which were detected at 195 nm. For radioactivity
monitoring, 2-ml effluent fractions were collected and the radioactivity was
measured by liquid scintilation counting.
    Delaney et al. (1980a) examined the use of HPLC for the isolation and se-
quencing of chondroitin sulphate oligosaccharides. Chondroitin-4-sulphate (from
whale cartilage) and chondroitin-6-sulphate (from shark cartilage) hyaluronidase
digests, containing tetra-, hexa-, octa- and decasaccharides, were separated by
low-pressure gel permeation chromatography (LP-GPC), reduced with sodium
[^{3}H]borohydride and purified by preparative paper chromatography and by HPLC on
an anion-exchange Partisil-10 SAX column (25 cm x 4.5 mm I.D.) (Whatman). The

tetra-, hexa-, octa- and decasaccharides were resolved by isocratic elution with 190, 260, 340 and 400 mM KH$_2$PO$_4$, respectively. The isolated homogeneous components obtained by repeated chromatography were used for the characterization of oligosaccharides. The HPLC migration of purified oligosaccharides was studied on the same SAX column using a linear gradient from 250 to 550 mM KH$_2$PO$_4$ at a flow-rate of 1 ml/min over a 20-min time interval, and then isocratic elution with the last buffer for additional 20 min. The isolated oligosaccharides were characterized in detail and could be sequenced.

Lee et al. (1980) described enzymic studies (using chondroitinases) or urinary isomeric chondroitin sulphates from patients with mucopolysaccharidosis and applied HPLC in these examinations. This method had considerable potential for chemical diagnostics. A Partisil-10 PAC (10 µm) column (25 cm x 4.6 mm I.D.) (Whatman) was employed. A 10-µl aliquot of each enzymic digest, equivalent to 0.05 ml of urine, was injected and chromatographed at 2 ml/min and 700 p.s.i. with a ternary solvent system of acetonitrile-methanol-0.5 M ammonium formate buffer (pH 4.8) (60:20:20, v/v). Separations were carried out isocratically at room temperature and were monitored by a Waters Assoc. Model 440 UV detector (254 nm). Changes in the aqueous ammonium formate buffer content had a significant effect on compound retention. Responses for unsaturated disaccharides derived from urinary chondroitin sulphates were linear from 100 ng to 10 µg injected, and good quantitation was obtained for 25 µl or less of injected samples.

Delaney et al. (1980b) examined the quantitation of sulphated disaccharides of heparin by HPLC. Heparin was converted by treatment with nitrous acid into sulphated disaccharides and the mixture was reduced with sodium [^{3}H]borohydride and purified by preparative paper electrophoresis and paper chromatography. Four purified saccharides were used as standards in the development of the HPLC procedure on a pre-packed Partisil-10 SAX anion-exchange column (25 cm x 4.5 mm I.D.) (Whatman), eluted isocratically with 40 mM KH$_2$PO$_4$ (pH 4.6) for 30 min (three monosulphated disaccharides were resolved); then a programmed convex gradient of 40-400 mM KH$_2$PO$_4$ was used over a 40-min interval (a disulphated disaccharide was eluted). Samples of 10 µl and a flow-rate of 1 ml/min were used. For the HPLC analysis of the disaccharide composition of heparin fractions, samples containing 25 mg/ml of heparin were prepared, deaminated, reduced, evaporated, redissolved in 210 µl of water and 10 µl or 20 µl were applied to the column for analysis. Fractions of 0.5 ml were collected and the recovery of all the ^{3}H-labelled products was determined by scintilation counting. Quantitative recoveries were obtained.

Lee and Tieckelman (1980) described two different HPLC separation methods for unsaturated disaccharides derived from heparan sulphate and heparin by enzymic treatment. In the first method a bonded cyanoamino polar Partisil 10 PAC column (25 cm x 4.6 mm I.D.) (Whatman) was eluted isocratically with a ternary solvent system [acetonitrile-methanol-0.5 M ammonium acetate (pH 6.5)]. Non-sulphated, monosulphated and disulphated disaccharides were eluted; the tri-sulphated disaccharide was retained more strongly and was eluted by increasing the aqueous ammonium acetate content in the mobile phase. In the second method (which was more satisfactory, especially for the rapid separation of disulphated and trisulphated disaccharides), ion-pair RPC was applied on a bonded C_{18} RPC column (25 cm x 4.6 mm I.D.), containing Partisil 10 ODS (10 μm) and pre-packed by Whatman, eluted with a mobile phase consisting of 0.005 M PIC-A reagent mixed with methanol. The contents of one bottle (15 ml) of pre-packed PIC-A reagent (tetrabutylammonium phosphate, purchased from Waters Assoc.) were mixed with 1 l of distilled water (pH 7.0). The solution of 10% of methanol and 90% of this PIC-A reagent gave a satisfactory result, suitable for quantitation. The flow-rate was 1 ml/min and the pressure 800 p.s.i., and UV detection at 254 nm was applied.

Lee et al. (1981) studied in detail the separation of reduced disaccharides derived from various glycosylaminoglycans by HPLC. Unsaturated disaccharides prepared by enzymic digestion of chondroitin sulphates, dermatan sulphate, heparan sulphate and heparin, followed by reduction with sodium borohydride, were separated by NPC using a Whatman Partisil-10 PAC column. Acetonitrile-methanol-0.5 M ammonium formate (pH 6.0 for chondroitin sulphate disaccharides, or pH 4.5 for heparan sulphate disaccharides) in varying ratios was used in order to select the optimal separation for isocratic elution at room temperature. UV detection (A_{254}) was used for monitoring. This method was well suited for the HPLC analysis of glycosaminoglycans, because the possibility of obtaining anomeric forms of unsaturated disaccharides was eliminated. The applicability of this method for the determinations of glycosylaminoglycans in biological samples was demonstrated. Rice et al. (1985) examined the HPLC of heparin-derived oligo-saccharides. Enzymically depolymerized heparin was separated into di-, tetra-, hexa-, octa- and decasaccharide mixtures by LP-GPC and the mixtures were resolved by a strong anion exchange 25 cm x 5 mm I.D. column (SAX, Phenomenex, CA, U.S.A.) using linear gradient elution with 0.012 M to 0.2 M NaCl. Detailed methodology was presented for the separation of large and highly sulphated oligosaccharides.

Van Eikeren and McLaughlin (1977) described the HPLC analysis of lysozyme-catalysed hydrolysis and transglycosylation of β(1→4)-linked N-acetyl-D-glucosamine oligomers. A μBondapak carbohydrate column (30 cm x 4 mm I.D.) (Waters Assoc.) was eluted with acetonitrile-water (70:30, v/v) at a flow-rate of 1.5 ml/min. The output of a DRI detector was coupled to a Houston Instruments recorder with an electronic integrator. ^{14}C-Labelled samples were counted with

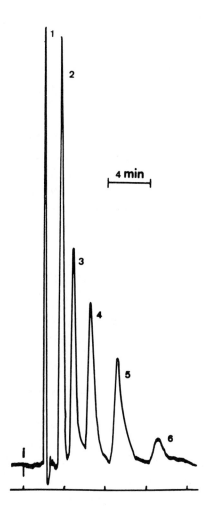

Fig. 11.13. Chromatographic separation of 15 µl of a standard calibration solution of N-acetyl-D-glucosamine (NAG) oligomers (each at 5 mg/ml) on a µBondapak
carbohydrate column using RI detector: (i) injection point; (1) water; (2) NAG
monomer; (3) NAG dimer; (4) NAG trimer; (5) NAG tetramer; (6) NAG pentamer.
Other chromatographic conditions are given in the text. (Reprinted from Van
Eikeren and McLaughlin, 1977.)

a Picker Nuclear Liquimat 220 liquid scintillation counter. An example of a
typical chromatographic separation is illustrated in Fig. 11.13. Hounsell et al.
(1985) dealt with the application of HPLC to the purification of oligosaccharides
containing neutral and acetamido sugars. Both RP-HPLC and anion-exchange HPLC
were discussed for the purification of oligosaccharides from human meconium,
milk and other sources.

11.5.3 Oligosaccharides in foodstuff raw materials and products

In addition to proteins, carbohydrates are also interesting components of soya beans. Havel et al. (1977) studied oligosaccharides released during the hydration of textured soya as determined by HPLC. Especially the amount of stachyose released from an extruded textured soya (Uni-Tex) during its hydration in boiling water was studied. Within the 25 min interval necessary for complete hydration, 75% of sugars originally present in the texturized product were released into the cooking water. The conditions of the chromatographic procedure were mentioned in Section 11.3.2 and an example of a chromatogram obtained was illustrated in Fig. 11.4. The determination of oligosaccharides in soya beans by HPLC using an internal standard (β-cyclodextrin) was published by Black and Bagley (1978). All the conditions were described in Section 11.3.2 and the results were illustrated in Fig. 11.3.

Wootton and Chaudhry (1981) studied the in vitro digestion of hydroxypropyl derivatives of wheat starch and described the analysis of its oligosaccharide fractions by HPLC. The products of partial digestion by porcine pancreatic α-amylase were analysed using a 30 cm x 4 mm I.D. μBondapak carbohydrate column (Waters Assoc.), eluted with acetonitrile-water (70:30, v/v), with RI detection. Oligosaccharides G_2 to G_{11} were well resolved. The separated peaks were quantitated by triangulation assuming similar RI properties for all the components. The detection limit based on maltose injections was 0.002 mg. Zygmunt (1982) evaluated the HPLC determination of mono- and oligosaccharides in pre-sweetened cereals and published the results of a wide collaborative study based on the work of 29 contributors. Chromatographic columns of C_{18} bonded phases eluted with acetonitrile-water mixtures were used. Eight samples consisting of six products were analysed in duplicate by HPLC and other standard methods and the results were tabulated and compared.

HPLC methods were also applied to the analysis of saccharides in molasses. Clarke and Brannan (1978) developed an HPLC method for the determination of sucrose and fructose and applied it to a variety of molasses samples; for the chromatographic conditions, see Section 11.3.2. The results were compared with those obtained by classical analytical methods. There is a problem in molasses analysis in separating a relatively large amount of saccharose from small amounts of other saccharide components, especially raffinose. Also, the resolution of fructose from psicose (formed during the industrial processing) requires special attention. How these problems were solved by Vrátný et al. (1979) using chromatography on DEAE-Spheron (cf., Section 11.3.3) is illustrated in Fig. 11.14.

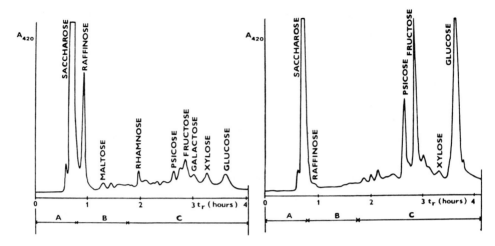

Fig. 11.14. Separations of saccharides in a 0.60-mg sample of sugar-beet molasses (left) and in a 0.30-mg sample of sugar-cane molasses (right). Medium basic DEAE-Spheron 300 (27 μm, 2.25 mequiv./g) in a 48.7 cm x 0.6 cm I.D. column equilibrated with buffer A was used in both instances, eluted in a stepwise manner with buffer A (0.03 M borate of pH 7.50), B (0.100 M borate of pH 8.85) and C (0.250 M borate of pH 8.88). Flow-rate, 50 ml/h; temperature, 60°C. (Reprinted from Vrátný et al., 1979.)

Mulcock et al. (1985) analysed saccharides in beet and examined an HPLC method. The concentrations of glucose, fructose and sucrose in beet extracts were determined with a total error of 2.7-10.5%. HPLC was found to be a suitable tool for sugar determinations and also allowed the exact calculation of the error.

11.5.4 Biomedical applications and methodological studies

Kuo and Yeung (1981) determined carbohydrates in urine by HPLC with optical activity detection. The conditions used were mentioned in Section 11.4.3. Sucrose, lactose, glucose, xylose, arabinose and fructose were quantitated in a 100-μl sample of human urine with a detectability of 100 ng. The reproducibility and reliability of this method should allow a better insight into the relationship between urinary sugars and physiological conditions.

Bienkowski and Conrad (1985) studied oligosaccharides formed by the treatment of heparin with nitrous acid. A mixture of the sodium [³H]borohydride-reduced oligosaccharides (from hog mucosa heparin treated with nitrous acid) was separated by GPC and the di- and tetrasaccharide pools were resolved by HPLC. In addition to already described components, new ones were found and their structures characterized. Apparently, the analysis of the deamidated mixture components could yield a chemical assay for the anticoagulant activity of heparin.

Ishii et al. (1985) described the HPLC analysis of changes in asparagine-linked oligosaccharides in regenerating rat liver. Detailed differences in the oligosaccharide chains between the regenerating liver and the controls were elucidated without the need for any isotope-labelling procedure. N-linked oligosaccharides were prepared from liver glycoproteins by glycopeptidase digestion, separated by affinity chromatography on a Con A-Sepharose column, and analysed by HPLC. The separation and identification of pyridylamino (PA) derivatives (cf., Section 11.4.2) was carried out by HPLC on a Shimadzu LC-3A column (250 mm x 4 mm I.D.) packed with TSK Gel LS 410 (5 μm, C_{18}) (Toyo Soda). The elution was performed at 50°C with 0.1 M phosphate buffer (pH 3.8) containing 0.1-1.0% 1-butanol at a flow-rate of 1 ml/min, with fluorescence detection (excitation 320 nm, emission 400 nm). An HPLC method was also used for the determination of monosaccharide components in hydrolysed oligosaccharides. High-mannose oligosaccharides $[(Man)_{5-9}GlcNAc_2]$ were well separated from complex oligosaccharides (bi-, tri- and tetra-antennary species with or without fucose). The recoveries of oligosaccharides were 90-100%. Differences in oligosaccharides in regenerating livers and controls were discussed in detail.

Some examples are important from the methodological point of view. Oshima et al. (1980) improved the separation of anomers of saccharides by HPLC on macro-reticular anion-exchange resins in the sulphate form (cf., Section 11.3.2). For example, after equilibration in 80% ethanol α- and β-anomers of D-glucose or D-galactose were well separated at 20°C (flow-rate 0.9 ml/min). The capacity factors of anomers of ten saccharides using the above conditions were tabulated.

From the methodological point of view, the perbenzoylation of oligosaccharides as a prerequisite for a sensitive analytical separation and quantitative determination is also important. Daniel et al. (1981a) described a method in which oligosaccharides were completely O-benzoylated without the concomitant N-benzoylation of acetamidodeoxyhexoses. The pre-column benzoylation allowed a quantitative determination of picomole amounts, because the absorbance at 233 nm of these derivatives is directly proportional to the number of benzoyl groups present. Both NPC (using a pellicular Zipax silica gel column) and RPC were examined, but the best results were obtained with a pre-packed Ultrasphere octyl column (25 cm x 4.6 mm I.D.) (Rainin Instruments, Woburn, MA, U.S.A.), eluted with an acetonitrile-water linear gradient (80 to 100% acetonitrile) at a rate of 2 ml/min. Benzoylated samples were dissolved in acetonitrile and aliquots (20 μl) were injected. Monitoring was via the UV absorbance in a variable-wavelength detector connected to an Autolabs System 1 computing integrator (Spectra-Physics, Santa Clara, CA, U.S.A.) and a strip-chart recorder.

An excellent separation of oligosaccharides containing up to ten sugar residues present in mannosidosis urine, or maltooligosaccharides containing up to fifteen sugar residues present in Karo syrup, were achieved in an analysis time of 30 min. In order to prevent the separation of anomers, reduction of complex samples prior to analysis was advisable. Examples of separations of various sugar types were given and the wide application possibilities of this method in clinical diagnosis were discussed.

Perbenzoylation procedures were as follows (Daniel et al., 1981a).

(a) With benzoic anhydride. Samples containing <1 μmol of total sugars in a screw-capped tube (100 mm x 13 mm I.D.) were dried in vacuo in the presence of P_2O_5 and benzoylated for up to 4 h at 37°C with 0.5 ml of pyridine containing 10% of benzoic anhydride and 5% of 4-dimethylaminopyridine as catalyst. Samples containing ≤1 nmol were benzoylated in silanized tubes with 100 μl of the reagent. The excess of reagents was removed by a modified Daniel's (1979) method. Each sample was then diluted with water (4.5 ml), shaken in a vortex mixer and applied to a column of Sep-Pak C_{18} (Waters Assoc.). The eluate was collected in the original tube and reapplied twice to the Sep-Pak column (to ensure maximal bonding), which was then washed with 10% (v/v) aqueous pyridine (10 ml), followed by distilled water (5 ml). The benzoylated oligosaccharides were eluted with acetonitrile (15 ml) into a 50-ml round-bottomed flask and evaporated by rotary evaporation. Each Sep-Pak column was re-used up to five times.

(b) With benzoyl chloride. The oligosaccharides were benzoylated with 10% (v/v) benzoyl chloride in pyridine (0.5 ml) for 16 h at 37°C. The benzoylated products were separated from the excess of reagents by solvent partition (Daniel, 1979).

The perbenzoylation method was reviewed and many biomedical applications were given and discussed by Daniel et al. (1981b).

11.6 COMMENTS ON LITERATURE ON THE LIQUID CHROMATOGRAPHY OF POLYSACCHARIDES AND OLIGOSACCHARIDES

Many summarizing treatments have been written on the separation of saccharides and it is hardly possible to list all of them. In this section only references directly available to the author will be given. Danishefsky et al. (1970) wrote a general introduction to polysaccharide chemistry, and Pazur (1970) dealt with the chemistry of oligosaccharides.

Čapek and Staněk (1975b) reviewed the separation of polysaccharides and the column liquid chromatography of mono- and oligosaccharides (1975a). Richter and Woelk (1977) dealt with HPLC as a method for the analysis of starch hydrolysates.

Samuelson (1978) reviewed the chromatography of oligosaccharides and related compounds on ion-exchange resins. Kennedy and Fox (1980) explained the fully automated ion-exchange chromatographic analysis of neutral mono- and oligosaccharides. Bundle et al. (1980) reviewed the possibilities of applications of preparative medium- and high-pressure liquid chromatography in oligosaccharide synthesis.

More recent summarizing papers have focused predominantly on HPLC methods. Verhaar and Kuster (1981) described and discussed the LC of sugars on silica-based stationary phases; column materials, eluent systems, detection methods and sample pre-treatment were dealt with. Brobst and Scobell (1982) explained modern chromatographic methods for the analysis of carbohydrate mixtures and commented on the separation of lower oligosaccharides both by LC and GC. Ivie (1982) discussed the use of HPLC for the separation and identification of sugars. Robards and Whitelaw (1986) have written a detailed review on the chromatography of monosaccharides and disaccharides, based on 364 references. Hanai (1986) reviewed newer advances in the liquid chromatography of carbohydrates.

REFERENCES

Aitzetmüller, K., *J. Chromatogr.*, 156 (1978) 354-358.
Aitzetmüller, K., *Chromatographia*, 13 (1980) 432-436; *C.A.*, 93 (1980) 130 728z.
Aitzetmüller, K., Böhrs, M. and Arzberger, E., *J. High Resolut. Chromatogr.*
   Chromatogr. Commun., 2 (1979) 589-590; *C.A.*, 92 (1980) 56 893k.
Alcock, N.J., Eckers, Ch., Games, D.E., Games, M.P.L., Lant, M.S., McDowall, M.A.,
   Rossiter, M., Smith, R.W., Westwood, S.A. and Wong, H.-Y., *J. Chromatogr.*,
   251 (1982) 165-174.
Anderson, D.M.W., Hendrie, A. and Munro, A.C., *J. Chromatogr.*, 44 (1969) 178-179.
Baenziger, J.U. and Natowicz, M., *Anal. Biochem.*, 112 (1981) 357-361.
Barker, P.E., Hatt, B.W. and Holding, S.R., *J. Chromatogr.*, 174 (1979) 143-151.
Barker, P.E., Hatt, B.W. and Vlachogiannis, G.J., *J. Chromatogr.*, 208 (1981)
   74-77.
Barker, S.A., Hatt, B.W. and Somers, P.J., *Carbohydr. Res.*, 11 (1969) 355-362.
Barth, H.G., *J. Liq. Chromatogr.*, 3 (1980) 1481-1496; *C.A.*, 93 (1980) 237 068m.
Barth, H.G. and Regnier, F.E., *J. Chromatogr.*, 192 (1980) 275-293.
Barth, H.G. and Regnier, F.E., *Methods Carbohydr. Chem.*, in press.
Barth, H.G. and Smith, D.A., *J. Chromatogr.*, 206 (1981) 410-415.
Bathgate, G.N., *J. Chromatogr.*, 47 (1970) 92-96.
Bergh, M.L.E., Koppen, P. and Van Den Eijnden, D.H., *Carbohydr. Res.*, 94 (1981)
   225-229; *C.A.*, 95 (1981) 111 108x.
Bergh, M.L.E., Koppen, P.L. and Van Den Eijnden, D.H., *Biochem. J.*, 201 (1982)
   411-415; *C.A.*, 96 (1982) 176 805n.
Bienkowski, M.J. and Conrad, H.E., *Semin. Thromb. Hemostasis*, 11 (1985) 86-88;
   C.A., 103 (1985) 105 248k.
Black, L.T. and Bagley, E.B., *J. Am. Oil Chem. Soc.*, 55 (1978) 228-232; *C.A.*,
   88 (1978) 119 483s.
Bock, W., Anger, H., Kohn, R., Malovikova, A., Dongowski, G. and Friebe, R.,
   Angew. Makromol. Chem., 64 (1977) 133-146; *C.A.*, 87 (1977) 169 520b.
Böhme, W., *Perkin-Elmer Chromatogr. Newsl.*, 8 (1980) 38.
Bonn, G., *J. Chromatogr.*, 322 (1985) 411-424.

Brobst, K.M. and Scobell, H.D., *Starch/Stärke*, 34 (1982) 117-121.
Brown, W., *J. Chromatogr.*, 53 (1970) 572-575.
Bundle, D.R., Iversen, T. and Josephson, S., *Am. Lab. (Fairfield, Conn.)*, 12 (1980) 93-94, 96-98; *C.A.*, 93 (1980) 210 683w.
Buytenhuys, F.A. and Van der Maeden, F.P.B., *J. Chromatogr.*, 149 (1978) 489-500.
Čapek, K. and Staněk, J., Jr., in Deyl, Z., Macek, K. and Janák, J. (Editors), *Liquid Column Chromatography (Journal of Chromatography Library*, Vol. 3), Elsevier, Amsterdam, 1975a, pp. 465-522.
Čapek, K. and Staněk, J., Jr., in Deyl, Z., Macek, K. and Janák, J. (Editors), *Liquid Column Chromatography (Journal of Chromatography Library*, Vol. 3), Elsevier, Amsterdam, 1975b, pp. 523-528.
Carlsson, B. and Samuelson, O., *Anal. Chim. Acta*, 49 (1970) 247-254.
Charles, D.F. and Meads, P.F., *Methods Carbohydr. Chem.*, 1 (1962) 520-524; *C.A.*, 58 (1963) 12 652f.
Charlesworth, J.M., *Anal. Chem.*, 50 (1978) 1414-1420.
Cheetham, N.W.H. and Teng, G., *J. Chromatogr.*, 336 (1984) 161-172.
Cheng, S.-L., *T'ai-wan T'ang Yeh Yen Chiu So Yen Chiu Hui Pao*, 90 (1980) 55-66; *C.A.*, 95 (1981) 76 264w.
Churms, S.C., *Adv. Carbohydr. Chem. Biochem.*, 25 (1970) 13-51.
Chytilová, Z., Mikeš, O., Farkaš, J. and Štrop, P., *J. Chromatogr.*, 153 (1978) 37-48.
Clarke, M.A. and Brannan, M.A., *Proc. Tech. Sess. Cane Sugar Refin. Res.*, (1978) 136-148; *C.A.*, 91 (1979) 59 138h.
Coles, E., Reinhold, V.N. and Carr, S.A., *Carbohydr. Res.*, 139 (1985) 1-11; *C.A.*, 103 (1985) 101 350q.
D'Amboise, M., Noël, D. and Hanai, T., *Carbohydr. Res.*, 79 (1980) 1-10; *C.A.*, 93 (1980) 18 604e.
Daniel, P.F., *J. Chromatogr.*, 176 (1979) 260-263.
Daniel, P.F., De Feudis, D.F., Lott, I.T. and McCluer, R.H., *Carbohydr. Res.*, 97 (1981a) 161-180; *C.A.*, 96 (1982) 158 482a.
Daniel, P.F., Lott, I.T. and McCluer, R.H., in Hawk, G.L., Champlin, P.B., Hutton, R.F. and Mol, C. (Editors), *Biological/Biomedical Applications of Liquid Chromatography 3, (Chromatographic Science Series*, Vol. 18), Marcel Dekker, New York, 1981b, pp. 363-382; *C.A.*, 96 (1982) 100 328p.
Danishefsky, I., Whistler, R.L. and Bettelheim, F.A., in Pigman, W., Horton, D. and Herp, A. (Editors), *The Carbohydrates. Chemistry and Biochemistry*, Vol. II A, Academic Press, New York, 1970, pp. 375-412.
Delaney, S.R., Conrad, H.E. and Glaser, J.H., *Anal. Biochem.*, 108 (1980a) 25-34; *C.A.*, 93 (1980) 234 213u.
Delaney, S.R., Leger, M. and Conrad, H.E., *Anal. Biochem.*, 106 (1980b) 253-261.
Djordjevic, S.P., Batley, M. and Redmont, J.W., *J. Chromatogr.*, 254 (1986) 507-510.
Donald, A.S.R., *J. Chromatogr.*, 134 (1977) 199-203.
Dreher, T.W., Hawthorne, D.B. and Grant, B.R., *J. Chromatogr.*, 174 (1979) 443-446.
Dudman, W.F. and Bishop, C.T., *Can. J. Chem.*, 46 (1968) 3079-3084.
Dutot, G., Biou, D., Durand, G. and Pays, M., *Feuill. Biol.*, 22 (1981) 101-104; *C.A.*, 95 (1981) 217 186m.
Escott, R.E.A. and Taylor, A.F., *J. High Resolut. Chromatogr. Chromatogr. Commun.*, 8 (1985) 290-292; *C.A.*, 103 (1985) 215 656z.
Fishman, M.L., Pfeffer, P.E., Doner, L.W., Barford, R.A. and Hoagland, P.D., *paper presented at the International Symposium on Column Liquid Chromatography, Philadelphia, PA, USA, June 6-11*, 1982.
Folch, J., Lees, M. and Stanley, G.H.S., *J. Biol. Chem.*, 226 (1957) 497-503.
Fonknechten, G., Bazard, D., Flayeux, R. and Moll, M., *Bios (Nancy)*, 11 (1980) 60-65; *C.A.*, 93 (1980) 163 746g.
Games, D.E., *Biomed. Mass Spectrom.*, 8 (1981) 454-462.
Games, D.E., Hirter, P., Kuhnz, W., Lewis, E., Weerasinghe, N.C.A. and Westwood, S.A., *J. Chromatogr.*, 203 (1981) 131-138.

Games, D.E. and Lewis, E., *Biomed. Mass Spectrom.*, 7 (1980) 433-436.
Gandelman, M.S. and Birks, J.W., *J. Chromatogr.*, 242 (1982) 21-31.
Gum, E.K., Jr. and Brown, R.D., Jr., *Anal. Biochem.*, 82 (1977) 372-375.
Gurkin, M. and Patel, V., *Am. Lab. (Fairfield, Conn.)*, 14 (1982) 64-73; *C.A.*, 96 (1982) 138 998v.
Hanai, T., *Adv. Chromatogr.*, 25 (1986) 279-307; *C.A.*, 104 (1986) 236 551w.
Hase, S., Hara, S. and Matsushima, Y., *J. Biochem. (Tokyo)*, 85 (1979) 217-220.
Hase, S., Ikenaka, T. and Matsushima, Y., *J. Biochem. (Tokyo)*, 90 (1981a) 407-414; *C.A.*, 95 (1981) 93 213w.
Hase, S., Ikenaka, T. and Matsushima, Y., *J. Biochem. (Tokyo)*, 90 (1981b) 1275-1279; *C.A.*, 96 (1982) 16 820k.
Hashimoto, T., Sasaki, H., Aiura, M. and Kato, Y., *J. Polym. Sci., Polym. Phys. Ed.*, 16 (1978) 1789-1800; *C.A.*, 89 (1978) 215 941f.
Havel, E., Tweeten, T.N., Seib, P.A., Wetzel, D.L., Liang, Y.T. and Smith, O.B., *J. Food Sci.*, 42 (1977) 666-668; *C.A.*, 87 (1977) 20 699j.
Hendrix, D.L., Lee, R.E., Baust, J.G. and James, H., *J. Chromatogr.*, 210 (1981) 45-53.
Heyraud, A. and Rinaudo, M., *J. Liq. Chromatogr.*, 3 (1980) 721-739; *C.A.*, 93 (1980) 106 496b.
Himmel, M.E. and Squire, P.G., *J. Chromatogr.*, 210 (1981) 443-452.
Hobbs, J.S. and Lawrence, J.G., *J. Sci. Food Agric.*, 23 (1972a) 45-51; *C.A.*, 76 (1972) 139 147r.
Hobbs, J.S. and Lawrence, J.G., *J. Chromatogr.*, 72 (1972b) 311-318.
Hokse, H., *J. Chromatogr.*, 189 (1980) 98-100.
Honda, S., *Anal. Biochem.*, 140 (1984) 1-47.
Honda, S., Matsuda, Y., Takahashi, M., Kakehi, K., Honda, A., Ganno, S. and Ito, M., *J. Pharmacobio-Dyn.*, 3 (1980) S-11, S-31; *C.A.*, 93 (1980) 163 927s.
Hostomská-Chytilová, Z., Mikeš, O., Vrátný, P. and Smrž, M., *J. Chromatogr.*, 235 (1982) 229-236.
Hough, L., Jones, J.V.S. and Wusteman, P., *Carbohydr. Res.*, 21 (1972) 9-17.
Hounsell, E.F., Jones, N.J. and Stoll, M.S., *Biochem. Soc. Trans.*, 13 (1985) 1061-1064; *C.A.*, 103 (1985) 156 641r.
Ishii, I., Takahashi, N., Kato, S., Akamatsu, N. and Kawazoe, Y., *J. Chromatogr.*, 345 (1985) 134-139.
Ivie, K.F., *Sugar Azucar*, 77 (1982) 80-81, 86-87; *C.A.*, 96 (1982) 183 126v.
Jandera, P. and Churáček, J., *J. Chromatogr.*, 98 (1974) 55-104.
Johnson, S. and Samuelson, O., *Anal. Chim. Acta*, 36 (1966) 1-11.
Jolley, R.L., Warren, K.S., Scott, Ch.D., Jainchill, J.L. and Freeman, M.L., *Am. J. Clin. Pathol.*, 53 (1970) 793-802; *C.A.*, 73 (1970) 63 009n.
Kainuma, K., Nakakuki, T. and Ogawa, T., *J. Chromatogr.*, 212 (1981) 126-131.
Kato, T. and Kinoshita, T., *Anal. Biochem.*, 106 (1980) 238-243.
Katz, S. and Pitt, W.W., Jr., *Anal. Lett.*, 5 (1972) 177-185; *C.A.*, 77 (1972) 28 437m.
Kennedy, J.F., *Biochem. Soc. Trans.*, 2 (1974) 54-64.
Kennedy, J.F. and Fox, J.E., *Carbohydr. Res.*, 54 (1977) 13-21; *C.A.*, 87 (1977) 15 531u.
Kennedy, J.F. and Fox, J.E., *Methods Carbohydr. Chem.*, 8 (1980) 3-12; *C.A.*, 93 (1980) 230 282t.
Kennedy, J.F., Fox, J.E. and Skirrow, J.C., *Starch/Stärke*, 32 (1980) 309-316; *C.A.*, 93 (1980) 188 088g.
Kesler, R.B., *Anal. Chem.*, 39 (1967) 1416-1422.
Khym, J.X. and Zill, L.P., *J. Am. Chem. Soc.*, 73 (1951) 2399-2340.
Kim, C.J., Hamielec, A.E. and Benedek, A., *J. Liq. Chromatogr.*, 5 (1982) 425-441; *C.A.*, 96 (1982) 200 049s.
Kuo, J.C. and Yeung, E.S., *J. Chromatogr.*, 223 (1981) 321-329.
Ladisch, M.R., Huebner, A.L. and Tsao, G.T., *J. Chromatogr.*, 147 (1978) 185-193.
Ladisch, M.R. and Tsao, G.T., *J. Chromatogr.*, 166 (1978) 85-100.
Lawrence, J.G., *Scan*, 5 (1974) 19-24; *C.A.*, 95 (1976) 71 766v.

Lee, G.J.-L. and Tieckelmann, H., *J. Chromatogr.*, 195 (1980) 402-406.
Lee, G.J.-L. and Tieckelmann, H., *J. Chromatogr.*, 222 (1981) 23-31.
Lee, G.J.-L., Evans, J.E., Tieckelmann, H., Dulaney, J.T. and Naylor, E.W.,
   Clin. Chim. Acta, 104 (1980) 65-75.
Lee, G.J.-L., Liu, D.-W., Pav, J.W. and Tieckelmann, H., *J. Chromatogr.*, 212
   (1981) 65-73.
Lever, M., *Anal. Biochem.*, 47 (1972) 273-279.
Majors, R.E., *J. Chromatogr. Sci.*, 15 (1977) 334-351; *C.A.*, 88 (1978) 44 499r.
Marx-Figini, M. and Soubelet, O., *Polym. Bull.*, 6 (1982) 501-508; *C.A.*, 96 (1982)
   204 488j.
McGinnis, G.D., Prince, S. and Lowrimore, J., *J. Carbohydr. Chem.*, 5 (1986) 83-97.
Mellis, S.J. and Baenziger, J.U., *Anal. Biochem.*, 114 (1981) 276-280; *C.A.*, 95
   (1981) 76 279e.
Mencer, H.J. and Grubisic-Gallot, Z., *J. Chromatogr.*, 241 (1982) 213-216.
Meuser, F., Klingler, R.W. and Niediek, E.A., *Getreide Mehl Brot*, 33 (1979)
   295-299; *C.A.*, 92 (1980) 74 461x.
Mikeš, O., Hostomská, Z., Štrop, P., Smrž, M., Slováková, S., Vrátný, P., Rexová,
   L'., Kolář, J. and Čoupek, J., *J. Chromatogr.*, 440 (1988) 287-304.
Mino, Y., Tsutsui, S. and Ota, N., *Chem. Pharm. Bull.*, 33 (1985) 3503-3506.
Moore, S., Spackman, D.H. and Stein, W.H., *Anal. Chem.*, 30 (1958) 1185-1190.
Mopper, K., *Anal. Biochem.*, 85 (1978) 528-532.
Mopper, K., Dawson, R., Liebezeit, G. and Hansen, H.-P., *Anal. Chem.*, 52 (1980)
   2018-2022; *C.A.*, 93 (1980) 179 073w.
Morris, H.R., Dell, A., Etienne, T. and Taylor, G.W., *Dev. Biochem.*, 10 (Front.
   Protein Chem.) (1980) 193-209; *C.A.*, 95 (1981) 20 660e.
Mrochek, J.E., Dinsmore, S.R. and Waalkes, T.P., *Clin. Chem.*, (Winston-Salem,
   NC), 21 (1975) 1314-1322; *C.A.*, 83 (1975) 128 376h.
Mulcock, A.P., Moore, S., Barnes, F. and Hickey, B., *Int. Sugar J.*, 87 (1985)
   203-207; *C.A.*, 104 (1986) 111 710s.
Natowicz, M. and Baenziger, J.U., *Anal. Biochem.*, 105 (1980) 159-164.
Noël, D., Hansi, T. and D'Amboise, M., *J. Liq. Chromatogr.*, 2 (1979) 1325-1336;
   C.A., 92 (1980) 60 767q.
Ohms, J.I., Zec, J., Benson, J.V., Jr. and Patterson, J.A., *Anal. Biochem.*, 20
   (1967) 51-57; *C.A.*, 67 (1967) 40 867c.
Okada, M., Uejima, O. and Nakakuki, T., *Shokuhin Sangyo Senta Gijutsu Kenkyu
   Hokoku*, 8 (1985) 61-73; *C.A.*, 103 (1985) 179 890k.
Orth, P. and Engelhardt, H., *Chromatographia*, 15 (1982) 91-96; *C.A.*, 96 (1982)
   200 005z.
Oshima, R., Kurosu, Y. and Kumanotani, J., *J. Chromatogr.*, 179 (1979) 376-380.
Oshima, R., Takai, N. and Kumanotani, J., *J. Chromatogr.*, 192 (1980) 452-456.
Otter, G.E., Popplewell, J.A. and Taylor, L., *J. Chromatogr.*, 49 (1970) 462-468.
Pazur, J.H., in Pigman, W., Horton, D. and Herp, A. (Editors), *The Carbohydrates.
   Chemistry and Biochemistry*, Vol. II A, Academic Press, New York, 1970, pp.
   68-137.
Rajakylä, E., *J. Chromatogr.*, 353 (1986) 1-12.
Rice, K.G., Kim, Y.S., Grant, A.C., Merchant, Z.M. and Linhart, R.J., *Anal.
   Biochem.*, 150 (1985) 325-331.
Richter, K. and Woelk, H.U., *Starch/Stärke*, 29 (1977) 273-277; *C.A.*, 87 (1977)
   153 670q.
Robards, K. and Whitelaw, M., *J. Chromatogr.*, 373 (1986) 81-110.
Samuelson, O., *Methods Carbohydr. Chem.*, 6 (1972) 65-75; *C.A.*, 77 (1972) 42 879p.
Samuelson, O., *Adv. Chromatogr.*, 16 (1978) 113-150.
Samuelson, O. and Strömberg, H., *Carbohydr. Res.*, 3 (1966) 89-96.
Samuelson, O. and Swenson, B., *Anal. Chim. Acta*, 28 (1963) 426-432.
Samuelson, O. and Thede, L., *J. Chromatogr.*, 30 (1967) 556-565.
Scobell, H.D. and Brobst, K.M., *J. Chromatogr.*, 212 (1981) 51-64.
Strubert, W. and Hovermann, W., *GIT Fachz.*, 22 (1978) 615-623; *C.A.*, 89 (1978)
   130 071x.
Supelco, *Supelco Rep.*, 5, No. 2 (1986) 6-7.

Torii, M., Alberto, B.P., Tanaka, S., Tsumuraya, Y., Misaki, A. and Sawai, T., *J. Biochem. (Tokyo)*, 85 (1979) 883-886; *C.A.*, (1979) 164 157r.

Torii, M. and Sakakibara, K., *J. Chromatogr.*, 96 (1974) 255-257.

Vandevelde, M.C. and Fenyo, J.C., *Carbohydr. Polym.*, 5 (1985) 251-273; *C.A.*, 103 (1985) 106 578e.

Van Eikeren, P. and McLaughlin, H., *Anal. Biochem.*, 77 (1977) 513-522.

Verhaar, L.A.Th. and Kuster, B.F.M., *J. Chromatogr.*, 220 (1981) 313-328.

Vivilecchia, R.V., Lighbody, B.G., Thimot, N.Z. and Quinn, H.M., *J. Chromatogr. Sci.*, 15 (1977) 424-433; *C.A.*, 88 (1978) 7565p.

Vrátný, P., Brinkman, U.A.Th. and Frei, R.W., *Anal. Chem.*, 57 (1985) 224-229.

Vrátný, P., Čoupek, J., Vozka, S. and Hostomská, Z., *J. Chromatogr.*, 254 (1983) 143-155.

Vrátný, P., Mikeš, O., Farkaš, J., Štrop, P., Čopíková, J. and Nejepínská, K., *J. Chromatogr.*, 180 (1979) 39-44.

Vrátný, P., Mikeš, O., Štrop, P., Čoupek, J., Rexová-Benková, L'. and Chadimová, D., *J. Chromatogr.*, 257 (1983) 23-35.

Vrátný, P., Ouhrabková, J. and Čopíková, J., *J. Chromatogr.*, 191 (1980) 313-317.

Wada, H., Makino, K., Takeuchi, T., Hatano, H. and Noguchi, K., *J. Chromatogr.*, 320 (1985) 369-377.

Walborg, E.F., Jr. and Kondo, L.E., *Anal. Biochem.*, 37 (1970) 320-329.

Walborg, E.F. and Lantz, R.S., *Anal. Biochem.*, 22 (1968) 123-133.

Wells, G.B. and Lester, R.L., *Anal. Biochem.*, 97 (1979) 184-190.

Wheals, B.B. and White, P.C., *J. Chromatogr.*, 176 (1979) 421-426.

White, C.A., Corran, P.H. and Kennedy, J.F., *Carbohydr. Res.*, 87 (1980) 165-173.

White, E.H. and Roswell, D.F., *Acc. Chem. Res.*, 3 (1970) 54-62; *C.A.*, 72 (1970) 84 346q.

Wooton, M. and Chaudhry, M.A., *Starch/Stärke*, 33 (1981) 200-202; *C.A.*, 95 (1981) 60 022e.

Yeung, E.S., Steenhoek, L.E., Woodruff, S.D. and Kuo, J.C., *Anal. Chem.*, 52 (1980) 1399-1402.

Zmrhal, Z.J., Heathcote, J.G. and Washington, R.J., in Deyl, Z., Macek, K. and Janák, J. (Editors), *Liquid Column Chromatography. A Survey of Modern Techniques and Applications* (*Journal of Chromatography Library*, Vol. 3), Elsevier, Amsterdam, 1975, pp. 666-711.

Zsadon, B., Otta, K.H., Tüdös, F. and Szejtli, J., *J. Chromatogr.*, 172 (1979) 490-492.

Zygmunt, L.C., *J. Assoc. Off. Anal. Chem.*, 65 (1982) 256-264; *C.A.*, 96 (1982) 179 545p.

Chapter 12

COMPOUND BIOPOLYMERS AND BIOOLIGOMERS

In Chapters 7 - 11 of Volume B, separations of simple polymeric or oligomeric substances were treated, i.e., substances composed of repeated units of the same type, e.g., amino acids → peptides or proteins, nucleotides → oligonucleotides, polynucleotides or nucleic acids, simple sugars → oligosaccharides or polysaccharides. In this chapter the rapid chromatographic separation of natural oligomeric or polymeric compounds containing also important molecular moieties of a different type will be dealt with, such as nucleoprotein complexes, glycolipids, glycopeptides and glycoside oligomeric derivatives. In addition, separations of several natural complex substances that are not well known are discussed.

12.1 VIRUSES, BACTERIOPHAGES AND RIBOSOMES

Viruses, phages and ribosomes or their subunits are extremely high-molecular-weight, complex particles, differing substantially in their molecular size and shape, and therefore size exclusion chromatography (SEC) was the first chromatographic approach to separate them from each other and from various lower molecular weight substances such as proteins, peptides, oligonucleotides and sugars. In order to facilitate their rapid separation, highly macroporous and rigid support materials had to be found. Haller (1965) was the first to propose and develop controlled-pore glass (CPG) with pore diameters ranging from 100 to 1700 Å, suitable for this purpose. Fig. 1.1 in the introductory part of Volume A illustrated the success of this effort. The advantage of this support was not only that it separated compound biopolymers quickly, but also that it could be sterilized and cleaned with strong acids without adversely affecting their pore diameters. Unfortunately, the polar surface of CPG (and also of porous silica) was shown to sorb irreversibly and denature some sensitive biological substances. Therefore, methods were sought for protecting the active surface of CPG or silica either by covalent bonding (e.g., using a Glycophase layer, reviewed by Regnier et al., 1977), or by adsorbed protecting layers (described in Section 12.1.1).
  However, during the further development of the rapid and selective isolation of viruses or their components, other separation modes [ion-exchange chromato-

graphy (IEC), reversed-phase chromatography (RPC), affinity chromatography etc.] were also tested and successfully used. Some of these methods will be reviewed in the following sections.

12.1.1 Size exclusion chromatography

12.1.1.1 Plant viruses and ribosomes

The first experiments were undertaken with plant viruses. An example of their rapid mutual separation using CPG and only hydrostatic pressure is illustrated by Fig. 12.1, taken from the pioneering work of Haller (1965). Almost complete separation was achieved in less then 10 min. It is interesting that TM-virus appeared at an elution volume of 18 ml, which is the dead-space of the column. As the virus consists mainly of rods 3000 Å long and 150 Å in diameter (cf., Laufer, 1944), this indicates that the length of the virus prevents it from entering the pores. The same phenomenon was observed by Steere (1964) on agar gel columns and was explained as being due to the rotation motion of the rod-shaped virus in the solution.

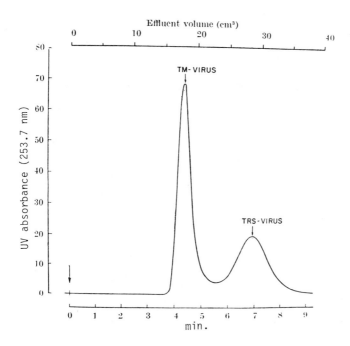

Fig. 12.1. Chromatographic separation of a mixture of tobacco mosaic virus (TM) and tobacco ring spot virus (TRS) on controlled-pore glass of average pore diameter 1700 Å. Bed dimensions: 50 cm x 1 cm. Sample: a mixture containing approximately 10^{11} particles of each virus in 0.06 ml of saline buffer. Eluent: 0.01 M phosphate buffer (pH 7.0) containing 0.85% NaCl. Flow by gravity was 5.2 ml/min cm^2. (Reprinted from Haller, 1965.)

TRS-virus, consisting of polyhedra of 260 Å diameter (Steere, 1956), readily
entered the 1700 Å pores of glass; the elution peak was close to 30 ml, typical
of admitted particles. Southern bean mosaic virus (SBMV) was also chromatographed (cf., Fig. 1.1 in Volume A); these particles had a diameter of 286 Å
(Leonard et al., 1953), did not enter the 260 Å pores of the glass used and
appeared at an elution volume of 18 ml, whereas the accompanying serum albumin
of $M_r = 7 \cdot 10^4$ (Phelps and Putnam, 1960), which was smaller than the pores,
was delayed and its position (slightly beyond 30 ml) was also influenced by the
charge effect at the particular pH. In a subsequent paper on virus isolation
with CPG, Haller (1967) described some other successful chromatographic experiments with plant viruses and bacteriophages (see the second half of this section).

However, Marcinka (1972), who also applied permeation chromatography on CPG
in the purification of plant viruses, found that red clover mottle virus,
tobacco mosaic virus, alfalfa mosaic virus and white clover mosaic virus were
adsorbed under certain conditions. Čech et al. (1977) tried to separate tobacco
mosaic virus on Spheron gel. This could be prepared in an extremely mcaroporous
form, and is a fully organic copolymer having no silanol adsorbing groups, that
could be used in high-performance liquid chromatography (HPLC).

Himmel and Squire (1981) described the HP-SEC of large biopolymers including
viruses and ribosomes on a TSK Gel G 5000 PW preparative column. This support
is also an organic polymer. It was used in the form of a 600 mm x 21.5 mm I.D.
column, connected with a protective pre-column. The eluent was 10 mM phosphate
buffer (pH 7), 100 mM in KCl, pumped at a flow-rate of 0.96 ml/min, and the
effluent was monitored by UV by detection at 280 nm. The average sample load
for calibration was 250 μl of a 3 mg/ml protein solution. In addition to a few
proteins and dextrans, the following viruses were chromatographed: southern bean
mosaic virus (SBMV, $M_r = 6.6 \cdot 10^6$; cf., Edsall, 1953), tomato bushy stunt virus
(TBSV, $M_r = 8.9 \cdot 10^6$; cf., Edsall, 1953), tobacco mosaic virus (TMV, $M_r = 39.4 \cdot 10^6$; cf., Caspar, 1963), and turnip yellow mosaic virus (TYMV, $M_r = 5.4 \cdot 10^6$; cf., Matthews, 1970).

The GPC of ribosomes and ribosome subunits (30S and 50S from *Escherichia coli*,
corresponding approximately to M_r $1.2 \cdot 10^6$ and $2.0 \cdot 10^6$, respectively) showed
co-elution at the void volume, and seemed to indicate that their size exceeded
that of the available pore openings. As the corresponding M_r values are not so
high, in the opinion of the cited authors it seems likely that the cause of this
phenomenon is the external size exclusion of these very large rods, whose length
(0.3 μm) is comparable to the interstitial space between the spherical packing
beads of radius ca. 9 μm. The cited authors have undertaken further detailed quantitative studies to test this hypothesis and the experiments are in progress. In
spite of these problems, the TSK Gel column chromatography may be useful as a pre-
paratory tool during the isolation and "clean-up" of ribosomes or their subunits.

Protein and virus samples were eluted as single symmetrical peaks with good resolution. GPC calibration constants (for dextrans) in this work were calculated both for molecular weights and for molecular radii and the data obtained are compared and discussed in relation to the other separated substances.

12.1.1.2 *Animal viruses and bacteriophages*

The first experiments on the rapid chromatography of bacteriophages and animal viruses were described by Haller (1967), who used CPG for the separation. MS_2 coliphage and Kilham rat virus were chromatographed. Purified preparations of ϕX-174, M13, M12, QB and T4 bacteriophages were obtained in large amounts by Gschwender et al. (1969), using a large-scale preparation on CPG. However, many investigators have found that some viruses adsorb so avidly to CPG that SEC cannot be applied. Hiatt et al. (1971) found some animal viruses (poliovirus, adenovirus, vesicular exanthema virus and viruses of vaccinia, yellow fever and rabies) that belonged to this adsorbed virus category. The adsorption of polio-virus was reduced (or eliminated) by pre-treatment of the glass with haemoglobin solutions followed by autoclaving (to denature haemoglobin in situ), but this technique was not found to be universal, e.g., it did not prevent the adsorption of rabies virus. The cited authors examined many techniques for preventing the adsorption of viruses and ultimately found that low concentrations of poly(ethylene oxide), a polyether of $M \approx 100\ 000^*$, prevented the adsorption of rabies virus to porous glass of average diameter 1250 Å (CPG 10-1250, Corning Glass Works, Corning, NY, U.S.A.).

Application procedure. The polyether can either be added to the eluent at a concentration of 0.04% or used to pre-condition the column by passage of one void volume of 0.4% solution, followed by five or more void volumes of distilled water or buffered salt solution.

An elution pattern was published by Hiatt et al. (1971) of mouse-adapted rabies virus (fixed strain PV_{12}) after passage over a 450 mm x 11 mm I.D. column of powdered CPG (porosity 1250 Å) which was preconditioned with poly(ethylene oxide), as mentioned above. The charge was 0.6 ml of 20% mouse brain suspension. The eluent (isotonic phosphate-buffered saline solution, pH 7.3) was pumped at a flow-rate of 0.8 ml/min.

According to Hiatt et al. (1971), the effect of poly(ethylene oxide) can be explained as follows. Polyethers have a strong affinity for complex formation by hydrogen bonding (Smith et al., 1959; Rösch, 1963). It seems probable that

---

*Polyox WSR IV-10, Union Carbide, New York, U.S.A. Poly(ethylene oxide) of M_r = 200 000 was also effective in preventing the adsorption. Polyethers of lower M_r were not tested; such compounds would be required for CPG of smaller pore size to ensure penetration.

multiple hydrogen bonds between the polyether and the electronegative oxygen atoms in the $-SiO_2-$ repeating structure of the glass account for the adsorption of poly(ethylene oxide). By competing for binding positions on the glass surface, poly(ethylene oxide) thus effectively blocks the adsorption of rabies virus.

According to Regnier et al. (1977), apparently polyethylene glycol (PEG) polymers hydrogen bond to surface silanol groups of CPG, decreasing the charge on the surface and preventing the adsorption of viruses and proteins. However, this is unfortunately a reversible process and PEG elutes from columns during continuous use. In spite of this, Darling et al. (1977) described the rapid purification of an RNA tumour virus and proteins by HP-SEC on bead columns made of CPG treated with PEG. Avian myeloblastosis virus (AMV) and hamster melanoma virus (HaMV) were purified from plasma proteins and tissue culture media components. The purified HaMV was still infectious, and the AMV-associated RNA-directed DNA polymerase ("reversed transcriptase") showed an 1100-fold purification of the virus after one column treatment. The time required for column purification was 5 min and electron microscopy of the purified virus showed intact particles.

The CPG (80-120 or 120-200 mesh) for this treatment was purchased from Analabs (North Haven, CT, U.S.A.) or Electro-Nucleonics (Fairfield, NJ, U.S.A.). The undesirable adsorption of CPG was greatly reduced by a modification of the method of Hawk et al. (1972), consisting in pre-treatment with PEG solution.

Pre-treatment procedure (Darling et al., 1977). A 100-g amount of CPG beads was added to 500 ml of degassed solution of 3% PEG 20M in distilled water. The suspension was maintained under vacuum to ensure that air held in the pores was removed to allow PEG to enter. The suspension was shaken occasionally to free any air bubbles adsorbed on the beads. After 30 min the suspension was connected to a vacuum line overnight. The excess of PEG was removed from the beads by repeated washing with distilled water until foam was no longer observed. The beads were air dried under vacuum (oven drying had to be avoided!).

Stainless-steel columns 1, 2 or 3 m long and 1/4 in. O.D. or 4.5 mm I.D. were used for chromatography. All columns were coiled, with a coil diameter of 30 cm. A simple dry-packing procedure and equilibration of the columns with eluent were applied. Separations were carried out at ambient temperature (for AMV) or at $0^{\circ}C$ (for HaMV). For example, for the separation of AMV and added alkaline phosphatase, a 2 m x 1/4 in. O.D. CPG-10-1250 pre-treated glass bead column was eluted with Tris buffer (0.01 M, pH 8.3) at a flow-rate 30 ml/h with a back-pressure of 100 p.s.i. Fractions of 0.5 ml were collected.

Eluting virions were detected (1) by assay for detergent-requiring polymerase activity versus protein concentration, (2) by a difference in UV absorption $(A_{254} - A_{280})$ and (3) by electron microscopy. Routinely, only the UV difference spectrum was checked, as the position of the virus peak, once determined, did not vary by more than 0.1 ml, which was within experimental error. If it was desired to retain the infectivity of the viruses, the radiation damage caused by UV light had to be avoided in repeated experiments. The UV detector did not need to be employed, as the peaks positions were reproducible.

Bhown et al. (1981) studied the purification and characterization of the gag gene products of avian-type C retroviruses by HPLC. A rapid chromatographic separation of translational products of this gene using volatile solvents was described. The viral structural components p27 (M_r = 27 000), p19 (M_r = 19 000) and p15 (M_r = 15 000), which are the fragments (split by a proteinase) of the primary translational product, i.e., the precursor protein Pr 76 (M_r = 76 000), were purified in milligram amounts nearly to homogeneity and in high yields. For the purification of protein components of the disrupted virus a series of four ^{125}I gel permeation columns (Waters Assoc., 30 cm x 0.78 cm I.D.) were used, eluted with acetic acid-propanol-highly purified water (20:15:65) at a flow-rate of 0.2 ml/min. The load was 1 - 2 mg of the total protein dissolved in a volume not exceeding 100 µl. The effluent was monitored spectrophotometrically at 280 nm. The peaks were collected manually, lyophilized and further charac-terized.

12.1.2 Other chromatographic modes

In the last few years other methods have been applied to the chromatography of detergent-disrupted virus components, especially of proteins, in addition to SEC. For example, Henderson et al. (1981) have shown that RP-HPLC could be used to separate proteins in murine leukaemia virus. Kårsnäs et al. (1983) described the purification of surface proteins of bovine viral diarrhoea virus (BVDV) in detergent-containing buffers by fast protein liquid chromatography (FPLC). BVDV is an RNA virus of the genus *Pestivirus*, family Togaviridae. First, the virus was purified by *Crotalaria juncea* lectin chromatography and the glycoprotein peplomers were dissociated from the virion by treatment with 1% Berol 172 (a non-ionic detergent, purchased from Berol Scandinavia, Stenungsund, Sweden). In a second *Crotalaria* lectin affinity chromatographic step the peplomers were re-chromatographed and the glycoconjugates carrying terminal galactose were isolated. For further purification of these glycoproteins, FPLC was employed. In order to overcome the tendency of hydrophobic proteins to aggregate, and with the low UV transparency of most conventional detergents which were able to suppress the aggregation, Berol 172 was applied as a suitable eluent component.

For IEC in the FPLC system, a Mono-Q column (Pharmacia, Uppsala, Sweden) was used and eluted with 0.05 M Tris-HCl buffer containing 0.2% of Berol 172. After 20 ml of isocratic elution a linear gradient of 1 M NaCl was applied up to 60% (reached in ca. 40 ml) and then isocratic elution was continued at this NaCl concentration up to the final 50 ml. Virus-infected and uninfected materials were compared. Proteins of the infected material were eluted above 24-26% NaCl in one preparation line, and above 28% NaCl in the other. Re-chromatography of the pooled fractions using 0.02 M piperazine-HCl buffer (pH 6) + 0.2% of Berol 172 with a linear gradient of salt concentration was successfully applied and led to single peaks. The influence of the critical micellar concentration (CMC) of the detergent in the eluent on the chromatographic and detection processes was also investigated.

The experiments showed that FPLC on ion-exchange columns in detergent-containing buffers can be used effectively for the separation of protein components of virus surfaces.

Phelan and Cohen (1983) studied gradient optimization principles in RP-HPLC and the separation of influenza virus components. Major proteins of A/Bangkok 1/79 x 73 influenza virus were chromatographed. The purity of the isolated proteins, their yields and their reactivity with monoclonal antibodies were discussed. These proteins are important, because several studies of the influenza virus, its components and their roles in host infection and immune responses depended on the isolated viral components.

Theoretical considerations were discussed at the beginning, considering the dependence of the resolution, R_s, on various gradient parameters. The virus was purified by sucrose gradient centrifugation and dissociated in 8 M guanidine-HCl and 2 mM dithiothreitol (DTT) to a final protein concentration of 3.9 mg/ml for chromatography. A 25 cm x 0.46 cm I.D. Aquapore RP-300 column (CO3-10A; Rainin Instrument, Woburn, MA, U.S.A.) was used for the chromatography. For the separation of dissolved influenza X-73 virus, material containing 78 µg of total virus protein was chromatographed using a linear gradient of 0.05% trifluoro-acetic acid (TFA) in water to 0.05% TFA in acetonitrile over 33 min and a flow-rate of 1.0 ml/min. In another experiment, a gradient from 5% to 75% acetonitrile in 0.05% TFA was used. The influence of gradient time was examined in detailed experiments.

For detection, measurement of the absorbance at 220 nm was applied. In order to measure the protein concentration of virus and column eluates, the samples were hydrolysed in 5 M NaOH, neutralized with 5 M HCl, diluted with 1 M borate buffer of pH 10.5 (containing 0.1% of Triton X-100, 0.1% of 2-mercaptoethanol and 1% of methanol), mixed with an equal volume of borate buffer containing

0.08% (w/v) of o-phthalaldehyde (Pierce, Rockford, IL, U.S.A.) and the fluorescence excited at 340 nm was measured at 455 nm. Calculations were based on
the measurement of standards. Column regeneration to improve its properties
after prolonged use was also described. The experiments showed that dissolution
of X-73 in 8 M guanidine hydrochloride and 2 mM DTT followed by RP-HPLC using
a gradient of acetonitrile in 0.05% TFA separated the major proteins of this
influenza virus. Although the total protein recovery was only ca. 20%, their
purity was ≥88%.

A series of papers were published by Van der Zee, Welling and co-workers,
describing various methods for the isolation and purification of proteins from
detergent-extracted Sendai virus. In the first paper Van der Zee et al. (1983)
described their purification by RPC.

Sendai virus is a paramyxovirus, ranging in size from 150 to 250 nm (Welling
et al., 1983). In addition to the internal nucleoprotein (NP) and polymerase
protein (P) in the nucleocapsis, there are three lipid-envelope-associated
proteins (HN, F and M) in Sendai virus (Scheid and Choppin, 1974, 1977). According to Chopin et al. (1981), NH protein (M_r = 66 000) is responsible for
haemagglutination and neuraminidase activities, and F (fusion) protein is
involved in cell fusion, virus penetration and haemolysis. This last mentioned
protein is present in the form of an inactive precursor F_o, from which it is
generated by proteolytic cleavage. F protein consists of two disulphide-linked
polypeptide chains, F_1 (M_r = 50 000 - 52 000) and F_2 (M_r = 13 500). Finally,
M (matrix) protein of M_r = 38 000 is important for virus assembly and viral
budding from the cell membrane. The envelope-associated proteins of Sendai virus
and their separation were the main interest of the group concerned.

Van der Zee et al. (1983) disrupted the virions of Sendai virus by the
addition of 2% (w/v) (final concentration) Triton X-100 for 30 min at room
temperature. The detergent to viral protein ratio was 3.3 (w/v). The detergent
was removed from the supernatant using Amberlite XAD-2, and at the same time
the proteins were reduced by the addition of dithiothreitol (DTT) to a final
concentration 20 mM (Amberlite XAD-2 was obtained from Serva, Heidelberg, F.R.G.,
and DTT from Sigma, St. Louis, MO, U.S.A.). Then the proteins were subjected to
RP-HPLC. Two types of column were employed: (1) a 300 mm x 4.6 mm I.D. column
was slurry packed with Nucleosil 10 C_{18} (Machery, Nagel & Co., Düren, F.R.G.),
and (2) a 40 mm x 3.2 mm column containing Supelcosil LC-318 (Supelco, Bellefonte,
PA, U.S.A.). The pore sizes were 10 and 30 nm, respectively.

Proteins containing extracts were chromatographed at a flow-rate of 1 ml/min
with a linear gradient from 10 to 60% of solvent B in A for 24 min. Solvent A
was 12 mM HCl in triply distilled water and B was 12 mM HCl in ethanol-1-butanol

(4:1, v/v). Both the organic solvents were obtained from E. Merck (Darmstadt, F.R.G.). The amount of proteins to be analysed was very small (10-50 µg). The purity of the isolated proteins was checked electrophoretically using SDS-PAGE. Protein F_2 was isolated in relatively large amounts (the yield was about 100%), whereas the recoveries of three other proteins (F_1, HN and M) were relatively low (from less than 5% to 50%). The amino acid analysis of F_2 was presented.

Welling et al. (1983) examined the anion-exchange chromatography of Sendai virus F protein in the presence of Triton X-100. Purified virions (46.5 mg/ml in 5 mM Tris-HCl, pH 7.23) were disrupted with Triton X-100 (purchased from BDH, Poole, U.K.) at a detergent to viral protein ratio of 0.9 (w/w) and dialysed against 0.02 M sodium phosphate buffer (pH 7.2). The retained material (containing mainly F and NH protein) was subjected to anion-exchange chromatography on a 50 mm x 5 mm I.D. Mono Q HR 5/5 column (Pharmacia, Uppsala, Sweden) and eluted with a gradient from 0.15 to 1.5 M sodium chloride in 0.02 M sodium phosphate (pH 7.2) containing 0.1% (v/v) of Triton X-100; the flow-rate was 1 ml/min. For detection the absorbance at 260 nm was measured (owing to the presence of the aromatic ring in Triton, which did not allow the detection at 280 nm) and the peak fractions were collected manually in low-protein-adsorption (Minisorb) tubes (Nunc, Roskilde, Denmark). To remove most of Triton the tubes were covered with dialysis-membrane tubing and dialysed overnight at 4°C against water with Bio-Beads SM 2 (Bio-Rad Labs., Richmond, CA, U.S.A.; cf., Volsky and Loyter, 1978). The dialysed fractions were reduced and checked by SDS-PAGE.

The detergent extract contained F and HN proteins. HN was not retained by the column, and elution with a salt gradient resulted in several peaks, containing mainly or only F protein. As Brij 35 detergent did not absorb at 280 nm, the same chromatography was performed with 0.1% (v/v) Brij in the eluent. A similar separation resulted, although some tailing of peaks was observed. The ion-exchange procedure described was found to be a rapid method for the purification of viral proteins in the presence of non-ionic detergents.

In continuation of these studies, Welling et al. (1984) described the isolation of detergent-extracted Sendai virus proteins by gel-filtration, ion-exchange and reversed-phase HPLC and the effect of immunological activity. Welling et al. (1985) combined size exclusion and reversed-phase HPLC of a detergent extract of Sendai virus, and Van der Zee and Welling (1985) studied the detection of Sendai virus proteins by RP-HPLC combined with immuno-chromatography.

Heukeshoven and Dernick (1985) described the characterization of a solvent system for the separation of water-insoluble poliovirus proteins by RP-HPLC, and Ricard and Sturman (1985) isolated the subunits of the coronavirus envelope glycoprotein E2 by hydroxyapatite HPLC.

12.2 GLYCOSPHINGOLIPIDS

12.2.1 Introduction

   Glycosphingolipids are relatively complicated substances of great biological importance and therefore a concise chemical (biochemical) introduction will be given before a description of their chromatographic separations. It is known (Lehninger, 1978a) that cell coats of higher organisms consist of three main groups of compounds: (1) glycosphingolipids, (2) acid mucopolysaccharides and (3) glycoproteins. The first group are membrane components and belong to the complex lipids called sphingolipids. They contain long-chain aliphatic hydroxy-amino alcohols, saturated or unsaturated, or a related base as their backbone, e.g., sphingosine (4-sphingenine), $CH_3(CH_2)_{12}CH=CHCH(OH)CH(NH_2)CH_2OH$. This is the first building-block component. More than 30 different long-chain amino alcohols have been found in sphingolipids of various biological species. In sphingolipids, the amino group of sphingosine (or other amino alcohol) forms an amide bond with a long aliphatic fatty acid (saturated or unsaturated, containing 18 - 24 carbon atoms), e.g., oleic acid, $CH_3(CH_2)_7CH=CH(CH_2)_7COOH$; the fatty acid is the second building-block component. Such amides are called ceramides and their structure can be illustrated by an example:

$$CH_3(CH_2)_{12}CH=CHCH(OH)\underset{\underset{\displaystyle CH_3(CH_2)_7CH=CH(CH_2)_7CO}{\overset{\displaystyle |}{NH}}}{CHCH_2OH}$$

Ceramide

Ceramides create a molecular basis for the construction of sphingolipids. The third characteristic building-block component of sphingolipids is a polar head group, which may be very large and complex. In relatively simple sphingomyelins (which are not oligomers and will not be treated in this book) it contains phosphorylcholine; this component confers a zwitterionic nature on the resulting complex compound.

   If sphingolipids contain one or more sugar residues in the position of their polar head, they are called neutral glycosphingolipids.

   The simplest species of this group are cerebrosides, which contain a mono-saccharide bound in a β-glycosidic linkage to the hydroxy group of ceramide. An example of galactocerebroside [containing galactose (Gal) and lignoceric acid (C_{24})] is illustrated by the formula

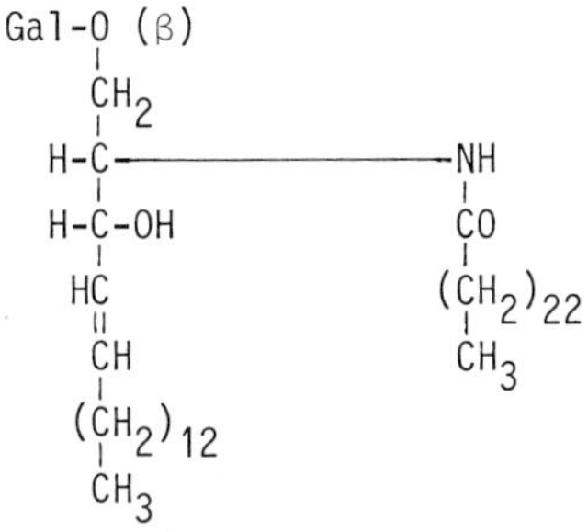

```
Gal-O (β)
    |
    CH₂
    |
 H-C——————————NH
    |             |
 H-C-OH          CO
    |             |
    HC          (CH₂)₂₂
    ‖             |
    CH           CH₃
    |
  (CH₂)₁₂
    |
    CH₃
```

Galactocerebroside

If two galactose units are bound to create galactosylceramide, the structure of such a dihexoside may be symbolized as

$$\text{Gal1} \xrightarrow{\beta} \text{4Gal1} \xrightarrow{\beta} \text{Ceramide}$$

A compound containing one glucose (Glc) unit bound to ceramide is called gluco-cerebroside. Two or more galactose units containing glucosylceramides are symbolized as illustrated below by the example of tetrahexoside, containing terminal galactosamine (GalN) in the N-acetylated form (NAc):

$$\text{GalNAc1} \xrightarrow{\beta} \text{3Gal1} \xrightarrow{\beta} \text{4Gal1} \xrightarrow{\beta} \text{4Glc1} \xrightarrow{\beta} \text{Ceramide}$$

These are examples of so-called neutral glycosphingolipids. Di-, tri- and tetrahexosides belong to biooligomers and will be dealt with in the Section 12.2.

Sulphate esters of galactocerebrosides (at position 3 of Gal) are also found in nature as components of brain tissue; they are called sulphatides.

The most complex group of glycosphingolipids are gangliosides, containing 2-10 or more carbohydrates. In their polar oligosaccharide head they contain one or more residues of sialic acid, i.e., N-acetylneuraminic acid (NeuAc) (the older designation is NANA) with the formula

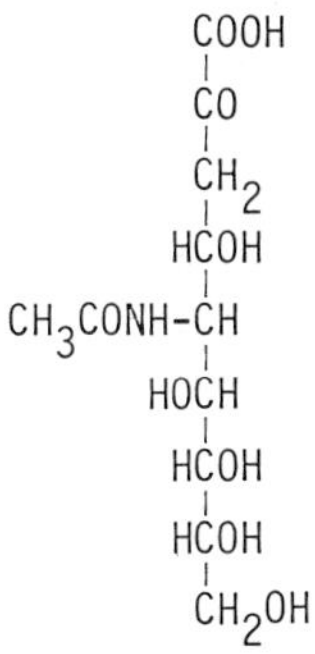

```
          COOH
           |
           CO
           |
           CH₂
           |
          HCOH
           |
CH₃CONH-CH
           |
          HOCH
           |
          HCOH
           |
          HCOH
           |
          CH₂OH
```

N-Acetylneuraminic acid (NeuAc) or sialic acid

[About 20 different sialic acids have been found in nature, but only a few have been detected in gangliosides (Ledeen and Yu, 1982). The above-mentioned NeuAc belong to the two major types.]

Most of the known gangliosides have Glc in the glycosidic linkage to ceramide. In the remaining part of the structure D-Gal or N-acetyl-D-GalN residues are often present. Nearly all vertebrate gangliosides are derived from lactosyl-ceramide and hence possess the glucosylceramide structure; they can be classified into ganglio and neolacto series [Wiegandt (1979); Leeden and Yu (1983)]. Exceptionally, other types have also been found. If we consider variances in the sialic acid type, about 50 individual gangliosides have been described so far, differing in the number and relative positions of the hexose and sialic acid residues (Leeden and Yu, 1982), and this number is increasing as new discoveries are being reported. A complicated example of a trisialoganglioside is represented by

$$
\begin{array}{cccccc}
& \beta & \beta & \beta & \beta & \\
3\mathrm{Gal}1 & \rightarrow 3\mathrm{GalNAc}1 \rightarrow & 4\mathrm{Gal}1 \rightarrow & 4\mathrm{Glc}1 \rightarrow & \mathrm{Ceramide} \\
\uparrow & & 3 & & \\
2\mathrm{NeuAc} & & \uparrow & & \\
& & 2\mathrm{NeuAc}8 & & \\
& & \uparrow & & \\
& & 2\mathrm{NeuAc} & &
\end{array}
$$

Even more complicated gangliosides have been found in tissues containing more sugar components, such as monosialomonofucooctahexaosylceramide.

Sphingolipids have been found in both plant and animal cells and are present in the largest amounts in brain and nerve tissues. The neutral glycosphingolipids play an important role in the cell surface (Lehninger, 1978a). Their non-polar tails penetrate into the lipid bilayer structure of cell membranes, whereas the polar heads protrude out from the surface and have various biological functions, e.g., they are responsible for blood-group specificity. In their positions on the cell surface the glycosphingolipids are concerned in organ and tissue specificity, in tissue immunity and cell-cell recognition processes (cancer cells have different glycosphingolipids from those in normal cells, which is important in cancer research and in clinical diagnostics). The accumulation of glyco-sphingolipids in cases of enzyme deficiency (genetic disorder in metabolism) leads to various lipid-storage diseases (glycosphingolipidoses; cf., Stanbury et al., 1978).

At present acidic glycosphingolipids (the gangliosides) are of great interest to biochemists and physicians as they are important membrane receptors (Svennerholm et al., 1980a; Hakomori, 1981; Fishman, 1982). They are components of the plasma membrane and contribute to the rich carbohydrate glycocylax, which determines the surface properties of cells (Ledeen and Yu, 1978, 1982; Wiegandt,

1982). In view of their localization in the plasma membrane outer leaflet, they participate in cell growth control and cell differentiation. Malignant transformation of cells triggers off major changes in their composition and therefore gangliosides can serve as tumour-associated markers (Hakomori and Kannagi, 1983). In addition, the gangliosides (abundant in nerve endings) have a function in the transmission of nerve impulses across synapses. Although originally discovered in the brain (especially the grey matter is very rich in gangliosides, where they represent 6% of the total lipids), these acidic glycosphingolipids have been found in other vertebrate tissues in smaller amounts. Several invertebrate gangliosides have also been identified.

In general, glycosphingolipids contain two long hydrophobic aliphatic hydrocarbon chains and are therefore soluble in organic solvents (e.g., chloroform or methanol), but they also contain a polar head. These are good properties for separations by means of partition chromatography in the normal-phase mode. Indeed, this technique has often been used for their fractionation. In some instance RPC has also been applied to their purification. Moreover, gangliosides contain ionogenic groups, allowing their simple isolation from neutral glycosphingolipids by anion-exchange chromatography and also their fractionation from each other. Both of these methods have often been used for the chromatographic separation of glycosphingolipids.

There are problems with the on-line detection of these substances if they are not modified with a suitable UV chromophore. Therefore, off-line detection has often been used for monitoring in column chromatography. A suitable and generally applicable method for the on-line detection of glycosphingolipids remains a problem.

12.2.2 Neutral glycosphingolipids

12.2.2.1 Unsubstituted natural products

HPLC has been applied for the analytical separation of glycosphingolipids (GSL) preferentially in the form of various derivatives, which will be discussed later in this section. However, the separation of GSL without derivatization has a great advantage, especially for preparations, because the separated lipids may be used directly in structural characterization and immunological analyses, and this is the main reason why it has been developed. Non-pressurized LC methods were often used in the beginning and various silica gel supports were applied as stationary phase for liquid-solid chromatography of GSL (Radin et al., 1956; Makita and Yamakawa, 1962; Gray, 1967; Vance and Sweeley, 1967; Siddiqui and McCluer, 1968; Puro, 1970; Hanahan et al., 1971; Ando and Yamakawa, 1973).

Pellicular supports have also been used for the separation of glycolipids, which allowed high-speed analyses (Stolyhwo and Privett, 1973; Sugita et al., 1974). Unfortunately, the low capacity of the last mentioned packing materials allowed only analytical applications.

Ando et al. (1976a) developed the high-performance preparative chromatography of lipids using totally porous silica (Iatrobeads), and applied it to the separation of molecular species of GSL. As the particle size was relatively large (60 μm), hydrostatic pressure was sufficient to achieve rapid separations. Ceramide dihexoside and ceramide trihexoside were both separated into two fractions of different molecular species, and two ceramide tetrasaccharides (globoside I and paragloboside) were also clearly separated. The separation of these molecular species was due to differences in their fatty acid and long-chain base compositions.

A sample of 400 mg of human erythrocyte glycolipid mixture was applied to a 120 cm x 1.7 cm I.D. column containing 140 g of Iatrobeads 6RS-8060 (Iatron Labs., Tokyo, Japan). The column was eluted with 1700 ml of a linear gradient of chloroform-methanol-water with the proportions changing from 83:16:0.5 to 55:42:3 (v/v) using two reservoirs. The column could be regenerated by washing with three column volumes of the starting solvent. In practice, an Iatrobeads column could be used for several chromatographic runs.

In another experiment, 100 mg of the fraction of human erythrocyte glycolipids (containing globoside I and paragloboside) were loaded on 85 g of Iatrobeads and the column was eluted with a convex gradient of chloroform-methanol-water with the proportions changing from 8:12:2 to 40:45:5 (v/v). The gradient was generated using three reservoirs in series containing the above eluent with proportions of 80:18:2, 55:42:3 and 40:55:5 (v/v).

The sugar content of each fraction was determined by the anthrone-sulphuric acid reaction, measuring the absorbance difference between 616 and 720 nm on a Hitachi Model 156 double-wavelength spectrometer. TLC was applied for the identification of the separated compounds. Also GLC (after methanolysis and trifluoroacetylation) and mass spectrometry were applied for detailed analysis of the fractions obtained. The physical properties of Iatrobeads were described in detail.

Watanabe and Arao (1981) described a new solvent system for the separation of neutral GSL. These glycolipids were extracted from erythrocyte membranes with hot ethanol and first purified by partitioning in chloroform-methanol (2:1). Acetylated glycolipids were further purified chromatographically on a Florisil column and neutral glycolipids were separated from gangliosides on an anion exchanger. Then the samples were applied to an HPLC column.

A stainless-steel column (500 mm x 4 mm I.D.) was slurry-packed with 10-µm Iatrobeads 6RS-8010 porous silica spheres (Iatron Labs.) using tetrabromoethane-tetrachloroethylene (60:40) at 400 kg/cm^2. The column was washed with iso-propanol-hexane (55:45) and equilibrated with the starting solvent (e.g., isopropanol-hexane-water, 55:44:1). For elution a linear gradient of isopropanol-hexane-water was used (the concentrations of hexane and water were varied, depending on the glycolipid composition). The gradient was initiated immediately after injection at a flow-rate of 2.0 ml/min. Every 0.5 or 1 min one effluent fraction was collected in the fraction collector. After use, the column was regenerated by washing with isopropanol-hexane (55:45) at a flow-rate of 1 ml/min for 60 min and could be used again without any loss of efficiency.

No on-line sample detection was applied. The fractions were analysed by TLC [pre-coated silica gel G plates, developed with chloroform-methanol-water (60:35:8 or 65:25:4); detection by spraying with orcinol reagent]. The glycolipid content in each fraction was determined by the anthrone-sulphuric acid reaction and calculated as globotetraosylceramide (cf., Suzuki et al., 1976).

GSL with mono- to dodeca- or tetrakisdecasaccharides were separated highly reproducibly within 60 min. The method was applied to preparative separations of highly complex glycolipids with blood group activity.

12.2.2.2 Per-O,N-benzoyl derivatives

The lack of a suitable on-line detection method is a disadvantage of the chromatography of unsubstituted glycosphingolipids (GSL). A simple derivatiza-tion method was sought that would introduce a suitable UV chromophore into the glycolipid and allow the utilization of a sensitive UV detector. In addition, the modifying group should be easily split off after the chromatographic separa-tion. This was believed to be a practical means of taking advantage of HPLC methods in glycolipid biochemistry.

Evans and McCluer (1972) reported a system for the separation of neutral glycolipids by HPLC of their benzoylated derivatives, prepared by reaction with benzoyl chloride, in a manner similar to that described by Acher and Kanfer (1972). They demonstrated that benzoylated cerebrosides could be readily separated into two fractions by HPLC on a silica gel column. It was originally believed that the derivatives prepared were only O-benzoyl derivatives and that the parent cerebrosides could be simply regenerated by catalytic deacylation with sodium methoxide in methanol. However, McCluer and Evans (1973) found that the reaction of cerebrosides containing non-hydroxy fatty acids with benzoyl chloride resulted in acylation of the amide in addition to the normal O-acylation (and mild alkali treatment of the N-diacyl derivative resulted in the formation of N-benzoylpsychosine). Cerebrosides containing hydroxy fatty acids did not create the undesired N-benzoyl derivatives. Derivatization with benzoic anhydride avoided amide acylation even in the presence of non-hydroxy fatty acid components; however, this last reaction was not sufficiently fast.

McCluer and Evans (1973) described the benzoylation of cerebrosides and derivatives that contained hydroxy and non-hydroxy fatty acids were isolated by HPLC with UV detection.

Benzoylation procedure. Cerebroside samples (0.1-1.0 mg) were benzoylated with 0.6 ml of benzoyl chloride-pyridine (1:5, v/v) for 1 h at 60°C (similar amounts were benzoylated with 0.6 ml of 5% benzoic anhydride in pyridine for 18 h at 110°C). Two procedures were utilized to purify the benzoylated products, as follows. (A) The reaction mixture was dried under a stream of nitrogen and the residue was dissolved in 5 ml of hexane and successively washed with 3 ml each of 95% methanol saturated with Na_2CO_3, 0.6 M HCl in 95% methanol and 95% methanol. The hexane layer was dried under a stream of nitrogen. (B) The reaction product was refluxed under methanol and chloroform was used for extraction; after successive washing of the lower layer, the solution was evaporated and dissolved in hexane.

For HPLC a 50 cm x 2.1 mm I.D. stainless-steel column was dry-packed with Zipax pellicular packing (Instrument Products Division, DuPont, Wilmington, DE, U.S.A.) and eluted with 0.13% methanol in pentane at a flow-rate of 1.5 ml/min.

In addition to Zipax, a MicroPak NH$_2$ column (microparticulate silica gel with a bonded polar phase consisting of 3-aminopropylsilane groups) was used for the separation of perbenzoylated galactosyl- and glucosylceramides (McCluer and Evans, 1976). Suzuki et al. (1976) also described the separation of molecular species of glucosylceramide by HPLC of their benzoyl derivatives.

Jungalwala et al. (1977) reported the determination of less than 1 nmol of perbenzoylated cerebrosides by HPLC using gradient elution. Both hydroxy and non-hydroxy fatty-acid-containing cerebrosides were analysed with UV detection at 230 or 280 nm. The quantitative range of the method was 0.5-10 nmol of cerebrosides. The detection limit for injected samples was about 1 pmol and the analysis time was less than 5 min. No preliminary purification of the cerebrosides from other lipids in brain extracts was necessary. The moisture-free cerebrosides were first benzoylated with 50 µl of 10% benzoyl chloride in pyridine and then separated on a 50 cm x 2.1 mm I.D. column of 27 µm Zipax porous layer silica beads. If gradient elution with 2.8-5.5% dioxane in hexane was applied, monitoring at 230 nm (i.e., at the absorption maximum of benzoylated cerebrosides) was possible. The column was regenerated by reversing the gradient for 1 min and equilibrating the column with 2.8% dioxane in hexane for 3-4 min. This procedure was at least 10 times more sensitive than isocratic elution with detection at 280 nm (McCluer and Evans, 1976). Jungalwala et al. (1977) also de-

scribed a variant of gradient elution with detection at 280 nm, in which a
linear gradient of 2-7% aqueous ethyl acetate in hexane was used for elution.
The column was regenerated to its original polarity by reversing the gradient
for 1 min and equilibrating the column with 2% aqueous ethyl acetate in hexane
for 3-4 min. The methods described are applicable to biological materials containing minute amounts of cerebrosides.

Similar improved methods for the quantitative analysis or microanalysis of
perbenzoylated neutral GSL were published by Ulman and McCluer (1977, 1978);
the first paper describes HPLC with monitoring at 280 nm and the second at
230 nm. An example of separations reported in the first paper is illustrated in
Fig. 12.2.

12.2.2.3 Per-O-benzoyl derivatives

The acylation of neutral glycosphingolipids (GSL) with benzoyl chloride was
reviewed above. A disadvantage of this procedure is that N-benzoylation occurred
in addition to the desired O-benzoylation, so that the recovery of the parent
GSL after mild alkaline hydrolysis was not possible, owing to the formation of
a mixture of products. Benzoylation with benzoic anhydride in pyridine did not
lead to the formation of N-benzoylated products, but the anhydride reaction was
sluggish, which is the reason why the benzoyl chloride method has been preferred.

Gross and McCluer (1980) described an HPLC analysis of neutral GSL per-O-
benzoyl derivatives. The method was based on the observation of Gupta et al.
(1977) that N,N-dimethyl-4-aminopyridine (DMAP) acted as a catalyst in the
acylation of phospholipids by fatty acid anhydrides. Therefore, Gross and McCluer
(1980) tried to acylate GSL with benzoic anhydride in the presence of DMAP.
Single products were formed and the parent GSL could be regenerated by mild
alkaline hydrolysis after chromatographic separation.

Per-O-benzoylation procedure (Gross and McCluer, 1980). Samples of 200 ng
each of mono- to tetraglycosylceramides were dried under a stream of nitrogen in
100 x 13 mm screw-capped tubes and desiccated in vacuo over P_2O_5 for at least
2 h. A freshly prepared solution of 20% (w/v) benzoic anhydride and 5% (w/v)
DMAP in pyridine (0.5 ml) was added. The tubes were briefly flushed with nitrogen,
capped tightly and incubated at 37°C for 4 h; the pyridine was then removed under
a stream of nitrogen. Hexane (3 ml) was added and the solution was washed four
times with 1.8 ml of 80% methanol saturated with sodium carbonate and twice with
80% methanol, as described by Ulman and McCluer (1978). The solvent was removed
under a stream of nitrogen and the benzoylated samples thus obtained were each
dissolved in an appropriate amount of CCl_4 for injection on to the HPLC column.

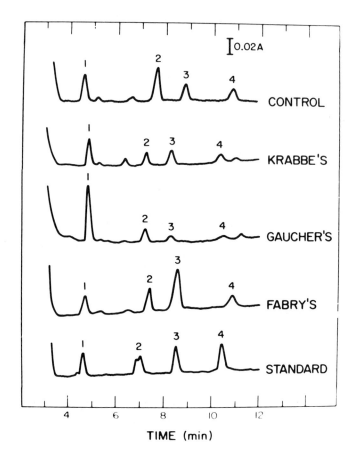

Fig. 12.2. HPLC analysis of plasma glycolipids from a healthy person and patients suffering from inborn errors of lipid metabolism. Glycolipid fractions from 1 ml of plasma were isolated, benzoylated and dissolved in 100 μl of CCl_4 and 40 μl were analysed on a pellicular silica gel (Zipax) column (50 cm x 2.1 mm I.D.) using a 10-min linear gradient of 2-17% ethyl acetate in hexane at a flow-rate of 2 ml/min. The absorbance at 280 nm was recorded. Peaks: 1 = Glc-Cer; 2 = Lac-Cer; 3 = Gal-Lac-Cer; 4 = globoside. The heights and the exact positions of the peaks are discussed in the original paper. On the right-hand side the diseases of the inborn errors of lipid metabolism are designated; for their characterization, see Table 14.4. (Reprinted from Ullman and McCluer, 1977.)

The HPLC column was a 50 cm x 2.1 mm I.D. stainless-steel tube packed with Zipax pellicular silica gel (DuPont, Wilmington, DE, U.S.A.). For elution a 13 min linear gradient from 2.5 to 25% dioxane in hexane was used, with a flow-rate of 2 ml/min. UV detection at 230 nm was applied.

Conversion of perbenzoylated products to the parent GSL (Gross and McCluer, 1980). The peaks obtained from HPLC were treated with 0.5 M methanolic sodium hydroxide solution for 1 h at 37°C. After partitioning in chloroform-methanol-water (8:4:3) the lipid products in the lower phase were examined by TLC.

Preparative and analytical HPLC of modified GSL using a similar method (per-O-benzoylation with benzoic anhydride catalysed by DMAP) was also described by McCluer and Ulman (1980).

12.2.2.4 O-Acetyl-N-p-nitrobenzoyl derivatives

Suzuki et al. (1977c) described the separation of molecular species of higher glycolipids by HPLC of their O-acetyl-N-p-nitrobenzoyl derivatives. The advantage of this modification of glycosphingolipids (GSL) for detection is the higher molar absorptivity at 254 nm compared with those of benzoyl derivatives. This labelling procedure can be easily performed on acetylated glycolipids. Suzuki et al. (1977c) separated O-acetyl-N-p-nitrobenzoyl derivatives of glucosyl-ceramide, lactosylceramide, trihexosylceramide, globoside and haematoside (N-acetyl type) into their molecular species on the basis of fatty acid components by HPLC using a reversed-phase column (μBondapak C_{18}). Acetonitrile was used as the eluent.

Yamazaki et al. (1979) examined the consecutive analysis of GSL on the basis of sugar and ceramide moieties by HPLC. A quantitative method for the determination of GSL in biological materials was described and applied to erythrocyte glycolipids. A crude lipid extract was separated into neutral and acidic fractions on DEAE-Sephadex. GSL fractions were obtained by acetylation and Florisil column chromatography. Acetylated GSL samples were N-p-nitrobenzoylated and chromatographed on a Zorbax SIL silica gel column. For elution an isopropanol gradient in hexane-chloroform was used. Derivatives were separated on the basis of their sugar chains. The peaks obtained were further chromatographed on a μBondapak C_{18} reversed-phase column on the basis of their lipid portions.

An improved technique for the separation of neutral GSL by HPLC was described by Suzuki et al. (1980). O-Acetyl-N-p-nitrobenzoyl derivatives of six neutral GSL (glucosylceramide, lactosylceramide, globotriaosylceramide, lactotriaosyl-ceramide, globotetraosylceramide and neolactotetraosylceramide) were separated. The recoveries of glucosylceramide and globotetraosylceramide for the derivatization procedure and chromatographic analysis were about 75%, and 1 nmol of glycolipid could be detected. The method was applied to the analysis of human erythrocyte neutral glycolipids and allowed the separation of certain glycolipids that contained equal numbers of sugar moieties. The derivatization technique was the same as that described by Yamazuki et al. (1979), including the purification of the derivatives on a Sephadex LH-20 column.

For the HPLC a 25 cm x 4.6 mm I.D. Zorbax SIL column was used and eluted at a flow-rate of 0.5 ml/min by programmed elution as follows: 5 min isocratic elution with 1% isopropanol in hexane-dichloroethane (2:1), 55 min with a linear gradient from 1 to 5% isopropanol in hexane-dichloroethane (2:1) and 10 min

isocratically with 5% isopropanol in hexane-dichloroethane (2:1). Before the
next analysis, the column was reactivated with the initial solvent for 15 min.

12.2.3 Acidic glycosphingolipids - gangliosides

Gangliosides are glycolipids and the composition of their components determines their solubility in water and organic solvents. The original method for
their isolation and purification (Folch et al., 1957) was based on partitioning
from chloroform-methanol (into which they were extracted from natural sources)
into the upper water-enriched phase, leaving behind the bulk of the other lipids.
This procedure was later modified by various workers, e.g. by Suzuki (1965); a
modification of this last-mentioned technique was described in detail in a
review by Ledeen and Yu (1982), where other approaches were also discussed.

Gangliosides contain one or more residues of sialic acid, which opens up the
possibility of ganglioside isolation using anion exchangers. The whole group of
gangliosides can be separated in this way from neutral glycosphingolipids, but
individual ganglioside subclasses (mono- to pentasialogangliosides) can be
fractionated according to the number of sialic acid residues.

Because of the hybrid lipophilic-hydrophilic character of ganglioside fractions, LSC on silica gel has often been used sucessfully for their fractionation.
In addition to column chromatography, TLC has also been applied for both analytical and semi-preparative separations.

All the column chromatographic separation techniques for gangliosides will be
reviewed in the following part of this Section. Most gangliosides have been
chromatographed in the underivatized form, which will be dealt with first.

12.2.3.1 Ion-exchange chromatography

In 1971 Winterbourn published the first paper on the separation of brain
gangliosides by column chromatography on DEAE-cellulose (Whatman DE-2), previously washed with ammonium acetate and methanol. Stepwise elution with gradually
increasing concentrations of ammonium acetate in methanol (0.022, 0.04, 0.1 and
0.25 M) was applied and the separation of gangliosides according to the number
of bound sialic acid residues was achieved. A group of gangliosides of human
myelin with sialosylgalactosyceramide (G_7) as a major component was isolated
from purified myelin using DEAE-Sephadex A-25 (Pharmacia) column chromatography
by Ledeen et al. (1973); this was the first application of DEAE-Sephadex for the
isolation of the total ganglioside group. Momoi et al. (1976) applied a concave
gradient of ammonium acetate in methanol for the elution of gangliosides from
a column of DEAE-Sephadex A-25, and these were further separated on silica gel.
Ivamori and Nagai (1981a,b) combined a separation on DEAE-Sepharose with

Iatrobeads 6RS-8060 chromatography. A detailed description of the gradient
elution of gangliosides from a DEAE-Sephadex A-25 column was described and
discussed by Ledeen and Yu (1982). Comparing the separation of gangliosides on
DEAE-Sephadex, QAE-Sephadex and DEAE-Sepharose, Ivamori and Nagai (1978a) found
that the last material gave the best separation and yields. All these methods
were based on the low-pressure chromatography.

The application of DEAE-silica gel as an ion exchanger for the isolation of
glycolipids (Kundu and Roy, 1978) can be considered as the first step towards
the development of the HPLC of gangliosides. A relatively rapid and quantitative
method (but non-pressurized, with only gravitational elution) was presented for
the isolation of gangliosides and neutral glycosphingolipids from animal tissues
and cells. The advantages of this method in comparison with ion exchangers based
on a carbohydrate matrix were: (1) faster flow-rate; (2) faster equilibration
with the starting buffer; (3) easier regeneration; (4) greater economy; and (5)
less susceptibility to microbial attack. Kundu and Roy (1978) described in
detail the preparation of the DEAE derivatives of porous silica gel (pore
diameter 200 nm; 200-400 mesh; surface area 150 m^2/g; pore volume 1.68 ml/g).
γ-Glycidoxypropyltrimethoxysilane and N,N-diethylethanolamine were used for
synthesis. The product had the formula

$$\text{Silica gel-O-}\underset{\underset{OCH_3}{|}}{\overset{\overset{OCH_3}{|}}{Si}}\text{-}(CH_2)_3\text{-O-CH}_2\text{-}\underset{\underset{O-(CH_2)_2-N}{|}}{\overset{}{CH}}\text{-CH}_2 \begin{matrix} {}\\ \diagup C_2H_5 \\ \diagdown C_2H_5 \end{matrix}$$

and was used in a 30 cm x 1.5 cm I.D. column, which was equilibrated with 0.2 M
sodium acetate in methanol and thoroughly washed with methanol, and then washed
with at least three bed volumes of chloroform-methanol-water(30:60:8, v/v) before
use. A 10-ml volume of beef brain total lipid extract was diluted to 50 ml with
chloroform-methanol-water (30:60:8, v/v) and allowed to flow through the column
(10 g of DEAE-silica gel) under gravity. After 50 ml had been added, the column
was eluted with 150 ml of the same solvent mixture. The combined eluates from
this first fraction consisted of the uncharged and zwitterionic lipids. The
second fraction, containing the acidic lipids, was eluted with 200 ml of chloro-
form-methanol-0.8 M sodium acetate (30:60:8, v/v). The total elution time was
approximately 5 h.

Kundu et al. (1979) also developed the DEAE-silica gel and DEAE-controlled-
pore glass (CPG) chromatography of gangliosides into a method that allowed the
separation of gangliosides according to the number of sialic acid residues into
mono-, di-, tri- and tetrasialogangliosides. The DEAE-silica gel was laboratory-
prepared as described above. DEAE-CPG was obtained from Electro-Nucleonics

(Fairfield, NJ, U.S.A.). The sorbents were compared with DEAE-Sephadex and the best results were obtained with DEAE-silica gel. A 60 cm x 2 cm I.D. column containing 40 mg of ganglioside mixture was eluted with methanol, 0.2 M ammonium acetate in methanol and 0.5 M ammonium acetate in methanol. Volumes of 200 ml of each solvent were connected in series with each other through a gradient mixer. The flow-rate was 1.5 ml/min. Fractions of 15 ml of the effluent were collected and aliquots of 500 µl were used for sialic acid assay by the resorcinol method (Svennerholm, 1957; Miettiner et al., 1959; Suzuki, 1964). The gangliosides were isolated after the removal of ammonium acetate by dialysis against cold water. DEAE-silica gel and DEAE-CPG as ion exchangers for the isolation of glycolipids and their application to the separation of gangliosides were described in detail by Kundu (1981a). Kundu (1981b) also described in detail techniques for the TLC of neutral glycosphingolipids and gangliosides; these techniques have often been used for the evaluation of fractions obtained by IEC.

Fredman et al. (1980) tried to combine the good separation ability of DEAE-dextran for gangliosides with the rigidity of silica gel, suitable for HPLC, and developed a new type of anion-exchange support, Spherosil-DEAE-dextran, which consists of porous glass beads (Spherosil silica gel) covered with cross-linked DEAE-dextran. The invention of this combined ion exchanger was described earlier by Tayot et al. (1978). In principle, 1 kg of Spherosil was poured into 2.3 l of a 7.5% aqueous solution of DEAE-dextran at pH 11.5 and then dried for 15 h at 80°C. DEAE-dextran was cross-linked with 1,4-butanediol diglycidyl ether (Aldrich, Milwaukee, WI, U.S.A.). The resin (200 ml) was freed from fines by decantation and then slurried into a glass column (3.0 cm I.D.) and converted into the acetate form by thorough washing with 2.0 M sodium acetate solution (five bed volumes). It was then rinsed with five bed volumes of distilled water and one bed volume of methanol. Before use one bed volume (200 ml) of chloroform-methanol-water (60:30:4.5, v/v) was passed through the column.

Dialysed brain gangliosides (corresponding to 1 mmol of neuraminic acid), dissolved in 200 ml of chloroform-methanol-water (66:30:4.5, v/v) were placed on the column and filtered at a flow-rate not exceeding 0.5 ml/min. The column was rinsed with 200 ml of the same chloroform-methanol-water mixture and 200 ml of methanol. The gangliosides were then eluted stepwise with (i) 1000 ml (five volumes) of 0.02 M, (ii) 2000 ml (ten volumes) of 0.10 M, (iii) 1600 ml (eight volumes) of 0.25 M, (iv) 1000 ml (five volumes) of 0.50 M and (v) 600 ml (three volumes) of 1.00 M potassium acetate in methanol. Mono-, di-, tri-, tetra- and pentasialoganglioside fractions were separated. The eluates were evaporated, dialysed, and their further purification was achieved by chromatography on silica gel or by TLC. This method has advantages over the previous preparative ion-exchange chromatography fractionation. The method was also described by Fredman (1980).

A very rapid method for the separation of gangliosides by anion-exchange chromatography on strongly basic Mono Q resin (9.8 µm, volume of the pre-packed bed = 1 ml) in combination with FPLC equipment (Pharmacia, Uppsala, Sweden) was presented by Månsson et al. (1985). The gangliosides were separated into mono-, di-, tri- and tetrasialoganglioside fractions by a discontinuous gradient of potassium acetate in methanol. The separation was completed in a volume of 50 ml (25 min). The Mono Q column was washed before use with 10 ml of methanol, 50 ml of 1 M potassium acetate in methanol and 10 ml of methanol at a flow-rate of 1 ml/min. A ganglioside sample dissolved in chloroform-methanol (1:2, v/v) was introduced on to the column at a flow-rate of 1 ml/min. The gangliosides were eluted with a stepwise gradient from 0 to 225 mM potassium acetate in methanol at a flow-rate of 2 ml/min. Fractions of 1 ml were collected and aliquots were assayed by HPTLC with 1-propanol-0.25% aqueous potassium chloride (3:1, v/v). The pooled fractions were dialysed against water and further assayed. After each analysis the column was washed with 5 ml of 1 M potassium acetate in methanol and 20 ml of methanol. Very good and rapid separations of gangliosides were achieved on Mono Q, but the application of this strongly basic anion exchanger had some disadvantages: part of the tetrasialoganglioside GQ1b was eluted among the monosialogangliosides, probably owing to lactonization of some of the sialic acid residues, and a certain non-specific adsorption effect also appeared.

Šmíd et al. (1986a) applied the medium-basic anion exchanger DEAE-Spheron to the separation of gangliosides. Linear gradient elution using ammonium acetate in methanol resulted in the complete separation of mono- to pentasialogangliosides in 35 min. The separation obtained, illustrated in Fig. 12.3, did not suffer from the undesirable catalytic activity of the anion exchanger on the sensitive gangliosides. Fractions were assayed by TLC and the technique of "ganglioside mapping" was employed (Ivamori and Nagai, 1978a) for their detailed identification. The rapid preparative separation of gangliosides on medium-basic anion exchangers was described by Šmíd et al. (1986b).

12.2.3.2 Chromatography on silica gel and controlled-pore glass

Ion-exchange chromatography permits only the group separation of gangliosides according to the number of sialic acid residues, and these groups usually contain more than one ganglioside species. Unfortunately, the present methods of IEC are not able to fractionate these groups into pure individual gangliosides and a subsequent separation is necessary. Some chromatographic methods allowing the direct fractionation of gangliosides (without any pre-fractionation by IEC) have been investigated. However, the combination of IEC with LSC (silica gel chromatography) was used most often and became almost customary.

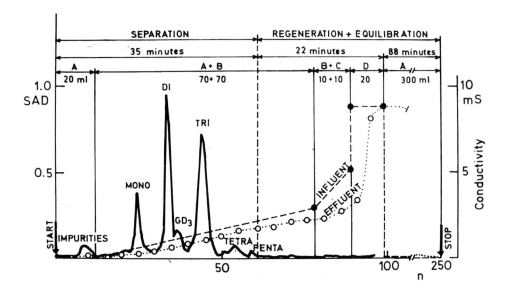

Fig. 12.3. Separation of gangliosides according to the number of sialic acid residues on the medium-basic anion exchanger Spheron 1000-DEAE (17 μm). The diagram illustrates the whole cycle of repeated accelerated IEC. The column (25 cm x 0.8 cm I.D.) was equilibrated with 2 M ammonium acetate in methanol and washed with methanol. The ganglioside sample (30 mg) was suspended in a mixture of 400 μl of water and 400 μl of methanol, cooled and applied using the under-layering method. Eluents: A = methanol; B = 0.2 M ammonium acetate in methanol; C = 0.5 M ammonium acetate in methanol; D = 2 M ammonium acetate in methanol. Fractions of 2 ml were taken at 35-s intervals. Temperature, 25°C; flow-rate, 3.4 ml/min (i.e., 6.8 ml/cm^2 min); pressure, 13 atm. After washing with methanol, the column was ready for the next sample. Mono to Penta refer to the number of sialic acid residues in ganglioside fractions. GD$_3$ is a disialoganglioside. SAD = relative colorimetric value (580 nm) of sialic acid detection (Miettinen and Takki-Luukkainen, 1959). The measurement of electrical conductivity (in milli-siemens, mS) was found to be very useful for monitoring both the chromatography and the regeneration and equilibration. (Reprinted from Šmíd et al., 1986.)

Momoi et al. (1976) used DEAE-Sephadex for the pre-fractionation and the fractions obtained were further chromatographed on totally porous spheres, Iatrobeads 6RS-8060 (d_p 60 μm) (Iatron Labs., Tokyo, Japan). Using this system, brain gangliosides GM1, GD1a, GD1b and GT1 were obtained in high purity and in milligram ammounts. The principle of the procedure is as follows. Sluries of 130 g of Iatrobeads in mixtures of chloroform-methanol in various proportions (depending on the content of the various gangliosides in fractions I-V from IEC) were poured into 120 cm x 1.7 cm I.D. columns and left overnight. Then solutions of 100 mg of the lipid fractions in small amounts of the appropriate chloroform-methanol-water mixtures were applied to the column. Each column was eluted with a linear gradient of methanol and water in chloroform-methanol-water (the composition being specified for individual IEC fractions) at a flow-rate of 0.8-1.0 ml/min, and fractions of 20 ml were collected. The components of fraction II were not separated on Iatrobeads, but re-chromatographed on a DEAE-Sephadex column. Nagai and Ivamori (1980) also further separated the fractions of mono- to tetrasialogangliosides (originating from chromatography on DEAE-Sepharose) using an Iatrobeads column and obtained individual gangliosides.

Kundu and Scott (1982) described a rapid separation of gangliosides by HPLC. First they employed IEC on DEAE-silica gel (Kundu, 1981), in which the sample was fractionated into groups according to the number of sialic acid residues. Then individual ganglioside species were obtained by chromatography on a 25 cm x 4.6 mm I.D. column of porous silica gel (5-μm Zorbax Sil, obtained from DuPont, Wilmington, DE, U.S.A.), which was eluted by programmed gradient elution with a mixture with an increasing water content and a decreasing hexane content. The column was equilibrated with solvent A (see below) prior to injection of the sample. A mildly sonicated ganglioside sample (up to 500 μg) dissolved in 50 μl of isopropanol-hexane-water was injected. A linear gradient of isopropanol-hexane-water from 55:42:3 (solvent A) to 55:25:20 (solvent B) was used during a period of 2 h. The flow-rate was 0.5 ml/min. Fractions of 1 ml were obtained with a fraction collector and assayed by TLC. The column was regenerated after use with the initial solvent for 15 min. Major ganglioside components of human erythrocytes and beef brain were separated and the procedure was found to be highly reproducible.

Nakamura et al. (1983) studied gangliosides of hog skeletal muscle. Mono-sialoganglioside and disialoganglioside groups were isolated on a DEAE-Sephadex A-25 column , using a linear gradient of sodium acetate in methanol for elution, and purified to homogeneity of the individual fractions by Iatrobeads column chromatography. For the fractionation of about 3 g of the monosialoganglioside fraction, a 130 cm x 1.4 cm I.D. column containing 45 g of support was eluted with a linear gradient system prepared with 1.2 l each of chloroform-methanol-7.5 M ammonia solution (70:50:3 and 36:65:3, v/v). The gangliosides in the effluent were analysed by TLC. The monosialoganglioside fraction was separated into six portions and seven gangliosides were further purified by successive Iatrobeads column chromatography using various solvent systems to obtain homogeneous preparations. Approximately 500 mg of disialogangliosides were chro-matographed on a 130 cm x 1.2 cm I.D. Iatrobeads column with a linear gradient of 1.5 l each of chloroform-methanol-7.5 M ammonia solution from 55:45:4 to 35:65:4 (v/v). TLC was used for the effluent analyses. Three pooled fractions were further purified by successive Iatrobeads column chromatography and five species were isolated. All the fractions obtained were examined using various chemical and biochemical modification reactions in order to determine their chemical composition and structure.

Several workers have tried to separate ganglioside mixtures in one chromatographic step. Tjaden et al. (1977) described the HPLC of glycosphingolipids
(with special reference to gangliosides) within 40 min using 9 μm LiChrosorb
Si 60 (E. Merck, Darmstadt, F.R.G.) as the stationary phase in a 25 cm x 2.8 mm
I.D. column. Artificial mixtures of six gangliosides and four neutral glycolipids were analysed. For gangliosides the eluent was chloroform-methanol-aqueous
HCl (60:35:4); the final HCl concentration was 0.01 M. For detection a moving
vire was applied with a flame ionization detector. In addition to three unknown
peaks, six known gangliosides were identified. In a separate chromatography of
neutral glycosphingolipids, one unknown peak appeared in addition to four known
species, when neutral chloroform-methanol (3:1) was used for elution. Also here
the same universal detector was employed for monitoring. Tjaden et al. (1977)
recommended the method for both analytical and micropreparative purposes.

Watanabe and Tomono (1984) developed a "one-step" fractionation of neutral
and acidic glycosphingolipids by HPLC. They connected one column of DEAE-
derivatized controlled-pore glass (CPG) serially with two columns of underivatized CPG. A mixture of gangliosides and neutral glycosphingolipids was loaded
on the DEAE-CPG column and washed using gradient elution based on chloroform-
methanol-water with increasing methanol and water contents, followed by a second
gradient with an increasing amount of lithium acetate solution. In the first
gradient elution neutral glycolipids (mono- to hexaglycosylceramides) were
separated within 80 min, and in the second mono- to tetrasialogangliosides were
separated within 60 min.

Controlled-pore glass (CPG-10) of pore diameter 75 Å, 200-400 mesh, and
DEAE-CPG of pore diameter 170 Å, 200-400 mesh, were purchased from Electro-
Nucleonics (Fairfield, NJ, U.S.A.). The first stainless-steel column (250 mm x
2.6 mm I.D.) was packed with DEAE-CPG. The ion exchanger was converted into the
acetate from by washing with 0.1 M lithium acetate in solvent D (see below) at
a flow-rate of 1 ml/min for 60 min, followed by washing with solvents C and B
for 30 min each. The first column was connected with a second column (500 mm x
2.6 mm I.D.) packed with CPG-10, which was further connected to a third column
identical with the second. The following mixtures were used as mobile phases:
(A) chloroform-methanol (95:5, v/v); (B) chloroform-methanol-water (90:10:0.5,
v/v); (C) chloroform-methanol-water (10:80:10, v/v); and (D) chloroform-methanol-
1 M lithium acetate (10:80:10, v/v). Glycolipid mixtures were loaded on to the
first column and eluted with a linear gradient of solvents B and C, followed by
gradient elution from solvent C to D (for optimal proportions, see the original
paper). The effluent was collected in a fraction collector and each fraction
(or every second fraction) was examined by TLC or using resorcinol reagent
(Suzuki, 1964).

A very good separation was obtained using this one-step fractionation method. This technique was applied to the determination of bovine brain gangliosides and neutral and acidic glycolipids isolated from rat kidney. The procedure was found to be highly reproducible.

12.2.3.3 *Application of reversed-phase and normal-phase chromatography to the separation of gangliosides*

In addition to IEC and LSC on silica gel, RPC and NPC have also been applied to separations of gangliosides. A great advantage of these methods is the possibility of a simple on-line detection of gangliosides in the effluent using the continuous measurement of the UV absorbance at low wavelengths. Kundu and Suzuki (1981) developed a simple micro-scale method for the isolation of gangliosides by RPC. A ganglioside fraction obtained by chromatography of the total lipid extract on DEAE-Sephadex (or DEAE-silica gel) was subjected to alkaline hydrolysis and salts and other non-lipid contaminants were removed by RPC on a Sep-Pak C_{18} cartridge (Waters Assoc., Milford, MA, U.S.A.). The purified gangliosides were then obtained by chromatography on a small Iatrobeads (6RS-8060) or Unisil (200-325 mesh) column (obtained from Iatron Labs., Tokyo, Japan, and Clarkson Chemical, Williamsport, PA, U.S.A., respectively).

The residues of evaporated fractions after the IEC of gangliosides from human brain white matter and human erythrocytes were treated with 15 ml of 0.1 M sodium hydroxide in methanol and incubated at $37^{o}C$ for 2 h to destroy the alkali-labile acidic phospholipids. After evaporation, the aqueous solution was neutralized with 0.5 M HCl and diluted to a concentration of 0.1 M salt and passed through a Sep-Pak C_{18} cartridge at a flow-rate of 1 ml/min. [The cartridge was pre-washed with 25 ml of chloroform-methanol (1:2), 25 ml of methanol and 50 ml of water before use.] The cartridge was washed with 25 ml of water and the gangliosides were then eluted with 5 ml of methanol followed by 25 ml of chloroform-methanol (1:2). The cartridge could be used again after washing with methanol and equilibration with water. Sulphatides and other coloured impurities were removed later by chromatography on Iatrobeads columns. In these experiments RPC on a C_{18} cartridge replaced a dialysis procedure in order to remove the salts and low-molecular-weight impurities. A quantitative recovery of gangliosides was obtained.

Sonnino et al. (1984) described the HPLC preparation of muscular species of GM1 and GD1a gangliosides with a homogeneous long-chain base composition (C_{18} or C_{20} sphingosine and C_{18} or C_{20} sphinganine, each in its natural *erythro* or unnatural *threo* form). An analytical or semi-preparative procedure was based on the application of HP-RPC under conditions suitable for resolving ganglioside species on the basis of (i) length, (ii) double bond occurrence and (iii) C-3 isomerism of the long-chain base.

The gangliosides were purified by Silica gel 100 (70-230 mesh, ASTM) column chromatography and structurally characterized as described by Ghidoni et al. (1980). After oxidation with 2,3-dichloro-5,6-dicyanobenzoquinone (DDQ) and reduction with $NaBH_4$, the *erythro* form was partially converted into the *threo* form.

For the analytical separation, 1-10 μg of ganglioside, dissolved in 25 μl of redistilled water, was applied to an RP Spherisorb S5 OD S2 column (250 mm × 4.5 mm I.D.). The temperature was 18-20°C, the eluent was acetonitrile-5 mM sodium phosphate buffer (pH 7), in volume ratios of 7:3 for GM1 and 3:2 for GD1a, and the flow-rate was 1.0 ml/min. One complete analysis took about 40 min for GM1 and 20 min for GD1a. For semi-preparative experiments the same support was used in a 250 mm × 10 mm I.D. column (Phase Separations, Queensferry, U.K.) using the above conditions with the exception of the load (up to 5 mg of natural or DDQ/$NaBH_4$-treated gangliosides, dissolved in 25 μl of redistilled water) and flow-rate (7.5 ml/min). The elution was monitored by UV detection at 195 nm. The effluent was automatically collected on a computer programmed fraction collector on the basis of the UV signals. A complete cycle of analysis took 60 min for GM1 and 40 min for GD1. Natural GM1 and GD1a were separated by the HPLC preparative technique into four fractions, the composition of which was clarified using detailed chemical studies and further HPLC analyses.

In one working day four or five cycles as described above could be performed using a 30-min wash with the elution solvent between cycles, so that a total of 20-25 mg of gangliosides per day could be processed.

Gazzotti et al. (1984a) examined the analytical and preparative HPLC of gangliosides using an RP LiChrosorb RP-8 or μBondapak RP-18 column and a mixture of acetonitrile and phosphate buffer at fixed or varying volume ratios. The HPLC separation resolved all common gangliosides into four molecular species containing C_{18}-sphingosine, C_{18}-sphinganine, C_{20}-sphingosine and C_{20}-sphinganine. Both the sensitivity and the precision of the method were high (detection limit 0.1 nmol, relative standard deviation less than 10%). The semi-preparative method allowed the preparation of 100-mg amounts of gangliosides of each molecular species in 2-4 days in a fully automated process, starting from single gangliosides (such as GM1 and GD1a). In the preparative method acetonitrile-phosphate buffer-tetrahydrofuran was used as the eluent and the addition of the corresponding radioactive tracer was required; this procedure, applied to GM1, was devised for processing up to 50 mg of ganglioside per analysis.

LiChrosorb RP-8 (5 μm) was used in the analytical column (150 mm × 4.6 mm I.D.) (Merck, Darmstadt, F.R.G.). The chromatography of the total ganglioside mixture was carried out at 18-20°C with acetonitrile-5 mM sodium phosphate buffer (pH 7.0) in a volume ratio of 1:1 for 8 min, then the ratio was altered continuously to

3:2 during the following 6 min and finally kept at 3:2 until the end of the run. The flow-rate was 0.5 ml/min and UV detection at 195 nm (absorption maximum for gangliosides) was applied. A complete analysis took about 45 min. Analytical check of the purity of individual gangliosides were carried out under isocratic conditions with suitable mobile phase compositions.

In an automatic semi-preparative method a 10-μm μBondapak RP-18 column (250 mm × 10 mm I.D.) (Water Assoc., Milford, MA, U.S.A.) was used at 18-20°C. The eluent was acetonitrile-5 mM sodium phosphate buffer (pH 7.0) in a volume ratio of 7:3 for GM1 and 3:2 for GD1a. The flow-rate was 3.0 ml/min and UV detection at 195 nm was applied. The automation of the process was described. For fully preparative separations a 10-μm μBondapak column (250 mm × 10 mm I.D.) was used and eluted at 18-20°C with acetonitrile-5 mM sodium phosphate buffer (pH 7.0)-tetrahydrofuran (9:7:4, v/v) at a flow-rate of 7.5 ml/min. A 50 mg amount of GM1 was applied, mixed with 0.5 μCi of the tritiated compound, dissolved in 250 μl of redistilled water. The elution profile was monitored with a computer-assisted HPLC radioactivity monitor (Berthold, Model LB 503) equipped with a 120-μl solid scintillator cell.

In addition to RPC, NPC was also applied to the rapid fractionation of gangliosides. Gazzotti et al. (1985) described the analytical and semi-preparative NP-HPLC separation of underivatized brain ganglioside mixtures into individual components. Gangliosides were applied to a LiChrosorb-NH$_2$ column and eluted with acetonitrile-phosphate buffer mixtures. For on-line monitoring, flow-through detection at 215 nm was applied. Mono- and oligosialogangliosides were separated in one step in a total elution time of less than 90 min with high reproducibility (1-2% relative standard deviations of retention times).

A 0.1 - 20-nmol portion of pure gangliosides, or 1 - 50-nmol portions (as lipid-bound sialic acid) of calf-brain gangliosides were dissolved in 10 μl of redistilled water and applied to an analytical column (250 mm × 4 mm I.D.) containing 7-μm LiChrosorb-NH$_2$ (Merck, Darmstadt, F.R.G.), which was eluted at 20°C with the following solvent mixtures: (A) acetonitrile-5 mM phosphate buffer (pH 5.6) (83:17); (B) acetonitrile-20 mM phosphate buffer (pH 5.6) (1:1). A computer-controlled gradient programme was applied: 7 min elution with solvent A, 53 min with a linear gradient from A to B (66:34), 20 min with a linear gradient from A to B (66:34) to A to B (36:64). The flow-rate was 1 ml/min, and a complete analysis took 80 min.

For preparative HPLC, a 1-5 mg portion of ganglioside mixture (calculated as sialic acid) was dissolved in 100 μl of water and applied to a 250 mm × 25 mm I.D. column of LiChrosorb-NH$_2$ and eluted with the above-described gradient programme. The flow-rate was 39 ml/min. Fractions of 20 ml were collected au-

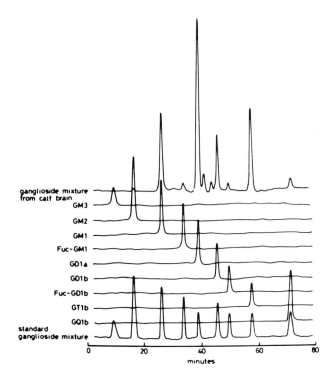

Fig. 12.4. Application of the NP-HPLC method to the separation of gangliosides present in a calf-brain ganglioside mixture. In the middle part of the figure analytical runs of the individual isolated gangliosides are illustrated. For comparison a profile of a standard mixture is given below. Chromatographic procedures are described in the text. (Reprinted from Gazzotti et al., 1985.)

tomatically, and 0.1-ml aliquots were dried and assayed by the HPTLC method described. Pooled fractions of purified gangliosides were submitted to the described analytical HPLC methods.

The following species were separated from a mixture of calf-brain gangliosides (in order of elution): GM3, GM2, GM1, Fuc-GM1, GD1a, GD1b, Fuc-GD1b, GT1b and GQ1b.

Monitoring at 215 nm represents only 60% absorbance compared with the maximum absorption of gangliosides at 195 nm. However, the use of this longer wavelength allowed the baseline variations usual at 195 nm (if a gradient was used) due to the solvent absorption to be avoided. The purity of each separated ganglioside was at least 99%. Fractionation of gangliosides as described by Gazzotti et al. (1985) (cf., Fig. 12.4) seems to give the most complete separation of these underivatized acid glycosphingolipids described so far.

12.2.3.4 *Separation of derivatized gangliosides*

The derivatization of gangliosides with a UV chromophore opened up the way to the simple continuous monitoring of their separation even in LSC on silica gel. In experiments with neutral glycosphingolipids, perbenzoylation was found to be a suitable derivatization method (see Section 12.2.2). Such a modification was believed to allow the transformation of the old LSC into a modern HPLC technique, because on-line monitoring of the effluent was possible.

Bremer et al. (1979) described the quantitative analysis of monosialogan-gliosides by HPLC of their perbenzoyl derivatives. The method was very sensitive owing to the high molar absorptivity of the benzoyl derivatives at 230 nm. It was used to determine the GM3 content of human plasma.

As little as a 3-nmol sample of a monosialoganglioside fraction prepared from biological material by a preliminary fractionation (on a DEAE-Sephadex A-25 column, dialysis or using a Sep-Pak C_{18} cartridge, or on a Unisil silicic acid column) was dried and converted into the perbenzoyl derivative by reaction with 0.1 ml of 10% benzoyl chloride in pyridine at $60^{o}C$ for 1 h. The product was purified by silicic acid chromatography and analysed by HPLC on a 50 cm x 2.1 mm I.D. LiChrospher Si 4000 column. The most satisfactory eluent was an 18-min linear gradient from 7 to 23% dioxane in hexane, or a 10-min gradient from 10 to 25% dioxane in hexane. The flow-rate was 2 ml/min and UV detection at 230 nm was applied; the detector response was found to be proportional to the amount of monosialoganglioside present. As little as 1 pmol of ganglioside could be conveniently estimated. After each gradient run was completed, a short reversed gradient (3 min) was used and a flow-through of the first solvent for 8 min returned the column to the original state, ready for further chromatography.

Lee et al. (1982) developed the HPLC of long-chain neutral glycosphingolipids and gangliosides. Perbenzoyl derivatives of both these types of glycolipids were chromatographed in a single HPLC system, consisting of Zipax stationary phase and a linear gradient of dioxane in hexane as the mobile phase. Twenty-four glycosphingolipids containing 1-10 sugars and 1 or 2 sialic acid residues were studied using this advantageous system. Gangliosides present in human leucocytes were evaluated.

A newer version of the quantitative analysis of more complex brain ganglio-sides by HPLC of their perbenzoyl derivatives was published by Ulman and McCluer (1985).

Traylor (1983) examined the HPLC resolution of *p*-nitrobenzyloxyamine deriva-tives of brain gangliosides, which were originally isolated from the natural material by the partitioning method of Folch et al. (1957) in the version of Suzuki (1965). The reversed-phase mode was used in these experiments with modified gangliosides. This derivatization involves ozonation and cleavage of the ceramide double bond, followed by oxime formation to the nascent aldehyde. Ozonolysis at -70°C was performed with a Supelco Microozonizer at an oxygen flow-rate of 10 ml/min. Ozonation required about 2 min per milligram of ganglioside. Triphenylphosphine was added to cleave the ozonide selectively to the aldehyde (the fuchsin test for aldehydes was routinely positive within 1 min). A molar ratio of p-nitrobenzyloxyamine to the sum of ganglioside-aldehyde and long-chain fatty acid aldehyde of 1:3:1 was routinely used. The reaction mixture was incubated in methanol for 15 min at 40-45°C. Immediately following the derivatization, chloroform was added to the reaction mixture, the solvent composition was adjusted to methanol-chloroform (7:2) and the mixture was applied to a 25 cm x 2 cm I.D. DEAE-Sephadex A-25 column (Ledeen et al., 1973); the ion exchanger was first washed four times with chloroform-methanol-0.8 M sodium acetate (30:60:8) and three times with chloroform-methanol-water (30:60:9), and then poured in the form of a suspension in methanol-chloroform (7:2) into a glass column. The above-mentioned reaction mixture was applied to the column at a flow-rate of 1 ml/min and the column was eluted with 300 ml of methanol-chloroform (7:2). The labelled gangliosides were eluted with 200 ml of methanol-chloroform (7:2) that had been made 0.2 M in ammonium acetate; the effluent was monitored at 254 nm and the fractions were assayed by the resorcinol method of Svennerholm (1957). Dialyzed samples were lyophilized.

For HPLC a 30 cm x 2 mm I.D. reversed-phase 10-μm μBondapak C_{18} column (Waters Assoc.) was used and eluted at ambient temperature at a flow-rate of 1 ml/min. Solvent programming began with the initial solvent composition methanol-water (50:50), which was held constant for 1 min and then linearly adjusted during 15 min to methanol-water (70:30). Peak areas were quantitated by a Columbia Scientific Industries (Austin, TX, U.S.A.) Model CRS-208 integrator.

Individual gangliosides were collected and identified by TLC and GC of their monosaccharides. Quantitative results were obtained in this method, together with a marked increase in sensitivity in comparison with conventional analytical methods.

Nakabayashi et al. (1984) studied the analysis and quantitation of gangliosides as p-bromophenacyl derivatives by HPLC. This method was found to be highly sensitive, and involves a simple and specific derivatization of the carboxylic group. The molar absorptivity of the derivatives at maximum absorption (261 nm) was about 23 000. The acidic fractions from IEC pre-fractionation on DEAE-Sephadex could be directly derivatized and the reaction mixtures directly injected into the HPLC equipment without any preliminary purification. Both normal- and reversed-phase modes could be applied for the separation of ganglioside mixtures.

Derivatization of gangliosides (Nakabayashi et al., 1984). Ganglioside, containing 1-15 μg of sialic acid, was dried in vacuo in a PTFE-lined glass tube and reacted with 20 μl of *p*-bromophenacyl bromide (Gasukuro Kogeyo, Tokyo, Japan) in dimethylformamide (10 mg/ml) at 60°C for 1 h. When the reaction was completed, the solution was immediately checked by TLC and 5 μl of the reaction mixture was directly injected on to an HPLC column.

For HPLC in the NPC mode a 25 cm x 4.6 mm I.D. column packed with silica gel (5-6 μm, Zorbax SIL, DuPont) was used and eluted according to the following solvent programme: 1 min isocratic elution with solvent mixture A [isopropanol-*n*-hexane-water (50:49:1, v/v)], then from A to B [the same components (55:35:10, v/v)], at a rate of 2%/min. For HPLC in the RPC mode a 15 cm x 4.6 mm I.D. column packed with octasilylated silica gel (5-6 μm Zorbax C_8, DuPont) was used and eluted with acetonitrile-methanol (4:1, v/v). UV detection at 261 nm (the absorbance maximum of *p*-bromophenacyl derivatives) was applied and the peak areas were calculated with an integrator (Shimadzu C-R1A).

Monosialogangliosides GM4, GM3, GM2 and GM1 were well separated using the normal-phase mode. The calibration graph was linear up to 100 μg of ganglioside sialic acid; the detection limit was about 10 ng. The reversed-phase mode was also successfully applied to analyses of gangliosides.

12.2.3.5 Sulphatides

Sulphatides also belong to acid glycosphingolipids. From the chemical point of view they are sulphate esters of galactocerebrosides and create glycolipid components of brain and other tissues. Desulphated sulphatides are often treated together with cerebrosides. RPC of sulphatides was described by Yahara et al. (1980).

Shimomura and Kishimoto (1983) presented and improved a procedure for the quantitative determination and characterization of sulphatides in rat kidney and brain by RP-HPLC. A significant improvement was made in the desulphation step. A solution of trifluoroacetic acid in ethyl acetate was used for solvolysis. The revised method was used to determine the levels of cerebrosides and sulphatides in rat kidney.

12.2.4 Special separation techniques

In this section several miscellaneous techniques will be mentioned that are not based on liquid column chromatography, but which may be of interest for those investigators who are going to separate glycolipids by HPLC methods.

Young et al. (1979) described an improved method for the covalent attachment of glycosphingolipids to solid supports and macromolecules. The olefinic bond of the sphingosine moiety was oxidized to a carboxyl group, which could be coupled to glass beads, agarose gel, proteins and polyacrylic hydrazide polymers. Solid supports and macromolecules prepared in this way may become useful for a variety of bioaffinity studies in cell biology and immunology.

Pick et al. (1984a,b) used a method for overpressured TLC (Tyihak et al., 1979), designated OPTLC, and used it for the class fractionation of neutral and acidic glycosphingolipids.

Otsuka and Yamakawa (1981) and Otsuka et al. (1983) described the application of droplet counter-current chromatography (DCC) for the separation of acidic glycosphingolipids from brain and neutral glycosphingolipids from mammalian erythrocytes (blood group substances).

12.2.5 Applications of HPLC of glycosphingolipids

12.2.5.1 Discoveries of novel glycosphingolipids

The investigation of glycosphingolipids is a very systematic, thorough and progressive process: nearly every year brings knew knowledge in this field and new species are discovered and structurally characterized; many of them are gangliosides.

Ando et al. (1976b) described the existence and structure of glucosamyl-lactosylceramide [lacto-N-triose(II)ceramide, amino CTH-I] in human erythrocyte membranes as a possible precursor of paragloboside and group-active glycolipids; high-speed Iatrobeads chromatography was used for its purification. Ando and Yu (1977) reported the isolation and structural characterization of a trisialo-ganglioside, GT1a, from human brain; it represented 0.6% of the total ganglio-sides. Also here Iatrobeads chromatography was applied as a final fractionation and purification step of the trisialoganglioside fraction obtained by IEC. Ivamori and Nagai (1978b) presented results of the isolation and characteriza-tion of a ganglioside (monosialosylpentahexaosylceramide) from human brain. Using a combination of IEC and Iatrobeads chromatography, approximately 2.1 μmol of an unknown ganglioside were obtained from 1 kg of human brain (this amount comprised 0.09% of the total lipid-bound sialic acid in the brain). The charac-terization procedures were described.

Ivamori and Nagai (1981c) studied monosialogangliosides of rabbit skeletal muscle and characterized a novel N-acetylneuraminosyl lacto-N-noroctaosyl ceramide. After IEC (DEAE-Sephadex, DEAE-Sepharose) the monosialoganglioside fraction was further fractionated by LCS on Silica gel 60 and Iatrobeads columns. As in the above-mentioned papers, also here the purity of the obtained fractions was examined by TLC. The newly isolated ganglioside represented 5.1% of the monosialoganglioside fraction. Chien and Hogan (1983) described a novel pentasialohexaosyl ganglioside of the globo series purified from chicken muscle. This monosialosyl ganglioside was isolated from a tetrahydrofuran-aqueous KCl extract of chicken pectoral muscle by a combination of IEC (DEAE-Sephadex) and silicic acid (Bio-Sil A) chromatography. The fractions were checked by TLC.

12.2.5.2 *Glycosphingolipids in research on and diagnostics of cancer*

Glycosphingolipids participate in cell recognition processes and their changes are closely connected with the process of tumour growth. This is the reason why they are intensively studied also from the medical point of view and high-speed chromatographic methods for glycosphingolipids can not only contribute to research on cancer, but they can be also used as rapid diagnostic methods in clinical biochemistry.

Hakomori et al. (1982) described the common structure of fucosyllactosamino-lipids accumulating in human adenocarcinomas and their possible absence in normal tissue. Two major glycolipids accumulating in human primary liver carcinoma (absent in normal liver) were characterized as lacto-N-fucopentaosyl(III)ceramide and difucosyllacto-N-norhexaosylceramide. The tissue extract [in isopropanol-hexane-water (55:25:20)] was partitioned four times (according to Folch et al., 1951, as modified by Svennerholm, 1963) and then separated by successive chromatography on DEAE-Sephadex (Yu and Ledeen, 1972). Neutral glycolipids were purified by three stages of HPLC: (i) low-pressure chromatography on an Iatrobeads 6RS-8060 column, (ii) re-chromatography of the selected fractions on an Iatrobeads 6RS-8060 column and (iii) repeated HPLC purification; for details see the original paper. Chemical characterization of the isolated species was described and its comparison with other related glycolipids was discussed in connection with oncogenesis.

Holmes and Hakomori (1982) isolated and characterized a new fucoganglioside accumulated in precancerous rat liver and in hepatoma induced by N-2-acetyl-aminofluorene. This novel ganglioside was found to be accumulated in the liver of rats fed with N-2-acetylaminofluorene before the development of hepatoma. This new fucoganglioside was absent in normal rat liver and in livers of rats fed with non-hepatitic carcinogens.

The animals were labelled in vivo with [^{3}H]fucose. The crude gangliosides obtained after tissue extraction, Folch partitioning (for citations see above in the preceding paper) and IEC on DEAE-Sephadex A-25 were further chromato-graphed on an Iatrobeads (6RS-8010) column (50 cm x 1 cm I.D.), eluting with a gradient system of isopropanol-hexane-water from 55:40:45 to 55:30:15 over 150 min at a flow-rate of 1 ml/min. The eluted gangliosides were detected both by liquid scintillation counting and by TLC. Fractions containing [^{3}H]fucosyl gangliosides were pooled and purified by HPLC on an analytical Iatrobeads

6RS-8010 column (80 cm x 4 mm I.D.) using the gradient described above and a flow-rate of 0.5 ml/min. The biochemical and chemical degradation of the novel ganglioside in order to establish the structure was described.

In addition to the mentioned species, also another (second) fucoganglioside was detected in a smaller amount in the precancerous liver and in hepatoma in vivo, but not in control tissue. The results indicated that the synthesis of new fucolipids was induced at an early stage during the process of carcinogenesis; this could be a unique marker for the diagnosis of hepatoma and its premalignancy.

Magnani et al. (1982) proved that monoclonal antibody-defined antigen associated with gastrointestinal cancer is a ganglioside containing sialylated lacto-N-fucopentaose(II). Monoclonal antibodies were produced by hybridomas obtained from a mouse immunized with colorectal carcinoma cells. These antibodies bind specifically to human gastrointestinal cancer cells. The antigen corresponsing to this antibody was isolated and chemically determined. It appeared to be a mono-sialoganglioside and its structure was proposed. The ganglioside antigen was isolated from human adenocarcinoma cell line SW 1116. The extract was first fractionated by IEC on DEAE-Sepharose CL-6B and the monosialoganglioside fraction was further separated on a Bio-Sil HA (-325 mesh) column (100 cm x 0.9 cm I.D.) using concave gradient elution from chloroform-methanol (4:1, v/v) to methanol-0.2% calcium chloride solution (50:3, v/v). Fractions containing antigen and exhibiting only one band in TLC (migrating between G_{M1} and G_{D1}) gave 300 µg of the monosialoganglioside, which was chemically characterized and examined by immunoassays.

Westrick et al. (1983a) studied gangliosides of human acute leukaemia cells (from cells of eight patients: four lymphoblastic and four non-lymphoblastic) and found great differences in the absolute amounts of gangliosides (they were lower than in normal cells), a simplified ganglioside pattern and reduced amounts of N-acetylneuraminosyllactotriaosylceramide. In one instance (a patient with acute non-lymphoblastic leukaemia) a disialylated ganglioside (GD_3) was proved, which was not found in normal leukocytes. TLC and HPLC methods in combination with glycosidase treatments were used for the identification of gangliosides. Westric et al. (1983b) described the isolation and characterization of gangliosides from chronic myelogenous leukaemia cells. Similarly to the above, they found differences in the patterns (lower total content of gangliosides). The major gangliosides were G_{M3}, a sialosylparagloboside, and another ganglioside of the reported structure (previously designated as an "i" active compound). No other gangliosides containing more than one sialic acid residue were detected in chronic myelogenous leukaemia glycosphingolipids. Folch partitioning (mentioned at the beginning of this section), DEAE-Sephadex chromatography and Bio-Sil A or Florisil chromatography were used for the isolation, and TLC, HPLC and biochemical and chemical methods were used for the degradation and identification of these glycosphingolipids. Differences between chronic myelogenous leukaemia and acute leukaemia were discussed.

Blaszczyk et al. (1984) studied a foetal glycolipid expressed on adenocarcinomas of the colon. Using IEC, HPLC and TLC they reported the detection and characterization of tumour-associated glycolipids from colorectal carcinoma and meconium, which resemble the gastrointestinal cancer antigen described by Magnani et al. (1981). These glycolipids contain the known X-determinant or stage-specific embryonic antigen-1, modified by a single sialo residue. Although the X-determinant was present in the normal colon, the sialylated-X glycolipids were absent and found only in adenocarcinomas. These monosialo-X glycolipid antigen might serve as potential tumour markers. Gonwa et al. (1984) reported the inhibition of mitogen- and antigen-induced lymphocyte activation by human leukaemia cell gangliosides. The number of viable cells was not reduced and the inhibition was due to blast formation. Three (in vitro) inhibiting gangliosides were isolated, which were not unique to leukaemia cells, but their concentration was increased in patients with cancer. Siddiqui et al. (1984) studied the differential expression of ganglioside G_{D3} by human leukocytes and leukaemia cells (both acute and chronic). Among myeloid cells acute leukaemia cells were positive for G_{D3}, whereas chronic leukaemia cells and normal neutrophils did not have detectable G_{D3}. All lymphocytic leukaemia cells (chronic and acute) contained G_{D3}, which were not detected in normal lymphocytes. The procedure was as follows. The upper phase from Folch partitioning was chromatographed on an Iatrobeads 6RS-8010) column using a linear gradient of 2-propanol-hexane-water from 55:35:10 to 55:30:15 at 2 ml/min for 150 min. The fractions were analysed on HPTLC plates, using detection with resorcinol reagent or by immunostaining.

12.2.5.3 Miscellaneous biomedical applications

The analytical chromatographic separation of glycosphingolipids has been used for the diagnostics of some inherited anomalies (enzyme deficiency diseases). Ullman et al. (1980) described the application of HPLC to the study of sphingolipidoses. Perbenzoylated sphingolipids, prepared from plasma from 32 Gaucher (β-glucosidase deficiency) and 6 Fabry (α-galactosidase deficiency) patients by solvent partition, were chromatographically separated on silicic acid or reversed-phase columns. Chromatographic analysis of sphingolipids provided useful supportive information for diagnoses because affected individuals were shown to possess increased circulating concentrations of pathognomic sphingolipid. Strasberg et al. (1983) also studied the HPLC analysis of perbenzoylated neutral glycolipids as an aid in the diagnosis of lysozomal storage disease. Concentrations of four glycolipids (GL1a, GL2a, GL3a and GL4a) were determined in normal plasma and compared with the concentrations in plasma from patients with Gaucher, Krabbe, Fabry, Sandhoff and Tay-Sachs diseases and also with hypocholesterolaemia. An Ultrasphere silica column (5 μm) eluted with a gradient of isopropanol in hexane proved useful for the analyses (see also Fig. 12.2).

Nagai and Ivamori (1980b) examined brain and thymus gangliosides, their molecular diversity, their biological implications and a dynamic annular model for their function in cell surface membranes. Iatrobeads column chromatography and ganglioside mapping were used. Gravotta and Maccioni (1985) studied gangliosides and sialoglycoproteins in coated vesicles from bovine brain.

Uemura et al. (1978) described the characterization of major glycolipids in bovine erythrocyte membrane; a combination of IEC (DEAE-Sephadex A-25) and chromatography on a mixed silicic acid-Hyflo Super Cell (3:1) column were used for the fractionation of neutral and acidic glycosphingolipids. Yamakawa et al. (1980) reported the analysis of red blood cell glycolipids by HPLC of acetylated and perbenzoylated derivatives; the derivatives of erythrocyte glycolipids were separated on a silica gel column on the basis of their sugar chain and the fractions were analysed by RPC on the basis of their lipid portions.

Suzuki et al. (1977a) used perbenzoylated derivatives of neutral glyco-sphingolipids for the micro-scale determination of seminolipids by HPLC and studied their application to the determination of the seminolipid content of boar spermatozoa. The decrease in the seminolipid content of testes of rats with vitamin A deficiency was also determined by HPLC using similar methods (Suzuki et al., 1977b).

Chien Jaw-Long and Hogan (1980) studied glycosphingolipids of skeletal muscle. Kosai et al. (1982) described a convenient method for the large-scale prepara-tion of sialo-G_{M1} ganglioside from brain. Gazzotti et al. (1984b) reported the preparation of tritiated molecular forms of gangliosides with a homogeneous long-chain base composition; gangliosides were labelled at C-6 of the terminal galactose or N-acetylgalactosamine or at C-3 of the long-chain base, and using an HPLC procedure eight different molecular species were prepared from each labelled ganglioside, which were further characterized. The reversed-phase mode according to Sonnino et al. (1984) was applied, employing the experimental conditions described in Section 12.2.3 (NPC and RPC). For radioactivity counting a 120-μl solid scintillator cell (Berthold LB 503 radioactivity monitor) was used). A computer-assisted automatic fraction collector (Gilson 201) was pro-grammed to separate peaks on the basis of UV or radioactivity signals.

12.2.6 Comments on literature

In addition to the references given in the introductory Section 12.2.1 the following citations are important. Glycolipids were reviewed by McKibbin (1970). Weigandt (1980) wrote introductory remarks on the chemical structure of gangliosides, including the classification of glycosphingolipids. The proposed pathways for the ganglioside biosynthesis were discussed by Yu and Ando (1980). Svennerholm et al. (1980b) edited symposium proceedings in which much useful information on gangliosides can be found. Macher and Sweeley (1978) dealt with the structure, biological sources and properties of glycosphingolipids. Ledeen and Yu (1982) have written a methodological review on the structure, isolation and analysis of gangliosides (including glycolipid classification). Kanfer and Hakomori (1983) have written a book on sphingolipid biochemistry. Eberendu et al. (1985) presented a review of chromatographic techniques for the analysis of glycolipids and phospholipids, based on 125 references. McCluer et al. (1986) have written a modern review on the HPLC of glycosphingolipids and phospholipids. Nagai et al. (1986) described the procedure of gene transfer as a novel approach to the gene-controlled mechanism of the cellular expression of glycosphingolipids.

The nomenclature of glycolipids is sometimes complicated and several recommendations have been made by individual authors, and proposals accepted by the official nomenclature commissions have also been published. A practical shortened ganglioside nomenclature was proposed by Svennerholm (1963, 1964), and the IUPAC-IUB Commissions (1977, 1978) recommended a unified nomenclature for lipids.

12.3 GLYCOPEPTIDES AND PEPTIDES OF BACTERIAL CELL WALLS, IMMUNOMODULATORS AND SYNTHETIC VACCINES

12.3.1 Introduction

The polysaccharides, peptidoglycans and proteins on the surface of bacterial cell walls give the main impulse for the immune response of macroorganisms attacked by microbes. The detailed mechanism of the immune response process is very complicated, but two general types of compounds can be classified on the bacterial cell walls, which are the most important to start with: (i) antigens, against which antibodies are synthesized by the macroorganism, and (ii) immunomodulators, modifying the immune response and the activity of antigens. In Sections 12.3.1-12.3.3 the second type of compounds will be dealt with, and the first type will be discussed in the Section 12.3.4. In research on both types of compounds chromatographic methods have played an important role.

Gram-positive and Gram-negative bacterial cell walls have a common feature, viz., the rigid structural framework consisting of parallel polysaccharide chains covalently cross-linked by peptide chains, and this framework constitutes 10-50% or even more of the weight of the cell wall. We shall not comment here on the accessory components, which are different in the two types of bacterial cell, but will focus our attention on the main macromolecular framework, a hetero-polymer, which is called peptidoglycan or murein (murus means wall in Latin). The peptidoglycan forms a completely continuous covalent structure around the cell. For example, Gram-positive bacteria are encased in up to 20 layers of cross-linked peptidoglycan (Lehninger, 1978b). The basic recurring unit in peptidoglycans (creating long polysaccharide chains) is a disaccharide of N-acetyl-D-glucosamine and N-acetylmuramic acid in a β(1→4) linkage, to the carboxyl group of which a peptide is attached by means of the N-terminus. The peptidoglycan cross-linking in the *Staphylococcus aureus* cell wall is given below as an example (according to Strominger and Ghuysen, 1967; see also peptido-glycan biosynthesis, dealt with by Strominger et al. 1967).

A tetrapeptide L-Ala-D-isoGln-L-Lys-D-Ala is joined by a pentaglycine bridge bound with its N-terminus to the carboxyl group of D-Ala in one chain and by its C-terminus to the ε-amino group of Lys in the other chain. In the cell walls of other bacteria similar but not identical peptide cross-linking patterns have been found [for surveys, see Rogers and Perkins (1968), Schleifer and Kandler (1972), Leive (1973) and Ježek (1986)].

It was found that for the immunization of the macroorganism (i.e., for the prevention of infection) the presence of intact bacterial cells is not necessary; some parts of cells were sufficient, especially bacterial cell walls and their oligomeric components. Peptidoglycan and their fragments showed in both in vitro and in vivo tests about 40 types of biological activity (Kotani et al., 1983; Adam and Lederer, 1984; Lederer, 1986; Ježek, 1986), such as modulation (potentiation or inhibition) of the immune response, pyrogenicity, stimulation of the natural non-specific resistence against microbial, viral and parasite infections and growth of tumours and modulation of cells participating in the process of the natural or induced immunity. Some of these biological activities did not seem to have any direct connection with the immune response process, e.g., the influence on the increase in the duration of slow-wave sleep (somnogenic activity). Many papers have been published describing the biological properties of various peptidoglycan fragments and their analogues. The turning-point in research on these compounds was the period 1974-75, when Adam et al. (1974), Ellouz et al. (1974) and independently Kotani et al. (1975) proved that the minimum size fragment showing high immunoadjuvant activity is N-acetylmuramyl-L-alanyl-D-isoglutamin, often designated muramyl dipeptide (MDP) or MDP-Pasteur in the literature, owing to its discovery in the Pasteur Institute in Paris.

(D)   MDP

Adjuvant activity is the stimulation of the production of antibodies against an antigen injected simultaneously with the adjuvant glycopeptide. Many experiments have been undertaken to substitute chemically or modify this basic active unit (MDP) in order to prepare analogues with various biological activities; for reviews see, e.g., Lederer (1986) and Ježek (1986).

Synthetic glycopeptides represent a very important contribution to the endeavour of discovering the best immunomodulators. The first peptide fragments of bacterial cell walls (the lysine pentapeptides) were synthesized by Garg et al. (1962), the first lysine glycopeptides (containing N-acetylmuramic acid) were prepared by Lanzilotti et al. (1964) and the first glycopeptides containing diaminopimelic acid were prepared by Arendt et al. (1974). The synthetic approach required the working out of some auxiliary synthetic methods, for instance for isoglutamine or isoasparagine derivatives (e.g., Straka and Zaoral, 1977) or for removal of protecting groups (e.g., Zaoral et al., 1978). Immunoadjuvant activity and structure specificity of some synthetic glycopeptides were studied (e.g., Mašek et al., 1978, 1979; Adam and Lederer, 1984; Lederer, 1986). Ovchinnikov et al. (1979) prepared the longest oligosaccharide glycopeptide (the tetra-saccharide with two bound dipeptides L-Ala-D-iGln) and the longest peptide bound to disaccharide (i.e., the pentapeptide) using semi-synthetic methods. The largest lysine glycopeptides [bis(N-Ac-Mur)tridecapeptides with a pentaglycine or penta-L-alanine bridge moiety] were synthesized by Zaoral et al. (1980) and tris(N-Ac-Mur)octadecapeptide by Ježek et al. (1987). Up to now about 1000 synthetic glycopeptides have been reported, and most of them have been patented (cf., e.g., Lefrancier and Lederer, 1981; Straka 1983; Ježek, 1986).

In addition to glycopeptides and peptides from bacterial cell walls, a number of other natural peptides (and their synthetic analogues) also show important activities in the immune response process [cf., e.g., the review (460 references) by Leclers and Vogel (1986)]. Similar activity was found, e.g., in the natural tetrapeptide tuftsin (Thr-Lys-Pro-Arg), which is a phagocytosis-stimulating factor isolated from the γ-globulin fraction of human blood serum and activates all functions of phagocytic cells (Najar and Nishioka, 1970; Najar et al., 1981). Tuftsin binds to specific receptors on phagocytic cells (Wleklik et al., 1986). In addition to tuftsin and its analogues, the natural pentapeptide proctolin (Arg-Tyr-Leu-Pro-Thr), a neuropeptide from the hindgut of the cockroach, *Periplaneta americana* L., also shows a positive activity on restoration of human PMN leukocytes defected by acute lymphoblastic leukaemia (Konopinska et al., 1986). In contrast, Orefici et al. (1986) found that peptides are not necessary in the modulation of all the immune responses: polysaccharides secreted by streptococci (streptococcal exopolysaccharides alone) could induce a variety of effector cells in the peritoneum of normal experimental mice, including natural killer lymphocytes showing cytotoxic activity against some tumour cell targets.

12.3.2 Chromatography of glycopeptide immunomodulators

The conditions for the chromatography of glycopeptides are limited to the
pH range of stability and also by the solubility of these substances. N-Acetylmuramyl peptides are readily soluble in water and in lower alcohols. They are
very sensitive to alkaline solutions, e.g., 0.05 M ammonia solution (at 37^{0}C)
can split off lactyl or lactopeptide in 1 h, and aqueous ammonia (pH 12) will
destroy the muramyl peptide immediately (for citations, see Ježek, 1986). In
a mildly acidic solution (pH 2-4) muramyl peptides are relatively stable, whereas
in strongly acid media muramic acid is decomposed. For monitoring in column
liquid chromatography, measurement of the UV absorbance at 200-210 nm may be
used. In preparative separations TLC is often applied for monitoring the purity
of fractions, in addition to HPLC.

In the study of natural glycopeptides, IEC has been used for the separation
of peptidoglycan fragments. Bacteriolytic peptidases are capable of breaking
some peptides that interlink the peptidoglycan strands of microbial cell walls
and the wall material is thus transferred into solution. Ion exchangers were
used to fractionate such a partial hydrolysate (cf., review by Mikeš and Šebesta,
1979). Tipper et al. (1967) split the peptide cross-bridges that interlink the
peptidoglycan strands of *Staphylococcus aureus* cell walls by means of exo-
peptidases produced by *Myxobacter*, and the digest was chromatographed on a
23 cm x 3 cm I.D. ECTEOLA-cellulose (Bio-Rad Cellex E) column using brief
isocratic elution with water and then a linear gradient of LiCl concentration.
The elution rate was 1 ml/min and the fraction volume was 15 ml. The desalted
fractions were further fractionated on Sephadex and CM-cellulose. On the basis
of the study of isolated polysaccharides and peptides, the structure of peptidoglycan was proposed.

Lefrancier et al. (1977) also used IEC (AG1 X-2 resin from Bio-Rad Labs.,
Richmond, CA, U.S.A. and Amberlite IRA-68 resin from Serva, Heidelberg, F.R.G.)
for the fractionation of synthetic glycopeptides [N-acetylmuramyl-L-Ala-D-iGln,
i.e., muramyl dipeptide (MDP) and analogues] in addition to preparative silica
gel column chromatography. Zaoral et al. (1978), in synthetic experiments on
the glycopeptide series, purified 270 mg of the crude intermediate N-acetyl-D-
glucosamine by chromatography on a silica gel (60-120 µm) column (60 cm x 2.5 cm
I.D.) eluted with 1-butanol-acetic acid-water (4:1:1) at a flow-rate of 50 ml/h
and with a fraction volume of 10 ml. The pure product was eluted in fractions
40-56 and, after evaporation at reduced pressure, a yield of 203 mg (73%) was
obtained. The same chromatographic method was also used for the purification
of other intermediates. In an additional contribution to the glycopeptide series

(Zaoral et al., 1980), some intermediates were purified by GPC on a 100 cm x 1 cm I.D. column of Sephadex in 1 M acetic acid at a flow-rate of 6 ml/h, monitored by TLC. No apparent loss due to decomposition of the products was observed in these experiments.

Lefrancier and Lederer (1981) found HPLC of MDP to be a promising method for various analytical purposes (to check its chemical or even stereochemical homogeneity, to study its stability or to quantify its dosage forms). Spherisorb ODS and LiChrosorb-NH$_2$ columns eluted with ammonium acetate-acetonitrile mixtures were found to be very effective even for the separation of isomers (see later). Phillips et al. (1984) studied the synthesis and fast atom bombardment-mass spectrometry of MDP and for the HPLC of glycopeptide samples used a Varian (Palo Alto, CA, U.S.A.) MicroPak AX-10 column (30 cm x 4 mm I.D.), eluted at 50°C with acetonitrile-triethylammonium trifluoroacetate (pH 3.1) (2:3) at 1 ml/min; the pressure was 9.6 MPa and UV detection at 206 nm was applied.

If the results of glycopeptide HPLC are to be interpreted, it is necessary to realize that the free muramyl peptides contain a half-acetal hydroxy group and are therefore subject to mutarotation. Because the half-time of the mutarotation is one order of magnitude longer than the chromatographic process, it is possible to employ this technique for the separation of individual anomers in the pyranoid or furanoid forms (e.g., see Ježek, 1981, 1986).

Halls et al. (1980) studied the HPLC of MDP and some derivatives on a Spherisororb ODS (5 µm) column (25 cm x 3 cm I.D.) using 5 mM ammonium acetate (pH 2.5)-acetonitrile (199:1, v/v) as eluent and found that this approach allowed the separation of α- and β-anomers in 6 min. This time was short enough to allow (with fairly good accuracy) the study of the ratio of the two anomers and its variation with time as a result of mutarotation (which proceeds in a time measurable in hours). Ježek (1981) and Buděšínský et al. (1982) examined in detail the cyclization and mutarotation processes, which influence the separation of individual components of glycopeptides: double, triple or quadruple peaks of the same component were often found and the isolated individual split again into the original forms if there was enough time to reach equilibrium in the solution.

Ježek (1981) studied the synthesis of glycopeptides of bacterial cell walls. Synthetically prepared MDP contained four substances, which were preparatively separated on a 25 cm x 0.4 cm I.D. column of Separon Si C$_{18}$ (10 µm) (Laboratory Instruments Works, Prague, Czechoslovakia), eluted with 0.01 M phosphate buffer (pH 5.0) containing 5% methanol at a flow-rate of 1.5 ml/min and with UV detection at 210 nm. Two pairs of anomers were obtained; the first two isomers were the main products and the other two were impurities. The fractions were identified as follows: 1 (k' = 4.7) = β-pyranoid isomer; 2 (k' = 8.5) = α-pyranoid;

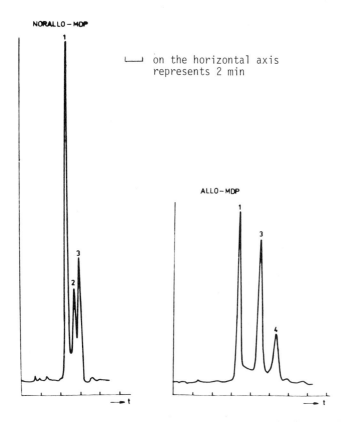

Fig. 12.5. Chromatography of synthetic glycopeptides by means of HP-RPC (for conditions, see text). Left: separation of nor-*allo*-MDP (i.e., N-Ac-Nor-*allo*-Mur-Ala-iGln); 1 = β-pyranoid isomer; 2 = α-furanoid form; 3 = α-pyranoid; 4 = β-furanoid (present in only a small amount; not numbered). Right: separation of *allo*-MDP (i.e., N-Ac-*allo*-Mur-Ala-D-iGln); 1 = β-pyranoid isomer; 2 = β-furanoid form (present in only a small amount in some experiments owing to rapid isomerization; not-numbered); 3 = α-pyranoid; 4 = α-furanoid. (Reprinted from Ježek, 1981; see also Buděšínský et al., 1982.)

3 (k' = 9.8) = β-pyranoid, isomuramic isomer; and 4 (k' = 15.8) = α-pyranoid, isomuramic isomer. Owing to the mutarotation process, the freshly prepared solution of *allo*-MDP contained four peaks before equilibrium and three peaks after equilibrium had been achieved.

Similar experiments with other synthetic glycopeptides (N-Ac-galacto-Mur-L-Ala-D-iGln, N-Ac-nor-*allo*-Mur-L-Ala-D-iGln and N-Ac-*allo*-Mur-L-Ala-D-iGln) provided a mixture of at least three forms in each instance (cf., Fig. 12.5). All the isolated individual components 1, 2, 3 or 1, 3, 4 after standing at an appropriate pH yielded a mixture of components of the original composition, i.e., containing α- and β-anomers in the pyranoid and furanoid form.

Lefrancier and Lederer (1981) used HPLC of MDP on a Spherisorb ODS (5 μm) column eluted with $5 \cdot 10^{-3}$ M ammonium acetate (pH 2.5)-acetonitrile (995:5) and separated α- and β-anomers; this method was able to follow their mutarotation. Using a LiChrosorb-NH$_2$ column eluted with $5 \cdot 10^{-3}$ M ammonium acetate (pH 3)-acetonitrile (10:90) the two anomers of MDP could be separated from the two peaks of its diastereoisomer containing a D-Ala residue. Lebl and Gut (1983) used the data from HPLC of muramyldipeptide anomers [MDP was prepared by Zaoral et al. (1978)] for the calculation of the rate constant of the reversible anomerization reaction. They employed a Zorbax ODS column (DuPont, Wilmington, DE, U.S.A.) eluted with 0.1% trifluoroacetic acid-1% methanol at 34°C. For the determination of the rate constant with an isolated anomer, 0.5 mg of MDP was separated at a flow-rate of 1.5 ml/min. The time of elution of the anomer peak that was collected was taken as zero. The anomer solution was kept at 34°C and samples were taken at several time intervals and analysed at a flow-rate of 6.8 ml/min. The equilibrium constant of the reaction was determined at the same flow-rate with a sample of MDP dissolved in the mobile phase equilibrated at 34°C for 2 days. The rate constant for β- to α-anomer conversion was found to be $k_2 = 3.25 \cdot 10^{-2}$ min^{-1} or $3.45 \cdot 10^{-2}$ min^{-1} at 34°C. An advantage of the method is the need for only two chromatographic experiments for the determination and no strict requirements with regard to the purity of the studied compounds.

12.3.3 Examples of prepared immunomodulators

A series of papers were published on studies of the biological activity of synthetic subunits of *Streptococcus* peptidoglycan. Rotta et al. (1979) studied the pyrogenic and thrombocytolytic activity of several analogues and found that synthetic glycopeptides displayed both activities comparable to those of natural peptidoglycans. Rotta et al. (1983) examined the relationship between peptidoglycan subunits and series of analogues and the fever effect and induction of tolerance to pyrogenicity. The homogeneity of the tested peptides was checked by TLC, NMR, HPLC and amino acid analyses. Structure-function relationships were discussed. Pekárek et al. (1985) described a relationship between peptidoglycan subunit and analogue structures and adjuvant activity in cell-mediated immunity. Eight compounds (homogeneous according to TLC, paper electrophoresis and HPLC) were tested; the examined compounds were three desmuramyl peptides, MDP-polylysine, *allo*- and nor-*allo*-MDP, [Abu]-nor-MDP and MDP. Their structure-function relationships were discussed. Rotta et al. (1986) studied the pyrogenicity and hypersensitivity inducing activities of streptococcal peptidoglycan and its synthetic analogues. Ovalbumin and M25 *Streptococcus* extract were

used as antigens, and six glycopeptides (from glycodipeptide up to glycotrideca-
peptide) as immunoadjuvants. Rýc et al. (1986) described the effect of
Streptococcus peptidoglycan and its synthetic analogues on rabbit blood cells.
Four linear and one cross-linked desmuramylpeptide, four linear and one cross-
linked muramylpeptide and twelve other analogues were investigated. Morphological
and ultrastructural changes in platelets were followed. Pekárek et al. (1986)
examined experimental allergic encephalomyelities (EAE) and the production of
lymphokins in animals stimulated by *Streptococcus* peptidoglycan and synthetic
analogues. Thirty peptides (both desmuramyl and muramyl derivatives, some of
them cross-linked) were tested; ten showed immunoadjuvant activity and twenty
were inactive.

Ivanov et al. (1981) studied the synthesis, structure and biological properties of glycopeptides containing the N-Ac-Glc-(β1$\rightarrow$4)-N-Ac-Mur-disaccharide unit.
The antitumour activity of eleven synthetic peptides and the adjuvant activity
of eight synthetic peptides were tested. The energetically preferred conformation
of the disaccharide dipeptide N-Ac-Glc-N-Ac-Mur-L-Ala-D-iGln (GMDP) was discussed.
Ivanov (1986) described studies on bacterial glycopeptides (analogues of di- and
tetrasaccharide muramylpeptides found in blastolysin, the antitumour preparation
isolated from *Lactobacillus bulgaricus* cells). Nine synthetic glycopeptides were
prepared and their immunomodulating properties were examined, both in antitumour
activity and experimental bacterial infections.

Zaoral et al. (1982) described the preparation and some biological properties
of three MDP analogues and Krchňák et al. (1983) those of nor-Mur-N-Ac-α-
aminobutyryl-D-iGln-Lys-Lys-Lys, which was prepared by solid-phase synthesis.
Preparative HPLC was performed on a 25 cm x 2 cm I.D. column of Separon R
(Laboratory Instrument Works, Prague, Czechoslovakia) using methanol-water-
trifluoroacetic acid (5:95:0.2) for elution and UV detection at 210 nm. The
compound forms a mixture of α- and β-anomers. Mašek and Flegl (1983) prepared
an unusual type of hydrophobic desmuramyldipeptide, L-Ala-D-iGln-adamantylamide,
which showed increased delayed hypersensitivity.

During the last 10 years, Kotani and co-workers have examined a series of
immunoadjuvant glycopeptides and their application forms and activities [see,
e.g., the paper by Kotani et al. (1977) in which the effect of administration
of glycopeptides with liposomes was reported]. Kotani et al. (1986) also studied
synthetic immunomodulators mimicking bacterial cell surface components. A description of the acyl derivatives of muramylpeptides and low toxic lipid A
analogues as possible adjuvants for vaccines was presented. 6-O-(2-Tetradecyl-
hexadecanoyl)-Mur-N-Ac-L-Ala-D-iGln (the so-called B 30 MDP), N^{α}-(Mur-N-Ac)-N^{α}-
methyl-L-Ala-D-iGln-N^{ε}-stearoyl-L-Lys and Mur-N-Ac-L-Ala-D-Gln α-*n*-butyl ester

(Murabutide; cf., Lefrancier et al., 1982) were prepared and tested in various forms, and also in the form of virosome influenza vaccine. Masihi and Lange (1986) reported the stimulation of non-specific resistance against respiratory infections by immunomodulators. Eleven MDP analogues were found to be effective (when combined with 6,6'-trehalose dimycolate) against influenza virus or *Mycobacterium tuberculosis* infection.

Ježek et al. (1986, 1987) continued the synthesis of peptides and glyco-peptides of bacterial cell walls. By means of solid-phase synthesis and synthesis in solution, 21 tetra- to octapeptides and 4 glycopeptides [N-acetylmuramylhexa-peptide to tris(N-acetylmuramyl)octadecapeptide] were prepared and tested for immunoadjuvant activity and pyrogenicity. For the preparative separation of glycopeptides (30-50 mg) a 250 mm x 9 mm I.D. column of 10-μm Partisil ODS-2 (Whatman, Clifton, NJ, U.S.A.) was used, eluted with a gradient system of 0.1% trifluoroacetic acid solution (or 0.05 M NaH$_2$PO$_4$-0.05 M H$_3$PO$_4$) and methanol, the methanol concentration being increased at 1%/min. For analytical checking of purity, a Spherisorb ODS (5 μm) column (250 mm x 4.6 mm I.D.) (Phase Separa-tions, Queensferry, U.K.) was applied, using the same mobile phases. Two of the prepared peptides were pyrogenic. Tris(N-acetylmuramyl)octapeptide showed immunoadjuvant activity comparable to MDP and was not pyrogenic.

Farkaš et al. (1986, 1987) described the synthesis of a new disaccharide analogue of MDP, O-(2-acetamido-2-deoxy-β-D-glucopyranosyl)-(1→4)-N-acetyl-nor-muramoyl-L-aminobutyryl-D-isoglutamine:

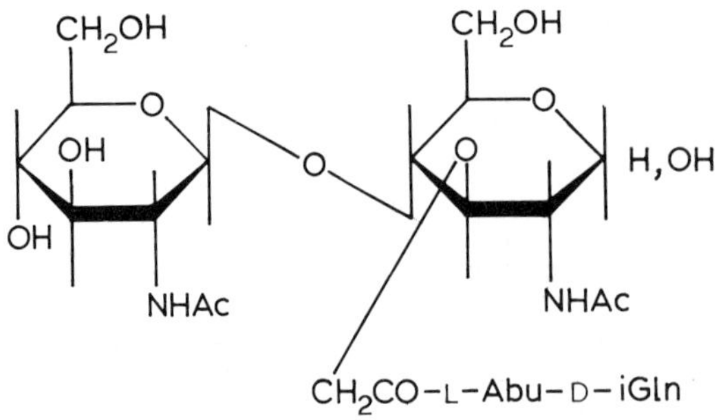

Some intermediates were purified by preparative column chromatography on silica gel (30-60 μm) using acetone-chloroform (1:9, v/v), chloroform-ethyl acetate (2:1, v/v), ethyl acetate-toluene (2:1, v/v) or chloroform-methanol (50:1 or 25:1, v/v), and checked by RPC on C$_{18}$ silica gel (10-15 μm) in methanol-water (3:1 or 97:3, v/v) or 60% methanol. The final product was chromatographed by RP-HPLC on a 25 cm x 0.4 cm I.D. column of C_{18} silica (Laboratory Instrument Works, Prague, Czechoslovakia) using water-methanol (97:3, v/v) for elution. The immunoadjuvant activity of the final compound was determined in guinea pigs by the delayed-type hypersensitivity assay using ovalbumin as antigen. The disaccharide, when applied with Freund's incomplete adjuvant, produced a 3.4-fold higher immunoadjuvant activity than MDP in control experiments. The undesirable pyrogenic effect of the product, unlike that in MDP, was very low (cf., Rotta et al., 1986).

12.3.4 Synthetic vaccines

12.3.4.1 Introduction

Not only peptidoglycan macromolecules on the bacterial cell surface but also protein macromolecules do not need to be in the intact macromolecular form to be able to elicit the immune response in the attacked macroorganism. Fragments of proteins are sufficient to act as antigens. Antigenic determinants (so-called protective epitopes) are relatively low-molecular-weight segments of antigens and therefore the possibility of preparing synthetic vaccines is opened up if the antigenic determinants of the natural antigens are reliably determined (Shinnic et al., 1983; Arnon, 1984). Many studies on the identification of the epitopes have been published. The antigenic determinants are believed to be present in the accessible hydrophilic areas on the surface of protein antigens, which can be found by theoretical prediction.

Hopp and Woods (1981) published a method for the prediction of antigenic determinants from amino acid sequences, assigning a hydrophilicity value for each amino acid, repetitively averaging these values along the chain (a table of hydrophilicity values of amino acids was published). Kyte and Doolitle (1982) presented a simple method for displaying the hydropathic character of proteins; a computer program was written that evaluated the hydrophilicity and hydrophobicity of proteins along the chain. However, the choice of peptides to be synthesized is not always straightforward (Van Regenmortel, 1986), as many antigenic determinants of proteins seem to be conformational (depending on the three-dimensional folding of the native molecule) rather than sequentional (conditioned only by the amino acid sequence of the isolated peptide segment, not influenced by the total structure of the protein). Different strategies have been developed for searching for the appropriate antigenic sites (Atasi, 1984; Weijer et al., 1986). Also the atomic mobility sites in the three-dimensional X-ray structure of proteins (Tainer et al., 1984) and their segmental mobility (Westhoff et al., 1984) were correlated with the location of antigenic determinants and their reactivity.

In addition to theoretical methods, experimental techniques have also been proposed for searching for determinants. Jemmerson and Paterson (1986b) presented a method for mapping epitopes on a protein antigen by partial proteolysis of the antigen-antibody complexes, in which chromatography played an important role. An antibody bound to a protein antigen showed a steric hindrance effect and decreased the rate of proteolytic cleavage of the antigen in regions involved in the antibody-antigen contact. RP-HPLC of the partial digest has made it possible to identify the contact area according to the relative amount (peak heights) and composition of the isolated peptides of the antigen, provided that the primary structure of the antigen was known and that a protease digest of the single antibody and the single antigen were chromatographed in preliminary experiments in order to localize their peptide peaks in the chromatogram. Cytochrome c and monoclonal antibody against it were used in these experiments. A 30-min tryptic digestion was found to be optimal. For HPLC a 250 mm x 4.6 mm I.D. column of C_{18} Spherisorb ODS-1 (Custom LC, Houston, TX, U.S.A.) was eluted with a linear gradient from 0 to 70% acetonitrile (0.1% TFA) during 90 min at a flow-rate of 1 ml/min. UV detection at 214 nm was applied. Collected peaks of the antigen were hydrolysed and analysed in an amino acid analyser. The applicability of this method was discussed in detail. Another, but very expensive and time-consuming, experimental method for determining the contact area between the antibody and antigen is X-ray crystallography of the antibody-antigen complex (Amit et al., 1985), from which the epitopes could be derived. This method is very precise but it cannot be generally used.

The theoretically proposed or practically found antigen determinants are the starting point for the peptide synthesis (Shinnick et al., 1983; Walter, 1986). However, so far the purified peptides corresponding to protective epitopes often possess poor immunogenicity, "at least partly because of the removal of components carrying adjuvant activity of the original vaccines (Kotani et al., 1986); hence the development of new types of vaccines requires studies of chemically well defined immunoadjuvants, which effectively potentiate the immunogenicity of protective epitopes". Also, the effect of additivity or synergism caused by a large number of determinants on the same natural macromolecular antigen contributes to the stronger antigenicity of natural antigens in comparison with single synthetic peptides.

The chromatography of synthetically prepared peptides for vaccination does not differ from the separation techniques for any other simple peptides, which were described in detail in Chapter 11, and therefore in the following part of this section only a few examples of synthetic vaccines will be presented.

12.3.4.2 Examples of efforts to prepare synthetic vaccines

First, synthetic vaccines against bacterial infections will be considered. Synthetic streptococcal M-protein vaccines were chosen because of their great medical importance. The usual vaccination against streptococcal infections with killed streptococci (or with a crude extract) is not possible owing to serious complications with rheumatic fever (Massel et al., 1969). The only known factor of the virulence and antigenicity of *Streptococcus pyogenes* is the M-protein, emanating as fibrils from the surface of streptococcal cells (Fox, 1974). The single-chain protein molecule is present as a dimer and has the form of an α-helical coiled-coil structure (Phillips et al., 1981). The isolated M-protein is an unsuitable antigen for vaccination owing to immune cross-reactions leading to acute rheumatic fever (Dale and Beachey, 1985). It has become apparent that the whole M-protein molecule is not required to evoke protective immunity and that various extracts containing only polypeptide fragments of M-protein retain type-specific protective immunogenicity [for references see Beachey et al. (1986) and Kühnemund et al. (1986)]. A solution to the vaccination problem might be seen in the preparation of synthetic peptide vaccines (as short as possible) that could retain antigenic determinants for the production of protective antibodies, but which would not contain segments responsible for the toxic cross-reactions (Beachey and Stollerman, 1971). Therefore, experiments with further cleavage of the fragments continued and yielded peptides containing as few as twelve amino acid residues that retained protective epitopes (Beachey et al., 1981).

These problems were discussed in detail by Beachey et al. (1986), who carried out experiments on the immune responses in rabbits immunized with synthetic streptococcal M-protein peptides (M5 and M6, containing 20 amino acid residues from the N-terminus of M-protein, and M24, containing two 35-residue peptides) conjugated to polylysine or tetanus toxoid. The location and synthesis of protective epitopes of type 5 streptococcal M-protein were studied by Seyer et al. (1986). Six synthetic peptides were prepared from M5 region 1-35 and tested. The segment SM5 (1-35) was immunogenic in rabbits and evoked opsonic antibodies against type 5 streptococci, that did not negatively cross-react with human tissues. All of the protective epitopes resided between amino acid residues 14 and 26. Kühnemund et al. (1986) studied types 1 and 12 of group A streptococci. The M1 and M12 proteins were prepared using phage-associated lysin (PAL) extraction and affinity chromatography. The maps of tryptic digests of samples of the two proteins were compared by two-dimensional TLC peptide mapping and nine common spots were found. A mild pepsin extraction to isolate M1 protein led to a protein fragment of M_r = 20 000, which (following affinity chromatography on immobilized fibrinogen) was identified as the N-terminal part of M-protein. Morávek et al.

(1986) further studied this purified fragment of M1 protein, which represents the exposed area of the molecule on the surface fibrils of streptococcal cells, which seems to be very important for differentiation of the individual serological types. The sequence of 39 amino acid residues was determined and the homology with the N-terminus of M5, M6 and M24 proteins was discussed, and also with the internal homology in repeated areas of M24 protein. All the mentioned results provide a great impulse to efforts for the preparation of synthetic vaccines.

In the second part of this section, synthetic antiviral vaccines (Brown, 1984) will be mentioned and the examples will be focused on influenza viruses. Synthetic vaccines against influenza viruses, which are of great public-health importance, are a great challenge for organic chemists synthesizing peptides. Their efforts are directed to the preparation of segments mimicking viral surface proteins which elicit the production of antibodies in the attacked macroorganism. There are two surface glycoproteins in influenza virus, haemagglutinin and neuramidinase. The first protein is the major viral surface antigen and will be dealt with later (cf., Müller et al., 1982; Shapira et al., 1984; Wabuke-Bunoti, 1984a,b; Hamšíková et al., 1986, 1987). It is strain-specific and can be seen by electron microscopy to form spikes protruding from the spherical lipid envelope in the form of a trimer; the monomer consists of two polypeptide chains denoted HA-1 and HA-2. A concise chronological review of viral antigen examinations will illustrate the research efforts that have been made and reasons for the selection of particular peptide sequences for experiments.

As early as in 1964 Laver reported structural studies on the protein subunits from three strains of influenza virus. Brand and Skehel (1972) described crystalline antigen from the influenza virus envelope. Jackson et al. (1979) studied the antigenic determinant of influenza virus in detail and described the immunogenicity of fragments isolated from haemagglutinin of A/Memphis/72 virus. The cyanogen bromide cleavage fragment (designated CN-1), prepared from the heavy-chain subunit HA-1, was responsible for immunogenic activity, but it was too large for synthesis because it contained 170 amino acid residues. Wilson et al. (1981) presented the three-dimensional structure of the haemagglutinin membrane glycoprotein determined by X-ray crystallography at 3 Å resolution. Wiley et al. (1981) identified the structures of the antibody-binding sites of Hong Kong influenza haemagglutinin and their involvement in the antigenic variations. Ward (1981) reviewed the structure of the influenza virus haemagglutinin; it was apparent that the fragment Ser 91 to Leu 108 of CN-1 (from HA-1) contained two Pro and three Tyr residues (important for antigenicity) and a folder corner in the X-ray structure; this was suggesting because there was a chance that this area might be hopeful for synthetic efforts. Müller et al. (1982) reported an anti-influenza response achieved by immunization with a synthetic conjugate containing the mentioned amino acid sequence. Atassi and Kurisaki (1984) published a novel approach to the location of the continuous protein antigenic sites by comprehensive synthetic surface scanning: antibody and T-cell activity to several influenza haemagglutinin synthetic sites. Arnon and Shapira (1984) dealt with anti-influenza synthetic vaccines and reported that at least twelve H3N2 influenza virus strains contained the above-mentioned sequence Ser 91 to Leu 108, so that it is very common in influenza viruses. Synthetic peptides corresponding to the antigenic site of haemagglutinin of various length were described by Shapira et al. (1984), and Wabuke-Bunoti et al. (1984a,b) studied cytolytic T-lymphocytes in relation to antibody responses to synthetic peptides of influenza virus haemagglutinin. Shapira et al. (1985) described a synthetic vaccine against influenza with built-in adjuvant (HA fragment 91-108 was conjugated to tetanus toxoid and then to MDP-Lys) and reported protective activity in mice immunized with such a system.

Bláha et al. (1986) synthesized by a solid-phase method haemagglutinin HA-2 fragments Ser 91 to Leu 108 (corresponding to H3N2-type haemagglutinin) in which Cys 97 was replaced with Ala 97 (peptide I) or Met 97 (peptide II), and HA-2 (1-13) (peptide III), and HA-1 (185-200) (peptide IV). Peptide I was identical with the Müller et al.'s (1982) peptide. For the purification of I, II and IV a column of Partisil ODS-2 or Separon Si C_{18} was used, eluted with a gradient 30-80% of methanol-water (+0.1% trifluoroacetic acid). All peptides were checked by elemental and amino acid analysis, Edman degradation and peptide mapping and behaved as homogeneous in TLC and analytical HPLC. With the exception of III the peptides were conjugated to tetanus toxoid and mixed with Freund's adjuvant. Hamšíková et al. (1986, 1987) studied the immunogenic properties of the above-mentioned synthetic peptides I and II. The induced antibodies reacted with synthetic peptides (coupled to bovine serum albumin) and with the haemagglutinins of the homotypic H3 viruses, but not with the haemagglutinins of the heterotypic influenza A viruses (when using purified virions as antigens, marked cross-reactivity was observed). The petide II conjugates appeared to be a more potent immunogen then I (probably owing to the presence of Met). None of the antibodies raised against the peptide conjugates displayed any activity in the haemagglutinin-inhibition neutralization or complement-fixation tests. In addition, in mice immunized with the same preparation, intranasally administrated H3N2 virus replicated to the same extent as in the control animals; these findings did not agree with the observation of Shapira et al. (1985). The data obtained suggest that the region corresponding to Ser 91-Leu 108 is not immunogenic under conditions of natural infection or in comparison with immunization with more complex

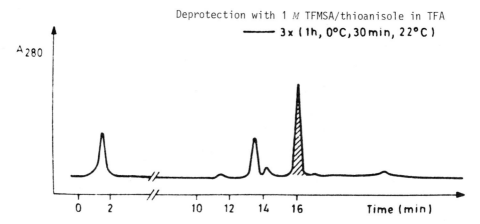

Fig. 12.6. HPLC of deprotected [Tyr]^{161}-VP$_1$(141-161) peptide preparation. For chromatographic conditions, see text. TSMFA = trifluoromethanesulphonic acid. (Reprinted from Dölling et al., 1986.)

virus material. It can be seen from this short history that the preparation of an effective synthetic vaccine will not be easy.

Dölling et al. (1986) described the synthesis of peptide sequences belonging to foot-and-mouth disease virus (FMDV) using a biphasic solvent system. The major immunogenic region of FMDV is located on the viral capsid protein VP$_1$ in the amino acid sequence 141-160 (Pfaff et al., 1982) with the accent on the Leu residues 148 and 151 of the immunogenic epitope 146-152 (Geysen et al., 1984). So far several attempts to prepare peptides from this region have been published. For references, see Dölling et al. (1986), who used a special solution synthesis for the preparation of VP$_1$ fragment 141-160, C-terminally extended by a tyrosine residue Tyr^{161} in order to allow specific coupling to proteins. The detailed strategy was discussed. For analyses of deprotected peptides a 5-μm Vydac C$_4$ (300 Å) column was used (cf., Fig. 12.6), eluted with a gradient of 2%/min of B in 0-40 min at a flow-rate of 1 ml/min; solutions: A = 0.08% TFA in water - n-propanol (9:1); B = 0.05% TFA in water - n-propanol (1:1).

The Tyr^{161} peptide was linked with human serum albumin using bis-diazotized benzidine. In preliminary experiments, the peptide showed good reactivity with an antiserum directed toward the complete virus, but immunization with the albumin-conjugate (peptide to protein ratio = 8:1) gave no virus neutralizing antibodies.

Krchňák and Malý (1986) used modern technology to prepare and test synthetic antigen fragments in which the isolation of the real intact-protein antigen was not necessary. The principle of this approach can be illustrated by the following steps: (1) determination of the DNA sequence expressing the antigen; (2) selection of the corresponding protein fragment(s) based on theoretical translation;

(3) synthetsis of the selected peptide(s); (4) preparation of hapten-carrier conjugate; (5) immunization of experimental animals; (6) titration of the antibody; (7) isolation of the peptide-specific antibody (may be omitted); and (8) study of the interaction of the antibody with the protein. Krchňák and Malý (1986) used this method to study the antigenic properties of oncoprotein of avian myeloblastosis virus (AMV). AMV is a member of the avian retroviruses, which are deffective for replication and cause leukaemia. The AMV oncogens are expressed in the transformed cell in the form of subgenomic spliced mRNA (which is also present in the virion) and the product of its translation is a protein of M_r = 48 000 (Boyle et al., 1983; Klempnauer et al., 1983; Malý and Krchňák, 1984).

Krchňák and Malý (1986) started from the published nucleotide sequence of the transforming gene (Rushlow et al., 1982) and selected the fragment 92-110, expected to be the most hopeful. The nonadecapeptide was synthesized by the Merrifield solid-phase method. The product was purified on Bio-Gel P-4 and by HPLC. The homogeneous preparation was coupled to keyhole limpet haemocyanin as the carrier, emulsified in complete Freund's adjuvant and administered intra-cutaneously to rabbits. High titres were obtained. The detailed immunological assays were discussed.

Mach et al. (1986) studied antibodies against a synthetic decapeptide pre-cipitating phosphokinase (measured by enzymic activity) expressed by src gene, which had been originally detected in Rous sarcoma virus. Oncogenes (i.e., genes responsible for the induction and maintenance of the transformed state of a cell) are intensively studied, because they give rise to tumours not only in model systems, but even in humans. The detection of protein products expressed by pathological genes permits their easier investigation and specific antibodies can be simply used for this purpose.

Owing to difficulties in the preparation of a potent specific antibody against the products of src gene, new ways were sought and fragments of phosphokinase (M_r = 60 000) were believed to be useful for the synthesis of antigens. Because the amino acid sequence of pp60^{src} was known, Mach et al. selected a surface hydrophilic domain, which could be characterized by a decapeptide containing a Tyr residue. This decapeptide was prepared by solid-phase synthesis and purified to 99% purity by HPLC. The peptide was coupled to serum albumin as a carrier and used for immunization of rabbits using Freund's complete adjuvant. All the immunized rabbits produced antibodies against the synthetic peptide in titres from 1:5000 to 1:10 000. The antibody against the synthetic peptide reacted even with native protein pp60^{src}. The amino acid sequence of the decapeptide used was compared with sequences of four other published heptapeptides (corresponding to different regions of the protein) eliciting antibodies which reacted with native pp60^{src}.

12.3.5 Comments on literature

First the literature relating to glycopeptide adjuvants will be summarized.
Kotani et al. (1982) dealt with the non-specific and antigen-specific stimulation of host defence mechanism by liophilic derivatives of muramyl dipeptides,
Leclerc et al. (1983) reviewed the potential use of synthetic muramyl peptides
as immunoregulating molecules, Straka (1983) the preparation of fragments of
bacterial cell wall peptidoglycan and analogues (in Czech; 280 references),
Adam and Lederer (1984) muramyl peptides, immunomodulators, sleep factors and
vitamins and Takada and Kotani (1985) immunopharmaceutical activities of synthetic muramyl peptides. Ježek (1986) (in Czech, 234 references) dealt with
glycopeptides of bacterial cell walls from the synthetic point of view. Lederer
(1986) presented an opening lecture, "Muramyl Peptides and Their Use in Vaccines",
at a Symposium and its expanded version contained 133 references. Books on this
theme have also been published by Friedman et al. (1981) on immunomodulation by
bacteria and by Adam (1985) on synthetic adjuvants, and Zaoral et al. (1986)
edited Symposium Proceedings on synthetic immunomodulators and vaccines.

Secondly, a survey of the literature relating to the problem of the prediction of peptide antigenic determinants will be given, because of its importance
with regard to synthetic approaches. Hopp and Woods (1981) described the prediction of protein antigenic determinants from amino acid sequences, and Kyte and
Doolitle (1982) a simple method for displaying the hydropathic character of
proteins. Sutcliffe et al. (1983) dealt with antibodies that reacted with
predetermined sites on proteins, and Berzofsky (1985) intrinsic and extrinsic
factors in protein antigenic structure. Tainer et al. (1985) reviewed the atomic
mobility component of protein antigenicity, and Jemmerson and Paterson (1986a)
the mapping of antigenic sites on proteins with implications for the design of
synthetic vaccines.

Thirdly, several reviews on synthetic vaccines will be summarized. Shinnick
et al. (1983) dealt with synthetic peptide immunogens as vaccines, Benjamin et
al. (1984) have written a reappraisal of the antigenic structure of proteins,
Arnon (1984) reviewed synthetic vaccines, Brown (1984) presented a specialized
review on synthetic antiviral vaccines and Walter (1986) commented the production and use of antibodies against synthetic peptides. Rowland (1986) reviewed
the synthetic antigen approach to the production of vaccines and a new generation of antiviral vaccines was discussed by Bittle (1986).

12.4 MISCELLANEOUS GLYCOPEPTIDES

12.4.1 *Viral glycopeptides*

Hollaway et al. (1979) described a procedure for the rapid separation of
Rauscher murine leukaemia virus type I and II glycopeptides, which consisted of
one to three amino acid residues and a large oligosaccharide chain. The desalted
glycosylated components obtained from radiolabelled virions or from gp70 viral
glycoprotein were dissolved in solvent A, i.e., in 0.1% acetic acid (pH 3.2) or
0.1% phosphoric acid (pH 2.85) and subjected to IP-RPC using a µBondapak C_{18}
column (30 cm x 4 mm I.D.) and brief isocratic elution, followed by a linear
gradient up to 30% acetonitrile (solvent B) at ambient temperature and a flow-
rate of 2 ml/min. The $[^3H]$glucosamine or $[^3H]$mannose employed for labelling were
detected by measurement of their decompositions per minute. The isolated components could be re-chromatographed using a similar procedure. The utilization
of an ion-pair technique eliminated the necessity for derivatization of the
primary amino group and hence the subsequent analysis of the sample was simple.

Basak and Compans (1981a) dealt with variations of glycosylation sites in
H_1N_1 strains of influenza virus. The glycosylated sites of haemagglutinin
glycoprotein of series of H_1N_1 strains (after pronase digestion) were compared
by GPC of glycopeptides on a 115 cm x 1 cm I.D. Bio-Gel P-6 column, and also
(after tryptic digestion) by mapping using RP-HPLC on a Waters µBondapak C_{18}
column, eluted with a 0.1% phosphoric acid-*n*-propanol (70:30) gradient. HPLC was
found to be effective for the detection of changes in glycosylation sites.
Basak and Compans (1981b) published an improved separation of glycosylated
tryptic peptides of haemagglutinin (HA) glycoprotein of influenza virus. For
the RPC a Waters µBondapak C_{18} column (30 cm x 3.9 mm I.D.) was eluted using a
gradient from 0.1% phosphoric acid (solvent A) to either acetonitrile (containing 0.1% of phosphoric acid)-0.1% phosphoric acid (40:60) or *n*-propanol
(containing 0.1% of phosphoric acid)-0.1% phosphoric acid (30:70) at room
temperature and a flow-rate of 2 ml/min; the gradient time was 150 min. Radiolabelled fractions were mixed with Scintiverse (Fisher Scientific) and counted
in a Beckman scintillation counter. Excellent separation was achieved with a
recovery of 90-95%. When reduction and alkylation were carried out before
tryptic digestion of HA glycoprotein, eight tryptic peptide classes originating
from HA glycoprotein of the A/USSS-R/90/77 virus strain were perfectly resolved.

Kemp et al. (1981) described the IP-RPC of viral tryptic glycoproteins from
influenza A/WSN(H_0N_1) and mink cell focus (MCF)-inducing (MCF-247) murine
leukaemia virus. The chromatography was developed for mapping purposes. Hydrophilic ion pairing was accomplished using 0.1% phosphoric acid on µBondapak and

LiChrosorb RP-18 ODS columns. Solvent A was 0.1% phosphoric acid (pH 2.85) and solvent B was 60% acetonitrile in solvent A. A brief isocratic elution (10 min) was followed by a linear gradient starting from 0% acetonitrile to 40% aceto-nitrile at a flow-rate of 2 ml/min, or starting from 12.5% acetonitrile to the final concentration of 35% at a flow-rate of 1 ml/min. The glycopeptides originating from [^{3}H]mannose-labelled cells and the radioactivity of 1-min fractions were determined by liquid scintillation countings. The chromatogram of the total HA tryptic glycopeptides was compared with those of HA$_1$ and HA$_2$ tryptic glycopeptides, prepared from haemagglutinin previously separated into HA$_1$ and HA$_2$ glycoproteins, which differed in electrophoresis. The recovery of glycopeptides was in the range 70-80%.

12.4.2 Glycopeptides indicating lysosomal storage disorders

Calatroni and Tira (1976) reported the isolation of acidic glycopeptides by means of an anion-exchange resin (Dowex 1) and the application of this method to some cases of glycosphingolipidosis or mucolipidosis.

Ng Ying-Kin and Wolfe (1980) described the HPLC of oligosaccharides and glycopeptides accumulating in lysosomal storage disorders. These types of com-pounds are excreted in the urine of a number of lysosomal storage disorders. The diseases studied were mannosidosis, fucosidosis, G$_{M1}$-gangliosidosis, G$_{M2}$-gan-gliosidosis variant 0, sialidosis, and aspartylglucosaminuria. A table was published giving the structures of oligosaccharides and glycopeptides accumulat-ing in tissues and excreted in the urine of patients suffering from the above diseases. A 30 cm x 4 mm I.D. µBondapak/carbohydrate column was isocratically eluted with acetonitrile-1% acetic acid (60:40, v/v) for sialic acid-free glycopeptides and acetonitrile-0.1 M sodium acetate-acetic acid buffer (pH 5.6) (54:45, v/v) for sialyl oligosaccharides and glycopeptides. Refractive index measurement was used for monitoring. A large detector response in the first 3 min represented the low-molecular-weight material present in the urine. The chromatographic separation could be completed in 20 min.

Lou et al. (1980) reported a split-stream cation-exchange chromatographic method for isolating glycopeptides from biological fluids, using Aminex A-7 in a carbohydrate analyser and citrate buffers for elution. Both ninhydrin and orcinol reaction patterns could be obtained.

12.5 MISCELLANEOUS POLYMERIC AND OLIGOMERIC SUBSTANCES

12.5.1 Complex carbohydrates

The chromatography of polysaccharide-protein complexes was reviewed by
Juřicová and Deyl (1975); glycosaminoglycans (mucopolysaccharides), glycoproteins
and glycopeptides were dealt with. The separation of dolichylpyrophosphoryl
oligosaccharides by LC was studied by Wells et al. (1981). Fourteen compounds of
this type (precursors of the asparagine-linked oligosaccharides in proteins)
were separated on base-treated (Smith and Lester, 1974) silicic acid Porasil
A-60 (37-75 μm) (Waters Assoc., Milford, MA, U.S.A.), packed into six 100 cm x
0.32 cm I.D. columns linked in series. In some instances a 3.5 cm x 0.32 cm I.D.
pre-column was used. Up to 500 μl of sample dissolved in chloroform-methanol-
water (10:10:3) or in chloroform-methanol-concentrated ammonia solution-water
(16:16:1:4) was applied to the column equilibrated with the initial solvent,
which was then pumped for 5 min prior to the gradient. The flow-rate was 2.0
ml/min and pressures of 1000 and 2000 p.s.i. were applied.

For the separation of dolichylpyrophosphoryl-N,N'-diacetylchitobiose, a
programmed non-linear gradient was used, formed by mixing solvent A (chloroform-
methanol-concentrated ammonia solution, 65:29:6) and solvent B (chloroform-
methanol-concentrated ammonia solution-water, 43:43:6:8). The total gradient
time was 60 min, at which the solvent composition was 65% A - 35% B, and this
mobile phase was pumped isocratically for a further 30 min. The gradient defined
the composition at t min representing a percentage of solvent B of $35(t/60)^{1/3}$.
Separation of more complicated dolichylpyrophosphoryl oligosaccharides was
carried out with a programmed non-linear gradient formed with solvent A and
solvent C (chloroform-methanol-concentrated ammonia solution-water, 40:42:6:12).
The gradient was run for 120 min from 0 to 100% solvent C, followed by pumping
solvent C for 75 min isocratically. In this instance the solvent composition
during the gradient at t min was represented by a percentage of solvent C of
$100(t/120)^{1/2}$. The complex oligosaccharides thus resolved retained their function as substrates in enzyme-catalysed reactions.

The chromatographic procedures described for the first time made available
many of these single intermediates for further study. Wells et al. (1982)
reviewed the resolution of dolichylpyrophosphoryl oligosaccharides by HPLC.

Roughley and Mort (1985) described the resolution of cartilage proteoglycan
and its proteolytic degradation products by HPLC.

12.5.2 Pteroyl oligoglutamates

Pteroyl oligo-γ-L-glutamates (PtGlu$_n$) are folic acid (FA) derivatives and have the general formula

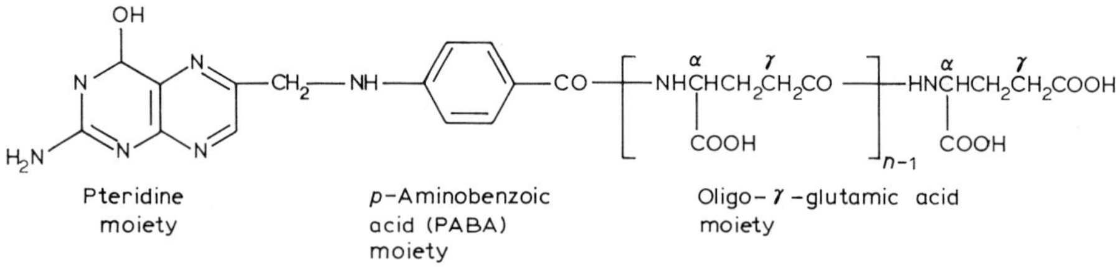

| Pteridine moiety | p-Aminobenzoic acid (PABA) moiety | Oligo-γ-glutamic acid moiety |

(for the parent FA $n=$ 1 and for Pt-oligoglutamates $n=$ 2-9), and have important biological functions as coenzymes essential in the synthesis of proteins and nucleic acids (Baugh and Drumendieck, 1971) and other functions. Because the FA derivatives are acids or oligoacids, IEC on microparticulate bonded phases was used for their rapid separation first (Reed and Archer, 1976; Stout et al., 1976). Naturally, the retention of PtGlu$_n$ in anion-exchange chromatography depends on the number of carboxyl groups, being highest with species containing most Glu residues.

Bush et al. (1979) studied the retention behaviour of PtGlu$_n$ in RPC on an ODS-silica column and found this method to be very suitable for good resolution and rapid analysis. When the carboxylic groups are largely undissociated (at pH 2), the retention of PtGlu$_n$ increases with the number of Glu residues and the elution order parallels that in IEC. At sufficiently high pH (e.g., at pH 4.5) the carboxilic groups are dissociated and the elution order is reversed (cf., Fig. 12.7). The logarithm of the capacity factor is linearly dependent (with the exception of FA) on the number of Glu residues over a wide range of eluent pH (cf., Fig. 4 in the original paper). As can be seen from Fig. 12.8, at pH 6 (at which the carboxyl groups are almost completely dissociated) the retention decreases with increasing number of Glu residues, so that both methods (IEC and RPC) complement each other. In general, the efficiency of IEC with bonded phases appears to be higher than that of RPC at low pH on ODS with the same particle size and column dimensions. IEC is recommended when the elution

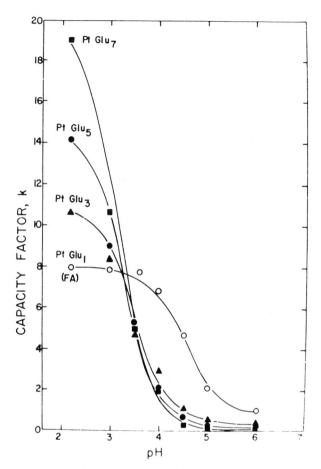

Fig. 12.7. Capacity factors of pteroyl-γ-glutamates as a function of the eluent pH in RPC. Column, 5-µm Partisil ODS-2; eluent, 0.1 M phosphate buffer containing 6% (v/v) acetonitrile; temperature, 45°C. (Reprinted from Bush et al., 1979.)

order of increasing numbers of Glu residues is desired. RPC is eminently suitable for separations with a reversed elution order, as at high eluent pH adequate resolution can be obtained by isocratic or (with more complex mixtures) gradient elution.

Cashmore et al. (1980) reviewed the separation of PtGlu$_n$ by HPLC. Both RPC and IEC methods were described in detail and compared. The conditions for the IEC of PtGlu$_n$ and PABA-oligo-γ-L-glutamates were as follows: a 25 cm x 4.6 mm I.D. column of 10-µm Partisil 10 SAX (Whatman, Clifton, NJ, U.S.A.), eluted with a linear gradient from 0.01 to 1.0 M sodium phosphate (pH 3.3); temperature, 50°C; inlet pressure at the start, 300 lb/in^2; flow-rate, 0.8 ml/min; sample, 3-10 µg of each component in 0.1 M phosphate buffer (pH 3.0); detection, UV (280 nm); and time required, 50 min.

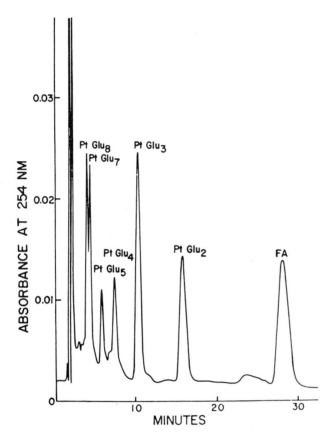

Fig. 12.8. RPC of pteroyl-γ-glutamates (PtGlu$_n$) obtained by isocratic elution at pH 6. Column, Spherisorb ODS (250 mm x 4.6 mm I.D.); eluent, 0.1 M phosphate buffer (pH 6) containing 1% (v/v) acetonitrile; flow-rate, 1.5 ml/min; inlet pressure, 11.42 MN/m^2; temperature, 45°C; sample size, ca. 10 μg of each component. FA = Folic acid. (Reprinted from Bush et al., 1979.)

Eto and Krumdieck (1982) studied the determination of three different pools of reduced one-carbon substituted folates and described the RP-HPLC of the azo dye derivatives of p-aminobenzoyl poly-γ-glutamates (AzoGlu$_n$) and its application to the study of unlabelled endogenous PtGlu$_n$ of rat liver. Picomole amounts of the AzoGlu$_n$ could be separated and quantitated. Azo dye derivatives were prepared from the corresponding PABA-γ-glutamates (containing 1-7 Glu residues) by a Braton-Marshall procedure slightly modified by Eto and Krumdieck (1980) and were purified by Bio-Gel P-2 polyacrylamide gel chromatography.

All the analyses were performed on a 25 cm x 0.46 cm I.D. column packed with 5-μm Spherisorb S5 ODS (carbon load approximately 7%) (Applied Science Division, Milton Roy, Philadelphia, PA, U.S.A.). Before sample application the column was

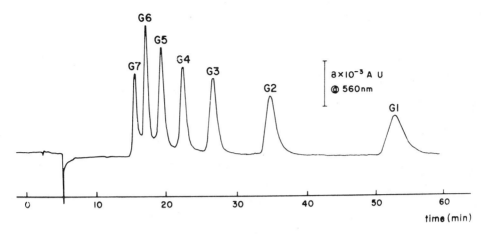

Fig. 12.9. Separation of synthetic azo-p-aminobenzoyl poly-γ-glutamates (AzoGlu$_n$) by RP-HPLC. For conditions, see text. G1-G7 indicate the number of glutamyl residues. (Reprinted from Eto and Krumdieck, 1982.)

routinely washed with 30 ml of solvent B (1-propanol-acetic acid-water, 10:7.5: 82.5) followed by 30 ml of solvent A (acetic acid-water, 7.5:92.5). The lyophilized AzoGlu$_n$ were dissolved in 1.0 ml of solvent A and 20-70 µl were injected. Immediately thereafter the concentration of 1-propanol in the mobile phase was increased to 10% (v/v) within 3 min, switching from solvent A to B employing a steep convex gradient. Isocratic elution followed at a flow-rate of 1 ml/min. UV detection at 560 nm was applied. The resolution was completed within 1 h. In the case of pulse labelling with radioactive FA (carried out by injecting rats intraperitoneally with 1.0 ml of a solution of 0.24 µg of [^{3}H]FA (25 µCi) in physiologic saline), the radioactivity of the isolated products was monitored with a Beckman LS-250 liquid scintillation spectrometer using Scinti-Verse (Fisher Scientific) scintillation fluid. An example of the HPLC elution pattern is illustrated in Fig. 12.9.

12.5.3 Oligomers containing nucleobases

Fiedler (1981a) described the quantitation of nikkomycins in biological fluids by IP-RPC. These antibiotics are produced by *Streptomyces tendae* and are interesting from an agricultural point of view because they possess high and selective antifungal and insecticidal activity (they interfere in the biosynthesis of chitin cell walls). Nikkomycins are nucleoside-peptide antibiotics, similar to polyoxins. Using a 120 mm x 4.6 mm I.D. column packed with 5-µm LiChrosorb RP-8, equipped with a 40 mm x 4.6 mm I.D. pre-column (Knauer, West

Berlin, F.R.G.), Fiedler was able to determine five nikkomycin components in
the filtrate of the fermentation broth in which they were cultured.

Both types of elution were used for effective separation. For isocratic elution the mobile phase employed contained 80 mM ammonium formate buffer (pH 4.7)
and 1 mM heptanesulphonic acid. The flow-rate was 1 ml/min and the pressure was
100 bar. For gradient elution, solvent A was 30 mM ammonium formate (pH 3.75)
containing 1 mM heptanesulphonic acid and solvent B was methanol containing
30 mM ammonium formate (pH 3.75) and 1 mM heptanesulphonic acid. The flow-rate
was 1 ml/min. UV detection at 290 nm was applied. A 10-μl volume of clear
supernatant of the fermentation broth was injected. When using gradient elution,
only a short equilibration time under initial conditions was necessary to achieve
reproducible results. In isocratic elution the column had to be equilibrated for
3 days after the regeneration with methanol in order to obtain reproducible
results; about 300 injections could be performed before regeneration of the
column and when using a pre-column about 3000 injections could be effected
(however, it was necessary to change the pre-column after 500 injections). The
advantages of HPLC of nikkomycins were obvious.

Van Haastert (1981) studied the HPLC of nucleobases, nucleosides and nucleotides, and in this cited second paper of the series the mobile phase composition for the separation of charged solutes by IEC was described. He examined
the application of a cation exchanger (Partisil 10 SCX) and an anion exchanger
(Partisil 10 SAX) and concluded that these ion exchangers also had reversed-
phase and normal-phase properties; their occurrence was determined by the
polarity of the mobile phase. In addition to various tested compounds,
S-adenoxyl-L-methionine (SAM) was chromatographed on a 250 mm x 4.6 mm I.D.
column of Partisil 10 SCX, using as the mobile phase 1.1 M ammonium acetate
(pH 5.6) and a flow-rate of 2 ml/min. SAM was separated into two peaks (stereoisomers at the sulphonium S$^+$ atom), SAM I = (-)-S-adenysyl-L-methionine and
SAM II = (+)-S-adenosyl-L-methionine. The two peaks were interconvertible by
heating. The twin peaks appeared only if reversed-phase and cation-exchange
properties of the support were present simultaneously, i.e., in the mixed-mode
chromatography process. An increase in buffer concentration or the addition of
propanol shifted the peak of SAM in the opposite direction.

Yip and Albarella (1985) described the use of HPLC in the preparation of
flavin adenine dinucleotide conjugates, and Ehrat et al. (1985), in the context
of research on neonucleoproteins, described the preparation and HPLC characterization of succinyl-lysozyme-diaminooctylpolycytidylic acid and realted polycytidylic acid conjugates.

12.5.4 Miscellaneous substances

Species of complex glycosides, reviewed by Courtois and Percheron (1970), also belong to biooligomers, and examples of their separation and detection will be mentioned in this section. Erni and Frei (1977) described a comparison of reversed-phase and partition or adsorption HPLC of some digitalis glycosides. Digitalis glycosides of the cardenolide groups are important drugs for the treatment of heart diseases. The authors attempted to find a simple isocratic system for a good separation with sufficient reproducibility. RPC could offer some advantages with regard to sample preparation of pharmaceutical formulations. Digitalis glycosides of the C series contain (on the hydroxy group in position 3 of the steroid skeleton) chains of repeated digitose or acetyldigitose residues, sometimes bound to glucose. A chromatogram of this group of compounds is illustrated in Fig. 12.10. The detection limits for digitalis glycosides in the abovementioned paper varied between 10 and 100 ng per injection and permitted the analysis of by-products even in low-dosage pharmaceutical formulations. The reproducibility for repetitive chromatograms was about 1% (relative standard deviation); significantly better results were obtained with automatic injections.

In Nibbering's (1982) review on ionization methods with emphasis on liquid chromatography-mass spectrometry, an analysis of oligomeric purpureaglycoside A (M_r = 926) was mentioned in addition to other glycosides, according to papers by Bruins (1980a,b).

Saito and Hayano (1979) examined the application of HP-GPC to humic substances in marine sediments. Humic and fulvic acids were extracted with 0.1 M $Na_4P_2O_7$-0.1 M NaOH solution from the marine sediment obtained from Sagami Bay (Japan) at a depth of 88 m and 11 km from the shore. The humic acids were preliminarily purified and injected (in volumes of 0.025-0.1 ml) into a TSK Gel G 3000 SW column (60 cm x 0.75 cm I.D.). Pure water and 0.1 M NaCl were used for elution at flow-rates of 0.5 and 1.0 ml/min. The pressure for a 1.0 ml/min flow-rate was 30 atm. UV detection at 254 nm was applied. The void volume (V_o) and the total effective column volume ($V_o + V_i$) were determined using Blue Dextran 2000 and acetone, respectively. In water the coulombic repulsion forces shifted the chromatographed compounds to the void volume, but in 0.1 M NaCl a normal chromatogram was obtained containing the eluted compounds between V_o and $V_o + V_i$. Part of the humic acid was of the higher molecular weight type and required a more porous support.

Melander et al. (1979) studied mobile phase effects in RPC and described changes in the conformation and retention of oligo(ethylene glycol) derivatives with temperature and eluent composition. Zorbax ODS and LiChrosorb RP-8 columns

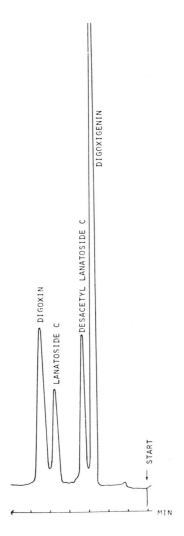

Fig. 12.10. Separation of the C series of digitalis glycosides by RPC. A 30 cm x 3.5 mm I.D. column of 10-μm Nucleosil C$_{18}$ (Machery, Nagel & Co., Düren, F.R.G.) was eluted with a 40% solution of acetonitrile-dioxan (1:1) in water; flow-rate, 1.3 ml/min; pressure, 140 bar; injection, 25-μl loop; detection, 220 nm. (Reprinted from Erni and Frei, 1977.)

were used. The results contributed to the verification of various theoretical approaches to chromatography and could serve as a model for studying the chromatographic behaviour of biological substances such as peptides and oligonucleotides.

Fiedler (1981b) described the preparative-scale HPLC of ferricrocin, a microbial product. The cyclic hexapeptide (containing three acetylated N^{δ}-hydroxyornithine residues, capable of binding one Fe^{3+} ion in a complex bond) is a metabolic product of *Aspergillus viridi-mutans* and was isolated from the fermentation broth by means of XAD-2 adsorption. It is a member of a broad and interesting family of natural hydroxamates (growth factors and antibiotics) (cf., review by Mikeš and Turková, 1964) and close to the structure of the antibiotic albomycin (cf., Turková et al. 1964, 1965) or grisein (Turková et al., 1966), which contains a pyrimidine moiety in addition to cyclic hexapeptide with three hydroxamate residues and one iron atom in a complex bond.

Ferricrocin could be purified to a high degree of homogeneity by means of RPC on a C_8 support. Columns 250 mm x 16 or 4.6 mm I.D. packed with 7-μm LiChrosorb RP-8 eluted with water-acetonitrile (9:1) at a linear flow-rate of 12 cm/min and ambient temperature were used for separation. UV detection at 270 nm was applied. The capacity of the separation was reached with a sample weight of 20 mg per gram of stationary phase.

REFERENCES

Acher, A.J. and Kanfer, J.N., *J. Lipid Res.*, 13 (1972) 139-142.
Adam, A., *Synthetic Adjuvants*, Wiley, New York, 1985, 239 pp.
Adam, A., Ciorbaru, R., Ellouz, F., Petit, J.-F. and Lederer, E., *Biochem. Biophys. Res. Commun.*, 56 (1974) 561-567.
Adam, A. and Lederer, E., *Med. Res. Rev.*, 4 (1984) 111-152.
Amit, A.G., Mariuzza, R.A., Phillips, S.E.V. and Poljak, R.J., *Nature (London)*, 313 (1985) 156-158.
Ando, S., Isobe, M. and Nagai, Y., *Biochim. Biophys. Acta*, 424 (1976a) 98-105.
Ando, S., Kon, K., Isobe, M., Nagai, Y. and Yamakawa, T., *J. Biochem. (Tokyo)*, 79 (1976b) 625-632.
Ando, S. and Yu, R.K., *J. Biol. Chem.*, 252 (1977) 6247-6250.
Ando, S. and Yamakawa, T., *J. Biochem. (Tokyo)*, 73 (1973) 387-396.
Arendt, A., Kolodziejczyk, A. and Sokolowska, T., *Rocz. Chem.*, 47 (1973) 1087-1089 (in English); 48 (1974) 1921-1927 (in English); *C.A.*, 79 1973) 115 878r; 82 (1975) 125 597z.
Arnon, R., in Roitt, I.M. (Editor), *Immune Intervention, Vol. 1, New Trends in Vaccines*, Academic Press, London, 1984, pp. 93-116.
Arnon, R. and Shapira, M., in Chanock, R.M. and Lerner, R.A. (Editors), *Modern Approaches to Vaccines*, Cold Spring Harbour Laboratory, New York, 1984, pp. 109-113.
Atassi, M.Z., *Eur. J. Biochem.*, 145 (1984) 1-20.
Atassi, M.Z. and Kurisaki, J., *Immunol. Commun.*, 13 (1984) 539-551.
Basak, S. and Compans, R.W., in Nayak, D.P. (Editor), *Genetic Variation Among Influenza Viruses, (ICN-UCLA Symposia on Molecular and Cellular Biology*, Vol. 21) Academic Press, New York, 1981a, pp. 253-262.
Basak, S. and Compans, R.W., *J. High Resolut. Chromatogr. Chromatogr. Commun.*, 4 (1981b) 302-304; *C.A.*, 95 (1981) 76 261t.
Baugh, C.M. and Krumdieck, C.L., *Ann. NY Acad. Sci.*, 186 (1971) 7-28; *C.A.*, 76 (1972) 69 195d.
Beachey, E.H., Seyer, J.M. and Dale, J.B., in Zaoral, M., Havlas, Z., Mikeš, O. and Procházka, Ž. (Editors), *Synthetic Immunomodulators and Vaccines*, Institute of Organic Chemistry and Biochemistry, Czechoslovak Academy of Science, Prague, 1986, pp. 189-196.

B366

Beachey, E.H., Seyer, J.M., Dale, J.B., Simpson, W.A. and Kang, A.H., *Nature (London)*, 292 (1984) 457-459.
Beachey, E.H. and Stollerman, G.H., *J. Exp. Med.*, 134 (1971) 351-362.
Benjamin, D.C., Berzofsky, D.A., East, I.J., Gurd, F.R.N., Hannum, C., Leach, S.J., Margoliash, F., Michael, J.G., Miller, A., Prager, E.M., Reichlin, M., Sercarz, E.E., Smith-Gill, S.J., Todd, P.E.E. and Wilson, A.C., *Annu. Rev. Immunol.*, 2 (1984) 67-101.
Berzofsky, J.A., *Science (Washington, D.C.)*, 229 (1985) 932-940.
Bhown, A.S., Bennett, J.C., Mole, J.E. and Hunter, E., *Anal. Biochem.*, 112 (1981) 128-134.
Bittle, J.L., *Vet. Clin. North Am.: Large Anim. Pract.*, 16 (1986) 1247-1257.
Bláha, I., Ježek, J., Zaoral, M., Hamšíková, E., Závadová, H. and Vonka, V., in Zaoral, M., Havlas, Z., Mikeš, O. and Procházka, Ž. (Editors), *Synthetic Immunomodulators and Vaccines*, Institute of Organic Chemistry and Biochemistry, Czechoslovak Academy of Science, Prague, 1986, pp. 203-207.
Blaszczyk, M., Ross, A.H., Ernst, C.S., Marchisio, M., Atkinson, B.F., Pak, K.Y., Steplewski, Z. and Koprowski, H., *Int. J. Cancer*, 33 (1984) 313-318.
Boyle, W.J., Lipsick, J.S., Reddy, E.P. and Baluda, M.A., *Proc. Natl. Acad. Sci. U.S.A.*, 80 (1983) 2834-2838.
Brand, C.M. and Skehel, J.J., *Nature (London)*, 238 (1972) 841-866.
Bremer, E.G., Gross, S.K. and McCluer, R.H., *J. Lipid Res.*, 20 (1979) 1028-1035; *C.A.*, 92 (1979) 054 407f.
Brown, F., *Annu. Rev. Microbiol.*, 38 (1984) 221-235.
Bruins, A.P., *Adv. Mass Spectrom.*, 8 (1980a) 246-254.
Bruins, A.P., *Anal. Chem.*, 52 (1980b) 605-607.
Buděšínský, M., Ježek, J., Krchňák, V., Lebl, M. and Zaoral, M., in Bláha, K. and Maloň, P. (Editors), *Peptides 1982 (Proceeding of 17th European Peptide Symposium, Prague, August 29th-September 3rd, 1982)*, Walter de Gruyter, Berlin, New York, 1983, pp. 305-310.
Bush, B.T., Frenz, J.H., Melander, W.R., Horváth, Cs., Cashmore, A.R., Dreyer, R.N., Knipe, J.O., Coward, J.K. and Bertino, J.R., *J. Chromatogr.*, 168 (1979) 343-353.
Calatroni, A. and Tira, M.E., *Clin. Chim. Acta*, 71 (1976) 137-141; *C.A.*, 85 (1976) 155 932j.
Cashmore, A.R., Dreyer, R.N., Horváth, Cs., Knipe, J.O., Coward, J.K. and Bertino, J.R., *Methods Enzymol.*, 66 (1980) 459-468.
Caspar, C., *Adv. Protein Chem.*, 18 (1963) 37-121.
Čech, M., Jelínková, M. and Čoupek, J., *J. Chromatogr.*, 135 (1977) 435-440.
Chien, J.-L. and Hogan, E.L., *ACS Symp. Ser.*, No. 128 (1980) 135-148; *C.A.*, 93 (1980) 109 351m.
Chien, J.-L. and Hogan, E.L., *J. Biol. Chem.*, 258 (1983) 10727-10730.
Choppin, P.W., Richardson, C.D., Merz, D.C., Hall, W.W. and Scheid, A., *J. Infect. Dis.*, 143 (1981) 352-363; *C.A.*, 95 (1981) 20 971g.
Courtois, J.E. and Percheron, F., in Pigman, W., Horton, D. and Herp, A. (Editors), *The Carbohydrates. Chemistry and Biochemistry*, Vol. IIA, Academic Press, New York, 1970, pp. 213-240.
Dale, J.B. and Beachey, E.H., *J. Exp. Med.*, 161 (1985) 113-122.
Darling, T., Alpert, J., Russel, P., Albert, D.M. and Reid, T.W., *J. Chromatogr.*, 131 (1977) 383-390.
Dölling, R., Winter, R., Schmidt, M., Furkert, J., Otto, H.-A., Liebermann, H. and Bienert, M., in Zaoral, M., Havlas, Z., Mikeš, O. and Procházka, Ž. (Editors), *Synthetic Immunomodulators and Vaccines*, Institute of Organic Chemistry and Biochemistry, Czechoslovak Academy of Science, Prague, 1986, pp. 218-227.
Eberendu, A.R.N. and Venables, B.J., *LC, Liq. Chromatogr. HPLC Mag.*, 3 (1985) 424-432; *C.A.*, 103 (1985) 100 985p.
Edsall, J., in Neurath, H. and Bailey, K. (Editors), *The Proteins*, Academic Press, New York, 1st ed., 1953, p. 549.

Ehrat, M., Cecchini, D.J. and Giese, R.W., *J. Chromatogr.*, 326 (1985) 311-320.
Ellouz, F., Adam, A., Ciorbaru, R. and Lederer, E., *Biochem. Biophys. Res. Commun.*, 59 (1974) 1317-1325.
Erni, F. and Frei, R.W., *J. Chromatogr.*, 130 (1977) 169-180.
Eto, I. and Krumdieck, C.L., *Anal. Biochem.*, 109 (1980) 167-184.
Eto, I. and Krumdieck, C.L., *Anal. Biochem.*, 120 (1982) 323-329; *C.A.*, 96 (1982) 196 006f.
Evans, J.E. and McCluer, R.H., *Biochim. Biophys. Acta*, 270 (1972) 565-569.
Farkaš, J., Ledvina, M., Brokeš, J., Ježek, J., Zajíček, J. and Zaoral, M., in Zaoral, M., Havlas, Z., Mikeš, O. and Procházka, Ž. (Editors), *Synthetic Immunomodulators and Vaccines*, Institute of Organic Chemistry and Biochemistry, Czechoslovak Academy of Science, Prague, 1986, pp. 123-128.
Farkaš, J., Ledvina, M., Brokeš, J., Ježek, J., Zajíček, J. and Zaoral, M., *Carbohydr. Res.*, 163 (1987) 63-72.
Fiedler, H.P., *J. Chromatogr.*, 204 (1981a) 313-318.
Fiedler, H.P., *J. Chromatogr.*, 209 (1981b) 103-106.
Fishman, P.W., *J. Membr. Biol.*, 69 (1982) 85-98.
Folch, J., Arsove, S. and Meath, J.A., *J. Biol. Chem.*, 191 (1951) 819-831.
Folch, J., Lees, M. and Sloane Stanley, G.H., *J. Biol. Chem.*, 226 (1957) 497-509.
Fox, E.N., *Bacteriol. Rev.*, 38 (1974) 57-86.
Fredman, P., in Svennerholm, L., Mandel, P., Dreyfus, H. and Urban, P.-F. (Editors), *Structure and Function of Gangliosides [Proceedings of Symposium on Structure and Function of Gangliosides, Strasbourg (Bischenberg), April 23rd-27th, 1979]; Adv. Exp. Med. Biol.*, 125 (1980) 23-31, Plenum Press, New York, London, 1980; *C.A.*, 93 (1980) 109 896t.
Fredman, P., Nilsson, O., Tayot, J.L. and Svennerholm, L., *Biochim. Biophys. Acta*, 618 (1980) 42-52; *C.A.*, 93 (1980) 21 825h.
Friedman, H., Klein, T.W., Szentivanyi, A. (Editors), *Immunomodulation by Bacteria and Their Products*, Plenum Press, New York, 1981, 308 pp.; *C.A.*, 97 (1982) B 37 435z.
Garg, G.H., Khosla, M.C. and Anand, N., *J. Sci. Ind. Res. B*, 21 (1962) 286-287.
Gazzotti, G., Sonnino, S. and Ghidoni, R., *J. Chromatogr.*, 348 (1985) 371-378.
Gazzotti, G., Sonnino, S., Ghidoni, R., Kirscher, G. and Tettamanti, G., *J. Neurosci. Res.*, 12 (1984a) 179-192.
Gazzotti, G., Sonnino, S., Ghidoni, R., Orlando, P. and Tettamanti, G., *Glycoconjugate J.*, 1 (1984b) 111-121.
Geysen, H.M., Meloen, R.H. and Barteling, S.I., *Proc. Natl. Acad. Sci. U.S.A.*, 81 (1984) 3998-4002.
Ghidoni, R., Sonnino, S., Tettamanti, G., Baumann, G., Reuter, G. and Schauer, R., *J. Biol. Chem.*, 255 (1980) 6990-6995.
Gonwa, T.A., Westrick, M.A. and Macher, B.A., *Cancer Res.*, 44 (1984) 3467-3470.
Gravotta, D. and Maccioni, H.J.F., *Biochem. J.*, 225 (1985) 713-721.
Gray, G.M., *Biochim. Biophys. Acta*, 144 (1967) 501-510.
Gross, S.K. and McCluer, R.H., *Anal. Biochem.*, 102 (1980) 429-433.
Gschwender, H.G., Haller, W. and Hofschneider, P.H., *Biochim. Biophys. Acta*, 190 (1969) 460-469.
Gupta, C.M., Radhakrishnan, R. and Khorana, H.G., *Proc. Natl. Acad. Sci. U.S.A.*, 74 (1977) 4315-4319.
Hakomori, S., *Annu. Rev. Biochem.*, 50 (1981) 733-764; *C.A.*, 95 (1981) 110236a.
Hakomori, S. and Kannagi, R., *J. Natl. Cancer Inst.*, 71 (1983) 231-251; *C.A.*, 99 (1983) 137 551u.
Hakomori, S.-i., Nudelman, E., Kannagi, R. and Levery, S.B., *Biochem. Biophys. Res. Commun.*, 109 (1982) 35-44.
Haller, W., *Nature (London)*, 206 (1965) 693-696.
Haller, W., *Virology*, 33 (1967) 740-743.
Halls, T.D.J., Raju, M.S., Wenkert, E., Zuber, M., Lefrancier, P. and Lederer, E., *Carbohydr. Res.*, 81 (1980) 173-176.
Hamšíková, E., Závadová, H., Zaoral, M., Ježek, J., Bláha, I. and Vonka, V., *J. Gen. Virol.*, 68 (1987) 2249-2252.

Hamšíková, E., Závadová, H., Zaoral, M., Ježek, J. and Vonka, V., in Zaoral, M., Havlas, Z., Mikeš, O. and Procházka, Ž. (Editors), *Synthetic Immunomodulators and Vaccines*, Institute of Organic Chemistry and Biochemistry, Czechoslovak Academy of Science, Prague, 1986, pp. 208-217.

Hanahan, D.J., Ekholm, J.E. and Benson, B., *Biochim. Biophys. Acta*, 231 (1971) 343-359.

Hawk, G.L., Cameron, J.A. and DuFault, L., *Prep. Biochem.*, 2 (1972) 193-203; *C.A.*, 77 (1972) 58 423w.

Henderson, L.E., Sowder, R. and Oroszlan, S., in Liu, T.-Y., Schechter, A.N., Heirikson, R.L. and Condliffe, P.G. (Editors), *Chemical Synthesis and Sequencing of Peptides and Proteins*, Elsevier, Amsterdam, 1981, pp. 251-260.

Heukeshoven, J. and Dernick, R., *J. Chromatogr.*, 326 (1985) 91-101.

Hiatt, C.W., Shelokov, A., Rosenthal, E.J. and Galimore, J.M., *J. Chromatogr.*, 56 (1971) 362-364.

Himmel, M.E. and Squire, P.G., *J. Chromatogr.*, 210 (1981) 443-452.

Hollaway, W.L., Kemp, M.C., Bhown, A.S., Compans, R.W. and Bennett, J.C., *J. High Resolut. Chromatogr. Chromatogr. Commun.*, 2 (1979) 149-151; *C.A.*, 91 (1979) 153 702j.

Holmes, E.H. and Hakomori, S.-I., *J. Biol. Chem.*, 257 (1982) 7698-7703.

Hopp, T.P. and Woods, K.R., *Proc. Natl. Acad. Sci. U.S.A.*, 78 (1981) 3824-3828.

IUPAC-IUB Commision on Biochemical Nomenclature, *The Nomenclature of Lipids*, Lipids, 12 (1977) 455-468; *C.A.*, 87 (1977) 147 491r; *J. Lipid Res.*, 19 (1978) 114-128; *J. Biol. Chem.*, 257 (1982) 3347-3351.

Ivanov, V.T. in Zaoral, M., Havlas, Z., Mikeš, O. and Procházka, Ž. (Editors), *Synthetic Immunomodulators and Vaccines*, Institute of Organic Chemistry and Biochemistry, Czechoslovak Academy of Science, Prague, 1986, pp. 174-188.

Ivanov, V.T., Rostovtseva, L.I., Andronova, T.M., Sorokina, I.B., Mal'kova, V.P., Gavrilov, Yu.D., Noskov, F.S. and Fridman, E.A., in Brundfeld, K. (Editor), *Peptides 80. Proceedings of 16th European Peptide Symposium (Helsingör, 1980)*, Scriptor, Copenhagen, 1981, pp. 494-500.

Iwamory, M. and Nagai, Y., *Biochim. Biophys. Acta*, 528 (1978a) 257-267.

Iwamori, M. and Nagai, Y., *J. Biochem. (Tokyo)*, 84 (1978b) 1601-1608.

Iwamori, M. and Nagai, Y., *Biochim. Biophys. Acta*, 665 (1981a) 205-213.

Iwamori, M. and Nagai, Y., *Biochim. Biophys. Acta*, 665 (1981b) 214-220.

Iwamori, M. and Nagai, Y., *J. Biochem. (Tokyo)*, 89 (1981c) 1253-1264.

Jackson, D.C., Brown, L.E., White, D.O., Dopheide, T.A.A. and Ward, C.W., *J. Immunol.*, 123 (1979) 2610-2617.

Jemmerson, R. and Paterson, Y., *Bio Techniques*, 4 (1986a) 18-30.

Jemmerson, R. and Paterson, Y., *Science (Washington, D.C.)*, 232 (1986b) 1001-1004.

Ježek, J., *Synthesa Glycopeptidů Bakteriálních Stěn (in Czech; Synthesis of Glycopeptides of Bacterial Cell Walls)*, Dissertation Thesis for CSc Degree, Institute of Organic Chemistry and Biochemistry, Czechoslovak Academy of Science, Prague, 1981, 113 pp.

Ježek, J., *Glykopeptidy Bakteriálních Stěn (Glycopeptides of Bacterial Cell Walls; a review in Czech containing 234 references)*, *Chem. Listy*, 80 (1986) 337-365.

Ježek, J., Straka, R., Krchňák, V., Ryba, M., Rotta, J., Mayer, P. and Zaoral, M., *Collect. Czech. Chem. Commun.*, 52 (1987) 1609-1624.

Ježek, J., Zaoral, M., Straka, R., Rotta, J. and Krchňák, V., in Zaoral, M., Havlas, Z., Mikeš, O. and Procházka, Ž. (Editors), *Synthetic Immunomodulators and Vaccines*, Institute of Organic Chemistry and Biochemistry, Czechoslovak Academy of Science, Prague, 1986, pp. 112-122.

Jungalwala, F.B., Hayes, L. and McCluer, R.H., *J. Lipid Res.*, 18 (1977) 285-292.

Juřicová, M. and Deyl, Z., in Deyl, Z., Macek, K. and Janák, J., *Liquid Column Chromatography (Journal of Chromatography Library, Vol. 3)*, Elsevier, Amsterdam, 1975, pp. 529-542.

Kanfer, J.N. and Hakomori, S.-I., *Sphingolipid Biochemistry (Handbook of Lipid Research, Vol. 3)*, Plenum Press, New York, London, 1983, 485 pp.

Kårsnäs, P., Moreno-Lopez, J. and Kristiansen, T., *J. Chromatogr.*, 266 (1983) 643-649.
Kasai, N., Sillerud, L.O. and Yu, R., *Lipids*, 17 (1982) 107-110; *C.A.*, 96 (1982) 158 520m.
Kemp, M.C., Hollaway, W.L., Prestidge, R.L., Bennett, J.C. and Compans, R.W., *J. Liq. Chromatogr.*, 4 (1981) 587-598; *C.A.*, 94 (1981) 187 950r.
Klempnauer, K.-H., Ramsay, G., Bishop, J.M., Moscivici, M., Moscivici, C., McGrath, J.P. and Levinson, A.D., *Cell*, 33 (1983) 345-355.
Konopińska, D., Sobótka, W., Kazanowska, B. and Bogusławska-Javorska, J., in Zaoral, M., Havlas, Z., Mikeš, O. and Procházka, Ž. (Editors), *Synthetic Immunomodulators and Vaccines*, Institute of Organic Chemistry and Biochemistry, Czechoslovak Academy of Science, Prague, 1986, pp. 85-89.
Kotani, S., Azuma, I., Takada, H., Tsujimoto, M. and Yamamura, Y., *Adv. Exp. Med. Biol.*, 166 (1983) 117-158.
Kotani, S., Kinoshita, F., Morisaki, I., Shimono, T., Okunaga, T., Takada, H., Tsujimoto, M., Watanabe, Y., Kato, K., Shiba, T., Kusumoto, S. and Okada, S., *Biken J.*, 20 (1977) 95-103.
Kotani, S., Takada, H., Tsujimoto, M., Kubo, T., Ogawa, T., Azuma, I., Ogawa, H., Matsumoto, K., Siddiqui, W.A., Tanaka, A., Nagao, S., Kohashi, O., Kanoh, S., Shiba, T. and Kusumoto, S., in Jeljaszewicz, J., Pulverer, G. and Roszkowski, W. (Editors), *Bacteria and Cancer*, Academic Press, London, 1982, pp. 67-107.
Kotani, S., Tsujimoto, M., Ogawa, T., Nerome, K., Ooya, A., Takahashi, T., Goto, Y., Shiba, T., Kusumoto, S. and Shimamoto, T., in Zaoral, M., Havlas, Z., Mikeš, O. and Procházka, Ž. (Editors), *Synthetic Immunomodulators and Vaccines*, Institute of Organic Chemistry and Biochemistry, Czechoslovak Academy of Science, Prague, 1986, pp. 40-64.
Kotani, S., Watanabe, Y., Kinoshita, F., Shimono, T., Morisaki, I., Shiba, T., Kusumoto, S., Tarumi, Y. and Ikenaka, K., *Biken J.*, 18 (1975) 105-111.
Krchňák, V., Ježek, J. and Zaoral, M., *Collect. Czech. Chem. Commun.*, 48 (1983) 2079-2081.
Krchňák, V. and Malý, A., in Zaoral, M., Havlas, Z., Mikeš, O. and Procházka, Ž. (Editors), *Synthetic Immunomodulators and Vaccines*, Institute of Organic Chemistry and Biochemistry, Czechoslovak Academy of Science, Prague, 1986, pp. 228-232.
Kühnemund, O., Morávek, L. and Havlíček, J., in Zaoral, M., Havlas, Z., Mikeš, O. and Procházka, Ž. (Editors), *Synthetic Immunomodulators and Vaccines*, Institute of Organic Chemistry and Biochemistry, Czechoslovak Academy of Science, Prague, 1986, pp. 287-299.
Kundu, S.K., *Methods Enzymol.*, Part D, 72 (1981a) 174-185.
Kundu, S.K., *Methods Enzymol.*, Part D, 72 (1981b) 185-204.
Kundu, S.K., Chakravarty, S.K., Roy, S.K., Roy, A.K., *J. Chromatogr.*, 170 (1979) 65-72.
Kundu, S.K. and Roy, S.K., *J. Lipid Res.*, 19 (1978) 390-395; *C.A.*, 88 (1978) 185 541k.
Kundu, S.K. and Scott, D.D., *J. Chromatogr.*, 232 (1982) 19-27.
Kundu, S.K. and Suzuki, A., *J. Chromatogr.*, 224 (1981) 249-256.
Kyte, J. and Doolitle, R.F., *J. Mol. Biol.*, 157 (1982) 105-132; *C.A.*, 97 (1982) 68 044a.
Lanzilotti, A.E., Benz, E. and Goldman, L., *J. Am. Chem. Soc.*, 86 (1964) 1880-1881.
Lauffer, M.A., *J. Am. Chem. Soc.*, 66 (1944) 1188-1194.
Laver, W.G., *J. Mol. Biol.*, 9 (1964) 109-124.
Lebl, M. and Gut, V., *J. Chromatogr.*, 260 (1983) 478-482.
Leclerc, C., Morin, A. and Chedid, L., in Thompson, R.A. and Rose, N.R. (Editors), *Recent Advances in Clinical Immunology*, Churchill Livingstone, Edinburgh, 1983, pp. 187-204.
Leclerc, C. and Vogel, F.R., *CRC Crit. Rev. Ther. Drug Carrier Syst.*, 2 (1986) 353-406.

Ledeen, R.W., Yu, R.K. and Eng, L.F., *J. Neurochem.*, 21 (1973) 829-839.
Ledeen, R.W. and Yu, R.K., in Marks, N. and Rodnight, R. (Editors), *Research in Methods in Neurochemistry*, Plenum, New York, 1978, pp. 371-410.
Ledeen, R.W. and Yu, R.K., *Methods Enzymol.*, 83 (1982) 139-191.
Lee, W.M.F., Westrick, M.A. and Macher, B.A., *Biochim. Biophys. Acta*, 712 (1982) 498-504; *C.A.*, 98 (1983) 68 198z.
Lederer, E., in Zaoral, M., Havlas, Z., Mikeš, O. and Procházka, Ž. (Editors), *Synthetic Immunomodulators and Vaccines*, Institute of Organic Chemistry and Biochemistry, Czechoslovak Academy of Science, Prague, 1986, pp. 3-39.
Lefrancier, P., Choay, J., Derrien, M. and Lederman, I., *Int. J. Pept. Protein Res.*, 9 (1977) 249-257.
Lefrancier, P., Derrien, M., Jamet, X., Choay, J., Lederer, E., Audibert, F., Parant, M., Parant, F. and Chedid, L., *J. Med. Chem.*, 25 (1982) 87-90; *C.A.*, 96 (1982) 471m.
Lefrancier, P. and Lederer, E., *Fortschr. Chem. Org. Naturst.*, 40 (1981) 1-47 (in English).
Lehninger, A.L., *Biochemistry. The Molecular Basis of Cell Structure and Function*, Worth, New York, 2nd ed., 1978a, pp. 272 and 292-295.
Lehninger, A.L., *Biochemistry. The Molecular Basis of Cell Structure and Function*, Worth, New York, 2nd., 1978b, p. 269.
Leive, L. (Editor), *Bacterial Membranes and Walls*, Marcel Dekker, New York, 1973, 495 pp.; *C.A.*, 85 (1976) 2445c.
Leonard, B.R., Jr., Anderegg, J.W., Shulman, S., Kaesberg, P. and Beeman, W.W., *Biochim. Biophys. Acta*, 12 (1953) 499-507.
Lou, M.F., Morrison, W.H., Dueber, Ch., *Biochem. Med.*, 23 (1980) 47-54; *C.A.*, 93 (1980) 3219j.
Mach, O., Grófová, M., Korec, E., Krchňák, V. and Černá, H., in Zaoral, M., Havlas, Z., Mikeš, O. and Procházka, Ž. (Editors), *Synthetic Immunomodulators and Vaccines*, Institute of Organic Chemistry and Biochemistry, Czechoslovak Academy of Science, Prague, 1986, pp. 233-235.
Macher, B.A. and Sweeley, C.C., *Methods Enzymol.*, 50 (1978) 236-251.
Magnani, J.L., Brockhaus, M., Smith, D.F., Ginsburg, V., Blaszcyk, M., Mitchell, K.F., Steplewski, Z. and Kaprowski, H., *Science (Washington, D.C.)*, 212 (1981) 53-55.
Magnani, J.L., Nilsson, B., Brockhaus, M., Zopf, D., Steplewski, Z., Koprowski, H. and Ginsburg, V., *J. Biol. Chem.*, 257 (1982) 14 365-14 369.
Makita, A. and Yamakawa, T., *J. Biochem. (Tokyo)*, 51 (1962) 124-133.
Malý, A. and Krchňák, V., *Folia Biol. (Prague)*, 30 (1984) 168-176 (in English).
Månsson, J.-E., Rosengren, B. and Svennerholm, L., *J. Chromatogr.*, 322 (1985) 465-472.
Marcinka, K., *Acta Virol.*, 16 (1972) 52-62.
Mašek, K. and Flegl, M., *Zentralbl. Pharm.*, 122 (1983) 127-129.
Mašek, K., Zaoral, M., Ježek, J. and Krchňák, V., *Experientia*, 35 (1979) 1397-1398.
Mašek, K., Zaoral, M., Ježek, J. and Straka, R., *Experientia*, 34 (1978) 1363.
Masihi, K.N. and Lange, W., in Zaoral, M., Havlas, Z., Mikeš, O. and Procházka, Ž. (Editors), *Synthetic Immunomodulators and Vaccines*, Institute of Organic Chemistry and Biochemistry, Czechoslovak Academy of Science, Prague, 1986, pp. 70-75.
Massell, B.F., Honikman, L.H. and Amezcua, J., *J. Am. Med. Assoc.*, 207 (1969) 1115-1119.
Matthews, R., *Turnip Yellow Mosaic Virus — Commonwealth Agricultural Bureaux and the Association of Applied Biologists, Descriptions of Plant Viruses*, No. 2, Culross, Coupar Angus, 1970.
McCluer, R.H. and Evans, J.E., *J. Lipid Res.*, 14 (1973) 611-617; *C.A.*, 80 (1974) 11 577w.
McCluer, R.H. and Evans, J.E., *J. Lipid Res.*, 17 (1976) 412-418; *C.A.*, 85 (1976) 89 528d.

McCluer, R.H. and Ullman, M.D., *ACS Symp. Ser.*, No. 128 (1980) 1-13; *C.A.*, 93 (1980) 109 940c.

McCluer, R.H., Ullman, M.D. and Jungalwala, F.B., *Adv. Chromatogr.*, 25 (1986) 309-353.

McKibbin, J.M., in Pigman, W., Horton, D. and Herp, A. (Editors), *The Carbohydrates (Chemistry and Biochemistry)*, Vol. IIB, Academic Press, New York, 1970, pp. 711-738.

Melander, W.R., Nahum, A. and Horváth, Cs., *J. Chromatogr.*, 185 (1979) 129-152.

Miettinen, T. and Takki-Luukkainen, T.T., *Acta Chem. Scand.*, 13 (1959) 856-858.

Mikeš, O. and Šebesta, K., in Mikeš, O. (Editor), *Chromatographic and Allied Methods*, Ellis Horwood, Chichester, New York, 1979, pp. 315-316.

Mikeš, O. and Turková, J., *Chem. Listy*, 58 (1964) 65-123; *C.A.*, 60 (1964) 7662h.

Momoi, T., Ando, S. and Nagai, Y., *Biochim. Biophys. Acta*, 441 (1976) 488-497.

Morávek, L., Kühnemund, O., Havlíček, J., Kopecký, P. and Pavlík, M., *FEBS Lett.*, 208 (1986) 435-438.

Müller, G.M., Shapira, M. and Arnon, R., *Proc. Natl. Acad. Sci. U.S.A.*, 79 (1982) 569-573.

Nagai, Y. and Iwamori, M., in Svennerholm, L., Dreyfus, H. and Urban, P.-F. (Editors), *Structure and Function of Gangliosides*, Plenum, New York, 1980a, pp. 13-21; *C.A.*, 93 (1980) 3211a.

Nagai, Y. and Iwamori, M., *Molecular and Cellular Biochemistry*, 29 (1980b) 81-90; *C.A.*, 93 (1980) 90 262j.

Nagai, Y., Nakaishi, H. and Sanai, Y., *Chem. Phys. Lipids*, 42 (1986) 91-103.

Najjar, V.A., Chaudnuri, M.K., Konopińska, D., Beck, B.D., Layne, P.P. and Linehan, L., in Hersh, E.M., Chirigos, M.A. and Mastrangelo, M.J. (Editors), *Augmenting Agents in Cancer Therapy*, Raven Press, New York, 1981, pp. 459-478.

Najjar, V.A. and Nishioka, K., *Nature (London)*, 228 (1970) 672-673.

Nakabayashi, H., Iwamori, M. and Nagai, Y., *J. Biochem. (Tokyo)*, 96 (1984) 977-984.

Nakamura, K., Nagashima, M., Sekine, M., Igarashi, M., Ariga, T., Atsumi, T., Miyatake, T., Suzuki, A. and Yamakawa, T., *Biochim. Biophys. Acta*, 752 (1983) 291-300.

Ng Ying-Kin, N.M.K. and Wolfe, L.S., *Anal. Biochem.*, 102 (1980) 213-219.

Nibbering, N.M.M., *J. Chromatogr.*, 251 (1982) 93-104.

Orefici, G., Tissi, L., Scaringi, L., Mosci, F., Pellegrini, G., Orsatti, R. and Marconi, P., in Zaoral, M., Havlas, Z., Mikeš, O. and Procházka, Ž. (Editors), *Synthetic Immunomodulators and Vaccines*, Institute of Oragnic Chemistry and Biochemistry, Czechoslovak Academy of Science, Prague, 1986, pp. 99-111.

Otsuka, H., Suzuki, A. and Yamakawa, T., *J. Biochem. (Tokyo)*, 94 (1983) 2035-2041.

Otsuka, H. and Yamakawa, T., *J. Biochem. (Tokyo)*, 90 (1981) 247-254.

Ovchinnikov, Yu.A., Ivanov, V.T., Rostovstseva, L.I., Andronova, T.M., Sorokina, I.B. and Malkova, V.P., *Ger. Offen.*, 28 47 608 (1979).

Pekárek, J., Rotta, J., Rýc, M., Zaoral, M., Straka, R. and Ježek, J., in Zaoral, M., Havlas, Z., Mikeš, O. and Procházka, Ž. (Editors), *Synthetic Immunomodulators and Vaccines*, Institute of Organic Chemistry and Biochemistry, Czechoslovak Academy of Science, Prague, 1986, pp. 153-158.

Pekárek, J., Rotta, J., Zaoral, M., Krchňák, V., Straka, R., Ježek, J. and Rýc, M., *Exp. Cell Biol.*, 53 (1985) 260-264.

Pfaff, E., Mussgay, M., Böhm, H.O., Schulz, G.E. and Schaller, H.E., *EMBO J.*, 1 (1982) 869-874; *C.A.*, 97 (1982) 160 801t.

Phelan, M.A. and Cohen, K., *J. Chromatogr.*, 266 (1983) 55-66.

Phelps, R.A. and Putnam, F.W., *The Plasma Proteins*, Vol. 1, Academic Press, New York, 1960, pp. 143 and 158.

Phillips, G.N., Jr., Flicker, P.F., Cohen, C., Manjula, B.N. and Fischetti, V.A., *Proc. Natl. Acad. Sci. U.S.A.*, 78 (1981) 4689-4693.

Phillips, L.R., Nishimura, O. and Fraser, B.A., *Carbohydr. Res.*, 132 (1984) 275-286.

B372

Pick, J., Vajda, J., Anh-Tuan, N., Leisztner, L. and Hollan, S.R., *J. Liq. Chromatogr.*, 7 (1984a) 2777-2791.
Pick, J., Vajda, J. and Leisztner, L., *J. Liq. Chromatogr.*, 7 (1984b) 2759-2776.
Puro, K., *Acta Chem. Scand.*, 24 (1970) 13-22.
Radin, N.S., Brown, J.R. and Lavin, F.B., *J. Biol. Chem.*, 219 (1956) 977-983.
Reed, L.S. and Archer, M.S., *J. Chromatogr.*, 121 (1976) 100-103.
Regnier, F.E., Gooding, K.M. and Chang, S.M., in Hercules, D., Hieftje, G., Snyder, L.R. and Evenson, M.A. (Editors), *Contemporary Topics of Analysis in Clinical Chemistry*, Vol. 1, Plenum, New York, 1977, pp. 1-48; *C.A.*, 87 (1977) 196 576f.
Ricard, C.S. and Sturman, L.S., *J. Chromatogr.*, 326 (1985) 191-197.
Rogers, H.J. and Perkins, H.R., *Cell Walls and Membranes*, Spon, London, 1968, 436 pp.
Rösch, M., *Fette Seifen Anstrichm.*, 65 (1963) 223-227; *C.A.*, 59 (1963) 848c.
Rotta, J., Rýc, M., Zaoral, M., Straka, R., Ježek, J. and Krchňák, V., in Zaoral, M., Havlas, Z., Mikeš, O. and Procházka, Ž. (Editors), *Synthetic Immunomodulators and Vaccines*, Institute of Organic Chemistry and Biochemistry, Czechoslovak Academy of Science, Prague, 1986, pp. 129-140.
Rotta, J., Rýc, K., Mašek, K. and Zaoral, M., *Exp. Cell Biol.*, 47 (1979) 258-268.
Rotta, J., Zaoral, M., Rýc, M., Straka, R. and Ježek, J., *Exp. Cell Biol.*, 51 (1983) 29-38.
Roughley, P.J. and Mort, J.S., *Anal. Biochem.*, 149 (1985) 136-141.
Rowlands, D.J., in Silver, S. (Editor), *Biotechnology: Potentials and Limitations (Dahlem Konferenzen 1986)*, Springer-Verlag, Berlin, Heidelberg, New York, Tokyo, 1986, pp. 139-154.
Rushlow, K.E., Lautenberger, J.A., Baluda, M.A., Perbal, B., Chirikian, J.G. and Reddy, E.P., *Science (Washington, D.C.)*, 216 (1982) 1421-1423.
Rýc, M., Rotta, J., Zaoral, M., Straka, R., Ježek, J., Krchňák, V., Farkaš, J., Pokorný, J. and Hříbalová, V., in Zaoral, M., Havlas, Z., Mikeš, O. and Procházka, Ž. (Editors), *Synthetic Immunomodulators and Vaccines*, Institute of Organic Chemistry and Biochemistry, Czechoslovak Academy of Science, Prague, 1986, pp. 141-152.
Saito, Y. and Hayano, S., *J. Chromatogr.*, 177 (1979) 390-392.
Scheid, A. and Choppin, P.W., *Virology*, 57 (1974) 475-490.
Scheid, A. and Choppin, P.W., *Virology*, 80 (1977) 54-66.
Schleifer, K.H. and Kandler, O., *Bacteriol. Rev.*, 36 (1972) 407-477.
Seyer, J.M., Beachey, E.H. and Dale, J.B., in Zaoral, M., Havlas, Z., Mikeš, O. and Procházka, Ž. (Editors), *Synthetic Immunomodulators and Vaccines*, Institute of Organic Chemistry and Biochemistry, Czechoslovak Academy of Science, Prague, 1986, pp. 197-202.
Shapira, M., Jibson, M., Müller, G.M. and Arnon, R., *Proc. Natl. Acad. Sci. U.S.A.*, 81 (1984) 2461-2465.
Shapira, M., Jolivet, M. and Arnon, R., *Int. J. Immunopharmacol.*, 7 (1985) 719-723.
Shimomura, K. and Kishimoto, Y., *Biochim. Biophys. Acta*, 754 (1983) 93-100.
Shinnick, T.M., Sutcliffe, J.G., Green, N. and Lerner, R.A., *Annu. Rev. Microbiol.*, 37 (1983) 425-446.
Siddiqui, B., Buehler, J., DeGregorio, M.W. and Macher, B.A., *Cancer Res.*, 44 (1984) 5262-5265.
Siddiqui, B.S. and McCluer, R.H., *J. Lipid Res.*, 9 (1968) 366-370.
Šmíd, F., Bradová, V., Mikeš, O. and Sedláčková, J., *J. Chromatogr.*, 377 (1986a) 69-78.
Šmíd, F., Mikeš, O., Bradová, V. and Ledvinová, J., in Tuček, S., Štípek, S., Šťastný, F. and Křivánek, J. (Editors), *Molecular Basis of Neural Function (Abstracts of the Sixth General Meeting of the European Society for Neurochemistry, Czechoslovakia, Prague, September 1-6, 1986)*, European Society for Neurochemistry, Prague, 1986b, R 12, p. 374.
Smith, K.L., Winslow, A.E. and Petersen, D.E., *Ind. Eng. Chem.*, 51 (1959) 1361-1364; *C.A.*, 54 (1960) 4529c.

Smith, S.W. and Lester, R.L., *J. Biol. Chem.*, 249 (1974) 3395-3405.
Sonnino, S., Ghidoni, R., Gazzotti, G., Kirschner, G., Galli, G. and Tettamanti, G., *J. Lipid Res.*, 25 (1984) 620-628.
Stanbury, J.B., Wyngaarden, J.B. and Fredrickson, D.S. (Editors), *The Metabolic Basis of Inherited Diseases (Inherited Variations and Metabolic Abnormality)*, McGraw-Hill, New York, 4th ed., 1978, pp. 544-916.
Steere, R.L., *Plant Virology*, University of Florida Press, 1964, p. 211.
Steere, R.L., *Phytopathology*, 46 (1956) 60-69; *C.A.*, 50 (1956) 7235c.
Stolyhwo, A. and Privett, O.S., *J. Chromatogr. Sci.*, 11 (1973) 20-25.
Stout, R.W., Cashmore, A.R., Coward, J.V., Horváth, C.G. and Bertino, J.R., *Anal. Biochem.*, 71 (1976) 119-124.
Straka, R., *Česk. Farm.*, 32 (1983) 167-181; *C.A.*, 99 (1983) 140 316b.
Straka, R. and Zaoral, M., *Collect. Czech. Chem. Commun.*, 42 (1977) 560-563.
Strasberg, P.M., Warren, I., Skomorowski, M.A. and Lowden, J.A., *Clin. Chim. Acta*, 132 (1983) 29-41.
Strominger, J.L., Izaki, K., Matsuhashi, M. and Tipper, D.J., *Fed. Proc., Fed. Am. Soc. Exp. Biol.*, 26 (1967) 9-22.
Strominger, L. and Ghuysen, M., *Science (Washington, D.C.)*, 156 (1967) 213-221.
Sugita, M., Iwamori, M., Evans, J., McCluer, R.H., Dulaney, J.T. and Moser, H. W., *J. Lipid Res.*, 15 (1974) 223-226.
Sutcliffe, J.G., Shinnick, T.M., Green, N. and Lerner, R.A., *Science (Washington, D.C.)*, 219 (1983) 660-666.
Suzuki, A., Handa, S., Ishizuka, I. and Yamakawa, T., *J. Biochem. (Tokyo)*, 81 (1977a) 127-134.
Suzuki, A., Handa, S. and Yamakawa, T., *J. Biochem. (Tokyo)*, 80 (1976) 1181-1183.
Suzuki, A., Handa, S. and Yamakawa, T., *J. Biochem. (Tokyo)*, 82 (1977c) 1185-1187; *C.A.*, 87 (1977) 196 600j.
Suzuki, A., Kundu, S.K. and Marcus, D.M., *J. Lipid Res.*, 21 (1980) 473-477.
Suzuki, A., Sato, M., Handa, S., Muta, Y. and Yamakawa, T., *J. Biochem. (Tokyo)*, 82 (1977b) 461-467.
Suzuki, K., *Life Sci.*, 3 (1964) 1227-1233.
Suzuki, K., *J. Neurochem.*, 12 (1965) 629-638; *C.A.*, 63 (1965) 10 399h.
Svennerholm, L., *Biochim. Biophys. Acta*, 24 (1957) 604-611.
Svennerholm, L., *Acta Chem. Scand.*, 17 (1963a) 239-250.
Svennerholm, L., *J. Neurochem.*, 10 (1963b) 613-623; *C.A.*, 60 (1964) 839f.
Svennerholm, L., *J. Lipid Res.*, 5 (1964) 145-155.
Svennerholm, L., *Methods Carbohydr. Chem.*, 4 (1972) 464-474.
Svennerholm, L., in Svennerholm, L., Mandel, P., Dreyfus, H. and Urban, P.-F. (Editors), *Structure and Function of Gangliosides [Proceedings of Symposium on Structure and Function of Gangliosides, Strasbourg (Bischenberg) France, April 23rd-27th, 1979]*, Plenum, New York, London, 1980, p. 11.
Svennerholm, L., Fredman, P., Elwing, H., Molmbren, J. and Strannegard, O., *ACS Symp. Ser.*, No. 128 (1980a) 373-390.
Svennerholm, L., Mandel, P., Dreyfus, H. and Urban, P.-F. (Editors), *Structure and Function of Gangliosides [Proceedings of Symposium on Structure and Function of Gangliosides, Strasbourg (Bischenberg) France, April 23rd-27th, 1979]*, Plenum, New York, London, 1980b, 571 pp.
Tainer, J.A., Getzoff, E.D., Alexander, H., Houghten, R.A., Olson, A.J., Lerner, R.A. and Hendrickson, W.A., *Nature (London)*, 312 (1984) 127-134.
Tainer, J.A., Getzoff, E.D., Paterson, Y., Olson, A.J. and Lerner, R.A., *Annu. Rev. Immunol.*, 3 (1985) 501-535.
Takada, H. and Kotani, S., in Stewart-Tull, D.E.S. and Davies, M. (Editors), *Immunology of the Bacterial Cell Envelope*, Wiley, Chichester, 1985, pp. 119-152.
Tayot, J.-L., Tardy, M., Gattel, P., Plan, R. and Roumiantzeff, M., in Epton, R. (Editor), *Chromatography of Synthetic and Biological Polymers*, Vol. 2, Ellis Horwood, Chichester, 1978, pp. 95-110.
Tipper, D.J., Strominger, J.L. and Ensign, J.C., *Biochemistry*, 6 (1967) 906-920.

Tjaden, U.R., Krol, J.H., Van Hoeven, R.P., Oomen-Meulemans, E.P.M. and Emmelot, P., *J. Chromatogr.*, 136 (1977) 233-243.
Traylor, T.D., Koontz, D.A. and Hogan, E.L., *J. Chromatogr.*, 272 (1983) 9-20.
Turková, J., Mikeš, O. and Šorm, F., *Collect. Czech. Chem. Commun.*, 31 (1966) 2444-2455.
Turková, J., Mikeš, O., Šraml, J., Knessl, O. and Šorm, F., *Antibiotiki*, 9 (1964) 506-516 (in Russian); *Fed. Proc.*, *Fed. Am. Soc. Exp. Biol.*, 24 (Transl. Suppl.) (1965) T725-T730 (in English).
Tyihak, E., Mincsovics, E. and Kalász, H., *J. Chromatogr.*, 174 (1979) 75-81.
Uemura, K.-i., Yuzawa, M. and Taketomi, T., *J. Biochem.*, 83 (1978) 463-471.
Ullman, M.D. and McCluer, R.H., *J. Lipid Res.*, 18 (1977) 371-378.
Ullman, M.D. and McCluer, R.H., *J. Lipid Res.*, 19 (1978) 910-913.
Ullman, M.D. and McCluer, R.H., *J. Lipid Res.*, 26 (1985) 501-506.
Ullman, M.D., Pyeritz, R.E., Moser, H.W., Wenger, D.A. and Kolodny, E.H., *Clin. Chem.*, 26 (1980) 1499-1503; *C.A.*, 93 (1980) 165 577b.
Vance, D.E. and Sweeley, C.C., *J. Lipid Res.*, 8 (1967) 621-630.
Van der Zee, R. and Welling, G.W., *J. Chromatogr.*, 327 (1985) 377-380.
Van der Zee, R., Welling-Wester, S. and Welling, G.W., *J. Chromatogr.*, 266 (1983) 577-584.
Van Haastert, P.J.M., *J. Chromatogr.*, 210 (1981) 241-254.
Van Regenmortel, M.H.V., *Trends Biochem. Sci.*, 11 (1986) 36-39.
Volsky, D.J. and Loyter, A., *FEBS Lett.*, 92 (1978) 190-194; *C.A.*, 89 (1978) 175 843s.
Wabuke-Bunoti, M.A.N., Taku, A., Garmen, N. and Fan, D.P., *J. Immunol.*, 133 (1984b) 2186-2193.
Wabuke-Bunoti, M.A.N., Taku, A., Fan, D.P., Kent, S. and Webster, R.G., *J. Immunol.*, 133 (1984a) 2194-2201.
Walter, G., *J. Immunol. Methods*, 88 (1986) 149-161.
Ward, C.W., *Curr. Top. Microbiol. Immunol.*, 94/95 (1981) 1-74.
Watanabe, K. and Arao, Y., *J. Lipid Res.*, 22 (1981) 1020-1024.
Watanabe, K. and Tomono, Y., *Anal. Biochem.*, 139 (1984) 367-372.
Weijer, W.J., Welling, G.V. and Welling-Wester, S., in Brown, F., Chanock, R.M. and Lerner, R.A. (Editors), *Vaccines 86. New Approaches to Immunization. Developing Vaccines Against Parasitic, Bacterial and Viral Diseases (Proceedings of Conference, 1985)*, Cold Spring Harbor Laboratory, Cold Spring Harbor, 1986, pp. 71-77.
Welling, G.W., Groen, G., Slopsema, K. and Welling-Wester, S., *J. Chromatogr.*, 326 (1985) 173-178.
Welling, G.W., Groen, G. and Welling-Wester, S., *J. Chromatogr.*, 266 (1983) 629-632.
Welling, G.W., Nijmeier, J.R.L., Van der Zee, R., Groen, G., Wilterding, J.B. and Welling-Wester, S., *J. Chromatogr.*, 297 (1984) 101-109.
Wells, G.B., Turco, S.J., Hanson, B.A. and Lester, R.L., *Anal. Biochem.*, 110 (1981) 397-406.
Wells, G.B., Turco, S.J., Hanson, B.A. and Lester, R.L., *Methods Enzymol.*, Part D, 83 (1982) 137-139.
Westhof, F., Altschum, D., Moras, D., Bloomer, A.C., Mondragon, A., Klug, A. and Van Regenmortel, M.H.V., *Nature (London)*, 311 (1984) 123-126.
Westrick, M.A., Lee, W.M.F., Goff, B. and Macher, B.A., *Biochim. Biophys. Acta*, 750 (1983a) 141-148.
Westrick, M.A., Lee, W.M.F. and Macher, B.A., *Cancer Res.*, 43 (1983b) 5890-5894.
Wiegandt, H., in Svennerholm, L., Mandel, P., Dreyfus, H. and Urban, P.-F. (Editors), *Structure and Function of Gangliosides [Proceedings of Symposium on Structure and Function of Gangliosides, Strasbourg (Bischenberg), April 23rd-28th, 1979]*, Plenum, New York, London, 1980, pp. 3-10.
Wiegandt, H., *Adv. Neurochem.*, 4 (1982) 149-223.
Wiely, D.C., Wilson, I.A. and Skehel, J.J., *Nature (London)*, 289 (1981) 373-378.
Wilson, I.A., Skehel, J.J. and Wiley, D.C., *Nature (London)*, 289 (1981) 366-373.

Winterbourn, Ch.C., *J. Neurochem.*, 18 (1971) 1153-1155.

Wleklik, M., Bump, N.J., Lee, J. and Najjar, V.A., in Zaoral, M., Havlas, Z., Mikeš, O. and Procházka, Ž. (Editors), *Synthetic Immunomodulators and Vaccines*, Institute of Organic Chemistry and Biochemistry, Czechoslovak Academy of Science, Prague, 1986, pp. 76-84.

Yahara, S., Kishimoto, Y. and Poduslo, J., *ACS Symp. Ser.*, No. 128 (1980) 15-33; *C.A.*, 93 (1980) 109 941d.

Yamakawa, T., Handa, S., Yamazaki, T. and Suzuki, A., *Proceedings of the 27th International Congress of Pure and Applied Chemistry*, 1980, 351-358; *C.A.*, 93 (1980) 109 942e.

Yamazaki, T., Suzuki, A., Handa, S. and Yamakawa, T., *J. Biochem. (Tokyo)*, 86 (1979) 803-809; *C.A.*, 91 (1979) 188 987h.

Yip, K.F. and Albarella, J.P., *J. Chromatogr.*, 326 (1985) 301-310.

Young, W.W., Jr., Laine, R.A. and Hakamori, S., *J. Lipid Res.*, 20 (1979) 275-278.

Yu, R.K. and Ando, S., in Svennerholm, L., Mandel, P., Dreyfus, H. and Urban, P.-F. (Editors), *Structure and Function of Gangliosides [Proceedings of the Symposium on Structure and Function of Gangliosides, Strasbourg (Bischenberg), April 23rd-28th, 1979]*, Plenum, New York, London, 1980, pp. 40-42.

Yu, R.K. and Ledeen, R.W., *J. Lipid Res.*, 13 (1972) 680-686.

Zaoral, M., Havlas, Z., Mikeš, O. and Procházka, Ž. (Editors), *Synthetic Immunomodulators and Vaccines (Proceedings of International Symposium, October 14th-18th, 1985, Třeboň, Czechoslovakia)*, Institute of Organic Chemistry and Biochemistry, Czechoslovak Academy of Science, Prague, 1986, 353 pp.

Zaoral, M., Ježek, J., Krchňák, V. and Straka, R., *Collect. Czech. Chem. Commun.*, 45 (1980) 1424-1446.

Zaoral, M., Ježek, J. and Rotta, J., *Collect. Czech. Chem. Commun.*, 47 (1982) 2989-2995.

Zaoral, M., Ježek, J., Straka, R. and Mašek, K., *Collect. Czech. Chem. Commun.*, 43 (1978) 1797-1802.

Chapter 13

APPLICATIONS OF HIGH-PERFORMANCE LIQUID CHROMATOGRAPHIC METHODS IN STRUCTURAL
STUDIES OF BIOPOLYMERS AND BIOOLIGOMERS

In the preceding part of this Volume B, the main topic was a description of
the separation of various compounds. In this chapter, the applications of chromatographic methods to the determination of the structures of polymeric and
oligomeric natural substances are reviewed. Of course, the actual technique in
this field is the chromatographic separation process, but the overall purpose, to
which the experimental approach is subordinated, is to determine chemical
structures, and this may bring some special requirements into the separation
processes.

First, high-performance liquid chromatographic (HPLC) techniques were applied
in the amino acid sequencing of proteins and peptides. Much work has been
published on this topic and therefore this area will be dealt with preferentially.
Second, the sequencing of nucleotides in nucleic acids, polynucleotides and
oligonucleotides will be mentioned briefly, because HPLC is not the main separation technique in this area. Third, sequencing experiments in efforts to determine the structures of oligosaccharides will be briefly discussed.

13.1 PROTEINS AND PEPTIDES

13.1.1 Introduction

Column chromatographic techniques have been used as standard separation
methods throughout the development of procedures for structural studies of
proteins; the determination of the complete covalent structure of bovine trypsinogen can be mentioned here as an example (Mikeš et al., 1966a,b, 1967). The
classical ion-exchange columns and tedious separation processes were gradually
replaced by modern rapid and very efficient methods. HPLC techniques are used
(i) for the purification of the starting protein or higher peptide, the structure of which is to be studied, (ii) for the separation and isolation of
fragmental peptides, arising from partial hydrolysis or specific splitting of
the protein or polypeptide studied, and (iii) for the separation of suitable
amino acid derivatives, both from the total hydrolysates for analytical purposes
and for the analytical monitoring of amino acids gradually split from the

terminal parts of the proteins and peptides. All the techniques are specialized
for the particular purpose. Especially the application of HPLC in the last two
areas have contributed much to the development of protein and peptide sequencing
to the present very advanced form.

13.1.2 Isolation of the starting proteins (polypeptides) for sequential studies

The separation of proteins was described in detail in Chapter 7 and the
separation of polypeptides in Chapter 9, so only brief comments and examples are
sufficient here. The requirements for a protein (polypeptide) isolated for
sequential studies are not as strict as in the isolation and characterization
of an unknown enzyme or protein (where the activity must be maximal and no contaminants are allowed). For sequential studies a purity of 95% is often sufficient. The analytical data for terminal amino acids of polypeptide contaminants
should be roughly one order of magnitude lower than that for the terminal amino
acids of the protein studied, to permit a clear distinction between the examined
polypeptide chain and ballast polypeptides.

Henderson et al. (1982) described the reversed-phase HPLC (RP-HPLC) of proteins and peptides and its impact on protein microsequencing. Met-enkephalin,
Leu-enkephalin, ribonuclease, insulin, lysozyme, bovine serum albumin and
ovalbumin were separated on μBondapak C_{18} and μBondapak Phenyl columns using a
linear gradient from 0.05% trifluoroacetic acid (TFA) in water to 60% (v/v)
acetonitrile containing 0.05% TFA over 60 min at a flow-rate of 1.0 ml/min. By
replacing the acetonitrile by a gradient with methanol (0 to 100%), similar
results were obtained. In another experiment Rauscher murine leukaemia virus
(17 mg) was disrupted and solubilized in 10 ml of 6 M guanidinium chloride at
pH 2.0 (TFA) and injected on to a 30 cm x 0.8 cm I.D. column of μBondapak C_{18}.
Using gradients with 1-propanol-0.05% TFA and acetonitrile-0.05% TFA nine defined
viral components were eluted.

Kamp et al. (1984) reported the application of HPLC in the separation of
ribosomal proteins of different organisms. Size exclusion chromatography (SEC)
(on a TSK Gel 2000 SW column), ion-exchange chromatography (IEC) and reversed-
phase chromatography (RPC) were described, employing new column packing materials,
different gradient systems and preparative columns. In addition to other methods
of characterization, the purity of isolated proteins was verified by direct
micro-sequencing. From *Escherichia coli*, 15S and 23L proteins were isolated
in sequencer purity by employing propanol gradients in combination with Vydac
TR-RP or Ultrapore RPSC columns. 6S proteins and 16L proteins were purified
by IEC (on a TSK IEX-530 CM column). Kamp (1985b) reviewed the HPLC of proteins
for sequential studies in a practical detailed manual.

Meloun et al. (1985c) described the HPLC separation of three variants of
unmodified acidic acrosin inhibitors from bull seminal plasma, using a 25 cm x
4.6 mm I.D. Vydac TP-RP 10 column and elution with a linear gradient from 20 to
40% acetonitrile in 40 min. The carboxymethylated derivatives of the isolated
inhibitors were compared by HPLC fingerprinting (cf. Section 13.1.4).

13.1.3 Separation and isolation of protein fragments and peptides from partial hydrolysates

13.1.3.1 Products of chemical cleavage

Specific chemical cleavage of polypeptide chains is usually used to prepare
small numbers of larger fragments of a protein. As a rule, there are no problems
with peak capacity. Often the greatest problem is to find conditions under which
the unnaturally split fragments are soluble in order to facilitate their chromatographic separation. Cyanogen bromide (CNBr) splitting behind methionine
residues (Gross, 1967) is used for the specific fragmentation of polypeptide
chains. Because there is only a very limited number of Met residues in proteins,
relatively large polypeptide fragments are usually obtained in this way and
chromatographic methods are most often used for their isolation from the digest.
In this way, the primary structure of protein can readily be established. Several examples will be presented.

Hughes et al. (1979) studied the application of the HPLC of peptides for
microsequence analysis. In addition to the separation of peptides from tryptic
digests, the chromatography of 53 nmol of CNBr peptides from the aminoethylated
α-chain of human haemoglobin was described. A LiChrosorb RP-18 (10 μm) column
(25 cm x 4.6 mm I.D.) was eluted with a gradient (at 0.7 ml/min) of buffers (A)
1.0 M pyridine acetate (pH 5.5) and (B) 1.0 M pyridine acetate (pH 5.5), 60% (v/v)
1-propanol, and 5% of the effluent was removed for detection with fluorescamine.
The fractions obtained were suitable for sequencing using the DABITC method
(see later). Stoming et al. (1979) reported the separation of $^A\gamma$ and $^G\gamma$ peptides
from the CNBr fragment located in the C-terminal area of human foetal haemoglobin.
Two μBondapak C_{18} columns were joined in series and eluted with a linear gradient
of 10-25% acetonitrile-0.01 M ammonium acetate over a period of 40 min at a
flow-rate of 1.0 ml/min. The absorbance at 280 nm was used for monitoring.

Takagaki et al. (1980) studied the amino acid sequence of the membraneous
segment of rabbit liver cytochrome b_5. A mixture of N-bromosuccinimide (NBS)
fragments (cleaved after tryptophan residues) were separated on a 30 cm x 0.39 cm
I.D. column of μBondapak C_{18} using a system of solvents (A) 88% (w/w) formic
acid in water and (B) 88% (w/w) formic acid in ethanol. A linear gradient from
20% (v/v) to 100% (v/v) of solvent B was used for elution. The separation of

CNBr fragments (on Sephadex LH-60) and *Staphylococcus aureus*-V8-protease digest
(using RP-HPLC) was also described. Hermodson and Mahoney (1981) also reported
the separation of CNBr peptides of human haemoglobin α-chain on a LiChrosorb
C_8 (5 μm) column (25 cm x 4.6 mm I.D.) eluted with a gradient of 0.1% (v/v)
trifluoroacetic acid (TFA) in water and 0.1% TFA (v/v) in n-propanol, prior to
sequencing.

Wilson et al. (1981) compared buffers and detection systems for the HPLC of
peptide mixtures. CNBr fragments from aminoethylated α-chain of human haemoglobin
were well separated on a LiChrosorb RP-18 (10 μm) column (25 cm x 4.6 mm I.D.)
using a gradient from 0.1% H_3PO_4-0.01 M $NaClO_4$ to 60% (v/v) 2-propanol in the
same initial buffer. Other examples of the separation of CNBr fragments were
also described. Common buffer systems for gradient RP-HPLC on RP-8 or -18
packings in connection with suitable detection systems were compiled and discussed. Frantíková et al. (1984) subjected S-sulphonated fractions of a CNBr
digest originating from the N-terminal region of human haemopexin, and preliminarily separated by SEC, to RP-HPLC on Separon-SI-C_{18} using a gradient of
1-propanol in 0.05% TFA. The fractions obtained were suitable for further structural studies.

Prestidge (1984) briefly reviewed the separation of protein fragments by HPLC.

13.1.3.2 Enzymic partial hydrolysates

The enzymic digests of polypeptide chains are often more complex than the
products of specific chemical cleavage. They may consist of many components,
which are usually readily soluble, because they were cleaved from the parent
proteins in a natural way. The reason for the large number of peptides is that
not all proteolytic enzymes are strictly specific for cleavage after or before
only one type of amino acid residue and, moreover, residues sensitive to specific
enzymic attack occur relatively often in proteins. Claims of chromatographic
efficiency in the separation of such a complex mixture are therefore excessive.

Many papers have been published describing the identification of haemoglobin
variants in which chromatography of enzymic digests of the proteins studied was
often used. Schroeder et al. (1979) applied RP-HPLC in the separation of tryptic
peptides of haemoglobin E. A μBondapak C_{18} (10 μm) column (30 cm x 3.9 mm I.D.)
was equilibrated with solution A (0.01 M ammonium acetate, pH 6.07) and eluted
with a gradient of 50 ml of A and 50 ml of B (40% acetonitrile in A). The flow-
rate was 1 ml/min and the pressure 1500 p.s.i., the work was carried out at room
temperature and the absorbance at 220 nm was used for monitoring. Fractions of
0.5 ml were collected. Wilson et al. (1979) separated by HPLC tryptic peptides
of normal α-, β-, γ- and δ-chains of human haemoglobins A, F and A_2, and of
abnormal chains of 25 variants. In addition, the HPLC separation of chymotryptic

peptides of the oxidized core of the normal α-chain was described. A Waters µBondapak C_{18} column and solvents (I) 0.01 M ammonium acetate (pH 7) and (II) 50% acetonitrile in I were used in a gradient elution programme at room temperature with a flow-rate of 1.5 ml/min and monitoring of the absorbance at 220 nm.

Schroeder et al. (1980) briefly reviewed experiments on the HPLC of haemoglobin tryptic peptides and Schroeder et al. (1981) presented more detailed information on the application of HPLC in the identification of human haemoglobin variants. In addition to Waters µBondapak C_{18}, Altex Ultrasphere ODS and DuPont Zorbax TMS columns were also used and an acetonitrile-phosphate buffer gradient for the elution of tryptic peptides.

Yi-Tao et al. (1982) described the identification of G-Philadelphia and Matsue-Oki haemoglobin variants in a black infant using HPLC and microsequencing. Tryptic and thermolytic digests were separated employing the conditions presented by Wilson et al. (1979). Yi-Tao et al. (1983) subsequently applied HPLC and microsequencing to the structural analysis of human haemoglobin variants and described the structural abnormalities of eleven α-chain and seven β-chain variants present in 37 Chinese families and 1 black family; tryptic and thermolytic digests were investigated.

Investigations of the primary structures of immunoglobulins and Bence-Jones proteins are further examples of the utilization of sequential studies of separated peptides originating from enzymic digests. Kratzin et al. (1980) examined the structure of monoclonal L-chain of κ-type, subgroup I, Bence-Jones protein Wes (M_r = 23 000). The tryptic digest was separated on a Zorbax C_8 (25 cm x 4.6 mm I.D.) column at 60°C by means of a gradient of volatile buffers (A) 50 mM ammonium acetate (pH 6.0) and (B) 60% acetonitrile in buffer A. In other experiments isocratic elution was employed with 25 mM ammonium acetate (pH 6 or 3, adjusted by means of TFA), the absorbance at 220 nnm being monitored. Of 22 peptides obtained, 18 were homogeneous and could be directly used for sequencing. Yang et al. (1981) studied in detail the chromatography and rechromatography of L-chain of κ-type, subgroup I, Bence-Jones protein Den (M_r = 23 000). An ODS-Hypersil (5 µm) column (25 cm x 4.6 mm I.D.) was eluted at 60°C at a flow-rate of 1.9 ml/min using three buffer systems: I, (A) 25 mM ammonium acetate (pH 6.0) and (B) 40% 50 mM ammonium acetate (pH 6.0) + 60% acetonitrile; II, (A) 5 mM KH_2PO_4-K_2HPO_4 (pH 6) and (B) 40% buffer A + 60% acetonitrile; and III, (A) water of pH 2.15 (adjusted with TFA) and (B) 40% buffer A + 60% acetonitrile. Linear gradients from 0 to 70% in 40 or 60 min were used with all three buffer systems and the absorbance at 220 nm was monitored. From the primary separation fourteen peptides were isolated in analytically pure form, and the remaining fragments were completely purified after one rechromatography.

The use of two-dimensional HPLC and chemical modification in sequence analysis
was studied by Takahashi et al. (1983), who described the complete amino acid
sequence of the λ-light chain of human immunoglobulin D. A first separation on
a macroreticular cation exchanger was employed, followed by RPC. For the first
step a Hitachi Gel 3013C (5-7 μm) column (25 cm x 0.41 cm I.D.) was eluted at
$70^{\circ}C$ using a gradient of (A) water and (B) 0.4 M ammonia + 0.005 M acetic acid,
containing 50% (v/v) acetonitrile and 25% (v/v) 1-propanol, adjusted to pH 6.2
with methanesulphonic acid. Aliquots dissolved in 0.1% TFA or 4 M guanidinium
chloride were applied to the RP column (SynChropak RP-P, 25 cm x 0.41 cm I.D.,
from SynChrom) and eluted for 40 min at a flow-rate of 0.7 ml/min with a linear
gradient from 0.1% TFA to 40% 1-propanol containing 0.1% TFA. In order to obtain
overlapping of the tryptic peptides, chemical modification of the light chain
and RPC were found to be the most reliable method and contributed much to the
determination of the primary structure of the light chain.

Examples of the separation of enzymic digests from miscellaneous proteins
will now be briefly reviewed. IP-RP-HPLC of tryptic peptides from plasma amyloid
P-component on a μBondapak C_{18} column using a volatile buffer gradient system
of 5-60% acetonitrile in 0.1% TFA and detection at 216 nm was described by
Anderson et al. (1981); the peptides obtained were suitable for immediate amino
acid sequence analysis. Hobbs et al. (1981) reported the analysis of rat caseins.
Both dephosphorylated and phosphorylated tryptic peptides were separated using
μBondapak Fatty Acid Analysis columns (30 cm x 0.39 cm I.D.), eluted at a flow-
rate 2.0 ml/min with a linear gradient from water-16 mM H_3PO_4 to 50% acetonitrile + 50% water-16 mM H_3PO_4 in 60 min; the temperature was $18^{\circ}C$, the sample
loading 150 μg and for detection the absorbance at 210 nm was measured. Yuan et
al. (1982) compared two systems for the preparation of samples by means of RPC
for microsequence analysis. One system used TFA-acetonitrile solvents with
absorbance monitoring at 206 nm and the other employed pyridine-formate and
pyridine acetate buffers with 1-propanol for elution and post-column fluorescence
detection. Each system was tested with RP-8, RP-18 and alkylphenyl supports. In
most applications, separation with TFA-acetonitrile in conjuction with an alkylphenyl column was the fastest and most efficient. A Waters 30 cm x 3.9 mm I.D.
μBondapak Phenyl (10 μm) column was eluted with a linear gradient from 0 to
50% B generated by solvents (A) 0.09% TFA and (B) TFA-water-acetonitrile (0.09:
9.91:90). Tryptic peptides from horse heart cytochrome c, [^{14}C]carboxymethylated
pig testes cytochrome P-450 and human plasma fibronectin were separated in
addition to synthetic opioid peptides.

Jonák et al. (1984) studied histidine residues in ribosomal elongation factor
EF-Tu from *Escherichia coli* protected by aminoacyl-tRNA against photooxidation.
The separation of tryptic digest of EF-Tu was performed on a 25 cm x 0.8 cm I.D.

column of Nucleosil ODS (5 μm) using 0.1% TFA, a gradient of ethanol at 50°C and a flow-rate of 3 ml/min, with absorbance monitoring at 220 nm. Tryptic digests of a complex of EF-Tu with GTP and with aminoacyl-tRNA, both non-irradiated and irradiated (in the presence of rose bengal dye to destroy histidine residues) were separated. The positions of the two sensitive histidine residues (characterizing the binding site of EF-Tu versus aminoacyl-tRNA) were located in the primary structure. Jonák et al. (1986) described the structural homology in the binding site for aminoacyl-tRNA between elongation factors EF-Tu from *Bacillus stearothermophilus* and *Escherichia coli*. To identify the binding site, the factors were labelled with $[^{14}C]$Tos-PheCH$_2$Cl and digested with trypsin and the radioactive peptides were separated on a CGC Separon Six C$_{18}$ (5 μm) column (15 cm x 0.32 cm I.D.) (Laboratory Instruments Works, Prague, Czechoslovakia), using elution with a linear gradient of methanol at a flow-rate of 0.5 ml/min; solvent (A) was 0.1% TFA in doubly distilled water and solvent (B) was 90% ethanol containing 0.1% TFA. Concentrations of 33% and 100% of B were achieved in 15 and 60 min, respectively. Detection at 220 nm was used and the fractions were collected manually.

Meloun et al. (1985b) prepared a tryptic digest of a large CNBr fragment (226 amino acid residues) of thermostable proteinase thermitase from *Thermoactinomyces vulgaris* in order to obtain peptides suitable for sequence analysis. A LiChrosorb RP-2 (7 μm) column (25 cm x 4.6 mm I.D.) was eluted with a system (I) consisting of solvents (A) 0.1% TFA in doubly distilled water and (B) methanol containing 0.1% TFA. Another system (II) also used for the separation employed solvents (A) 0.05 M ammonium acetate (pH 6) and (B) methanol. Linear gradients were used and a 90% concentration of B was achieved in 60 min at a flow-rate of 2 ml/min. Detection was at 280 nm, with manual collection of fractions. The sequential data obtained contributed to the determination of the complete primary structure of this interesting thermostable enzyme (Meloun et al., 1985a).

Various natural proteolytic processes can be followed by the HPLC separation of peptides formed and by their sequential analyses. Kehl et al. (1981) reported the analysis of human fibrinopeptides formed during coagulation of plasma, in which fibrinogen is cleaved by thrombin, releasing fibrin and low-molecular-mass fibrinopeptides. For chromatographic separation and quantitation, Hibar RT 250-4 and LiChrosorb RP-18 (5 μm) RPC columns, purchased from Merck (Darmstadt, F.R.G.) were employed. A gradient elution system consisting of (A) 0.025 M ammonium acetate with orthophosphoric acid (pH 6.0) and (B) 0.05 M ammonium acetate with orthophosphoric acid (pH 6.0), mixed with an equal volume of acetonitrile, was used. Programmed elution started with 12% B-88% A and the final concentration of 28% B-72% A was reached in 30 min. The flow-rate was 1.5 ml/min and the temperature 40°C, with detection at 210 nm. Seidah et al. (1983) studied the enzymic maturation of pro-opiomelanocortin by anterior pituitary granules. For the RP-HPLC of the released peptides an Altex 5-μm ODS column (25 cm x 0.39 cm I.D.) was used. Two buffer systems were employed: (I) solvent A, 0.13% (v/v) hepta-fluorobutyric acid (HFBA) in water and B, 0.13% (v/v) HFBA in 1-propanol. After equilibration elution of the sample was carried out with a linear gradient from 25% to 55% B in 150 min at 0.5 ml/min; fractions of 0.5 ml per tube were col-lected automatically. (II) Solvent A, 0.1% (v/v) TFA in water and B, 0.1% (v/v) TFA in 1-propanol. After equilibration, the sample was eluted with a linear gradient from 0% to 70% B in 140 min at 0.5 ml/min. For SEC of the maturation products Waters Protein Analysis columns (30 cm x 0.78 cm I.D.), one I-60, two I-125 and one I-250 connected in series in the given order, were isocratically eluted with a buffer of 0.2 M triethylamine phosphate + 6 M guanidinium chloride (pH 3) at a flow-rate of 1 ml/min.

Yang et al. (1979) examined a special application of HPLC in studies of protein structure, viz., the determination of disulphide bond pairing. The method is similar to diagonal paper chromatography and diagonal paper electrophoresis (Mikeš and Holeyšovský, 1957, 1958; Brown and Hartley, 1963, 1966; Hartley, 1970). Fractions of peptides from the first chromatography were collected and any S-S bonds broken. The fractions were then reinjected to identify (by their different chromatographic mobilities) the individual peptides that were linked by the S-S bonds. Peptides from the enzymic digest of lysozyme were labelled with fluorescamine and identified as described. Thannhauser et al. (1985) published a more advanced technique for the two-dimensional determination of disulphide pairing employing a continuous-flow sensitive disulphide detection system [using the reaction with disodium 2-nitro-5-thiosulphobenzoate (NTSB) and monitoring the reaction product 2-nitro-5-thiobenzoic acid (NTB) at 410 nm] and demonstrated its application to peptide mapping of bovine pancreatic ribonuclease A by RP-HPLC. The disulphide-containing peptides (obtained by RP-HPLC from partial enzymic digests of ribonuclease using the above-mentioned specific detection, and quantitatively analysed) were reduced with a solution of 100 mM sodium sulphite and 40 mM S-sulphocysteine at pH 8.0. The reaction products were rechromatographed and two peptide-containing peaks were obtained in each case. The positions of the peaks could be compared with the peptide map of the fully sulphonated protein and the identification could be confirmed by amino acid analysis. For HPLC a 30 cm x 4.6 mm I.D. RP C_{18} column was eluted using the solvent system A (0.09% TFA in water) and B (0.09% TFA in acetonitrile); a linear gradient from 2% to 35% B in 1 h was employed at a flow-rate of 2 ml/min. Detection at 210 nm was employed. The sensitive disulphide detection system (detection limit 5 pmol) was described in detail.

The practical procedures of the HPLC separation of peptides for the sequential studies (obtained both by chemical and enzymic digestions) were reviewed by Kamp (1985a).

13.1.4 Peptide mapping ("fingerprinting")

In many instances, the preparative or semipreparative chromatographic isolation of peptides cleaved from proteins or polypeptides does not need to be the final step in structural studies of proteins. A simple comparison of chromatographic positions (elution volumes) of peptides prepared from closely related (homologous) proteins may often be sufficient to identify easily a peptide in which the compared primary structures differ. The isolation and sequential characterization of such a single peptide (or of a few peptides) can help in clarifying the difference between primary structures of the homologous proteins without any need for the laborious isolation and structural characterization of all the cleaved peptides, the structures of which may be known in advance from the homologous proteins used for comparison. In other instances, the process of releasing peptides during digestion may be interesting and important and can be followed by repeating the measurement of chromatograms performed using identical separation conditions.

Ingram (1958) studied abnormal human haemoglobins by comparison of known normal and unknown sickle-cell anaemia species. Tryptic digests of both the proteins were subjected to a combination of two separation processes: in the first step paper electrophoresis was utilized (this was the so-called "first dimension") and then, after drying (without any detection), the same sheet of paper was subjected to paper chromatography in the perpendicular direction (the "second dimension"). The separated peptides were revealed by ninhydrin detection and the spots obtained were found to be spread in a characteristic way over the area of the paper. This experiment showed that the two proteins differed in only one peptide, which was isolated and characterized. Ingram named this technique "fingerprinting" and it has often been used subsequently. The method was reviewed by Bennett (1967) and described with methodological details, and the term "peptide maps" was used for the designation of the results. Paper chromatography and electrophoresis could be replaced by repeating chromatographic runs on peptides on an automatic amino acid analyser; the use of such analysers and other automatic equipment to monitor peptide separations by column methods was reviewed in detail by Hill and Delaney (1967) and named "peptide mapping". The automated microprocedures for peptide separations (using amino acid analysers, ion-exchange columns and volatile buffers) were reviewed by Herman and Vanaman (1977) and the detection of peptides by fluorescence methods (by means of fluorescamine and o-phthalaldehyde) for these purposes was reviewed by Lai (1977). In the following part of this section, HPLC peptide mapping (finger-printing) will be dealt with.

Van der Rest et al. (1977) compared the human collagen "fingerprints" pro-duced by clostridiopeptidase A digestion and HPLC. A DC-4A resin column (part of a Durrum D-500 amino acid analyser; Durrum Instruments, Palo Alto, CA, U.S.A.) and Durrum sodium citrate buffer system were used to produce a discontinuous pH and ionic strength gradient. For monitoring, ninhydrin detection at 590 nm was employed. Type I, II and III collagens and the reconstitution of type I collagen were studied. Hancock et al. (1978) described the application of RP-HPLC to peptide mapping of proteins. Waters μBondapak Fatty Acid and μBondapak C_{18} (10 μm) columns (30 cm x 4 mm I.D.) were used. Acetonitrile-water mixtures (each containing 0.1% phosphoric acid) were used for elution. Thermolysin digests of acyl carrier protein derivatives were studied and other examples of application of this method were presented. The application of analytical peptide mapping by HPLC to the study of intestinal calcium binding proteins was examined by Fullmer and Wasserman (1979). All the separations were performed on a μBondapak C_{18} RP (10 μm) column (30 cm x 4 mm I.D.). A linear gradient from 0.1% ortho-phosphoric acid (solvent A) to acetonitrile (solvent B) at a constant flow-rate of 2 ml/min with column pressures varying from 500 to 1000 p.s.i. was used for elution. Tryptic digests were chromatographed. The time course of the digestion is illustrated in Fig. 13.1. The complete amino acid sequence of the three forms of bovine calcium binding protein and a very efficient procedure for peptide mapping by HPLC were also published by Fullmer and Wasserman (1980).

The mapping and isolation of large peptide fragments from bovine neurophysins and biosynthetic neurophysin-containing species by HPLC were described by Chaiken and Hough (1980). Tryptic fragments of performic acid-oxidized bovine neurophysin I and II were fractionated by RP-HPLC using γ-cyanopropyl-bonded phase columns (Zorbax CN). Elution was effected using a gradient from triethylammonium phosphate buffer to mixtures of this buffer with increasing proportions of acetonitrile. The effectiveness of HPLC mapping in detecting small amounts of neurophysin-related substances was demonstrated by a map obtained for ^{35}S-con-taining tryptic peptides derived from neurophysin-containing cell-free transla-tion products. The elution profile of the label was accompanied by that of an internal control of authentic neurophysin II peptides observed by monitoring at 215 nm. The correspondence of the major peaks established clearly the identity of the species.

Takahashi et al. (1981) examined an analytical and preparative method for peptide separation by HPLC on a macroreticular anion-exchange resin; the system could also be used for peptide mapping. Diaion CDR-10 (5-7 μm) was obtained from

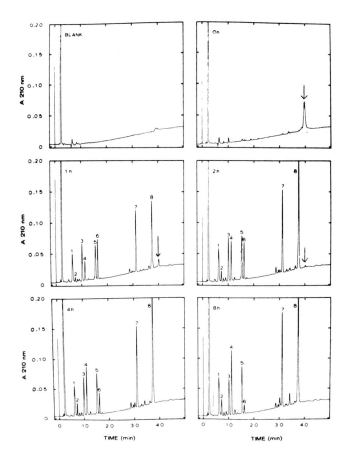

Fig. 13.1. Time course of the tryptic [TPCK (tosylphenylalanyl chloromethyl ketone)-treated] digestion of bovine intestinal calcium-binding proteins. All incubations were carried out in 0.2 M N-ethylmorpholinoacetate buffer (pH 8.1) with a trypsin:protein ratio of 1:100 (w/w). The protein concentration was approximately 1 μg/μl. A 20-μl volume (20 μg, 2 nmol) was removed from the incubation mixture at the time indicated, mixed with 5 μl of acetic acid and injected directly. Chromatographic conditions are described in the text. The blank contained buffer, trypsin and acetic acid only. A zero time control sample was removed immediately following the addition of trypsin and acidified. An arrow indicates the position of undigested protein. Note the disappearance of peak 6 with a coincidental increase of peaks 2 and 4 with time. (Reprinted from Fullmer and Wasserman, 1979.)

Mitsubishi Kasei Industry (Tokyo, Japan) and packed into a 50 cm x 0.4 cm I.D. column. A convex gradient from distilled water to 0.25 M methanesulphonic acid containing 50% (v/v) acetonitrile and 25% (v/v) isopropanol (adjusted to pH 8.2 with aqueous ammonia) was used for elution at a flow-rate of 0.89 ml/min and a starting temperature of 22°C, which was increased to 70°C in 22 min and then was maintained at this value until the end of the run. During the temperature programme the pressure dropped from 125 to 88 kg/cm^2. Tryptic digests of

calmodulin, bovine brain 14-3-2 protein and amyloid Bence-Jones protein NIG-51
were chromatographed. Prestidge et al. (1981) described an approach to the
semisynthesis of acyl carrier protein and confirmed that peptide mapping by RP-
HPLC was an extremely sensitive technique for detecting slight differences
between closely related proteins. Böhlen and Kleeman (1981) and Kleeman (1981)
also examined the analytical and preparative mapping of complex peptide mixtures
by RP-HPLC and found this approach to be very promising both for mixtures
containing large amounts of small peptides (generated, e.g., by enzymic digests)
and for simpler mixtures of large peptides (obtained, e.g., by treatment with
CNBr).

Bishop et al. (1981) studied the peptide mapping of tryptic digests of a
haemoglobin variants by hydrophilic ion-pair RP-HPLC using acetonitrile-water-
orthophosphoric acid (0.1%) elution systems. A μBondapak C_{18} (10 μm) column
(30 cm x 4 mm I.D.) was eluted at flow-rates of 1.5, 1.7 or 3.0 ml/min at $22^{\circ}C$.
The sample size varied from 50 to 250 μl and detection was at 230 or 254 nm.
The method should be particularly useful for examining variants that exhibit
minor sequence differences, such as Gly→Ala replacement. Abercrombie et al.
(1982, 1983) examined the use of RP-HPLC in structural studies of neuropeptides,
photolabelled derivatives and biosynthetic precursors. Cyanopropylsilyl packings
were found to be efficient and convenient for the separation of major tryptic
peptides of performic acid-oxidized or reduced and alkylated neurophysins.
Using a peptide mapping system the site of modification of a photoaffinity-
labelled derivative of bovine neurophysin II was identified. A DuPont Zorbax CN
HPLC column (25 cm x 0.46 cm I.D.), previously equilibrated with triethylammonium
phosphate (TEAP), was eluted with a linear gradient of acetonitrile mixed with
TEAP starting at 0% acetonitrile at 0 min and going to 40% acetonitrile at 60 min
at a flow-rate of 0.8 ml/min. Ageing of the columns was observed after 50-100
gradient runs.

Minasian et al. (1983) published a comparative study of the separation of the
tryptic peptides of the β-chain of normal and abnormal haemoglobins (Hb) by
RP-HPLC, and peptide mapping of aminoethylated Hb β-chain was found to be a
rapid and extremely versatile technique. Hearn et al. (1983a) described the
separation and assignment of tryptic peptides of human growth hormone (hGH) and
the 20 kdalton hGH variant. Here also chromatographic mapping procedures were
used successfully. Vensel et al. (1983) studied two-dimensional peptide mapping
by RP column chromatography and applied it to the sequence determination of
cytochrome c from a wild type and a mutant of the butterfly *Pieris brassicae*.
An initial fractionation at pH 7.2 with 100 mM potassium phosphate, followed
by chromatography with 0.1% TFA, was applied to the separation of chymotryptic
digests. The single residue difference was easily detected and identified using

this method. Meloun et al. (1985c) identified the structural differences in
acidic acrosin inhibitors from bull seminal plasma not only by TLC fingerprinting,
but also by peptide mapping examination of tryptic digests of carboxymethylated
acrosin inhibitors BUSI I A, B1 and B2. A 25 cm x 4.6 mm I.D. LiChrosorb RP-2
column was eluted with 0.1% TFA and after 10 min with a linear gradient from 0
to 60% methanol in 60 min.

13.1.4.1 Comments on literature
Hearn (1980) reviewed the use of RP-HPLC for the structural mapping of polypeptides and proteins. Fullmer and Wasserman (1982) dealt with the application
of HPLC to peptide mapping studies. Schroeder (1984) reviewed protein identification by peptide mapping. Kamp (1985a) described in detail practical guidelines
for peptide separation, including "fingerprinting" (mapping).

13.1.5 Protein (peptide) sequencing

As the HPLC of amino acid derivatives, permitting their efficient separation
and sensitive detection, is closely connected with the problems of amino acid
sequencing, the principles of these procedures will be briefly reviewed to
illustrate the purpose of HPLC methods, dealt with in Section 13.1.7.

13.1.5.1 Principle of the Edman sequencing method
The Edman (1949) method is the most important procedure for the determination
of amino acid sequences in proteins and peptides. It starts from the N-terminal
part of the peptide chain, which reacts easily and quantitatively (in mildly
alkaline medium) with phenyl isothiocyanate (PITC) to produce a phenylthio-
carbamoyl peptide (PTC-peptide). Anhydrous acids cleave the PTC-peptide and the
anilino-thiazolinone derivative of the N-terminal amino acid (ATZ-amino acid)
is split off from the remaining part of the peptide chain. After extraction of
the ATZ derivative from the reaction mixture, the shorter remaining peptide
chain is ready for the next sequencing step. The extracted ATZ-amino acid
derivative is then converted using an aqueous acid into the phenylthiohydantoin
(PTH) derivative, which can be used for chromatographic identification. The
order of reactions can be illustrated by the following scheme:

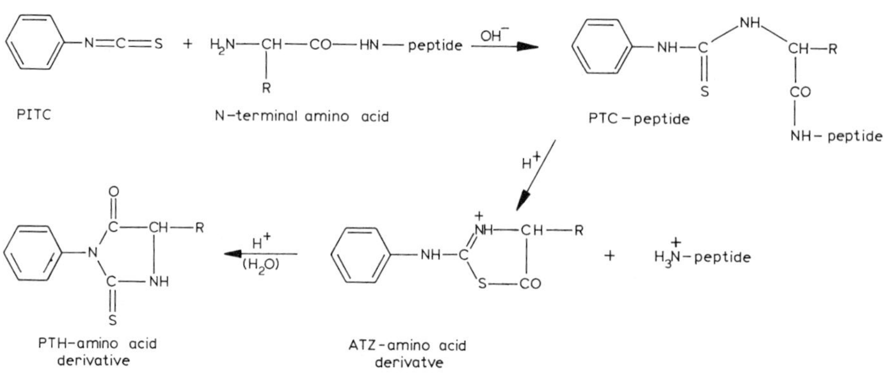

PITC   N-terminal amino acid   PTC-peptide

PTH-amino acid
derivative

ATZ-amino acid
derivatve

Edman (1956) studied the mechanism of the phenyl isothiocyanate degradation of peptides, and later Edman and Begg (1967) automated this process with the development of a protein sequenator, which was successfully applied in the determination of the primary structures of many proteins and polypeptides. This type of sequencing method cannot be directly used when the N-terminal amino acid is blocked, for instance if a pyroglutamyl residue is present at the N-terminus [such as in the β-chain of relaxin (Schwabe and McDonald, 1977)]. The Edman degradation method is not the only sequential method for the study of the primary structure of proteins [a survey of various chemical methods was compiled by Pavlík (1986)], but in every instance it is the most important and most often used procedure. The split-off PTH-amino acids (or other amino acid derivatives used for characterization) are identified by TLC, GC and nowadays mostly by HPLC; these methods will be reviewed in detail in Section 13.1.7.

13.1.5.2 Manual liquid-phase sequencing

The original Edman method (1949) was elaborated as a manual procedure. In spite of automation by means of sequenators of various types, the manual method is still important, because it is cheap and simple. In modern versions it may also be relatively rapid and can compete with expensive automated equipment. However, it is laborious and the yields of the products in the individual steps are not as high as in sequenators. The yield is very important in repetitive reactions, because it is decisive for the length of the peptide chain, which can be determined. Therefore, manual methods are often used for sequencing of shorter peptides and for orientation experiments.

Tarr (1975) described a general procedure for the manual sequencing of small amounts of peptides, based on the classical Edman cycle. Chang et al. (1976) examined the application of 4-(N,N-dimethylamino)azobenzene-4'-isothiocyanate (DABITC) as a chromophoric Edman reagent (structurally related to the pH indicator Methyl Orange) for protein sequence analysis, which after coupling, splitting off, and conversion, produces the 4-(N,N-dimethylaminoazobenzene)-4'-thiohydantoin (DABTH) derivative of the amino acid; this product is red in acidic media on a polyamide sheet in TLC and can also be used in HPLC identification. These compounds have the following formulae:

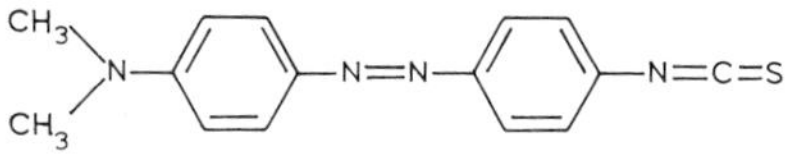

DABITC

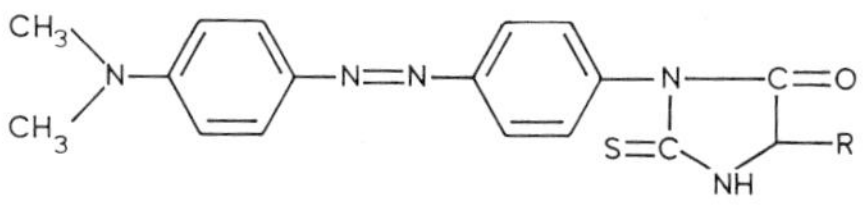

DABTH derivative of amino acid

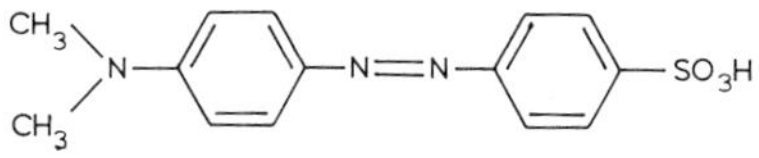

Methyl Orange

Pre-column derivatization of peptides with DABITC was described by Chang (1981a) (cf., Section 9.4.3.1) and this labelling of DABTC-peptides can also be employed for peptide sequencing.

Improved manual sequencing methods were reviewed in detail by Tarr (1977). Chang et al. (1978) developed a microsequence analysis of peptides and proteins using a DABITC-PITC double coupling method: the sample first reacts with DABITC and then with PITC in order to achieve quantitative coupling after the cleavage and conversion reaction. The coloured DABTH-amino acid derivatives are identified by TLC and also by HPLC (Chang, 1981c). Manual batchwise sequencing methods [modifications of Tarr's (1975) method] were reviewed by Tarr (1982). Practical procedures for modern manual sequencing methods were described in detail by Kuhn and Crabb (1985) and by Tarr (1986). Meloun et al. (1986) discussed the homology of bovine spleen cathepsin B with other SH-proteinases and clarified the corresponding parts of cathepsin B by both automatic Edman degradation and manual DABITC methods applied to peptides from complete tryptic digest, CNBr digest and limited proteolysis.

13.1.5.3 Automated liquid-phase sequencing (spin cup sequencer)

This most often used method was first developed by Edman and Begg (1967) and operates on the principle of the PITC degradion scheme. The automatic process of the formation of the PTC-derivative of a protein and the splitting off of the terminal amino acid as the thiazoline ATZ derivative proceeds in a continuously spinning cylindrical cup. The protein solution is spread as a thin film over the inner surface of the lower part of the cup and evaporated. Reagents are pumped in so as to flow over the protein from the bottom to the small trough in the inner part at the top of the spinning cup; they are sucked from this trough to undergo further processing. The thiazolinones (ATZ derivatives) are converted into the corresponding PTH derivatives in a separate operation in special test-tubes and identified by TLC.

The sequence of 60 amino acids from the N-terminus of humpback whale myoglobin was established in this way. The degradation proceeded at a rate of 15 cycles in 24 h with a yield in individual cycles in excess of 98%. An amount of 0.25 μmol of protein was required for the analysis.

Many improvements have been published that have brought automatic sequencing to its present high level and some of them are discussed here. Niall (1969) described the sequential analysis of proteins and peptides in a commercial Beckman-Spinco sequenator. The application of 0.1 M Quadrol [N,N,N',N'-tetrakis-(2-hydroxypropyl)ethylenediamine non-volatile buffer] in the microsequence analysis of proteins and peptides was examined by Brauer et al. (1975); this method permitted extensive degradation in the absence of a carrier protein. Wittmann-Liebold et al. (1976) constructed a device coupled to a modified sequenator for the automatic conversion of anilinothiazolinones (ATZ) into PTH-amino acids. Hollaway et al. (1979) reviewed the application of HPLC in struc- tural studies of proteins, Henschen-Edman and Lottspeich (1980) dealt with aspects of automated and microscale sequencing, including the HPLC of PTH amino acids. Van Beeumen et al. (1980) developed an isocratic liquid chromatographic method for all common PTH-amino acids, suitable for sequencing purposes.

Bhown et al. (1981) presented an improved procedure for high-sensitivity microsequencing involving the use of aminoethylaminopropyl glass beads (for the in situ purification of Quadrol) in a Beckman sequencer and an Ultrasphere ODS column for PTH-amino acid identification. The design and operation of a com- pletely automatic Beckman microsequencer were described by Rodrigues et al. (1984). A system was built for the automatic conversion of ATZ derivatives extracted from the spinning cup with subsequent on-line HPLC separation of the PTH derivatives. The auto-convertor auto-sampler system was controlled by a tape programmer or microprocessor and operated by transfer of the sample from the conversion vial into an HPLC injection loop by nitrogen pressure. Sub-nanomole amounts of proteins could be sequenced. The complete primary structure of thermitase from *Thermoactimyces vulgaris* and the structural features of this subtilisin-type proteinase were published by Meloun et al. (1985a); both an automated Beckman sequencer and a manual DABITC method were used for sequencing.

Ashman (1985) described in detail the use of on-line HPLC for PTH-amino acid identification. The on-line detection of amino acid derivatives released by the automated Edman degradation of polypeptides was also dealt with by Wittmann-Liebold and Ashman (1985).

A modification of the on-line principle of the analysis of PTH-amino acids during the sequential degradation of proteins and peptides using optical sensor-controlled injection was developed by Pavlík et al. (1988).

13.1.5.4 Solid-phase sequencing

A certain degree of solubility of a sample polypeptide in the reagent solutions was a major and unpleasant problem in liquid-phase sequencing. Solid-phase sequencing is characterized by immobilization of the sample polypeptide on to a solid insoluble carrier (an organic copolymer similar to the carrier for Merrifield peptide synthesis, or modified glass spheres) at the C-terminus. Modified amino acids are gradually split off from the free N-terminus of the polypeptide chain using the Edman method. Solid-phase sequencing can be operated both manually and by means of a sequenator. Such equipment consists of a short column containing the immobilized sample to be analysed, through which the reagent solutions are slowly filtered. Because the immobilization yield is often low and the method is laborious, it is often used only for the sequencing of shorter peptides, which leads to problems in liquid-phase sequencing owing to their high solubility. To increase the sensitivity of the method, a combined DABITC-PITC method was used for the identification of split-off amino acids.

Laursen (1971) was the first to develop the solid-phase Edman degradation of peptides by means of an automatic peptide sequencer. An improved peptide resin was used for immobilization, a complete cycle required 2 h and the degradation yields averaged 94% per cycle. An amount of 0.1 µmol of the insulin B-chain was degraded through 18 cycles and then the identification of PTH derivatives became ambiguous. The automated conversion of ATZ- to PTH-amino acid derivatives was also applied in solid-phase instruments (Bridgen, 1977). Wilson and Hughes (1980) described the use of DABITC in automatic sequencing and DABTH identification by HPLC. Sequencing was described in both the liquid phase (in the presence of Polybrene) and solid phase, following attachments via Lys residues to p-phenylene diisothiocyanate-glass, or via the lactone formed after Trp cleavage by BNPS-skatole or Tyr or His cleavage by N-bromosuccinimide to 3-aminopropyltriethoxy-silane glass. The methods of immobilization on to the solid carriers were explained by Laursen and Machleidt (1980). Fully automated solid-phase sequencing with on-line identification of PTHs by HPLC was described by Machleidt and Hofer (1980).

Hughes et al. (1981) discussed methods for the isolation of proteins in amino acid analyses with the exclusive use of standard HPLC equipment. As an example of microsequence analysis a series of RP-HPLC analyses from the solid-phase degradation of 200 pmol of the basic proteinase inhibitor from the bovine pancreas were illustrated by chromatograms. Kamp et al. (1984), in their extensive work on ribosomal proteins, used (in addition to liquid-phase sequencing of larger proteins) also solid-phase sequencing, both manual and automatic. The proteins were attached to p-phenylene diisothiocyanate-activated amino glass by the DABITC-PITC double coupling method (Chang et al., 1978; Salnikow et al., 1981). The released DABTH derivatives of amino acids were identified by TLC or isocratic HPLC. Automated solid-phase micro-sequencing using DABITC on-column immobilization of proteins was reviewed with practical details by Salnikow (1985). A rapid solid-phase protein microsequencer was described by Walker et al. (1986) that was designed to determine amino acids sequences with sub-nanomole amounts of protein. The degradation cycle took only 24 min, i.e. it was more than twice as fast as any other sequencer, and the strongly polar amino acids (such as phosphoserine) could be recovered quantitatively and identified without difficulty. This solid-phase sequencer was found to be suitable also for strongly hydrophobic proteins (proteolipids from membranes). The construction of this equipment and examples of applications were demonstrated on subunits of mito-chondrial F_1-ATPase, on a protein isolated from mouse gap junctions and on mitochondrial phosphate-transport protein.

13.1.5.5 Gas-liquid solid-phase sequencing

One of the latest principles of sequencers is a gas-liquid solid-phase peptide and protein sequenator, described by Hewick et al. (1981). The sample is applied on a membrane composed of glass fibres, which is kept in a special glass chamber. After wetting with PITC solution, the membrane is "washed" by a stream of an inert gas saturated with vapour of trimethylamine or triethylamine. After the formation of the PTC derivative, vapour of trifluoroacetic acid is applied in the same way as before. The ATZ derivatives formed are extracted into the con-version vessel and PTH derivatives of amino acids are automatically injected into the connected HPLC equipment. Various improvements of this system were described (e.g., Hawke et al., 1985). The gas-phase sequencing of peptides and proteins was reviewed in detail and in a practical manner by Reimann and Wittmann-Liebold (1985).

13.1.5.6 C-Terminal sequencing

In addition to the immobilization of the peptide chain using the binding reaction at the C-terminus (N-terminus is sequenced), immobilization via the N-terminus was also used in experiments with sequencing starting from the C-terminus. Because the C-terminal sequencing approach has not yet been developed into a routinely applied method and is not used very often (in spite of its potential importance in protein chemistry), it is mentioned here only briefly, without a detailed discussion.

Stark et al. (1968) described the sequential degradation of peptides from their carboxyl termini using the reaction with ammonium thiocyanate (rhodanide) in an activated solvent containing acetic anhydride. Peptides react under mild conditions to form peptidylhydantoins, which can be cleaved in aqueous solution with acetohydroxamate to form the thiohydantoin derivative of the original C-terminal amino acid. The carboxyl is freed from the shorter acetylated peptide, in which a new amino acid residue (in the original peptide the last but one residue) becomes terminal:

```
Peptide-CO-HN-CH-COOH + NH4NCS  ----------->  Peptide-CO-N------CH-R
              |                                           |       |
              R                                         S=C      C=0
                                                          \      /
                                                           N
                                                           H

(Original C-terminal    (Ammonium          (Peptidyl thiohydantoin)
 amino acid)             thiocyanate)

                      HN------CH-R
                       |       |                +       Peptide-COOH
                      S=C     C=0
                        \     /
                         N
                         H
                   (2-Thiohydantoin derivative       (Shorter peptide)
                    of C-terminal amino acid)
```

Some amino acids (proline, aspartic acid) are not removed in this way.

Meuth et al. (1982) investigated the stepwise sequence determination from the carboxyl terminus of peptides (also in the solid-phase version) and employed the above-mentioned method developed by Stark et al. The two steps of the degradation were studied using various model peptides. In addition to other methods, RP-HPLC was also used for the rapid and sensitive identification of C-terminal residues. The amino acid 2-thiohydantoins were separated on a 30 cm x 4 mm I.D. column of MicroPak MCH-10, eluted with a linear gradient formed by mixing acetonitrile solution in sodium acetate buffer (pH 3.7) with the same buffer.

The solid-phase COOH-terminal sequential degradation was reviewed by Dabre (1977). Pavlík (1986) discussed a series of miscellaneous methods published on C-terminal sequencing, including their detailed chemistry.

Chang et al. (1981b) described the application of dimethylaminoazobenzene-sulphonyl (DABS) derivatives of amino acids in another type of C-terminal sequence analysis of polypeptides at the picomole level: amino acids were split off from the C-terminus of the peptide chain by carboxypeptidase and the DABS derivatives were sensitively identified by RP-HPLC.

13.1.5.7 Dipeptide splitting-off method

In addition to the determination of the amino acid sequence in polypeptides by means of stepwise degradation methods, other methods of systematic sequencing were also investigated. One is the application of dipeptidyl aminopeptidase I, which splits the peptide chain into dipeptides. If this enzymic digestion is coupled with one step of an Edman degradation, a suitable method may be created for sequential studies of oligopeptides, because overlapping sets of dipeptides can be prepared in this way.

McDonald et al. (1969) presented new observations on the substrate specificity of cathepsin C (dipeptidyl aminopeptidase I), including the degradation of β-corticotropin and other peptide hormones. McDonald et al. (1972) reviewed in detail the preparation and specificity of dipeptidyl aminopeptidase I, and Callahan et al. (1972) covered methods of sequencing of peptides with dipeptidyl aminopeptidase I. For this approach the efficient chromatographic separation of mixtures of dipeptides is important. Dizdaroglu and Simic (1980) described the separation of underivatized dipeptides by HPLC on a bifunctional weak anion-exchange bonded phase. The mixture of dipeptides was eluted from a Varian MicroPak AX-10 column (30 cm x 0.4 cm I.D.) by an isocratic method using various mixtures of 0.01 M triethylammonium acetate (TEAA) buffer and acetonitrile. TEAA buffers allowed detection at 200-220 nm. Complex mixtures of dipeptides were well separated. This approach was briefly reviewed by Dizdaroglu (1981).

13.1.5.8 Comments on literature

In addition to the literature cited above, much detailed and useful information on techniques important for amino acid sequencing in proteins was given by Hirs and Timasheff (1977). Birr (1980) edited the Proceedings of a conference on solid-phase methods of sequencing, Liu et al. (1981) the Proceedings of a conference on the chemical synthesis and sequencing of peptides and proteins and Lottspeich et al. (1981) a book on HPLC in protein and peptide chemistry, in which amino acid sequence analysis was also dealt with. Elzinga (1982) edited a book on methods in protein sequence analysis and Wittmann-Liebold (1982) evaluated the current status of protein sequencing. Wittmann-Liebold (1983) also dealt with the advanced automatic microsequencing of proteins and peptides. Tschesche (1983, 1985) edited books on modern methods in protein chemistry, which consist of review articles. Bhown and Bennett (1984) reviewed the use of HPLC in protein sequencing and Shively (1987) edited a practical manual on the microcharacterization of polypeptides.

*13.1.6 Application of mass spectrometry in combination with HPLC to peptide
identification and sequencing*

The combination of liquid chromatography with mass spectrometry (LC-MS) is
an independent and interesting branch in the spreading tree of peptide sequencing
methods. Chromatographic fractions can be detected and identified by mass
spectrometry (MS). Ions of fragments of peptide chains can be arranged in order
of their exact molecular weights and the amino acid sequence can be derived from
these data. However, not all laboratories can afford to develop this principle.
Modern MS instruments, required for good results, are very expensive and a high
level of technical expertise is demanded for their operation. In addition, the
interpretation of the results is complicated by the presence of numerous side-
products formed in the derivatization process. Only relatively short peptides
have been studied using this technique. Several examples of work of this type
will be mentioned.

Novotný et al. (1972) investigated the uniformity and species-specific
features of the N-terminal amino acid sequence of porcine immunoglobulin λ-chains.
The N-terminal homoserine-containing nonapeptide was isolated from the CNBr
digest of the λ-chain, purified by GPC, O,N-perbenzoylated according to Wilkas
and Lederer (1968), examined by MS and compared with the amino acid sequence of
homologous proteins. Dawkins et al. (1978) investigated polypeptide sequencing
by LC-MS. An automatic system utilizing a liquid chromatograph-mass spectrometer-
computer system was tested using simple model oligopeptides. Nasimov et al.
(1980) studied the HPLC-MS of 5-substituted 3-[*p*-(3'-phenylindenonyl)phenyl]-2-
thiohydantoin (ITH) amino acid derivatives (cf., Section 13.1.7) for sequence
analysis of proteins. A method was described employing RP-HPLC and chemical
ionization MS, which followed sequence Edman degradation by using 2-*p*-isothio-
cyanophenyl-3-phenylindone; the sensitivity of the automated degradation was
50-100 pmol. RP-HPLC was carried out on a 25 cm x 4.6 mm I.D. Zorbax ODS column
eluted with 0.01 *M* sodium acetate solution (pH 5.6) using an acetonitrile
gradient. A standard mixture of twenty amino acids was separated into seventeen
fractions in 18 min; ITH-Leu and -Ile were unresolved. The relative standard
deviation was 1% and the limit of sensitivity of UV detection was 5 pmol (according to the authors, this is lower than for PTH-amino acids).

Yu et al. (1981) examined the analysis of N-acetyl-N,O,S-permethylated
peptides by combined LC-MS using a moving belt interface. Quadrupole MS was
employed and examples were given of the LC-MS of N-acetyl-permethylated Leu-
enkephalin and a mixture of derivatized peptides. It was shown that C-methylated
peptides can be separated by LC and identified by MS. In all instances sequence
information was obtainable by this approach. A Varian 5000 HPLC instrument and

a Finnigan 4000 mass spectrometer equipped with a Finnigan LC-MS moving-belt interface were connected to facilitate the on-line procedure. For HPLC a Supelcosil LC-8 column (150 mm x 4.6 mm I.D.) was used, eluted with a linear gradient from 5% to 70% acetonitrile in water in 20 min at a flow-rate of 1 ml/min. The amount of sample of permethylated peptides was about 1 μg in 5 μl of acetonitrile. A stream of hot nitrogen was passed through the HPLC effluent to create an aerosol spray for deposition of the effluent on the belt.

Baranowski et al. (1982) identified the enzymically produced peptide fragments of angiotensin I (generated by dipeptidyl carboxypeptidase, an angiotensin-converting enzyme) by LC-MS. The peptide fragments were derivatized with fluorescamine, separated by RP-HPLC and His-Leu derivatives were identified by MS. A Partisil ODS column (25 cm x 0.45 cm I.D.) was used for the separation and acetonitrile diluted 9:1 with 1 M acetic acid (final pH 3.5) for isocratic elution. In another experiment 38% acetonitrile in 0.1 M ammonium acetate solution (pH 4.0) was used for isocratic elution. For MS a Varian MAT 112S mass spectrometer was used. Samples were applied in an off-line procedure.

Desiderio and Stout (1984) reviewed field desorption mass spectrometry (FDMS) and reported recent data obtained with an off-line combination of HPLC and FDMS (to measure the levels of neuropeptides in biological tissue) and also the use of FD and FAB (fast atom bombardment)-MS techniques to obtain amino acid sequence information on biologically active peptides.

13.1.7 Separation of amino acid derivatives important for protein (peptide) sequencing and HPLC amino acid analyses

Phenylthiohydantoin (PTH) derivatives of amino acids (cf., Section 13.1.5) can be identified by TLC, GLC (some of them after trimethylsilylation to obtain volatile products) or after back-hydrolysis to the free parent amino acids, followed by their chromatography. Nowadays, HPLC is the most often used method for the identification and quantification of PTH-amino acids for the purpose of protein (peptide) sequencing. In special cases the HPLC of PTH-amino acids is also used for the quantitative analysis of total hydrolysates of peptides. However, not only PTH derivatives are employed in protein (peptide) sequencing or for amino acid analyses, and other suitable coloured (or fluorescent) derivatives are also used. Examples of such procedures will be described in the following Sections.

13.1.7.1 PTH-amino acids

Many papers have been published on this topic and examples are summarized in Table 13.1. NPC on silica was the first approach to the HPLC of PTH-amino acids.

TABLE 13.1

HPLC SEPARATION OF PTH-AMINO ACIDS: A CHRONOLOGICAL SURVEY OF EXAMPLES

Type	Packing material; column dimensions (cm × mm I.D.)	Elution program; elution solutions	Temperature; flow-rate; pressure; total time	Detection; (absorbance, nm); sensitivity or amount applied	Notes (AA = amino acid)	Ref.
NPC	Silica (100 × 1.8)	Isocratic: (I) heptane-CHCl$_3$, (50:50, v/v); (II) 100% CHCl$_3$; (III) 3% CH$_3$OH in CHCl$_3$	25°C; 1 ml/min; 1800 p.s.i.	A254; 25 µg	PTH derivatives were divided into 3 groups according to polarity and differently eluted	Graffeo et. al. (1973)
NPC	Mercosorb SI 60 (20 or 30 × 3)	Isocratic: (I) CH$_2$Cl$_2$-tert.-butanol-dimethyl sulphoxide (e.g., 500:4:0.4, v/v); (II) CH$_2$Cl$_2$-dimethyl sulphoxide-water (e.g., 80:15:2, v/v)	Ambient; 250 bar; 7–10 min	A260; 5 nmol	Hydrophilic and hydrophobic derivatives were eluted separately	Frank and Strubert (1973)
RPC	C$_{18}$ Corasil (3 ft. × 2 mm)	Gradient B into A at 2%/min: (A) 0.01 M sodium acetate (pH 3.8)-acetonitrile (95:5); (B) acetonitrile	Ambient; 0.8 ml/min; 700 p.s.i.; 30 min	subnano-mole range	Common PTH-AA separated	Zimmerman et al. (1973)
NPC	Silica (25 × 2.1)	Concave gradient from hexane-CH$_3$OH-propanol (3980:9:11) to CH$_3$OH-propanol (9:11)	1000 p.s.i.; 40 min	A254; 2–5 nmol	All PTH-AA except Arg and His resolved	Matthews et al. (1975)

(Continued on p. B400)

TABLE 13.1 (continued)

Type	Packing material; column dimensions (cm × mm I.D.)	Elution program; elution solutions	Temperature; flow-rate; pressure; total time	Detection; (absorbance, nm); sensitivity or amount applied	Notes (AA = amino acid)	Ref.
RPC	Zorbax ODS (25 × 2)	Linear gradient of B into A at 1%/min: (A) acetonitrile-0.01 M sodium acetate, pH 5.0 (5:95, v/v); (B) acetonitrile.	50°C; 0.5 ml/min; 2000 p.s.i.; 20 min	A254; 5 pmol	All PTH-AA eluted; His and Arg unresolved	Zimmerman et al. (1976)
		Isocratic elution: 50% 0.1 M sodium acetate (pH 5.0) + 50% acetonitrile	50°C; 0.5 ml/min; 2000 p.s.i.; 16 min	A254; 5 pmol	PTH-His and -Arg resolved separately	
RPC	Zorbax ODS (22 × 4.6)	Gradient of B into A according to given time scale: (A) 0.01 M sodium acetate (pH 4.5); (B) acetonitrile	62°C; 1 ml/min; 375 p.s.i.; 20 min	A254; 1-2 nmol	20 PTH-AA separated	Zimmerman et al. (1977)
RPC	LiChrosorb RP-8 (5 μm) (two 15 × 3.9 columns in series)	Isocratic: acetonitrile-0.01 M sodium acetate, pH 4.6 (25:75, v/v)	50°C; 0.9 ml/min; 3000 p.s.i.; 75 min	A254; 1 nmol	21 PTH-AA separated	Abrahamsson et al. (1978)
Special	Val-Ala-Pro immobilized on Partisil-10 (30 × 2.1)	Isocratic: 1% citric acid in water, pH ≈ 2.5	1 ml/min; 66 min	A254	13 PTH-AA separated	Fong and Grushka (1978)
RPC	Hypersil ODS or Spherisorb ODS, both 5 μm (23 × 4.6)	Gradient from 27% to 65% B in 55 min: (A) 0.01 M sodium acetate in water; (B) 0.01 M sodium acetate in CH3OH	55 min	A254	24 PTH-AA derivatives separated	Østvold et al. (1978)

RPC	μBondapak C$_{18}$ (30 × 4)	Concave gradient running from 5% to 100% B in 40 min: (A) 8 mM diethylenetriamine (DETA), 20 mM in dichloroacetic acid (DCA) (pH 4.2); (B) buffer A containing 60% (v/v) acetonitrile	Ambient; 4.0 ml/min; 36 min	$A268$; 10 nmol	16 PTH-AA resolved (Pro + Val and Tyr + Phe + norLeu gave common peaks)	Annan (1979, 1981)
NPC	LiChrosorb Si 60 (5 μm) (15 × 4.8)	Gradient: (I) heptane; (II) CH$_2$Cl$_2$-C$_2$H$_5$OH-H$_2$O (89:7:10:0.3, v/v). For 6 min isocratic 10% II; then gradient from 10% to 91% II in 25 min, finally isocratic 91% II.	100 ml/h; 40-50 bar; 35 min	$A254$	13 PTH-AA separated, others in mixtures	Caude (1979)
RPC	μBondapak C$_{18}$ (30 × 4)	Linear gradient from 95% A + B to 55% A + 45% B in 15 min, then isocratic: (A) water-CH$_3$OH-CH$_3$COOH-acetone (900:100:2.5:0.05, v/v) (pH 4.1); (B) water-CH$_3$OH-CH$_3$COOH (100:900:0.25, v/v).	2.5 ml/min; 20 min	$A254$	20 PTH-AA eluted, 17 well separated	Hollaway et al. (1979)
RPC	2 μBondapak C$_{18}$ columns	Gradient from 5% to 90% B in 45 min: (A) CH$_3$CN-0.01 M sodium acetate, pH 4 (10:90); (B) CH$_3$CN- same buffer (90:10)	2 ml/min; 40 min	$A254$	15 PTH-AA separated (part of a review)	Schaettle (1979)
RPC	R Sil C$_{18}$ HL (10 μm) (30 × 4.6)	Isocratic: 36% CH$_3$CN in 0.01 M sodium acetate, pH 4.5	64°C; 34 min		22 PTH-AA separated	Van Beeumen et al. (1980)

(Continued on p. B402)

TABLE 13.1 (continued)

Type	Packing material; column dimensions (cm × mm I.D.)	Elution program; elution solutions	Temperature; flow-rate; pressure; total time	Detection; (absorbance, nm); sensitivity or amount applied	Notes (AA = amino acid)	Ref.
RPC	Ultrasphere ODS (5 μm) (25 × 4.6)	Gradient from 0 to 40% B in 20 min, then isocratic to 30 min, then gradient from 40% to 0% B in 5 min: (A) 5% tetrahydrofuran (THF) in 5.25 mM acetate, pH 5.15; (B) 90% CH_3CN, 10% THF.	45°C; 1.3 ml/min; 35 min	$A254$; 0.5–1 nmol	21 PTH-AA separated	Cooke and Olsen (1980)
RPC	μBondapak C_{18} (30 × 3.9)	Concave gradient from 13 to 34% of ethanol in ammonium acetate, pH 5.1 (defined by conductivity, 25 μS, corresponding to about 0.1 mM) in 10 min, then isocratically to 24 min, then stepwise to final 13% up to 30 min.	37°C; 1000 p.s.i.; 1.1 ml/min; 30 min	$A254$; 2 nmol	21 PTH-AA well separated (cf., Fig. 13.2)	Fohlman et al. (1980)
RPC	μBondapak C_{18} (30 × 3.9)	Linear gradient with 0.01 M sodium acetate in 10–48% ethanol	1 ml/min; 43 min	$A254$, $A267$ or $A280$; 50 pmol	Comparison of various packing materials and chromatographic conditions; 25 derivatives were chromatographed	Greibrokk et al. (1980)

	Column	Mobile phase	Conditions	Detection	Results	Reference
RPC	Bondapak C$_{18}$ (30 × 4)	Gradient: 5% to 45% B: (A) 900 ml water + 2.5 ml acetic acid, adjusted to pH 4.4 (1 M NaOH), + 100 ml CH$_3$OH and 0.05 ml acetone; (B) 100 ml water, 900 ml CH$_3$OH, 0.25 ml acetic acid	Ambient; 2 ml/min; 30 min	A254; 0.5–1 nmol	17 PTH-AA resolved. Three chromatographic systems compared with particular reference to Glu and Asp derivatives	Harris et al. (1980)
RPC	μBondapak Phenyl-alkyl (30 × 4)	Linear gradient from 20% to 55% B over 60 min: (A) 0.085% propionic acid, pH 3.8; (B) 90% CH$_3$OH	1.0 ml/min; 30 min	A268; 0.5–1 nmol	21 PTH-AA derivatives resolved	Henderson et al. (1980)
RPC	μBondapak C$_{18}$ (30 × 4.6)	Isocratic: 67% 0.01 M sodium acetate, pH 5.4 with acetic acid, + 33% acetonitrile + 5% dichloroethane	62°C; 1 ml/min; 23 min	A254; 2 pmol	18 peaks resolved	Henschen-Edman and Lottspeich (1980)
RPC	Hibar 250-4 LiChrosorb RP-18 (5 μm) (25 × 4.6)	Isocratic: 68.5% 0.01 M sodium acetate, pH 5.2 + 31.5% acetonitrile + 0.5% dichloroethane	62°C; 1.5 ml/min; 16 min	A254; 3 pmol	18 PTH-AA resolved	Lottspeich (1980)
RPC	A tandem of two columns: DuPont Zorbax ODS followed by Merck Hibar RP-18 (25 × 4.6)	Isocratic: 68% 0.035 M sodium acetate, pH 4.32 + 32% acetonitrile	52°C; 2 ml/min	A269	23 PTH-AA resolved with automatic "peak-picking"	Rose and Schwartz (1980)

(Continued on p. B404)

TABLE 13.1 (continued)

Type	Packing material; column dimensions (cm × mm I.D.)	Elution program; elution solutions	Temperature; flow-rate; pressure; total time	Detection; (absorbance, nm); sensitivity or amount applied	Notes (AA = amino acid)	Ref.
RPC	Spherisorb 5S ODS (25 × 4.6)	Up to 4 min isocratic 10% B, then gradual linear gradient to 30% B in 5 min, linear gradient to 35% B in 7 min, reversed gradient to 10% B in 1 min and isocratic elution (10% B) for 3 min: (A) 33 mM sodium acetate, pH 5.1; (B) C_2H_5OH	62°C; 2 ml/min; 95 atm; 20 min	$A254$; 1 nmol	All commonly encountered PTH derivatives separated. Column packed in the laboratory. The consumption of cheap and non-toxic C_2H_5OH is 8.6 ml per analysis. After 300 analyses washing with C_2H_5OH is necessary	Sottrup-Jensen et al. (1980)
RPC	Zorbax ODS	Quasi-isocratic system containing CH_3CN	20-30 min	$A254$		Zalut and Harris (1980)
RPC	Ultrasphere ODS (5 μm) (15 × 4.6)	Programme: 2 min isocratic, 77% A + 23% B, 5 min linear gradient to 47% A + 53% B, then isocratic 13 min: (A) 0.04 M sodium acetate, pH 3.72, containing 50 μl/l of acetone; (B) 100% CH_3OH, containing 250 μl/l of CH_3COOH	1.5 ml/min; 20 min	$A254$; 400-500 pmol	Standard mixture of 17 PTH derivatives chromatographed	Bhown et al. (1981)
RPC	Zorbax ODS (15 × 4.6)	Stepwise: (I) 0.04 M sodium acetate, pH 5.0-acetonitrile (85:15, v/v), 5 min (II) same components (58:42, v/v), 7 min	62°C; 1 ml/min; 435 p.s.i.; 12 min	$A254$; 10-20 nmol	All common PTH derivatives separated; cf. Fig. 13.3	Bledsoe and Pisano (1981)

RPC	Si 100 RP C$_{18}$ (30 cm length)	Isocratic: acetonitrile–water (45:55, v/v) + 0.5% H$_3$PO$_4$	1.9 ml/min; 90 bar; 8 min		12 PTH-AA separated as illustration in broader study	Engelhardt (1981)
RPC	PRP-1 (polystyrene copolymer, 10 μm) (15 × 4.1)	Gradient elutions: acetonitrile mixed in various proportions with different 0.01 M buffers of various pH (1.6, 5.5, 6.1, 11.0)	1 ml/min		Group separation achieved in the frame of a broader study of PRP-1 applications	Iskanderani and Pietrzyk (1981)
RPC	Nucleosil 10 C$_{18}$ (30 × 4.6)	Linear gradient: methanol–water 0–100% in 15 min. Isocratic elutions using several mixtures; the best was 42% tetrahydrofuran in 1 mM perfluorooctanoic acid	1.5 ml/min; 15 min	A_{254}	As an illustration in a broader study 8 PTH-AA were perfectly separated	Schoenmakers et al. (1981)
RPC	Ultrasphere ODS (5 μm) (25 × 4.6)	Isocratic: 48% acetonitrile 0.065 M in acetic acid, adjusted to pH 4.54 with NH$_3$	55°C; flow-rate varied, 0.5 (0.5 min), 1.0 (4.5 min), 2.0 (3.5 min) and 0.5 ml/min (0.5 min); max. 18 MPa; 6–9 min	A_{269}; 100 pmol	All common PTH-AA separated; cf. Fig. 13.4	Tarr (1981)

(Continued on p. B406)

TABLE 13.1 (continued)

Type	Packing material; column dimensions (cm × mm I.D.)	Elution program; elution solutions	Temperature; flow-rate; pressure; total time	Detection; (absorbance, nm); sensitivity or amount applied	Notes (AA = amino acid)	Ref.
RPC	Ultrasphere ODS (5 µm) (25 × 4.6)	Stepwise: (I) 25% (v/v) acetonitrile in 0.05 M ammonium + triethylammonium acetate, pH 4.5, prepared according to special requirements; (II) 50% (v/v) acetonitrile in 0.05 M ammonium acetate. Step I to II followed the appearance of PTH-Gly	50°C; 1 ml/min; 1600 p.s.i.; 12 min	A_{254}; 250 pmol	Excellent separation of all common PTH-AA. Molar absorptivities of PTH-AA summarized	Black and Coon (1982)
RPC	Ultrasphere ODS	Gradient from 7 to 95% B in a described time sequence: (A) 10% acetonitrile + 90% trifluoroacetate-acetate buffer (33 mM-2 mM; pH 5.6); (B) 75% acetonitrile + 25 % trifluoro-acetate buffer (35 mM, pH 3.6)	48°C; 1.3 ml/min; 17 min	A_{254}; 200–300 pmol	Common PTH-AA resolved. Two supports and three buffer systems compared. The given procedure was the best	Hawke et al. (1982)
RPC	Ultrasphere ODS (25 cm × 4.6)	Isocratic: 0.01 M sodium acetate (pH 4.9)-acetonitrile (62.2:37.8). Gradient 27% to 48.6% B in 12 min, then immediately to 30% and held for 1 min, then gradient to 49.8% in 10 min and held for 2 min, then returned to 27%	55°C; 1 ml/min; 30 min. 55°C; 1 ml/min; 25 min	A_{254}	Common PTH-AA separated. Two supports compared. Optimization considerations	Noyes (1983)

	Column	Solvent system	Conditions	Detection	Notes	Reference
RPC	Microbore-1 column packed with nitrile alkylsilane bonded phase PTH AA-4 (15 × 1)	Gradient elution, ternary solvent system A-B-C according to a timing sequence: (A) 0.01 M sodium acetate, pH 5.0; (B) methanol; (C) acetonitrile	30°C; 50 µl/min	A_{270}; 0.5-1 pmol	20 PTH-AA resolved	Cunico et al. (1984)
RPC	Ultrasphere ODS (25 × 4.6)	Linear gradient system of acetonitrile in sodium acetate buffer (0.03 M, pH 4.2): 25% acetonitrile (1 min), 25-55% acetonitrile (8 min), 55-60% acetonitrile (6 min), 60% acetonitrile (7 min)	35°C; 1.2 ml/min; 17 min	A_{269}	22 PTH-AA separated. A novel method for the determination of Ser was described (the yield of Ser after Edman degradation was 70-80%)	Pavlík and Kostka (1985)

B408

However, it was difficult to separate all PTH-derivatives in a single chromatographic run; the hydrophobic and hydrophilic derivatives had to be separated
separately and strongly polar derivatives caused difficulties similar to those
in GLC. RPC brought the final solution for the separation of PTH-amino acids.
It can be readily seen from Table 13.1 how the persistent endeavours of a group
of researchers led gradually to very good present-day separations.

TABLE 13.2

ONE- AND THREE-LETTER SYMBOLS FOR AMINO ACIDS

IUPAC-IUB Commission (1968); see also Table 4 in Mikeš (1976).

Symbol	Abbreviation	Amino acid
A	Ala	Alanine
B	Asx	Aspartic acid or asparagine
C	Cys	Cysteine
D	Asp	Aspartic acid
E	Glu	Glutamic acid
F	Phe	Phenylalanine
G	Gly	Glycine
H	His	Histidine
I	Ile	Isoleucine
K	Lys	Lysine
L	Leu	Leucine
M	Met	Methionine
N	Asn	Asparagine
P	Pro	Proline
Q	Gln	Glutamine
R	Arg	Arginine
S	Ser	Serine
T	Thr	Threonine
V	Val	Valine
W	Trp	Tryptophan
X		Undetermined or atypical amino acid
Y	Tyr	Tyrosine
Z	Glx	Glutamic acid or glutamine or pyrrolidone carboxylic acid
-		Deletion

Three examples of RPC are illustrated here. The results of the application
of gradient elution according to Fohlman et al. (1980) are shown in Fig. 13.2
and a stepwise elution (according to Bledsoe and Pisano, 1981) in Fig. 13.3.
However, isocratic elution is now often recommended for monitoring the Edman
procedure when carrying out protein sequencing often and in series, in spite
of the fact that the gradient elution facilitates better resolution. There are
two reasons for this popularity of isocratic elution: (i) a constant baseline
is achieved more simply than in gradient chromatography and (ii) the once

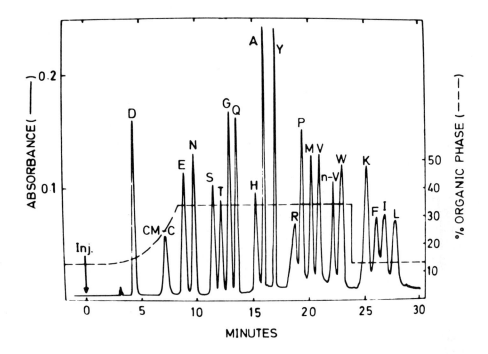

Fig. 13.2. Separation of 21 PTH-amino acids on a Waters μBondapak C₁₈ column using a concave ethanol gradient in ammonium acetate (pH 5.1). For chromato-graphic conditions see Table 13.1. The one-letter symbols for native amino acids (see Table 13.2) are used for designation; CM-C=PTH-carboxymethylcysteine; n-V= PTH-norvaline (internal standard). (Reprinted from Fohlman et al., 1980.)

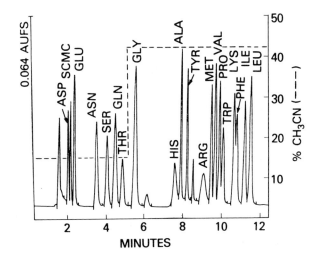

Fig. 13.3. Stepwise elution of PTH-amino acids on a Zorbax ODS column using acetonitrile in 0.04 M sodium acetate (pH 5.0). For conditions, see Table 13.1. (Reprinted from Bledsoe and Pisano, 1981.)

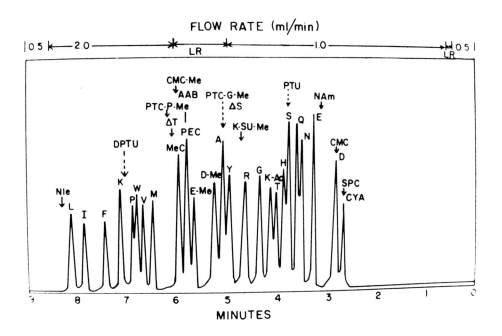

Fig. 13.4. Isocratic separation of common PTH-amino acids on an Ultrasphere ODS column plus 40-mm guard column of 10 μm ODS (both at 55°C) in 48% acetonitrile in 0.065 M ammonium acetate (pH 4.54). For detailed conditions see Table 13.1. LR = Linear ramp between adjacent flow-rates. Peaks are identified by single-letter symbols (see Table 13.2). Other abbreviations: CYA = cysteic acid; SPC = sulphopropyl-Cys; CMC = carboxymethyl-Cys; NAm = nicotinamide; PTU = phenylthio-urea; K-Ac = ε-acetyl-Lys; K-SU-Me = ε-succinyl-Lys methyl ester; PTC-G-Me = methyl ester of PTC-Gly (an intermediate in the conversion of sequencing product to PTH-Gly); ΔS = dehydro-Ser; D-Me = methyl ester of Asp; E-Me = methyl ester of Glu; PEC = pyridylethyl-Cys; AAB = α-aminobutyric acid; MeC = S-methyl-Cys; CMC-Me = methyl ester of CMC; ΔT = dehydro-Thr; PTC-P.Me = methyl ester of PTC-Pro; DPTU = diphenylthiourea; K = ε-PTC-Lys, Nle = norleucine. (Reprinted from Tarr, 1981.)

equilibrated and tested column can be used directly for a greater number of repeated analyses. An example of isocratic elution according to Tarr (1981) is illustrated in Fig. 13.4.

A ternary solvent system (50% aqueous acetate buffer solution - 25-50% acetonitrile - 0-25% DMSO, THF or DMF) for the separation of 3-phenyl-2-thio-hydantoin-amino acids by RPC was examined by Spatz and Roggendorf (1979). Watson and Carr (1979) described a simplex algorithm for the optimization of gradient elution HPLC and illustrated it with the separation of PTH-amino acids.

Annan (1981) and Schlesinger (1984) reviewed the separation and analysis of PTH-amino acids by HPLC.

13.1.7.2 Coloured and fluorescent derivatives of amino acids

The amino group of the N-terminal amino acid of a protein or peptide has often been used as a suitable reactive position for labelling in sequence studies. The labelled amino acid was identified in the total hydrolysate of the protein (peptide) using chromatographic methods. In this way the N-terminal amino acid can be determined very easily. One of the oldest reagents of this type used for the determination of protein sequences (Sanger, 1945) was 2,4-dinitrofluoro-benzene (DNFB), which labelled the amino acid by the formation of a dinitrophenyl (DNP) derivative:

DNFB   DNP–derivative

The yellow DNP derivatives of amino acids were usually identified by various forms of partition chromatography, such as paper chromatography, and later by TLC. In the further development of protein (peptide) sequencing, this reagent was gradually replaced with other more advanced reagents, so that only a few workers tried to separate DNP-amino acids by HPLC, which appeared in sequencing practice after the main period of application of DNFB reagent was over. Therefore, only one example of the HPLC of DNP-amino acids will be presented here.

Iskandarani and Pietrzyk (1981) examined the RP-HPLC of DNP-amino acids using PRP-1 (polystyrene-divinylbenzene copolymer, 10 µm) packed into a 15 cm x 4.1 mm I.D. column and eluted with an acidic (pH 1.6) or alkaline (pH 11.0) solvent system. The best results were obtained with an alkaline solvent system containing 0.01 M phosphate buffer with sodium chloride to give an ionic strength of 0.1 M, and a flow-rate of 1 ml/min. The solvent contained 12% acetonitrile and was applied isocratically for 20 min, followed by a linear gradient to 25% aceto-nitrile for an additional 40 min. Fourteen DNP derivatives were resolved.

Later, DNFB was replaced with 5-dimethylaminonaphthalene-1-sulphonyl chloride (dansyl chloride, DNS-Cl), described in the Section 6.6.2 (Volume A), which has often been used for the identification of amino acids by TLC. The DNS derivatives are easily formed by the reaction of DNS-Cl with amino groups. The main advantage of DNS derivatives is the sensitive fluorescence detection, but in laboratories where this type of detector is not available UV detection can also be used although with lower sensitivity.

Cooke et al. (1978) tried to use metal ions for selective separations in HPLC (cf., Section 4.4.6) and also examined the separation of DNS-amino acids by C_{12}-dien-Zn^{II} chromatography. Using 10^{-3} M zinc sulphate 1% ammonium acetate and acetonitrile-water (36:65), eight DNS-amino acids were separated in 35 min. Kobayashi and Imai (1980) examined the determination of fluorescent compounds by HPLC with chemiluminescence detection (cf., Section 6.6.3); the system bis(2,4,6-trichlorophenyl) oxalate (TCPO) - hydrogen peroxide was applied for monitoring the effluent after the chromatography of DNS-amino acids. A 30 cm x 4 mm I.D. column of µBondapak C_{18} was maintained at 35°C and eluted isocratically with 0.05 M Tris-HCl buffer (pH 7.7)-acetonitrile (16:9, v/v) at a flow-rate of 0.18 ml/min. A mixture of 5 mM TPCO in ethyl acetate (0.51 ml/min) and 0.5 M hydrogen peroxide in acetone (1.2 ml/min) was continuously added in the front of the mixing coil and mixed with the effluent. A detector (Schoeffel FS-970LC fluoromonitor) containing a 3x3x7 mm flow cell was adopted with the light source off. A mixture of DNS derivatives of Glu, Ala, Met and nor-Leu was separated in 30 min as an example. The sensitivity was in the femtomole range and the relationship between peak height and amount was linear up to more than 50 fmol. Engelhardt (1981) mentioned the enantiomeric separation of DNS-amino acids using a chiral bonded phase. The eluent contained no chiral component [acetonitrile-buffer (3:7, v/v); buffer, ammonium acetate-0.1 M ammonia solution, 10^{-5} M Cu^{2+}, pH 7.8; flow-rate 1 ml/min; pressure 20 bar]. The DNS derivatives of L-Ser, D-Ser, L-Met, D-Met, L-Phe and D-Phe were well separated in one run. The retention mechanism was based on ligand exchange (Davankov, 1980).

Iskandarani and Pietrzyk (1981), in their studies on the application of the polystyrene-divinylbenzene copolymer PRP-1 in RP-HPLC, also examined the separation of DNS derivatives using strongly acidic (pH 1.75), mildly acidic (pH 3.6) and basic (pH 11.0) mobile phases. Using mildly acidic conditions [0.01 M phosphate buffer, (adjusted with sodium chloride to an ionic strength of 0.1 M)-acetonitrile (80:20)], a flow-rate of 1 ml/min and a 15 cm x 4.1 mm I.D. column, they resolved twelve DNS-amino acids into eight peaks; a brief isocratic elution (20% acetonitrile for 12 min) was followed by a linear gradient to 50% acetonitrile in an additional 20 min. Tapuhi et al. (1981a,b) published a study on the dansylation of amino acids for HPLC analysis (cf., Section 13.1.8). As an example the chromatographic separation of five DNS-amino acids, dansylsulphonic acid (DNS-OH), dansylamide (DNS-NH$_2$) and dansylethylamide (DNS-NHEt) (148 pmol of each) was illustrated by a chromatogram. A 15 cm x 4.6 mm I.D. column of C_8 Supelcosil (5 µm) was eluted at a flow-rate of 2 ml/min with methanol-water (42:58, v/v) containing 0.6% (v/v) glacial acetic acid and 0.008% (v/v) triethylamine. Fluorescence detection with an excitation wavelength of 250 nm and a 470-nm emission filter was used for monitoring.

Simmons and Meisenberg (1983) described the separation of DNS-amino acid amides by HPLC. This method is important for the structural analysis of natural amidated peptides, from which the C-terminal amides can be released enzymically. A 25 cm x 4.6 mm I.D. column packed with Ultrasphere ODS (5 μm) was eluted at a flow-rate of 1 ml/min using the following gradient programme: 0 min, 20% B; 8 min, 20% B; 20 min, 35% B; 30 min, 40% B; 40 min, 80% B; and 45 min, 80% B; solvent A was 5 mM sodium phosphate (pH 7.0) and B was acetonitrile. An amount of 10 nmol of each DNS-amino acid amide was applied and detection was at 254 nm; monitoring at 206 and 275 nm was also applied. DNS-amino acid amides can be simply separated from mono-DNS-amino acids, the products of amino acids normally present in excess in biological smaples.

Dimethylaminoazobenzene thiohydantoins (DABTH) are another group of coloured derivatives of amino acids that are very useful for the sequencing of proteins and peptides (cf., Section 13.1.5). They have often been separated by TLC. Chang et al. (1980) reported the analysis of DABTH-amino acids by HPLC. A Knauer (West Berlin, F.R.G.) RP-8 (7 μm) column was eluted at 40°C using a linear gradient from 60% to 85% B in 20 min, followed by isocratic elution (85% B) for an additional 12 min. However, all DABTH-amino acids of the standard mixture were eluted in the first part of the programme within the gradient. Solvent A was 8 mM acetate (pH 5) and B was 90% methanol. The flow-rate was 1.5 ml/min and detection was by absorbance measurement in the visible region at 436 nm. The molar absorptivity is 34 000 at 420 nm, so the sensitivity is about twice that for PTH derivatives of amino acids at 254 nm (in this instance the molar absorptivity is only 16 000 at 269 nm). The sensitivity of detection of DABTH-amino acids could be further increased by employing acidic solvents (then the molar absorptivity is 47 000). As little as 10 pmol of a DABTH-amino acid can be detected against a stable baseline, so this labelling is also used in the form of a pre-column derivatization for amino acid analyses of total hydrolysates of peptides (cf., Section 13.1.8). The described method proved very useful especially in sequencing using the double-coupling (DABITC-PITC) method (cf., Section 13.1.5); the repetitive yields for the manual solid-phase sequential DABITC-PITC degradation of various protein or peptide species was 91-99.4%.

In the manual method for the N-terminal sequence analysis of polypeptides at the picomole level (cf., Section 13.1.5), Chang (1981c) used a Zorbax ODS column (DuPont) and samples of 5-25 μl of DABTH derivatives were usually injected. A standard mixture of 19 amino acid derivatives (10 or 5 pmol of each) was eluted using the following programme: a linear gradient from 40% to 70% B in 10 min, then 70% B isocratically to 15 min (total time), then stepwise to 80% B and finally 80% B isocratically to 30 min (total time). Solvent A was

35 mM acetate buffer (pH 5.0) and B was acetonitrile. The flow-rate was 1 ml/min
at 22°C. The methods described allowed the complete sequence of peptides and
the N-terminal sequence of proteins to be sensitively determined without use
of an automatic sequenator or any radioactive materials.

Mancheva et al. (1981) described the HPLC separation of coloured diphenylindenonylthiohydantoin derivatives of amino acids suitable for sequencing
studies. 2-p-Isothiocyanophenyl-3-phenylindenone or diphenylindenonyl isothiocyanate (DIITC) was synthesized by Ivanov et al. (1967) and proposed as a
suitable reagent for sequencing of peptides (Ivanov and Mancheva, 1973a).

2-p-Isothiocyanophenyl-
-3-phenylindenone

Diphenylindenonylthiohydantoin derivative
of amino acid
(ITH amino acid)

Using this reagent, an N-terminal amino acid was identified by TLC as a coloured
diphenylindenonylthiohydantoin (ITH) derivative (Ivanov and Mancheva, 1973b) or
as a fluorescent isobenzofuran derivative after additional treatment with sodium
ethoxide (Ivanov and Mancheva, 1975). Nasimov et al. (1980) used this reagent
for sequencing purposes and separated and identified ITH derivatives by means of
HPLC in combination with mass spectrometry (cf., Section 13.1.6). Mancheva et
al. (1981) described both gradient and isocratic elution of ITH derivatives.

The following conditions were used for gradient elution: a Knauer LiChrosorb
RP-18 column (25 cm x 0.46 cm I.D.) was isocratically eluted with a 44% solution
of A in B for 3 min and then a linear gradient from 44% to 94% A in B was used
at 2%/min with a flow-rate of 2 ml/min at 24°C. Solvent A was acetonitrile and
B was 0.01 M sodium acetate solution (pH 4.80). A sample of 10 µl contained
50 pmol of each ITH. Detection was at 267 nm. A mixture of 21 ITH-amino acids
was resolved in 28 min, with the exception of two common peaks for Ile + Leu
and Trp + Met derivatives.

Two systems for isocratic elution using the same column were proposed, both
resolving ITH derivatives Leu and Ile. One system resolved Trp and Met, but not
Tyr + Ala + His, and the other system did not resolve two pairs, Met + Trp and
Pro + Val. The conditions used for isocratic elution with the first system were
an A-B (44:56) mixture at a flow-rate 4 ml/min and 24°C; solvent A was acetonitrile-ethyl acetate (78:22) and B was 0.01 M sodium acetate solution (pH 4.8).

The conditions for the second isocratic system were an A-B (48:52) mixture at a flow-rate of 2 ml/min and 62°C; solvent A was acetonitrile-ethyl acetate (85:15) and B was 0.01 M sodium acetate solution (pH 4.8). The total time taken was 20 and 22 min, respectively.

Applying sensitive amino acid analysis to the C-terminal sequence analysis of peptides (cf., Section 13.1.5), Chang et al. (1981b) determined amino acids split from polypeptides by carboxypeptidase by means of RP-HPLC of 4-(NN-dimethyl-amino)azobenzene-4'-sulphonyl (DABSYL) derivatives. It is known that dimethyl-aminoazobenzenesulphonyl chloride (DABS-Cl) reacts with the amino group of an amino acid to form a DABS derivative (Lin and Chang, 1975). The excess of DABS-Cl used can be quantitatively hydrolysed to its corresponding sulphonic acid (Methyl Orange) and can be well separated from all the DABS-amino acids.

$$CH_3\!-\!N(CH_3)\!-\!C_6H_4\!-\!N\!=\!N\!-\!C_6H_4\!-\!SO_2Cl \xrightarrow{RNH_2} CH_3\!-\!N(CH_3)\!-\!C_6H_4\!-\!N\!=\!N\!-\!C_6H_4\!-\!SO_2\!-\!NHR$$

DABS-Cl   DABS–derivative of amino acid

$$\xrightarrow{H_2O} CH_3\!-\!N(CH_3)\!-\!C_6H_4\!-\!N\!=\!N\!-\!C_6H_4\!-\!SO_3H$$

Methyl Orange

The DABS derivatives can be well chromatographed by RPC. Chang et al. (1981b) used a Zorbax ODS column (DuPont). The injected sample size was 5-25 µl, containing 3-5 pmol of each DABSYL derivative. A linear gradient from 20% to 70% B in 25 min was followed by isocratic elution at 75% B for an additional 7 min. The flow-rate was 1.2 ml/min and the temperature was 22°C. Solvent A was 50 mM acetate buffer (pH 4.13) and B was acetonitrile. A mixture of sixteen DABS-amino acid derivatives were well resolved, but aspartic acid and serine derivatives were not resolved and for their separation solvent A was 28 mM phosphate buffer (pH 7.2) and B was acetonitrile. A linear gradient from 20% to 43% B in 15 min was followed by isocratic elution to 20 min (total time), then a steep gradient to 65% B to 25 min (total time) was followed by isocratic elution (65% B) to the final 30 min. Aspartic acid and serine derivatives were well resolved, but the separation of other DABS-amino acids was not as good as with the first system.

Chang et al. (1982) described another chromatographic system allowing a complete baseline separation of DABS-amino acids. A Merck LiChrosorb RP-18 column was maintained at 50°C and eluted at a flow-rate 1 ml/min by gradients of 12-37% B (0-22 min), 37-60% B (22-35 min), kept at 60% B (35-40 min), and then a reversed gradient from 60% to 12% B (40-45 min) followed. Solvent A was phosphate buffer (12 mM, pH 6.5), containing 4% (v/v) of dimethylformamide, and B was acetonitrile containing 4% (v/v) of dimethylformamide. Twenty DABS derivatives were well resolved.

13.1.7.3 Comments on literature

The application of HPLC to the separation of amino acids was reviewed by Engelhardt (1981); the review included PTH and DNS derivatives. HPLC of DNS-amino acids was reviewed by Wilkinson (1984) and HPLC of DABSYL-amino acids by Lin (1984), including a description of the procedure for the synthesis of DABS-Cl.

13.1.8 Quantitative analysis of total protein hydrolysates using HPLC of derivatized amino acids

The first and fundamental prerequisite when studying the primary structure of a protein is the determination of its molecular weight and amino acid composition. An automatic recording amino acid analyser was developed by Spackman et al. (1958) and was based on IEC and ninhydrin detection of free amino acids, products of the total hydrolysis of the protein or peptide studied. This classical approach to amino acid analysis and its various modifications has been reviewed elsewhere (Zmrhal et al., 1975; Hare, 1977; Benson, 1977). Our attention will be focused predominantly on the pre-column derivatization of amino acids to permit the application of HPLC equipment for the analysis. For laboratories concerned with protein chemistry, an efficient and specialized amino acid analyser is an indispensable tool if larger proteins are to be analysed. However, in laboratories with other directions of research, which do not possess an amino acid analyzer, but which are equipped with a good modern liquid chromatograph, the quantitative amino acid analysis of peptides is also possible, and this approach will be dealt with in the following part of this section. Aspects covered will be the o-phthaldialdehyde method and the chromatography of DNS derivatives, PTH derivatives and other suitable derivatives of amino acids.

13.1.8.1 o-Phthaldialdehyde (OPA) method

The principle of the OPA reaction was explained in Section 6.6.2. Roth (1971) described fluorescence reactions of amino acids: o-phthaldialdehyde (also called o-phthalaldehyde, OPA), which is a non-fluorescent reagent, reacts with the amino group of an amino acid in an alkaline medium in the presence of a reductive thiol-containing reagent (2-mercaptoethanol) to give rise to strongly fluorescent compounds (λ_{ex} = 340 nm, λ_{em} = 455 nm). This permitted the fluorimetric assay of amino acids down to the nanomole level without any heating of the reaction mixture, and the fluorescence could be measured as soon as 5 min after mixing the reagents. The reaction was well suited for the automatic determination of amino acids, but proline and hydroxyproline were not detected. Relative fluorescence intensities of individual amino acids were published. Benson and Hare (1975) studied the fluorogenic detection of primary amines in the picomole range and compared the OPA method with the fluorescamine method (cf., Section 6.6.2) and ninhydrin detection. Simons and Johnson (1978) examined in detail the reaction of OPA and thiols with primary amines, especially the formation of 1-alkyl(and aryl)-thio-2-alkylisoindoles. Svedas et al. (1980) studied the kinetics of the interaction of amino acids with OPA and the spectrometric assay of the reaction products.

The HPLC determination of amino acids in the picomole range using the OPA method was described by Hill et al. (1979). Standard amino acids (5 pmol of each) derivatized with OPA-ethanethiol were chromatographed on a μBondapak C_{18} column (30 cm x 3.9 mm I.D.) and eluted using a linear gradient programme from 9% to 49% B into A of 40 min duration, with 5 min isocratic held at 15 min after initiation; the flow-rate was 2 ml/min. Solvent A was 0.0125 M Na_2HPO_4 (pH 7.2) and B was acetonitrile. For fluorescence detection the excitation wavelength was 229 nm and for emission a 470 nm cut-off filter was used. The sensitivity was 5 pmol and the analysis time was 45 min. Threonine and lysine derivatives were not resolved. Hodgin (1979) used (after a 1-min derivatization period with OPA reagent) a ternary gradient for the elution of reaction products in the RP mode from a 15 cm x 5 mm I.D. column of Microsil C_8. The composition of the solvent, containing methanol, tetrahydrofuran and 0.02 M potassium acetate, varied with time. The complete analysis time, including the derivatization step, was less then 40 min and the average minimum detectable amount was about 40 pg. The RP-LC analysis of amino acids after reaction with OPA was described by Gardner and Miller (1980), who used the following procedure:

OPA reagent (concise recipe). A solution of 3 g of boric acid in 90 ml of water was adjusted to pH 10.5 with potassium hydroxide. In a separate container a solution of 0.05 g of OPA (Aldrich) in 1 ml of ethanol was mixed with 0.05 ml of 2-mercaptoethanol. The two solutions were mixed, diluted to 100 ml and 0.3 ml of a 30% solution of Brij was added (to enhance the fluorescence of lysine; cf., Benson and Hare, 1975).

Derivatization. Within 3-4 min before injection, the amino acids (dry or dissolved in water) were mixed with OPA reagent to form fluorescent derivatives.

Chromatography. A Varian MicroPak MCH-5 RP column (30 cm x 4 mm I.D.) was eluted stepwise at a flow-rate 1 ml/min by solutions A (0-14 min) and B (14-40 min), where A = methanol-water (17:192) and B = methanol-0.05 M KH_2PO_4 (pH 4.5)

(3:2). In some experiments a gradient system was also employed. For fluorescence detection the excitation wavelength was 340 nm and the emission wavelength 455 nm.

Eighteen amino acid derivatives were resolved and detected at the picomole level. The method was simple and allowed the measurement or confirmation of the presence of low levels of amino acids in aqueous solutions. This technique was demonstrated by measuring low concentrations of amino acids released from zooplankters stressed by contaminants.

Larsen and West (1981) used the following procedures for quantitative analysis by the OPA-HPLC method:

OPA derivatization reagent (summarized recipe). Ethanethiol (6 µl) (Aldrich), 25 µl of Brij solution (Pierce) (10 mg/ml in water) and 1.25 ml of 0.5 M Na_2HPO_4 (pH 10.3) were added to 50 µl of a 0.745 M ethanolic solution of OPA (Sigma) (100 mg/ml) and the solution was diluted to 1 l with water. A fresh (opaque) solution was prepared daily.

Derivatization procedure. A small sample (10-20 µl) of amino acid standard solution (containing 1 mg/ml of amino acid in 0.1 M hydrochloric acid) was placed in a 500-µl conical reaction vial. An aliquot of OPA reagent equal to twice the volume of the amino acid standard solution was added and the mixture was allowed to react for 1 min. An aliquot (2-10 µl) of the reaction mixture was injected into the HPLC system.

Chromatography. A mixture of 8.33 µmol of each amino acid after reaction for 10 min with OPA reagent was applied on a 30 cm x 0.39 cm I.D. µBondapak Fatty Acid Analysis column (Waters) and eluted with a linear gradient from 0 to 38% B over 38 min, then from 38 to 70% over 10 min; the flow-rate was 2.0 ml/min at ambient temperature. Solution A was 0.01875 M triethylamine-acetic acid and B was acetonitrile. Fifteen amino acids were resolved [the order of peaks was Asp, Glu, (His), Ser, His, Thr, Gly, Ala, Arg, Tyr, Val, Met, Ile, Leu, Phe, Lys]. Complete analysis required 46 min, with an injection to injection time of 56 min. The lower limit of quantitation (using the equipment described) was 100 pmol; the detection limit was 25 pmol. Application to a typical peptide hydrolysate was described.

Hughes et al. (1982) studied amino acid analysis using standard HPLC equipment, which was modified (with the addition of an interface) to an amino acid analyser. Detection was effected with either ninhydrin or OPA. The separation time was 45 min. Good reproducibility of chromatography permitted peak-height or peak-area measurements of amino acids in protein hydrolysates and quantification in the range 10 pmol to 25 nmol. For details of the construction of the apparatus and the composition of the solutions used, the original paper should be consulted. Cunico and Schlabach (1983) compared the ninhydrin and OPA post-column detection techniques for the HPLC of free amino acids after IEC on strong cation exchangers in the Na$^+$ or Li$^+$ form. The amino acid detection limit with OPA was near 5 pmol for amino acids with primary amino groups and 100 pmol for secondary amino groups (proline). The detection limit with ninhydrin was near 100 pmol.

Jones and Gillian (1983) studied the OPA pre-column derivatization and HPLC of polypeptide hydrolysates and physiological fluids using 3-μm RP column packings and a methanol gradient in 0.1 M aqueous sodium acetate (pH 7.2) for elution. Complete separation was achieved with a relative deviation of ±1.5% and a detection limit of less then 100 fmol. The analysis time for peptide hydrolysates was 18 min and for 48-component physiological fluids less then 50 min. Various commercial columns of different dimensions were tested. As an example, a 23-component mixture produced by OPA derivatization of amino acids was resolved on a Microsorb C$_{18}$ column (10 cm x 4.6 mm I.D.) eluted with a complicated gradient system of the above components at a flow-rate of 1.6 ml/min in 19 min.

Reagent solution. Jones and Gillian used commercial Fluoraldehyde OPA Reagent Solution (OPA in borate buffer containing 2-mercaptoethanol and also Brij 35), which is stable for a least 6 months if stored at 4°C and can be used directly (this complete reagent is obtainable from Pierce, Rockford, IL, U.S.A.). It can be replaced by a solution prepared by dissolving 50 mg of OPA in 1.25 ml of absolute methanol followed by the addition of 50 μl of 2-meracptoethanol and 11.2 ml of 0.4 M sodium borate (pH 9.5); however, this solution is stable for only 1-2 weeks.

Derivatization procedure. Aliquots of 5-10 μl of standards or unknown sample were mixed with 5 μl of the fluoroaldehyde reagent. After 1 min, 20-100 μl of 0.1 M sodium acetate solution (pH 7.0) were added, the solution was mixed and a 20-μl sample was subjected to analysis.

Total acid hydrolysates of peptides. These were prepared as follows. Peptides (50-500 μmol) were dried in a hydrolysis tube, the tube was sealed under vacuum after the addition of constant-boiling hydrochloric acid (6 M) and then heated at 110°C for 22 h. After hydrolysis, the acid was removed by lyophilization and the resulting residue dissolved in 25-50 μl of water. [Aliquots (10 μl) were used for the OPA derivatization, described above.] The original paper also describes the total enzymic hydrolysis of peptides by means of aminopeptidase.

Preparation of biological samples. Serum, urine or cerebrospinal fluid samples (25 μl) were mixed vigorously with acetonitrile (75 μl). Following centrifugation at 10 000 g for 2 min, 10-μl aliquots were subjected to the OPA derivatization procedure described above and analysed.

Ashman and Bosserhoff (1985) described in detail amino acid analysis by HPLC and pre-column derivatization using the OPA method (their book is in fact a comprehensive practical manual). As an example, conditions for the separation of 100 pmol of a sixteen component standard reaction mixture will be given. A 25 cm x 4.5 mm I.D. column filled with Hypersil ODS (5 μm) was eluted with a complicated gradient prepared by mixing methanol-tetrahydrofuran (100:3) with 12.5 mM Na_2HPO_4 (pH 7.2). A stock OPA solution consisted of 56 mg of OPA dissolved in 1 ml of methanol + 56 μl of mercaptoethanol. The reaction buffer was 1 M potassium borate (pH 10.4).

13.1.8.2 Chromatography of dimethylaminonaphthalenesulphonyl (DNS) amino acid derivatives

The principle of the reaction of DNS chloride with amino groups was explained in Section 6.6.2. Bayer et al. (1976) studied the separation of DNS-amino acids by HPLC on silica gel (LiChrosorb Si 60, 5 μm) columns (50 cm x 0.3 cm I.D.). A dual column system was applied. The first column was eluted in the normal-phase mode with a gradient from benzene-pyridine-acetic acid (50:5:0.5, v/v) to pyridine-acetic acid (30:30, v/v) at 65°C and a flow-rate 1 ml/min. The second colum was eluted isocratically with benzene-pyridine-acetic acid-methanol (100: 50:0.5:50, v/v). Twenty-one DNS derivatives were resolved using relative fluorescence measurement for detection. RP separation on a LiChrosorb RP-8 column (50 cm x 0.3 cm I.D.) was also tested using a methanol gradient in 0.01 M sodium phosphate. Hsu and Currie (1978) used the HPLC of DNS derivatives for the rapid amino acid analysis of hydrolysates of peptides with not too long chains. A 25 cm x 4.6 mm I.D. column of 10-μm Partisil PAC was eluted isocratically with acetonitrile-water-acetic acid (30:70:1 or 50:50:1) at a flow-rate of 1 ml/min with detection at 254 nm. Hydrolysates of a deca- or penta-peptide were analysed and the corresponding DNS-amino acids resolved. The analysis time was 30-45 min. A histidine tetrapeptide was analysed on a 3 ft. x 1/8 in. Vydac polar phase column, eluted isocratically with acetonitrile-water-acetic acid at a flow-rate of 1 ml/min.

Procedures. Hydrolyses were carried out in 6 M hydrochloric acid (200 μl) at 120°C for 20-24 h in sealed Pyrex tubes (50 mm x 6 mm O.D.). Hydrolysates were lyophilized, adjusted to pH 8.5 with $NaHCO_3$ buffer (200 μl) and treated with a 10-20-fold molar excess of DNS-Cl in acetone, incubated at 40°C for 1 h, lyophilized and dissolved in 1.0 ml of 50% acetonitrile in water or 50% methanol in water prior to LC.

Olson et al. (1979) examined the separation of DNS-amino acids on an RP-8 column eluted with a sodium acetate-phosphate-acetonitrile gradient. Fluorescence detection was used with sensitivity of >1000 pmol. The analysis time was 58 min.

Tapuhi et al. (1981a) dealt with practical considerations in the chiral separation of DNS-amino acids by RP-HPLC using metal chelate additives. They also
tested the HPLC separation of conventional amino acid mixtures by means of a
DNA procedure. A Hypersil C_{18} (5 μm) column (15 cm x 4.6 mm I.D.) was eluted
with a 40-min linear gradient from 14% to 100% B. Solution A was 5% methanol-3%
tetrahydrofuran-0.57% acetic acid-0.088% triethylamine and solution B was 70%
methanol-3% tetrahydrofuran-0.57% acetic acid-0.088% triethylamine. All the DNS-
amino acids were well separated. A UV monitor (254 nm) (picomole sensitivity)
or a spectrofluoromonitor (femtomole sensitivity) was used for detection.
Relative standard deviations of 0.5-1.5% for individual DNS-amino acids were
reported.

Tapuhi et al. (1981b) published a detailed study of the dansylation of amino
acids for HPLC analysis. Derivatization conditions were examined (time, temperature); the yield appeared to be independent of the ratio of DNS chloride
to amino acids over a 1000-fold range. A comparative study of liquid chromatography of DNS derivatives, gas chromatography and amino acid analyser methods
for amino acid analysis was published. For the HPLC of DNS-amino acids a 15 cm x
4.6 mm I.D. Supelcosil C_8 (5 μm) column was isocratically eluted with methanol-
water (42:58, v/v) containing 0.6 vol.-% of glacial acetic acid and 0.008% of
triethylamine. For fluorescence measurement the excitation wavelength was 250 nm
and a 470-nm emission filter was employed. Symmetrical chromatographic peaks
were obtained in all instances and the yield was determined by comparing the
peak heights of DNS-amino acids from the reaction with those of standards.
However, the maximum yield of the reaction varied for individual amino acids
from 90 to 113% and the time to reach the maximum yield at ambient temperature
also varied (e.g. several minutes for Pro, 20 min for Val and more than 100 min
for Asn). Great differences were found also at 95°C.

The optimum conditions for derivatization were as follows. The reactant
solution consisted of pure DNS-Cl (free from hydrolysis products) dissolved in
specially purified* acetonitrile (1.5 mg/ml, 5.56 mM). Amino acids for analysis
were dissolved in 40 mM lithium carbonate buffer (pH 9.5, adjusted with hydrochloric acid) to a concentration of 0.001-1 mM, 2 ml of this solution were pre-
heated to 60°C in a special Pierce Reacti-Therm heating module and 1 ml of the
DNS-Cl solution was added. After brief shaking, the reaction vial was maintained
at 60 ± 0.03°C for 5 min. The reaction was then terminated by adding 100 μl of
2% ethylamine solution. The reaction vials were wrapped with aluminium foil to
exclude light.

---

*HPLC-grade acetonitrile was distilled over DNS-Cl in order to remove potential
 interferences.

De Jong et al. (1982) also studied amino acid analysis by HPLC. They briefly
reviewed recent methodology for amino acid analysis utilizing HPLC equipment,
and evaluated the usefulness of pre-column DNS derivatization. They used a
Brownlee Aquapore RP-300 column (25 cm x 4.6 mm I.D.), eluted with a gradient
of methyl ethyl ketone - 2-propanol at 55°C. Measurement of relative fluorescence
was used for detection. DNS-amino acids at the level of hundreds of picomoles
could be routinely analysed. The method was also compared with ninhydrin detection.

13.1.8.3 *Chromatography of phenylthiocarbamyl (PTC) and phenylthiohydantoin (PTH) derivatives of amino acids*

The principle of this derivatization was explained in Section 13.1.5. One of
the first experiments on the RP-HPLC of PTH-amino acid mixtures for total
hydrolysate amino acid analyses was carried out by Haag and Langer (1974).
Employing stepwise elution with water-acetonitrile-propanol mixtures, the separation of a mixture of fourteen PTH derivatives was accomplished in 40 min; a
Corasil C_{18} column was applied. A gradient system was used by Margolies and
Brauer (1978) in an attempt to separate 22 PTH amino acid derivatives. Useful
references to various experiments on the HPLC separation of PTH derivatives
of amino acids can be found in Table 13.1.

HPLC analysis by means of PTH-amino acid derivatives was thoroughly studied
by Greibrokk et al. (1980). Four C_{18}-reversed phase packings were tested
(Spherisorb ODS, μBondapak C_{18}, Hypersil ODS and Spherisorb XOA 600 Normaton C_{18})
together with different mobile phases. The modification of amino acids to PTC
and PTH derivatives (according to Rosmus and Deyl, 1972) was described in detail.
The optimum chromatographic conditions are given in Table 13.1. The realistic
detection limit was 50 pmol of each amino acid. The speed and sensitivity of
this HPLC method competed favourably with post-column ninhydrin detection in
classical amino acid analysers. Because of the time-consuming preparation of
the derivatives, Greibrokk et al. (1980) considered the PTH method to be of
interest to laboratories without a sensitive amino acid analyser (but with HPLC
equipment) who occasionally need to carry out amino acid analyses and peptide/
protein sequencing, because the same analytical procedure could be used.

Heinrikson and Meredith (1984) examined amino acid analysis by RP-HPLC using
pre-column derivatization with PITC to obtain PTC derivatives (cf., Section
13.1.5). Phenylthiocarbamylation of amino acids proceeded smoothly in 5-10 min
at room temperature. PTC derivatives were dissolved in 0.05 M ammonium acetate
(pH 6.8) and injected on to a 5-μm octyl (C_8) or octadecyl (C_{18}) column (25 cm x
0.46 cm I.D.); both types operated successfully. Three solvents were used for
programmed elution at 52°C: (A) 0.05 M ammonium acetate (pH 6.8); (B) 0.1 M

acetate (pH 6.8) in methanol-water (80:20, v/v) or 0.1 M acetate (pH 6.8) in
acetonitrile-methanol-water (44:10:46); and (C) acetonitrile-water (70:30, v/v).
Four systems using programmed elution with these mixtures were tested and described. Detection at 254 nm was used. All amino acids (except proline) were
converted quantitatively into PTC derivatives (without conversion to PTH
derivatives), which were chromatographically separated in amounts of 500-600 pmol
each. The results were comparable in sensitivity and precision to those obtained
by state-of-the-art IEC analysers, and the equipment need not be dedicated to
a single purpose. The method was verified by the analysis of reduced and carboxymethylated lysozyme (0.0015 mg = pmol of protein was used) and on other proteins
and peptides.

Bidlingmeyer et al. (1984) also described a rapid analysis of amino acids
employing pre-column phenylthiocarbamide (PTC) derivatization. Using UV detection
at 254 nm and a 12-min analysis time, a sensitivity of 1 pmol was obtained in
resolving eighteen derivatives. The following procedures were applied.

Hydrolysis of proteins and peptides (summarized recipe). A solution containing
0.1-5 µg of the sample was dried under vacuum in a 50 mm x 6 mm tube, 200 µl of
constant-boiling hydrochloric acid (roughly 6 M) containing 1% (v/v) phenol were
added, the tube was sealed under vacuum and the sample was hydrolysed at 150°C
for 1 h or at 108°C for 24 h. [In the original paper, the application of a
Pico-Tag Workstation (Waters) for all the analytical operations was described.]

Derivatization (summarized recipe). Samples containing up to 25 µmol of each
amino acid in 50 mm x 6 mm tubes were dried under vacuum and re-dried after
adding 10-20 µl of ethanol-water-triethylamine (TEA) (2:2:1). The derivatization
reagent was prepared fresh daily and consisted of ethanol-TEA-water-phenyl
isothiocyanate (PITC) (7:1:1:1, v/v) (the PITC was stored at -20°C under nitrogen); 300 µl of the reagent were sufficient for twelve samples. PTC-amino acids
were formed by adding 20 µl of the reagent to the dried sample and kept sealed
in a special vacuum vial for 20 min at room temperature. The PTC derivatives
were not converted into PTH derivatives (cf., Section 13.1.5).

Chromatography. An application-specified Pico-Tag column (15 cm x 3.9 mm I.D.)
containing high-efficiency bonded-phase was eluted with a convex gradient from
10% to 51% B in 10 min, followed by a steep washing programme to 100% B. Solution
A was 0.14 M sodium acetate solution containing 0.5 ml/l of TEA and titrated
to pH 6.35 with glacial acetic acid and B was 60% acetonitrile in water.

13.1.8.4 Chromatography of miscellaneous derivatives of amino acids

In addition to the above methods, various other pre-column derivatizations
of amino acids have been used to facilitate their HPLC analysis. In 1968 Ghosh
and Whitehouse described 7-chloro-4-nitrobenzo-2-oxa-1,3-diazole (NBD-Cl) as a

fluorigenic reagent for amino acids and other amines. The principle of the
reaction was explained in Section 6.6.2. Watanabe and Imai (1982) examined pre-
column labelling for the HPLC of amino acids with the corresponding fluoro
derivative of this compound and its application to protein hydrolysates. The
fluorigenic reagent 7-fluoro-4-nitrobenzyl-2-oxo-1,3-diazol (NBD-F) reacted with
amines very easily in the same reaction scheme as for the chloro derivative.

Derivatization was very simple and proceeded in ethanol-0.1 M phosphate buffer
(pH 8.0) (50:50) at 60°C for 1 min.

Chromatography. A 1.5-μg aliquot of derivatized protein hydrolysate was
analysed by RPC on a μBondapak C$_{18}$ (10 μm) column (30 cm x 3.9 mm I.D.). Elution
was carried out as follows: (i) an isocratic procedure with the solution A
[methanol-tetrahydrofuran (THF) in 0.1 M phosphate buffer (pH 6) (3.75:1.6:94.65)]
for 24 min; (ii) a linear gradient from 100% A to 100% B [methanol-THF in 0.1 M
phosphate buffer (pH 6) (25:15:60)] over 30 min; (iii) isocratic elution with B
for 6 min; and (iv) isocratic elution with solvent C [methanol-water (40:60)]
for 12 min. Fluorescence detection was employed (excitation wavelength 470 nm,
emission wavelength 530 nm). The detection limit for each amino acid was 10 fmol.

Chang et al. (1983) described a complete separation of dimethylaminoazobenzene-
sulphonyl (DABS)-amino acids (cf., Chang et al., 1981b). Amino acid analysis
with low nanogram amounts of polypeptide could be achieved after hydrolysis and
derivatization with dimethylaminoazobenzenesulphonyl chloride (DABS-Cl). For
formulae and chemistry, see the Section 13.1.5. A complete baseline separation
of all DABS-amino acids at the 1-2 pmol level was possible. For the chromatographic conditions for the separation of DABS-amino acids, see Section 13.1.7.
The accuracy of this system was demonstrated by the composition analysis of
two immunoglobulin chains (214 amino acid residues) with differences at only
three amino acid residue positions.

Einarsson et al. (1983) studied the determination of amino acids with
9-fluoromethyl chloroformate (FMOC) and RP-HPLC. The reaction with the amino
group proceeded according to

CH$_2$OCOCl   CH$_2$OCOHNR

FMOC-chloride   FMOC-derivative

The reaction of amino acids with FMOC was very rapid (30 s) under mild conditions.

Derivatization (summarized recipe). The reagent was prepared by dissolving 155 mg of FMOC (Chemical Dynamics, South Plainfield, NJ, U.S.A.) in 40 ml of acetone to give a concentration of 15 mM. To 0.4 ml of sample were added 0.1 ml of 1 M borate buffer, (pH 6.2) and 0.5 ml of the reagent. After 40 s the mixture was extracted twice with 2 ml of pentane. The aqueous solution was then ready for injection of an aliquot.

Chromatography. A Shandon Hypersil ODS (3 µm) column (12.5 cm x 4.6 mm I.D.) was used, eluted with a gradient from acetonitrile-methanol-acetic acid buffer (pH 4.20) (10:40:50) to acetonitrile-acetic acid buffer (50:50) over 9 min. The acetic acid buffer was prepared by adding 3 ml of acetic acid and 1 ml of triethylamine to 1 l of water. The flow-rate was 1.3-2 ml/min. Fluorescence detection with an excitation wavelength of 250 nm and an emission wavelength of 320 nm was used. Twenty amino acid derivatives were separated with quantification by peak-height measurement. The detection limit was in the low femtomole range. The relative standard deviation was 2.4-6.4%. The detector response for most amino acids was linear in the range 0.1-50 µmol.

13.1.8.5 Comments on literature

In addition to the references cited above, some other useful citations to the description of chromatographic methods used in the determination of the structures of proteins can be found in Chapter 7. Rosmus and Deyl (1971, 1972) reviewed very thoroughly chromatographic methods in the analysis of protein structures, namely methods for the identification of N-terminal amino acids in peptides and proteins. Birr (1980) edited the Proceedings of a conference on methods in peptide and protein sequence analysis. Hearn et al. (1983b) edited a book on the HPLC of proteins and peptides. Hancock and Harding (1984) reviewed separation conditions for the HPLC of underivatized and derivatized amino acids and Bhown and Bennett (1984) the use of HPLC in protein sequencing. Henschen et al. (1985) edited a book on HPLC in biochemistry, in which methods for the separations important in protein structure determination were also considered. Current developments in the stepwise Edman degradation of peptides and proteins were reviewed by Han et al. (1985).

The results of studies on the amino acid sequencing in proteins, summarized in a condensed form from the total literature, were collected in an *Atlas of Protein Sequence and Structure* (Dayhoff et al., 1978).

B426

13.2 NUCLEIC ACIDS, POLYNUCLEOTIDES AND OLIGONUCLEOTIDES

A knowledge of the structure of nucleic acids is extremely valuable in present-day science. Access to the genetic information encoded in the base sequence of DNA provides data that are of central importance to molecular biology, life sciences and many corresponding applied sciences. Not long ago, examinations of the primary structures of proteins were the main source of biochemical information on gene expression. Today, direct sequencing of nucleotides in nucleic acids prevails. The establishment of the nucleotide sequence in a nucleic acid (i.e., so-called nucleic acid sequencing) has developed so quickly that it is now much easier then protein sequencing.

This is the reason why some workers anticipate a decrease in the importance of amino acid sequence determinations, e.g., Malcom (1978) in an article "The decline and fall of protein chemistry?" According to Malcom, the zenith of protein sequencing reached in six papers published in 1978 (cf., Fowler and Zabin, 1978), where the complete amino acid sequence of β-galactosidase was described; the molecular weight of this enzyme was 116 349 and it contained 1021 amino acid residues. Even in this paper the sequence determination was aided by analysis of CNBr-peptides obtained from a polypeptide fragment produced by a lacZ termination mutant strain. An intense expansion of molecular genetics methods was predicted, including nucleic acid sequencing. Malcom (1978) wrote, "It is still easier to purify a particular protein than to purify and characterize a particular segment of DNA, but latter is becoming easier all the time".

13.2.1 Nucleic acid sequencing

With respect to the great importance of nucleotide sequencing for the life sciences, these methods must be mentioned here. However, nucleic acid sequencing is based predominantly on electrophoretic separation methods, and HPLC is applied only for sequencing of not too long oligonucleotides, mainly in connection with synthetic procedures, for checking the correctness of the sequences prepared. Therefore, for the sequencing of nucleic acids and longer fragments, reference will be made to the specialized literature (cf., Section 13.2.3).

The longest sequence published [172 282 base pairs (bp)] was that determined by Baer et al. (1984) for the DNA of the B 95-8 Epstein-Barr virus genome. According to Martin and Davies (1986), the total number of sequences reported to international data banks was about $7 \cdot 10^6$ bp. The present total or that in the near future will be about 10^7 nucleotides, but most of this genetic information belongs to procaryonts. Because the first DNA sequencing was described in 1977, this total (the result of 10 years effort) does not represent any major progress, as the value of 10^7 is only about three times higher than the genetic information on the bacterium *Escherichia coli*, especially in comparison with the large genomes of eucaryonts, which are almost unknown. A great lack of information is being felt with regard to human genome (containing about $3 \cdot 10^9$ bp), the sequencing of which may represent a turning point in cancer research (Dulbecco, 1986), among others.

13.2.2 HPLC in the study of oligonucleotide and polynucleotide structures

In this section, the sequential analysis of synthetic oligonucleotides will be dealt with, together with examples of the application of HPLC in the determination of the chain length of oligonucleotides, and in the analysis of mRNA terminal cap structures. The former type of method is based on IEC and the latter on RPC.

13.2.2.1 Manual methods

Van Boom and De Rooij (1977) reported a sequence analysis of synthetic oligonucleotides by anion-exchange HPLC. The method was based on determining the amounts of 5'-mononucleotides and nucleosides sequentially liberated from the 3'-ends of the oligonucleotides by the action of snake venom phosphodiesterase. Samples were withdrawn from the enzymic digest at appropriate time intervals and the enzyme was deactivated by the addition of 0.2 M KH_2PO_4 (pH 3.5). The released amounts were chromatographed (after the addition of internal standards for quantification) on Permaphase AAX at room temperature by isocratic elution. A 1 m x 2.1 mm I.D. column was eluted with 0.005 M KH_2PO_4 (pH 4.5) and detection was at 254 nm. This method permitted the unambigous direct establishment of the sequence of oligonucleotides of synthetically prepared RNA and DNA fragments up to a length of six units without any recourse to prior ^{32}P labelling of the substrate.

Komiya and Takemura (1979) described the nucleotide sequence of 5S ribosomal RNA from rainbow trout (*Salmo gairdnerii*) liver, using complete and partial pancreatic RNase A and Rnase T1 digestion, followed by low-pressure DEAE-cellulose chromatography with a linear gradient of sodium chloride in 7 M urea-0.01 M Tris-HCl buffer (pH 7.5) for elution. Komiya et al. (1979; cf. also 1978) developed the above RNase digest approach for HPLC application and described methods for the sequencing of oligoribonucleotides and RNA fragments, using UV monitoring of the effluent for identification of the components. For chromatography AS-Pellionex SAX (at ambient temperature) or AL-Pellionex WAX (at 50°C) was used and the peaks in the effluent were rapidly scanned (5 s per 100 nm) with a Union-Giken SM 303 high-speed UV spectromonitor, connected directly to the outlet of the UV detector (254 nm), so that any peak could be immediately identified by its spectrum under a continuous solvent flow. For elution, concentration gradients of H_3PO_4-KH_2PO_4 (pH 3.3-4.2) or sodium chloride in 0.002 M Tris-sulphuric acid (pH 7.6) were used. If necessary, methanol (10-20%) or urea (5-7 M) was added to facilitate the elution of oligonucleotides. The determination of retention volumes and comparison of the measured spectra with those of standards allowed quantitative and qualitative analyses of oligonucleotides to be achieved at the same time. The method was rapid, taking only 10-30 min for one chromatographic run and required less than 0.01 A_{260} unit of sample per nucleotide material in one peak. It should be particularly useful in institutions where facilities for radioisotope experiments are absent or poor.

13.2.2.2 Automated procedure

De Rooij et al. (1979) reported an automated sequence analysis of synthetic DNA fragments by anion-exchange HPLC. The previously described method [Van Boom and De Rooij (1977); see the initial paragraph of this section] was automated by regular pumping of a small proportion (12 µl) of the oligonucleotide partial hydrolysate (enzyme-substrate digest mixture) from a thermostated syringe, at intervals of 12.5 min, into a 10-µl sample loop, which was then applied to a 50 cm x 2.1 mm I.D. column of Permaphase AAX (DuPont) and eluted isocratically with 0.0025 M KH_2PO_4 buffer (pH 4.5) at a flow-rate of 1 ml/min, 20°C and a pressure of 500 p.s.i. Peak areas were recorded with an LDC Model 308 computing integrator. The reproducibility of the method was good, so that products obtained after the partial digestion could be identified from their retention times. Owing to the very good reproducibility of the injections and the accuracy of the product quantitation, no internal standards were necessary (average standard deviation $\sigma = 0.9\%$). For example, the composition of the synthetic decanucleotide d-T-($CG_2A_3T_3$) and the 5'-terminal nucleoside were derived from the molar fractions of the total digest: dT (1.0), pdC (0.9), pdT (3.3), pdA (2.9) and pdG (1.9).

The sequence of this decanucleotide was derived as follows: after 20 injections of the partial digest (total time 4 h), the recorded peak areas were divided by the molar absorptivities (Van Boom and De Rooij, 1977) of the corresponding nucleoside 5'-phosphates, and the values obtained (peak area/ε_{254}) were plotted against time. Curves for the times of the appearance of molar fractions of the nucleoside-5'-phosphates were constructed in this way. The total sequence d-TpApTpCpApApGpTpTpG was derived from the reversed order of half-times of appearance of the corresponding nucleoside-5-phosphates (cf., Fig. 13.5).

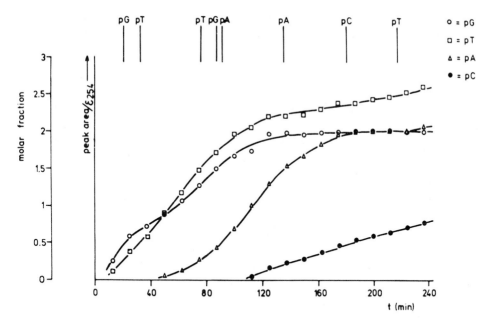

Fig. 13.5. Time curves of the appearance of the nucleoside 5'-phosphates (pdG, pdT, pdA and pdC) from phosphodiesterase action on the synthetic decamer d-TpApTpCpApApGpTpTpG, plotted according to the data obtained by the automatic sequencing analysis. The vertical bars headed by the symbols for the nucleoside 5'-phosphates represent the half-times of appearance of the corresponding nucleoside 5'-phosphates. (Reprinted from De Rooij et al., 1979.)

13.2.2.3 Other contributions of HPLC to the determination of structures of oligonucleotides and nucleic acids

Crowther et al. (1981) studied the HPLC of oligonucleotides. During the development of synthetic techniques for these oligomers, a novel method for the determination of oligonucleotide chain length (as a prerequisite for the sequencing) was also reported, using diesterase cleavage and IP-RP-HPLC. Any terminal phosphate was first removed by means of alkaline phosphatase. The second enzyme used for the total hydrolysis (snake venom diesterase) hydrolysed the phosphate linkages on the 3'-side: a single nucleoside (the terminal monomer) and 5'-monophosphate nucleosides were produced. When this mixture was chromatographed under conditions for the separation of nucleosides and monophosphate nucleotides, the molar fraction of nucleotides to nucleosides gave the chain length, absolute base composition, and leading nucleotide of the oligomer. A 100 mm x 4.6 mm I.D. column packed with Partisil 5 C_8 (Whatman) was eluted isocratically using 9% methanol in 0.05 M KH_2PO_4, and 15 mM tetrabutylammonium iodide was added as a hetaeron. Chains composed of 5-10 monomers could be determined rapidly and accurately using only nanomoles of the solute.

B430

Albers et al. (1981) reported the analysis of mRNA 5'-terminal cap structures and internal N^6-methyladenosine by RP-HPLC. A simple procedure for the determination of the complete methylation profile of a mRNA molecule in a single chromatographic separation was described. The mRNA was selectively hydrolysed to its component nucleosides, leaving its cap structures intact. The hydrolysis products were separated on an ODS RP column using a mobile phase containing acetonitrile and ammonium formate (a weak ion-pairing reagent). [Methyl-^{3}H]-poly(A)-containing mRNA was used to demonstrate the efficiency of the procedure. Three RP columns were employed.

One separation experiment will be mentioned here as an example: a mixture of 5'-terminal cap 0, cap 1,2'-O-methyl-nucleosides and an internal standard, N^6-methyladenosine, were separated on a 25 cm C_{18} Partisil PXS 5/25 ODS (Whatman) column, eluted isocratically, stepwise and finally using a gradient of mixtures of acetonitrille and 1 M ammonium formate. The sample contained 0.1-0.2 nmol of each component and was injected in a volume of 4 μl. Detection at 254 nm was used. A good separation of fifteen components was achieved within 3.5 h.

13.2.3 Comments on literature

Brownlee (1972) wrote a book on RNA sequencing. Maxam and Gilbert (1980) reviewed and described in detail the sequencing of end-labelled DNA with base-specific chemical cleavage and Smith (1980), in an analogous way, DNA sequence analysis by primed synthesis. Pačes (1982a) reviewed the determination of the primary structure of nucleic acids (both DNA and RNA). Pačes (1982b) also dealt with molecular cloning for the analysis of DNA sequences and Bankier (1984) advances in dideoxy sequencing (Sanger's method). Wu et al. (1984) reviewed in detail and in the form of a practical manual the purification and sequence analysis of synthetic deoxyribonucleotides. A detailed manual on RNA sequencing was prepared by Digweed et al. (1985). Bilofsky et al. (1986) reported on the GenBank genetic sequence data bank and Hamm and Cameron (1986) on the EMBL (European Molecular Biology Laboratory) data library. Martin and Davies (1986) reviewed progress and prospects in automated DNA sequencing. Academic Press (1987) published the third annual, new and complete eight-volume compendium containing nucleic acid sequences included in EMBL and GenBank genetic data banks (GenBank is a registered trade mark of the U.S. Department of Health and Human Services). On 4270 pages, 11 413 sequences are registered, containing 8 442 357 bases. This compendium was compiled for those without computer access, who need information on which genes have already been published and of which sequences are composed. The data were classified into 13 Code Sections according to organisms, and the Database Directory can help in finding the required information easily.

13.3 POLYSACCHARIDES, OLIGOSACCHARIDES, COMPLEX DERIVATIVES AND GLYCOCONJUGATES

13.3.1 *Introduction*

Carbohydrate polymers and oligomers (both simple and complex), and especially glycoconjugates, i.e. glycoproteins and glycolipids, play many important roles in living nature (e.g., glycoconjugates participate significantly in the cell-cell recognition process and in blood group specificity), without all of these activities being perfectly known yet. Therefore, the elucidation of the exact structure of these polymeric or complex compounds and of the oligosaccharide moieties of glyconjugates is a very important task in chemistry, biochemistry and the life sciences, because the complete chemical structure is the generally acknowledged basis for all other functional considerations. In this section, methods for the determination of the sequence and branching of sugars in all oligomeric and polymeric compounds will be dealt with briefly, applications of HPLC in this field being emphasized. In general, methods contributing to struc-tural research can be divided into three groups: (a) auxiliary analyses in order to determine quantitatively the composition of monosaccharides in the total hydrolysate of the studied compound, and to supply other important analytical data; (b) chemical methods for the investigation of oligosaccharide sequencing and branching, including derivatization of saccharide chains and their hydrolysis; and (c) biochemical methods for the specific splitting of oligosaccharide chains by suitable enzymes at defined positions. The various methods will be briefly reviewed in this order.

13.3.1.1 *Analytical methods*

The quantitative analysis of neutral sugars in hydrolysates can be best achieved by automated anion-exchange chromatography in borate buffers (Spiro, 1966a; Lee, 1972a), and for the determination of amino sugars and hexose aminitols cation-exchange chromatography on amino acid analysers has been elaborated (Lee, 1972a; Spiro, 1972). However, the old spectrophotometric methods for the determination of glucosamine and chondrosamine (Elson and Morgan, 1933), or N-acetylglucosamine and N-acetylchondrosamine (Morgan and Elson, 1934), including N-acetylated amino sugars liberated enzymically (Reissig et al., 1955), are sometimes used in structural studies. Warren (1959) described the determina-tion of sialic acid liberated from a complex by neuramidinase using the thio-barbituric acid method, after any O-acetyl groups had been removed. The deter-mination of sialic acid by means of an amino acid analyser was described by Liu (1972). Some reviews cited in Section 11.6 may be also consulated for newer analytical references.

13.3.1.2 Chemical approach

Among the chemical techniques for structural studies of complex carbohydrates
(McNail et al., 1982), one of the earliest degradation methods involved acid-
catalysed partial hydrolysis into oligosaccharide fragments, which were separated,
purified, analysed and further sequenced. Measurement of the timed release of
monosaccharides during the mild acid hydrolysis provides information on the
sugar sequence (Spiro, 1966b, 1972). The resulting sequence of the original
chain could be deduced from the structures of overlapping fragments. Another
important method (Li and Li, 1977) is periodate oxidation, in the form of serial
Smith degradation (Spiro, 1966b, 1972) and β-elimination (Svensson, 1978); when
the carbohydrate units are linked to the peptide by O-glycosidic bonds involving
serine or threonine, they can be obtained in the form of reduced oligosaccharides
(after β-elimination by mild alkaline treatment in the presence of sodium borohydride). An especially important chemical method is methylation or permethylation (Hakomori, 1964; Lindberg, 1972; Lindberg and Lönngren, 1978). Application
of this method to N-acetylated glycopeptides by means of methyl iodide and
catalysis by the methylsulphinyl carbanion in dimethyl sulphoxide can provide
important structural information about their carbohydrate units concerning the
linkages between sugar residues and their extent of branching (Spiro, 1972).
O-Methylated monosaccharides can be obtained from the hydrolysates of the
modified oligosaccharide chains, and isolated and characterized by means of IEC,
GLC, TLC or even paper chromatography (PC). Their detailed analysis permits
the following conclusions: tetra-O-methyl ethers indicate a terminal non-reducing
position, tri-O-methyl ethers are obtained from singly substituted internally
located sugar residues and di-O-methyl derivatives are formed from monosaccharides located at branch positions. From methylated hexosamines, O-methyl
ethers of N-methylamino sugars are formed after acid hydrolysis, and can be
identified chromatographically.

McNeil et al. (1982) reviewed in detail the structural analysis of complex
carbohydrates using HPLC and mass spectrometry (MS), with particular regard to
the HPLC separation of peralkylated oligosaccharide alditols. They considered
the sequencing of complex carbohydrates by partial hydrolysis of the peralkylated
complex followed by conversion of the resulting mixture of partially alkylated
oligosaccharides into differently peralkylated oligosaccharide alditols. The
preparation of peralkylated oligosaccharide alditols by partial hydrolysis,
reduction and ethylation of permethylated complex carbohydrates was described
(in addition to the preparation of diagnostic oligosaccharide fragments by
other methods) and the results obtained by the HPLC of peralkylated alditols
on C_{18} columns eluted with acetonitrile-water mixtures using refractive index
detection were discussed. MS detection was also discussed, and the analysis of

obtained fractions by chemical ionization (CI) MS, GLC-MS, direct probe MS and high-resolution proton magnetic resonance spectroscopy (^{1}H NMR) was described. In addition, the determination of the anomeric configuration of the glycosyl linkages of peralkylated oligosaccharide alditols by HPLC retention times, determination of ring forms (pyranosyl, furanosyl) of the glycosyl residues, determination of labile points of attachement of complex carbohydrate and sequence summation of the analytical results into the structure of the original complex carbohydrate were illustrated with examples.

Physico-chemical methods applied to structural studies of saccharides (e.g., for the determination of the configuration of anomeric protons) involve mainly NMR and IR spectroscopy. Barker et al. (1982) described the ^{13}C NMR spectroscopy of complex carbohydrates and Dabrowski et al. (1982) the analysis of glyco-sphingolipids by ^{1}H NMR spectroscopy, which also seems to be a powerful tool for sequencing some types of complex carbohydrates (McNeil et al., 1982).

13.3.1.3 Biochemical approach

A final approach to the structural elucidation of poly- and oligosaccharides (both single and complex) and of glycoconjugates is the application of bio-chemical techniques. They employ specific enzymes (glycosidases) to split the glycosidic bonds. Two classes of these enzymes must be distinguished: (i) exo-glycosidases and (ii) endoglycosidases.

13.3.1.4 Exoglycosidases

Exoglycosidases are hydrolases that cleave monosaccharide units from the non-reducing end of oligosaccharides (both single and in glycoconjugates). Two kinds of specificities are decisive for the cleavage of a substrate: (a) glycone specificity and (b) aglycone specificity, e.g., a β-galactosidase distinguishes (Kobata, 1979)

The glycone specificity is very high and is directed to the sugar moiety, including its anomeric configuration. Owing to the strict stereochemical specificity (and because a very small amount of substrate is sufficient for analysis),
the specific glycosidases are often used to determine the anomeric configuration.
However, the glycone specificity is not absolute: β-N-acetylhexosaminidases exist which cleave both β-N-AcGlc and β-N-AcGal linkages. Because exoglycosidases
cleave only monosaccharide residues located at the non-reducing terminal, they
can be used for the gradual sequencing of oligosaccharide chains (the amount
of the released monosaccharide can be measured by GC or LC). If a complete
structure of an oligosaccharide moiety composed of different sugars is to be
established, it is necessary to submit it to sequential digestion by a number
of different enzymes (see later, Table 13.3).

*General experimental technique for application of glycosidases (Li and Li,
1977).* A 0.05-0.5 μmol amount of the substrate is dissolved in 100-500 μl of
an appropriate buffer and incubated with a specific enzyme at 37°C for 4-20 h.
At the end of the reaction, an aliquot of the digest is analysed for the free
saccharide liberated (the rate of liberation of a monosaccharide unit from
different complex carbohydrates varies considerably, depending on the nature
of the aglycone moieties. It is therefore essential to assess the extent of
hydrolysis several times during incubation).

A very sensitive and convenient method is now available (Kobata, 1979) for
the sequencing of oligosaccharides: the chain is first reduced with $NaB[^3H]_4$
and the tritium-labelled sample is subjected to enzyme digestion. The reaction
products are analysed by chromatography (e.g., by GPC, PC, or other methods)
and, because the tritium label is located at the reducing end of the chain, the
recovered part of the chain can be used in the next digestion. For the sequencing
of a dodecasaccharide by this approach, as little as 10 μg of the sample was
sufficient (cf., Section 13.3.2).

The aglycone specificity can influence the catalytic ability of the enzyme
very substantially and the splitting may be suppressed for stereochemical reasons
and with respect to the binding site of the enzyme. Therefore, even if the
expected monosaccharide is not released by a specific exoglycosidase, it does
not necessarily mean that the sugar is not located in the non-reducing terminus.
The specifities of glycosidases of the same type but of different origin
(isolated from different sources) often vary in their substrate requirements.

13.3.1.5 Endoglycosidases
Endoglycosidases are enzymes that recognize oligosaccharides as specific
glycones and therefore they can act in the middle portion of an oligosaccharide
chain and can also release oligosaccharide moieties from glycoproteins and

TABLE 13.3

ENZYMES USEFUL FOR STRUCTURAL STUDIES OF OLIGO- AND POLYSACCHARIDES, COMPLEX CARBOHYDRATES AND GLYCOCONJUGATES (DEGRADING ENZYMES)

Sources: Ginsburg (1972, 1978, 1982), Spiro (1972) and Kobata (1979). Glycosidases compiled by Spiro (1972) as useful in the study of carbohydrate units of glycoproteins are designated here GP, exoglycosidases summarized by Kobata (1979) as useful in the structural examination of glycoconjugates are designated here EX and endoglycosidases reviewed by Kobata (1979) are designated here EN. This table does not include enzymes used solely for diagnostic purposes (such as the examination of genetic mucopolysaccharide storage disorders or sphingolipidoses). Enzyme classification numbers (E.C.) are incorporated only if they were given in the cited original papers or reviews.

Name	Designation	Source	Ref.
N-Acetylgalactosamine deacetylase	EX;E.C. 3.2.1.49	*Clostridium tertium*	Marcus (1972)
α-N-Acetylgalactosaminidase	EX;E.C. 3.2.1.49	*Aspergillus niger*	McDonald and Bahl (1972)
	EX;GP; E.C. 3.2.1.49	Beef liver	Weissman (1972b)
	GP	*Clostridium perfringens*	McGuier et al. (1972)
	GP	*Lumbricus terrestris*	Buddecke et al. (1969)
		Porcine liver	Weissmann and Hinrichsen (1969)
endo-α-N-Acetylgalactosaminidase	EN;E.C. 3.2.1.97	*Diplococcus pneumoniae*	Endo and Kobata (1976)
	EN;E.C. 3.2.1.97		Umemoto et al. (1977)
			Kobata and Tasaki (1978)
α-N-Acetylglucosaminidase	GP	Pig liver	Weissmann (1972a)
β-N-Acetylglucosaminidase	GP	*Aspergillus niger*	Bahl and Agraval (1972)
		Hen oviduct	Tarentino and Maley (1972a)
		Phaseolus vulgaris	Agraval and Bahl (1972)
β-N-Acetylglucos(or hexos)aminidase	EX;GP	*Clostridium perfringens*	McGuire et al. (1972)
endo-β-N-Acetylglucosaminidase	EN;E.C. 3.2.1.96	Hen oviduct	Tarentino and Maley (1978)
	EN	*Streptomyces plicatus*	Tarentino et al. (1978)

(Continued on p. B436)

TABLE 13.3 (continued)

Name	Designation	Source	Ref.
endo-β-N-Acetylglucosaminidases C_I and C_{II}		*Clostridium perfringens*	Kobata (1978)
	EN;E.C. 3.2.1.96	Fig	Ogata et al. (1977)
	EN;E.C. 3.2.1.96	Fig	Chien et al. (1977)
	EN;E.C. 3.2.1.96	Mamalian kidney, liver and spleen	Nishigaki et al. (1974)
endo-β-N-Acetylglucosaminidase D	EN;E.C. 3.2.1.96	*Diplococcus pneumoniae*	Muramatsu (1978)
endo-β-N-Acetylglucosaminidase H	EN;E.C. 3.2.1.96	*Streptomyces griseus*	Arakawa and Muramatsu (1974)
endo-β-N-Acetylglucosaminidase L		*Streptomyces plicatus*	Trimble et al. (1982)
β-N-Acetylhexos(or glucos)amini-dase	EX;GP;E.C. 3.2.1.52	*Diplococcus pneumoniae*	Hughes and Jeanloz (1964b)
	GP	Jack bean meal	Li and Li (1972a)
Amylase		*Pseudomonas stutzeri*	Robyt and Ackerman (1972)
Bacteriophage-induced capsular polysaccharide depolymerase		*Aerobacter aerogenes*	Yurewicz et al. (1972)
Chondroitinases	Special enzyme	*Flavobacterium heparinum* and *Proteus vulgaris*	Suzuki (1972a)
Chondrosulphatases		*Proteus vulgaris*	Suzuki (1972b)
α-L-Fucosidase	EX;E.C. 3.2.1.51	Almond emulsin	Kobata (1982)
	EX;E.C. 3.2.1.51	*Aspergillus niger*	Bahl (1972)
1,2-α-L-Fucosidase	EX;GP;E.C. 3.2.1.51	*Clostridium perfringens*	Aminoff (1972)
	GP	Rat epidermis	Bahl (1970)
α-Galactosidase	EX;GP;E.C. 3.2.1.22	*Aspergillus niger*	Bahl and Agraval (1972)
	EX;GP;E.C. 3.2.1.22	Coffee bean	Courtius and Petek (1966)
	EX;E.C. 3.2.1.22	Fig	Li and Li (1972b)
		Pinto beans (*Phaseolus vulgaris*)	Agraval and Bahl (1972)
β-Galactosidase	EX;GP;E.C. 3.2.1.23	*Aspergillus niger*	Bahl and Agraval (1972)
	EX;GP;E.C. 3.2.1.23	*Clostridium perfringens*	McGuire et al. (1972)
	EX;GP;E.C. 3.2.1.23	*Diplococcus pneumoniae*	Hughes and Jeanloz (1964a)
	GP	*Escherichia coli*	Spiro (1962)
		Human liver	Meisler (1972)
	EX;GP;E.C. 3.2.1.23	Jack bean meal	Li and Li (1972a)
	GP	Pinto beans (*Phaseolus vulgaris*)	Agraval and Bahl (1972)
		Sweet almond emulsin	Lee (1972b)

Enzyme	Source	Code	Reference
endo-β-galactosidase	*Diplococcus pneumoniae*	EN	Kobata and Tasaki (1978)
	Escherichia freundii	EN	Fukuda and Matsumura (1976)
	Flavobacterium keratolyticus		Kitamikado et al. (1982)
Glucoamylase	*Aspergillus niger*		Pazur (1972)
α-Glucosidase	Rat liver lysozomes		Brown et al. (1972)
β-Glucosidase	Pinto bean (*Phaseolus vulgaris*)		Agraval and Bahl (1972)
	Sweet almond emulsin		Lee (1972b)
Glycosyl asparaginase (4-L-aspartylglycosamine aminohydrolase)	Hen oviduct		Tarentino and Maly (1972b)
	Hog kidney		Kohno and Yamashina (1972)
α-Glucuronidase	*Helix pomatia*	GP	Spiro and Bhoyroo (1971)
β-Glucuronidase	Rat liver lysozomes	E.C. 3.2.1.31	Stahl and Touster (1972)
Heparinase and heparitinase (eliminases)	Flavobacteria		Linker and Hovingh (1972)
α-Fucosidase	*Aspergillus niger*	GP	Weissmann and Hinrichsen (1969)
α-L-Fucosidase	*Bacillus fulminans*	EX;E.C. 3.2.1.51	Kochibe (1973)
	Charania lampas	EX;E.C. 3.2.1.51	Iijima and Egami (1971)
	Turbo cornutus	EX;E.C. 3.2.1.51	Iijima et al. (1971)
	Almond emulsin	EX;E.C. 3.2.1.51	Yoshima et al. (1979)
α-L-Fucosidase I	*Aspergillus niger*	EX;E.C. 3.2.1.24	Swaminathan et al. (1972)
1,2-α-Mannosidase			Matta and Bahl (1972)
	Hen oviduct		Sukeno et al. (1972)
	Hog kidney		Okumura and Yamashina (1972)
	Jack bean meal	EX;GP;E.C. 3.2.1.24	Li and Li (1972a)
	Pine apple	EX;E.C. 3.2.1.24	Li and Lee (1972)
(Mannopyranosidase)	Pinto beans (*Phaseolus vulgaris*)	GP	Agraval and Bahl (1972)
	Sweet almond emulsin	GP	Lee (1972b)
	Turbo cornutus	EX;GP;E.C. 3.2.1.24	Muramatsu and Egami (1967)

(Continued on p. B438)

B438

TABLE 13.3 (continued)

Name	Designation	Source	Ref.
β-Mannosidase (Mannopyranosidase)	EX;GP;E.C. 3.2.1.25	Hen oviduct	Sukeno et al. (1972)
	EX;E.C. 3.2.1.25	Pine apple	Li and Lee (1972)
	EX;E.C. 3.2.1.25	Snail	Sugahara and Yamashina (1972)
	EX;E.C. 3.2.1.25	*Turbo cornutus*	Muramatsu and Egami (1967)
Neuramidinase (sialidase)	EX;E.C. 3.2.1.18	*Arthrobacter ureafaciens*	Uchida et al. (1974)
	EX;GP;E.C. 3.2.1.18	*Clostridium perfringens*	Cassidy et al. (1965)
	GP	*Clostridium perfringens, Vibrio cholerae,* influenza virus	Cuatrecasas (1972)
	GP	*Diplococcus pneumoniae*	Hughes and Jeanloz (1964a)
	EX;E.C. 3.2.1.18	Myxoviruses (influenza virus)	Drzeniek et al. (1966)
		Rat heart	Tallman and Brady (1972)
	EX;GP;E.C. 3.2.1.18	*Vibrio cholerae (Vibrio comma)*	Ada et al. (1961)
Phosphomannanase Trehalase		*Bacillus circulans*	McLellan (1972)
		Myxobacterium smegmatis and *Streptomyces hydroscopicus*	Patterson et al. (1972)
β-D-Xylosidase		*Bacillus pumilus*	Kersten-Hilderson et al. (1982)
		Penicillium wortmanni	Deleyn et al. (1982)
β-Xylosidase	GP;E.C. 3.2.1.37	*Charonia lampas*	Fukuda et al. (1969); Yasuda et al. (1970)

glycolipids; in this way they can substantially contribute to the structural studies of glycoconjugates. According to Kobata (1979), endoglycosidases "are opening a new age in the structural study of glycoconjugates", because they release the oligosaccharide fragment amenable to further investigation. Examples of their applications were presented by Kobata (1979) and the problem of possible contamination of enzymes was discussed.

A large number of enzymes applied both in the sequencing of simple oligosaccharides and in the degradation of complex carbohydrates are listed in Table 13.3. The detailed specificity of particular enzymes (in addition to that obvious from its name) cannot be explained in a simple form suitable for tabulation, because various limiting conditions and many exceptions must be considered, and therefore the cited papers should be consulted for details; the same applies to the experimental details (concentrations, pH, quality of buffers, temperature, time intervals, etc.).

The combination of biochemical and chemical approaches, i.e., the application of splitting by glycosidases with methylation analysis, offers (according to Li and Li, 1977) the most powerful tool available for elucidating the structures of heterosaccharide chains.

In the following sections a more specialized approach to research on the structure of poly- and oligosaccharidic chains in (a) simple and complex carbohydrates, (b) glycolipids and (c) glycoproteins will be discussed and illustrated by examples. The general techniques will differ for particular glycoconjugates (Li and Li, 1977): when glycosphingolipids are treated with glycosidases, the resulting glycolipid with a shorter saccharide chain can be detected easily (e.g., by TLC), whereas when glycoproteins are digested, the liberated sugar is assayed more easily than the glycoprotein residue.

13.3.2 *Oligosaccharides, polysaccharides, complex carbohydrates*

13.3.2.1 *Oligosaccharides*

According to Pazur (1970), the following information is required for the total elucidation of the structure of an oligosaccharide: (1) molecular weight, M_r; (2) identity of monomeric residues; (3) sequence of monomeric residues; (4) positions of the monomeric residues involved in the glycosidic bonds, (5) type of ring structure; and (6) stereochemical configuration of the glycosidic bond. Some of these requirements can be covered by the application of the method developed by Hase et al. (1978, 1979c), viz., structural analyses of oligosaccharides by tagging of the reducing end sugars with a fluorescent compound (cf., Section 11.4.2). A potential aldehyde group was combined with 2-aminopyridine by means of reductive amination with sodium cyanoborohydride giving a fluorescent

2-aminopyridine (PA) derivative, which could be separated from inorganic salts and the remaining 2-aminopyridine by sorption on Dowex 50 cation-exchange column and elution with dilute ammonia. The reaction scheme can be illustrated as follows:

| Oligosaccharide | 2- Aminopyridine | | PA-derivative |

The procedure for the combination of a sugar specimen with 2-aminopyridine (Hase et al., 1978) is as follows. About 1 mg of a sugar was dissolved in 20 µl of water and 80 µl of reagent, made by mixing 83 mg of 2-aminopyridine, 32 mg of sodium cyanoborohydride, 37 µl of acetic acid acid and 0.33 ml of methanol, were added. The solution was heated in a sealed tube at 75°C for 3-7 h and then the reaction mixture was mixed with 1.3 ml of Dowex 50-X2 (H$^+$) resin. The resin was washed with water and eluted with 0.6 M aqueous ammonia and the eluate was evaporated to dryness (the yield was 80-90%).

The sugar derivative can be used for the determination of (1) the degree of polymerization, (2) the sequence of sugar units, (3) the linkage points of the sugar units and (4) the preparation of "fingerprints". This derivatization method was combined with TLC or two-dimensional paper electrophoresis, and the modified substances were detected by fluorescence under UV light.

Hase et al. (1979b) applied the above method to the sequence analysis and the determination of linkage points of sugar units as follows. The PA derivative of an oligosaccharide was partially hydrolysed, the products were separated by paper electrophoresis and the fluorescent compounds were eluted and analysed by TLC. In order to determine the linkage points, the PA derivatives were permethylated and then partially hydrolysed. The newly produced hydroxyl groups were methylated with [^{2}H]methyl iodide. After the separation of neutral and basic components by passage through a Dowex 50 (H$^+$) column, the fluorescent components (now labelled with [^{2}H$_3$]methyl groups) were separated by TLC, eluted, hydrolysed and converted into alditol acetates. The linkage points were determined by GC-MS.

Later, Hase et al. (1981b) modified this method for the determination of the molecular weights of the oligosaccharides. The PA derivative of interest was mixed with a small amount of PA derivatives of standard oligosaccharides of different M_r and subjected to paper electrophoresis. Sections of paper corresponding to standard oligosaccharides were extracted with water and each extract was analysed by HPLC with fluorescence spectrometric detection. The

chromatographic details were described in Section 11.4.2. The M_r of the PA
derivative was calculated by comparing its mobility with those of PA derivatives
of internal standards detected together with the sample. This method was applied
to the determination of the M_r of the sugar moieties of Taka-amylase A and quail
ovomucoid.

Yamashita et al. (1982) described a method for the analysis of oligosaccharides by gel filtration; a relatively simple and rapid method for the separation
and characterization of oligosaccharides composed of 8-20 monosaccharides was
developed. The equipment consisted of a water-jacketted thick-walled column
(2 m x 2 cm I.D.) or of two water-jacketted columns (1 m x 2 cm I.D.) connected
by Teflon tubing (0.5 mm I.D.). The column was packed with Bio-Gel P-4 (under
400 mesh) and kept at $55^{\circ}C$ in order to minimize the possible interaction of
Bio-Gel beads with the oligosaccharides. The column was fitted at the top with
an adaptor fitting and connected to a refractive index (RI) detector. The bottom
of the column was connected through a universal injector to a high-pressure pump.
All connections were made with Teflon tubing (0.5 mm I.D.).

Oligosaccharides could be released from natural sources (serum glycoproteins)
by hydrazinolysis or by *endo*-β-N-acetylglucosaminidase digestion, and had to
be labelled with tritium by reduction with NaB^3H_4. An aliquot of the radioactive
oligosaccharide ($5 \cdot 10^4$-$50 \cdot 10^4$ cpm) was mixed with 0.4 ml of the non-radioactive
standard glucose oligomer solution*, applied on the column and eluted with pre-
warmed ($55^{\circ}C$) deaerated, distilled water at a flow-rate of 1 ml/min. Standard
glucose oligomers were detected by the RI detector. The effluent was fractionated
into 1.0-4.0 ml per tube. The radioactivity in each tube was determined by
liquid scintillation counting on aliquots of samples. The position of the radioactive oligosaccharide could then be determined accurately by superimposing
the radioactivity curve on that obtained by RI monitoring and was expressed in
glucose units. As this value was a constant for all oligosaccharides, Yamashita
et al. (1982) presented a large table of many natural oligosaccharide fragments
characterized by their structure and glucose units and discussed the collected
data in detail.

The oligosaccharide labelled on its reducing terminus could be then gradually
digested with various exoglycosidases and the remaining radioactive fragment
(characterized by glucose unit value) could be digested and chromatographed
again. From the data obtained the monosaccharide sequence and branching could
be derived. This approach was illustrated by an example of sequencing of an
oligosaccharide released from human transferrin.

---

*Standard glucose oligomers could be obtained as follows: 100 mg of dextran (MW
  200 000) were heated in 1 ml of 1 M hydrochloric acid for 4 h at 100ºC. The
  solution was lyophilized and redissolved in distilled water to give a 10 mg/ml
  solution.

13.3.2.2 Polysaccharides

In this section several newer examples of the application of the above-mentioned methods to the elucidation of the structure of higher molecular weight carbohydrates will be presented. Valent et al. (1980) reported a general and sensitive chemical method for the sequencing of glycosyl residues of complex carbohydrates, which employed fractionation of complex mixtures of peralkylated oligosaccharides by RP-HPLC. The isolated di-, tri- and (in some instances) tetrasaccharides generated by successive partial acid hydrolysis, reduction and ethylation of a permethylated complex carbohydrate, were unambiguously identified by GC-MS. From the overlapping structures obtained the complete oligosaccharide sequences may be pieced together and combined into the glycosyl sequence of the complex carbohydrate. This approach was illustrated on examples using lichenan from *Cetraria islandica*, gum xanthan and cell-wall xyloglucan nonasaccharide, secreted by sycamore cells into the culture medium. The experimental approach was described in detail and thoroughly discussed. For chromatography a pre-packed Zorbax ODS column (25 cm x 4.6 mm I.D.), equipped with a Whatman Co:Pell ODS guard column (7 cm x 2 mm I.D.) were used and eluted with 50% (v/v) acetonitrile-water mixtures. The flow-rate was 0.5 ml/min and the pressure was 400 lb/in.^{2}. Fractions of 0.5 ml were collected. An RI detector was used.

Taylor et al. (1985) described the application of HPLC to a study of the branching of dextrans. The lengths of the side-chains in dextrans were examined by enzymic hydrolysis and LC. The linear (IM_n) and branched (B_n) oligosaccharide products of endodextranase activity were separated by LC. The branched fractions obtained from *Leuconostoc mesenteroides* NRRL B-512 (F) dextran were resolved into components and treated with (1→6)-α-D-glucan glucohydrolase. From a knowledge of the specificity of both the dextranases together with results of methylation analysis, 3^3-α-isomaltosylisomaltosaccharides and 3^3-α-D-glucosylisomaltosaccharides were identified. In addition to LC, paper chromatography (PC) was used for the separation.

For LC a Dextropak cartridge (10 cm x 0.8 cm) inserted into radial-compression separating system (Waters Assoc.) was employed using water as solvent. A differential refractometer was used for monitoring. Branched oligosaccharides were separated at a constant flow-rate of 2 ml/min; for a complete separation of glucose, IM_2 and IM_3 from higher oligosaccharides the flow-rate was linearly increased from 0.5 to 3 ml/min during the first 10 min of the operating programme. The determination of B_n-2 oligosaccharides and B_4 (3^3-α-D-glucosylisomaltotriose) in the final products indicated the proportion of glucosyl side-chains in dextrans.

Hara et al. (1986) studied a (1→6) branched (1→3)-β-D-glucan (T-3-G) from an aqueous extract of *Dictiophora indusiata* Fisch. The water-soluble glucan T-3-G was isolated from a hot-water extract of the fruit bodies of *Dictiophora indusiata* Fisch, and was found to be homogeneous by ultracentrifugal analysis, electrophoresis and GPC ($M_r \approx 5.1 \cdot 10^5$). In order to study the structure, SEC of enzymic degradation products on a Bio-Gel P-2 column (98 cm x 1.5 cm I.D.) eluted with water and monitored by sugar assay (measurement of absorbance at 490 nm) was used to elucidate the structure, in addition to other methods. The separated disaccharides were characterized by PC and GLC (of the volatile derivatives), and the M_r was derived from GPC on Bio-Gel P-2. The data obtained from all the examinations led to the conclusion that the glucan T-3-G has a main chain composed of (1→3)-β-D-glucopyranosyl residues, and has as side-chains two single β-(1→6)-linked D-glucopyranosyl groups attached, on average, to every fifth sugar residue of the main chain; the branching of T-3-G occurs regularly at O-6 of the β-(1→3)-linked backbone. The conformation of the polysaccharide chains was discussed.

Melton et al. (1986) examined the structural characterization of an oligosaccharide isolated from pectic polysaccharide rhamnogalacturonan II (RG-II). This is a structurally complex acidic pectic (D-galactosyluronic acid-rich) polysaccharide that is present in the primary (growing) cell walls of higher plants. The isolation and structural characterization of 23 oligosaccharide fragments of the residue of RG-II that remained after removal of hepta- and disaccharides by partial acid hydrolysis were discribed. The carboxyl groups of RG-II were first dideutero-reduced and the carboxyl-reduced polysaccharide was per-O-methylated and then fragmented by partial hydrolysis with acid. The derivatized oligosaccharides were reduced, to afford a mixture of partially O-methylated oligoglycosylalditols, which were then per-O-methylated. For their isolation Sep-Pak cartridge was used and eluted repeatedly first with water, then with acetonitrile-water (3:17), acetonitrile (100%) and finally with dichloromethane. The 100% acetonitrile fractions contained per-O-methylated oligoglycosylalditols. These were further separated on an Altex Ultrasphere ODS (5 μm) column (24 cm x 4.6 mm I.D.) (protected by a guard column), eluted with two linear gradients of acetonitrile-water (first to 3:17, second to 4:1). The fractions were characterized by chemical ionization mass spectrometry (CI-MS), fast atom bombardment (FAB) MS and ^{1}H NMR analysis. It was shown that seventeen of the oligosaccharides isolated from RG-II were parts of a single heptasaccharide.

The fine structure of (1→3),(1→4)-β-D-glucan from *Zea* shoot cell walls was studied by Kato and Nevins (1986). The β-D-glucan was hydrolysed by *Bacillus subtilis* β-D-glucan 4-glucanohydrolase, and the oligosaccharides were characterized by methylation analysis of the fragments and by methylation analysis of the secondary fragments, generated by the treatment of isolated oligosaccharides with *Streptomyces* QM B814 cellulase. The first generation of fragments were fractionated by GPC on a Bio-Gel P-2 column (45 cm x 3.5 cm I.D.), eluted with water, and for the second generation of fragments a 150 cm x 1.5 cm I.D. Bio-Gel P-2 column was used, also eluted with water. In both instances detection at 490 nm was employed in assays for carbohydrates. The results of the detailed analysis of the fractions obtained demonstrated that the native polysaccharide consisted mainly of cellotriosyl and cellotetraosyl residues joined by single (1→3) linkage. Details of the structure were discussed.

O'Neill et al. (1986a) described the structure of the extracellular polysaccharide produced by the bacterium *Alcaligenes* (ATCC 31 555) species. This anionic polysaccharide contains L-mannose, L-rhamnose, D-glucose and D-glucuronic acid. For the resolution of its structure analysis of methylated and methylated carboxyl-reduced polysaccharide and partial acid hydrolysis and base-catalysed β-elimination were applied. Series of oligosaccharides were obtained, which were isolated as their alkylated alditol derivatives by RP-HPLC and characterized by FAB MS, EI-MS and ^{1}H NMR spectroscopy. For isolation Sep-Pak cartridges were used, eluted with water and 20% aqueous acetonitrile; the methylated oligosaccharide alditols were eluted with 60% aqueous acetonitrile. For RP-HPLC a 25 cm x 4.6 mm I.D. Zorbax ODS column and a Brownlee C_{18} guard column were eluted isocratically with 60% aqueous acetonitrile. A DRI detector was used for monitoring and fractions were collected manually. Dry residues were dissolved in acetone and applied to FAB MS or EI-MS. Results of all the analyses led to the proposal of a repeated pentasaccharide unit as the fundamental characteristic of the studied polysaccharide.

O'Neill et al. (1986b) examined the extracellular gelling polysaccharide produced by *Enterobacter* (NCIB 11 870) species. Quantitative analysis of the monosaccharide composition was followed by periodate oxidation linkage analysis using Smith degradation (Dutton and Merrifield, 1982) and methylation analysis of methylated and methylated carboxyl-reduced polysaccharides and bacteriophage-induced enzymic depolymerization. Oligosaccharides from the dialysed and desalted enzymic digest were isolated by means of preparative PC and characterized by MS. The reduced (NaB^{2}H$_4$) and methylated pentasaccharides were isolated from Sep-Pak C_{18} cartridges (Waters Assoc.) by elution with 60% aqueous acetonitrile and purified by RP-HPLC on a 25 cm x 4.6 mm I.D. Zorbax ODS column by isocratic elution with 60% aqueous acetonitrile according to O'Neill and Selvendran (1985); DRI measurement was used for monitoring. The fractions obtained were characterized by FAB MS, direct insertion EI-MS and ^{1}H NMR spectroscopy. From the results the following repeating tetrasaccharide unit was proposed for characterizing the structure of the studied polysaccharide:

→4)-α-D-GlcpA-(1→3)-α-L-Fucp-(1→3)-β-Glcp-(1→

$$\overbrace{4} \atop \uparrow \atop \underbrace{1}$$

β-D-Glcp

Mueller-Harvey et al. (1986) studied the linkage of p-cumaroyl and feruloyl groups to cell-wall polysaccharides of barley straw. The cell walls were treated with *Oxyporus* "cellulase" (a mixture of polysaccharide hydrolases) and compounds were released containing p-coumaroyl and feruloyl groups bound to carbohydrates. For their isolation HPLC was employed, using 25 cm x 1 cm or 0.5 cm I.D. columns of Spherisorb 5 ODS-1 (Phase Separations) eluted with methanol-water (28:72 or 22:78) at 2, 3 or 6 ml/min. The eluates were monitored at 250, 280 or 325 nm. The isolated fractions were characterized by analysis of products released by acid hydrolysis, enzymic "cellulase" hydrolysis, methylation analysis, NMR spectroscopy and FAB MS. Two repeated compounds were identified as O-[5-O-(*trans*-p-coumaroyl)-α-L-arabinofuranosyl]-(1→3)-O-β-D-xylopyranosyl-(1→4)-D-xylopyranose and O-[5-O-(*trans*-feruloyl)-α-L-arabinofuranosyl]-(1→3)-O-β-D-xylopyranosyl-(1→4)-D-xylopyranose.

The structure of the exopolysaccharide from *Rhizobium* sp. strain ANU 280 (a derivative of the broad host-range strain NGR 234) was investigated by Djordjevic et al. (1986). The fragments generated by partial acid hydrolysis were fractionated by sequential IEC and GPC. For IEC a 40 cm x 1.5 cm I.D. column of DEAE-Sephadex A25 was used, eluted isocratically with 25 mM Tris-HCl (pH 7.5) and with a convex gradient formed by the same buffer and 100 mM sodium chloride in the same buffer. For GPC a 95 cm x 1.5 cm I.D. column of Bio-Gel P-2 was eluted with 200 mM trimethylammonium formate (pH 3.3). Aliquots of the fractions were analysed for hexoses and uronic acid. The structures of the fragments obtained were assigned by Smith degradation, glycosidase treatment and ^{13}C NMR spectroscopy. Extensive overlap of structures between fragments, together with the results of periodate oxidation and spectrophotometric analysis permitted the assignment of the complicated nonasaccharide unit and terminal group.

13.3.2.3 Complex carbohydrates

Delaney et al. (1980a) described the quantification of sulphated disaccharides of heparin by HPLC. Heparin was converted into sulphated disaccharides by treatment with nitrous acid. The mixture obtained was reduced with sodium [^{3}H]borohydride and the disaccharides were purified by paper electrophoresis and PC. Four disaccharides were isolated, structurally characterized and used as standards for HPLC on a Partisil-10 SAX anion-exchange column. The chromato-

graphic conditions were described in Section 11.5.2. It was demonstrated that
this HPLC method gave reproducible and quantitative results in direct assays
of aliquots of $[^3H]$ borohydride-reduced heparin deamination mixtures. The four
sulphated disaccharides formed when heparin was treated with nitrous acid
represented the primary units from which the heparin sequences are constructed.
The characterization of heparin sequences requires quantitation of the relative
amounts of these disaccharides and determination of the order in which they
occur along the polymer. This paper was one of the first contributions to such
complex work.

In subsequent work, Delaney et al. (1980b) examined an HPLC approach for the
isolation and sequencing of higher chondroitin sulphate oligosaccharides (tetra-,
hexa-, octa- and decasaccharides). Conditions for the separation of products
of hyaluronidase digests of chondroitin-4-sulphate (of whale cartilage) and
chondroitin-6-sulphate (of shark cartilage) were reported in Section 11.5.2.
Each chromatographically isolated major peak was purified to homogeneity and
characterized for reducing and non-reducing terminal saccharides, for internal
disaccharides and for the degree of polymerization. The chondroitinase digestion converted the oligosaccharides into disaccharides, which were analysed to
determine the relative amounts of 4- and 6-sulphated disaccharides in each
position of the disaccharide, and further characterized. Methods suitable for
deriving the saccharide sequence were described and discussed.

Lee et al. (1981) also described an HPLC separation of reduced disaccharides
derived from glycosylaminoglycans. Reduced unsaturated disaccharides were
derived from enzymic digestion followed by reduction with sodium borohydride of
chondroitin sulphates, dermatan sulphate, heparan sulphate and heparin. The
chromatographic conditions were described in Section 11.5.2. The applicability
of the method to the determination of glycosaminoglycans in biological samples
was demonstrated in the original paper.

In subsequent sections the determination of the structure of the saccharide
moieties in the glycoconjugates will be dealt with. This is a generic name
(Kobata, 1979) for compounds in which sugars are covalently linked to either
lipids or proteins.

13.3.3 Glycolipids

In the examination of the chemical structure of saccharide moieties of glycolipids, HPLC is not the main separation technique. TLC is used most often,
because the stepwise-degraded glycolipids can be simply detected and characterized by a change in mobility in comparison with suitable standard substances,
and because more samples can be examined simultaneously on the same sheet or

plate. Therefore, only a very brief treatment will be given here of the methods
used.

Both chemical and biochemical methods have been applied to the elucidation
of the monosaccharide sequences and branching of oligosaccharide chains in
glycolipids. The isolation and characterization of glycosphingolipids using
chemical methods was reviewed by Esselman et al. (1972), who considered
methanolysis (with 0.75 M methanolic hydrochloric acid) to be the first useful
step in the structural characterization of these substances (methanolysis was
studied in detail by Chambers and Clamp, 1971). The products were sphingolipid
bases and their O-methyl derivatives, fatty acid methyl esters, and methyl
glycosides. These components could be analysed by GLC (methylglycosides after
trimethylsilylation).

The method for cleaving the carbohydrate portion from the lipid is ozonolysis
(according to Wiegandt and Bashang, 1965). The ozonolysis yield, which is a
relatively simple operation as described by Esselman et al. (1972), is about
80%, and can also be performed on a small scale, requiring 1 mg of lipid. This
review also describes the permethylation of glycosphingolipids and their partial
acid hydrolysis.

The chemical release of oligosaccharides from acetylated glycolipids using
sodium metaperiodate and osmium tetraoxide treatment and alkaline degradation
was described by Hakomori and Siddiqui (1972). The degradation can easily by
controlled and few milligrams of glycolipid are sufficient; when combined with
^{3}H labelling, as little as 100 µg are sufficient. This method cannot be applied
to glycolipids containing an O-acyl group.

Improved procedures for the chemical modification and degradation of glycosphingolipids were published by MacDonald et al. (1980); treatment of the
resulting glycolipid aldehyde (or ketone) after periodic acid and osmium
tetroxide oxidation (or with sodium methoxide and other modifications) resulted
in the release of the intact oligosaccharide in high yield.

The chemical methods were used by Smirnova et al. (1982) for a study of the
branched disialoglycolipids from the starfish *Asterias amurensis*. On the basis
of total and partial acid hydrolysis, methanolysis, methylation, periodate and
chromium trioxide oxidation, the main sialolipid was identified as 8-O-methyl-N-
glycolylneuraminosyl-(2→6)-[8-O-methyl-N-glycolylneuraminosyl-(2→3)]-N-acetyl-
galactosaminyl-(β1→3)-galactosyl-(β1→4)glucosyl-(β1→1)-ceramide.

Biochemical methods (enzymic degradation of gangliosides) were reviewed by
Li and Li (1980), who also discussed the function of protein activators (con-
ditioning the activity of mammalian exoglycosidases). The release of oligosaccharides from glycosphingolipids by *endo*-β-galactosidase of *Escherichia
freundii* was reported by Fukuda et al. (1978) [see also comments by Kobata

(1979)]. The structure of a new glycolipid antigen isolated from human erythrocyte membranes reacting with antibodies directed to globo-N-tetraosyl-ceramide (globoside) was reported by Kannagi et al. (1982). The reactivity to the antibody was eliminated by treatment with *endo*-β-galactosidase (*Escherichia freundii*). One of the fractions isolated by repeated HPLC was examined structurally by means of *endo*-β-galactosidase, exoglycosidases, methylation analyses and direct probe MS. This glycolipid might represent the second blood group P antigen belonging to the lacto series.

Another contemporary general biochemical method for the resolution of structures of saccharide moieties in glycosphingolipids is the application of exoglycosidases. Successive treatments lead to stepwise splitting of individual saccharides from the chain terminus and these structural changes are reflected in differences in the mobilities of the glycolipid residues containing shorter saccharide chains. They can easily be detected by means of TLC. Li and Li (1977) recommended the addition of detergents to the reaction mixtures, which enhances the exoglycosidase digestion of many glycolipids. They gave examples of structural studies of a globoside, a brain disialosylpentahexosylceramide and a tissue ganglioside. Watanabe and Hakomori (1979) studied gangliosides of human erythrocytes and described a novel ganglioside with a unique N-acetylneuro-aminosyl-(2→3)-N-acetylgalactosamine structure. Successive exoglycosidase treatment was used, in addition to methylation analysis and direct probe MS of permethylated intact and desialylated glycolipid.

Another example was published by Slomiany et al. (1980). Complex fucolipids with the hog gastric mucosa structure of ceramide eicosahexoside were studied using partial acid hydrolysis, sequential degradation with specific exoglyco-sidases, oxidation with periodate and chromium trioxide and permethylation analysis. Chien and Hogan (1980) characterized two gangliosides of the para-globoside series from chicken skeletal muscle. Using specific enzymic hydrolysis and methylation analysis the following structures were determined: (A) NeuAc-α-(2→3)-Gal-β-(1→4)-GlcNAc-β-(1→3)-Gal-β-(1→4)-GlcCer; and (B) NeuAc-α-(2→3)-Gal-β-(1→4)GlcNAc-β-(1→3)-Gal-β-(1→4)-GlcNAc-β-(1→3)-Gal-β-(1→4)GlcCer. Other examples of newer structural studies can be found in the elucidation of the structure of isolated novel gangliosides (for citations see Section 12.2.5), such as gangliosides from human erythrocytes (Kundu et al., 1983).

Takamizawa et al. (1986) studied gangliosides of bovine buttermilk and isolated and characterized a novel monosialoganglioside with a new branching structure. Here also degradation with exoglycosidases and methylation analysis were employed and, in combination with TLC and GLC of partially methylated aldohexitol acetates, led to a new core structure with this branched oligosaccharide chain:

Gal β1→4GlcNAcβ1
                    ↘ 6
                      ⌐Galβ1→4Glcβ1→Cer
                    ↗ 3
NeuAcα2→6Galβ1→4GlcNAcβ1

13.3.4 Glycopeptides and glycoproteins

The importance of the structural determination of the glycoprotein saccharide
moieties has been already mentioned in Section 13.3.1. The characterization of
carbohydrate units of glycoproteins was reviewed by Spiro (1966b); procedures
for the separation of glycopeptides from glycoproteins, analyses of monosaccharide and amino acid portions and the characterization of the glycopeptide
bond were described. The structure of glycoproteins was also reviewed by Spiro
(1970).

13.3.4.1 Analytical methods

The preliminary investigation consists in acquiring the necessary analytical
data for the glycoprotein studied. Among the first requirements is the determination of monosaccharidic units, of which the oligosaccharide chains are
composed.

Release of neutral sugars (Spiro, 1972). A dilute solution of a glycoprotein
(1-5 mg/ml) is hydrolysed in 1-2 M hydrochloric acid (or sulphuric acid) for
4-6 h in a sealed tube at 100°C. The conditions for optimal release will vary
with the protein under study and depend on the stability of the glycosidic
linkages and the resistance to acidic degradation of the released sugars. The
free neutral sugars can be determined by borate complex chromatography (for
citations on general analytical methods see Section 13.3.1).

Release of hexosamines and hexosaminitols (Spiro 1977). The glycoprotein is
hydrolysed in 4 M hydrochloric acid in a sealed tube at 100°C for 6 h. In most
instances it is convenient to separate first the amino sugars from neutral
sugars and some of amino acids and peptides by sorption and elution from a small
column of Dowex 50 (H^{+}); glucosamine and galactosamine can be separated and
analysed by means of an amino acid analyser.

One of the most important tasks in the study of a glycoprotein is the
characterization of the bond by which the carbohydrate moiety is attached to the
peptide chain, i.e., the identification of the sugar and amino acid involved
in the glycopeptide bond. Three types of linkages were found in glycoproteins
(Spiro, 1972): (i) the glycosamine bond, which involves N-acetylglucosamine
(GlcNAc) and the amide group of asparagine (Asn); (ii) the alkali-labile
O-glycosidic bond to the hydroxyamino acids serine (Ser) and threonine (Thr)

[which may involve N-acetylgalactosamine (GalNAc), galactose (Gal), xylose (Xyl) or mannose (Man)]; and (iii) the alkali-stable O-glycosidic bond of Gal or arabinose (Ara) to hydroxylysine (Hyl).

The first type (i) of glycosamine bond should be isolated from the partial enzymic* or acid** hydrolysate of the glycoprotein "in natura" and chromatographically characterized by comparing with a standard. A peptide as short as possible containing this glycosamine bond is desired, which is why repeated digestion is often applied. Asn-linked oligosaccharides can be chromatographed, e.g., on an Aminex-H (Bio-Rad) column (57 cm x 0.9 cm I.D.) at 53°C using elution with 0.1 M sodium citrate buffer (pH 2.80) containing 3% propanol (the sample buffer should have a slightly lower pH of 2.2) (Tarentino et al., 1970). For the determination of the second type of glycosidic bond (ii), mild alkaline treatment is used (incubation with 0.05-0.5 M sodium hydroxide solution at $0-45^{\circ}$C for 15-216 h). Such a treatment is accompanied by a rapid decline in the analytical values for Ser or Thr and results in a β-elimination process in which reducing products are formed. If no reducing agent is added, it is not possible to identify the sugar component participating in the bond. If the alkaline treatment is performed in the presence of sodium borohydride (0.15-1.0 M), the reduced oligosaccharides will give (after acid hydrolysis) the corresponding alditol (galactitol, galactosaminitol, xylitol or mannitol). The resulting amino sugar alcohol can be identified and determined in an amino acid analyser (in citrate-borate buffers). The third type of glycosamine bond (iii) can be elucidated by the isolation of short Hyl-glycopeptides (see below). If there is evidence for the presence of two or more bond types, the corresponding glycopeptides should be resolved and studied separately.

Special analytical procedures have been developed that are very useful for the investigation of glycoproteins. Carlson (1972) described an assay for N-acetylgalactosaminitol, used when N-acetylgalactosamine (GalNAc) linked glycosidically to Ser (or Thr) is to be determined. Alkali treatment releases GalNAc by β-elimination with concomitant formation of dehydroalanine from Ser (and α-aminocrotonic acid from Thr). Degradation of GalNAc is controlled by the reduction of its reducing group with sodium borohydride. The N-acetylgalactosaminitol and also N-acetylhexosamines formed are acid hydrolysed into galactosaminitol and amino sugars. The resulting free amino groups are N-acetylated with [^{14}C]acetic anhydride and the labelled derivatives are separated by paper electrophoresis and quantified by liquid scintillation counting. Another assay for following alkali-catalysed β-elimination, which releases carbohydrate chains

---

*e.g., pronase followed by glycosidases.
**e.g., 2 M hydrochloric acid for 12-20 min at 100°C.

linked O-glycosidically to Ser or Thr in glycoproteins, was described by
Plantner and Carlson (1972): the unsaturated amino acids that are formed in this
process absorb in the UV region at 240 nm and the extent of β-elimination can
be determined by the increase in absorbance at this wavelength. The concentration of both pyruvate and α-ketobutyrate (which are products of acid hydrolysis
of the olefinic amino acid residues) can be measured enzymically be means of
lactic acid dehydrogenase coupled with NADH.

According to Spiro (1972), the usual conditions for achieving β-elimination
are incubation with 0.1 M sodium hydroxide solution at 37°C for 48-72 h, but
the optimal conditions should be established separately for each glycoprotein
or glycopeptide. When β-elimination is carried out in the presence of sodium
sulphite, the unsaturated products formed are converted into their sulphonyl
derivatives (Ser → cysteic acid; Thr → α-amino-β-sulphonylbutyric acid). As
these strong acids cannot be simply determined on a cation-exchange amino acid
analyser, Spiro (1972) developed a chromatographic procedure using the anion
exchanger Dowex 1, which separates these two components clearly from an acid
hydrolysate. The bond between Gal and Hyl (5-O-β-D-galactopyranosylhydroxy-
lysine) is relatively very stable against alkali treatment and therefore this
unit and a short disaccharide terminated by this bond, Gal-Hyl (such as
2-O-α-D-glucosylgalactose disaccharide), can be split after alkaline hydrolysis
(which cleaves all peptide bonds linked to Hyl) and determined using an amino
acid analyser (Spiro, 1972).

A specific quantitation by HPLC of protein (lysine)-bound glucose in human
serum albumin (non-enzymically glycosylated) and in other glycoproteins was
developed by Schleicher and Wieland (1981). The protein was hydrolysed in 6 M
hydrochloric acid at 65°C for 18 h to yield furosine [ε-N-(2-furoylmethyl)-L-
lysine], a specific degradation product of the fructose-lysine bond. Furosine
was then separated by HPLC and quantitated against a prepared standard. The
method was used in biochemical diagnostics to determine the glycosyl-albumin
in diabetic patients (100 μl or less of serum was sufficient) and also for the
determination of other glycoproteins.

13.3.4.2 Determination of the structure of the oligosaccharide moiety

Studies involving periodate oxidation, methylation and graded acid hydrolysis
are more easily performed and interpreted for glycopeptides than intact proteins.
In addition, glycopeptides are more accessible to the degradation by glycosidases
than intact glycoproteins. Therefore (according to Spiro, 1972), as the first
step in the elucidation of the structure of the carbohydrate units of a glycoprotein and the nature of its glycopeptide bond, it is useful to digest the
protein moiety extensively with a protease of broad specificity (such as pronase

from *Streptomyces griseus*). To achieve maximum proteolysis, prolonged digestion is carried out at $37^{\circ}C$ at a concentration of 25 mg/ml of the glycoprotein with several additions of the enzyme (up to 2% of the substrate weight). In order to prevent bacterial growth, several drops of toluene are added to the glycoprotein mixture.

Glycopeptides in the proteinase digest can be fractionated by means of GPC or IEC. Examples of the separation of glycopeptides from the pronase digest of glomerular basement membrane on an 82 cm x 2.1 cm I.D. column of Sephadex G-25 eluted with 0.1 M pyridine acetate buffer (pH 5.0), monitored with anthrone detection and ninhydrin, and the separation of glycopeptides of the pronase digest of human thyroglobulin on a Dowex 50-X2 column (120 cm x 1.9 cm I.D.) eluted with a linear gradient from 1 mM pyridine formate (pH 3.0) to 0.155 M pyridine formate, monitored by anthrone and resorcinol (cf., Chapter 11), were included in the review by Spiro (1972).

Oligosaccharide fragments can also be prepared from glycoproteins by the action of endoglycosidase or by treatment with anhydrous hydrazine (Mizuochi et al., 1978). Natowicz and Baenziger (1980) presented examples of structures of such glycoprotein-derived oligosaccharides which were used for sequence studies. Partial acid hydrolysis of glycoproteins also leads to oligosaccharides suitable for this purpose, in addition to β-elimination using mild alkaline treatment.

The first and simplest methods for the elucidation of the structure of individual oligosaccharides included the measurement of the timed release of monosaccharides during mild acid hydrolysis, together with periodate oxidation and Smith degradation (Spiro, 1966b). Some more advanced methods of periodate oxidation and methylation treatment were reviewed by Spiro (1972). Examples of structural studies of glycoproteins were presented by Li and Li (1972).

Biochemical (enzymic) methods using exoglycosidases and endoglycosidases for the sequencing of oligosaccharides were dealt with in Section 13.3.1. Natowicz and Baenziger (1980) published an example of the sequential degradation of an oligosaccharide prepared from ceruloplasmin by hydrazinolysis. The ^{3}H-labelled reduced oligosaccharide was gradually treated with (A) neuraminidase, (B) neuraminidase + β-glycosidase, (C) neuraminidase + β-galactosidase + β-N-acetyl-glucosaminidase and (D) the above three enzymes + endoglycosidase. The products were centrifuged, mixed with internal standards and chromatographed on two 100 cm x 0.6 cm I.D. columns of Bio-Gel P-4 (-400 mesh) connected in series using water ($55^{\circ}C$) for elution. Very good separation of fragments was obtained, which opened way for structural characterization.

Hase et al. (1979a, 1981a) applied a method based on tagging the reducing end of oligosaccharides with a fluorescent compound to obtain PA derivatives (cf., Section 13.3.1) to the study of glycoproteins. The saccharides obtained by deamination with nitrous acid (or by hydrazinolysis) of oligosaccharide portions of the glycoproteins fetuin, taka-amylase A and ovalbumin were examined in this way. The fingerprints (obtained by two-dimensional paper electrophoresis) were characteristic of the chemical structures of the original oligosaccharides.

Hase et al. (1982) modified this method by employing column chromatography and applied it to the identification of the trimannosylchitobiose structure in sugar moieties of ovomucoid isolated from the egg white of Japanese quail. The ovomucoid contained mannose, galactose, glucosamine and sialic acid (molar ratio 8.6:1.3:6.6:0.5). Sugar chains were liberated from the polypeptide portion by hydrazinolysis, N-acetylated and fractionated on a Bio-Gel P-2 column (150 cm x 2 cm I.D.) using elution with water, and aliquots of the fractions were used for the detection of sugars by the orcinol-sulphuric acid method with absorbance measurement at 425 nm. Three fractions were obtained, P_1-P_3. Fraction P_3 (with the lowest M_r) was coupled with 2-aminopyridine and chromatographed on a Bio-Gel P-2 column (150 cm x 2.4 mm I.D.) using 0.02 M acetic acid for elution. The main fluorescent fraction had the sugar composition $Man_3GlcNAc_1$-PA-$GlcNAc_1$ and was subjected to exoglycanase digestion and M_r determination. The sequence of enzymes employed was α-mannosidase, β-mannosidase and β-N-acetyl-hexosaminidase, using digestion in 0.05 M sodium acetate (pH 4.6) at 37^0C for 16 h (or 40 h with the last-mentioned enzyme) in the presence of toluene. After every step the digest was separated using HPLC and the isolated products were subjected to the next digestion. For HPLC a column (300 mm x 4 mm I.D.) packed with TSK Gel LS 410 (5 μm, C_{18} RP) (Toyo Soda, Tokyo, Japan) was eluted with 0.2 M ammonium acetate (pH 4) containing 0.1% 1-butanol at a flow-rate of 1.6 ml/min. Fluorescence detection was applied with an excitation wavelength of 320 nm and an emission wavelength of 400 nm. The pyridylaminated reducing ends of the sugar moieties were also submitted to chemical studies. The fluorescent PA derivatives of major oligosaccharides indicated the following structure for the major sugar chains of Japanese quail ovomucoid:

$$Man\alpha 1 \searrow_6$$
$${}_3 Man\beta 1 \to 4GlcNAc\beta 1 \to 4GlcNAc$$
$$Man\alpha 1 \nearrow^3$$

Tang and Williams (1985) reported an improved method for the LC of 1-deoxy-1-(2-pyridylamino)alditol derivatives of oligosaccharides and its application to structural studies of the carbohydrate moieties of glycoproteins. The alditol derivatives were prepared by reductive amination. An amine-modified silica column was used to separated the components of a mixture of the pyridylamino derivatives of oligosaccharides from mono- to dodecasaccharide in 20 min. The reductive amination of reducing saccharides with 2-aminopyridine (Hase et al., 1981a) was considered to be a useful method for the introduction of a fluorescent label that enabled 0.1-pmol amounts to be detected by LC with a fluorescent detector. However, the yields of the products were low. Therefore, Tang and Williams (1985) developed a modified, more effective procedure and applied it to the analysis of oligosaccharides derived from glycoproteins. The products of the reductive amination were separated from the excess of inorganic salts and 2-aminopyridine by preparative paper electrophoresis. For the LC of PA derivatives a Hypersil (5 μm) column (25 cm x 0.49 cm I.D.) modified with 1,6-diaminohexane was employed and for elution a linear gradient from 80% to 60% aqueous acetonitrile containing 0.1% of 1,6-diaminohexane gave an excellent separation of the carbohydrate derivatives. The flow-rate was 2 ml/min and a Kratos FS LC fluorimeter was used for monitoring (excitation at 230 nm, emission above 340 nm using a cut-off filter).

An example of a procedure for a small-scale reductive amination (Tang and Williams, 1985) is as follows. A mixture of maltooligosaccharides (1 mg, from starch hydrolysate) was reductively aminated with 2-aminopyridine (5 mg) and sodium cyanoborohydride (1.5 mg) in N,N-dimethylformamide (0.2 ml) containing 10% of glacial acetic acid at 95°C for 3 h. The products were isolated by preparative PC and a solution of the products in water was used for LC.

Baudyš et al. (1982) described the characterization of the sulphated glyco-peptide of chicken pepsinogen, isolated from the S-sulphonated enzymogen, digested with TPCK-trypsin and (following SEC on Sephadex G-25) with chymotrypsin. High-voltage paper electrophoresis led to the isolation of three acidic glyco-peptides of identical amino acid sequence Val-Ser-Thr-*Asn*-Glu-Thr-Val-Tyr, differing in the composition of the sugar moiety, which was attached to the peptide chain via asparagine *Asn*. The localization of the oligosaccharide moiety in the primary structure of chicken pepsinogen was determined and discussed with regard to the homologous areas of other acid proteinases. Schmidt et al. (1987) studied the structure of a sulphated glycopeptide Val-Ser-Thr-*Asn*-Glu-Thr, derived from the same region of S-sulphonated chicken pepsinogen by thermolysin cleavage. The carbohydrate composition of the most abundant species was $Man_3GlcNAc_7$, containing five sulphate groups. The oligopeptide was studied in ^2H_2O by one- and two-dimensional NMR at 300 and 500 MHz. Peptide resonances were completely assigned from two-dimensional shift-correlated spectra (COSY). The pattern of Man H1, H2 and H3 chemical shifts and other examinations indicated that the glycopeptide was a derivative belonging to the tetraantennary complex class and can be expressed by

```
              (7)
        HO₃S-GlcNAcβ1                                              COOH
                     ↘                                              |
                      4                                            Thr
  HO₃S-GlcNAcβ(1→2)Manα1                                           |
        (5)           (4)   ↘                                      Glu
                            3  (3)      (2)        (1)             |
                             Manβ(1→4)GlcNAcβ(1→4)GlcAc-Asn
                            6                                      |
        (5')         (4')    ↗                                     Thr
  HO₃S-GlcNAcβ(1→2)Manα1                                           |
                 4 6                                               Ser
               ↗ ↖                                                 |
     HO₃S-GlcNAcβ1     1βGlcNAc-SO₃H                               Val
          (10)            (7')                                     |
                                                                  NH₂
```

[the saccharide numbering in parentheses follows the recommendation of Vliegenthart et al. (1981)]. This is an example of a modern physico-chemical approach to the determination of a glycopeptide moiety of a protein.

13.3.5 Comments on literature

The literature on the chemistry of polymeric and oligomeric carbohydrates and derivatives is very rich. Pigman and Horton (1970, 1972, 1980) edited a multi-volume book on the chemistry and biochemistry of carbohydrates, and Neufeld and Ginsburg (1966) and Ginsburg (1972, 1978, 1982) a series of books on methods for the study of complex carbohydrates. Lindberg (1972) reviewed the methylation analysis of polysaccharides, Lindberg and Lönngren (1978) the methylation analysis of complex carbohydrates (general procedure and application to sequence analysis), McNeil et al. (1982) the structural analysis of complex carbohydrates using HPLC and MS and Barker et al. (1982) the ^{13}C NMR analysis of complex carbohydrates.

Li and Li (1977) reviewed the use of enzymes in the elucidation of the structures of glycoconjugates, Kobata (1979) the use of endo- and exoglycosidases for structural studied of glycoconjugates, Li and Li (1980) the general principles of applying the structural analysis of glycoconjugates by using glycosidases and ^{13}carbon NMR spectroscopy, and Vliegenthart (1980) the high-resolution 1H NMR spectroscopy of carbohydrate structures.

Esselman et al. (1972) dealt with methods for the isolation and characterization of glycosphingolipids, Hakomori and Siddiqui (1972) the release of oligo-saccharides from glycolipids, Macher and Sweeley (1978) the structure, biological sources and properties of glycosphingolipids, Li et al. (1980a) the enzymic degradation of gangliosides, Dabrowski et al. (1982) the analysis of glyco-sphingolipids by high-resolution ^{1}H NMR spectroscopy and Hakomori (1983) the chemistry of and research methods for glycosphingolipids.

Spiro (1966b) reviewed the characterization of carbohydrate units of glycoproteins, Spiro (1970) the structure, metabolism and biology of glycoproteins,
Spiro (1972) studies of the carbohydrates of glycoproteins, Carlson (1972)
assays for N-acetylgalactosaminitol and Plantner and Carlson (1972) assays for
olefinic amino acids (products of the β-elimination reaction in glycoproteins).

REFERENCES

Abercrombie, D.M., Hough, C.J., Seeman, J.R., Brownstein, M.J., Gainer, H.,
    Russell, J.T. and Chaiken, I.M., *Anal. Biochem.*, 125 (1982) 395-405.
Abercrombie, D.M., Hough, C.J., Seeman, J.R., Brownstein, M.J., Gainer, H.,
    Russell, J.T. and Chaiken, I.M., in Hearn, M.T.W., Regnier, F.E. and Wehr,
    C.T. (Editors), *High-Performance Liquid Chromatography of Proteins and
    Peptides* (Proceedings of the First International Symposium), Academic Press,
    New York, 1983, pp. 129-138.
Abrahamsson, M., Gröningsson, K. and Castensson, S., *J. Chromatogr.*, 154 (1978)
    313-317.
Academic Press, *Nucleotide Sequences 1986/1987* (eight volumes), Harcourt Brace
    Jovanovich International, Orlando, FL, 1987, 4270 pp.
Ada, G.L., French, E.L. and Lind, P.E., *J. Gen. Microbiol.*, 24 (1961) 409-421;
    C.A., 55 (1961) 22 485b.
Agraval, K.M.L. and Bahl, O.P., *Methods Enzymol.*, 28 (1972) 720-728.
Albers, R.J., Coffin, B. and Rottman, F.M., *Anal. Biochem.*, 113 (1981) 118-123;
    C.A., 95 (1981) 32 268t.
Aminoff, D., *Methods Enzymol.*, 28 (1972) 763-769.
Anderson, J.K., Hollaway, W.L. and Mole, J.E., *J. High Resolut. Chromatogr.
    Chromatogr. Commun.*, 4 (1981) 417-418; *C.A.*, 95 (1981) 146 309r.
Annan, W.D., *J. Chromatogr.*, 173 (1979) 194-197.
Annan, W.D., in Rattenbury, J.M. (Editor), *Amino Acid Analysis* (Proceedings of
    Symposium, 1979) Ellis Horwood, Chichester, 1981, pp. 66-70; *C.A.*, 95 (1981)
    111 058f.
Arakawa, M. and Muramatsu, T., *J. Biochem. (Tokyo)*, 76 (1974) 307-317.
Ashman, K., in Lectures Edited for *International Conference on Novel Techniques
    in Protein Micro-Sequence Analysis* (October 1985, Berlin-Dahlem; Wittmann-
    Liebold, B., Erdmann, V. and Salnikov, J., Organizers), Max-Planck Institute
    for Molecular Genetics, Berlin-Dahlem, 1985, pp. 1-16.
Ashman, K. and Bosserhoff, A., *Modern Methods in Protein Chemistry*, Vol. 2,
    Walter de Gruyter, Berlin, New York, 1985, pp. 155-171.
Baer, R., Bankier, A.T., Biggin, M.D., Deininger, P.L., Farrell, P.J., Gibson,
    T.J., Hatfull, G., Hudson, G.S., Satchwell, S.C., Séguin, C., Tuffnell, P.S.
    and Barrell, B.G., *Nature (London)*, 310 (1984) 207-211.
Bahl, O.P., *J. Biol. Chem.*, 245 (1970) 299-304.
Bahl, O.P., *Methods Enzymol.*, 28 (1972) 738-743.
Bahl, O.P. and Agraval, K.M.L., *Methods Enzymol.*, 28 (1972) 728-734.
Bankier, A.T., *Bio Techniques*, 2 (1984) 72-77.
Baranowski, R., Westenfelder, C. and Currie, B.L., *Anal. Chem.*, 121 (1982) 97-102.
Barker, R., Nunez, H.A., Rosevear, P. and Serianni, A.S., *Methods Enzymol.*, 83
    (1982) 58-69.
Baudyš, M., Kostka, V., Grüner, K. and Pohl, J., *Collect. Czech. Chem. Commun.*,
    47 (1982) 709-718.
Bayer, E., Grom, E., Kaltenegger, B. and Uhman, R., *Anal. Chem.*, 48 (1976)
    1106-1109.
Bennett, J.C., *Methods Enzymol.*, 11 (1967) 330-339.
Benson, J.R., *Methods Enzymol.*, 47 (1977) 19-31.
Benson, J.R. and Hare, P.E., *Proc. Natl. Acad. Sci. U.S.A.*, 72 (1975) 619-622.

Bhown, A.S. and Bennett, J.C., in Hancock, W.S. (Editor), *Handbook on Use of HPLC for the Separation of Amino Acids, Peptides and Proteins*, Vol. II, CRC Press, Boca Raton, FL, 1984, pp. 267-278.

Bhown, A.S., Mole, J.E. and Bennett, J.C., *Anal. Biochem.*, 110 (1981) 355-359; *C.A.*, 94 (1981) 152 798v.

Bidligmeyer, B.A., Cohen, S.A. and Tarvin, T.L., *J. Chromatogr.*, 336 (1984) 93-104.

Bilofsky, H.S., Burks, C., Fickett, J.W., Goad, W.B., Lewitter, F.I., Rindone, W.P., Swindell, C.D. and Tung, C.S., *Nucleic Acids Res.*, 14 (1986) 1-14; *C.A.*, 104 (1986) 129 078b.

Birr, C. (Editor), *Methods in Peptide and Protein Sequence Analysis (Proceedings of 3rd International Conference on Solid Phase Methods in Protein Sequence Analysis, Heidelberg, 1979)*, Elsevier, Amsterdam, 1980, 531 pp.; *C.A.*, 94 (1981) B 27 011g.

Bishop, C.A., Hancock, W.S., Brennan, S.O., Carrel, R.W. and Hearn, M.T.W., *J. Liq. Chromatogr.*, 4 (1981) 599-612.

Black, S.D. and Coon, M.J., *Anal. Biochem.*, 121 (1982) 281-285; *C.A.*, 96 (1982) 196 026n.

Bledsoe, M. and Pisano, J.J., in Liu, T.Y., Schechter, A.N., Heinrikson, R. and Condliffe, P.G. (Editors), *Chemical Syntheses and Sequencing of Peptides and Proteins* (Developments in Biochemistry 1981), Elsevier, Amsterdam, 1981, pp. 245-249; *C.A.*, 96 (1982) 200123m.

Böhlen, P. and Kleeman, G., *J. Chromatogr.*, 205 (1981) 65-75.

Brauer, A.W., Margoliash, M.N. and Haber, E., *Biochemistry*, 14 (1975) 3029-3035.

Bridgen, J., *Methods Enzymol.*, 47 (1977) 385-391.

Brown, D.H., Illingworth Brown, B. and Jeffrey, P.L., *Methods Enzymol.*, 28 (1972) 805-813.

Brown, J.R. and Hartley, B.S., *Biochem. J.*, 89 (1963) 59P-60P; 101 (1966) 214-228.

Brownlee, G.G., *Determination of Sequences in RNA*, North Holland, Amsterdam, New York, 1972, 265 pp.; *C.A.*, 81 (1974) B 46655t.

Buddecke, E., Schauer, H., Werries, E. and Gottschalk, A., *Biochem. Biophys. Res. Commun.*, 34 (1969) 517-521.

Callahan, P.X., McDonald, J.K. and Ellis, S., *Methods Enzymol.*, 25 (1972) 282-298.

Carlson, D.N., *Methods Enzymol.*, 28 (1972) 43-45.

Cassidy, J.T., Jourdian, G.W. and Roseman, S., *J. Biol. Chem.*, 240 (1965) 3501-3506.

Caude, M., *Feuill. Biol.*, 106 (1979) 39-51; *C.A.*, 90 (1979) 199 683m.

Chaiken, I.M. and Hough, Ch.J., *Anal. Biochem.*, 107 (1980) 11-16; *C.A.*, 93 (1980) 163 749k.

Chambers, R.E. and Clamp, J.R., *Biochem. J.*, 125 (1971) 1009-1018.

Chang, J.-Y., *Biochem. J.*, 199 (1981a) 537-545; *C.A.*, 96 (1982) 31 034a.

Chang, J.-Y., *Biochem. J.*, 199 (1981c) 557-564; *C.A.*, 96 (1982) 48 482t.

Chang, J.Y., Brauer, D. and Wittmann-Liebold, B., *FEBS Lett.*, 93 (1978) 205-214.

Chang, J.Y., Creaser, E.H. and Bentley, K.W., *Biochem. J.*, 153 (1976) 607-611.

Chang, J.Y., Knecht, R. and Braun, D.G., *Biochem. J.*, 199 (1981b) 547-555.

Chang, J.Y., Knecht, R. and Braun, D.G., *Biochem. J.*, 203 (1982) 803-806; *Methods Enzymol.*, 91 (1983) 41-48.

Chang, J.Y., Lehmann, A. andWittmann-Liebold, B., *Anal. Biochem.*, 102 (1980) 380-383; *C.A.*, 92 (1980) 193 702m.

Chien, J.L. and Hogan, E.L., *Biochim. Biophys. Acta*, 620 (1980) 454-461.

Chien, S.-F., Weinburg, R., Li, S.-C. and Li, Y.-T., *Biochem. Biophys. Res. Commun.*, 76 (1977) 317-323.

Cooke, N.H.C. and Olsen, K., *J. Chromatogr. Sci.*, 18 (1980) 1-12.

Cooke, N.H.C., Viavattene, R.L., Eksteen, R., Wong, W.S., Davies, G. and Karger, B.L., *J. Chromatogr.*, 149 (1978) 391-415.

Courtois, J.E. and Petek, F., *Methods Enzymol.*, 8 (1966) 565-571.

Crowther, J.B., Jones, R. and Hartwick, R.A., *J. Chromatogr.*, 217 (1981) 479-490.

Cuatrecasas, P., *Methods Enzymol.*, 28 (1972) 879-902.
Cunico, R.L. and Schlabach, T., *J. Chromatogr.*, 266 (1983) 461-470.
Cunico, R.L., Simpson, R., Correia, L. and Wehr, C.T., *J. Chromatogr.*, 336 (1984) 105-113.
Dabrowski, J., Hanfland, P. and Egge, H., *Methods Enzymol.*, 83 (1982) 69-86.
Darbre, A., *Methods Enzymol.*, 47 (1977) 357-369.
Davankov, V.A., *Adv. Chromatogr.*, 18 (1980) 139-195.
Dawkins, B.G., Arpino, P.J. and McLafferty, F.W., *Biomed. Mass Spectrom.*, 5 (1978) 1-6.
Dayhoff, M.O., Hunt, L.T., Barker, W.G., Schwartz, R.M. and Orcutt, B.C., *Protein Segment Dictionary 78 from the Atlas of Protein Sequence and Structure*, Vol. 5, Suppl. 1, 2 and 3, National Biomedical Research Foundation, Georgetown University Medical Center, Washington, DC, 1978, 470 pp.
De Jong, C., Hughes, G.J., Van Wieringen, E. and Wilson, K.J., *J. Chromatogr.*, 241 (1982) 345-359.
Delaney, S.R., Conrad, H.E. and Glaser, J.H., *Anal. Biochem.*, 108 (1980b) 25-34.
Delaney, S.R., Leger, M. and Conrad, H.E., *Anal. Biochem.*, 106 (1980a) 253-261.
Deleyn, F., Claeyssens, M. and De Bruyne, C.K., *Methods Enzymol.*, 83 (1982) 639-644.
De Rooij, J.F.M., Bloemhoff, W. and Van Boom, J.H., *J. Chromatogr.*, 177 (1979) 380-384.
Desiderio, D.M. and Stout, Ch.B., in Hancock, W.S. (Editor), *Handbook on Use of HPLC for the Separation of Amino Acids, Peptides and Proteins*, Vol. II, CRC Press, Boca Raton, FL, 1984, pp. 197-211.
Digweed, M., Pieler, T. and Erdman, V.A., *Texts for Advanced FEBS Course 1985* (organized by Wittmann-Liebold, B., Salnikov, J. and Erdmann, V.), *Micro-Sequence Analysis of Proteins*, Max-Planck Institut fûr Molekulare Genetik, Technische Universität Berlin and Frei Universität Berlin, West Berlin, 1985, pp. 1-18.
Dizdaroglu, M., *Varian Instruments at Work*, Varian, Palo Alto, CA, 1981, LC-128, pp. 1-6.
Dizdaroglu, M. and Simic, M., *J. Chromatogr.*, 195 (1980) 119-126.
Djordjevic, S.P., Rolfe, B.G., Batley, M. and Redmond, J.W., *Carbohydr. Res.*, 148 (1986) 87-99.
Drzeniek, R., Seto, J.T. and Rott, R., *Biochim. Biophys. Acta*, 128 (1966) 547-558.
Dulbecco, R., *Science (Washington, D.C.)*, 231 (1986) 1055-1056.
Dutton, G.G.S. and Merrifield, E.H., *Carbohydr. Res.*, 105 (1982) 189-203.
Edman, P., *Acta Chem. Scand.*, 4 (1949) 283-293.
Edman, P., *Acta Chem. Scand.*, 10 (1956) 761-768.
Edman, P. and Begg, G., *Eur. J. Biochem.*, 1 (1967) 80-91.
Einarsson, S., Josefsson, B. and Lagerkvist, S., *J. Chromatogr.*, 282 (1983) 609-618.
Elson, L.A. and Morgan, W.T.J., *Biochem. J.*, 27 (1933) 1824-1828.
Elzinga, M. (Editor), *Methods in Protein Sequence Analysis*, Humana Press, Clifton, NJ, 1982, 589 pp.; *C.A.*, 97 (1982) B 195 376k.
Endo, Y. and Kobata, A., *J. Biochem. (Tokyo)*, 80 (1976) 1-8.
Engelhardt, H., in Lottspeich, F., Henschen, A. and Hupe, K.-P. (Editors), *High Performance Chromatography in Protein and Peptide Chemistry*, Walter de Gruyter, Berlin, 1981, pp. 55-69.
Esselman, W.J., Laine, R.A. and Sweeley, C.C., *Methods Enzymol.*, 28 (1972) 140-156.
Fohlman, J., Rask, L. and Peterson, P.A., *Anal. Biochem.*, 106 (1980) 22-26.
Fong, G.W.-K. and Grushka, E., *Anal. Chem.*, 50 (1978) 1154-1161.
Fowler, A.V. and Zabin, I., *J. Biol. Chem.*, 253 (1978) 5521-5525 (for complete information see also pp. 5484-5520 and refs. 1-4 on p. 5522).
Frank, G. and Strubert, W., *Chromatographia*, 6 (1973) 522-524.
Frantíková, V. Borvák, J., Kluh, I. and Morávek, L., *FEBS Lett.*, 178 (1984) 213-216.

Fukuda, M., Muramatsu, T. and Egami, F., *J. Biochem. (Tokyo)*, 65 (1969) 191-199.
Fukuda, M.N. and Matsumura, G., *J. Biol. Chem.*, 251 (1976) 6218-6225.
Fukuda, M.N., Watanabe, K. and Hakomori, S.-I., *J. Biol. Chem.*, 253 (1978) 6814-6819.
Fullmer, C.S. and Wasserman, R.H., *J. Biol. Chem.*, 254 (1979) 7208-7212.
Fullmer, C.S. and Wasserman, R.H., *Dev. Biochem.*, 14 (*Calcium Binding Proteins: Struct. Funct.*), Elsevier, Amsterdam, 1980, 363-370; *C.A.*, 96 (1982) 30 198v.
Fullmer, C.S. and Wasserman, R.H., in Hawk, G.L., Champlin, P.B., Hutton, F.R. and Mol, C. (Editors), *Biological/Biomedical Applications of Liquid Chromatography, III (Chromatographic Science Series*, Vol. 18), Marcel Dekker, Basle, 1982, pp. 175-196; *C.A.*, 96 (1982) 100 325k.
Gardner, W.S. and Miller III, W.H., *Anal. Biochem.*, 101 (1980) 61-65.
Ghosh, P.B. and Whitehouse, M.H., *Biochem. J.*, 108 (1968) 155-156.
Ginsburg, V. (Editor), *Methods Enzymol. (Complex Carbohydrates, Part B)*, 28 (1972) 1-1057.
Ginsburg, V. (Editor), *Methods Enzymol. (Complex Carbohydrates, Part C)*, 50 (1978) 1-628.
Ginsburg, V. (Editor), *Methods Enzymol. (Complex Carbohydrates, Part D)*, 83 (1982) 1-684.
Graffeo, A.P., Haag, A. and Karger, B.L., *Anal. Lett.*, 6 (1973) 505-511.
Greibrokk, T., Jensen, E. and Ostvold, G., *J. Liq. Chromatogr.*, 3 (1980) 1277-1298; *C.A.*, 93 (1980) 217 371b.
Gross, E., *Methods Enzymol.*, 11 (1967) 238-255.
Haag, A. and Langer, K., *Chromatographia*, 7 (1974) 659-662.
Hakomori, S., *J. Biochem. (Tokyo)*, 55 (1964) 205-208.
Hakomori, S.-I., in Kanfer, J.N. and Hakomori, S.-I. (Editors), *Sphingolipid Biochemistry (Handbook of Lipid Research*, Vol. 3), Plenum Press, New York, London, 1983, pp. 1-165.
Hakomori, S.-I. and Siddiqui, B., *Methods Enzymol.*, 28 (1972) 156-159.
Hamm, G.H. and Cameron, G.N., *Nucleic Acids Res.*, 14 (1986) 5-9.
Han, K.-K., Belaiche, D., Moreau, O. and Briand, G., *Int. J. Biochem.*, 17 (1985) 429-445.
Hancock, W.S., Bishop, C.A., Prestidge, R.L. and Hearn, M.T.W., *Anal. Biochem.*, 89 (1978) 203-212.
Hancock, W.S. and Harding, D.R.K., in Hancock, W.S. (Editor), *CRC Handbook of HPLC for the Separation of Amino Acids, Peptides and Proteins*, Vol. I, CRC Press, Boca Raton, FL, 1984, pp. 235-262.
Hara, C., Kiho, T. and Ukai, S., *Carbohydr. Res.*, 145 (1986) 237-246.
Hare, P.E., *Methods Enzymol.*, 47 (1977) 3-18.
Harris, J.U., Robinson, D. and Johnson, A.J., *Anal. Biochem.*, 105 (1980) 239-245; *C.A.*, 93 (1980) 91 298n.
Hartley, B.S., *Biochem. J.*, 19 (1970) 805-822.
Hase, S., Hara, S. and Matsushima, Y., *J. Biochem. (Tokyo)*, 85 (1979c) 217-220.
Hase, S., Ikenaka, T. and Matsushima, Y., *Biochem. Biophys. Res. Commun.*, 85 (1978) 257-263.
Hase, S., Ikenaka, T. and Matsushima, Y., *J. Biochem. (Tokyo)*, 85 (1979a) 989-994; *C.A.*, 90 (1979) 182 648f.
Hase, S., Ikenaka, T. and Matsushima, Y., *J. Biochem. (Tokyo)*, 85 (1979b) 995-1002; *C.A.*, 90 (1979) 182 765s.
Hase, S., Ikenaka, T. and Matsuchima, Y., *J. Biochem. (Tokyo)*, 90 (1981a) 407-414.
Hase, S., Ikenaka, T. and Matsushima, Y., *J. Biochem. (Tokyo)*, 90 (1981b) 1275-1279.
Hase, S., Okawa, K. and Ikenaka, T., *J. Biochem. (Tokyo)*, 91 (1982) 735-737; *C.A.*, 96 (1982) 138 282g.
Hawke, D.H., Harris, D.C. and Shively, J.E., *Anal. Biochem.*, 147 (1985) 315-330; *C.A.*, 103 (1985) 67 611p.
Hawke, D., Yuan, P.-M. and Shively, J.E., *Anal. Biochem.*, 120 (1982) 302-311.
Hearn, M.T.W., *J. Liq. Chromatogr.*, 3 (1980) 1255-1276; *C.A.*, 93 (1980) 217 319r.

Hearn, M.T.W., Grego, B. and Chapman, G.E., *J. Liq. Chromatogr.*, 6 (1983a) 215-228; *C.A.*, 98 (1983) 119840n.
Hearn, M.T.W., Regnier, F.E. and Wehr, C.T. (Editors), *The HPLC of Proteins and Peptides*, Academic Press, Orlando, FL, 1983b, 267 pp.; *C.A.*, 99 (1983) B 118 839c.
Heinrikson, R.L. and Meredith, S.C., *Anal. Biochem.*, 136 (1984) 65-74.
Henderson, L.E., Copeland, T.D. and Oroszlan, S., *Anal. Biochem.*, 102 (1980) 1-7; *C.A.*, 92 (1980) 106 688v.
Henderson, L., Sowder, R. and Oroszlan, S., in Elzinga, M. (Editor), *Methods in Protein Sequence Analysis*, Humana Press, Clifton, NJ, 1982, pp. 409-416.
Henschen, A., Hupe, K.-P., Lottspeich, F. and Voelter, W., *High-Performance Liquid Chromatography in Biochemistry*, Verlag Chemie, Weinheim, 1985, 638 pp.; *C.A.*, 104 (1986) B 48 306x.
Henschen-Edman, A. and Lottspeich, F., in Birr, C. (Editor), *Methods in Peptide and Protein Sequence Analysis (Proceedings of 3rd International Conference on Solid Phase Methods in Protein Sequence Analysis, Heidelberg, 1979)*, Elsevier, Amsterdam, 1980, pp. 105-114; *C.A.*, 94 (1981) 43 661z.
Herman, A.C. and Vanaman, T.C., *Methods Enzymol.*, 47 (1977) 220-236.
Hermodson, M. and Mahoney, W.C., in Liu, T.Y., Schechter, A.N., Heinrikson, R. and Condliffe, P.G. (Editors), *Chemical Synthesis and Sequencing of Peptides and Proteins* (Developments in Biochemistry 1981), Elsevier, Amsterdam, 1981, pp. 119-130; *C.A.*, 96 (1982) 177 211c.
Hewick, R.M., Hunkapiller, M.V., Hood, L.E. and Dreyer, W.J., *J. Biol. Chem.*, 256 (1981) 7990-8005.
Hill, D.W., Walters, F.H., Wilson, T.D. and Stuart, J.D., *Anal. Chem.*, 51 (1979) 1338-1341.
Hill, R.L. and Delaney, R., *Methods Enzymol.*, 11 (1967) 339-351.
Hirs, C.H.W. and Timasheff, S.N. (Editors), *Methods Enzymol (Enzyme Structure, Part E)*, 47 (1977) 3-498.
Hobbs, A.A., Grego, B., Smith, M.G. and Hearn, M.T.W., *J. Liq. Chromatogr.*, 4 (1981) 651-659; *C.A.*, 94 (1981) 187 953u.
Hodgin, J.C., *J. Liq. Chromatogr.*, 2 (1979) 1047-1059; *C.A.*, 91 (1979) 206 635y.
Hollaway, W.L., Bhown, A.S., Mole, J.E. and Bennett, J.C., *Chromatogr. Sci.*, 10 (1979) 163-176; *C.A.*, 90 (1979) 147 915u.
Hsu, K.-T. and Currie, B.L., *J. Chromatogr.*, 166 (1978) 555-561.
Hughes, R.C. and Jeanloz, R.W., *Biochemistry*, 3 (1964a) 1535-1543; *C.A.*, 61 (1964) 14 957c.
Hughes, R.C. and Jeanloz, R.W., *Biochemistry*, 3 (1964b) 1543-1548; *C.A.*, 61 (1964) 14 957c.
Hughes, G.J. and Winterhalter, K.H., *J. Chromatogr.*, 235 (1982) 417-426.
Hughes, G.J., Winterhalter, K.H. and Wilson, K.J., *FEBS Lett.*, 108 (1979) 81-86.
Hughes, G.J., Winterhalter, K.H. and Wilson, K.J., in Lottspeich, F., Henschen, A and Hupe, K.P. (Editors), *High Performance Chromatography in Protein and Peptide Chemistry*, Walter de Gruyter, Berlin, New York, 1981, pp. 175-191; *C.A.*, 97 (1982) 52 011b.
Iijima, Y. and Egami, F., *J. Biochem. (Tokyo)*, 70 (1971) 75-78.
Iijima, Y., Muramatsu, T. and Egami, F., *Arch. Biochem. Biophys.*, 145 (1971) 50-54.
Ingram, V.M., *Biochim. Biophys. Acta*, 28 (1958) 539-545.
Iskandarani, Z. and Pietrzyk, D.J., *Anal. Chem.*, 53 (1981) 489-495.
IUPAC-IUB Commision on Biochemical Nomenclature, *J. Biol. Chem.*, 243 (1968) 3557-3559.
Ivanov, Ch.P. and Mancheva, I.N., *Anal. Biochem.*, 53 (1973a) 420-430.
Ivanov, Ch.P. and Mancheva, I.N., *J. Chromatogr.*, 75 (1973b) 129-132.
Ivanov, Ch.P. and Mancheva, I.N., *C.R. Acad. Bulg. Sci. (Dokl. Bolg. Akad. Nauk)*, 28 (1975) 1399-1402; *C.A.*, 84 (1976) 56 142z.
Ivanov, Ch.P., Mancheva, I.N. and Gaitandzhiev, S., *C.R. Acad. Bulg. Sci. (Dokl. Bolg. Akad. Nauk)*, 20 (1967) 799-802; *C.A.*, 68 (1968) 29 455s.

Jonák, J., Petersen, T.E., Meloun, B. and Rychlík, I., *Eur. J. Biochem.*, 144 (1984) 295-303.

Jonák, J., Pokorná, K., Meloun, B. and Karas, K., *Eur. J. Biochem.*, 154 (1986) 355-362.

Jones, B.N. and Gilligan, J.P., *J. Chromatogr.*, 266 (1983) 471-482.

Kamp, R.M., *Separation of Peptides*, in Lectures edited for *International Conference on Novel Techniques in Protein Micro-Sequence Analysis* (October 1985, Berlin-Dahlem; Wittmann-Liebold, B., Erdmann, V. and Salnikov, J., Organizers), Max-Planck Institute for Molecular Genetics, Berlin-Dahlem, 1985a, pp. 1-14.

Kamp, R.M., *High Performance Liquid Chromatography of Proteins*, in Lectures edited for *International Conference on Novel Techniques in Protein Micro-Sequence Analyses* (October 1985, Berlin-Dahlem; Wittmann-Liebold, B., Erdmann, V. and Salnikov, J., Organizers), Max-Planck Institute for Molecular Genetics, Berlin-Dahlem, 1985b, pp. 1-16.

Kamp, R.M., Bosserhoff, A., Kamp, D. and Wittmann-Liebold, B., *J. Chromatogr.*, 317 (1984) 181-192.

Kannagi, R., Fukuda, M.N. and Hakomori, S., *J. Biol. Chem.*, 257 (1982) 4438-4442.

Kato, Y. and Nevins, D.J., *Carbohydr. Res.*, 147 (1986) 69-85.

Kehl, M., Lottspeich, F. and Henschen, A., *Hoppe-Seylers' Z. Physiol. Chem.*, 362 (1981) 1661-1664; *C.A.*, 96 (1982) 82 121j.

Kersters-Hilderson, H., Claeyssens, M., Van Doorslaer, E., Saman, E. and De Bruyne, C.K., *Methods Enzymol.*, 83 (1982) 631-639.

Kitamikado, M., Ito, M. and Li, Y.-T., *Methods Enzymol.*, 83 (1982) 619-625.

Kobata, A., *Methods Enzymol.*, 50 (1978) 567-574.

Kobata, A., *Anal. Biochem.*, 100 (1979) 1-14.

Kobata, A., *Methods Enzymol.*, 83 (1982) 625-631.

Kobata, A. and Takasaki, S., *Methods Enzymol.*, 50 (1978) 560-567.

Kobayashi, S. and Imai, K., *Anal. Chem.*, 52 (1980) 424-427.

Kochibe, N., *J. Biochem. (Tokyo)*, 74 (1973) 1141-1149.

Kohno, M. and Yamashina, I., *Methods Enzymol.*, 28 (1972) 786-792.

Komiya, H., Nashikawa, K., Ogawa, K. and Takemura, S., *Nucleic Acids Res.*, Spec. Publ. No. 5 (1978) 467-470.

Komiya, H., Nishikawa, K., Ogawa, K. and Takemura, S., *J. Biochem. (Tokyo)*, 86 (1979) 1081-1088; *C.A.*, 91 (1979) 206 621r.

Komiya, H. and Takemura, S., *J. Biochem. (Tokyo)*, 86 (1979) 1067-1080.

Kratzin, H., Yang, Ch.-Y., Krusche, J.U. and Hilschmann, N., *Hoppe-Seylers' Z. Physiol. Chem.*, 361 (1980) 1591-1598; *C.A.*, 94 (1981) 63 482v.

Kuhn, C.C. and Crabb, J.W., *Modern Manual Sequencing Methods*, in Lectures edited for *International Conference on Novel Techniques in Protein Micro-Sequence Analysis* (October 1985, Berlin-Dahlem; Wittmann-Liebold, B., Erdmann, V. and Salnikov, J., Organizers), Max-Planck Institute for Molecular Genetics, Berlin-Dahlem, 1985, pp. 1-17.

Kundu, S.K., Samuelsson, B.E., Pascher, I. and Marcus, D.M., *J. Biol. Chem.*, 258 (1983) 13857-13866.

Lai, C.Y., *Methods Enzymol.*, 47 (1977) 236-240.

Larsen, B.R. and West, F.G., *J. Chromatogr. Sci.*, 19 (1981) 259-265.

Laursen, R.A., *Eur. J. Biochem.*, 20 (1971) 89-102.

Laursen, R.A. and Machleidt, W., *Methods Biochem. Anal.*, 26 (1980) 201-284; *C.A.*, 93 (1980) 163 671d.

Lee, G.J.-L., Liu, D.-W., Pav, J.W. and Tieckelmann, H., *J. Chromatogr.*, 212 (1981) 65-73.

Lee, Y.C., *Methods Enzymol.*, 28 (1972a) 63-73.

Lee, Y.C., *Methods Enzymol.*, 28 (1972b) 699-702.

Li, Y.-T., King, M.-J. and Li, S.-Ch., in Svennerholm, L., Mandel, P., Dreyfus, H. and Urban, P.-F. (Editors), *Structure and Function of Gangliosides*, Plenum Press, New York, 1980a, pp. 93-104.

Li, Y.T., Koerner, T.A.W. and Li, S.Ch., *Proceedings of 27th International congress on Pure and Applied Chemistry*, 1980b, pp. 199-204; *C.A.*, 93 (1980) 109 890m.

Li, Y.-T. and Lee, Y.-C., *J. Biol. Chem.*, 247 (1972) 3677-3683.
Li, Y.-T. and Li, S.-Ch., *Methods Enzymol.*, 28 (1972a) 702-713.
Li, Y.-T. and Li, S.-Ch., *Methods Enzymol.*, 28 (1972b) 714-720.
Li, Y.-T. and Li, S.-Ch., *Glycoconjugates*, 1 (1977) 51-67.
Li, Y.-T., Nakagawa, H., Kitamikado, M. and Li, S.-Ch., *Methods Enzymol.*, 83 (1982) 610-619.
Lin, J.-K., in Hancock, W.S. (Editor), *CRC Handbook of HPLC for the Separation of Amino Acids, Peptides and Proteins*, Vol. I, CRC Press, Boca Raton, FL, 1984, pp. 359-366.
Lin, J.K. and Chang, J.-Y., *Anal. Chem.*, 47 (1975) 1634-1638.
Lindberg, B., *Methods Enzymol.*, 28 (1972) 178-195.
Lindberg, B. and Lönngren, J., *Methods Enzymol.*, 50 (1978) 3-33.
Linker, A. and Hovingh, P., *Methods Enzymol.*, 28 (1972) 902-911.
Liu, T.-Y., *Methods Enzymol.*, 28 (1972) 48-54.
Liu, T.-Y., Schechter, A.N., Heinrikson, R.L. and Condliffe, P.G. (Editors), *Chemical Synthesis and Sequencing of Peptides and Proteins (Proceedings of International Conference, May 8-9, 1980, Bethesda, MD, U.S.A.) (Developments in Biochemistry*, Vol. 17), Elsevier, Amsterdam, New York, 1981, 274 pp.; *C.A.*, 96 (1982) B 181 620c.
Lottspeich, F., *Hoppe-Seyler's Z. Physiol. Chem.*, 361 (1980) 1829-1834; *C.A.*, 95 (1981) 2770m.
Lottspeich, F., Henschen, A. and Huppe, K.-P. (Editors), *High Performance Chromatography in Protein and Peptide Chemistry*, Walter de Gruyter, Berlin, New York, 1981, 388 pp.
MacDonald, D.L., Patt, L.M. and Hakomori, S., *J. Lipid Res.*, 21 (1980) 642-645.
Macher, B.A. and Sweeley, Ch.S., *Methods Enzymol.*, 50 (1978) 236-251.
Machleidt, W. and Hofer, H., in Birr, C. (Editor), *Methods in Peptide and Protein Sequence Analysis (Proceedings of the 3rd International Conference on Solid Phase Methods in Protein Sequence Analysis, Heidelberg, 1979)*, Elsevier, Amsterdam, 1980, pp. 35-47; *C.A.*, 94 (1981) 1681a.
Malcom, A.D.B., *Nature (London)*, 275 (1978) 90-91.
Mancheva, I.N., Nikolov, R.N. and Pfletschinger, J., *J. Chromatogr.*, 213 (1981) 99-103.
Marcus, D.M., *Methods Enzymol.*, 28 (1972) 967-970.
Margolies, M.N. and Brauer, A., *J. Chromatogr.*, 148 (1978) 429-439.
Martin, W.J. and Davies, R.W., *Bio/Technology*, 4 (1986) 890-895.
Matta, K.L. and Bahl, O.P., *Methods Enzymol.*, 28 (1972) 749-755.
Matthews, E.W., Byfield, P.G.H. and MacIntire, I., *J. Chromatogr.*, 110 (1975) 369-373.
Maxam, A.M. and Gilbert, W., *Methods Enzymol.*, 65 (1980) 499-560.
McDonald, J.K., Callahan, P.X. and Ellis, S., *Methods Enzymol.*, 25 (1972) 272-281.
McDonald, J.K., Zeitman, B.B. and Reilly, T.J., *J. Biol. Chem.*, 244 (1969) 2693-2709.
McDonald, M.J. and Bahl, O.P., *Methods Enzymol.*, 28 (1972) 734-738.
McGuire, E.J., Chipowsky, S. and Roseman, S., *Methods Enzymol.*, 28 (1972) 755-763.
McLellan, W.L., *Methods Enzymol.*, 28 (1972) 921-925.
McNeil, M., Darvill, A.G., Åman, P., Franzén, L.-E. and Albersheim, P., *Methods Enzymol.*, 83 (1982) 3-45.
Meisler, M., *Methods Enzymol.*, 28 (1972) 820-824.
Meloun, B., Baudyš, M., Kostka, V., Hausdorf, G., Frömmel, C. and Höhne, W.E., *FEBS Lett.*, 183 (1985a) 195-200.
Meloun, B., Baudyš, M., Pavlík, M., Kostka, V., Hausdorf, G. and Höhne, W.E., *Collect. Czech. Chem. Commun.*, 50 (1985b) 885-896.
Meloun, B., Jonáková, V. and Henschen, A., *Biol. Chem. Hoppe-Seyler*, 366 (1985c) 1155-1160.
Meloun, B., Pohl, J., Baudyš, M. and Kostka, V., *Biol. Chem. Hoppe-Seyler*, 367 (1986) (Supplement, Abstracts, 17th FEBS Meeting, 1986, West Berlin), p. 365, Abstr. FRI 06.03.50.

Melton, L.D., McNeil, M., Darvill, A.G., Albersheim, P. and Dell, A., *Carbohydr. Res.*, 146 (1986) 279-305.

Meuth, J.L., Harris, D.E., Dwulet, F.E., Crowl-Powers, M.L. and Gurd, F.R.N., *Biochemistry*, 21 (1982) 3750-3757.

Mikeš, O., in Fox, J.L., Deyl, Z. and Blažej, A. (Editors), *Protein Sequence and Evolution (Proceedings of International Symposium, Smolenice and Kočovce, 1975)*, Marcel Dekker, New York, 1976, p. 290.

Mikeš, O. and Holeyšovský, V., *Chem. Listy*, 51 (1957) 1497-1500; *Collect. Czech. Chem. Commun.*, 23 (1958) 524-528; *C.A.*, 51 (1957) 16 635e.

Mikeš, O., Holeyšovský, V., Tomášek, V. and Šorm, F., *Biochem. Biophys. Res. Commun.*, 24 (1966a) 346-351.

Mikeš, O., Tomášek, V., Holeyšovský, V. and Šorm, F., *Biochim. Biophys. Acta*, 117 (1966b) 281-284.

Mikeš, O., Tomášek, V., Holeyšovský, V. and Šorm, F., *Collect. Czech. Chem. Commun.*, 32 (1967) 655-677; *C.A.*, 66 (1967) 55 746r.

Minasian, E., Sharma, R.S., Leach, S.J., Grego, B. and Hearn, M.T.W., *J. Liq. Chromatogr.*, 6 (1983) 199-213; *C.A.*, 98 (1983) 194 367s.

Mizuochi, T., Yonemasu, K. and Yamashita, K., *J. Biol. Chem.*, 253 (1978) 7404-7409.

Morgan, W.T.J.and Elson, L.A., *Biochem. J.*, 38 (1934) 988-995.

Mueller-Harvey, I., Hartley, R.D., Harris, P.J. and Curzon, E.D., *Carbohydr. Res.*, 148 (1986) 71-85.

Muramatsu, T., *Methods Enzymol.*, 50 (1978) 555-559.

Muramatsu, T. and Egami, F., *J. Biochem. (Tokyo)*, 62 (1967) 700-709.

Nasimov, I.V., Levina, N.B., Shemyakin, V.V., Rosynov, B.V., Bogdanova, I.A. and Merimson, V.G., in Birr, C. (Editor), *Methods in Peptide and Protein Sequence Analysis (Proceedings of 3rd International Conference on Solid Phase Methods in Protein Sequence Analysis, Heidelberg, 1979)*, Elsevier, Amsterdam, 1980, pp. 475-483; *C.A.*, 94 (1981) 43 666e.

Natowicz, M. and Baenziger, J.U., *Anal. Biochem.*, 105 (1980) 159-164.

Neufeld, E.F. and Ginsburg, V. (Editors), *Methods Enzymol. (Complex Carbohydrates, Part A)*, 8 (1966).

Niall, H.D., *Fractions*, No. 2 (1969) 1-10 (Beckman Instruments, Fullerton, CA).

Nishigaki, M., Muramatsu, T. and Kobata, A., *Biochem. Biophys. Res. Commun.*, 59 (1974) 638-645.

Novotný, J., Dolejš, L. and Franěk, F., *Eur. J. Biochem.*, 31 (1972) 277-289.

Noyes, C.M., *J. Chromatogr.*, 266 (1983) 451-460.

Ogata, A.M., Muramatsu, T. and Kobata, A., *J. Biochem. (Tokyo)*, 82 (1977) 611-614.

Okumura, T. and Yamashina, I., *Methods Enzymol.*, 28 (1972) 792-796.

Olson, D.C., Schmidt, G.J. and Slavin, W., *Chromatogr. Newsl.*, 7 (1979) 22-25.

O'Neill, M.A., Morris, V.J., Selvendran, R.R., Sutherland, I.W. and Taylor, I.T., *Carbohydr. Res.*, 148 (1986b) 63-69.

O'Neill, M.A. and Selvendran, R.R., *Carbohydr. Res.*, 145 (1985) 45-58.

O'Neill, M.A., Selvendran, R.R., Morris, V.J. and Eagles, J., *Carbohydr. Res.*, 147 (1986a) 295-313.

Østvold, G., Jensen, E. and Greibrokk, T., *Medd. Nor. Farm. Selsk.*, 40 (1978) 173-179.

Pačes, V., *Chem. Listy*, 76 (1982a) 176-199; *C.A.*, 96 (1982) 158 406d.

Pačes, V., *Biol. Listy*, 47 (1982b) 293-298; *C.A.*, 98 (1983) 120 440p.

Patterson, B.W., Ferguson, A.H., Matula, M. and Elbein, A.D., *Methods Enzymol.*, 28 (1972) 996-1000.

Pavlík, M., *Zhodnocení Chemických Metod a Činidel Navržených k Postupnému Odbourávání Polypeptidových Řetězců (Reseršní Práce ke Kandidátskému Minimu) (Evaluation of Chemical Methods and Reagents Proposed for Gradual Degradation of Polypeptide Chains)* (compilation part of CSc Thesis), Institute of Organic Chemistry and Biochemistry, Czechoslovak Academy of Sciences, Prague, 1986, 52 pp.

Pavlík, M., Jehnička, J. and Kostka, V., *J. Chromatogr.*, (1988) submitted for publication.

Pavlík, M. and Kostka, V., *Anal. Biochem.*, 151 (1985) 520-525.
Pazur, J.H., in Pigman, W. and Horton, D. (Editors), *The Carbohydrates (Chemistry and Biochemistry)*, Academic Press, New York, 1970, pp. 69-137.
Pazur, J.H., *Methods Enzymol.*, 28 (1972) 931-934.
Pigman, W. and Horton, D. (Editors), *The Carbohydrates (Chemistry and Biochemistry)*, Academic Press, New York, Vol IA, 2nd ed., 1972, pp. 1-642; Vol. IB, 1980, pp. 644-1627; Vol. IIA, 1970, pp. 1-469; Vol. IIB, 1970, pp. 471-853.
Plantner, J.J. and Carlson, D.M., *Methods Enzymol.*, 28 (1972) 46-48.
Prestidge, R.L., in Hancock, W.S. (Editor), *CRC Handbook of HPLC for the Separation of Amino Acids, Peptides and Proteins*, Vol. II, CRC Press, Boca Raton, FL, 1984, pp. 279-285.
Prestidge, R.L., Harding, D.R.K., Moore, C.H. and Hancock, W.S., *Bioorg. Chem.*, 10 (1981) 277-282.
Reimann, F. and Wittmann-Liebold, B., *Gas Phase Sequencing of Peptides and Proteins*, in Lectures for *International Conference on Novel Techniques in Protein Micro-Sequence Analysis* (October 1985, Berlin-Dahlem; Wittmann-Liebold, B., Erdmann, V. and Salnikov, J., Organizers), Max-Planck Institute for Molecular Genetics, Berlin-Dahlem, 1985, pp. 1-10.
Reissig, J.L., Strominger, J.L. and Leloir, L.F., *J. Biol. Chem.*, 217 (1955) 959-966.
Robyt, J.F. and Ackerman, R.J., *Methods Enzymol.*, 28 (1972) 925-930.
Rodriguez, H., Kohr, W.J. and Harking, R.N., *Anal. Biochem.*, 140 (1984) 538-547; *C.A.*, 101 (1984) 86 538v.
Rose, S.M. and Schwartz, B.D., *Anal. Biochem.*, 107 (1980) 206-213; *C.A.*, 93 (1980) 163 751e.
Rosmus, J. and Deyl, Z., *Chromatogr. Rev.*, 13 (1971) 163-302.
Rosmus, J. and Deyl, Z., *J. Chromatogr.*, 70 (1972) 221-339.
Roth, M., *Anal. Chem.*, 43 (1971) 880-882.
Salnikov, J., *Automated Solid-Phase Microsequencing using DABITC On-Column-Immobilization of Proteins*, in Lectures edited for *International Conference on Novel Techniques in Protein Micro-Sequence Analysis* (October 1985, Berlin-Dahlem; Wittmann-Liebold, B., Erdmann, V. and Salnikov, J., Organizers), Max-Planck Institute for Molecular Genetics, Berlin-Dahlem, 1985, pp. 1-14.
Salnikov, J., Lehmann, A. and Wittmann-Liebold, B., *Anal. Biochem.*, 117 (1981) 433-442.
Sanger, F., *Biochem. J.*, 39 (1945) 507-515; *C.A.*, 40 (1946) 5399[3].
Schaettle, E., *Lab. Praxis*, 2 (1978) 36-38, 40; *C.A.*, 91 (1979) 35 023r.
Schleicher, E. and Wieland, O.H., *J. Clin. Chem. Clin. Biochem.*, 19 (1981) 81-87; *C.A.*, 94 (1981) 152 805v.
Schlesinger, D.H., in Hancock, W.S. (Editor), *CRC Handbook of HPLC for the Separation of Amino Acids, Peptides and Proteins*, Vol. I, CRC Press, Boca Raton, FL, 1984, pp. 367-379.
Schmidt, P.G., Baudyš, M., Tang, J. and Kostka, V., *Biochemistry*, (1987) submitted for publication.
Schoenmakers, P.J., Billiet, H.A.H. and De Galan, L., *J. Chromatogr.*, 205 (1981) 13-30.
Schroeder, W.A., in Hancock, W.S. (Editor), *CRC Handbook of HPLC for the Separation of Amino Acids, Peptides and Proteins*, Vol. II, CRC Press, Boca Raton, FL, 1984, pp. 287-300.
Schroeder, W.A., Shelton, J.B. and Shelton, J.R., *Hemoglobin*, 4 (1980) 551-559; *C.A.*, 94 (1981) 1667a.
Schroeder, W.A., Shelton, J.B., Shelton, J.R. and Powars, D., *J. Chromatogr.*, 174 (1979) 385-392.
Schroeder, W.A., Shelton, J.B. and Shelton, J.R., *Prog. Clin. Biol. Res.*, 60 (Advances in Hemoglobin Analysis) (1981) 1-22; *C.A.*, 95 (1981) 146 235p.
Schwabe, Ch. and McDonald, J.K., *Biochem. Biophys. Res. Commun.*, 74 (1977) 1501-1504.

Seidah, N.G., Pélaprat, D., Rochemont, J., Lambelin, P., Dennis, M., Chan, J.S.D., Hamelin, J., Lazure, C. and Chrétien, M., *J. Chromatogr.*, 266 (1983) 213-224.

Shively, J.E. (Editor), *Microcharacterization of Polypeptides: A Practical Manual*, Humana Press, Clifton, NJ, 1987, in press.

Simmons, W.H. and Meisenberg, G., *J. Chromatogr.*, 266 (1983) 483-489.

Simons, S.S., Jr. and Johnson, D.F., *J. Org. Chem.*, 43 (1978) 2886-2891.

Slomiany, B.L., Slomiany, A., Galicki, N.I. and Kojima, K., *Eur. J. Biochem.*, 113 (1980) 27-32.

Smirnova, G.P., Glukhoded, I.S. and Kochetkov, N.K., *Bioorg. Khim.*, 8 (1982) 971-979; *C.A.*, 97 (1982) 184 843y.

Smith, A.J.H., *Methods Enzymol.*, 65 (1980) 560-580; *C.A.*, 93 (1980) 91 387r.

Sottrup-Jensen, L., Petersen, T.E. and Magnusson, S., *Anal. Biochem.*, 107 (1980) 456-460.

Spackman, D.H., Stein, W.H. and Moore, S., *Anal. Chem.*, 30 (1958) 1190-1206.

Spatz, R. and Roggendorf, E., *Fresenius' Z. Anal. Chem.*, 299 (1979) 267-270. *C.A.*, 92 (1980) 87 596v.

Spiro, R.G., *J. Biol. Chem.*, 237 (1962) 646-652.

Spiro, R.G., *Methods Enzymol.*, 8 (1966a) 3-26; *C.A.*, 68 (1968) 10 103d.

Spiro, R.G., *Methods Enzymol.*, 8 (1966b) 26-52; *C.A.*, 68 (1968) 26 849n.

Spiro, R.G., *Annu. Rev. Biochem.*, 39 (1970) 599-638.

Spiro, R.G., *Methods Enzymol.*, 28 (1972) 3-43.

Spiro, R.G. and Bhoyroo, V.D., *Fed. Proc. Fed. Am. Soc. Exp. Biol.*, 30 (1971) 1223.

Stahl, P.D. and Touster, O., *Methods Enzymol.*, 28 (1972) 814-819.

Stark, G.R., *Biochemistry*, 7 (1968) 1796-1807; *C.A.*, 69 (1968) 106p.

Stoming, T.A., Garver, F.A., Gangarosa, M.A., Harrisson, J.M. and Huisman, T.H.J., *Anal. Biochem.*, 96 (1979) 113-117.

Sugahara, K. and Yamashina, I., *Methods Enzymol.*, 28 (1972) 769-772.

Sukeno, T., Tarentino, A.L., Plummer, T.H., Jr. and Maley, F., *Methods Enzymol.*, 28 (1972) 777-782.

Suzuki, S., *Methods Enzymol.*, 28 (1972a) 911-917.

Suzuki, S., *Methods Enzymol.*, 28 (1972b) 917-921.

Svedas, V.-J.K., Galaev, I.J., Borisov, I.L. and Berezin, I.V., *Anal. Biochem.*, 101 (1980) 188-195.

Svensson, S., *Methods Enzymol.*, 50 (1978) 33-38.

Swaminathan, N., Matta, K.L. and Bahl, O.M., *Methods Enzymol.*, 28 (1972) 744-749.

Takagaki, Y., Gerber, G.E., Nihei, K. and Khorana, H.G., *J. Biol. Chem.*, 255 (1980) 1536-1541; *C.A.*, 92 (1980) 159 913w.

Takahashi, N., Isobe, T., Kasai, H., Seta, K. and Okuyama, T., *Anal. Biochem.*, 115 (1981) 181-187.

Takahashi, N., Takahashi, Y. and Putnam, F.W., *J. Chromatogr.*, 266 (1983) 511-522.

Takamizawa, K., Iwamori, M., Mutai, M. and Nagai, Y., *J. Biol. Chem.*, 261 (1986) 5625-5630.

Tallman, J.F. and Brady, R.O., *Methods Enzymol.*, 28 (1972) 825-829.

Tang, P.W. and Williams, J.M., *Carbohydr. Res.*, 136 (1985) 259-271.

Tapuhi, Y., Miller, N. and Karger, B.L., *J. Chromatogr.*, 205 (1981a) 325-337.

Tapuhi, Y., Schmidt, D.E., Lindner, W. and Karger, P.L., *Anal. Biochem.*, 115 (1981b) 123-129.

Tarentino, A.L. and Maley, F., *Methods Enzymol.*, 28 (1972a) 772-776.

Tarentino, A.L. and Maley, F., *Methods Enzymol.*, 28 (1972b) 782-786.

Tarentino, A.L. and Maley, F., *Methods Enzymol.*, 50 (1978) 580-584.

Tarentino, A., Plummer, T.A., Jr. and Maley, F., *J. Biol. Chem.*, 245 (1970) 4150-4157.

Tarentino, A.L., Trimble, R.B. and Maley, F., *Methods Enzymol.*, 50 (1978) 574-580.

Tarr, G.E., *Anal. Biochem.*, 63 (1975) 361-370.

Tarr, G.E., *Anal. Biochem.*, 111 (1981) 27-32; *C.A.*, 94 (1981) 170 381h.

Tarr, G.E., *Methods Enzymol.*, 47 (1977) 335-357.
Tarr, G.E., in Elzinga, M. (Editor), *Methods in Protein Sequence Analysis*, Humana Press, Clifton, NJ, 1982, p. 223-232; *C.A.*, 98 (1983) 194 543w.
Tarr, G.E., in Shively, J.E. (Editor), *Microcharacterization of Polypeptides: A Practical Manual*, Humana Press, Clifton, NJ, 1987, in press.
Taylor, C., Cheetham, N.W.H. and Walker, G.J., *Carbohydr. Res.*, 137 (1985) 1-12.
Thannhauser, T.W., McWherter, Ch.A. and Scheraga, H.A., *Anal. Biochem.*, 149 (1985) 322-330.
Trimble, R.B., Tarentino, A.L., Aumick, G.E. and Maley, F., *Methods Enzymol.*, 83 (1982) 603-610.
Tschesche, H. (Editor), *Modern Methods in Protein Chemistry, Review Articles*, Walter de Gruyter, Berlin, Vol. 1, 1983, 463 pp.; *C.A.*, 100 (1984) B 2469s; Vol. 2, 1985, 434 pp.; *C.A.*, 104 (1986) B 65 381k.
Uchida, Y., Tsukada, Y. and Sugimori, T., *Biochim. Biophys. Acta*, 350 (1974) 425-431.
Umemoto, J., Bhavanandan, V.P. and Davidson, E.A., *J. Biol. Chem.*, 252 (1977) 8609-8614.
Valent, B.S., Darvill, A.G., McNeil, M., Robertsen, B.K. and Albersheim, P., *Carbohydr. Res.*, 79 (1980) 165-192; *C.A.*, 92 (1980) 215 666g.
Van Beeumen, J., Van Damme, J., Tempst, P. and DeLey, J., in Birr, C. (Editor), *Methods in Peptide and Protein Sequence Analysis (Proceedings of 3rd International Conference on Solid Phase Methods in Protein Sequence Analysis, Heidelberg*, Elsevier, Amsterdam, 1980, pp. 503-506; *C.A.*, 94 (1981) 60 838e.
Van Boom, J.H. and De Rooij, J.F.M., *J. Chromatogr.*, 131 (1977) 169-177.
Van der Rest, M., Cole, W.G. and Glorieux, F.H., *Biochem. J.*, 161 (1977) 527-534.
Vensel, W.H., Fujita, V.S., Tarr, G.E., Margoliash, E. and Kayser, H., *J. Chromatogr.*, 266 (1983) 491-500.
Vliegenthart, J.F.G., in Svennerholm, L., Mandel, P., Dreyfus, H. and Urban, P.-F. (Editors), *Structure and Function of Gangliosides (Advances in Experimental Medicine and Biology*, Vol. 125), Plenum Press, New York, London, 1980, pp. 77-91.
Vliegenthart, J.F.G., Van Halbeck, H. and Dorland, V., *Pure Appl. Chem.*, 53 (1981) 45-77.
Walker, J.E., Fearnley, I.M. and Blows, R.A., *Biochem. J.*, 237 (1986) 73-84.
Warren, L., *J. Biol. Chem.*, 233 (1959) 1971-1974.
Watanabe, K. and Hakomori, S., *Biochemistry*, 18 (1979) 5502-5504.
Watanabe, Y. and Imai, K., *J. Chromatogr.*, 239 (1982) 723-732.
Watson, M.W. and Carr, P.W., *Anal. Chem.*, 51 (1979) 1835-1842.
Weissmann, B., *Methods Enzymol.*, 28 (1972a) 796-800.
Weissmann, B., *Methods Enzymol.*, 28 (1972b) 801-805.
Weissmann, B. and Hinrichsen, D.F., *Biochemistry*, 8 (1969) 2034-2043; *C.A.*, 71 (1969) 9777g.
Wiegandt, H. and Bashang, G., *Z. Naturforsch., Teil B*, 20 (1965) 164-166; *C.A.*, 62 (1965) 14 984b.
Wilkas, E. and Lederer, E., *Tetrahedron Lett.*, (1968) 3089-3092.
Wilkinson, J.M., in Hancock, W.S. (Editor), *CRC Handbook of HPLC for the Separation of Amino Acids, Peptides and Proteins*, Vol. I, CRC Press, Boca Raton, FL, 1984, pp. 339-350.
Wilson, J.B., Lam, H., Pravatmuang, P. and Huisman, T.H.J., *J. Chromatogr.*, 179 (1979) 271-290.
Wilson, K.J., Honegger, A. and Hughes, G.J., *Biochem. J.*, 199 (1981) 43-51; *C.A.*, 96 (1982) 65 063z.
Wilson, K.J. and Hughes, G.J., in Birr, C. (Editor), *Methods in Peptide and Protein Sequence Analysis (Proceedings of 3rd International Conference on Solid Phase Methods in Protein Sequence Analysis, Heidelberg, 1979)*, Elsevier, Amsterdam, 1980, pp. 491-501; *C.A.*, 94 (1981) 12 335s.
Wittmann-Liebold, B., in Elzinga, M. (Editor), *Methods in Protein Sequence Analysis*, Humana Press, Clifton, NJ, 1982, pp. 27-63.
Wittmann-Liebold, B., in Tschesche, H. (Editor), *Modern Methods in Protein Chemistry*, Walter de Gruyter, Berlin, New York, 1983, pp. 267-302.

Wittmann-Liebold, B. and Ashman, K., in Tschesche, H. (Editor), *Modern Methods in Protein Chemistry*, Walter de Gruyter, Berlin, 1985, pp. 303-327.

Wittmann-Liebold, B., Graffunder, H. and Kohls, H., *Anal. Biochem.*, 75 (1976) 621-633.

Wu, R., Wu, N.-H., Hanna, Z., Georges, F. and Narang, S., in Gait, M.J. (Editor), *Oligonucleotide Synthesis (A Practical Approach)*, IRL Press, Oxford, 1984, pp. 135-151.

Yamashita, K., Mizuochi, T. and Kobata, A., *Methods Enzymol.*, 83 (1982) 105-126.

Yang, Ch.-Y., Pauly, E., Kratzin, H. and Hilschmann, N., *Hoppe-Seylers' Z. Physiol. Chem.*, 362 (1981) 1131-1146; *C.A.*, 95 (1981) 201 816f.

Yang, H.S., Studebaker, J.F. and Parravano, C., *Chromatogr. Sci.*, 10 (1979) 247-260; *C.A.*, 90 (1979) 147 919y.

Yasuda, Y., Takahashi, N. and Murachi, T., *Biochemistry*, 9 (1970) 25-32; *C.A.*, 72 (1970) 63 041w.

Yi-Tao, Z., Headlee, M.E., Henson, J., Lam, H., Wilson, J.B. and Huisman, T.H.J., *Biochim. Biophys. Acta*, 707 (1982) 206-212; *C.A.*, 97 (1982) 179 691d.

Yi-Tao, Z., Shu-Zheng, H., Reynolds, A., Lam, H., Webber, B.B., Wilson, J.B. and Huisman, T.H.J., *Sci. Sin., Ser. B, (Engl. Ed.)*, 26 (1983) 836-849; *C.A.*, 99 (1983) 191 035s.

Yoshima, H., Takasaki, S., Ito, S.M. and Kobata, A., *Arch. Biochem. Biophys.*, 194 (1979) 394-398.

Yuan, P.M., Pande, H., Clark, B.R. and Shively, J.E., *Anal. Biochem.*, 120 (1982) 289-301; *C.A.*, 96 (1982) 177 210b.

Yurewicz, E.C., Chalambor, M.A. and Heath, E.C., *Methods Enzymol.*, 28 (1972) 990-996.

Yu, T.J., Schwartz, H., Giese, R.W., Karger, B.L. and Vouros, P., *J. Chromatogr.*, 218 (1981) 519-533.

Zalut, C. and Harris, H.W., *J. Biochem. Biophys. Methods*, 2 (1980) 155-161; *C.A.*, 92 (1980) 176 736h.

Zimmerman, C.L., Appella, E. and Pisano, J.J., *Anal. Biochem.*, 75 (1976) 77-85.

Zimmerman, C.L., Appella, E. and Pisano, J.J., *Anal. Biochem.*, 77 (1977) 569-573.

Zimmerman, C.L., Pisano, J.J. and Appella, E., *Biochem. Biophys. Res. Commun.*, 55 (1973) 1220-1224.

Zmrhal, Z.J., Heathcote, J.G. and Washington, R.J., in Deyl, Z., Macek, K. and Janák, J. (Editors), *Liquid Column Chromatography*, Elsevier, Amsterdam, 1975, pp. 665-711.

Chapter 14

REGISTER OF BIOPOLYMERS, BIOOLIGOMERS AND OLIGOMERIC FRAGMENTS SEPARATED BY
HPLC OR MPLC METHODS

   This Register contains lists of separated substances based on data from all
the available sources, i.e. both from original papers and from reviews or
lectures. References in the Register refer to the full-title Bibliography
(Chapter 15).
   The Register consists of four tables. The first (Table 14.1) is the Contents
to the Register, the main second Table (Table 14.2) lists the chromatographed
substances, the third (Table 14.3) brings the key to the symbols used for
packings. The list of the producers and/or distributors of packings is given
in Table 4.10 in Part A. As many papers cited in the Register are important
from the point of view of medical (diagnostical) applications, the fourth
Table (Table 14.4) is added at the end of the main Register, summarizing this
type of papers. An Appendix (Table 14.5) collecting the newer references of
the Bibliographic Addendum (Chapter 15) was later attached to the main Register.

TABLE 14.1

CONTENTS

TABLE 14.2

CHROMATOGRAPHED SUBSTANCES

Key to abbreviations in the "mode of chromatography" column: AC = affinity; AD = adsorption; BA = bioaffinity; HI = hydrophobic interaction; IE = ion-exchange; IM = immuno-affinity; IP-RP = ion-pair reversed-phase; LE = ligand-exchange; MM = mixed mode; n.d. = not defined or no data available; NP = normal phase; RP = reversed-phase; SE = size (steric) exclusion, or gel permeation.

Substances	Mode of chromato-graphy	Packings	References
14.2.1 ENZYMES, ZYMOGENS			
14.2.1.1 Alcohol dehydrogenases			
Alcohol dehydrogenase	RP	Aquapore RP 300, LiChrosorb RP-18	741
	SE	Separon HEMA 300 Glc, TSK Gels SW	116,181,470
—, horse liver	AC	Cibacron blue/Silica	422
—, —	BA	AMP-Silica	396,522
—, sheep liver	SE	µBondagel E-linear	261
—, yeast	IE	Polyanion-SI	568
14.2.1.2 Aldolases			
Aldolase	SE	Glycophase/CPG	32
—	SE	Agarose-12%	282,283
—	SE	Fractogel TSK HW-55F,	27,198,216,
		TSK Gels SW, Toyopearl HW-55F	579
—	SE	Diol-modified silica	583,584
—	SE	AP/LiChrospher, PAP/LiChrosorb	155,158,625
—	AD	µBondapak C_18	596
—, rabbit	SE	LiChrosorb Diol, SynChropak	562,583,607
		GPC 100	

Aldolase-SDS (sodium dodecylsulphate) complexes	SE	CPG	178
14.2.1.3 Alkaline phosphatases			
Alkaline phosphatase, beef liver, calf intestinal	IE	DEAE-Glycophase/CPG	103,468,469, 562,564,602, 606,701
—, calf intestinal	AC	T.-dyes/Silica	646,647
—, *Escherichia coli*	SE	PEG/CPG	125
—, human placenta	SE	TSK Gel 3000 SW	368
14.2.1.4 Amylases			
Amylase	SE	SynChropak GPC 100	562
Amylases and isoamylases (in normal human serum and of patients with acute pancreatitis and mumps)	SE	I-125 PC	170
α-Amylase, *Bacillus subtilis*	IE	Phospho-Spheron	468,469
α-Amylase crude, hog pancreatic	HI	Spheron 300	658
Macroamylase, see section 14.2.1.20			
14.2.1.5 Carbonic anhydrases			
Carbonic anhydrase	RP IE	Mono Q, SAX	269,568
—	RP	Aquapore RP 300, LiChrosorb RP-18, Partisil ODS, Supelcosil LC 18	30,741
—	SE	SynChropak GPC 100	562
14.2.1.6 Catalases			
Catalase	IE	CPG	485
—	HI, RP	Nucleosil C18, μBondapak C18	490,596
—	SE	AP/LiChrosphere, Diol/Silica, LiChrosorb Diol, PSG-1000, SynChropak RPC 100	562,583,584, 625,633
—	SE	Agarose-12%, PEG/CPG, Spheron 1000	72,125,282, 283
—	SE	Fractogel TSK HW-55F, Toyopearl HW-55F, TSK Gels SW	181,198,216, 470,579
—	SE	Amide Si-1000, PAP/LiChrosorb	155,158

Table 14.2 (continued)

Substances	Mode of chromatography	Packings	References
14.2.1.7 Cellulolytic enzymes			
Cellulases, *Penicillium funiculosum*	SE	LiChrosorb Diol	88
Cellulolytic enzyme complexes (technical) of *Trichoderma viride-reesei* (Onozuka R-10) and of *Aspergillus niger*	IE	DEAE- and Phospho-Spheron 300, Spheron 1000-DEAE, -TEAE, -S, -C, and -Phosphate	297,468,469, 473
Cellulolytic enzyme complex of *Trichoderma viride-reesei* mutant	IE	Spheron 1000-DEAE, -Phosphate	298
Cellulolytic enzymes of *Trichoderma viride-reesei*	IE	DEAE-Silica, Mono Q	495,568
14.2.1.8 Chymotrypsin, chymotrypsinogen			
α-Chymotrypsin A	IE	AP/Silochrom	154
—	IE	SP-Glycophase/CPG, Mono S	104,564,568
—	RP	ODS-HC/Sil X, Partisil ODS, Syn-Chropak RP-P, μBondapak C18, μBondapak CN	678
—, bovine, from pancreatic extract	SE	TSK Gel 3000 SW	680
α-Chymotrypsin and chymotrypsinogen A, bovine	BA	Cbz-Gly-D-Phe-Spheron 300	72
—	BA	Soy bean inhibitor/LiChrospher	328,329
—	HI	PEG/CPG	102
—	IP-RP	Nucleosil CN	20
—	SE	LiChrosorb Diol	607,701
—	SE	TSK Gels PW, I-125 PC	246,272
Chymotrypsinogen A, bovine	IE	QAE-Glycophase/CPG, Zipax SCX	31,104
—	RP	LiChrosorb RP-8, μBondapak C18, μBondapak Phenylalkyl	273,320,596
—	HI	Diol-C8/Silica, Partisil C-8, Spheron 300	518,658
—	IE	CM-, Phospho-, Sulpho-Spheron 300, Spheron 1000-Phosphate	72, 467,470, 473,475,476
—	SE	Glycophase/CPG	32
—	SE	I-125 PC, SynChropak GPC 100	317,562

Substance	Method	Column	Ref.
——	SE	Diol/Silica, LiChrosorb Diol	88,582-584
——	SE	Agarose-12%, PAP/LiChrosorb	158,282,283
——	SE	TSK Gels SW	27,427,579,694
——, acetylated, succinylated	SE	LiChrosorb Diol	607
——, denaturated (8 M urea, 6 M guanidinium chloride, 0.1% SDS - sodium dodecylsulphate)	SE	Glyceryl/CPG, CPG	63,178
Chymotrypsinogen A, B	SE	Fractogel TSK HW-55F	216
Chymotrypsinogen B, bovine	SE	Toyopearl HW-55F	198

14.2.1.9 Creatine kinases (phosphokinases)

Substance	Method	Column	Ref.
Creatine kinase	RP	Aquapore RP 300, LiChrosorb RP-18	741
—— (BB), rabbit	IE	DEAE-Glycophase/CPG	604
—— isoenzymes	IE	DEAE- and QAE-Glycophase/CPG	101,103,104,129,327,562,602,606
——	IE	SynChropak AX 300, Vydac Pelli AX	74,380
——, bovine	IE	DEAE-Glycophase/CPG	564
——, human serum profiling	IE	DEAE-Glycophase/CPG	604,605
——, human serum, after acute myocardial infarction or severe accident	IE	DEAE-Glycophase/CPG, MP-AX, PEI-LiChrospher	562,563,601,701

14.2.1.10 Cytochromes c and cytochrome P-450

Substance	Method	Column	Ref.
Cytochrome c	HI, RP	Hypersil ODS	470,520
——	IE	AP/Silochrom	154
——	IE	CM-Glycophase, CM-Polyamide, Mono S	269,320,568
——	IP-RP	Nucleosil C18, Nucleosil CN	19,490
——	RP	Aquapore RP 300, CNP-Silica, DPh-Silica, LiChrosorb RP-8, -18, LiChrospher C8, Octyl-Silica, Ultrasphere octyl	320,415,741
——	RP	RP-8, μ-CN/Silica	142,574
——	SE	Amide/Si, PAP/LiChrosorb	155,158
——	SE	Diol/Silica, LiChrosorb Diol	88,583,584,607,701

Table 14.2 (continued)

Substances	Mode of chromatography	Packings	References
—	SE	Glyceryl/CPG, Glycophase/CPG	120,564
—	SE	Glycophase/CPG, Glycophase/ LiChrospher, Glycophase/Partisil	103
—	SE	I-125 PC	272,317,573
—	SE	Porasil DX, Spheron 1000	633,716
—	SE	SynChropak GPC 100	142,562
—	SE	TSK Gels PW	8,240,246
—	SE	TSK Gels SW	8,181,269, 281,336,427, 579,680,694
—, isoproteins	SE	Fractogel TSK HW-55F, -65F, Micro-Pak TSK SW Gels, Toyopearl HW-55F	198,216,727
—, bovine, horse heart	HI,RP	μBondapak C$_{18}$, μBondapak Phenyl	46,596,757
[14C]Cytochrome c	HI	Partisil ODS	513
Cytochrome c, trinitrophenylated- and nitrobenzodiazol-derivative	IE	Amberlite CG 50, Bio Rex 70, Hydroxylapatite	707
—, denatured (in 8 M urea, 6 M guanidinium chloride and 0.1% SDS)	SE	Glyceryl/CPG, TSK Gels SW	63,337,339
—, SDS-complexes	SE	CPG	178
Cytochrome P-450, multiple forms, of solubilized rabbit hepatic microsomes from liver, kidney cortex and intestinal mucose (with and without drug treatment)	IE	Anpack AX	387

14.2.1.11 Glucose oxidases

Substances	Mode of chromatography	Packings	References
Glucose oxidase	SE	Separon HEMA 300-Glc	116
—, fungal, commercial	BA	Concanavalin A/Silica	73
—, technical, *Aspergillus niger*	IE	DEAE-Spheron 300	474,562
—, —	IE	Sulpho-Spheron	72,476

14.2.1.12 Hexokinases

Substances	Mode of chromatography	Packings	References
Hexokinase, crude yeast extract	AC	Cibacron blue/Silica, T.-dyes/ Silica	422,646,647

Sample	Method	Column / Support	References
— isoenzymes, rat liver	IE	SynChropak AX 300	10,470
— , rat liver and testicular tissue extracts	IE	DEAE-Glycophase/CPG	606,701

14.2.1.13 Lactate dehydrogenases

Lactate dehydrogenase

Sample	Method	Column / Support	References
	SE	MicroPak TSK Gels SW	727
	SE	TSK Gels-SW, -PW	8,269
	IE	MP-AX, PEI-LiChrospher	601,701
— isoenzyme profiles, rat kidney, rat muscle, human spermatozoa (LD$_1$ to LD$_5$, LD$_{x2}$, LD$_{x3}$), human serum normal and after acute myocardial infarction or after severe accident	IE	DEAE-Glycophase/CPG	103,129,381, 562,564,602, 603,619,701
— isoenzymes, commercial, in spleen and heart extract, bovine heart and muscle, human serum			422,646,647
— — (H$_4$,M$_4$), pig heart and rabbit muscle	AC	Cibacron blue/Silica, T.-dyes/Silica	396,522
— — (H$_4$,M$_4$), pig heart and rabbit muscle	BA	AMP/Silica	447
— — , human serum	IE	IEX 525 QAE	185,186,562, 563,604-606, 701
— — — chromatography profiling, from patients with elevated LDH-activity, with myocardial infarction, with liver or muscle damage or with necrotizing pancreatitis	IE	DEAE-Glycophase/CPG	
— — , in addition to LD$_1$-LD$_5$ also LD$_3$	IE	SynChropak AX 300	743
— — , in human serum of an accident victim with broken hip	IE	DEPA-column	701
— — , in normal human serum and after myocardial infarction	IE	SynChropak AX 300	470
— — , LD$_1$ to LD$_5$	IE	DEAE- and QAE-Glycophase/CPG	101
— — , rat kidney	IE	SynChropak AX 300	10

14.2.1.14 Lysozymes

Lysozyme

Sample	Method	Column / Support	References
—	HI, RP	Aquapore RP 300, LiChrosorb RP-18, Spheron 300	658,741
—	HI, RP	Diol-C$_8$/Silica, Hypersil ODS, Partisil C-8	470,518,520
—	IE	AP/Silochrom	154
—	IE	CM-Polyamide	269
—	IE	CM-Spheron 300	467,470,475
—	IE	Phospho-Spheron 300, Spheron Phosphate 1000	72,473,476

B480

Table 14.2 (continued)

Substances	Mode of chromatography	Packings	References
—	RP	Radial-Pak A/C18, μBondapak Alkylphenyl, μBondapak C18, μBondapak Phenyl	262,273,274
—	SE	Fractogel TSK HW-55F, I-125 PC, LiChrosorb Diol, MicroPak, Shodex OH Pak B-804, SynChropak GPC-100, TSK Gels SW	216,436,547
—, chicken = egg white = hen egg	SE	LiChrosorb Diol	582,607,701
	SE	TSK Gels SW	368,579
	SE	Glyceryl/CPG, Porasil-DX (Carbowax), Separon HEMA 300 Glc	116,542,633
—, egg white	RP	Nucleosil CN, Partisil ODS, Supelcosil LC-18	19,30
—, succinylated	SE	LiChrosorb Diol	607
—, testing of Spheron ion-exchanger hydrophobicity	IE	DEAE-Spheron 300	477
14.2.1.15 Pectolytic (pectic) enzymes			
Pectic enzymes, technical (Leozym - *Aspergillus niger*)	IE	DEAE-Spheron 300	474
— — (Leozym - *Aspergillus niger*)	IE	Spheron ion-exchangers, all types (1000-C, -DEAE, -Phosphate, -S, -TEAE)	566
— — (Pectinex ultra)	IE	Spheron 1000-S	470
— — (Pectinex ultra - *Aspergillus niger*, Rohament P)	IE	All types of commercial Spheron ion exchangers	471
— — (Pectinex ultra, Rohament P)	IE	Spheron 1000-DEAE, Spheron 1000-C	468, 469
14.2.1.16 Pepsin, pepsinogen			
Pepsin	BA	Separon-INHIB	691,692
—	SE	I-125 PC	272
—	SE	TSK Gels SW, TSK Gels PW	8,368,579

	Method	Column	References
—, fresh and stored	SE	MicroPak TSK Gels SW	727
—, pepsinogen, porcine	SE	LiChrosorb Diol	607,701
14.2.1.17 Peroxidase			
Peroxidase, horse radish	AD	Silicone/CPG	486
—	SE	I-125 PC	272
—	SE	PEG/CPG	125
—	SE	TSK Gels PW	246
—	SE	TSK Gels SW	181,336,470
14.2.1.18 Ribonuclease			
Ribonuclease A, bovine pancreas	HI, RP	Hypersil ODS	470,520
—	IE	Mono S, Zipax SCX	31,568
—	IP-RP	Nucleosil CN	19
—	RP	LiChrosorb RP-8	320,585
—	RP	µBondapak C_{18}, µBondapak Phenyl, µBondapak Phenylalkyl	273,274
—	SE	Agarose-12%	282,283
—	SE	Glycophase/CPG	32
—	SE	I-125 PC	317,573
—	SE	LiChrosorb Diol	607,701
—	SE	TSK Gels PW	246
—	SE	TSK Gels SW	27,317,336, 579,694
Ribonuclease, bovine pancreas extract	AC	Cibacron blue/Silica	422
—, in 6 M guanidinium chloride	SE	TSK Gels SW	339
Ribonuclease s-peptide	IE	MicroPak AX-10	146
14.2.1.19 Trypsin, trypsinogen			
Trypsin, bovine	IE	DEAE-Glycophase/CPG	102,104,470
—	IP-RP, RP	ODS-HC/Sil X-1, SynChropak RP-P, µBondapak C_{18}, µBondapak CN, µBondapak f.a.	233,565,678
—	SE	Glyceryl/CPG, Spheron 300 Glc, Spheron 1000	116,542,716
—, acetylated	SE	LiChrosorb Diol	607

Table 14.2 (continued)

Substances	Mode of chromatography	Packings	References
— partly autolyzed, α- and β-trypsin	BA	Separon-INHIB, Soy bean-inhibitor/ Lichrospher	116,328
—, trypsinogen	SE	Lichrosorb Diol	607,701
—, α- and β-trypsin, crude trypsin	HI, IP-RP	Nucleosil CN, Spheron 300	20,657
Trypsinogen	SE	I-125 PC, PAP/Lichrospher, TSK Gels SW	158,272,579
14.2.1.20 Various enzymes			
Acetylcholin esterase, erythrocyte	SE	Glyceryl/CPG	120
Acyl carrier protein	IP-RP	µBondapak Alkylphenyl, µBondapak C18, µBondapak f.a.	229,232
Adenylate kinase	IP-RP	µBondapak f.a.	233
—, thermolysin digestion	SE	TSK Gel 3000 SW	269
Aldehyde dehydrogenase, sheep liver	IP-RP, RP	µBondapak Alkylphenyl,	226,229
— — —	SE	µBondagel E-linear	261
Amino acid oxidase, isoenzymes	IE	Mono P	568
Amyloglucosidase	SE	Diol/Silica	584
Arylsulfatase isoenzymes in sera of patients with colorectal cancer, in human urine, in limpet	IE	DEAE-Glycophase/CPG	75,563
Aspergillopeptidase technical, *Aspergillus oryzae* (Amylorizin P1 OX)	BA	Separon-INHIB	692
Aspergillopeptidase technical (crude protease), *Aspergillus sojae*	IE	CM-Spheron 300, DEAE-Spheron 300	474,562
Carboxypeptidase G2, *Pseudomonas* species	AC	T.-dyes/Silica	646,647
Citrate synthetase, from bacterial mutant cells	IE	Mono Q	568
DNA-polymerase, from HeLa cells extract	SE	Toyopearl HW-55F	198
DNA-polymerase (RNA-directed, "reversed transcriptase"), avian myeloblastosis virus-associated	SE	PEG/CPG	125
Elastase	IP-RP	Nucleosil CN	20

Table 14.2 (continued)

Substances	Mode of chromatography	Packings	References
Urease, soy beans, crude	SE	Spheron 1000	72
Urease, commercial	SE	TSK Gel 3000 SW	341
14.2.2 GLYCOLIPIDS			
14.2.2.1 Gangliosides			
Beef brain and human erythrocyte mono- to tetrasialo-gangliosides	IE	DEAE/CPG, DEAE/Sil	384,385
Brain gangliosides GM1, GD1a, GD1b, GT1	AD	Iatrobeads	494
Brain human gangliosides GD1a, GD1b, GM1, GQ, GT1b, p-nitro-benzoyloxyamine derivatives	RP	μBondapak C18	687
Brain human or bovine gangliosides GD1a, GD1b, GM1-GM3, GT1	AD	LiChrosorb Si 60	679
Ganglioside micelles	SE	Glyceryl/CPG	120
Leukocyte gangliosides GA1, GA3, GM1, GM3, GM4	AD	Zipax	411
Monosialoganglioside standards – perbenzoylated derivatives GM1 to GM4, gangliosides from human plasma and liver, GM3	AD	LiChrosphere Si 4000	77
Mono- to tetrasialogangliosides	AD	Iatrobeads	504,505
Mono- to pentasialogangliosides in human infant forebrain and cerebellum	IE	DEAE-Dextran/Spherosil	173,174
Tay-Sachs ganglioside, GM2	AD	LiChrosorb Si 60	679
14.2.2.2 Neutral glycosphingolipids (cerebrosides)			
Beef brain and human erythrocyte glycosphingolipids	IE	DEAE/Sil	385
Ceramide derivatives, O-acetyl-N-p-nitrobenzoylated: glucosyl- (Glc-Cer), lactosyl- (Gal-Glc-Cer), trihexosyl-ceramides, globoside, hematoside (N-acetyl type)	RP	μBondapak C18	661
——, O-acetyl-N-p-nitrobenzoylated: glucosyl-, lactosyl-, lactotriaosyl-, globotriaosyl-, globotetraosyl-, neolacto-tetraosyl-ceramide	AD	Zorbax SIL	662

Compound	Mode	Column/Sorbent	Ref.
—, perbenzoylated: glucosyl-, lactosyl (Lac-Cer), galactosyllactosyl- (Gal-Lac-Cer), trihexosides, globoside (GalNAc-Gal-Lac-Cer), hematoside (Neu-NAc-Lac-Cer)	AD	Zipax	695,697
—, perbenzoylated: glucosyl-, lactosyl-, globotriaosyl-, globotetraosyl-ceramide, globoside	AD	Zipax	212,696
Ceramide dihexoside, trihexoside, tetrasaccharides (globoside I and paragloboside), phytosphingosine, sphingodienine	AD	Iatrobeads	16
—, from human plasma	AD	Zipax	324
	IE	DEAE/CPG, DEAE/Sil	384
Ceramide trihexosides, globoside, paragloboside, in human erythrocytes			
Cerebrosides, benzoylated	AD	Zipax	324, 449
—, benzoylated: beef brain standard, calf brain lipid extract, rat brain lipid extract (both with hydroxy fatty acids and nonhydroxy fatty acids), human plasma, new-born mouse dorsal root ganglion cells	AD	Zipax	324
—, benzoylated, brain, with hydroxy and nonhydroxy fatty acids, N-stearoyl-gluco-cerebrosides	AD	Zipax	449
—, calf brain, benzoylated	AD	Zipax	450
—, N-acetylpsychosine, psychosine (N-benzoyl derivative) as identification product	AD	Zipax	449,696
Erythrocyte glycolipids, after acetylation and N-p-nitro-benzoylation	NP,RP	Zorbax SIL, µBondapak C$_{18}$	751
—, human	AD	Zipax	695
Erythrocyte glycosphingolipids, after acetylation and N-p-nitrobenzoylation; cytolipin R, globoside I	NP,RP	Zorbax SIL, µBondapak C$_{18}$	749,751
—, human	AD	Silicic acid	16
Forssman globoside, paragloboside	AD	Zipax	411
Galactolactoceramide (Gal-Lac-Cer), globoside	AD	Silica-Pel	1
Galactosylceramides, benzoylated, mouse brain extract, myelin	AD	Zipax	450
Glycosphingolipids neutral GL1a, GL2a, GL3, GL4	AD	LiChrosorb Si 60	679
Hematosides (gangliosides GM3), dog erythrocytes	AD	LiChrosorb Si 60	679
Plasma glycosphingolipids of adult and juvenile patients with Gaucher's disease (β-glucosidase deficiency)	AD	Zipax	696
—, (perbenzoylated) from patients with Gaucher's disease, Fabry's disease (α-galactosidase deficiency) and Krabbe's disease (galactocerebrosidase deficiency)	AD	Zipax	695,697
Urine sediment sphingolipids, perbenzoylated	AD	Zipax	697

Table 14.2 (continued)

Substances	Mode of chromatography	Packings	References
14.2.3. MISCELLANEOUS SUBSTANCES			
Dolichylpyrophosphoryl oligosaccharides (all 14 known lipid-linked oligosaccharides were separated in the form of tritiated derivatives); I, GlcNAc; II, GlcNAc2; III, Man1GlcNAc2; IV, Man2GlcNAc2; V, Man3GlcNAc2; VI, Man4Glc-NAc2; VII, Man5GlcNAc2; VIII, Man6GlcNAc2; IX, Man7GlcNAc2; X, Man8GlcNAc2; XI, Man9GlcNAc2; XII, Glc1Man9GlcNAc2; XIII, Glc2Man9GlcNAc2; XIV, Glc3Man9GlcNAc2	NP	Porasil A-60	730,731
Glycopeptides: mono-, di- and octa-saccharide glycopeptides (4 species) from liver or urine of normal persons and of patients with lysozomal storage disorders affecting the catabolism of glycoproteins (fucosidosis, sialidosis and aspartylglucosaminuria)	NP	μBondapak Carbohydrate	511
Humic (and fulvic) acid from marine sediment	SE	TSK Gel 3000 SW	593
Nucleoside-amino acids: (+)S-adenosyl-L-methionine, (-)S-adenosyl-L-methionine	IE, RP	Partisil 10 SCX	712
Proteolipids (microsomal) from *Neurospora crassa*	NP	LiChrosorb	430
14.2.4. NUCLEIC ACIDS			
14.2.4.1 Deoxyribonucleic acids (DNAs)			
Deoxyribonucleic acid (DNA)	SE	AP/LiChrospher	625
DNA	SE	SynChropak GPC	562
—, calf thymus	AD	Silicone/CPG	487
— —, digestion by deoxyribonuclease – time study	IE	DEAE-Spheron 300	72,476
— —, calf thymus fragments (digestion by bovine spleen deoxyribonuclease II)	SE	Glycophase/LiChrospher	564
	SE	Glycophase/LiChrospher	103
—, digestion by deoxyribonuclease	SE	Glycophase/LiChrospher	470
— restriction fragments (21 species)	RP	RPC-5	280
—, salmon sperm	SE	Glycophase/CPG	564,565

14.2.4.2 Ribonucleic acids (RNAs)

Description	Technique	Support	Ref.
m-RNA (9S), rabbit globin, 32pCp radiolabeled	IP-RP	μBondapak C18	510
Poly(A)-RNA (mRNA from *Bombyx mori* silkworm)	SE	Fractogel TSK HW 65F	216
RNA, *Escherichia coli*, 32pCp radiolabeled	IP-RP	μBondapak C18	510
r-RNA, *Escherichia coli*, crude	AD	Silicone/CPG	487
Yeast crude commercial RNA	SE	Spheron 1000	72

14.2.4.3 Transfer ribonucleic acids (tRNAs)

Description	Technique	Support	Ref.
Rapid resolution of tRNAs and ribosomal RNAs	RP	RP supports	344
tRNA, *Escherichia coli*	HI	Spherisorb	285
— [3H]Tyr-isoacceptors, separation from mixed uncharged RNAs	BA	Lectin-agarose	642
— isoacceptors, aminoacylated with [14C]-labeled amino acids: Ser- (from *Escherichia coli*), Leu- and Tyr- (*Klebsia aerogenes*), Phe- (*Rhodospirillum rubrum*)	RP	RPC-5	141
—, rabbit liver extract, Tyr- and N-formyl-Met-tRNA standards	IE, SE	MicroPak MAX, MicroPak TSK Gels SW	725,727
—, 4S, yeast, 32pCp radiolabeled	IP-RP	μBondapak C18	510
tRNAs, Asn- and His-unsubstituted, queuosine-containing, [3H]Asn-, [3H]His- labeled derivatives	IE, RP	Aminex A-28, RPC-5	642
—, bovine liver	AD	Silicone/CPG	487
—, bovine liver, fractionation, tRNA^{Ser}	AD, IE	Silicone/CPG, TMA-Silicone/CPG	508
— [14C]Leu- and Tyr-isoacceptors	RP	RPC-5	725
—, *Escherichia coli*: Ala I-, Ala II-, Arg I-, Arg II-, Asn-, Asp-, fMet1-, fMet2-, Glu-, His I-, His II-, Ile I-, Ile II-, Leu I- to Leu V-, Phe-, Tyr-, Val I-, Val II-	RP	RPC-1 to RPC-4	345
—, *Escherichia coli*: Glu-tRNA, [14C]Leu-tRNA, tRNA^{Phe} and Phe-tRNA	IE,RP	Aminex A-28, RPC-5	644,725
—, *Escherichia coli*, mixed: tRNA^{Glu}, [14C]Leu-tRNA isoacceptors, [3H]Lys-tRNA isoacceptors	IE, RP	Aminex A-28, RPC-5	642
—, human, synthesized in early (^{3}H) and late (^{14}C) stages of proliferating lymphocytes	IE, RP	Aminex A-28, RPC-5	644
—, liver, separation of [14C]Lys-tRNAs from other (uncharged) tRNAs	AC, RP	CMB-cellulose, RPC	642
—, mammalian and bacterial, unsubstituted, separation of queuine-containing tRNAs from others	AC, RP	CMB-cellulose, RPC	642
—, mammalian, separation of Asp-tRNA from others	BA	Lectin-agarose	642

B488

Table 14.2 (continued)

Substances	Mode of chromatography	Packings	References
—, mixture of species	SE	Glycophase/CPG	564,565
—, mouse plasma tumor cells ([³H]Tyr-) and mouse liver ([¹⁴C]Tyr-)	IE, RP	Aminex A-28, RPC-5	644
—, rat liver, Lys-tRNA	IE	Aminex A-25, A-28, DA-X12-11, HA-X8, RPC-5	369
—, separation of Lys-tRNA from uncharded [³H or ¹⁴C] non-queuine - tRNAs (Glu-, Gly-, Lys-, Phe-) and queuine-tRNAs (Asn-, Asp-, His-, Tyr-)	AC	RP-boronate	643
Tyr- and N-formyl-Met-tRNA	IE	DEAP/Silica	2

14.2.5 OLIGONUCLEOTIDES

14.2.5.1 Oligodeoxyribonucleotides

Substances	Mode of chromatography	Packings	References
Deoxydinucleotide monophosphates, all 16 possible species and 15 available deoxydinucleotides	IE	Partisil-10 SAX	143
Dodeca- to hexadecadeoxyadenylates, oligodeoxyadenylates	IE	PEI-MAX	540
Mixed oligonucleotides d(CACCA)rC, d(CATTCACCA)rC, containing 3-35 bases d(CCATTCACCA)rC	IE	PEI-PE-MAX	436
Mixture of oligodeoxyribonucleotides	IE	SynChropak AX 300	10
	RP, SE	TSK Gel 3000 SW, μBondapak C₁₈	752
Oligodeoxyadenylic acids (2 to 6 or 2 to 7 bases)	IE	Micropak NH₂	22
Oligothymidylic acids	IP-RP	LiChrosorb RP-18	649
— and other oligonucleotides (TpT, TpTp, TpTpTp, CpGp, ApAp, ApT)	IE, RP, SE	DEAE TSK 2000 SW, TSK Gel 3000 SW, μBondapak C₁₈	752
— (2-4, 2-7 or 2-10 bases)	RP	Bondapak C₁₈/Porasil B, μBondapak C₁₈	180
Protected oligonucleotides for polynucleotide synthesis: (a) di- to tetranucleotides with free 5'-phosphate end group (17 species), (b) oligonucleotides carrying 5'-mono-methyoxytrityl end groups (19 species), (c) 2-(p-trityl-phenyl)sulfonyl-ethyl-protected mono- to trinucleotides (11 species), (d) unprotected oligonucleotides with 5'-hydroxyl end groups (20 species)			

Compound	Method	Column/Support	Ref.
Purine oligonucleotides from depyridinated DNA (herring sperm)	RP	Nucleosil C_{18}	617
Pyrimidine deoxytetranucleoside triphosphates, all sequence isomers of (dC_3,dT), (dC_2,dT_2) and (dC,dT_3)	IE	Partisil-10 SAX	144
— oligodeoxyribonucleotides (di- to deca-)	IE	DEAE-Spheron 300	476
— pentanucleotide tetraphosphates from depurinated herring sperm DNA (terminal dephosphorylated); all sequence isomers of (dC,dT_4), (dC_2,dT_3), (dC_3,dT_2)	IE	MicroPak AX-10	148
Synthetic 2',-deoxyribohexanucleotides, both unprotected d-CpCpApApGpC and protected DMTrCBzpCAnpABzpABzpGiBupCAn and reaction components DMTrCBzpCAnpABzpABz and d-pGiBupCAnoAc	IE	Pellionex AL WAX	414
Synthetic deoxyribonucleoside diphosphates (partially protected) - field desorption mass spectrometry identification	n.d.	n.d.	598
Synthetic octa- and dodecanucleotides, deprotected products of the phosphotriester solid phase synthetic method	IE, RP	Partisil 10 SAX, μBondapak C_{18}	190
Synthetic oligodeoxyribonucleotides (di- to pentadeca-), 17 species	RP, SE	I-125 PC, LiChrosorb RP-8	491
Synthetic oligodeoxyribonucleotides, mixtures, derivatives from different stages of synthesis	IE, MM	MicroPak NH_2	22
Synthetic oligonucleotides, deprotected, CG dimer	IP-RP	Partisil 10 C_8	121
Synthetic protected and deprotected oligonucleotides, dimer CG, tetramer (CG)2 and hexamer (CG)3	NP, RP	Partisil ODS, Partisil 10 PAC	121
Trideca- to pentadecathymidylates	IE	PEI-MAX	540
Tyrosine suppresor t-RNA gene oligonucleotide synthetic segments corresponding to the terminal regions (tri-, penta-, hexa-, hepta-, octa-, nona- and undecanucleotides)	RP	μBondapak C_{18}, μBondapak C_{18}/Porasil B	42

14.2.5.2 Oligodeoxyribonucleotides prepared by enzymic cleavage

Compound	Method	Column/Support	Ref.
DNA, calf thymus, deoxytrinucleotide diphosphates from alkaline phosphatase digest	IE	Partisil 10 SAX	143
—, digestion by deoxyribonuclease	SE	Glycophase/LiChrospher Si 1000	470
—, salmon sperm; deoxyribonuclease digest, di- and tri-nucleotides	IE	LiChrosorb NH_2	722

14.2.5.3 Oligodeoxyribo- and oligoribonucleotides prepared by chemical or specific partial hydrolysis

Compound	Method	Column/Support	Ref.
DNA-acid digest deoxyoligonucleotides	IE	Dowex 1	641
DNA, *Bacillus subtilis*; Pyrimidine oligonucleotides from specific partial hydrolysate	IE	DEAE-Spheron 300	595

Table 14.2 (continued)

Substances	Mode of chromatography	Packings	References
—, herring sperm; oligoguanosine phosphates from partial hydrolysate, (dG)3 to (dG)8	RP	LiChrosorb RP-18	618
DNAs (readily available), preparation of oligonucleotides from partial specific hydrolysates	RP	Nucleosil C18	611
mRNA-based hydrolysis products	RP	Partisil ODS	121
Poly(A)ribonucleotides from poly(A) partial alkaline hydrolysate (A2 to A19)	IE	PEI-MAX	540
rRNA (*Artemia salina*) 5S, 5.8S and 28S alkali resistant oligonucleotides	IE	AL Pellionex WAX	703
14.2.5.4 Oligoribonucleotides (adenosine oligonucleotides)			
2'-5'-Adenine oligonucleotide inhibitors (2-5A) in interferon-treated EMC virus-infected mouse L-cells (ppA2'p5' A2'p5'A to ppp(A2'p)4A)	MM, RP	LiChrosorb NH2	735
Adenosine signal nucleotides, low-molecular-weight mediators in the regulation and coordination of intracellular reactions	RP	RP-NH2/LiChrosorb	66
Diadenosin monophosphate and triadenosin diphosphate	IE	MicroPak AX-10	153
Di- to nonaadenylic acid	IE	TMCA-silica	21
Oligoadenylates (A2'p)1-4A, (A2'p5')1-4A, and related; (A3'p5')1-4 and related; 5'mono- to triphosphorylated and nonphosphorylated oligomers from reticulocytes and chemically or enzymically synthesised; oligoadenylates from interferon-treated encephalomyocarditis virus infected L-cells	MM, RP	Hypersil C18, LiChrosorb NH2, μBondapak C18	96
— (Ap)1-11A	RP	ODS	453
— ApA to (Ap)4A	IP-RP	LiChrosorb RP-18	322,323
— ApA to (Ap)15A	IE	DMA/Silica	323
— ppp(A2'p)2-4A and core (A2'p)2-4A, detection in crude extracts from interferon-treated or EMC virus infected L-cells	RP	μBondapak C18	362
Oligoadenylic acids (2-10 bases)	IE	DEAE-TSK 2000 SW	752
—, 5'-triphosphorylated, (2'-5')-oligo A	MM, RP	LiChrosorb NH2, μBondapak C18	82

14.2.5.5 Oligoribonucleotides (guanosine oligonucleotides)

Diguanosine oligophosphates (tri- and tetra-) from cyst of brine shrimp (*Artemia salina*)	IE	AX	81
Guanosine oligonucleotides associated with elongation factor Tu (*Escherichia coli*)	RP	SRPA	67
— tetra- and pentaphosphates	IE	Micropak SAX	725

14.2.5.6 Oligoribonucleotides (uridine oligonucleotides)

Oligouridylic acids UpU to (Up)4U	IP-RP, SE	LiChrosorb Diol, LiChrosorb RP-18	322,323
— UpU to (Up)3U, UpU to (Up)4U, UpU to (Up)7U	IE, SE, RP	DMA/Silica, LiChrosorb Diol, LiChrosorb RP-8	699
— — UpU to (Up)8U	IE	DMA/Silica	323
— — UpU to (Up)11U	RP	ODS	453

14.2.5.7 Oligoribonucleotides (various)

Block copolymers with ApG primer: ApG(pA)0-6, ApG(pU)0-4 and ApG(pC)0-2	RP	ODS	453
Cap structured and related dinucleotides (22 species), dinucleotide di- to pentaphosphates, Gp3A synthesis reaction mixture	n.d.	Nucleosil SB, Partisil PAC	218
Cytidine oligonucleotides (Cp)1-19C	RP	ODS	453
Enzymic synthesis, oligonucleotides (A-)4G-(Ap)1 to (A-)4G-(Ap)21	IE, RP	RPC-5	640
Mixture of all 16 possible ribodinucleoside monophosphates	IE	Partisil 10 SAX	143
Mixture of synthetic oligonucleotides, containing 6-16 bases	IE	Polyanion SI HR	568
Nicotinadenin dinucleotides NADH, NAD+, NADP+	IE	Aminex A-25	723
Pyridine dinucleotides in cell extracts NADH, NAD+, NADPH, NADP+	RP	µBondapak C18	321
Ribodinucleotide monophosphates (16 species)	RP	ODS	453
Ultra-pure NADP from commercial product	IE	Polyanion SI HR	568

14.2.5.8 Oligoribonucleotides prepared by enzymic cleavage

Escherichia coli 5S rRNA enzymic digest oligoribonucleotides (pancreatic or T1 ribonuclease)	IE	AL Pellionex WAX	703,704
Oligonucleotides, snake venom hydrolysis products	RP	Partisil ODS	121

Table 14.2 (continued)

Substances	Mode of chromato-graphy	Packings	References
Poly(A,G)-ribonuclease (T$_1$) digest oligonucleotides, from (Ap)$_7$GP to (Ap)$_{22}$GP	IE, RP	RPC-5	640
Rainbow trout liver 5S rRNA ribonuclease T$_1$ or ribonuclease A digest (small oligonucleotides for sequence analysis)	IE	AL Pellionex WAX	365
Torulopsis (Candida) utilis 5S rRNA ribonuclease T$_1$ digest oligonucleotides	IE	AL Pellionex WAX	365
tRNA (glutamate), normal and bisulfite treated ribonuclease (pancreatic or T$_1$) digest oligonucleotides	IE, RP	RPC-5	641
Yeast RNA trinucleotide diphosphates from pancreatic ribonuclease digest	IE	Partisil 10 SAX	143
Yeast tRNA^{Phe} (both native and modified) ribonuclease T$_1$ digest oligonucleotides	IE, MM	Zorbax-NH$_2$	454

14.2.6. OLIGOSACCHARIDES

14.2.6.1 Acid oligosaccharides (mucopolysaccharide fragments and sialooligosaccharides)

Substances	Mode of chromato-graphy	Packings	References
Chondroitin-4-sulphate and chondroitin-6-sulphate di-saccharides (reduced saturated and nonsaturated); shark cartilage [^{3}H]-labeled chondroitin-6-sulphate, hyaluronidase digest di- to decasaccharides; whale cartilage [^{3}H]-labeled chondroitin-4-sulphate, hyaluronidase digest di- to tetra-saccharides	IE	Partisil SAX	127
Chondroitin sulphate chondroitinase digest, non-saturated disaccharides	MM, NP	Partisil PAC	406
Chondroitin sulphates A, B, C and glycosaminoglycans (muco-polysaccharides) substituted nonsaturated disaccharides resulting from chondroitinase digestion	AD	LiChrosorb SI-100, Partisil PXS	405,410
Chondroitin sulphates A (chondroitin-4-sulphate), B (dermatan sulphate) and C (chondroitin-6-sulphate), heparan-sulphate and heparin disaccharides resulting from chondroitinase digestion and reduction with borohydride	MM, NP	LiChrosorb NH$_2$, Partisil PAC	407,408

B493

Description	Method	Column/packing	Ref.
Heparan-sulphate and heparin nonsaturated disaccharides derived from heparinase digest	IP-RP, NP	Partisil ODS, Partisil PAC	409
Heparin (nitrous acid treated and reduced), [3H]-labeled sulphated disaccharides	IE	Partisil SAX	128
Oligosaccharides bearing zero to 4 sialic acid residues	IE	MicroPak AX-10	24
Sialooligosaccharides	MM	LiChrosorb-NH2	52,53

14.2.6.2 *Analyses of body fluid oligosaccharides*

Description	Method	Column/packing	Ref.
Acid oligosaccharides (reduced) from mucolipidosis I (sialidosis) urine	RP	LiChrosorb RP-8	124
Carbohydrates in human serum	IE	AX 2633	292
Disaccharides (lactose, sucrose) and trisaccharides in human urine	AD, RP	HPX-87 heavy metal, C_{18}RP	386
Neutral oligosaccharides (reduced and nonreduced per-O-benzoyl derivatives) from mannosidosis urine and from Sandhoff's disease urine	RP	LiChrosorb RP-8	124
Oligosaccharides (reduced and nonreduced per-O-benzoyl derivatives) from mannosidosis urine	NP, RP	Ultrasphere octyl, Zipax	123,124
Tri- to heptasaccharides and corresponding borohydride-reduced compounds (11 species) from liver or urine of normal persons and of patients with lysosomal storage disorders affecting the catabolism of glycoproteins: fucosidosis, G_{M1}-gangliosidosis, G_{M2}-gangliosidosis, mannosidosis and sialidosis	NP	μBondapak Carbohydrate	511
Unsaturated sulphated disaccharides resulting from enzymic digestion (chondroitinases) of urinary isomeric chondroitin sulphates (from patients with mucopolysaccharidoses and from normal individuals)	MM, NP	Partisil PAC	406, 410
Urinary nondialyzable carbohydrates from cancer patients (fucose/galactose ratios)	IE	AX 2633	292

14.2.6.3 *Cellulose fragments*

Description	Method	Column/packing	Ref.
Cellodextrins (cellobiose to cellohexaose) and also sucrose	IE	DEAE-Spheron 300, Spheron-TEAE 1000	296
— (cellobiose to cellohexaose)	NP	Aminex 50 W-X4 Ca^{2+}	390
— (cellobiose to celloheptaose) and also lactose, sucrose	NP	AG 50 W-X4 Ca^{2+}	389

Table 14.2 (continued)

Substances	Mode of chromato-graphy	Packings	References
—, (DP$_1$ to DP$_6$, in the case of DP$_4$-DP$_6$ partial resolution of α- and β-anomers)	RP	Separon Silica C$_{18}$	718
—, enzymic hydrolysates of celloheptaose	IE	DEAE-Spheron 300, Spheron-TEAE 1000	296
—, enzymic hydrolysis	NP	Aminex 50 W-X4 Ca^{2+}	390
Cellooligosaccharides reduced and normal (G$_2$, G$_2$H to G$_6$, G$_6$H)	NP	PXS-1025 PAC, μBondapak Carbohydrate	215
Cellulose (filter paper), enzymic hydrolysates (*Trichoderma viride-reesei* cellulases)	IE	DEAE-Spheron 300, Spheron-TEAE 1000	296
Cellulose (filter paper), enzymic hydrolysates (*Trichoderma viride-reesei* cellulases)	NP	Aminex 50 W-X4 Ca^{2+}	390
14.2.6.4 Chitin fragments and cell-wall derived oligosaccharides			
Cell-wall saccharide from *Klebsiella*-bacterium (tetrasaccharide and its oligomers, n = 2, 3, 6 and 12)	NP	μBondapak	596
Chitin-derived [^{3}H]-oligomers (NaB^{3}H$_4$ reduced, G$_2$ to G$_5$)	NP	MicroPak AX-5, AX-10, μBondapak Carbohydrate	461
Chitin partial hydrolysate	SE	BioGel P-4	509
N-Acetyl-D-glucosamine oligomers: (NAG)$_1$ to (NAG)$_5$; lysozyme-catalyzed hydrolysis and transglycosylation of NAG-oligomers	NP	μBondapak Carbohydrate	710
14.2.6.5 Cyclodextrins			
β-Cyclodextrin	NP	μBondapak Carbohydrate	62
—, partial hydrolysate	SE	Fractogel TSK HW-40S	216
Cyclodextrins	NP	Aminex 50 W-X4 Ca^{2+}	287
—	NP	Silica + Modifier I	6
Cyclodextrins α, β, γ (consisting of 6, 7 and 8 α-1-4 linked glucose units)	NP	μBondapak Carbohydrate	761

14.2.6.6 Fragments of dextrans, plant gums, and some other glycans

Dextran B-1355-S, acetolysate, tetrasaccharide-isomalto-dextranase digest fragments (4 species)	IE	LCR-3	685
— B-1355, trisaccharide fraction obtained by acetolysis; dextran B-1397, pentasaccharide mixture obtained by dextranase digest	IE	LCR-3	682
— B-1397 (*Arthrobacter*), G_2-dextranase digest oligo-saccharides (8 species)	IE	LCR-3	684
Dextran-derived [^{3}H]-oligomers (G_2 to G_{17}, NaB^3H_4 reduced)	NP	MicroPak AX-5, AX-10, μBondapak Carbohydrate	461
Dextran NRRL B-1355-S, acetolysed and deacetylated product	IE	LCR-3	683
— partial hydrolysates, [^{3}H]-oligomers (G_2 to G_{12}, NaB^3H_4 reduced)	SE	BioGel P-4	509
— T-40, partial hydrolysate	SE	Fractogel TSK HW-40S	216
Elsinan (glucan of *Elsinoe leucospila*), trisaccharides obtained by acid hydrolysis	IE	LCR-3	682
Isomaltooligosaccharides (from partial hydrolysate of dextran) in the form of pyridylaminoderivatives	RP	TSK Gel LS 410	244
Lichenan (ex *Cetraria islandica*) or xanthan (gum, commercial) – oligosaccharide fragments (partially methylated, partially ethylated)	RP	Zorbax ODS	702
Plant gum (sap of the lacquer tree *Rhus succedanea*), acid hydrolysate and also lactose and raffinose	MM, SE	H-Gel 3011-s (H^+), TSK Gel 212 (H^+)	383, 536
Xyloglucan nonasaccharide, prepared from xyloglucan secreted by sycamore cells culture (permethylated); xyloglucan nonasaccharide (permethylated); partial acid digest fragment-oligosaccharides (additionally ethylated)	RP	Zorbax ODS	702

14.2.6.7 Malto- and isomaltooligosaccharides

Isomaltooligosaccharides (n = 4–13), in the form of 2-aminopyridine derivatives	RP	TSK Gel LS 410 C$_{18}$	243
Isomaltose and oligosaccharides (n = 3–5), in the form of 2-aminopyridine derivatives	SE	TSK Gel 2000 PW	243
— (to isomaltohexaose)	IE	LCR-3	683
—, isomaltitol, isomaltotetraitol, isomaltohexaitol	IE	LCR-3	684
Maltodextrins (DP$_1$ to DP$_9$), in the case of DP$_3$ to DP$_7$ partial resolution of α- and β-anomers	RP	Separon Silica C$_{18}$	718

Table 14.2 (continued)

Substances	Mode of chromato-graphy	Packings	References
— (maltose to maltoheptaose)	NP	Aminex 50 W-X4 Ca^{2+}	390
Maltooligosaccharides	NP	Bondapak CX/Corasil, μBondapak Carbohydrate	570
— (G$_2$ to G$_{15}$ glucose units)	NP	Jascopak SN-01, μBondapak Carbohydrate	325
Maltose, maltooligosaccharides	NP	Aminex A6 Ca^{2+}, Aminex Q 15 S Ca^{2+}	570
Maltose to maltopentaose, maltosylsucrose, maltotriosyl-sucrose, maltooligosylsucrose	NP	μBondapak Carbohydrate	524
Maltotriose to maltopentaose	IE, NP	Aminex A-27, LiChrosorb NH$_2$, μBondapak	330
—	NP	Silica+Modifier I	6
Oligosaccharide standards (maltose to maltooctaose, dextran 10)	MM, NP	Aminex X6 Ag$^+$	626

14.2.6.8 Oligosaccharide moieties of glycoproteins and proteoglycans

Substances	Mode of chromato-graphy	Packings	References
α$_1$-Acid glycoprotein glycopeptides (3 species) in the form of pyridylaminoderivatives	RP	TSK Gel LS 410 C$_{18}$	244
Ceruloplasmin derived [^{3}H]-di- and tri-antennary desialyzed oligosaccharides, released by hydrazinolysis or by endo-glycosidases	NP	MicroPak AX-5, AX-10, μBondapak Carbohydrate	461
— (human) oligosaccharides	NP	MicroPak AX-10	24
— oligosaccharides, neuramidase-, β-galactosidase- and β-N-acetylglucosaminidase-digest fragments	SE	Bio-Gel P-4	509
Fetuin, endoglycosidase digest- or hydrazinolysate-oligo-saccharides, [^{3}H]-labeled after reduction with NaB^{3}H$_4$	SE	Bio-Gel P-4	509
-- sugar moiety	NP	MicroPak AX-10	24
β-Glucuronidase (human) oligosaccharides	RP	TSK Gel LS 410 C$_{18}$	243
Immunoglobulin IgG (bovine), endoglycosidase digest or	NP	MicroPak AX-10	24
hydrazinolysis-derived [^{3}H]-labeled oligosaccharides	SE	Bio-Gel P-4	509

Ovalbumin - endoglycosidase digest or hydrazinolysate, [³H]-labeled oligosaccharides	SE	Bio-Gel P-4	509
Ovalbumin glycopeptide V oligosaccharides	NP	MicroPak AX-10	24
Ovalbumin-derived high mannose [³H]-labeled oligosaccharides	NP	MicroPak AX-5, AX-10, μBondapak Carbohydrate	461
Orosomucoid oligosaccharides; orosomucoid-[³H]-tetraantennary desialyzed oligosaccharides released by hydrazinolysis or with endoglycosidases	NP	MicroPak AX-5, AX-10, μBondapak Carbohydrate	24,461
Ovomucoid (Japanese quail) enzymic digest, N-acetylated oligosaccharides also in the form of pyridylamino derivatives	RP	TSK Gel LS 410 C_{18}	243-245
α_1-Proteinase inhibitor, sugar moiety	RP	TSK Gel LS 410 C_{18}	243
Taka amylase, sugar moiety (in the form of pyridylamino derivatives)	RP	TSK Gel LS 410 C_{18}	243, 244

14.2.6.9 Oligosaccharides in food-stuff products and raw materials

Beers (Hungarian), carbohydrate analysis	NP	Excalibar NH_2, μBondapak Carbohydrate	686
Fruit juices, apple juice, sucrose	IE	DA X-4	497
—, various (apple, orange, cola drink), oligosaccharides	n.d.	Shodex S-801/S	138
Molasse analyses, sucrose in molasses	NP	PSX 10/25 PAC	109
— (sugar beet, sugar can)	IE	DEAE-Spheron 300	719
— (sugar can: psicose, raffinose, sucrose)	NP	Ostion LG KS	720
Presweetened cereals, analysis: sucrose and maltose in addition to fructose and glucose	RP	C18 Corasil	762
Soya protein product, soybean (defatted, deproteinized)	NP	Silica + Modifier I	6
Soybean oligosaccharides	NP	μBondapak Carbohydrate	62
Soy textured, released oligosaccharides: raffinose, stachyose, sucrose	NP	Partisil-10 PAC	248

(i) Syrups

Conversion syrup	MM, NP	Aminex HPX-87 Ca^{2+} or Ag^+	78
Corn syrup acid hydrolyzed (oligosaccharides DP1 to DP 15)	MM, NP	Aminex HPX-42 Ca^{2+}, Aminex 50W-X4 Ag^+, Aminex X6 Ag^+, HC-40 X5 Ag^+	626
— enzyme hydrolyzed (oligosaccharides DP1 to 6, DP1 to 15)	MM, NP	Aminex Q-15 S Ag^+, Aminex X4 Ag^+	626

Table 14.2 (continued)

Substances	Mode of chromatography	Packings	References
— — enzyme converted, oligosaccharides	MM, NP	Aminex A-7 Ag$^+$, Aminex X6 Ag$^+$	626
Fructose corn syrup oligosaccharides	MM, NP	Aminex A5 Ca^{2+}	78
Karo corn syrup, acetylated oligosaccharides (DP1 to DP15)	RP	Bondapak C$_{18}$/Corasil, Vydac RP OD	729
Karo syrup, reduced maltooligosaccharides	NP	Zipax	123,124
—, reduced and nonreduced maltooligosaccharides	RP	LiChrosorb RP-8, Ultrasphere octyl	123,124
Staley corn syrup oligosaccharides, sucrose	NP	Aminex HPX-42 Ca^{2+}, Aminex HPX-87 H$^+$	743
Syrup prepared from starch and sucrose by the action of cyclodextrin-glycosyltransferase; glucosylsucrose	NP	µBondapak Carbohydrate	524

14.2.6.10 Separation studies of various oligosaccharides
(i) Di- and trisaccharides

Substances	Mode of chromatography	Packings	References
Cellobiose, isomaltose, isomaltotriose, isopanose, lactose, maltose, maltotriose, panose, sucrose	IE	LC-R-3	348
Cellobiose, lactose, maltose, raffinose, sucrose, trehalose	MM, NP	Radial-Pak (TEPA)	275
Di- and trisaccharides	NP	MicroPak NH$_2$	436
Lactose anomers (α-lactose, β-lactose), raffinose, sucrose	IE, MM, NP	CDR-10 Cl$^-$	537
Lactose, maltose, raffinose	MM	TSK Gel LS 212 H$^+$	536
Lactose, maltose, sucrose	MM, NP	SCX Li$^+$	397
Lactose, raffinose, sucrose	MM, NP	TAP	534

(ii) Di-, tri- and tetrasaccharides

Substances	Mode of chromatography	Packings	References
Cellobiose, chitobiose, gentibiose, Glc-NAc-glc-myoinositol, isomaltose (also reduced), lactose (also reduced), lacto-N-tetraose, maltose (also reduced), maltotriose, malto-tetraose, raffinose, stachyose, trehalose	RP	Bondapak C$_{18}$/Corasil, Vydac RP OD	729
Cellobiose, lactose, maltose, melezitose, raffinose, stachyose, sucrose, trehalose	IE	Aminex A-15, DEAE-Spheron 300	107,719

Compounds	Mode	Stationary phase	Ref.
Isomaltose, lactose, maltose, raffinose, stachyose, sucrose	NP	µBondapak Carbohydrate	1,62,570
Lactose, maltose, melezitose, raffinose, stachyose	NP	Silica + Modifier I	6
(iii) Pentasaccharides and higher oligosaccharides			
Glucooligosaccharides	MM, NP	Aminex 50W-X4 Ca^{2+}	287
Higher oligosaccharides	MM, NP	Aminex HPX-42 Ag$^+$	78
Lacto-N-fucopentaose, maltopentaose	RP	Bondapak C$_{18}$/Corasil, Vydac RP OD	729
(iv) Modified oligosaccharides			
2-Aminopyridine derivatives: cellobiose, chitobiose, gentiose, lactose, laminaribiose, laminarioligosaccharides (n = 2–5), maltose	RP, SE	TSK Gel LS 410 C$_{18}$, TSK Gel 2000 PW	243
Dansyl hydrazones: lactose, maltose	AD, NP	LiChrosorb Si 60	152
α-Linked glucose disaccharides and their tetrahydroborate-reduced products: isomaltose and isomaltitol, kojibiose and kojibiitol, maltose and maltitol, nigerose and nigeritol	IE	LC-R-3	683
Per-O-benzoyl derivatives: cellobiose, gentibiose, isomaltose, isomaltotriose, kojibiose, maltose, maltotriose, nigerose, raffinose, trehalose	RP	Ultrasphere octyl	123
Reduced and de-N-acetylated: hexosamine containing di- and trisaccharides	NP	Locarte No 12	149
14.2.6.11 Starch- and amylose-derived oligosaccharides			
Amylose, partial acid hydrolysate, maltooligosaccharides (in the form of pyridylamino derivatives)	RP	TSK Gel LS 410 C$_{18}$	244
— (potato, III), partial acid hydrolysate, acetylated oligosaccharides DP2 to DP35	RP	Bondapak C$_{18}$/Corasil, Vydac RP OD	729
Starch-derived oligosaccharides	IE	LC-R-3	348
Starch hydrolysates	MM, NP	Aminex Q 15S Ca^{2+}	570
— —, glucooligosaccharides DP2 to DP20	NP	µBondapak Carbohydrate	570
— wheat (hydroxypropyl derivative), porcine pancreatic	MM, NP	Silica DMPA	732
α-amylase digest (G2 to G11 oligosaccharides)	NP	µBondapak Carbohydrate	744

Table 14.2 (continued)

Substances	Mode of chromatography	Packings	References
14.2.7 PEPTIDE ANTIBIOTICS			
Bacitracin A, B, B_1, B_2, C, D, E, F, F_1, F_2, G, X	RP	µBondapak C_{18}, µBondapak C_{18}/Corasil	689,690
Cerexin A, B_1–B_4, C, D_1–D_4	RP	Nucleosil C_{18}	638,639
Circulin A, B	RP	Hitachi Gel 3011	355
Colistin	RP	Nucleosil C_{18}	355
Colistin A, B, C	RP	Amberlite XAD-2, Hitachi Gel 3010, 3011, LiChrosorb RP 8	326,355
Colistin (Polymyxin E)	RP	µBondapak C_{18}	689
Cyclosporin A in plasma or serum; in urine	RP	RP 8	412,597
Gramicidin	HI	LiChrosorb RP-8	492
Gramicidin linear Val-A, Ile-A, B, Val-C, Ile-C	RP	Bondapak Phenyl/Porasil B, µBondapak Phenyl	363
Gramicidin Val-A, Ile-A, Val-B, Ile-B, Val-C, Ile-C	RP	Zorbax ODS	23
Nikkomycins (nucleoside-peptide antibiotics produced by *Streptomyces tendae*)	RP	LiChrosorb RP-8	164
Octapeptin A_1–A_4, B_1–B_4, C_1	RP	Nucleosil C_{18}	638,676
Polymyxin A_1, A_2, B_0, B_1, B_2, D_1, D_2, E_1, E_2, K_1, K_2, M_1, M_2, P_1, P_2	RP	Hitachi Gel 3010, 3011	355
— B_1–B_2			
— B_1–B_3, C_1, C_2, D_1, D_2, E_1 (Colistin A), E_2 (Colistin B), F_1–F_3, M_1, M_2, T_1, T_2	RP	LiChrosorb RP-8, µBondapak C_{18}	326,689
Tridecapaptin A_α, A_β, B_α–B_δ, C_{α_1}, C_{α_2}, C_{β_1}	RP	Nucleosil C_{18}	676
Tyrocidins	RP	Nucleosil C_{18}	638,639
	RP	Zorbax ODS	23
14.2.8 PEPTIDE ENZYME INHIBITORS			
Glyoxalase I inhibitors	SE	µBondagel E-125 + E-300	172
Renin inhibitor (octapeptide)	RP	Nucleosil C_{18}, µBondapak C_{18}	489, 596

Trypsin inhibitor (pancreatic, bovine) and [^{14}C]-methylated derivative	RP	Aquapore RP 300, LiChrosorb RP-18	741
— —, potato, crude extract	BA	Chymotrypsin-Spheron 300	72
— —, soybean	SE	SynChropak GPC-100	214,562
— —	SE	TSK Gels SW	317,579,694
— —	SE	I-125 PC	272,317
— —	IE	CM-Glycophase/CPG	104,564
— —	IE	Mono Q, SAX	269,568
— —	IP-RP	Nucleosil CN	20

14.2.9 PEPTIDE (PROTEIN) FRAGMENTS PREPARED BY CHEMICAL CLEAVAGE

14.2.9.1 Cyanogen bromide digest fragments

Aminopeptidase *Aeromonas*, C-terminal fragment	RP	LiChrosorb RP-18	738
Collagen I, II, III	RP	C_{18} Vydac TP 201	705
— — bovine	RP	LiChrosorb C_{18}	61
Factor B of the complement system	IP-RP	μBondapak C_{18}	291
Globin fetal, human	RP	LiChrosorb RP-8	435
— —, α-chain (44- and 65-amino acid residues)	RP	Vydac TP and six other silicas coated with SiMe$_2$ClC_8	538
— —, α-chain (fragments 1-32, 33-76, 77-141) and γ-chain (fragments 1-55, 56-133, 134-146)	RP	LiChrosorb C_8, Silica 2362 C_8 or C_{18}, Zorbax CN and C_{18}	539
Haemaglutinin of influenza virus (of bromelain released fragment and of the component polypeptide chains BHA$_1$ and BHA$_2$)	SE	I-125 PC	724
Haemoglobin *Dicrocoelium*, C-terminal fragment	RP	LiChrosorb RP-18	738
Haemoglobin human, α-chain, aminoethylated	RP	LiChrosorb RP-8 or RP-18	300,738
— —, A$_γ$, β, γ-chains	RP	LiChrosorb RP-8	277,435
— —, A$_γ$ and G$_γ$ globin chains	RP	μBondapak C_{18}	112
Haemoglobin human fetal, A$_γ$ and G$_γ$ globin chains (carboxy-terminal fragments) from an individual with A$_γ$-hereditary persistent hemoglobin heterozygosity)	RP	μBondapak C_{18}	655
Haemoglobin HbF thalassemic	IP-RP	μBondapak C_{18}	266
Histocompatibility antigens, H-2^a heavy chains	SE	I-125 PC	11
Lysozyme	SE	Glycophase/CPG	289
Mollusc globin	RP	LiChrosorb RP-18	301
M-structural protein from influenza virus	IP-RP	μBondapak C_{18}	291

Table 14.2 (continued)

Substances	Mode of chromatography	Packings	Reference
Myoglobin	IE	CM-Glycophase/CPG	289
—	RP	RP-18	70
—	SE	Glycophase/CPG	289
Pancreatic polypeptide, human (HPP)	RP	LiChrosorb RP-18	119
Parvalbumin rat (search for conditions of fragmentation)	RP	LiChrosorb RP-18	737
Serum albumin (separation of two fragments 299-329 and 124-298, in the case of the second fragment succinylated and non-substituted form were compared)	IE	CM-Spheron 300	467,470,475
Vasoactive intestinal peptide (VIP)	RP	LiChrosorb RP-18	119
14.2.9.2 N-Bromosuccinimide digest fragments			
(Apo)cytochrome b$_5$, rabbit liver	RP	μBondapak C$_{18}$	667
14.2.9.3 Peptides prepared using mild acid cleavage			
P 27 Avian oncovirus structural protein	IP-RP	μBondapak C$_{18}$	291
14.2.10 PEPTIDE (PROTEIN) FRAGMENTS PREPARED BY ENZYMIC CLEAVAGE			
(Proteinase used for cleavage is given in brackets)			
14.2.10.1 Human haemoglobin (Hb) variants (all cleaved by trypsin)			
Hb A and 10 abnormal Hb; Hb (normal) AEβ and AEδ; Hb oxidized (core)	RP	Ultrasphere ODS, Zorbax C$_8$, Zorbax CN, Zorbax ODS, Zorbax TMS, μBondapak C$_{18}$	622
Hb abnormal α-chain, α-thalassemia-2, one tryptic fragment cleaved by thermolysin	RP	μBondapak C$_{18}$	759
Hb abnormal α- and β-chains in 37 Chinese families, β-chains were aminoethylated (in addition to trypsin also chymotrypsin was used and some tryptic fragments were cleaved by thermolysin)	RP	μBondapak C$_{18}$	760
Hb A, C (β6 Gly→Lys) and J (β69 Gly→Asp)	IP-RP	μBondapak C$_{18}$	266

Hb A, F, A$_2$, normal α-, β-(aminoethylated), γ-, δ-chains and abnormal chains of 25 Hb-variants from persons suffering with haemoglobinopathies	RP	μBondapak C$_{18}$	736
Hb Cambridge	IP-RP	μBondapak Alkyl-Phenyl	265
Hb Chiapas α$_2$ (114 Pro→Arg)	RP	n.d.	308
Hb E	RP	μBondapak C$_{18}$	621
Hb Pasadena (β75 (E 19) Leu→Arg), non-modified and oxidized globin	RP	n.d.	318
Hb variants	RP	μBondapak C$_{18}$	58

14.2.10.2 *Various proteins*

Acid brain-specific S-100 protein (pepsin)	IE	DC-4A	517
Actin, carboxymethylated (pepsin)	IE	W-3 IER	319
Actinidin (trypsin)	IP-RP	μBondapak f.a.	233
Acyl carrier protein (thermolysin)	IP-RP	μBondapak f.a.	233
— —, both native and semisynthetic (thermolysin), peptide mapping	IP-RP	μBondapak f.a.	552
Alloantigen IA, murine, [^{3}H]- and [^{14}C]-labelled β-chain (trypsin)	RP	RC-C$_{18}$ (radial)	234
Amyloid protein NIG-51 (trypsin)	IE	Diaion CDR-10	671
(Val-4)-Angiotensin III (dipeptidylaminopeptidase)	IP-RP	μBondapak C$_{18}$	417
Apolipoprotein B (trypsin)	RP	RC-C$_{18}$ (radial)	234
Apomyoglobin, sperm-whale (trypsin)	RP	μBondapak C$_{18}$	100
Arabinose binding protein, normal and defective mutant (trypsin), peptide mapping	RP	LiChrosorb C$_8$	277
Bence-Jones protein "Den", isolated from urine of a plasmozytom-patient (trypsin), one of the tryptic fragments T$_{18}$ (chymotrypsin)	RP	ODS/Hypersil	753
Bence-Jones protein "Wes", monoclonal L-chain of κ-Type, subgroup I (trypsin): 18 species isolated in pure form for sequencing from 22 resulting peptides	RP	Zorbax C-8	370
Brain proteins, bovine, S-100b protein and 14-3-2 protein (trypsin)	IE	Diaion CDR-10	671
Calcium-binding proteins, bovine intestinal and chick intestinal (trypsin)	RP	μBondapak C$_{18}$	182
Calcium-binding protein, bovine intestinal, N-bromosuccinimid cleaved (trypsin)	RP	μBondapak C$_{18}$	183

Table 14.2 (continued)

Substances	Mode of chromato-graphy	Packings	References
Calmodulin, bovine brain (trypsin)	IE	Diaion CDR-10	671
Calmodulin (trypsin)	IE, MM	MicroPak AX-10	146
— , peptides 1-77, 78-148, 1-90, 1-106, 107-148	RP	μBondapak Phenyl	531
Casein α_1; β_1, β_2, rat, phosphorylated a dephosphorylated (trypsin)	RP	μBondapak f.a.	286
Collagen, human, I, II, III (clostridiopeptidase A), mapping, "fingerprints"	IE	DC-4A	706
Corticotropin-like intermediary lobe peptide (CLIP), Ser31-phosphorylated form and non-phosphorylated form (trypsin and pepsin)	RP	C18 Sep-Paks, μBondapak C18	84
(Apo)Cytochrome b5, rabbit liver (trypsin, S. aureus V8 protease)	RP	μBondapak C18	667
Cytochrome c, horse heart (trypsin)	IE, MM	MicroPak AX-10	146
—	RP	Ultrasphere RP-8, Zorbax ODS, μBondapak Phenyl	757
Cytochrome P-450, pig testes, [14C]-carboxymethylated (trypsin)	RP	μBondapak Phenyl	757
β-Endorphin, camel and rat (trypsin)	RP	LiChrosorb RP-18	586
— , human (trypsin)	RP	LiChrosorb RP-18	119
Fibronectin, human plasma (trypsin)	RP	μBondapak Phenyl	757
Globin β-chain, baboon (trypsin and thermolysin)	RP	Ultrasphere ODS, μBondapak C18	635
Globin, human (trypsin), soluble peptides	IE	DA and DC resins	76
Glucan cellobiohydrolase, Trichoderma viride-reesei (trypsin)	IE	Mono Q	568
Haemagglutinin glycoprotein, influenza virus (trypsin), [3H]-glucosamine and [14C]-glucosamine labelled glycopeptides of various virus strains, mapping	RP	μBondapak C18	38,39
Haemagglutinin, influenza virus, component glycopeptide from the BHA1 chain (Staphylococcus aureus V8-protease)	RP	Zorbax ODS	724
— , component polypeptide chains BHA1 and BHA2 (bromelain)	SE	I-125 PC	724

— —, component polypeptide chain BHA$_2$ of bromelain- released fragment (trypsin)	RP	Zorbax ODS	724
Haemoglobin (trypsin)	IE	SSC	51
Haemoglobin A (trypsin)	RP	Ultrasphere ODS, Zorbax C8, Zorbax CN, Zorbax ODS,	620
— —	IP-RP	μBondapak C$_{18}$	229
— —	RP	μBondapak C$_{18}$	458
Haemoglobin human α-chain, citraconylated and aminoethylated (timed trypsin digest)	RP	ODS; LiChrosorb RP-8 or RP-18	300
— —, oxidized core (chymotrypsin)	RP	μBondapak C$_{18}$	736
Haemoglobin E, α- and β-chains (trypsin)	RP	μBondapak C$_{18}$	621
Haemoglobin goldfish α-chain (trypsin); β-chain, amino- ethylated and maleylated (trypsin)	RP	LiChrosorb RP-8 or RP-18	300
Immunoglobulin G, human, λ-chain, carboxymethylated (trypsin)	RP	RC-C$_{18}$ (radial)	234
Immunoglobulin M, heavy chain (trypsin)	IP-RP	μBondapak C$_{18}$	291
Insulin A and B chains (staphylococcal proteinase)	RP	μBondapak C$_{18}$	100
Insulin B chain, oxidized (trypsin or chymotrypsin)	RP	μBondapak C$_{18}$	678
β-Lactoglobulin (chymotrypsin)	SE	TSK Gel 3000SW	181
Lysozyme, hen egg (trypsin)	HI	Spheron 300	658
— —	IE	SSC	51
— —	IP-RP	μBondapak C$_{18}$	291
— —	RP	Zorbax C$_8$, Zorbax ODS, Zorbax CN	724
— —	SE	I-125 PC	724
— —, oxidized (trypsin)	RP	RP-18	70
— —, reduced and alkylated (trypsin)	IE, MM	Micro-Pak AX-10	146
α-Melanocyte-stimulating hormone (α-MSH), mono-N-acetyl form, N,O-diacetyl form; γ$_3$-melanotropin (trypsin and chymotrypsin)	RP	C$_{18}$ Sep-Pak, μBondapak C$_{18}$	84,85
Mollusc globin cyanogen bromide fragment (trypsin)	RP	LiChrosorb RP 18	301
Myelin basic protein (trypsin)	RP	μBondapak C$_{18}$	574
— —, oxidized (trypsin)	RP	RP-18	70
Myoglobin (trypsin)	IE	CM-Glycophase/CPG	289
— —	RP	μBondapak C$_{18}$	289
— —, citraconylated (trypsin)	SE	I-125 PC	724

Table 14.2 (continued)

Substances	Mode of chromatography	Packings	References
Neurophysins, bovine and rat, Pro-neurophysin I = Pro-pre-sophysin and Pro-neurophysin II = Pro-oxyphysin, oxidized (trypsin)	MM, RP	Zorbax CN	3,98
Ovalbumin (trypsin)	RP	LiChrosorb RP-8	586
—, oxidized (trypsin)	RP	RP-18	70
Parvalbumin, chicken (trypsin), mapping and fragmentation	RP	LiChrosorb RP-18	737
—, rat muscle (trypsin), 11 peptides containing from 3 to 15 residues	RP	LiChrosorb RP-18	741
Phosphofructokinase, sheep heart (trypsin), mapping	IP-RP	μBondapak C_{18}	229
Phosphogluckinase, sheep heart (trypsin), mapping	IP-RP	μBondapak C_{18}	266
Plasma amyloid P-component (trypsin), 28 peptides	IP-RP	μBondapak C_{18}	14
Prolyl hydroxylase small subunit (trypsin); cross-reacting protein immunologically related to prolyl hydroxylase (trypsin)	RP	LiChrosorb RP-18	586
Pro-opiocortin, camel pituitary (trypsin)	RP	LiChrosorb RP-18	353,590
Protein A, Streptococcus aureus (trypsin)	IE	Mono S	568
Rauscher murine leukemia virus, p^{30} protein (endoprotease Lys-C)	RP	μBondapak C_{18}	274
Relaxin β-chain, porcine (trypsin), N-terminal octapeptide	RP	μBondapak C_{18}	623
Ribonuclease A, carboxymethylated (chymotrypsin)	IE	W-3 IER	319
Ribonuclease (pepsin and trypsin)	IE	DC-4A	517
— oxidized (trypsin)	IE	SSC	43,51
Ribonuclease S peptide (dipeptidylaminopeptidase)	IP-RP	μBondapak C_{18}	417
— (trypsin, pepsin)	HI, RP	LiChrosorb RP-8 and RP-18	492, 493
Serum albumin, bovine, methylated and oxidized (trypsin)	RP	RP-18	70
Thyroglobulin, human (trypsin)	RP	ODS	269
Thyroglobulin 19-S, sheep (trypsin)	IP-RP	μBondapak C_{18}	264
Thyrotropin, bovine (trypsin)	IP-RP	μBondapak C_{18}	264
Trypsin inhibitor, basic, pancreatic, [^{3}H]-carboxy-methylated (trypsin)	RP	LiChrosorb RP-18	738
Tooth pulp extract, before and after treatment with proteinases	RP	μBondapak C_{18}	133

Compound	Mode	Column	Ref.
Urotensin I, carp (trypsin)	RP	LiChrosorb RP-8	496
Viral proteins: (a) influenza A haemagglutinin glycoproteins (trypsin) polypeptides; (b) mink cell focus-inducing murine leukemia virus, envelope precursor proteins, [³H]-mannose labeled (trypsin)	IP-RP	LiChrosorb RP-18, μBondapak C_{18}	346

14.2.11 PEPTIDE (PROTEIN) HORMONES AND OTHER BIOGENIC PEPTIDES

14.2.11.1 Adrenocorticotropic hormone ACTH (corticotropin, melanotropin and allied peptides)

(i) ACTH, fragments and analogues

Compound	Mode	Column	Ref.
ACTH I, ACTH II, ACTH (1-39) calf, human and rat, ACTH (1-38), calf, ACTH (1-24), ACTH (1-18) analogue, ACTH (1-39) and ACTH (18-39), phosphorylated and nonphosphorylated	RP	Sep-Pak C_{18}, μBondapak C_{18}	45-49,84,86
— derivatives (5 analogues), ACTH human and porcine (21 analogues)	HI, RP	Nucleosil C_{18}	196,675
— human (1-18), (1-24), (1-39), (4-19), (5-10), (11-39), (12-15), (17-39), (18-39); ACTH porcine (1-39); ACTH sheep	IP-RP, RP	LiChrosorb RP-18, μBondapak C_{18}	56,90,257, 452,455,608, 629
— human and porcine (1-39) [^{125}I]-labeled; ACTH human (1-39); ACTH (1-13), (1-18), (1-24), (4-10), (5-10), (18-39), (34-39)	RP	Hypersil ODS, Sep-Pak C_{18}	514,515,520, 610
— [^{125}I]-iodinated or chloramine T-treated polypeptides	RP	μBondapak C_{18}	629
— (1-24) pentaacetate	IP-RP	μBondapak C_{18}/Corasil	229,232
—/β-LPH precursor	RP	Sep-Pak C_{18}, μBondapak C_{18}	48
— synthetic decapeptide with natural sequence of ACTH; ACTH (1-24), monitoring the coupling reaction in the solid phase synthesis	IE	DC-resin, H-Gel 3000, -a or -s	41,670

(ii) CLIP (corticotropin-like intermediary lobe peptide)

Compound	Mode	Column	Ref.
ACTH (18-39) calf, human, rat (CLIP$_1$, CLIP$_2$), Ser^{31}-phosphorylated and nonphosphorylated forms	RP	Sep-Pak C_{18}, μBondapak C_{18}	48,49,84,86, 257
CLIP, ACTH (18-39) human	RP	Hypersil ODS	515
— rat, ACTH (18-38), (18-39), various forms (nonglycosylated, nonphosphorylated, glycosylated, phosphorylated and phospho-glycosylated	RP	ODS	83

(iii) Gastrins

Compound	Mode	Column	Ref.
Gastrin (12-15)	IP-RP, RP	Bio-Rad ODS, LiChrosorb RP-18	455,456
— (ACTH 4-10)	IP-RP	LiChrosorb RP-18	608

Table 14.2 (continued)

Substances	Mode of chromatography	Packings	References
Gastrin I, human (17 amino acid residues); [^{125}I]-labeled human gastrin	HI, RP	Hypersil ODS, PRP 1, Sep-Pak C_{18}	520,571,610
Gastrin releasing peptide analogues, 5 species (1-13, 14-27, 18-27, 19-27, Ac-Asn 19-19-27)	RP	Supelcosil LC-18, Ultrasphere ODS	434
(iv) MSH (melanotropin, melanocyte stimulating hormone)			
α-MSH (ACTH 1-13)	RP	Sep-Pak C_{18}, µBondapak C_{18}	46,47,49,84, 90,257,452, 629
—	HI, IP-RP, RP	LiChrosorb RP-8, RP-18	455,492,493, 562,608
—	RP	B-R ODS	455,456
—	RP	Cosmosil C_{18}, Hypersil ODS	427,515,520
—	SE	TSK Gel 2000 SW	427
— desacetyl, monoacetyl, diacetyl	RP	Sep-Pak C_{18}, µBondapak C_{18}	49
β-MSH fragment, both unreacted with Bolton-Hunter reagent and the modified peptide	HI	LiChrosorb RP-8	492
β-MSH procine	RP	µBondapak C_{18}	629
γ-MSH	RP	Sep-Pak C_{18}, µBondapak C_{18}	49,85
— bovine	RP	µBondapak C_{18}	629
MSH, [^{125}I]-iodinated or chloramine T-treated polypeptides	RP	µBondapak C_{18}	629
(v) Various ACTH-related substances			
Corticotropin-immunoreactive compounds in human plasma, cerebrospinal fluid and urine (chromatography profiles)	IP-RP	LiChrosorb RP-18, ODS	608,609
Memory peptide (ACTH 4-10)	RP	Hypersil ODS	515
Synacthen (ACTH 1-24)	HI, RP	Hypersil ODS	515,520
—	RP	Sep-Pak C_{18}, µBondapak C_{18}	46,47
Tetracosactide	IP-RP	Cosmosil C_{18}	427
—	SE	TSK Gel 2000 SW	427

14.2.11.2 Angiotensins, angiotensinogen and saralasin

Description	Method	Column/support	Ref.
Angiotensins	IP-RP	Bondapak C18/Corasil, μBondapak, μBondapak C18 Phenyl/Porasil B, μBondapak f.a.	255
—; [125I]-iodinated or chloramine T-treated polypeptides	RP	μBondapak C18	629
— (5 species)	HI, RP	LiChrosorb RP-8, Radial Pak A/C18	262,327,492
—, synthetic	IE	Partisil SCX	555
—, synthetic fragments VIHPF, VVIHPF, YIHPF	IP-RP	μBondapak f.a.	231,232
(Segments are indicated in one-letter symbols of amino acids)			
Angiotensin I, DABTC- (dimethylaminoazobenzenthiocarbamoyl-) derivative	RP	μBondapak C18	100
—; I, separation of impurities	RP	ODS	440
—; I and II; [Asp1,Ile5]-angiotensins; tetradecapeptide (artificial renin substrate generating angiotensin I) and enzymically released tetrapeptide LVYS	RP	μBondapak C18	257,258,629, 681
—; I and II; [Ile5]-angiotensin II, [Ile5]-angiotensin II, [Val5]-angiotensin II, des[Asp1],[Ile5]-angiotensin II, separation of angiotensins from impurities; angiotensin I D-diastereoisomers (5 species), angiotensin II (truncated forms, 4 species), α- and β [Asp1, Val5]-angiotensin I	IP-RP, RP	LiChrosorb RP-18, ODS	361,441-443, 608
—; I, II, III	SE	TSK Gel 2000 SW	427
—; I, II, III and dipeptide HL split by angiotensin-converting enzyme (E.C. 3.4.15.1, rat lung dipeptidyl carboxypeptidase); analogues of angiotensins I, II, III, 12 species	MM, RP	MicroPak AX-10, Partisil ODS	26,145,146
—; I, II, III; iodinated derivatives, des[Asp1],[Ile18]-angiotensin II; two analogues of angiotensin III ([D-Pro6]- and [ThioPro6]-derivatives)	RP	Bio-Sil ODS, LiChrosorb RP-8, μBondapak Alkylphenyl	217,289,496
— II	HI, IP-RP, RP	Cosmosil C18, Hypersil ODS	427,520
— II	RP	B-R ODS, LiChrosorb RP-8, RP-18	455,496
— II	RP	μBondapak C18	90,130
Angiotensinogen (equine), protein N-terminal tetradecapeptide	IP-RP	ODS	361
Saralasin ([Sarcosine1, Ala8]-angiotensin II)	RP	Bio-Sil ODS	217

14.2.11.3 Bombesin, eledoisin, allied peptides and analogues

Description	Method	Column/support	Ref.
Bombesin	HI, RP	Hypersil ODS	520
—	HI, RP	PRP 1	571

Table 14.2 (continued)

Substances	Mode of chromatography	Packings	References
— analogues (20 species)	RP	MicroPak MCH-5 C_{18}	728
— nonmodified and diastereomer [D-His^{12}]	RP	Supelcosil LC-18, Ultrasphere ODS	434
Eledoisin-related peptide; peptide KFIGLM amide	RP	Bondapak C_{18},	575
	RP	B-R ODS, LiChrosorb RP-18, μBondapak C_{18}	130, 455
Eledoisin, separation of impurities	RP	ODS	441,442
Litorin	RP	Supelcosil LC-18, Ultrasphere ODS	434
Physalaemin	RP	B-R ODS, LiChrosorb RP-18	455
Ranatensin (undecapeptide from the skin of the frog *Rana pipens*)	RP	MicroPak MCH-10 C_{18}, -MCH-N-Cap C_{18}	726,728
14.2.11.4 Bradykinin and analogues			
Bradykinin	SE	TSK-Gel 2000 SW	427
—	RP	B-R ODS	455,458
—	RP	Cosmosil C_{18}, LiChrosorb RP-18	427,455
—	RP	μBondapak C_{18}	130,132,136
—, [Phe^8]-bradykinin, [Tyr^8]-bradykinin	RP	LiChrosorb RP-18, μBondapak C_{18}	629,737
—, bradykinin potentiator C, Lys-bradykinin, Met,Lys-bradykinin	MM	MicroPak AX-10	146
14.2.11.5 Calcitonins			
Calcitonin human	RP	ODS, Sep-Pak C_{18}, μBondapak C_{18}	46,47,717
— human lung tumor cell culture medium; human medullary thyroid carcinoma tissue culture; porcine; salmon	HI	Hypersil ODS, Partisil ODS	513
— human, salmon	HI, RP	Hypersil ODS	520
— human synthetic and rat, nonlabeled and [^{125}I]-labeled	IP-RP	μBondapak C_{18}	391
14.2.11.6 Factors inhibiting release of hormones (somatostatin and other inhibitory factors)			
C-LHIH, luteinizing hormone inhibiting factor (hormone) from porcine hypothalami	RP	Phenyl/Corasil, Poragel P-T	99

Compound / description	Mode	Column / packing	Ref.
D-GHIH, release of growth hormone inhibiting factor (hormone) from porcine hypothalami		Phenyl Porasil/Corasil, Poragel P-T	208
Factor inhibiting melanocyte-releasing hormone and metabolites of the factor, five N-alkylaminonaphthalene-sulfonyl derivatives	RP	µBondapak C_{18}, µBondapak Phenyl	302
Hypothalamic extract	HI	Hypersil APS, Hypersil ODS	513
MIF, factor inhibiting melanocyte stimulating hormone (natural) and synthetic analogue peptide (L-Pro-N-methyl-D-Leu-GlyNH$_2$) after derivatization with fluorescent 7-chloro-4-nitrobenzyl-2-oxa-1,3-diazole	RP	Spherosil ODS	373
PIF (or PIH), prolactin inhibiting factor (hormone) from porcine hypothalami	n.d.	Phenyl Porasil/Corasil, Poragel, Poragel PN, Poragel PS, Poragel P-T	208, 209

(i) Somatostatin and analogues

Compound / description	Mode	Column / packing	Ref.
Somatostatin	HI, IP-RP, RP	Cosmosil C_{18}, Hypersil ODS	427,520
—	IP-RP, RP	B-R ODS, LiChrosorb RP-8 and RP-18	5,455,456
—	MM	MicroPak AX-10	146
—	MM, SE	Glycophase G/CPG	165
—	RP	MicroPak-MCH C_{18}, -MCH-N-Cap C_{18}	728
—	RP	µBondapak C_{18}	46,130,257, 452,746
—	RP	µBondapak C_{18}, Sep-Pak C_{18}	47
—	RP	µCN	574
—	SE	TSK Gel 2000 SW	427
— and analogues	RP	µBondapak C_{18}	90
—, [D-Cys^{14}]-, [D-Trp^8]-somatostatin	RP	µBondapak C_{18}	575
— commercial, separation of impurities	RP	ODS	441,442
—, di-S-acetamidomethyldihydrosomatostatin	RP	PrepPak C_{18}	688
—, 6 highly active analogues: D$_8$ and L-5F-Trp^8-, D- and L-6F-Trp^8- and D- and L-5Br-Trp^8-derivatives; separation of diastereoisomers	RP	µBondapak $_{18}$	465
— synthetic; pancreatic small somatostatin from pigeon	RP	µBondapak C_{18}, µBondapak CN	577

Table 14.2 (continued)

Substances	Mode of chromatography	Packings	References
14.2.11.7 Glucagon			
Glucagon	IP-RP	Bondapak C$_{18}$/Corasil, Cosmosil C$_{18}$	232,427
—	IP-RP, RP	µBondapak C$_{18}$	56,90
—	MM, RP	MicroPak AX-10, MicroPak MCH 5	146,728
—	MM, SE	CPG, Glycophase G/CPG	165,178
—	RP	B-R ODS	455,456
—	RP	Hypersil ODS, Polygosil C$_{18}$	520,532
—	RP	LiChrosorb RP-18	56,455,608
—	SE	I-125 PC, TSK Gel 2000 SW	317,427
—, DABTC (dimethylaminoazobenzenthiocarbamoyl) derivative	RP	µBondapak C$_{18}$	100
—, porcine; glucagon and its radio-labeled oxidized and iodinated derivatives (tracers)	RP	µBondapak C$_{18}$	258,444
—, glucagon monocomponent	HI, RP	PRP 1, µCN	40,571
14.2.11.8 Insulin and allied proteins			
Insulin	IM	BE CB6	701
—	IP-RP	Nucleosil CN, µCN	19,574
—	MM, SE	Glycophase G/CPG	165
—	RP	MicroPak MCH C$_{18}$	728
—	SE	Fractogel TSK HW 55F, I-125 PC	216,317
—	SE	PEG/CPG	125
—	SE	SynChropak GPC-100	214
— and its radiolabeled iodinated derivatives (tracers)	RP	µBondapak C$_{18}$	444
— bovine, human, porcine; [Thr^{B30}]-bovine, synthetic	HI, RP	Nucleosil C$_{18}$	196,490,675
—, ovine, porcine, Novo	IP-RP, RP	Sep-Pak C$_{18}$, µBondapak C$_{18}$	46,47,56,229, 232,257,258, 266,273,274, 596
—, porcine; desamidoinsulin bovine, porcine	RP	Nucleosil C$_{18}$	665

Compound	Method	Stationary phase	References
— human, porcine	HI, RP	PRP 1, Radial-Pak A/C$_{18}$ Vydac TP coated	262,538,571
— (denatured), α-monocarbamoylated, β-monocarbamoylated, dicarbamoylated and tricarbamoylated insulin	IE	Polyanion SI HR 5/5	568
— in SDS (sodium dodecylsulphate), in 6 M G-HCl (guanidinium chloride); insulin porcine	SE	TSK Gels SW	337,339,427, 579
— in the presence of SDS	IE	Partisil SCX	427
—; porcine insulin	IP-RP, RP	B-R ODS, Cosmosil C$_{18}$	427,458
— —	IP-RP	μBondapak Alkylphenyl, μBondapak f.a.	229,232,267, 273,274
— —	RP	LiChrosorb RP-8 and RP-18	56,140,740
— —	RP	Supelcosil C$_{18}$	543
Insulin porcine natural and semisynthetic, insulin human synthetic, insulin derivatives A-21-monodesamido- and B-31- monoarginine insulin	RP	LiChrosorb RP-18, μBondapak C$_{18}$	11,257,258
Insulin A- and B-chains	SE	TSK Gels SW	694
— —, bovine insulin	RP	μBondapak C$_{18}$	100
— —, DABTC derivatives	IE	CM-Spheron, Phospho-Spheron	467,476
— oxidized, A- and B-chains	HI, RP	Diol C$_8$/Silica, Hypersil ODS	518,520
— B-chain fragment (22-27) prepared by solid phase synthesis	IP-RP	μBondapak C$_{18}$	266
— B-chain denatured in 6 M G-HCl, in 0.1% SDS, in 8 M urea	SE	CPG, Glyceryl/CPG	63,178
Insulin-like growth factor (IGF) human	RP	LiChrosorb RP-8	740
— proteins	RP	LiChrosorb RP-8	140
— potential by-products	RP	LiChrosorb RP-8	140
Proinsulin bovine, porcine	HI, IP-RP, RP	Nucleosil C$_{18}$, Supelcosil C$_{18}$, μBondapak C$_{18}$	196,229,266, 307,543,665
— C-peptide bovine, solid phase synthetic fragment (34→45), des-Pro^9-undecapeptide (major deletion product)	IP-RP, RP	LiChrosorb ODS, μBondapak C$_{18}$, Bondapak f.a.	266,267,307

14.2.11.9 Luteinizing hormone releasing factor LH-RF (hormone LH-RH), luliberin, LRF, LRH

Compound	Method	Stationary phase	References
LH-FSH releasing indole decapeptide hormone	RP	Spherisorb ODS	374
LH-RF	IP-RP, RP	B-R ODS, LiChrosorb RP-18	455,456,608
— —	RP	μBondapak C$_{18}$	46,130,452, 574
— —	RP	Phenyl/Corasil, Poragel PN, PS	238
— —	SE	TSK Gel 2000 SW	427
— and analogues	RP	μBondapak C$_{18}$	90

Table 14.2 (continued)

Substances	Mode of chromato-graphy	Packings	References
— and [D-His²]diastereoisomer	RP	Bondapak C18	575
— and 2 analogues ([Gly⁴, Phe³,⁵] and [Gly⁴, Phe⁵])	RP	Nucleosil C18	675
— enzymic digest (by impurities in bovine plasma or serum albumin)	RP	μBondapak C18	452
—, [³H]- and [³H]-Pro⁹-derivatives	RP	μBondapak C18, μBondapak CN, μCN	574,577
14.2.11.10 Lipotropins (LPH)			
ACTH/β-LPH precursor	RP	Sep-Pak C18, μBondapak C18	48
β-Lipotropin and fragments [β-LPH39-45, β-LPH61-65 (enkephalins); β-LPH61-76 (α-endorphin), β-LPH61-91 (β-endorphin)]	RP	Hypersil ODS	515
— human and β-LPH34-39 ovine	HI, RP	Hypersil ODS	520
— β-LPH61-91 fragments (23 species), LPH88-91	RP	Sep-Pak C18, μBondapak C18	48,420,452
— ovine and its pyroGlu-form	RP	Partisil ODS-2	750
— rat	IE	Partisil SCX	589
—; βs-LPH61-69, [Leu⁵]βs-LPH61-69	RP	LiChrosorb RP-18	589,590
γ-Lipotropin	RP	Sep-Pak C18, μBondapak C18	49,452
14.2.11.11 Neuropeptides			
(i) Endorphins			
α-Endorphin	MM, RP	Cosmosil C18, MicroPak AX-10, μBondapak C18	90,146,427
—	RP	B-R ODS, LiChrosorb RP-18, Micro-Pak MCH C18, MicroPak MCH-N-Cap	455,726,728
—	SE	TSK Gel 2000 SW	427
α-, β-, γ-Endorphin	RP	Radial-Pak A/C18	262
α-, β-, γ-, and δ-Endorphin	RP	Sep-Pak C18, μBondapak C18	49
α- and β-Endorphin, ovine	RP	Hypersil ODS	520
β-Endorphin	IP-RP, RP	Hypersil ODS, LiChrosorb RP-8 and RP-18, Sep-Pak C18, μBondapak C18	48,56,197, 452,514,608

Description	Method	Packing	Refs
— camel	RP	μBondapak Phenyl	122
— human	RP	C-8 ES, RP-8 Lobar, μBondapak C_{18}	46,189
— — and $[Leu^5]$-, $[D-Ala^2]$- and $[p-Cl-Phe^4]$-derivatives	RP	LiChrosorb RP-18	119
— —, ovine and porcine	RP	μBondapak C_{18}, μCN	574
— sheep	RP	μBondapak C_{18}	629
—, β-Leu- and β-Met-endorphin	RP	μBondapak C_{18}	257
Endorphin analogues (7 species); endorphins (α-, β-)	RP	Hypersil ODS, μBondapak Phenyl	122,515
Endorphin and related peptides (α-, des-Tyr-α-, β-, des-Tyr-β-, γ-, des-Tyr-γ-endorphin)	RP	μBondapak C_{18}	420
Endorphins	RP	LiChrosorb RP-18	353,733

(ii) Enkephalins

Description	Method	Packing	Refs
Enkephalin analogues (14 species)	RP	μBondapak Phenyl	122
— peptides (5 species)	RP	Amberlite XAD-4	548
— (endogenous), in an extract of guinea pig striatum after the reaction with 2-methoxy-2,4-diphenyl-3(2H)-furanone; the fluorophors are chromatographed	RP	LiChrosorb RP-18	733
$[^3H]$-Leu-enkephalin	HI	Partisil 10 ODS	513
Leu-enkephalin	IP-RP, RP	Cosmosil C_{18}, μBondapak Alkyl-phenyl	427,567
— amide (O-benzyl derivative)	IP-RP	Sep-Pak C_{18}, μBondapak Alkyl-phenyl, μBondapak Phenyl/Porasil B	229,266
—, produced by solid phase synthesis	IP-RP	Phenyl/Porasil B, μBondapak C_{18}, μBondapak C_{18} Corasil, μBondapak f.a.	255
— and Ala^2-Leu-enkephalin	RP	μBondapak C_{18}	134,135
— (in the presence of SDS)	IE	Partisil SCX	427
—, N-acetyl-N,O,S-permethylated derivative for mass spectra detection	RP	Supelcosil LC 8	756
— protected peptides used for the protease-catalyzed synthesis: Boc-Gly-, Boc-Tyr(Bzl)-, -Gly-N_2H_2Ph, -Leu-N_2H_2Ph, -Phe-N_2H_2Ph, -Phe-OEt-, -Tyr-N_2H_2Ph	NP	LiChroprep Si 60	382
Met-enkephalin	RP	B-R ODS, Hypersil ODS, Sep-Pak C_{18}, μBondapak C_{18}	47,458,514
—, $[D-Ala^2]$-derivative	IE, MM	MicroPak AX	146
—	IE, MM	MicroPak AX	743
— in rat and mouse brain tissue after decapitation or microwave irradiation	RP	Yanapak ODS-T	502

Table 14.2 (continued)

Substances	Mode of chromatography	Packings	References
Leu- and Met-enkephalins	HI, RP	Hypersil ODS, LiChrosorb RP-18, Yanapak ODS-T	502,515,520 608
——	IE	SCX	71,457
——	RP	B-R ODS, LiChrosorb RP-18	455,456
——	RP	μBondapak C_{18}	46,130,132, 136,257
——	RP	PrepPak-500 C_{18}, μBondapak C_{18}, μBondapak Phenyl, μBondapak Phenylalkyl	60,273,274
——, separation of impurities	SE	TSK Gel 2000 SW	427
—— and Ala^2-Leu-enkephalin (internal standard)	RP	ODS	441,442
—— and [D-Ala^2]-Met-enkephalin	RP	μBondapak C_{18}	673,748
	RP	MicroPak MCH C_{18}, MicroPak MCH-N-Cap, μBondapak C_{18}	726,728,745
—— DABTC derivatives	RP	μBondapak C_{18}	100
——, [^{125}I]-iodinated and chloramine T-treated polypeptides	RP	μBondapak C_{18}	629
—— and des-Tyr-(Met-enkephalin)	RP	μBondapak C_{18}	420
(iii) Neuropeptides in general			
Brain neuropeptides in canine thalamus tissue, extract, in canine spinal cord tissue	RP	μBondapak C_{18}	134,135
Casomorphin (β-), β-casomorphin (1-5)	IP-RP	Cosmosil C_{18}	427
	SE	TSK Gel 2000 SW	427
Delta sleep inducing peptide DSIP	MM	MicroPak AX	146
Dynorphin	RP	MicroPak AX, μBondapak C_{18}	132,136
Dynorphins [D- and L-Arg^6]	RP	LiChrosorb RP-18	737
Neuropeptides, mass spectrometry after HPLC	RP	μBondapak C_{18}	500
Opioid peptides, synthetic	RP	Ultrasphere RP-8	757
Opiate receptor binding material	IE	SCX	71
Peptide fraction from canine tooth pulp, from caudate nucleus tissue and from spinal cord tissue extract	RP	μBondapak C_{18}	132,133,136

Peptides in tooth pulp extract	RP	μBondapak C$_{18}$	448
— — (dog) and in caudate nuclei and hypothalami from dog brains	RP	μBondapak C$_{18}$	673,748
Pro-opiocortin and tryptic digest peptides (with opioid activity)	RP	LiChrosorb RP-18	590
Pro-opiocortin-protein (camel pituitary)	RP	EM LiChrosorb RP-8	353

(iv) Substance P

Substance P	IE	Partisil SCX	555
	IP-RP	B-R ODS, Cosmosil C$_{18}$	427,458
	RP	ODS	441,442
	RP	μBondapak C$_{18}$	130,132,136, 257,452
—, fragment P4-11 [^{3}H]-labeled	SE	TSK Gel 2000 SW	427
— non-labeled and [^{3}H]-labeled	HI, RP	Hypersil ODS	520
— sulfoxide and analogues [11-Met]- and [8-Tyr]-	RP	Hypersil ODS, μBondapak C$_{18}$	167,513
	RP	μBondapak C$_{18}$	167,650

14.2.11.12 Neurotensin

Neurotensin	HI, IP-RP, RP	Cosmosil C$_{18}$, Hypersil ODS, μBondapak C$_{18}$	90,130,427, 520
—	HI, MM	MicroPak AX 10, PRP 1	146,571
—	IP-RP, RP	B-R ODS, LiChrosorb RP-18	455,456,458, 608
—, DABTC derivative	SE	TSK Gel 2000 SW	427
—	RP	μBondapak C$_{18}$	100

14.2.11.13 Nonapeptides (pituitary hormones and analogues)

(i) Ornipressin

Ornipressin and other nonapeptides	AD, MM	Silica gel SI 60 RP-8	176
— — —	RP	RP-8	176,177
— — —	RP	Nucleosil C$_8$ and C$_{18}$, RP-8, Spherisorb S5 ODS	377,378

Table 14.2 (continued)

Substances	Mode of chromato-graphy	Packings	References
(ii) Oxytocin and analogues			
Oxytocin (-SH)	HI	Nucleosil C$_{18}$	196
Oxytocin (-SS-) and analogues (17 species), diastereoisomers (20 species)	RP	Separon SI-C-18	399,402
— and diastereoisomers [2-D-Tyr], [4-D-Gln], [2-D-Tyr-4-D-Gln]	RP	LiChrosorb RP-8, OD-silanized c.	379
— and 7 diastereoisomers [1-hemi-D-Cys]-, [2-D-Tyr]-, [4-D-Gln]-, [5-D-Asn]-, [6-hemi-D-Cys]-, [7-D-Pro]-, [8-D-Leu]- and specifically labeled diastereoisomers [1-hemi-D-(α-^{2}H)-Cys]-, [6-hemi-DL-(α-^{2}H)-Cys]-, [2-DL-(α-^{13}C)-Tyr]-, [8-DL-(2-^{13}C)-Leu]-, [3-DL-(2-^{13}C)-Leu]-, [8-D-(2-^{13}C)-Leu]-, [8-L-(2-^{13}C)-Leu]-, [3-D-(2-^{13}C)-Leu]- and [3-L-(2-^{13}C)-Leu]-oxytocin	RP	Partisil ODS, µBondapak C$_{18}$	394,395,713
— and other peptide hormones	AD, MM	Silica gel Si 60	176
—	HI RP	Nucleosil C$_8$ and C$_{18}$, RP-8, Spherisorb S5 ODS	177,196,377, 378,597
—	IE	Partisil SCX	555
—	IP-RP, RP	Cosmosil C$_{18}$, Hypersil ODS	427,520,728
—	IP-RP, RP	LiChrosorb RP-2, RP-8 and RP-18, Sep-Pak C$_{18}$, µBondapak C$_{18}$	49,56,64, 257,650
—	SE	TSK Gel 2000 SW	427
— and protected or nonprotected peptides formed during the synthesis (9 species)	RP	B-R ODS, LiChrosorb RP-8 and RP-18	455,496,503
— carba-analogues, oxidation products of deamino-1-carba-oxytocin; deamino-6-carba-analogues with modified Phe in p-position (11 species), synthetic intermediates of carba-analogues (12 species), carba-analogues both sulfides and sulfoxides (35 species)	RP	Separon SI-C-18	399-401
—, fluorescent Fluoram derivative and 16 fluorescent derivatives	RP	MicroPak MCH C$_{18}$, MicroPak MCH-N-Cap C$_{18}$, Partisil ODS	419

— in extract of anteria ganglia of marine mollusc *Aplysia californica*	RP	LiChrosorb RP-8	496
—, post-column fluorescence derivatization	RP	RP-8	176
(iii) Various nonapeptides			
Demoxytocin and other nonapeptides	RP	Nucleosil C_8 and C_{18}, RP-8, Spherisorb ODS	377,378
Felypressin and other nonapeptides	RP	Nucleosil C_8 and C_{18}, RP-8, Spherisorb ODS	377
Glumitocin and other nonapeptides	RP	LiChrosorb RP-2 and RP-8, μBondapak C_{18}	64
Isotocin and other nonapeptides	RP	LiChrosorb RP-2 and RP-8, μBondapak C_{18}	64
Lypressin and other nonapeptides	RP	Nucleosil C_8 and C_{18}, RP-8, Spherisorb ODS	377,378
Lysopressin	RP	μBondapak C_{18}	257
Mesotocin and other nonapeptides	RP	LiChrosorb RP-2 and RP-8	64
Nonapeptides and peptide derivatives containing sulphur (11 species)	RP	μBondapak C_{18}, Separon SI C-18	399
Nonapeptides (mixture); trace enrichment; separation	RP	LiChrosorb RP-2 and RP-8, Nucleosil C_8 and C_{18}, RP-8, Spherisorb ODS, μBondapak C_{18}	64,175,377, 378
Pitressin (posterior pituitary extract, crude Arg-vasopressin)	HI, RP	Hypersil ODS	520
Pituitaries rat, crude extract	SE	Agarose-12%	283
(iv) Vasopressin and analogues			
Vasopressin	IE	Partisil SCX	555
—, posterior pituitary [35S]-CysH-labeled	SE	SynChropak GPC-100	214
— and analogues (8 species); (L- or D-homoarginine = [Har], β-mercaptopropionic acid = [Mpa]): [D-Arg8]-, [L-Har8]-, [D-Har8]-, [Mpa1]-, [Mpa1,L-Har8]-, [Mpa1,D-Har8]-, [Mpa1,D-Arg8]-derivatives	RP	Nucleosil C_{18}	418
—, [Arg8]-vasopressin	HI, IP-RP, RP	Cosmosil C_{18}, Sep-Pak C_{18}, μBondapak C_{18}	49,90,427, 650
— [Arg8]	SE	TSK Gel 2000 SW	427

Table 14.2 (continued)

Substances	Mode of chromatography	Packings	References
— — and some of its diastereoisomers	RP	LiChrosorb RP-8 and RP-18, Spherisorb ODS, μBondapak C₁₈	65
— — in extract of anteria ganglia of marine mollusc	RP	LiChrosorb RP-8	496
Aplysia californica			
— and [Lys⁸]	RP	Hypersil ODS, LiChrosorb RP-2, RP-8 and RP-18, μBondapak C₁₈	56,64,496, 520
— —, DABTC derivatives	RP	μBondapak C₁₈	100
— carba-analogues (both sulfides and sulfoxides), 5 species	RP	Separon SI C-18	401
— [Lys⁸]	AD, RP	RP-8, Silica gel Si 60	176,177
— [¹²⁵I]-iodinated or chloramine T treated polypeptides	RP	μBondapak C₁₈	629
— —, post column fluorescence derivatization; [Arg⁸]-vaso-pressin, fluorescent Fluram derivative	RP	RP-8, Partisil ODS	176,419
(v) Vasotocin			
Vasotocin [Arg⁸]	HI, IP-RP, RP	Hypersil ODS, μBondapak C₁₈	520,650
— —	IP-RP, RP	LiChrosorb RP-8 and RP-18	496,608
— —, in extract of anteria ganglia of marine mollusc	RP	LiChrosorb RP-8	496
Aplysia californica			
14.2.11.14 Secretin and analogues			
Secretin	HI	PRP 1	571
—	RP	ODS, Polygosil C₁₈	532,717
— —, analogues, diastereoisomers [L-Ala⁴]- and [D-Ala⁴]-des-His¹-secretin; intermediate protected peptides for synthesis	RP	LiChrosorb RP-18	25
— porcine (synthetic), "VIP-secretin"	RP	Zorbax ODS	40
14.2.11.15 Thyrotropin TH, TSH and allied factors (hormones) TRF, TRH			
Thyroid stimulating hormone, TSH	RP	RadialPak A/C₁₈	262
— —, β-TSH fragment (solid phase synthesis)	IP-RP	Bondapak C₁₈/Corasil, Phenyl Porasil B, μBondapak C₁₈, μBondapak f.a.	255

Thyrotropin (thyrotropic hormone TH) bovine; ovine, α- and β-subunits	IP-RP, RP	ODS, μBondapak C_{18}	266,269
— crude, ovine	SE	PAP/LiChrosorb	156
— releasing factor (hormone) TRF, TRH	RP	B-R ODS, LiChrosorb RP-18	455,458
TRH and analogues: desamido-TRH, TRH-GlyNH$_2$, 3-methyl-His-TRH	IP-RP	μBondapak C_{18}	650
— and [D-His^2]-isomer	IE	DC-resin	41
— and [Phe^2]-analogue	IE	DC-1 A	715
— —, separation of diastereoisomers	RP	μBondapak C_{18}	90,575,578
— and 3,3-dimethyl-ProNH$_2$ analogue	RP	Spherisorb CN, Spherisorb ODS	306
— and free acid TRH-OH	RP	μBondapak C_{18}	452

14.2.11.16 Various peptides (proteins) with biogenic acitivity, not listed above

Antamanid (synthetic linear decapeptide)	IP-RP	Bondapak C_{18}/Corasil, Phenyl/ Porasil B, μBondapak C_{18}, μBondapak f.a.	231,232,255
Anticholecystokinin peptide	RP	μCN	40
Aprotinin	HI	Nucleosil C_{18}	196
Caerulein	RP	B-R ODS, LiChrosorb RP-18	455,456
Cholecystokinin CCK-8	RP	μBondapak C_{18}	452
Contraceptive polypeptide (structure TPRK) synthetic (One letter symbols of amino acids are used)	RP	Phenyl/Corasil, Poragel P-T	99
Epidermal growth factor EGF	RP	RP-8	545
Experimental allergic encephalitogenic peptide EAE and p-iodo-Phe^1-EAE	MM, RP	LiChrosorb RP-18, MicroPak AX-10	146,740
Growth hormone human	RP	Radial Pak A/C_{18}, RC-18 (radial), other details not available	206,262
Growth hormone, rat pituitaries crude extract	SE	Agarose-12%	283
Kentsin (structure TPRK)	RP	μBondapak C_{18}	746
Mellitin	HI, RP	Hypersil ODS	520
Motilin	HI	PRP 1	571
Neurohypophyseal peptide hormones	RP	LiChrosorb RP-2 and RP-8, μBondapak C_{18}	64
Neurophysin I and II (neurohypophyseal binding proteins for nonapeptides)	RP	RP-18, Sep-Pak C_{18}, Zorbax-CN, μBondapak C_{18}	49,98,200
Pancreatic polypeptide human (HPP) and bovine (BPP)	HI, RP	LiChrosorb RP-18, PRP-1	119,571

Table 14.2 (continued)

Substances	Mode of chromatography	Packings	References
Parathyrin (parathyroid hormone human native 1-84), parathyrin large fragments 1-34, synthetic, purification from side products	RP	Nucleosil C_{18}, Sep-Pak C_{18}, μBondapak C_{18}	50,354
Parathyroid hormone	SE	Fractogel TSK HW 55-F	216
— (structure NSILA)	IE	Partisil SCX	555
Peptide hormones, various (57 peptides including peptide hormones)	RP	μBondapak C_{18}	660
Prolactin	MM	MicroPak AX-10	146
Thymopoietin (immunoregulatory thymic hormone), active fragment (pentapeptide 32-36) and its tyrosine di- to tetrapeptide fragments	IP-RP	μBondapak C_{18}, μBondapak CN	677
Thymopoietin active fragment 32-36	SE	TSK Gel 2000 SW	427
Thymosin α_1	IE, RP	Partisil 10 SAX, μBondapak C_{18}	189
Tuftsin [a hormone-like tetrapeptide from leucophilic immunoglobulin G (IgG)]	IP-RP, RP	B-R ODS, Cosmosil C_{18}, LiChrosorb RP-18, W-ODS-2	13,427,455
— analogue (structure TKPPR) and reaction by-product (structure TKR)	SE	TSK Gel 2000 SW	427
	RP	LiChrosorb RP-18	737
Vasoactive intestinal peptide VIP, chicken and pork, natural	HI, RP	LiChrosorb RP-18, PRP-1, Zorbax ODS, μCN	40,119,571
Urotensins II_α, II_β and II_γ (carp); urotensin I (carp), tryptic digest peptides	RP	LiChrosorb RP-8	496

14.2.12 PEPTIDES

(One-letter symbols are often used for their sequence characterization)

14.2.12.1 Dipeptides

(i) Data for collections of dipeptides

| Dipeptides of various hydrophobicity (9 species) | RP | LiChrosorb C_8 | 372 |

Data for 11 peptide species	MM	TP$_1$/Partisil	168
Data for 11 species	MM	TP$_2$/Silica	350
Data for 14 species	RP	RP-18	426
Data for 15 species	IP-RP	μBondapak C$_{18}$	417
Data for 18 species	MM	TP$_3$/Partisil	169
Data for 20 species	RP	PRP-1	312
Data for 44 species	RP	Bondapak NH$_2$, Nucleosil CN, Phenyl-Sil-X	425
Data for 50 species	MM	MicroPak AX-10	147
Data for 53 species	RP	Spherisorb ODS	425
Data for 97 species	RP	ODS- Hypersil	425

(ii) Miscellaneous dipeptides

Alanine-based dipeptides (8 species)	RP	LiChrosorb C$_8$	372
Aspartame (Asp-Phe-OMe, artificial sweetener)	IE	SCX Zipax	171
Detection by ESR (electron spin resonance) spectroscopy after γ-irradiation (4 species)	IE	IEX-210-Sc	498,499
Diketopiperazin His-Pro	RP	μBondapak C$_{18}$	452
Phenylalanine (N-terminal) dipeptides (19 species)	RP	C$_{18}$-TMCS/Silica	576
Proline-containing dipeptides (PA, PG, PV, AP)	RP	LiChrosorb RP-18	459
Separation using chelate packings	NP, RP	DBS-Cd(II), DBS-Zn(II)	115
Various dipeptides	IE	DC-resin	41
—	IE	KU-2	546
—	IE	M-72	519
—	IP-RP	ODS	361
—	RP	Zorbax ODS	11
—	SE	Glycophase/CPG, I-125 PC, LiChrosorb Diol, Shodex OHpak B-804, SynChropak GPC 100, TSK Gels SW, μBondagel	547,562,564

14.2.12.2 Dipeptides, tripeptides and tetrapeptides

(i) Data for collections of di- and tripeptides (synthetic, non-protected)

Data for 4 species	AD	Partisil 5	192
Data for 5 species	RP	μBondapak Alkylphenyl	226

Table 14.2 (continued)

B524

Substances	Mode of chromatography	Packings	References
Data for 5 species	RP	PrepPak-500 C_{18}, μBondapak C_{18}	60
Data for 9 species	RP	PRP-1	312
Data for 12 species	IP-RP	μBondapak C_{18}	271
Data for 16 species	RP	Amberlite XAD-2, -4, -7	371
(ii) Miscellaneous peptides			
γ-Glutamyl di- and tripeptides (6 species)	IE	Aminex A-28	315
Glycin-based di- and tripeptides (14 species)	RP	LiChrosorb C_8	372
Hydrophobic peptides	IP-RP, RP	ODS Hypersil, Radial-Pak A/C_{18}, μBondapak C_{18}	229,258,262
(iii) Synthetic tetrapeptides			
CAGY (solid phase)	IP-RP	μBondapak C_{18}	265
GGFL, GGFM (enkephalin fragments)	RP	μBondapak C_{18}	132
GGPL, GGPM	RP	μBondapak C_{18}	136
LAGV (and corresponding deletion peptides from solid phase synthesis)	IE	AA-15	160
— (and 4 deletion peptides)	RP	Bio-Beads S-XI resin B, Styragel 1000 A, μBondapak C_{18}	424,553
LGGG	RP	PrepPak-500 C_{18}, μBondapak C_{18}	59
TYYF	IE	SCX	457
YGGF (enkephalin fragment)	IE	SCX, Zipax	71

14.2.12.3 Dipeptides up to higher peptides

(For tetra- and higher peptides sequence identification is given; *n* r. = number of residues)

(i) Data for collections of synthetic non-derivatized peptides

Substances	Mode of chromatography	Packings	References
Tri- to hexapeptides, data for 4 species, 4 r.: EHPG, 6 r.: EHYWKP	MM	Hydrogel IV	238

Peptides	Mode	Column	Ref.
Di- to pentapeptides, 7 species, 4 r.: LWMR, 5 r.: LWMRF	IP-RP, RP	µBondapak Alkylphenyl	226,229,230, 266
Di- to heptapeptides, 8 species, 4 r.: LWMR, MRFA, 5 r.: LGMRF, 7 r.: FSKLGDG	IP-RP	µBondapak Alkylphenyl, µBondapak f.a.	231
Tri- to pentapeptides, 8 species, 4 r.: GAGY, LWMR, 5 r.: LWMRF	IP-RP	µBondapak C_{18}	264
Tri- to hexapeptides, 8 species, 4 r.: LWMR, MRFA, 5 r.: LWMRF, YIHPF, 6 r.: VYIHPF	RP	Bondapak C_{18}/Corasil, Bondapak Phenyl/Corasil	228
Di- to decapeptides, 9 species, 4 r.: EHPG, 6 r.: EHYWKP, 8 r.: ESYGLRPG, 10 r.: EHWSYGLRPG (*i.e.* luteinizing hormone releasing factor), EHYSWGLRPG	RP	Phenyl/Corasil, Poragel PN, Poragel PS	238
Di- to hexapeptides, 11 species, 4 r.: FGFG, FGGF, VAAF, 6 r.: KFIGLM	RP	LiChrosorb RP-18	493
Di- to hexapeptides, 11 species, 4 r.: LWMR, MRFA, 5 r.: FLEEI, LWMRF, 6 r.: KETYSK, LWMRFA	MM	MicroPak AX-10	146
Tri- to pentapeptides, 11 species, 4 r.: LGGG, LWMR, MRFA, 5 r.: LWMRF	IP-RP	µBondapak C_{18}	241
Nonapeptides and peptide derivatives containing sulphur (11 species)	RP	Separon SI C-18	399
Di- to octadecapeptides, 14 species, 4 r.: LWMR, MRFA, 5 r.: LWMRF, 8 r.: FVQWLMDT, 18 r.: LESFLKSWLSALEQALKA (W = -Trp(CHO)-)	IP-RP	Radial-Pak C_{18}	549
Di- to heptapeptides, 15 species, 4 r.: LLVV, LWMR, 5 r.: AAYAA, 7 r.: MEHFRWG	RP	Nucleosil C-18, µBondapak C_{18}	489,596
Di- to hexapeptides, 22 species, 4 r.: FGFG, FGGF, VAAF, WADFamide, 6 r.: KFIGLM	HI	LiChrosorb RP-18	492
Di- to pentapeptides, 23 species, 4 r.: FFFF, GGGAamide, LWMR, MRFA, 5 r.: LWMRF	RP	µBondapak C_{18}	257
Di- to 32-residue peptides	RP	ODS	717
57 various peptides including peptide hormones	RP	µBondapak C_{18}	660
98 various peptides (from dipeptides to 31-residue peptide) both synthetic and from natural sources, 82 species are tetra- or higher peptides	RP	LiChrosorb RP-8 or RP-18	739

(ii) Various synthetic peptides

Peptides	Mode	Column	Ref.
KETYSK	MM	MicroPak AX-10	146
LWMR, LWMRF, MRFA	IP-RP	µBondapak C_{18}, µBondapak f.a.	232

Table 14.2 (continued)

Substances	Mode of chromatography	Packings	References
MRF, LWMR, synthetic 1-15 segment of human apolipoprotein C-I (p-iodoPhe in Phe-14 position)	MM	RC-C$_{18}$	236
RKDVY	RP	LiChrosorb RP-8 and RP-18	740

14.2.12.4 Homopeptides

Substances	Mode of chromatography	Packings	References
Alanine oligomers Ala$_2$ to Ala$_6$	HI, IE, RP	Amberlite XAD-2, -4, -7, LC-R-1, LiChrosorb RP-18, MicroPak AX-10, PRP-1	147,312,371, 492,535
Glycine oligomers Gly$_2$ to Gly$_6$	IE, SE	KU 2, LC-R-1, M-72, TSK Gels PW and SW	246,336,519, 535,546
Phenylalanine oligomers Phe$_2$ to Phe$_5$	HI, IP-RP,	B-R ODS, Hypersil ODS, LiChrosorb RP-18, μBondapak C$_{18}$	257,258,455, 492,520,608
Various homopeptides Ala$_5$, Gly$_3$, Leu$_3$, Val$_2$, Met$_2$, Phe$_3$, Trp$_2$, Tyr$_3$	RP	B-R ODS, LiChrosorb RP-18	455
— Trp$_2$, Tyr$_3$	MM, RP	Radial Pak A/C$_{18}$, TP/Partisil	168,262

14.2.12.5 Isomeric peptides

(i) Diastereoisomeric peptides

Substances	Mode of chromatography	Packings	References
Benzoyl-Phe-Ala-0-Me, benzoyl-Phe-Ala-0-Benzyl	AD, NP	Microporasil	203
Diketopiperazine isomers D-His-L-Pro, L-His-L-Pro	RP	Partisil ODS	306
Dipeptides, N-terminal-Phe dipeptide mixtures	RP	μBondapak C$_{18}$	575,578
—, data for 3 pairs: Phe-Ser, Phe-Ala, Leu-Val	RP	Amberlite XAD-4	548
—, 6 pairs	RP	ODS/Hypersil, Spherisorb ODS	425
—, 7 pairs	MM	MicroPak AX-10	147
—, 16 pairs	RP	PRP-1	312
—, 23 pairs	RP	LiChrosorb RP-8, μBondapak C$_{18}$	92
—, 27 pairs	RP	LiChrosorb C$_8$	372
Pyroglutamyl-histidyl-dimethylprolineamide diastereoisomers L-L-L, D-L-L, L-D-L, L-L-D	LE	Partisil ODS, Spherisorb-CN (Cu^{2+})	659

Peptide	Mode	Sorbent	Ref.
Tripeptides of Ala-Leu-Gly (2 diastereoisomers)	RP	Amberlite XAD-4	548
Trialanine diastereoisomers (3 species)	RP	PRP-1	312
Tripeptide diastereoisomers (6 species)	RP	LiChrosorb C_8	372
— — (8 species)	RP	LiChrosorb RP-8, μBondapak C_{18}	92
	IE	AA-15	349
Tetrapeptide diastereoisomers of Leu-Ala-Gly-Val and related peptides (9 species)	RP	Partisil ODS, Spherisorb CN, Spherisorb ODS	306
— of L-Tyr-D-Ala-Gly-L-MePheNH$(CH_2)_2NMe_2$; L-Ala^2 (enkephalin analogues)	LE	Spherisorb-CN(Cu^{2+})	306
Pentapeptide diastereoisomers of L-Tyr-D-Ala-Gly-L-Phe-L-Met; D-Met^5 (enkephalin analogues)	RP	Spherisorb ODS	306
— of L-Tyr-D-Ala-Gly-L-Phe-L-Met; L-Ala^2 and D-Met^5; and of L-Tyr-D-Ala-Gly-L-MePhe-D-Leu; D-MePhe^4 (enkephalin analogues)			

(ii) Sequence-isomeric peptides

Peptide	Mode	Sorbent	Ref.
Dipeptides of (Phe,Ser), testing of strongly acidic and strongly basic mobile phases	RP	Amberlite XAD-4	548
Dipeptides, data for 4 pairs	MM	TP$_1$/Partisil	168
—, 5 pairs	MM	TP$_2$/Silica	350
—, 12 pairs	MM	MicroPak AX-10	147

14.2.12.6 Modified peptides

Peptide	Mode	Sorbent	Ref.
DABTC dipeptides	RP	μBondapak C_{18}	100
DNP (dinitrophenyl) di- to tetrapeptides, checking of Merrifield peptide synthesis, 5 species	RP	μBondapak C_{18}	596
L-Met-dipeptides (Met-Ala, Ala-Met) complexed with palladium, Pd(II)-thioether (linear or intracyclic) peptide complexes	IP-RP	C_{18} Radial-Pak, RCSS, μBondapak C_{18}	392,393
N-acetyl-N,O,S-permethylated oligopeptides, mass spectrometry detection, 8 modified dipeptides, 6 tripeptides	RP	Supelcosil LC-8	756

14.2.12.7 Peptides of natural origin

(With the exception of peptide hormones, which are registered in 14.2.11)

Peptide	Mode	Sorbent	Ref.
Cerebrospinal fluid peptides (human)	RP	LiChrosorb RP-18	279
Drosophila funebris paragonia extract, 27-residue peptide, separation of Val- and Leu-form	RP	LiChrosorb RP-18	737

Table 14.2 (continued)

Substances	Mode of chromatography	Packings	References
Ferricrocin, metabolic product of *Aspergillus viridinutans*	RP	LiChrosorb RP-8	163
Fibrinopeptides A and B (bovine)	RP	μBondapak C_{18}	445
Fibrinopeptides A, AP, AY, B, des-Arg-B (human)	RP	LiChrosorb RP-18	343
Fibrinopeptides human and mammalian	RP	Micropak C_{18}	632
Folic acid derivatives, see lower pteroyl-derivatives			
Glutathione (γ-L-glutamyl-L-cysteinylglycine) in the whole blood	IE	Zipax SCX	554
—, SH form, in human blood, post-column derivatization with N-chlorodansylamide and fluorescence detection	IP-RP	μBondapak C_{18}	501
—, SS and SH forms, bis-N,N'-(γ-glutamyl)cystine, carboxy-methyl-γ-glutamylcystine, carboxymethylglutathione, γ-glutamylcysteic acid, γ-glutamylcysteine	IE	Aminex A-28	666
—, SS form	MM	MicroPak AX-10	146
	IE	Aminex A-7	421
Glycopeptides from biological fluids (urine), both ninhydrin and orcinol reaction patterns	NP	μBondapak Carbohydrate	511
— in urine; Rauscher murine leukemia virus (labeled with ^{3}H-glucosamine or ^{3}H-mannose)	RP	μBondapak C_{18}	290
Histidine-containing dipeptides anserine (β-alanyl-2-methyl-histidine), carnosine (β-alanylhistidine), homocarnosine, ophidine (β-alanyl-3-methylhistidine); carnosine and anserine in guinea pig brain	IE	Partisil 10 SCX	507
Histidine-containing dipeptides, carnosine (mouse olfactory bulb), anserine and carnosine (synthetic)	RP	Partisil 10/25 PXS	734
Pteroyl-oligo-γ-L-glutamates (2-8 glutamyl residues), (2 to 5, 7, 8 glutamic acid residues); p-aminobenzoyl-oligo-L-glutamates (2-4 glutamic acid residues)	IE, RP	Partisil ODS 2, Partisil 10 SAX, Spherisorb ODS	91,95
Pteroyl-oligo-γ-L-glutamates, azo-p-aminobenzoyl-γ-oligo-glutamates synthetic (Glu_1 to Glu_7), nonlabeled, derived from rat liver folates; non-substituted-, methylene-, methyl-, methenyl-, formyl- and formimino-derivatives	RP	Spherisorb ODS	161

Ribonuclease S-peptide	MM	MicroPak AX-10	146
Serum peptides (in normal human and uremic serum)	IP-RP, SE	Cosmosil, TSK Gel 2000 SW	428
Spleen-derived immunosuppressive peptide, bovine	RP	Prep-Pak 500/C$_{18}$, μBondapak C$_{18}$	413

14.2.12.8 Prediction of retention time of peptides

Prediction (based on amino acid composition and on relative lipophilicity of amino acid residues) for 23 and 25 peptide species	RP	ODS columns	455,456
Prediction for peptides containing 20 amino acid residues or less, based on retention of 100 peptide species	RP	ODS columns	458

14.2.12.9 Protected peptides

Boc (t-butyloxycarbonyl) protected isomeric dipeptides, 4 species	AD, NP	Silica (n.d.)	1
Boc protected tripeptides after treatment with TFA or HF, or TFA and KOH	IE	DC-1A	715
Di- and tripeptides, protected trifluoroacetyl-, -OMe, 12 species	MM	TP$_3$/Partisil	169
Diastereoisomeric hydrophobic di- and tripeptides, protected Z-(benzyloxycarbonyl-), -OMe	RP	Develosil ODS	388
Di- to hexapeptides, hydrophobic, protected Boc-, OMe, 23 species; sequential isomeric (Boc-, -OMe)-protected dipeptides (2 pairs), tripeptides (3 species), hexapeptides (3 species)	NP	μPorasil	506
Hexapeptide, acyl carrier protein-synthetic peptide 1-6, pentachlorophenyl ester derivative	RP	Bondapak Phenyl Porasil B silanized, Supelcosil LC8, μBondapak Alkylphenyl	552
Neurotoxin II (cobra Naja-naja), synthetic di- to 33-residue peptides, [Aoc-, Boc-, Z-, Tos-(4-toluenesulphonyl-), -OBzl-, -OBzh (-benzhydryloxy), -OBu^t derivatives], 10 species	SE	Sephadex LH 20, Si 40, PSM-60S, Silasorb 600, Silicagel L, Zorbax Sil	698
Pentapeptide Ac-Ser-Thr-Ile-Glu(OBzl pNO$_2$)-Arg(NO$_2$)OH	IP-RP	Bondapak Phenyl Porasil B silanized, μBondapak C$_{18}$	232
Tripeptide Z(OMe)-Gly-Ala-Phe-OBzl (Bzl = benzyl), D-Ala racemisation test	RP	μBondapak f.a. μBondapak C$_{18}$	356,357
Tripeptides derivatized, 7 species	AD, NP	Partisil 5	192

Table 14.2 (continued)

Substances	Mode of chromatography	Packings	References
Tri- to octapeptides [Boc-, Bu^t-(*tert*.-butyl), ONp-(*p*-nitrophenyl-), Acm- (acetamidomethyl), Z-, OBu^t (-tert-butyl ester), -OMe]-protected	AD, NP	Silica Gel 60	188
Tetrapeptide (GluOBzl)$_4$, hexapeptide Ac-Ser-Thr-Ile-Glu(OBzlpNO$_2$)-Asp(OBzlpNO$_2$)-Arg(NO$_2$)OH	IP, RP	µBondapak f.a.	231
Tetra- to heptapeptides (protected and non-protected)	MM	Partisil	57
Variety of synthetic peptides up to tetradecapeptide used for the synthesis of somatostatin [N-benzyloxycarbonyl-, N-2-(*p*-biphenylyl)-2-propyl-oxycarbonyl-, N-*t*-butyloxy-carbonyl-, O- and S-*t*-butyl- and S-acetamidomethyl-derivatives], 11 species	AD, NP	Silica Gel 60	187

14.2.13 POLYSACCHARIDES

14.2.13.1 Cellulosics (modified cellulose)

Substances	Mode of chromatography	Packings	References
Carboxymethyl cellulose	SE	SynChropak GPC	36,37
— hydroxyethyl cellulose	SE	SynChropak GPC	36
Cellulose nitrate (and fragments)	SE	LiChrospher CH-8	446
Hydroxyethyl cellulose	SE	SynChropak GPC	36,37

14.2.13.2 Dextrans (products of Leuconostoc mesenteroides, derivatives and fragments)

Substances	Mode of chromatography	Packings	References
Blue dextran (M_r 2·10^6), dextrans M_r 1·10^4 to 5·10^4	SE	Glycophase/CPG	564,565
[Carboxyl-^{14}C]-dextran (M_r 2·10^4), void volume indicator	SE	Bio-Gel P-4	509
Dextran	SE	TSK Gels PW	247
— B 161 D40 (M_r 7·10^4); T 2000	SE	Toyopearl HW 55S and 65S, TSK Gels 33	33
— B; D 4133; D 5251	SE	µBondagel E 300, µBondagel E linear	150
— standard T 2000, dextran average M_r 234 000	SE	TSK Gels SW and PW	338
— T 5000	SE	CPG, LiChrospher Si 100	464
Dextrans	SE	GC-Silica	126
— of various molecular weights	SE	GA/Silica, Glycophase/Silica	157

FITC-dextran 3 (M_r 2900)	IE	TSK Gels LS 212 (Na^+)	536
— 3 and FITC-dextran 20 (M_r 19 000)	SE	H-Gel 3011s (H^+)	383
Polydextran	SE	μBondagel	462

(i) Dextran standards

Dextrans 9.4 (M_r 9400), 40 (M_r 40 000), 90 (M_r 90 000), 500 (M_r 500 000)	SE	LiChrosorb Diol	583
— M_r 9700; 70 000; 500 000	SE	TSK Gel 5000 PW	281
— T 10, T 20, T 70, T 500	SE	μBondagel E 300, μBondagel E linear	150
— T 10, T 20, T 40, T 70, T 150, T 250, T 500, T 2000	SE	TSK Gels SW	336
— T 10, T 40, T 70, T 250	SE	TSK Gels SW	181
— T 10, T 40, T 70, T 150, T 250, X	SE	CPG	351
— T 10, T 110, T 2000	SE	Spheron 100 000	72

14.2.13.3 Miscellaneous polysaccharides

Amylopectin (degradation of maize starch by grinding)	SE	CPG, LiChrospher Si 100	464
Amylose (DP 200, DP 570, DP 1130, DP 2560)	SE	CPG, LiChrospher Si 100	464
Chondroitin sulphates (whole cartilage, pig skin and shark cartilage)	SE	TSK Gels SW	181
Gum arabic	SE	TSK Gels SW	181
Heparin sodium	SE	TSK Gels SW	181
Inulin	SE	TSK Gels SW	181
Pectin	SE	μBondagel E linear	166
Pectins amidated, high methoxy- or low methoxy-substituted	SE	SynChropak	35,37
Polysaccharides produced by *Xanthomonas manihotis*	SE	μBondagel	106
Pullulan	SE	μBondagel E linear, μBondagel E 300	150
Starch (maize) grinded and digested with pullulanase, β-amylase	SE	CPG, LiChrospher Si 100	464

14.2.14. PROTEINS (INTACT, LARGE FRAGMENTS, SUBUNITS)

(With the exception of enzymes and zymogens, which are registered in 14.2.1)

14.2.14.1 Alfalfa leaf proteins

Alfalfa leaf proteins	IE	Pel SBCX/Silica, Zipax SCX	27,31
— — —	SE	Glycophase/CPG, SEC-500	27,32

Table 14.2 (continued)

Substances	Mode of chromatography	Packings	References
14.2.14.2 Blood haemolyzates			
(*cf.* also the title "Haemoglobins")			
Blood haemolyzate	SE	TSK Gels SW	311
— from normal subjects and diabetic patients; glycosylated haemoglobins; HbA1c	IE	Bio Rex 70	213,721,740
—, haemoglobins HbA1, A2	IP-RP	µBondapak C_{18}	265
— (human), haemoglobin from adults and newborns, from individuals showing β-thalassemia trait and Hb disorders	IE	Bio-Rex 70, SynChropak AX 300	201,223,301
14.2.14.3 Calcium-binding proteins			
Calcineurin B	RP	µBondapak Phenyl	360,531
Calmodulin, *Acanthamoeba Castellani*	AC	Phenothiazine-Affigel	531
—, testis and *Acanthamoeba Castellani*	RP	µBondapak Phenyl	360,531
α-Lactalbumin	RP	µBondapak Phenyl	360,531
Myosin, light chains	RP	µBondapak Phenyl	531
Parvalbumin, carp, *Xenopus Laevis*	RP	µBondapak Phenyl	360,531
— chicken, rat	RP	Aquapore 300, LiChrosorb RP-18	737,741
Troponin C, sceletal muscle	RP	µBondapak Phenyl	360,531
14.2.14.4 Collagen and fragments			
Collagen α1, human	RP	CNP/Silica, DPh/Silica, EM Li-Chrosorb RP-8, LiChrospher C_8, Octyl/Silica	320,415
— α1, α2, lathyrogenic chickens		CNP/Silica, DPh/Silica, Octyl/Silica	415
— α1(I), α1(II), α1(III), α1(IV), α1(IV) BM, α2(IV) BM, α1-3(V), β(I), γ(I), fragments C, C1, 50 K, 7 S, 7 S coll; polypeptide chains of various origin: rat cartilage, calf skin, mouse tumour, EHS tumour, human placenta	SE	Separon HEMA 1000 Glc	137

— I, II, III, human	RP	CNP/Silica	162
— I, II, lathyric chick embryo cartilage	RP	CNP/Silica	162
— I, III, both non-oxided and after performic acid oxidation	SE	Separon HEMA 1000 Glc	429

14.2.14.5 Egg proteins

["Ovalbumin (egg albumin)" is registered under the separate title]

Avidin	SE	I-125 PC	272
Conalbumin II	IE	Mono Q	568
Conalbumin, egg white	SE	SynChropak GPC-100, TSK Gels SW	337,368
Egg white proteins, chicken	IE	CM-Spheron 300, Spheron Phosphate 1000	72,473, 475, 476
— —, chicken	SE	TSK Gel 3000 SW	181
Ovoglobulin, denatured in SDS solution	SE	TSK Gels SW	337
Ovomucoid, denatured in SDS solution	SE	TSK Gels SW	337

14.2.14.6 Ferritin, apoferritin

Apoferritin, monomer, dimer	SE	TSK Gel 3000 SW, TSK Gel 5000 PW	27,281
Ferritin	SE	Amide Si-100, -500, AP/LiChrosphere, PAP/LiChrosorb	155,158,625
—	SE	I-125 PC, SynChropak GPC-100	573,562
—	SE	Diol modified Silica, LiChrosorb Diol	88,583,584
—	SE	Fractogel TSK HW-55F, -65F, Toyopearl HW 55F	198,216
—	SE	Agarose-12%, Separon 300 Glc	116,284
—	SE	TSK Gels SW	427,579

14.2.14.7 Fibrinogen

Fibrinogen, human plasma	SE	Micropak TSK Gels SW, TSK Gels SW	8,680,727
—	HI	Octyl-Agarose-12%	285
—	SE	Fractogel TSK HW-55F, -65F	216
—	SE	SynChropak GPC 100	562
—	SE	TSK Gels PW	8,246

Table 14.2 (continued)

Substances	Mode of chromatography	Packings	References
14.2.14.8 Globulins			
("Immunoglobulins" are registered under the separate title)			
α-Globulin bovine; globulin bovine	SE	CPG, MicroPak TSK 3000 SW	436,481
— human plasma	SE	Micropak TSK Gels SW	727
—	HI	Octyl-Agarose-12%	285
α₂- and β₂-Globulins, human plasma	IE, SE	DEAE- and CM-Trisacryl, Ultrogel AcA 202	591
α-Globulins, β-globulins, γ-globulins	SE	TSK Gel 3000 SW	680
β₁A-globulin (C3c)	SE	TSK Gel 3000 SW	340
γ-Globulin	SE	TSK Gels SW and PW	8
—; (γ₁- to γ₃-)	SE	Fractogel TSK HW-55F, -65F, MicroPak TSK Gels SW, TSK Gels SW	216,336,727
— bovine, H- and L-chain	SE	TSK Gel SW	694
— human	IE	DEAE-Dextran/Spherosil	674
— —; bovine	SE	LiChrosorb Diol	607,701
— — denatured in SDS (sodium dodecylsulphate) and in 6 M G-HCl (guanidinium chloride) solution	SE	TSK Gels PW	240,246
	SE	TSK Gels SW	337,339
—, human serum	SE	TSK Gel 3000 SW	181
14.2.14.9 Haemoglobins, globins			
(*cf.* also the separate title "Blood haemolyzates")			
Globin human chains; α, β, βᴬ, Aᵧ I Aᵧ T Gᵧ chains; globin γ chains, human adult, cord blood and baboons adult, ab-normal baboon β-chain from Hb-papio B silent mutation	RP	LiChrosorb RP-8, Ultrasphere ODS, μBondapak C₁₈	301,634,635
Haemoglobin	AD	Silicone/CPG	486
—	HI	PEI-Glycophase/CPG	102

	Mode	Column	Ref.
—	IE	DEAE-Dextran/Spherosil	674
—	RP	Partisil 10 ODS	30
—	SE	LiChrosorb Diol, MicroPak TSK Gels SW, Spheron 1000, TSK Gel 5000 PW	281,436,701, 716,727
— A, A_2, F, human, α, β, $A_\gamma I$ $A_\gamma T$ G_γ chains, chains of normal adults or HPFH homozygotes or of various HPFH or thalassemia heterozygotes; hemoglobin of human embryonic types, synthesised *in vitro* cultures of haeme-induced K-562 cells, α, γ and ζ chains present in isolated Hb zones	RP	μBondapak C_{18}	305
— A_1, A_2 from haemolyzate	IP-RP	μBondapak C_{18}	265
—, adult and fetal	AC	T. dyes/Silica	646,647
— bovine	SE	CPG, LiChrosorb Diol, SynChropak GPC-100, TSK Gels SW	368,481,607
— F (G_γ and A_γ chains) from six newborns and parents and from patients with sickle cell anemia	RP	μBondapak C_{18}	636
— fetal calf, α- and γ-chain	RP	μBondapak C_{18}	94
— human, adult and fetal (A, C, F), separation of globin chains α, β, A_γ, G_γ, normal and abnormal, determination of β/γ ratio	RP	μBondapak C_{18}	110-113
— fetal, β^S, A_T, $A_\gamma I$, G_γ chains, haemoglobins of many patients with different type of haemoglobinopathies, especially with sickle cell anemia or HbS-HPFH condition	RP	μBondapak C_{18}	303
—, haemolyzates	SE	TSK Gels PW, TSK Gels SW	246,311
—, tritiated leucine labeled	RP	μBondapak C_{18}	113
— mixtures, variants A_1, A_2, C, F and S	IE	DEAE-, CM- and QAE-Glycophase/CPG	101, 103,104, 560,564
— rat, simetryn sulfoxide and iodoacetamide modified globins, α- and β-chains	RP	LiChrosorb RP-8 and RP-18	740
—, α- and β-chains, from normal persons and 9 variants from persons with haemoglobinopathies	RP	RP-18, Supelcosil C_{18}	544
Haemoglobins			
— A, A_0, A_1, A_2, C, F, S, blood haemolyzates of human individuals (newborn and adult) showing β-thalassemia trait, β^+- or A-β-thalassemia or other Hb-disorders (sickle cell trait), subjects in A^+ low F condition	IE	CPG, TSK Gel IEX-500	366,485
	IE	SynChropak AX 300	195,201,223, 304,562,563, 701

Table 14.2 (continued)

Substances	Mode of chromatography	Packings	References
—glycosylated, A1, A1a+b, A1c, AI, haemolyzates from blood of normal and diabetic patients	IE	Amberlite CG 50, Bio-Rex 70	4,213,301,556, 559,721,740
—, pre-FI-, FI-, after-FI-zones, post-FI-components, analyses of red cell haemolyzates (newborn, adults) with elevated HbF and various genetic abnormalities; haemoglobin modified (acetylated, glycosylated), haemoglobin F [14C]-glucose or [14C]-glucose-6-phosphate labeled, enzymatically acetylated	IE	Bio-Rex 70	4
14.2.14.10 Immunoglobulins			
Immunoglobulin A (IgA)	SE	TSK Gel 3000 SW	340
— E (IgE)	MM	MicroPak AX-10	146
— G (IgG)	IE	SynChropak AX 300	600
— G bovine	SE	TSK Gels SW	317,340
— G human	SE	TSK Gels SW	139
—	BA	Anti-IgI/LiChrospher	701
—	SE	CPG	222
—	SE	TSK Gel 3000 SW	364
—, light chains	IE	Partisil SCX	555
— M (IgM) human	SE	CPG	222
Immunoglobulins from human plasma (large scale preparation of IgG)	IE, SE	DEAE- and CM-Trisacryl, Ultrogel AcA 202	591,592
14.2.14.11 Lactalbumins			
α-Lactalbumin, bovine whey	RP	RP-8, μBondapak Phenyl	142,360,531
—	SE	PAP/LiChrosorb, SynChropak GPC-100, TSK Gels SW	139,142,158
—, β-lactalbumin	IE	Mono Q HR 5/5	568
14.2.14.12 Lactoglobulins			
Lactoglobulin	SE	TSK Gels SW	579

— reduced, carboxyamidomethylated, chromatographed in SDS (sodium dodecylsulphate) solution	SE	TSK Gel 3000 SW	669
β-Lactoglobulin			
— bovine	HI	Diol-C$_8$/Silica	518
— —	IE	Mono Q	568
— —	SE	I-125 PC, LiChrosorb Diol	272,607,701
— —	SE	TSK Gels PW	240,246
— —	SE	TSK Gels SW	139,181,336
— — whey	RP	Partisil ODS, RP-8	30,142
— —	SE	Fractogel TSK HW-55F, -65F, SynChropak GPC-100	142,216
— in 6 M guanidinium chloride	SE	TSK Gels SW	339
γ-Lactoglobulin, whey	RP	RP-8	142
—	SE	SynChropak GPC-100	142

14.2.14.13 Lipoproteins, apolipoproteins

Apolipoprotein components of human serum, very low density (VLDL) and high density (HDL) lipoproteins; C$_x$, C-I, C-II, C-III$_0$, C-III$_1$, C-III$_2$, C-III$_x$, A-I, A-II	AD(MM), IP-RP, RP	C$_{18}$-packings, Radial-Pak A C$_{18}$, Zorbax C$_8$, μBondapak Alkylphenyl	227,229,234-236,624
Apolipoproteins, human serum	SE	TSK Gel 3000 SW	269
Lipoprotein fraction of human serum, normal and pathological (hyperlipidemia), VLDL + chylomicron fraction, low density (LDL), HDL$_2$, HDL$_3$, very high density (VHDL) lipoproteins; choline containing phospholipids quantitation, measuring of cholesterol contents in normal group, in hyperlipidemic subjects and in HDL$_2$ and HDL$_3$ of patients with atherosclerotic heart disease and cirrhosis of liver; separation of serum albumin and ovalbumin from lipoproteins	SE	TSK Gels 5000 PW and 3000 SW, TSK Gels (combinations of PW and SW)	523, 525-530
Lipoproteins, human serum, chylomicron (ascitic), VLDL, LDL and HDL, HDL$_2$, HDL$_3$, determination of cholesterol content, various sera of normal persons and of patients with dislipoproteinemia, hyperlipidemia, primary biliary cirrhosis, acute liver hepatitis, liver cirrhosis etc.; lipoproteins of adult persons, of a fatty child and of babies; detection of triglycerides	SE	TSK Gels SW, TSK Gels PW and SW, TSK Gels PW	219,239,240,364
— in animal sera (bovine, dog, rat)	SE	TSK Gels 5000 PW and 3000 SW	530

Table 14.2 (continued)

Substances	Mode of chromatography	Packings	References
14.2.14.14 Milk and cheese proteins			
("Lactalbumins" and "Lactoglobulins" are registered under the separate titles)			
Casein bovine, α_s1, β	SE	TSK Gels SW	139
Cheese proteins	SE	TSK Gel 3000 SW, Ultropak 2135	242
Milk proteins, reconstituted low-heat skim milk powder casein micelles	SE	CPG (Carbowax)	342
Milk serum proteins, preparation and purification	IE, SE	Spherosil ion exchangers	479
Protease peptone (whey)	SE	SynChropak GPC-100	142
Skim milk proteins	SE	TSK Gels SW	139
Whey proteins	RP	RP-8	142
Whey proteins, mixture	SE	SynChropak GPC-100	142
14.2.14.15 Myoglobin, apomyoglobin			
Apomyoglobin, denatured in SDS complex, in 8 *M* urea or in 6 *M* G-HCl (guanidinium hydrochloride)	SE	CPG, Glyceryl/CPG	63,178
— sperm whale	RP	µBondapak Phenyl	757
Myoglobin	IE	DEAE- or QAE-Glycophase/CPG	102,104,564
	IE	PEI-Glycophase/CPG	102
	IP-RP, HI, RP	Hypersil ODS, Nucleosil CN	19,470,520
—	SE	Amide Si-100 and Si-500, Spheron 1000	155,716
—	SE	Fractogel TSK HW-55F, -65F, Toyopearl HW-55S, -55F, I-125 PC	216,331-335, 358,359 272,573
—	SE	Diol-modified silica, LiChrosorb Diol	583,584,701
—	SE	MicroPak TSK Gels SW	436,727
—	SE	Glycophase/Silica, PAP/LiChrosorb, PEG/CPG	125,126,158

— and [14C]-methylated myoglobin	SE	TSK Gels SW	336,579,680
— denatured in SDS or 6 M G-HCl solutions	SE	TSK Gels PW or SW	8,231,338
—	RP	Aquapore RP-300, LiChrosorb RP-18	741
— equine	SE	TSK Gels SW	337,339
—	HI	Diol-C_8/Silica	518
— heart, sperm whale	SE	LiChrosorb Diol, TSK Gels PW	246,607
—	SE	SynChropak GPC-100, TSK Gels SW	181,368,562

14.2.14.16 Ovalbumin (egg albumin)

Ovalbumin	AC	T.-dyes/Silica	646,647
—	IE	Mono Q	568
—	IE	Pel SBCX/Silica, Zipax SCX	27,31
—	IE	PEI/LiChrospher, SAX, SynChropak AX 300	269, 562,708, 709
—	RP	C_8-coated silica, RP-8	142,538
—	RP	Nucleosil C_{18}, μBondapak C_{18}, μBondapak f.a., μBondapak Phenyl, μBondapak Phenylalkyl	273,274,490
—	SE	Agarose-12%	282,283
—	SE	Amide Si-100, -500	155
—	SE	Diol modified silica, LiChrosorb Diol	584,607,701
—	SE	Fractogel TSK HW-55F, -65F, Toyopearl HW 55F	198,216
—	SE	Glyceryl/CPG, Glycophase/CPG	32,542
—	SE	Glycophase/Silica, PAP/LiChrosorb, PEG/CPG	125,126,158
—	SE	I-125 PC	272,317,573
—	SE	MicroPak TSK Gels SW, TSK Gels SW	27,181,317, 336,368,427, 579,694,727
—, [14C]-methylated ovalbumin	SE	SynChropak GPC-100	142,368,562
—, denatured in 8 M urea, 6 M G-HCl or 0.1% SDS	RP	Aquapore RP-300, LiChrosorb RP-18	741
	SE	CPG, Glyceryl/CPG, TSK Gels SW	63,178,337, 339
—, reduced, carboxyamidomethylated, chromatographed in SDS	SE	TSK Gel 3000 SW	669
—, separation from lipoproteins	SE	TSK Gels PW or combination PW with SW gels	246,528,529

Table 14.2 (continued)

Substances	Mode of chromato-graphy	Packings	References
14.2.14.17 Plasma proteins			
[*cf.* also "Serum proteins" and "Serum (plasma) albumin"]			
Plasma human	HI	Octyl-agarose-12%	285
	IE	DEAE-Spheron 300, Phospho-Spheron 300	72,476,477
—, preparative fractionation	IE, SE	DEAE- and CM-Trisacryl, Ultrogel AcA 202	591,592
—	SE	CPG, TSK Gels PW, μBondagel E_2	222,246,714
14.2.14.18 Serum (plasma) albumin			
Serum albumin	HI	PE-Glycophase/CPG	102
—	IE	AP/Silochrom, QAE-Glycophase/CPG, SynChropak AX 300	104,154,600
—	SE	Glycophase/CPG, Glycophase/Li-Chrospher, Glycophase/Partisil, LiChrospher SI-100, LiChrosorb Diol, PEG/CPG	103,125,701
—	SE	Micropak TSK Gels SW, TSK Gels PW, TSK Gels SW	8,219,727
— chicken	RP	Aquapore RP 300, LiChrosorb RP-18	741
—	SE	Glycophase/CPG, μBondagel	628
—, evaluation of binding parameters for two drugs (warfarin and furosemide)	SE	Agarose-12%	282,283
— bovine	AC	AMP/Silica, T.-dyes/Silica	422,522,646,647
—, monomer, dimer	AD	CPG, Silicone/CPG	483,486
—	HI, RP	C_8-coated silica, Hypersil ODS, Nucleosil C_{18}, μBondapak C_{18}	490,520,538,596
—	IE	CPG, Pel SBCX/Silica, Zipax SCX	27,31,485
—	IE	Mono Q, PEI/LiChrospher, Syn-Chropak AX 300	568,708,709

—	RP	CNP/Silica, DPh/Silica, Li-Chrosorb RP-8, LiChrospher C-8	320,415
—	RP	LiChrosorb RP-8, Methylsilica, Partisil ODS, RP-8, Supelcosil LC-18, μBondapak Phenyl	30,142,274
—	SE	Amide Si-100, PAP/LiChrosorb, Porasil DX (Carbowax)	155,158,633
—	SE	Diol modified silica, LiChrosorb Diol	88,582,584, 607
—	SE	Fractogel HW-55F, -65F, Toyopearl HW-55F, -55S	198,216,331-335,358,359
—	SE	CPG, Glyceryl/CPG	120,221,481
—	SE	I-125 PC	272,317,573
—	SE	MicroPak TSK 3000 SW, TSK Gels SW	27,139,181, 317,336,436, 579,694
—	SE	SynChropak GPC-100	142,214,562
—, adsorption on the packings and its prevention	SE	TSK Gels PW, TSK SW and PW Gels	246,281,338
—, binding of bilirubin	SE	CPG	482
—, denatured (in 8 M urea, 6 M guanidinium chloride or 0.1% sodium dodecylsulphate)	SE	SynChropak GPC-100	423
—, reduced, carboxyamidomethylated and chromatographed in the presence of SDS	SE	CPG, Glyceryl/CPG, TSK Gels SW	63,178,337, 339
—, separation from human serum albumin	SE	TSK Gel 3000 SW	669
— human	BA	Anti-HSA/Silica	396,522
—	HI	Aquapore RP-300, LiChrosorb RP-18, Octyl-agarose-12%, Spheron 300	285,658,741
—	IE	CM-Spheron 300, DEAE-Spheron 300, Phospho-Spheron 300, Spheron Phosphate 1000	72,467,473, 475-477
—	IE	DEAE-dextran/Spherosil	674
—, large scale preparative fractionation	IE, SE	DEAE- and CM-Trisacryl, Ultrogel AcA 202	591
—	SE	Glyceryl/CPG, Glycophase/CPG	542,564
—	SE	TSK Gel 3000 SW	340,364,680
—	SE	TSK Gels PW, TSK Gels SW and PW	240,528,529

Table 14.2 (continued)

Substances	Mode of chromato-graphy	Packings	References
— —; binding of bilirubin	SE	Separon 300 Glc; SynChropak GPC	116,423
— —, separation from bovine serum albumin	BA	Anti-HSA/Silica	396,522
— —, separation from human plasma	AC	T-dyes/Silica	592,644,647
— —, [99mTc]-derivative	SE	µBondagel E	742

14.2.14.19 Serum proteins

[*cf.* also the titles "Plasma proteins" and "Serum (plasma) albumin"]

Substances	Mode of chromato-graphy	Packings	References
Serum proteins, calf, fetal	NP	LiChrosorb Diol	585
— —	RP	LiChrosorb RP-8	585
— —, horse	SE	Spheron 100	72
— —, human	HI	Octyl-agarose-12%, Spheron 300	284,658
— —	IE	Agaropectin	284
— —	IE	DEAE-Glycophase/CPG	103,564
— —	SE	CPG, Glycophase/CPG	222,564,565
— —	SE	Fractogel TSK HW-55F, -65F,	116,216
		Separon 300 Glc	
— —	SE	Micropak TSK 3000 SW, TSK Gels SW	181,336,436, 680
— —, adult and neonatal	IE	PEI-Glycophase/CPG, PEI/LiChro-spher, SynChropak AX 300	10,102,423, 563,701
— —, dual-wavelength profiles	IE	SynChropak AX 300	600
— —, rabbit	AD	CPG, Silicone/CPG	483,484,486
— —	IE	CPG	485

14.2.14.20 Thyroglobulins

Substances	Mode of chromato-graphy	Packings	References
Thyroglobulin	SE	Agarose-12%	282-284
	SE	AP/LiChrospher, Diol modified silica	584,625
—	SE	Fractogel TSK HW-55F, -65F, Toyopearl HW 55F	198,216

—	SE	PSG-1000 (Carbowax)	633
—	SE	SynChropak GPC-100	214
—, bovine	SE	TSK Gels SW, TSK Gels PW	8,281,336
— denaturated in SDS or in 6 M G-HCl	SE	TSK Gels PW	240
— human	SE	TSK Gels SW	337,339
— 19S iodoprotein, human	SE	μBondagel E-linear	261
	RP	Radial-Pak A/C$_{18}$	262

14.2.14.21 *Transferrins*

Transferrin	IE	Mono Q, SynChropak AX 300	568,600
—, bovine, human	SE	LiChrosorb Diol	607,701
—, human serum	SE	TSK Gels SW	340,364,579, 680
— —, preparative fractionation	IE, SE	DEAE- and CM-Trisacryl, Ultrogel AcA 202	591

14.2.14.22 *Various proteins*

α$_1$-Acid glycoprotein	IE	SynChropak AX 300	600
α$_1$-, α$_2$ HS-glycoprotein	SE	TSK Gel 3000 SW	340
Actinidin	IP-RP	μBondapak f.a.	233
Albumen bovine, egg, human	SE	Diol-modified silica, LiChrosorb Diol	583
Algae proteins, crude extract from *Ceramium rubrum*	IE	DEAE-Agarose-20%	283
Antigen-specific T-cell-replacing factor	RP	LiChrosorb MPLC RP-18	11
— —	SE	I-125 PC	11
Antigens, tumor associated, fetal	SE	SynChropak GPC-100	214
—, histocompatibility-, pH-2, heavy chains	RP	LiChrosorb MPLC RP-18	11
—, H-2^a, heavy chains	RP	LiChrosorb MPLC RP-18	12
α-1 Antitrypsin, rat serum	IE	Mono Q	568
Aprotinin	SE	I-125 PC, TSK Gel 2000 SW	272,427
Bacterial proteins, partially purified *Escherichia coli* extract	HI	Octyl-Agarose-12%	283
Bilirubin-binding proteins in neonatal and adult serum	SE	SynChropak GPC-100	423,563,701
Ceruloplasmin degraded	SE	Ultropac 2135, TSK Gel 3000 SW	645
Chlorocruorin, sea worm *Potamilla leptochaeta*	SE	TSK Gel 5000 SW	281
Clupein	SE	TSK Gels SW	694
Cobra neurotoxin	RP	Hypersil ODS	470,520

Table 14.2 (continued)

Substances	Mode of chromatography	Packings	References
γ-Crystallin, bovine lens, crude	IE	Mono S	568
Erabutoxin a	SE	TSK Gels SW	694
Estradiol receptors, [³H]-labeled	SE	SynChropak GPC-100	214
Factor VIII	SE	CPG	439
Factor B (human serum, purified), factor B (enzymic cleavage products B_a, B_b), factor $C3_b$, factor D	SE	I-125 PC	516
Fibronectin, human plasma, glycosylated Con-A-Sepharose-binding large molecular weight tryptic fragments	RP	μBondapak Phenyl	757
Glycoproteins, beef thyroid plasma membrane	IP-RP	μBondapak C_{18}	265
— acid cf. above "Acid glycoproteins"			
Haptoglobin	SE	TSK Gel 3000 SW	340
Hemopexin	SE	TSK Gel 3000 SW	340
Histocompatibility antigens cf. above "Antigens"			
Histone H 4, calf thymus	SE	TSK Gels SW	694
Interferon, human fibroblast	HI, RP	Variously coated silica	179
—, human leukocyte	MM, NP	LiChrosorb Diol	587,588
—	RP	LiChrosorb RP	587,588
Keratin-based proteins, keratin biodegradation products	SE	Glycophase/CPG	28
Liver homogenate (extract), rat	IE	CM-Glycophase/CPG, DEAE-Glyco-phase/CPG	102,103,564
— — —	SE	Glycophase/CPG, Glycophase/Li-Chrospher, Glycophase/Partisil, TSK Gel 3000 SW	103,181
α2-Macroglobulin	SE	CPG, TSK Gel 3000 SW	222,340
Metallothionein	MM, SE	TSK Gel 3000 SW	663
β2-Microglobulin	RP	LiChrosorb MPLC RP-18	11,12
Microsomal proteins	AD	Silicone/CPG	486
Paramyosin, oyster, denatured in 8 M urea, 6 M G-HCl (guanidinium chloride) or in 0.1% SDS	SE	Glyceryl/CPG	63
Parotid protein mixture	IE	CPG	485
Phycocyanin	SE	Agarose-12%	284

Phycoerythrin	SE	Agarose-12%, μBondagel E linear	284,596
Placental globulins	IE	DEAE-Dextran/Spherosil	674
— homogenate	IE	Mono Q	568
Prealbumin	IE	SynChropak AX 300	600
Protein glues, industrial, intact and bacteria-degraded	SE	Glyceryl/CPG	470,542
Proteinoids ("thermal proteins"), thermal copolymer of Glu + Trp	SE	μBondagel E-125 + E-300	172
Proteins and polypeptides, elution data for 20 species	SE	PAP/LiChrosorb	156
Proteoglycan subunits, from bovine articular cartilage, from bovine nasal cartilage, from rat chondrosarcoma cartilage	SE	AP/LiChrospher	625
Ribonucleoproteins, rat liver; proteins of the 30-40S nuclear ribonucleoprotein particles	IE	SynChropak AX 300	700
Seed proteins, lectins, crude extract of *Vicia ervilia* seeds	AC	Mannose-agarose-12%	283
Staphylococcal α-toxin	AD	CPG	68
Thiamin-binding proteins (rat brain, rat heart, rat sciatic nerve, rat liver)	SE	TSK Gel 3000 SW	352
Thyroid-binding protein components, sheep, solubilized, crude	SE	PAP/LiChrosorb	156
Thyroid protein extract, human	SE	μBondagel E linear	261
Tropomyosin, rabbit, denatured in 8 M urea, 6 M G-HCl or in 0.1% SDS	SE	CPG, Glyceryl/CPG	63,178

14.2.14.23 *Viral proteins*

Avian myeloblastosis virus protein, solubilized	SE	PEG/CPG	125
— oncovirus P-27, structural protein	IP-RP	μBondapak C$_{18}$	291
Gross murine leukemia virus, structural protein (disrupted in 6 M G-HCl)	RP	μBondapak f.a., μBondapak Phenylalkyl I-125 PC	273
Precursor viral polyprotein (translation product of the gag gene of avian-type C retroviruses) and its cleaved products (viral structural components p27, p19 and p15)	SE		55
Rauscher murine leukemia virus protein	RP	μBondapak C$_{18}$, μBondapak Phenyl	274

14.2.15 VIRUSES

Avian myoblastosis virus (RNA tumour virus)	SE	PEG/CPG	125
Hamster melanoma virus (RNA tumour virus)	SE	PEG/CPG	125
Rabies virus (mouse-adapted fixed strain)	SE	PEO/CPG	278

Table 14.2 (continued)

Substances	Mode of chromato-graphy	Packings	References
Southern bean mosaic virus (SBMV)	SE	CPG, TSK Gel 5000 PW	221,281
Tobacco bushy stunt virus (TBSV)	SE	TSK Gel 5000 PW	281
Tobacco mosaic virus (TMV)	SE	CPG, TSK Gel 5000 PW	221,281
Tobacco ring spot virus (TRSV)	SE	CPG	221
Turnip yellow mosaic virus (TYMV)	SE	TSK Gel 5000 PW	281

TABLE 14.3

KEY TO THE SYMBOLS USED FOR PACKINGS LISTED IN THE REGISTER

The majority of the packings cited in the Register are commercial products. For more detailed descriptions see Chapter 4. Full names and addresses of producing or distributing firms are given in Table 4.10 of Part A. Packings designated "home made" were developed by authors cited in Section 14.2 or they were prepared in laboratories according to the references given in the cited paper. Further general information on "home made" packings can be found in the Table 4.11 of Part A. In some cases published "home made" packings or similar products are now also manufactured commercially; these cases are mentioned in the "Producer" column, if such information is available. d.n.a. means details not available.

Symbol for the packings	Concise characterization of the packings	Producer or distributor
AA-15	Sulphonated styrene-divinylbenzene resin (X8)	Beckman-Spinco
AG 50W	Sulphonated styrene-divinylbenzene resin (X4)	Bio Rad Labs.
Agaropectin	Natural cation-exchanger	Home made
Agarose-12%	Agarose 12%-crosslinked	Home made
AL Pellionex WAX	Weak pellicualr anion exchanger	Reeve Angel, Whatman
Amberlite CG 50	Weak microporous cation-exchange resin	Rohm and Haas

Name	Description	Manufacturer
Amberlite XAD	Polystyrene-divinylbenzene copolymer, which acts as a stationary phase for RP- or HI-chromatography	Mallinckrodt, Rohm and Haas
Amide/Si	Hydrophilic amide stationary phase on silica support	Home made
Aminex A-5, A-6, A-7, HPX-42, HPX-87, Q 15 S, 50 W	Sulphonic type of cation-exchange resins	Bio Rad Labs.
Aminex A-25, A-27, A-28	Quaternary nitrogen anion-exchange resins	Bio Rad Labs.
Aminex X 6	Experimental 6%-crosslinked strong cation-exchange resin	Bio Rad Labs
AMP/Silica	Adenosinmonophosphate-modified silica	Home made
Anpack AX	Anion-exchange resin	Anpack
Anti-HSA/Silica	Immobilized antibody, specific for human serum albumin	Home made
Anti-IgG/LiChrospher	Immobilized immunosorbent (anti-human immunoglobulin)	Home made
AP/LiChrospher	Hydrophilic amide stationary phase on silica	Home made
AP/Silochrom	Aminopropyl modified silica	Home made
Aquapore RP 300	Reversed-phase silica gel packings	Brownlee
AX	Anion-exchanger, d.n.a.	d.n.a.
AX 2633	Anion-exchange resin	Hitachi
BE CB6	Immobilized monoclonal immunosorbent for insulin	Home made
Bio-Beads S-X1	Styrene-divinylbenzene copolymer adsorbent	Bio Rad Labs.
Bio-Gel P-4	Polyacrylamide gel	Bio Rad Labs.
Bio Rex 70	Weakly acidic cation-exchange resin	Bio Rad Labs.
Bio-Sil ODS	Octadecyl silyl column	Bio Rad Labs.
Bondapak C$_{18}$/Corasil	Octadecyl silyl packings	Waters Assoc.
Bondapak Phenyl/Corasil	Hydrophobic packings with aromatic nucleus	Waters Assoc.
B-R ODS	Octadecyl silyl column	Bio Rad Labs.
Cbz-Gly-D-Phe-Spheron 300	Packings containing proteinase-inhibitor	Home made
CDR-10	Strongly basic macroreticular anion-exchange resin	Mitsubishi Kasei
C-8 ES	C-8 reversed phase column	ES Industries
Chymotrypsin/Spheron 300	Immobilized chymotrypsin	Home made
Cibacron blue/Silica	Immobilized Cibacron blue (triazine dye)	Home made
CMB-Cellulose	(Dihydroxyborylphenyl)carboxymethyl derivative of cellulose	Home made
CM-Glycophase/CPG	Carboxymethyl Glycophase bonded on controlled porosity glass, weakly acidic cation exchanger	Home made or Pierce
CM-Polyamide	Carboxymethyl derivative of polyamide packings	d.n.a.
CM-Spheron 300	Carboxymethyl derivative, weakly acidic macroporous fully synthetic organic cation exchanger	Home made or Laboratory Instruments
CM-Trisacryl	Carboxymethyl derivative, organic fully synthetic weakly acidic cation exchanger	Réactifs IBF

Table 14.3 (continued)

Symbol for the packings	Concise characterization of the packings	Producer or distributor
CNP/Silica	Cyanopropyl silica	Home made
Cosmosil C18	Octadecylsilane column	Nakarai
Concanavalin A/Silica	Concanavalin A immobilized on inorganic packings	Home made
C18 Corasil	Octadecyl silyl column	Waters Assoc.
C18 packings	Octadecyl silyl packings and columns	Kontron
CPG	Controlled porosity glass	Home made (originator Haller), Analabs, Electro-Nucleonics, Pierce or Sigma
CPG (Carbowax)	Desactivated controlled pore glass	Home made
CPG/uncoated	Controlled porosity glass without hydrophilic covering	Pierce Eurochemie
C18 Radial-Pak	Octadecyl silyl support, radial-packed column	Waters Assoc.
C18RP	Octadecyl silyl column	d.n.a.
C8 or C18/Silicas	Vydac TP and six other silicas, C8- or C18-coated	Home made
C18-TMSC/Silica	C18-Trimethylchlorosilane capped silica	Home made
C18-Vydac TP 201	Octadecyl silyl packings	The Separations Group
DA and DC resins	Anion- and cation exchangers, divinylbenzene-polystyrene copolymers	Dionex
DA X-4 to X-12	Polystyrene anion exchangers	Durrum
DBS-Cd(II)	Diamine bonded silica with Cd(II) in chelate complex	Home made
DBS-Zn(II)	C12-dien-loaded C8- and C18-silica columns with Zn(II)-chelating complex	Home made
DC-1A or -4A	Cation-exchange resins	Durrum
DC resins	Polystyrene-divinylbenzene strongly acidic cation-exchange resins	Durrum
DEAE-Agarose-20%	Diethylaminoethyl-agarose 20%-crosslinked, weakly basic macroporous anion exchanger	Home made
DEAE/CPG	Diethylaminoethyl derivative of controlled porosity glass, weakly basic anion exchanger	Electro-Nucleonics
DEAE-Dextran/Spherosil	Composite packings, Spherosil covered with crosslinked diethylaminoethyl-dextran, anion exchanger	Home made
DEAE-Glycophase/CPG	Weakly basic hydrophilic macroporous anion exchanger based on controlled porosity glass	Home made, Corning Glass or Pierce
DEAE-Sil	Diethylaminoethyl-derivative of silica gel	Home made

Name	Description	Source
DEAE/Silica	Diethylaminoethyl-phase bonded silica column	Applied Science
DEAE-Spheron 300	Medium basic macroporous anion exchanger, fully synthetic hydrophilic copolymer	Home made or Laboratory Instruments
DEAE-Trisacryl	Medium basic synthetic anion exchanger	Réactifs IBF
DEAE-TSK 2000 SW	Macroporous medium basic hydrophilic anion-exchanger with silica gel matrix	Toyo Soda
DEAP/Silica	Silica modified with N,N-diethylaminopropyltrimethoxy-silane and γ-glycidoxypropyltrimethoxysilane	Patent, ref. 2
DEPA-column	Diethyl-2,3-epoxypropylamine column	Home made
Develosil ODS	Octadecylsilyl support	d.n.a.
Diaion CDR-10	Macroreticular anion-exchange resin	Mistubishi Kasei
Diol-C₈/Silica	Octylsilane hydrophilic/hydrophobic hybrid gel	Home made
Diol/Silica	Angular silica modified with 1,2-epoxy-3-propoxy-propyltrimethoxysilane	Home made
DMA/Silica	Dimethylamino-modified silica anion exchanger	Home made
Dowex 1	Strongly basic polystyrene-divinylbenzene resin	Dow Chemical
DPh/Silica	Diphenyl-coated silica	Home made
Excalibar NH₂	d.n.a.	Applied Science
Fractogel TSK	Totally porous hydrophilic semi-rigid spherical gel manufactured from vinyl polymers. HW 40 to 75 designate fractionation ranges, S and F particle size	E.M. Science
GA/Silica	Glycinamide phase bonded on silica	Home made
GC/Silica	γ-Glycidoxypropyltrimethyoxysilane modified silica	Home made
Glyceryl/CPG	Glycerol bonded via propylsilane to controlled porosity glass	Home made or Electro-Nucleonics
Glycophase/CPG	Glyceryl-propylsilyl controlled pore glass	Home made, Corning Glass or Pierce
Glycophase/LiChrospher	LiChrospher coated with Glycophase layer	Home made
Glycophase/Partisil	Partisil coated with Glycophase layer	Home made
Glycophase/Silica	Glyceryl-propylsilane coated silica microspheres	Home made
HA-X8	Polystyrene anion-exchange resin	Hamilton
HC-40	Polystyrene cation-exchange resin (X5)	Hamilton
H-Gel 3000, 3010, 3011 or 3012, -a or -s	Gels aminated (-a) or sulphonated (-s)	Hitachi
HPX-87 heavy metal	Cation exchanger in the ionic form of heavy metal	Bio Rad Labs.
Hydrogel IV	Gel for size exclusion chromatography proving to have some degree of reversed-phase character	Waters Assoc.
Hydroxylapatite	Mineral packings	Bio Rad Labs.

Table 14.3 (continued)

Symbol for the packings	Concise characterization of the packings	Producer or distributor
Hypersil APS	Aminopropylsilyl silica, spherical porous packings	Shandon
Hypersil ODS	Octadecylsilyl silica support	Shandon
Iatrobeads	Totally porous silica gel spheres	Iatron
IEX-525-QAE	Strongly basic anion exchanger	Toyo Soda
IEX-210-SC	Strongly acidic cation exchanger	Toyo Soda
INHIB/Spheron	Cbz-Gly-(D-Phe)-Spheron 300 (immobilized proteinase inhibitor)	Home made
I-125 PC	Hydrophilic gel permeation matrix for protein analysis columns	Waters Assoc.
Jascopak SN-01	Packings for HPLC of oligosaccharides	Nihon Bunho
KU 2	Strongly acidic cation exchanger	d.n.a.
LC-R-1	Polystyrene sulphonated resin	Jeol
LC-R-3	Quaternary ammonium anion-exchange resin	Jeol
Lectin-agarose	Lectin immobilized to agarose	Home made
LiChroprep Si 60	Porous silica gel	Merck
LiChrosorb	Porous silica gel	Altex Scientific, Rainin
LiChrosorb C$_8$ or C$_{18}$	Octylsilyl or octadecylsilyl phases bonded on LiChrosorb	Ace Scientific, Rainin
LiChrosorb Diol	Glycerol coatings covalently bonded on LiChrosorb	Merck, Knauer, Rainin
LiChrosorb MPLC RP-18	Column for reversed-phase chromatography of proteins	Brownlee
LiChrosorb NH$_2$	Isopropylamine bonded phase (weak anion exchanger)	Altex Scientific, Merck
LiChrosorb RP-2, RP-8, RP-18	Ethyl-, octyl- or octadecylsilyl phases bonded to LiChrosorb	Ace Scientific, Altex Scientific, Brownlee, E.M. Laboratories, E.M. Science, Knauer, MCB, Merck, Rainin, Scientific Products
LiChrosorb Si 60, Si 100	Porous silica gel	Brownlee
LiChrospher CH-8 or C$_8$	Octylsilyl coated silica gel	Merck
LiChrospher Si 100, Si 4000	Porous silica gel	Merck
Locarte No 12	X8 Sulphonated polystyrene resin (cation exchanger)	Locarte
M-72	Sulphopolystyrene cation exchanger	d.n.a.
Mannose-agarose-12%	12%-Crosslinked agarose, substituted by D-mannose	Home made
M-AR 100	100 mesh Silicic acid column	Mallinckrodt
Methylsilica	Methylated silica	Home made

Name	Description	Manufacturer
MicroPak AX	Difunctional weak anion exchanger bonded phase, prepared on LiChrosorb Si-60 silica	Varian
MicroPak MAX	Silica-based medium anion exchanger	Varian
MicroPak MCH C18	Octadecyl hydrocarbon monolayer bonded on silica	Varian
MicroPak NH2	Aminoalkyl silica	Varian
MicroPak SAX	Strongly basic anion exchanger based on silica support	Varian
MicroPak TSK Gels SW	TSK Gels for size exclusion chromatography	Varian
Microporasil	Fine silica gel for partition and adsorption chromatography	Waters Assoc.
Mono P	Special anion exchanger for chromatofocusing	Pharmacia
Mono Q	Strongly basic (quaternary amino) anion exchanger for protein separation	Pharmacia
Mono S	Strongly acidic (sulphonated) cation exchanger for protein separation	Pharmacia
MP-AX	Microparticulate anion exchanger, polyethylenimine on silica	Home made
Nucleosil C8	Octyl phase bonded on silica	Machery-Nagel
Nucleosil C18	Octadecyl phase bonded on silica	Chrompak, Machery-Nagel
Nucleosil CN	Cyanogen phase bonded on silica	Machery-Nagel
Nucleosil SB	d.n.a.	Machery-Nagel
Octyl-agarose-12%	12%-Crosslinked agarose modified with octyl groups	Home made
Octyl/Silica	Silica support, to which octyl groups are covalently bonded	Home made
ODS	Microparticulate porous silica derivatized with octadecylsilyl groups	d.n.a.
ODS-HC/Sil-X-1	Column for reversed-phase chromatography	Perkin-Elmer
ODS/Hypersil	Hypersil derivatized with octadecylsilyl groups	Shandon
OD-silanized c.	Octadecylsilanyzed column	d.n.a.
ODS/Partisil	Partisil derivatized with octadecylsilyl groups	Whatman
ODS/Silica	Microparticulate porous silica fully derivatized with octadecylsilyl groups	Waters Assoc.
Ostion LG KS	Strongly acidic cation exchanger (sulphonated styrene-divinylbenzene copolymer)	United Chemical
PA-35	Cation exchanger, sulphonated styrene-divinylbenzene copolymer	Beckman
PAP/LiChrosorb	Propylamide phase bonded to LiChrosorb	Home made
Partisil	"Naked" silica gel	Whatman

Table 14.3 (continued)

Symbol for the packings	Concise characterization of the packings	Producer or distributor
Partisil C$_8$	Partisil 10 covered with octylsilyl groups	Whatman
Partisil ODS	Partisil derivatized with octadecyl groups	Whatman
Partisil PAC	Polar cyanoamide bonded phase	Altex Scientific, Whatman
Partisil PXS	Silica gel polar adsorbent	Brownlee, Whatman
Partisil SAX	Strong anion exchanger based on silica gel	Chrompak, Whatman
Partisil SCX	Silica gel containing covalently bonded sulphonic acid groups, strong cation exchanger	Whatman
PEG/CPG	Polyethyleneglycol derivative of controlled pore glass	Home made
PEI-Glycophase/CPG	Polyethyleneimine bound to Glycophase/CPG	Home made or Pierce
PEI-LiChrospher	LiChrospher Si coated with polyethyleneimine and crosslinked	Home made or SynChrom
PEI-MAX	Polyethyleneimine coated silica, medium basic anion exchanger	Home made or SynChrom
PEI-PE-MAX	Polyethyleneimine-pentaerythritol tetraglycidylether, crosslinked on silica, medium basic anion exchanger	Home made
Pellionex AL WAX	Pellicular weak anion exchanger	Whatman
Pel SBCX/Silica	Sulphobenzyl pellicular silica cation exchanger	Home made
PEO/CPG	Poly(ethylene oxide)-treated controlled porosity glass	Home made
Phenothiazine/Affigel	2-Chloro-10-(3-aminopropyl)-phenothiazine linked to Affigel 10	d.n.a.
Phenyl/Corasil	Phenyl reversed-phase bonded on Corasil	Waters Assoc.
Phenyl/Porasil B	Phenyl derivative reversed-phase packings	Waters Assoc.
Phenyl Porasil/Corasil	Phenyl derivative reversed-phase packings	Waters Assoc.
Phospho-Spheron 300	Medium acidic macroporous hydrophilic fully synthetic cation exchanger copolymer	Home made or Laboratory Instruments
Polyanion SI	Weak anion exchanger covered on silica surface	Pharmacia
Polygosil C$_{18}$	Octadecylsilyl-silica stationary phase	Machery-Nagel
Poragel PN	Styrene-vinylderivative copolymer	Waters Assoc.
Poragel PS	Polystyrene-type polymer, containing pyridine rings	Waters Assoc.
Poragel P-T	d.n.a.	Waters Assoc.
Porasil A-60	Base-treated silicic acid	Waters Assoc.
Porasil DX	Porous silica gel	Waters Assoc.
Porasil DX (Carbowax)	Porous silica gel desactivated with Carbowax	Home made or Waters Assoc.

Name	Description	Source
PrepPak C$_{18}$	Radially compressed cartridge containing octadecyl-silyl bonded phase	Waters Assoc.
PRP-1	Macroporous polystyrene-divinylbenzene copolymer	Hamilton
PSG-1000 (Carbowax)	Porous silica gel 1000, desactivated with Carbowax	Home made
PSM-60 S	Silica gel column for SEC	DuPont
PXS-10/25 PAC	Bonded cyano-amide type polar phase column for carbohydrate analysis	Whatman
QAE-Glycophase/CPG	Strongly basic anion exchanger immobilized on controlled porosity glass	Home made or Pierce Eurochemie
Radial-Pak A C$_{18}$	Radially compressed C$_{18}$-packings in flexible-wall cartridge	Waters Assoc.
Radial-Pak (TEPA)	Radial Pak silica column modified with tetraethylene pentamine	Waters Assoc.
RC-C 18 (radial)	Radially compressed C$_{18}$-packings	Waters Assoc.
RC SS	Radially compressed system, Bondapak C$_{18}$ in polyethylene cartridge	Waters Assoc.
RP 8 or RP 18	C$_8$- or C$_{18}$-reversed-phase packings	Altex-Beckman, Brownlee, E.M. Laboratories, Hewlet-Packard, Merck
RP 8 L	C$_8$-reversed-phase packings	Lobar
RP boronate	Polychlorotrifluoroethylene beads coated with N-(m-decanoyl)phenylboronic acid	Home made
RPC	Reversed-phase matrix immobilized on polychlorotrifluoroethylene beads	Home made
RPC$_1$ to RPC$_4$	Chromosorb W or G modified with quaternary ammonium compounds	Home made
RPC-5	Pellicular layer of trioctylammonium chloride adsorbed to inert perchlorotrifluoroethylene beads immobilized on the surface of powdered polychlorotrifluoroethylene	Home made
RP-NH$_2$/LiChrosorb	Modified reversed-phase packings	Merck
SAX	Strong anion exchanger	d.n.a.
SCX	Strongly acidic cation-exchange resin	DuPont
SCX/Zipax	Strongly acidic cation-exchange groups attached to Zipax silica	DuPont
SEC-500	Microparticulate silica	DuPont
Separon HEMA 300-Glc or 1000-Glc	Separon HEMA (*i.e.* hydroxyethylmethacrylate) 300 or 1000 with immobilized glucose	Home made or Laboratory Instruments

B554

Table 14.3 (continued)

Symbol for the packings	Concise characterization of the packings	Producer or distributor
Separon-INHIB	Separon HEMA, modified with ε-aminocaproyl-L-Phe-D-Phe-OMe (protease inhibitor)	Home made
Separon Si C_{18}	Spherical silica gel with bonded octadecylsilyl chains	Laboratory Instruments
Sephadex LH-20	Hydroxypropylated Sephadex G-25 (crosslinked poly-dextran), which can be used also in organic or mixed solvents for SEC, AD-, NP- and MM-chromatography	Pharmacia
Sep-Pak C_{18}	Octadecylsilyl silica cartridge	Waters Assoc.
Shodex OH Pak B 804	Packings for SEC	d.n.a.
Shodex S-801/S'	d.n.a.	Perkin-Elmer
Si 40 or 60	Silica gel	Merck
Silasorb 600	Granulated silica gel	Lachema
Silica 2362 C_8 or C_{18}	Silica for reversed-phase chromatography of proteins, now produced as RP-P	SynChrom
Silica DMPA	Silica column dynamically modified with different di- or polyamine modifiers	Home made
Silica Gel L	Granulated silica gel	Lachema
Silica + Modifier	Silica impregnated _in situ_ with HPLC Amine Modifier I	Home made
Silica-Pel	Pellicular silica gel Zipax	DuPont
Silicic acid	Silica in prepacked column	Mallinckrodt
Silicone/CPG	Silicone coated controlled porosity glass	Home made
Soy-bean inhibitor/LiChrospher	Inhibitor immobilized on aminosilanized silica	Home made
SP-Glycophase/CPG	Sulphopropylderivative of Glycophase on controlled porosity glass	Home made or Pierce
Spherisorb	Hexylsilica	Deeside
Spherisorb-CN (Cu^{2+})	Packings for ligand-exchange chromatography	Phase Separations
Spherisorb-CN (without Cu^{2+})	Packings for reversed-phase chromatography	Phase Separations
Spherisorb ODS	Octadecylsilane bonded phase	Milton Roy, Phase Separations, Reeve Angel, Spectra Physics
Spheron 300 or 1000	Macroporous fully synthetic glycolmethacrylate-(bis-glycolmethacrylate) copolymer	Lachema
Spheron 1000-C	Weakly acidic macroporous hydrophilic cation exchanger	Lachema
Spheron 1000-DEAE	Medium basic macroporous hydrophilic anion exchanger	Lachema

Name	Description	Source
Spheron 1000-Phosphate	Medium acidic macroporous hydrophilic cation exchanger	Lachema
Spheron 1000-S	Strongly acidic macroporous hydrophilic cation exchanger	Lachema
Spheron 1000-TEAE	Strongly basic macroporous hydrophilic anion exchanger	Lachema
Spherosil ion exchangers	DEA-, QMA-, C- and S-derivatives of silica gel Spherosil	Rhône-Poulenc
Spherosil ODS	Octadecylsilane bonded column	Spectra-Physics
SRPA	Aminoalkyl reversed-phase silica particles	Riedel de Haën
SSC	Sulphonated styrene copolymer X8, Type PA 35; strongly acidic cation exchanger	Beckman
Styragel 1000 A	Highly crosslinked polystyrene resin 10^3 Å	Waters Assoc.
Supelcosil C_{18}	Octadecyl phase bonded to silica	Supelco
Supelcosil LC-8	Octyl phase bonded to silica	Supelco State
Supelcosil LC-18	Octadecyl phase bonded to silica	Supelco
SynChropak AX 300	Microparticulate silica modified with polyethylene-imine, medium basic anion exchanger, macroporous	SynChrom
SynChropak CX 300	Microparticulate cation exchanger, carboxymethyl derivative, macroporous	SynChrom
SynChropak GPC	Hydrophilized macroporous silica for SEC of proteins, numbers 100-300 designate fractionation range	SynChrom
SynChropak RP-P	Column for reversed-phase chromatography	SynChrom
Sulpho-Spheron 300	Strongly acidic fully synthetic macroporous hydrophilic cation exchanger	Home made
TAP	Triamine phase bonded on LiChrosorb	Home made
T.-dyes/Silica	Triazine dyes (Cibacron blue and others) attached to silica	Home made
TMA-silicone/CPG	Siliconized controlled pore glass, coated with tri-alkylmethylammonium chloride	Home made
TMCA-silica	Trimethylcetylamine derivative of silica	Home made
Toyopearl HW, S, F	Semirigid hydrophilic vinyl polymer containing plenty of hydroxyl groups, prepared for SEC of biopolymers. HW 40-75 designate fractionation ranges, S and F particle size	Toyo Soda
TP$_1$/Partisil	Tripeptide phase (Val-Phe-Val) synthesized on silica gel	Home made
TP$_2$/Silica	Tripeptide (Val-Ala-Ser) bonded on silica gel	Home made
TP$_3$/Partisil	Tripeptide phase (Val-Ala-Pro) synthesized on silica gel	Home made
TSK Gels 2000-4000 SW	Hydrophilized macroporous silica gel for SEC of bio-oligomers an biopolymers	Varian, Toyo Soda

Table 14.3 (continued)

Symbol for the packings	Concise characterization of the packings	Producer or distributor
TSK Gels 3000-5000 PW	Macroporous copolymer modified for SEC of oligomers and polymers	Toyo Soda
TSK Gel LS 212	Cation exchanger	Toyo Soda
TSK Gel LS 410	Gel for reversed-phase chromatography	Toyo Soda
TSK Gel IEX-500	Ion exchangers; six products available, including strong and weak anion and cation exchangers	Toyo Soda
Ultrasphere octyl	Octylsilane bonded to silica	Altex Scientific, Rainin
Ultrasphere-ODS	Octadecylsilane bonded to silica	Altex Scientific
Ultrasphere RP-8	Octylderivative of silica for reversed phase chromatography	Altex Scientific
Ultrogel AcA	Gel for SEC; agarose + polyacrylamide matrix	LKB, Réactifs IBF
Ultropak 2135	Column packed with TSK Gel 3000 SW	LKB
Variously coated silica	Columns investigated carried cyanopropyl-, octyl-, octadecyl-, cyclohexyl-, phenylalkyl-, diphenyl- or diol-coatings	Home made
Vydac Pelli AX	High capacity pellicular anion exchanger	The Separation Group
Vydac RP OD	Reversed-phase octadecyl support	The Separation Group
Vydac TP coated	Vydac TP and six other silicas coated with n-octyldi-methylchlorosilane or with octyltrichlorosilane	Home made
W-3 IER	Amino acid analyzer ion-exchange resin	Beckman
W-ODS-2	Octadecylsilyl support	Whatman
Yanapak ODS-T	Octadecylsilyl support	Yanagimoto
Zipax	Pellicular silica gel, controlled surface porosity beads	DuPont
Zipax SCX	Pellicular strong cation exchanger	DuPont
Zorbax C8	Octylsilyl derivative of silica	DuPont
Zorbax CN	Cyanopropylsilyl derivative of silica	DuPont
Zorbax NH2	Aminopropyl silica	DuPont
Zorbax ODS	Octadecylsilyl derivative of silica	DuPont
Zorbax Sil	Totally porous silica gel	DuPont
Zorbax TMS	Trimethylsilyl derivative of silica	DuPont
μBondagel	Microparticulate porous silica gel	Waters Assoc.
μBondapak Alkylphenyl	Microparticulate silica with alkylphenyl surface coatings	Waters Assoc.
μBondapak AX	Amine-bearing columns for "normal phase" chromatography	Varian

μBondapak C18	Microparticulate silica covered with octadecylsilyl bonded phase	Waters Assoc.
μBondapak Carbohydrate	Microparticulate silica with an aminopropyl functional hydrophilic group for "normal phase" chromatography	Waters Assoc.
μBondapak C18/Corasil	Microparticulate Corasil with octadecylsilyl coatings	Waters Assoc.
μBondapak C18/Porasil	Microparticulate Porasil B with octadecylsilyl coatings	Waters Assoc.
μBondapak CN	Column for RPC based on cyano-derivative bonded phase	Waters Assoc.
μBondapak f.a.	Fatty acid analysis column	Waters Assoc.
μBondapak Phenyl	Microparticulate silica with phenylsilyl coatings	Waters Assoc.
μBondapak Phenyl/Corasil	Phenylsilyl coatings on microparticulate Corasil	Waters Assoc.
μBondapak Phenyl/Porasil	Microparticulate Porasil B with phenylsilyl coatings	Waters Assoc.
μCN/Silica	Cyanopropylsilane bonded to 10 μm silica	Waters Assoc.
μPorasil	Microparticulate porous silica	Waters Assoc.

TABLE 14.4

BIOMEDICAL APPLICATIONS OF HPLC (MPLC) OF BIOPOLYMERS AND BIOOLIGOMERS

This Table lists applications of rapid column chromatography of biopolymers and biooligomers in medicine, diagnoses and in metabolic or hereditary studies, or papers introducing new possibilities in these areas. Further chromatographic information on listed cases can be found in Table 14.2. References refer to the full-title Bibliography (Chapter 15).

Diseases, accidents, inherited variants, metabolic abnormalities, diagnoses	Chromatographed substances	Section of Table 14.2	References
14.4.1 Accident victims			
Broken hip	Lactate dehydrogenase isoenzymes in human serum	14.2.1.13	701
Severe accident	Creatine kinase isoenzyme in human serum	14.2.1.9	701
— —	Lactate dehydrogenase isoenzyme profile in victim serum in comparison with normal human serum	14.2.1.13	601

Table 14.4 (continued)

Diseases, accidents, inherited variants, metabolic abnormalities, diagnoses	Chromatographed substances	Section of Table 14.2	References
14.4.2 Biochemical analyses			
Biological fluids	Glutathione in blood	14.2.12.7	501,554
—	Glycopeptides in urine	14.2.12.7	421
—	Oligosaccharides in human serum	14.2.6.2	292
—	Oligosaccharides in urine	14.2.6.2	386
Enzymes	Arylsulphatase isoenzymes in human urine	14.2.1.20	563
—	Creatine kinase isoenzymes	14.2.1.9	74,101,103,104, 129,327,380,602, 604-606,741
—	Lactate dehydrogenase isoenzyme	14.2.1.13	10,101,103,129, 186,381,396,422, 447,470,522,564, 602,603,605,619, 646,647,743
Fibrinogens, human and mammalian	Fibrinopeptides	14.2.12.7	632
Microsomes (solubilized rabbit hepatic, from liver, kidney cortex and intestinal mucose) with or without preliminary drug-treatment	Cytochrome P-450 (multiple forms)	14.2.1.10	387
Triglycerides, selective enzymic detection	Serum lipoproteins	14.2.14.13	240
14.4.3 Cancer, carcinomas, tumours, oncoviruses, etc.			
Avian oncovirus, structural protein	Peptides prepared by acid cleavage	14.2.9.3	291
Cancer patients, urinary nondialyzable carbohydrates	Fucose/galactose ratios	14.2.6.2	292
Colorectal cancer patients	Arylsulphatase isoenzymes in sera	14.2.1.20	75
Medullary thyroid carcinoma, human, tissue culture	Calcitonin	14.2.11.5	513
Mink cell focus-inducing murine leukemia virus, envelope precursor protein	Tryptic digest peptides	14.2.10.2	346
Rauscher murine leukemia virus	Glycopeptides	14.2.12.7	290
Tumour lung, human, cell culture medium	Calcitonin	14.2.11.5	513
Tumour mouse, EHS tumour	Collagen polypeptide chains	14.2.14.4	137

14.4.4 Cardiac episodes (in the cardiatic care units), myocardial infarction

Acute myocardial infarction-serum in comparison with normal serum and preinfarction-serum	Creatine kinase isoenzymes (sequential profiles)	14.2.1.9	601,604
Isoenzyme profiling of normal serum and of a patient in cardiatic care unit	Creatine kinase isoenzymes	14.2.1.9	562,563,601
Isoenzyme profiling of normal human serum and of a patient with periodical chest pains on admission in the cardiatic care unit and 18 h later; consecutive serum profiling of the same patient after cardiac episode	Creatine kinase isoenzymes (MM and MB)	14.2.1.9	701
	Lactate dehydrogenase isoenzymes	14.2.1.13	604,701
Myocardial damage	Lactate dehydrogenase isoenzymes	14.2.1.13	606
Myocardial infarction-serum, serum with elevated LDH-activity, comparison with normal serum	Lactate dehydrogenase isoenzymes (LD_1 to LD_5) profiles	14.2.1.13	185,562,563,601

14.4.5 Diabetes mellitus

Blood haemolysates from normal subjects and diabetic patients	Haemoglobins, glycosylated haemoglobins	14.2.14.2, 14.2.14.9	213,301,556,599, 721,740

14.4.6 Haemoglobinopathies

14.4.6.1 Abnormal haemoglobins

Abnormal haemoglobins	Trypsin-, chymotrypsin- and thermolysin-digest peptides	14.2.10.1	760
	Trypsin-digest peptides	14.2.10.1	266,622,759
Hb-papio B silent mutation	Baboon globin β-chain	14.2.14.9	635
Haemoglobin E	Tryptic-digest peptides	14.2.10.1	621
Analytical	Determination of β to γ ratio of globin chains	14.2.14.9	111
	Haemoglobins, globin chains	14.2.14.9	30,94,101-104,110, 112,195,265,366,485, 486,560,564,634,636, 646,647,674,701,740

Table 14.4 (continued)

Diseases, accidents, inherited variants, metabolic abnormalities, diagnoses	Chromatographed substances	Section of Table 14.2	References
14.4.6.2 Different types of haemoglobinopathies, Hb-disorders			
Newborns and individuals with Hb-disorders	Haemoglobins from blood haemolysates	14.2.14.2	223
Patients with HbS-HPFH condition and other haemoglobinopathies	Haemoglobins	14.2.14.9	303-305
Subjects in A⁺ low F condition	Haemoglobins from haemolysates	14.2.14.2, 14.2.14.9	223,304
14.4.6.3 Genetics, hereditary studies			
Hereditary persistent Aᵧ haemoglobin heterozygosity	CNBr-fragments of haemoglobin	14.2.9.1	655
Various genetic abnormalities (homozygous hereditary persistence of fetal haemoglobin, elevated Hb F, a child with trisomy-D-13 and sicle cell trait, homozygous αβ-, β0- and β+-thalassemia, β-thalassemia heterozygotes or HPFH homozygotes or heterozygotes)	Haemoglobins	14.2.14.9	4,305
14.4.6.4 Haemoglobin variants			
Abnormal haemoglobins in 37 Chinese families	Haemoglobin α- and β-chains, enzymic-digest peptide fragments	14.2.10.1	760
Cambridge	Trypsin-digest peptide	14.2.10.1	265
Chiapas α₂	Trypsin-digest peptides	14.2.10.1	308
βNew York	Globin chain	14.2.14.9	113
Pasadena	Trypsin-digest peptides	14.2.10.1	318
Sickle cell anemia	Haemoglobins	14.2.14.9	4,223,303,304,636
Thalassemia	Haemoglobin (Hb F) CNBr-fragments	14.2.9.1	266
α-Thalassemia	Trypsin-digest peptides of haemo-globin α-chain	14.2.11.1	759
β-Thalassemia	Hb of blood haemolysates	14.2.14.2, 14.2.14.9	201,562,563
β-Thalassemia heterozygotes, homozygotes	Haemoglobins	14.2.14.9	4,305
β+- or A-β-thalassemia	Haemoglobins	14.2.14.2	233,301,304

14.4.7 Influenza

Influenza A haemagglutinin glycoproteins	Trypsin-digest peptides	14.2.10.2	346
M-structural protein from influenza virus	CNBr-fragment	14.2.9.1	291

14.4.8 Lathyrogenic chickens

	Collagen polypeptide chains	14.2.14.4	162,415

14.4.9 Lipoprotein metabolism abnormality

Acute liver hepatitis	Lipoproteins	14.2.14.3	240
Atherosclerotic heart disease, comparison with normal group	Cholesterol content in lipoproteins	14.2.14.13	526
Biochemical analysis of human serum lipoproteins, detection of triglycerides	Lipoproteins	14.2.14.13	240
Biochemical analysis of human serum lipoproteins, determination of cholesterol (incl. enzymic detection)	Lipoproteins	14.2.14.13	239,526,527,529
Biochemical analysis of human serum lipoproteins, phospholipids quantification	Choline containing lipoproteins	14.2.14.13	525
Cirrhosis of liver, comparison with control subjects	Lipoproteins, measurement of cholesterol content	14.2.14.13	527,528,530
—	Lipoproteins	14.2.14.13	240,528
Coronary heart disease	Lipoproteins	14.2.14.13	528
Dislipoproteinemia, comparison with normal subjects	Human serum lipoproteins	14.2.14.13	219
Hyperlipidemia, comparison with normal subjects	Lipoproteins, measuring of cholesterol content	14.2.14.13	239,526,529,530
—, serum of normal young women and of pathological case	Total lipoproteins	14.2.14.13	523
—, various human sera, adult persons, baby and fatty child	Lipoproteins	14.2.14.13	240,528

14.4.10 Liver damage

	Lactate dehydrogenase isoenzyme profiles	14.2.1.13	604,606

14.4.11 Lysosomal storage disorders, affecting the catabolism of glycoproteins

Mucolipidoses, Aspartylglucosaminuria (aspartyl-glycosamine amide hydrolase deficiency; aspartyl-2-deoxy-2-acetamidoglycosylamine accumulates	Glycopeptides in urine	14.2.12.7	511

Table 14.4 (continued)

Diseases, accidents, inherited variants, metabolic abnormalities, diagnoses	Chromatographed substances	Section of Table 14.2	References
—, Fucosidosis (α-L-fucosidase deficiency; fucose-containing sphingolipids and glyco-protein fragments accumulate)	Glycopeptides in urine Oligosaccharides in urine	14.2.12.7 14.2.6.2	511 511
—, Mannosidosis (α-mannosidase deficiency; mannose-containing oligosaccharides accumulate)	Oligosaccharides in urine	14.2.6.2	123,124,511
—, Mucolipidosis I (sialidosis)	Oligosaccharides in urine	14.2.6.2	124
Sialidoses (neuraminic acid oligosaccharides and neuraminic acid glycopeptides accumulate)	Glycopeptides in urine Oligosaccharides in urine	14.2.12.7 14.2.6.2	511 511
Sphingolipidoses, Fabry's disease (α-galacto-sidase deficiency; ceramide trihexoside accumulates)	Glycolipids in plasma	14.2.2.2	695,697
—, Gaucher's disease (β-glucosidase deficiency; glucocerebroside accumulates)	Glycolipids in plasma	14.2.2.2	695–697
—, GM1-, GM2-gangliosidosis (β-galactosidase, hexosaminidase deficiency; GM1-, GM2- ganglio-sides accumulate)	Oligosaccharides in urine	14.2.6.2	511
—, Krabbe's disease (β-galactosidase deficiency; galactocerebroside accumulates)	Glycolipids in plasma	14.2.2.2	695
—, Sandhoff's disease (hexosaminidase A,B deficiency; GM2-ganglioside, globoside accumulate)	Oligosaccharides in urine	14.2.6.2	123,124
—, Tay-Sachs' disease (hexosaminidase A deficiency; GM2-ganglioside accumulates)	Glycolipids	14.2.12.1	679
14.4.12 *Lysosomal storage disorders, affecting the catabolism of proteoglycans*			
Mucopolysaccharidoses	Oligosaccharides, enzymic (chondroitinase) digest of urinary isomeric chondroitin sulphates	14.2.6.1, 14.2.6.2	406,410
14.4.13 *Macroamylasemia*			
	Macroamylase in human serum	14.2.1.20	170

14.4.14 *Multiple myeloma*

—	Enzymic-digest peptides of Bence-Jones protein isolated from urine of a plasmozytom-patient	14.2.10.2	753
	Trypsin-digest peptides of Bence-Jones protein for amino acid sequencing	14.2.10.2	370

14.4.15 *Mumps*

Comparison of patient's serum and normal human serum · Amylases and isoamylases · 14.2.1.4 · 170

14.4.16 *Pancreatitis*

Acute pancreatitis; serum of a patient is compared with normal human serum · Amylases and isoamylases · 14.2.1.4 · 170

Necrotizing pancreatitis · Lactate dehydrogenase isoenzyme profile · 14.2.1.13 · 604

14.4.17 *Muscle damage* · Lactate dehydrogenase isoenzyme profile · 14.2.1.13 · 604,606

14.4.18 *Uremia*

Normal human and uremic serum · Serum peptides · 14.2.12.7 · 428

TABLE 14.5

APPENDIX

This Table contains the list of references corresponding to the full title Bibliography Addendum (Section 15.1), which collects more recent citations (1983-1986). In the majority of cases this Appendix presents group classification of the listed material. In addition to chromatographed substances it also contains references to methodically or theoretically oriented works and to international symposia organized on the topics described in this book.

Subject	References
Affinity chromatography	772, 801, 823, 934, 1019, 1020, 1118, 1167, 1203, 1248, 1292, 1297
—, immunoaffinity	1151-1153
Antibiotics	781, 803, 808
Antibodies	806, 828, 1006, 1146, 1179
Antigens	1040, 1054, 1076
Automation, robotics, computerization	784, 839, 964, 1066, 1265
Blood proteins, haemoglobins and globins	771, 796, 804, 812, 995, 1077, 1103, 1115, 1137, 1165, 1177, 1218, 1246, 1254
—, immunoglobulins, γ-globulins	837, 843, 890, 891, 900, 949, 1114, 1183, 1229
—, plasma and serum proteins	829, 875, 952, 1033, 1223, 1235, 1242, 1245
—, serum albumin	786, 818, 840, 1050, 1206, 1236
Carbohydrates	1156
Chromatofocusing	973, 974
Chromatographic special techniques	848, 1039, 1043
Detection methods, amino acids and peptides	845, 857, 879, 916, 1002, 1130, 1133, 1209
—, chemiluminescence and fluorescence	850, 966, 1096, 1260
—, derivatization	1166
—, potentiometric, electrochemical	766, 1011, 1209
—, radioactivity	1187, 1201
Enzymes	790, 807, 817, 827, 830, 836, 868, 887, 947, 970, 976, 978, 982, 1006, 1011, 1062, 1074, 1085, 1089, 1094, 1106, 1124, 1167, 1187, 1188, 1210, 1224, 1240, 1271, 1293, 1301

Table 14.5 (continued)

Subject	References
—, miscellaneous	770, 849, 863, 955, 986, 994, 1008, 1030, 1090, 1136, 1172, 1173
—, phosphopeptides	913, 1284
—, prediction of retention data	782, 915, 921, 922
Plasmin, plasminogen	984, 1194
Polysaccharides	799, 931, 1111
Preparative and large scale chromatography, chemical engineering	832, 993, 1079, 1176, 1202
Proteinases, peptidases, inhibitors	903, 912, 920, 985, 996, 1191, 1193, 1267, 1295
Proteins, cytochrome P-450	783, 886, 980, 1086
—, general aspects, separation and theory	768, 769, 811, 838, 841, 854, 871 – 873, 896, 903, 904, 906, 919, 926, 943, 948, 957, 979, 981, 1024, 1026, 1078, 1098, 1129, 1142, 1147, 1222, 1238, 1250, 1255, 1263, 1290, 1291
—, insulin	1051, 1163, 1275
—, milk	777, 792
—, miscellaneous	791, 800, 814, 819, 918, 935, 965, 1027, 1049, 1052, 1055, 1072, 1164, 1205, 1215, 1259, 1303
—, Rhodopsin	851, 1116
—, Toxins	1148, 1175, 1227, 1252, 1287
—, Urinary	1063, 1087
Protein sequencing and the determination of the covalent structure, separation of DNS-amino acids	948, 1197
— — —, separation of PTH derivatives of amino acids	764, 779, 844, 883, 999, 1037, 1069, 1131, 1144, 1145, 1185, 1196, 1219
— — —, strategy, peptide fractionation or mapping, sample preparation, CNBr fragments purification	880, 885, 916, 1000, 1001, 1003, 1056, 1070, 1073, 1083, 1099, 1100 – 1102, 1112, 1128, 1141, 1237, 1264
Proteoglycans	805, 1169
RPC techniques	861, 991, 992
Supports (syntheses, evaluations, analyses)	774, 825, 847, 906, 907, 958, 1015, 1032, 1036, 1064, 1068, 1104, 1170, 1226, 1256, 1257, 1266
Symposia	936, 937, 944, 945, 1031, 1158, 1249, 1258, 1273, 1302
Synthetic vaccines	797, 856, 932, 1058, 1113, 1302
Theoretical relationships	968, 991, 992, 1010, 1022, 1071
Viruses and viral proteins	815, 820, 954, 1013, 1149, 1161, 1262, 1276 – 1279, 1288

Chapter 15

BIBLIOGRAPHY

   This chapter consists of a list of full title references of original papers, reviews and books dealing with the HPLC and MPLC of proteins and peptides, nucleic acids and oligonucleotides, polysaccharides and oligosaccharides, glyco- lipids containing oligosaccharide chains and similar biooligomeric substances and/or describing chromatographic packings developed for the separation of such compounds. The English translation is given in parentheses after the title in the original language, if it differs from English. If not evident from the title an additional characteristic of a paper is briefly given in square brackets. For papers in a language other than English or for papers published in journals or publications not available in the majority of libraries the *Chemical Abstracts* (*C.A.*) references are also given. The citations are numbered in spite of the fact that they are listed alphabetically because they also represent references for Chapter 14 (Register of all biopolymers and their oligomeric fragments separated by HPLC or MPLC methods). Please note that an Addendum, listing the most recent publications, starts on p. B612.

1 Abbott, S.R., Practical aspects of normal-phase chromatography [review in- cluding besides others short peptides, short oligosaccharides, glycolipids], *J. Chromatogr. Sci.*, 18 (1980) 540-550.
2 Abbott, S.R., Mixed phase chromatographic compositions [for the separation of multisite biopolymers, specifically t-RNAs], *U.S. Pat.*, 4 298 500 (1981); *C.A.*, 96 (1982) 31 217n.
3 Abercombie, D.M., Hough, C.J., Seeman, J.R., Brownstein, M.J., Gainer, H., Russell, J.T. and Chaiken, I.M., Use of reverse-phase high-performance liquid chromatography in structural studies of neurophysins, photolabeled derivatives, and biosynthetic precursors, *Anal. Biochem.*, 125 (1982) 395-405.
4 Abraham, E.C., Cope, N.D., Braziel, N.N. and Huisman, T.H.J., On the chroma- tographic heterogeneity of human fetal hemoglobin, *Biochim. Biophys. Acta*, 577 (1979) 159-169.
5 Abrahamsson, M. and Groeningsson, K., High performance liquid chromatography of the tetrapeptide somatostatin, *J. Liq. Chromatogr.*, 3 (1980) 495-511.
6 Aitzetmueller, K., Boehrs, M. and Arzberger, E., Separation of higher sugars using HPLC Amine Modifier I, *J. High Resolut. Chromatogr., Chromatogr. Commun.*, 2 (1979) 589-590.
7 Alcock, N.J., Eckers, Ch., Games, D.E., Games, M.P.L., Lant, M.S., McDowall, M.A., Rossiter, M., Smith, R.W., Westwood, S.A. and Hee-Yen Wong, High- performance liquid chromatography - mass spectrometry with transport interfaces [review including also peptides and sugars], *J. Chromatogr.*, 251 (1982) 165-174.
8 Alfredson, T.V., Wehr, C.T., Tallman, L. and Klink, F., Evaluation of new microparticulate packings for aqueous steric exclusion chromatography [in- cluding proteins], *J. Liq. Chromatogr.*, 5 (1982) 489-524.

B568

9 Alpert, A.J., New materials and techniques for HPLC of proteins [PhD Thesis, Purdue University 1979], Diss. Abstr., Int. B, 41 (1980) 2145; C.A., 94 (1981) 79 538d.
10 Alpert, A.J. and Regnier, F.E., Preparation of a porous microparticulate anion-exchange chromatography support for proteins [also for the separation of oligonucleotides], J. Chromatogr., 185 (1979) 375-392.
11 Alvarez, V.L., Roitsch, C.A. and Henriksen, O., High-pressure liquid chromatography of proteins and peptides, Immunol. Methods, 2 (1981) 83-103; C.A., 95 (1981) 111 063d.
12 Alvarez, V.L., Roitsch, C.A. and Henriksen, O., Purification of H-2^a heavy chain and β_2-microglobulin by reverse-phase high-performance liquid chromatography, Anal. Biochem., 115 (1981) 353-358.
13 Amoscato, A.A., Babcock, G.F. and Nishioka, K., Analysis of contaminants in commercial preparations of the tetrapeptide tuftsin by high-performance liquid chromatography, J. Chromatogr., 205 (1981) 179-184.
14 Anderson, J.K., Hollaway, W.L. and Mole, J.E., Ion-pair reverse-phase HPLC of tryptic peptides from plasma amyloid P-component using a volatile buffer system, J. High Resolut. Chromatogr., Chromatogr. Commun., 4 (1981) 417-418.
15 Ando, J. and Ariji, T., Koatsu Ekitai Kuromatogurafui, Sono Seikagaku to Ikagakue no Oyo (High Pressure Liquid Chromatography, Its Use in Biochemistry and Medical Chemistry), Tokyo Kagaku Dojin, Tokyo, 1980, 410 pp; C.A., 94 (1981) 27 008m.
16 Ando, S., Isobe, M. and Nagai, Y., High performance preparative column chromatography of lipids using a new porous silica, Iatrobeads®. I. Separation and molecular species of sphingoglycolipids, Biochim. Biophys. Acta, 424 (1976) 98-105.
17 Anhalt, J.P., Antibiotics, in Marton, L.J. and Kabra, P.M. (Editors), Liquid Chromatography in Clinical Analysis, Humana Press, Clifton, NJ, 1981, pp. 163-184.
18 Aoshima, H., High performance liquid chromatography studies on protein: multiple forms of soybean lipoxygenase-1, Anal. Biochem., 95 (1979) 371-376.
19 Asakawa, N., Tsuno, M., Hattori, T., Ueyama, M., Shinoda, A. and Miyake, Y., High performance liquid chromatography of proteins. II. Separation of proteins by reversed phase high performance liquid chromatography [in Japanese], Yakugaku Zasshi, 101 (1981) 279-282; C.A., 95 (1981) 43 643r.
20 Asakawa, N., Tsuno, M., Hattori, T., Ueyama, M., Shinoda, A. and Miyake, Y., High performance liquid chromatography of proteins. III. Separation of proteins by reversed phase high performance liquid chromatography [in Japanese], Yakugaku Zasshi, 101 (1981) 708-712; C.A., 95 (1981) 164 872y.
21 Aukati, M.F., Bubenshchikova, S.N., Kagramanova, V.K. and Baratova, L.A., Ob ispol'zovanii novogo anionoobmennika dlya razdeleniya oligonukleotidov metodom zhidkostnoi khromatografii vysokogo davleniya (On the use of new anion-exchanger for separation of oligonucleotides by high pressure liquid chromatography), Vestn. Mosk. Univ., Ser. 2: Khim., 19 (1978) 350-352; C.A., 89 (1978) 159 647a.
22 Aukaty, M.F., Bubenschikova, S.N., Kagrammanova, V.K. and Baratova, L.A., Use of an amino-silica column for the high-performance liquid chromatographic analysis of synthetic oligodeoxynucleotides, J. Chromatogr., 137 (1977) 351-356.
23 Axelsen, K.S. and Vogelsang, S.H., High-performance liquid chromatographic analysis of gramicidin, a polypeptide antibiotic, J. Chromatogr., 140 (1977) 174-178.
24 Baenziger, J.U. and Natowicz, M., Rapid separation of anionic oligosaccharide species by high performance liquid chromatography, Anal. Biochem., 112 (1981) 357-361.

25 Bakkum, J.T.M., Beyerman, H.C., Hoogerhout, P., Olieman, C. and Voskamp, D., Reverse-phase high-performance liquid chromatography of protected peptides in the sequential synthesis of secretin and analogs, *Recl. Trav. Chim. Pays-Bas*, 96 (1977) 301-306; *C.A.*, 88 (1978) 105 770w.

26 Baranowski, R., Westenfelder, C. and Currie, B.L., Identification of enzymatically produced peptide fragments by liquid chromatography and mass spectrometry, *Anal. Chem.*, 121 (1982) 97-102.

27 Barford, R.A., High performance liquid chromatography: potential for protein separations, *Proc. Annu. Reciprocal Meat Conf.*, 33 (1981) 102-106; *C.A.*, 95 (1981) 38 240c.

28 Barford, R.A., Kupec, J. and Fishman, M.L., Monitoring keratin degradation in activated sludge by rapid gel permeation chromatography, *Water Pollut. Control Fed. J.*, May (1977) 764-767; *C.A.*, 87 (1977) 72 866f.

29 Barford, R.A., Olszewski, L.T., Saunders, D.H., Magidman, P. and Rothbart, H.L., Bound-monolayer cation exchangers for the analysis of biochemically significant molecular species, *J. Chromatogr. Sci.*, 12 (1974) 555-558.

30 Barford, R.A., Sliwinski, B.J., Breyer, A.C. and Rothbart, H.L., Mechanism of protein retention in reversed-phase high-performance liquid chromatography, *J. Chromatogr.*, 235 (1982) 281-288.

31 Barford, R.A., Sliwinski, B.J. and Rothbart, H.L., Interactions of proteins with bonded-phase ion exchangers, *J. Chromatogr.*, 185 (1979) 393-402.

32 Barford, R.A., Sliwinski, B.J. and Rothbart, H.L., Observations on the rapid size-exclusion chromatography of proteins, *Chromatographia*, 12 (1979) 285-288.

33 Barker, P.E., Hatt, B.W. and Vlachogiannis, G.J., Suitability of TSK-gel Toyopearl packing for the gel permeation chromatographic analysis of dextran, *J. Chromatogr.*, 208 (1981) 74-77.

34 Barth, H.G., A practical approach to steric exclusion chromatography of water-soluble polymers [review], *J. Chromatogr. Sci.*, 18 (1980) 409-429.

35 Barth, H.G., High performance gel permeation chromatography of pectins, *J. Liq. Chromatogr.*, 3 (1980) 1481-1496.

36 Barth, H.G. and Regnier, F.E., High-performance gel permeation chromatography of water-soluble cellulosics, *J. Chromatogr.*, 192 (1980) 275-293.

37 Barth, H.G. and Regnier, F.E., High-performance gel permeation chromatography of industrial gums; analysis of pectins and water-soluble cellulosics, *Methods Carbohydr. Chem.*, in press.

38 Basak, S. and Compans, R.W., Improved separation of glycosylated tryptic peptides by RP-HPLC, *J. High Resolut. Chromatogr., Chromatogr. Commun.*, 4 (1981) 302-304.

39 Basak, S. and Compans, R.W., Variation of glycosylation sites in H_1N_1 strains of influenza virus, in Nayak, D.P. (Editor), *Genetic Variation Among Influenza Viruses (ICN-UCLA Symposia on Molecular and Cellular Biology*, Vol. 21), Academic Press, New York, 1981, pp. 253-262.

40 Bataille, D., Besson, J., Gespach, C. and Rosselin, G., High-performance liquid chromatography of hormonal peptides: use of trifluoroacetic acid-diethylamine for their separation, in Rosselin, G., Fromageot, P. and Bonfils, S. (Editors), *Hormone Receptors in Digestion and Nutrition (Proc. 2nd Int. Symp. Horm. Recept. Dig. Tract Physiol.)*, Elsevier, Amsterdam, 1979, pp. 79-88; *C.A.*, 92 (1980) 106 667n.

41 Bauer, H., Buervenich, C., Fuchs, S., Kraas, E. and Voelter, W., High-pressure liquid chromatography of peptide hormones, in Gupta, D. and Voelter, W. (Editors), *Hypothalamic Hormones - Chemistry, Physiology and Clinical Applications (Proc. 2nd Eur. Colloq. Hypothal. Horm. 1976)*, Verlag Chemie, Weinheim, 1978, pp. 77-83; *C.A.*, 90 (1979) 99 353k.

42 Belagaje, R., Brown, E.L., Fritz, H.J., Lees, R.G. and Khorana, H.G., Studies on polynucleotides. CXLIX. Total syntheses of a tyrosine suppressor transfer RNA gene. XIV. Chemical syntheses of oligonucleotide segments corresponding to the terminal regions, *J. Biol. Chem.*, 254 (1979) 5765-5780.

B570

43 Bennett, D.J. and Creaser, E.H., Automated column peptide analysis, *Anal. Biochem.*, 37 (1970) 191-194.
44 Bennett, H.P.J., Browne, C.A., Brubaker, P.L. and Solomon, S., A comprehensive approach to the isolation and purification of peptide hormones using only reversed-phase liquid chromatography, in Hawk, G.L., Champlin, P.B., Hutton, R.F. and Mol, Ch. (Editors), *Biological/Biomedical Applications of Liquid Chromatography III (Chromatographic Science Series*, Vol. 18), Marcel Dekker, New York, Basle, 1982, pp. 197-210; *C.A.*, 96 (1982) 136 024a.
45 Bennett, H.P.J., Browne, C.A., Goltzman, D. and Solomon, S., Isolation of peptide hormones by reversed-phase high-pressure liquid chromatography, in Gross, E. and Meienhofer, J. (Editors), *Peptides: Structure and Biological Function (Proc. 6th Am. Pept. Symp.)*, Pierce, Co., Rockford, IL, 1979, pp. 121-124; *C.A.*, 93 (1980) 234 214v.
46 Bennett, H.P.J., Browne, C.A. and Solomon, S., The use of perfluorinated carboxylic acids in the reversed-phase HPLC of peptides, *J. Liq. Chromatogr.*, 3 (1980) 1353-1365.
47 Bennett, H.P.J., Browne, C.A. and Solomon, S., Purification of the two major forms of rat pituitary corticotropin using only reversed-phase liquid chromatography, *Biochemistry*, 20 (1981) 4530-4538.
48 Bennett, H.P.J., Browne, C.A. and Solomon, S., Biosynthesis of phosphorylated forms of corticotropin-related peptides, *Proc. Natl. Acad. Sci. U.S.A.*, 78 (1981) 4713-4717.
49 Bennett, H.P.J., Browne, C.A. and Solomon, S., Complete purification of pituitary peptides using reversed-phase HPLC alone, in Rich, D.H. and Gross, E. (Editors), *Peptides: Synthesis-Structure-Function (Proc. 7th Am. Pept. Symp.)*, Pierce Chemical Co., Rockford, IL, 1981, pp. 785-788; *C.A.*, 97 (1982) 17 280s.
50 Bennett, H.P.J., Solomon, S. and Goltzman, D., Isolation and analysis of human parathyrin in parathyroid tissue and plasma. Use of reversed-phase liquid chromatography, *Biochem. J.*, 197 (1981) 391-400.
51 Benson, J.V., Jr, Jones, R.T., Cormick, J. and Patterson, J.A., Accelerated automatic chromatographic analysis of peptides on a spherical resin, *Anal. Biochem.*, 16 (1966) 91-106.
52 Bergh, M.L.E., Koppen, P. and Van den Eijnden, D.H., High-pressure liquid chromatography of sialic acid-containing oligosaccharides, *Carbohydr. Res.*, 94 (1981) 225-229.
53 Bergh, M.L.E., Koppen, P.L. and Van den Eijnden, D.H., Specificity of ovine submaxillary-gland sialyltransferases. Application of high-pressure liquid chromatography in identification of sialo-oligosaccharide products, *Biochem. J.*, 201 (1982) 411-415.
54 Bethell, G.S., Ayers, J.S., Hearn, M.T.W. and Hancock, W.S., Investigation of the activation of various insoluble polysaccharides with 1,1'-carbonyl-diimidazole and of the properties of the activated matrices, *J. Chromatogr.*, 219 (1981) 361-372.
55 Bhown, A.S., Bennett, J.C., Mole, J.E. and Hunter, E., Purification and characterization of the gag gene products of avian-type C retroviruses by high-pressure liquid chromatography, *Anal. Biochem.*, 112 (1981) 128-134.
56 Biemond, M.E.F., Sipman, W.A. and Olivie, J., Quantitative determination of polypeptides by gradient elution high-pressure liquid chromatography. *J. Liq. Chromatogr.*, 2 (1979) 1407-1435.
57 Bij, K.E., Horváth, Cs., Melander, W.R. and Nahum, A., Surface silanols in silica-bonded hydrocarbonaceous stationary phases. II. Irregular retention behavior and effect of silanol masking [includes peptide separations], *J. Chromatogr.*, 203 (1981) 65-84.
58 Bishop, C.A., Hancock, W.S., Brennan, S.O., Carrel, R.W. and Hearn, M.T.W., High performance chromatography of amino acids, peptides and proteins. XXIII. Peptide mapping by hydrophilic ion-paired, reversed-phase high performance liquid chromatography for the characterization of the tryptic digest of haemoglobin variants, *J. Liq. Chromatogr.*, 4 (1981) 599-612.

59 Bishop, C.A., Harding, D.R.K., Mayer, L.J., Hancock, W.S. and Hearn, M.T.W., High-performance liquid chromatography of amino acids, peptides and proteins. XXI. The application of preparative reversed-phase high-performance liquid chromatography for the purification of a synthetic underivatised peptide, *J. Chromatogr.*, 192 (1980) 222-227.

60 Bishop, C.A., Mayer, L.J., Harding, D.R., Hancock, W.S. and Hearn, M.T.W., High-performance liquid chromatography of peptides and proteins. XXIV. The preparative separation of synthetic peptides on reversed-phase silica packed in radially compressed flexible-walled columns, *J. Liq. Chromatogr.*, 4 (1981) 661-680.

61 Black, C., Douglas, D.M. and Tanzer, M.L., Separation of cyanogen bromide peptides of collagen by means of high-performance liquid chromatography, *J. Chromatogr.*, 190 (1980) 393-400.

62 Black, L.T. and Bagley, E.B., Determination of oligosaccharides in soybeans by high pressure liquid chromatography using an internal standard, *J. Am. Oil Chem. Soc.*, 55 (1978) 228-232; *C.A.*, 88 (1978) 119 483s.

63 Blagrove, R.J. and Frenkel, M.J., Determination of protein molecular weights in denaturing solvents using glyceryl-CPG, *J. Chromatogr.*, 132 (1977) 399-404.

64 Blevins, D.D., Burke, M.F. and Hruby, V.J., Parameters affecting high performance liquid chromatographic separations of neurohypophyseal peptide hormones, *Anal. Chem.*, 52 (1980) 420-424.

65 Blevins, D.D., Burke, M.F., Hruby, V.J. and Larsen, B.R., Factors affecting the separation of arginine vasopressin peptide diastereoisomers by HPLC, *J. Liq. Chromatogr.*, 3 (1980) 1299-1318.

66 Block, W., Luestorff, J. and Pingoud, A., The use of HPLC for the analysis and determination of signal nucleotides derived from guanosine and adenosine, *Fresenius' Z. Anal. Chem.*, 311 (1982) 422.

67 Block, W. and Pingoud, A., The identification and analysis of nucleotides bound to the elongation factor Tu from Escherichia coli, *Anal. Biochem.*, 114 (1981) 112-117.

68 Bock, H.G., Skene, P., Fleischer, S., Cassidy, P. and Harshman, S., Adsorption chromatography [of proteins] on controlled pore glass with the use of chaotropic buffers, *Science (Washington, D.C.)*, 191 (1976) 380-383.

69 Boehlen, P., High-performance liquid chromatography in neuropeptide research [review], *Psychopharmacol. Bull.*, 15 (1979) 46-50; *C.A.*, 92 (1980) 18 175v.

70 Boehlen, P. and Kleeman, G., Analytical and preparative mapping of complex peptide mixtures by reversed-phase high-performance liquid chromatography, *J. Chromatogr.*, 205 (1981) 65-75.

71 Bohan, T.P. and Meek, J.L., Met-enkephalin: rapid separation from brain extracts using high-pressure liquid chromatography, and quantitation by binding assay, *Neurochem. Res.*, 3 (1978) 367-372; *C.A.*, 89 (1978) 102 978e.

72 Borák, J., Čaderský, I., Kiss, F., Smrž, M. and Víška, J., Poly(hydroxyethyl methacrylate) gels (Spheron®) [separation of proteins], in Epton, R. (Editor), *Chromatography of Synthetic and Biological Polymers, Vol. 1, Column Packings, GPC, GF and Gradient Elution*, Ellis Horwood, Chichester, 1978, pp. 91-108.

73 Borchert, A., Larsson, P.-O. and Mosbach, K., High-performance liquid affinity chromatography on silica-bound concanavalin A [of glycoproteins peroxidase and glucose oxidase], *J. Chromatogr.*, 244 (1982) 49-56.

74 Bostick, W.D., Denton, M.S. and Dinsmore, S.R., Liquid-chromatographic separation and on-line bioluminescence detection of creatine kinase isoenzymes, *Clin. Chem. (Winston-Salem, N.C.)*, 26 (1980) 712-717.

75 Bostick, W.D., Dinsmore, S.R., Mrochek, J.E. and Waalkes, T.P., Separation and analysis of arylsulfatase isoenzymes in body fluids of man, *Clin. Chem. (Winston-Salem, N.C.)*, 24 (1978) 1305-1316; *C.A.*, 89 (1978) 142 370g.

76 Bradshaw, R.A., Bates, O.J. and Benson, J.R., Peptide separations on substituted polystyrene resins. Effect of cross-linkage, *J. Chromatogr.*, 187 (1980) 27-33.

77 Bremer, E.G., Gross, S.K. and McCluer, R.H., Quantitative analysis of mono-sialogangliosides by high performance liquid chromatography of the perbenzoylated derivatives, *J. Lipid Res.*, 20 (1979) 1028-1035; *C.A.*, 92 (1979) 054 407f.

78 Brobst, K.M. and Scobell, H.D., Modern chromatographic methods for the analysis of carbohydrate mixtures [lower oligosaccharides], *Starch/Stärke*, 34 (1982) 117-121.

79 Brown, E. and Boschetti, E., The theory and development of new hydrophilic polymers for separating of active proteins [review in French], *Inf. Chim.*, 222 (1982) 131-134; *C.A.*, 96 (1982) 218 596c.

80 Brown, P.R., *High Pressure Liquid Chromatography (Biochemical and Biomedical Applications)*, Academic Press, New York, 1973, 202 pp.; *C.A.*, 79 (1973) 102 378c.

81 Brown, P.R., Latest developments in HPLC [review including diguanosine oligophosphates], *Chem. Aust.*, 45 (1978) 257-264; *C.A.*, 90 (1979) 50 699p.

82 Brown, R.E., Cayley, P.J. and Kerr, I.M., Analysis of (2'-5')-oligo(A) and related oligonucleotides by high-performance liquid chromatography, *Methods Enzymol.*, 79 (1981) 208-216.

83 Browne, C.A., Bennett, H.P.J. and Solomon, S., Biosynthesis of glycosylated and phosphorylated forms of corticotropin-like intermediary lobe peptide, in Rich, D.H. and Gross, E. (Editors), *Peptides: Synthesis, Structure, Function (Proc. of the 7th Amer. Pept. Symp.)*, Pierce Chemical Co., Rockford, IL, 1981, pp. 509-512.

84 Browne, C.A., Bennett, H.P.J. and Solomon, S., Isolation and characterization of corticotropin- and melanotropin-related peptides from the neurointer-mediary lobe of the rat pituitary by reversed-phase liquid chromatography, *Biochemistry*, 20 (1981) 4538-4546.

85 Browne, C.A., Bennett, H.P.J. and Solomon, S., The isolation and characteriza-tion of γ_3-melanotropin from the neurointermediary lobe of the rat pituitary, *Biochem. Biophys. Res. Commun.*, 100 (1981) 336-343.

86 Brubaker, P.L., Bennett, H.P.J., Baird, A.C. and Solomon, S., Isolation of $ACTH_{1-39}$, $ACTH_{1-38}$ and CLIP from the calf anterior pituitary, *Biochem. Biophys. Res. Commun.*, 96 (1981) 1441-1448.

87 Brynda, E., Štrop, P., Mikeš, F. and Kálal, J., Polarity of some chromato-graphic materials intended for the separation of biopolymers: spectroscopic label technique, *J. Chromatogr.*, 196 (1980) 39-50.

88 Buchholz, K., Goedelmann, B. and Molnar, I., High-performance liquid chro-matography of proteins: analytical applications [trypsin autolysis, cellulases], *J. Chromatogr.*, 238 (1982) 193-202.

89 Bundle, D.R., Iversen, T. and Josephson, S., Preparative medium and high pressure chromatography [in oligosaccharide synthesis; a review], *Am. Lab. (Fairfield, Conn.)*, 12 (1980) 93-94, 96-98; *C.A.*, 93 (1980) 210 683w.

90 Burgus, R. and Rivier, J., Use of high pressure liquid chromatography in the purification of peptides, *Proc. Eur. Pept. Symp.*, *14th, 1976*, (1976) 85-94.

91 Bush, B.T., Frenz, J.H., Melander, W.R., Horváth, Cs., Cashmore, A.R., Dryer, R.N., Knipe, J.O., Coward, J.K. and Bertino, J.R., Retention behavior of pteroyl-oligo-γ-L-glutamates in reversed-phase chromatography, *J. Chromatogr.*, 168 (1979) 343-353.

92 Cahill, W.R. Jr., Kroeff, E.P. and Pietrzyk, D.J., Applications of tert-butyloxycarbonyl-L-amino acid-N-hydroxysuccinimide esters in the chromato-graphic separation and determination of D,L-amino acids and diastereomeric dipeptides, *J. Liq. Chromatogr.*, 3 (1980) 1319-1334.

93 Calatroni, A. and Tira, M.E., Isolation of acidic glycopeptides from urine by means of anion-exchange resins, application to some cases of glyco-sphingolipidosis or mucolipidosis, *Clin. Chim. Acta*, 71 (1976) 137-141; *C.A.*, 85 (1976) 155 932j.

94 Canas, P.E. and Congote, L.F., Effects of cyclic nucleotides on hemoglobin syntheses in fetal calf liver cells in culture [HPLC of globin chains], *Can. J. Biochem.*, 60 (1982) 1-7.

95 Cashmore, A.R., Dreyer, R.N., Horváth, Cs., Knipe, J.O., Coward, J.K. and
   Bertino, J.R., Separation of pteroyl-oligo-γ-L-glutamates by high-performance liquid chromatography, *Methods Enzymol.*, 66 (1980) 459-468.
96 Cayley, P.J., Brown, R.E. and Kerr, I.M., High-performance liquid chromatography of 2'-5' oligoadenylates and related oligonucleotides [review],
   J. Liq. Chromatogr., 5 (1982) 2027-2039; *C.A.*, 98 (1983) 49 798d.
97 Cazes, J. and Delamare, X. (Editors), *Liquid Chromatography of Polymers and
   Related Materials. II.* [including dextrans and monitoring nitrocellulose
   production], Marcel Dekker, New York, Basle, 1980, 262 pp.
98 Chaiken, I.M. and Hough, Ch.J., Mapping and isolation of large peptide
   fragments from bovine neurophysins and biosynthetic neurophysin-containing
   species by high-performance liquid chromatography, *Anal. Biochem.*, 107
   (1980) 11-16.
99 Chang Ding, Greibrokk, T. and Bowers, C.Y., Differentiation of factor C-LHIH
   and the synthetic contraceptive polypeptide, H-Thr-Pro-Arg-Lys-OH,
   Biochim. Biophys. Res. Commun., 65 (1975) 1208-1213.
100 Chang Jui-Yoa, Isolation and characterization of polypeptide at the picomole
    level. Pre-column formation of peptide derivatives with dimethylaminoazo-
    benzene isothiocyanate, *Biochem. J.*, 199 (1981) 537-545.
101 Chang Shung-Ho, High speed anion exchange chromatography of proteins, *Diss.
    Abstr., Int. B*, 37 (1977) 3923-3924.
102 Chang Shung-Ho, Gooding, K.M. and Regnier, F.E., Use of oxiranes in the
    preparation of bonded phase supports [for the separation of proteins], *J.
    Chromatogr.*, 120 (1976) 321-333.
103 Chang Shung-Ho, Gooding, K.M. and Regnier, F.E., High performance liquid
    chromatography of proteins, *J. Chromatogr.*, 125 (1976) 103-114.
104 Chang Shung-Ho, Noel, R. and Regnier, F.E., High speed ion exchange chro-
    matography of proteins, *Anal. Chem.*, 48 (1976) 1839-1845.
105 Chang Shung-Ho and Regnier, F.E., Chromatographic supports and methods and
    apparatus for preparing them, *U.S. Pat.*, 4 029 583; *C.A.*, 87 (1977)
    98 467b.
106 Cheng, S.-L., Estimation of molecular weight distribution of microbial
    polysaccharide by high-performance gel permeation chromatography [in
    Chinese], *T'ai-wan T'ang Yeh Yen Chiu So Yen Chiu Hui Pao*, 90 (1980) 55-66;
    C.A., 95 (1981) 76 264w.
107 Chytilová, Z., Mikeš, O., Farkaš, J. and Štrop, P., Chromatography of
    sugars on DEAE-Spheron, *J. Chromatogr.*, 153 (1978) 37-48.
108 Clark, P.M.S. and Kricka, L.J., High-resolution analytical techniques for
    proteins and peptides and their applications in clinical chemistry
    [review], *Adv. Clin. Chem.*, 22 (1981) 247-296; *C.A.*, 96 (1982) 48 404u.
109 Clarke, M.A. and Brannan, M.A., Sugars in molasses, *Proc. Tech. Sess. Cane
    Sugar Refin. Res.*, (1978) 136-148; *C.A.*, 91 (1979) 59 138h.
110 Congote, L.F., Rapid procedure for globin chain analysis in blood samples
    of normal and β-thalassemic fetuses, *Blood*, 57 (1981) 353-360; *C.A.*, 94
    (1981) 135 229t.
111 Congote, L.F., Reversed-phase high pressure liquid chromatography of globin
    chains: its application for the prenatal diagnosis of β-thalassemia,
    Prog. Clin. Biol. Res., 60 (*Adv. Hemoglobin Anal.*, 1981) 39-52; *C.A.*, 95
    (1981) 128 574u.
112 Congote, L.F., Bennett, H.P.J. and Solomon, S., Rapid separation of the α,
    β, ^{G}γ and ^{A}γ human globin chains by reversed-phase high pressure liquid
    chromatography, *Biochem. Biophys. Res. Commun.*, 89 (1979) 851-858.
113 Congote, L.F. and Kendall, A.G., Rapid analysis of labeled globin chains
    without acetone precipitation or dialysis by high-pressure liquid chro-
    matography and ion exchange chromatography, *Anal. Biochem.*, 123 (1982)
    124-132; *C.A.*, 97 (1982) 123 231m.
114 Conrad, E.C., High pressure liquid chromatography "everything a food chemist
    wanted in an analytical technique and more" [review], in Charalambous, G.
    (Editor), *Liquid Chromatography Analysis of Food and Beverages*, Vol. 2,
    Academic Press, New York, 1979, pp. 237-254; *C.A.*, 91 (1979) 209 394e.

B574

115 Cooke, N.H.C., Viavattene, R.L., Eksteen, R., Wong, W.S., Davies, G. and
Karger, B.L., Use of metal ions for selective separations in high-performance liquid chromatography [including dipeptides], *J. Chromatogr.*,
149 (1978) 391-415.

116 Čoupek, J., Macroporous spherical hydroxyethyl methacrylate copolymers,
their properties, activation and use in high performance affinity chromatography [Spheron, Separon HEMA], in Gribnau, T.C.J., Visser, J. and
Nivard, R.J.F. (Editors), *Affinity Chromatography and Related Techniques*
(*Analytical Chemistry Symposia Series*, Vol. 9), Elsevier, Amsterdam, 1981,
pp. 165-179; *C.A.*, 96 (1982) 100 346t.

117 Čoupek, J., Gemeiner, P., Jirků, V., Kálal, J., Kubánek, V., Kuniak, L.,
Peška, J., Rexová, L'., Štamberg, J., Švec, F., Turková, J., Veruovič, B.
and Zemek, J., Výzkum a vývoj nosičů pro imobilisaci enzymů a buněk v
Československu (Study and development of supports for the immobilization
of enzymes and cells in Czechoslovakia) [review in Czech including supports
suitable also for HPLC of biopolymers], *Chem. Listy*, 75 (1981) 512-533;
C.A., 95 (1981) 37 762n.

118 Čoupek, J., Křiváková, M. and Pokorný, S., New hydrophilic materials for
chromatography: glycol methacrylates [Spheron], *J. Polym. Sci. Polym.*
Symp., 42 Part C (1973) 185-190.

119 Coy, D.H., Determination of purity of large synthetic peptides using HPLC,
in Hawk, G.L., Champlin, P.B., Hutton, R.F., Jordi, C.H. and Mol, C.
(Editors), *Biological/Biomedical Applications of Liquid Chromatography II*
(*Chromatographic Science Series*, Vol. 12), Marcel Dekker, New York, Basle,
1979, pp. 283-292; *C.A.*, 92 (1980) 37 022j.

120 Crone, H.D. and Dawson, R.M., Residual anionic properties of a covalently
substituted controlled-pore glass, glyceryl-CPG, *J. Chromatogr.*, 129
(1976) 91-96.

121 Crowther, J.B., Jones, R. and Hartwick, R.A., High-performance liquid chromatography of the oligonucleotides, *J. Chromatogr.*, 217 (1981) 479-490.

122 Currie, B.L., Chang Jaw-Kang and Cooley, R., High-performance liquid chromatography of enkephalin and endorphin peptide analogs, *J. Liq. Chromatogr.*,
3 (1980) 513-527.

123 Daniel, P.F., De Feudis, D.F., Lott, I.T. and McCluer, R.H., Quantitative
microanalysis of oligosaccharides by high-performance liquid chromatography, *Carbohydr. Res.*, 97 (1981) 161-180.

124 Daniel, P.F., Lott, I.T. and McCluer, R.H., Separation of perbenzoylated
oligosaccharides of biological origin, in Hawk, G.L., Champlin, P.B.,
Hutton, R.F. and Mol, C. (Editors), *Biological/Biomedical Applications*
of Liquid Chromatography III (*Chromatographic Science Series*, Vol. 18),
Marcel Dekker, New York, Basle, 1981, pp. 363-382; *C.A.*, 96 (1982) 100 328p.

125 Darling, T., Alpert, J., Russel, P., Albert, D.M. and Reid, T.W., Rapid
purification of an RNA tumor virus and proteins by high-performance
steric exclusion chromatography on porous glass bead columns, *J. Chro-*
matogr., 131 (1977) 383-390.

126 Dawkins, J.V. and Yeadon, G., Macromolecular separations by liquid exclusion
chromatography, *Faraday Symp. Chem. Soc.* [1980], 15 (1981) 127-138; *C.A.*,
96 (1982) 65 067d.

127 Delaney, S.R., Conrad, H.E. and Glaser, J.H., A high-performance liquid
chromatography approach for isolation and sequencing of chondroitin
sulfate oligosaccharides, *Anal. Biochem.*, 108 (1980) 25-34.

128 Delaney, S.R., Leger, M. and Conrad, H.E., Quantitation of the sulfated
disaccharides of heparin by high performance liquid chromatography, *Anal.*
Biochem., 106 (1980) 253-261.

129 Denton, M.S., Bostick, W.D., Dinsmore, S.R. and Mrochek, J.E., Continuously
referenced on-line monitoring of creatine kinase and lactate dehydrogenase
isoenzymes for use in clinical diagnostics, *Chromatogr. Sci.*, 12 (1979)
(*Biol. Med. Appl. Liq. Chromatogr. 2*) 165-191; *C.A.*, 92 (1980) 36 513h.

130 Desiderio, D.M. and Cunningham, M.D., Triethylamine formate [volatile] buffer for HPLC-field desorption mass spectrometry of oligopeptides, *J. Liq. Chromatogr.*, 4 (1981) 721-733.

131 Desiderio, D.M., Sabbatini, J.Z. and Stein, J.L., HPLC and field desorption mass spectrometry of hypothalamic oligopeptides, *Adv. Mass Spectrom.*, 8B (1980) 1298-1305; *C.A.*, 94 (1981) 43 563u.

132 Desiderio, D.M. and Stout, Ch.B., High performance liquid chromatography and field desorption mass spectrometry measurement of endogenous amounts of neuropeptides in biologic tissue, CRC Press, in press.

133 Desiderio, D.M. and Stout, Ch.B., High performance liquid chromatography and mass spectrometry of neuropeptides in biologic tissue, *Adv. Chromatogr.*, 22 (1983) 1-36.

134 Desiderio, D.M. and Yamada, S., Measurement of endogenous leucine enkephalin in canine thalamus by high-performance liquid chromatography and field desorption mass spectrometry, *J. Chromatogr.*, 239 (1982) 87-95.

135 Desiderio, D.M. and Yamada, S.Ch.B., FDMS measurement of picomole amounts of leucine enkephalin in canine spinal cord tissue, *Biomed. Mass Spectrom.*, 10 (1983) 358-362.

136 Desiderio, D.M., Yamada, S., Tanzer, F.S., Horton, J. and Trimble, J., High-performance liquid chromatographic and field desorption mass spectrometric measurement of picomole amounts of endogenous neuropeptides in biologic tissue, *J. Chromatogr.*, 217 (1981) 437-452.

137 Deyl, Z., Macek, K., Adam, M. and Horáková, M., High-performance gel permeation chromatography of collagens, *J. Chromatogr.*, 230 (1982) 409-414.

138 DiCesare, J.L. and Ettre, L.S., New ways to increase the specificity of detection in liquid chromatography [review including some low molecular weight sugars], *J. Chromatogr.*, 251 (1982) 1-16.

139 Dimenna, G.P. and Segall, H.J., High-performance gel-permeation chromatography of bovine skim milk proteins, *J. Liq. Chromatogr.*, 4 (1981) 639-649.

140 Dinner, A. and Lorenz, L., High performance liquid chromatographic determination of bovine insulin, *Anal. Chem.*, 51 (1979) 1872-1873.

141 Dion, R. and Cedergren, R.J., High-pressure liquid chromatography of tRNA, *J. Chromatogr.*, 152 (1978) 131-136.

142 Diosady, L.L., Bergen, I. and Harwalkar, V.R., High performance liquid chromatography of whey proteins, *Milchwissenschaft*, 35 (1980) 671-674; *C.A.*, 94 (1981) 14 083g.

143 Dizdaroglu, M. and Hermes, W., Separation of small DNA and RNA oligonucleotides by high-performance anion-exchange liquid chromatography, *J. Chromatogr.*, 171 (1979) 321-330.

144 Dizdaroglu, M., Hermes, W., Von Sontag, C. and Schott, H., Separation of the sequence isomers of pyrimidine deoxytetranucleoside triphosphates by high-performance ion-exchange liquid chromatography, *J. Chromatogr.*, 169 (1979) 429-435.

145 Dizdaroglu, M., Krutzsch, H.C. and Simic, M.G., Separation of angiotensin by high-performance liquid chromatography on a weak anion-exchange bonded phase, *Anal. Biochem.*, 123 (1982) 190-193.

146 Dizdaroglu, M., Krutzsch, H.C. and Simic, M.G., Separation of peptides by high-performance liquid chromatography on a weak anion-exchange bonded phase, *J. Chromatogr.*, 237 (1982) 417-428.

147 Dizdaroglu, M. and Simic, M.G., Separation of underivatized dipeptides by high-performance liquid chromatography on a weak anion-exchange bonded phase, *J. Chromatogr.*, 195 (1980) 119-126.

148 Dizdaroglu, M., Simic, M.G. and Schott, M., Separation and sequencing of the sequence isomers of pyrimidine deoxypentanucleoside tetraphosphates by high-performance liquid chromatography, *J. Chromatogr.*, 188 (1980) 273-279.

149 Donald, A.S.R., Separation of hexosamines, hexosaminitols and hexosamine-containing di- and trisaccharides on an amino acid analyser, *J. Chromatogr.*, 134 (1977) 199-203.

150 Dreher, T.W., Hawthorne, D.B. and Grant, B.R., Comparison of open-column and high-performance gel permeation chromatography in the separation and molecular-weight estimation of polysaccharides [dextrans, pullulan], *J. Chromatogr.*, 174 (1979) 443-446.

151 Dubin, P.L., Aqueous exclusion chromatography [general review including also biopolymers], *Sep. Purif. Methods*, 10 (1981) 287-313; *C.A.*, 96 (1982) 173 612m.

152 Dutot, G., Biou, D., Durand, G. and Pays, M., High-performance liquid chromatography of oligosaccharides in the form of dansylhydrazones [in French], *Feuill. Biol.*, 22 (1981) 101-104; *C.A.*, 95 (1981) 217 186m.

153 Edelson, E.H., Lawless, J.G., Wehr, C.T. and Abbott, S.R., Ion-exchange separation of nucleic acid constituents by high-performance liquid chromatography, *J. Chromatogr.*, 174 (1979) 409-419.

154 Eltekov, Yu.A., Kiselev, A.V., Khokhlova, T.D. and Nikitin, Yu.S., Adsorption and chromatography of proteins on chemically modified macroporous silica-aminosilochrom, *Chromatographia*, 6 (1973) 187-189.

155 Engelhardt, H., Application of HPLC to the separation of amino acids, proteins and their derivatives, in Lottspeich, F., Henschen, A. and Hupe, K.-P. (Editors), *High Performance Chromatography in Protein and Peptide Chemistry*, Walter de Gruyter, Berlin, 1981, pp. 55-69.

156 Engelhardt, H., Ahr, G. and Hearn, M.T.W., High-performance liquid chromatography of amino acids, peptides and proteins. XXXI. Experimental studies with a bonded N-acetylaminopropylsilica stationary phase for the aqueous high-performance exclusion chromatography of polypeptides and proteins, *J. Liq. Chromatogr.*, 4 (1981) 1361-1379.

157 Engelhardt, H. and Mathes, D., Chemically bonded stationary phases for aqueous high-performance exclusion chromatography [dextrans, proteins], *J. Chromatogr.*, 142 (1977) 311-320.

158 Engelhardt, H. and Mathes, D., High performance liquid chromatography of proteins using chemically-modified silica supports, *Chromatographia*, 14 (1981) 325-332.

159 Epton, R., *Chromatography of Synthetic and Biological Polymers. Vol. 1. Column Packings, GPC, GF and Gradient Elution*, (368 pp.). *Vol. 2. Hydrophobic, Ion-Exchange et Affinity Methods*, (353 pp.), Ellis Horwood, Chichester, 1978.

160 Erickson, B.W. and Prystowsky, M.B., Continuous-flow solid-phase peptide synthesis using an HPLC system, in Hawk, G.L., Champlin, P.B., Hutton, R.F., Jordi, H.C. and Mol, Ch. (Editors), *Biological/Biomedical Applications of Liquid Chromatography II. (Chromatographic Science Series*, Vol. 12), Marcel Dekker, New York, Basle, 1979, pp. 293-305; *C.A.*, 92 (1980) 164 280s.

161 Eto, I. and Krumdieck, C.L., Determination of three different pools of reduced one-carbon-substituted folates. III. Reversed-phase high-performance liquid chromatography of the azo dye derivatives of p-aminobenzoyl-poly-γ-glutamates and its application to the study of unlabeled endogenous pteroylpolyglutamates of rat liver, *Anal. Biochem.*, 120 (1982) 323-329.

162 Fallon, A., Lewis, R.V. and Gibson, K.D., Separation of the major species of interstitial collagen by reverse-phase high-performance liquid chromatography, *Anal. Biochem.*, 110 (1981) 318-322.

163 Fiedler, H.P., Preparative-scale high-performance liquid chromatography of ferricrocin, a microbial product, *J. Chromatogr.*, 209 (1981) 103-106.

164 Fiedler, H.P., Quantitation of nikkomycins in biological fluids by ion-pair reversed-phase high-performance liquid chromatography, *J. Chromatogr.*, 204 (1981) 313-318.

165 Fischer, L.J., Thies, R.L. and Charkowski, D., High performance liquid chromatographic separation of insulin, glucagon and somatostatin, *Anal. Chem.*, 50 (1978) 2143-2144.

166 Fishman, M.L., Pfeffer, P.E., Doner, L.W., Barford, R.A. and Hoagland, P.D., Characterization of pectin by high performance size exclusion chromatography (HPSEC) on E linear μBondagel. *Presented at the International Symposium on Column Liquid Chromatography Philadelphia, PA, June 6-11, 1982.*

167 Floor, E. and Leeman, S.E., Substance P sulfoxide: Separation from substance P [undecapeptide amide] by high-pressure liquid chromatography, biological and immunological activities and chemical reduction, *Anal. Biochem.*, 101 (1980) 498-503.

168 Fong, G. Wing-Kin and Grushka, E., High-pressure liquid chromatography of amino acids and dipeptides on a tripeptide bonded stationary phase, *J. Chromatogr.*, 142 (1977) 299-309.

169 Fong, G. Wing-Kin and Grushka, E., Effects of pH, ionic strength and organic modifier on the chromatographic behavior of amino acids and peptides using a bonded peptide stationary phase, *Anal. Chem.*, 50 (1978) 1154-1161.

170 Fourmy, D., Pradayrol, L., Bommelaer, G. and Ribet, A., Macroamylasémie: détection rapide par chromatographie liquide haute performance (Macroamylasemia: rapid detection by high-performance liquid chromatography) [macroamylase], *Gastroenterol. Clin. Biol.*, 6 (1982) 249-251; *C.A.*, 96 (1982) 176 658s.

171 Fox, L., Anthony, G.D. and Lau, E.P.K., High-performance liquid chromatographic determination of L-aspartyl-L-phenylalanine methyl ester in various food products and formulations [aspartame; sweetener], *J. Assoc. Off. Anal. Chem.*, 59 (1976) 1048-1050; *C.A.*, 86 (1977) 15 283u.

172 Fox, S.W., Syren, R.M. and Windsor, C.R., Thermal copoly(amino acids) as inhibitors of glyoxalase I, *Submolecular Biology and Cancer (Ciba Foundation Series,* 67) Excerpta Medica, 1979, pp. 175-193.

173 Fredman, P., Isolation and separation of gangliosides on a new form of glass bead ion exchanger, *Adv. Exp. Med. Biol.*, 125 (1980) 23-31; *C.A.*, 93 (1980) 109 896t.

174 Fredman, P., Nilsson, O., Tayot, J.L. and Svennerholm, L., Separation of gangliosides on a new type of anion-exchange resin, *Biochim. Biophys. Acta*, 618 (1980) 42-52.

175 Frei, R.W., Trace enrichment and chemical derivatization in liquid chromatography, problems and potential in environmental analysis [including some peptides], *Int. J. Environ. Anal. Chem.*, 5 (1978) 143-155.

176 Frei, R.W., Michel, L. and Santi, W., Post-column fluorescence derivatization of peptides. Problems and potential in high-performance liquid chromatography, *J. Chromatogr.*, 126 (1976) 665-677.

177 Frei, R.W., Michel, L. and Santi, W., New aspects of post-column derivatization in high-performance liquid chromatography [pharmaceutically important nona-peptides], *J. Chromatogr.*, 142 (1977) 261-270.

178 Frenkel, M.J. and Blagrove, R.J., Controlled pore glass chromatography of protein-sodium dodecyl sulphate complexes, *J. Chromatogr.*, 111 (1975) 397-402.

179 Friesen, H.-J., Stein, S. and Pestka, S., High performance hydrophobic (reverse phase) liquid chromatography of proteins: facts and artefacts examplified with human fibroplast interferon and other proteins, *International Symposium on Affinity Chromatography and Related Techniques Veldhoven, June 22-26, 1981, Abstracts*, B-55, Katholieke Universiteit, Nijmegen.

180 Fritz, H.J., Belagaje, R., Brown, E.L., Fritz, R.H., Jones, R.A., Lees, R.G. and Khorana, H.G., High-pressure liquid chromatography in polynucleotide synthesis (Studies on polynucleotides 146), *Biochemistry*, 17 (1978) 1257-1267.

181 Fukano, K., Komiya, K., Sasaki, H. and Hashimoto, T., Evaluation of new supports for high-pressure aqueous gel permeation chromatography: TSK-GEL SW type columns [proteins, enzymes, protein hydrolysis, oligosaccharides], *J. Chromatogr.*, 166 (1978) 47-54.

182 Fullmer, C.S. and Wasserman, R.H., Analytical peptide mapping by high-performance liquid chromatography. Application to intestinal calcium-binding proteins, *J. Biol. Chem.*, 254 (1979) 7208-7212.

183 Fullmer, C.S. and Wasserman, R.H., The amino acid sequence of bovine intestinal calcium-binding protein, in Siegel, F.L., Carafoli, E., Kretsinger, R.H., MacLennan, D.H. and Wasserman, R.H. (Editors), *Calcium Binding Proteins: Structure and Function* (*Developments in Biochemistry*, Vol. 14), Elsevier, Amsterdam, 1980, pp. 363-370; *C.A.*, 96 (1982) 30 198v.

184 Fullmer, C.S. and Wasserman, R.H., High performance liquid chromatography application to peptide mapping studies, in Hawk, G.L., Champlin, P.B., Hutton, R.F. and Mol, Ch. (Editors), *Biological/Biomedical Applications of Liquid Chromatography III* (*Chromatographic Science Series*, Vol. 18), Marcel Dekker, New York, Basle, 1982, pp. 175-196; *C.A.*, 96 (1982) 100 325k.

185 Fulton, J.A., Schlabach, T.D., Kerl, J.E., Toren, E.C. Jr. and Miller, A.R., Dual-detector-post-column reactor system for the detection of isoenzymes separated by high-performance liquid chromatography I. Description and theory, *J. Chromatogr.*, 175 (1979) 269-281.

186 Fulton, J.A., Schlabach, T.D., Kerl, J.E. and Toren, E.C. Jr., Dual-detector-post-column reactor system for the detection of isoenzymes separated by high-performance liquid chromatography II. Evaluation and application to lactate dehydrogenase isoenzymes, *J. Chromatogr.*, 175 (1979) 283-291.

187 Gabriel, T.F., Jimenez, M.H., Felix, A.M., Michalewsky, J. and Meienhofer, J., Purification of protected synthetic peptides by preparative high performance liquid chromatography on silica gel 60, *Int. J. Pept. Protein Res.*, 9 (1977) 129-136.

188 Gabriel, T.F., Michalewsky, J. and Meienhofer, J., Preparative high-performance liquid chromatography applied to peptide synthesis, *J. Chromatogr.*, 129 (1976) 287-293.

189 Gabriel, T.F., Michalewsky, J. and Meienhofer, J., Development of preparative scale high performance liquid chromatography for peptide and protein purification [review], in Eberle, A., Geiger, R. and Wieland, T. (Editors), *Perspectives in Peptide Chemistry*, S. Karger, Basle, 1981, pp. 195-206; *C.A.*, 94 (1981) 116 960h.

190 Gait, M.J., Singh, M., Sheppard, R.C., Edge, M.D., Greene, A.R., Heathcliffe, G.R., Atkinson, T.C., Newton, C.R. and Markham, A.F., Rapid synthesis of oligonucleotides IV. Improved solid phase synthesis of oligo deoxyribonu-cleotides through phosphotriester intermediates, *Nucleic Acids Res.*, 8 (1980) 1081-1096.

191 Games, D.E., Combined high performance liquid chromatography mass spectrometry [review including also peptides], *Biomed. Mass Spectrom.*, 8 (1981) 454-462.

192 Games, D.E., Eckers, Ch., Gower, J.L., Hirter, P., Knight, M.E., Lewis, E., Rao, K.R.N. and Weerasinghe, N.C., Combined HPLC-MS and its potential in clinical studies [review], *Clin. Res. Cent. Symp. (Harrow)*, No. 1 (Curr. Dev. Clin. Appl. HPLC, GC, MS) (1980) 97-118; *C.A.*, 94 (1981) 152 731t.

193 Games, D.E., Hirter, P., Kuhnz, W., Lewis, E., Weerasinghe, N.C.A. and Westwood, S.A., Studies of combined liquid chromatography-mass spectrometry with a moving-belt interface [discussion including peptides], *J. Chromatogr.*, 203 (1981) 131-138.

194 Games, D.E. and Lewis, E., Combined liquid chromatography mass spectrometry of glycosides, glucuronides, sugars and nucleosides, *Biomed. Mass Spectrom.*, 7 (1980) 433-436.

195 Gardiner, M.B., Carver, J., Abraham, B.L., Wilson, J.B. and Huisman, T.H.J., Further studies on the quantitation of the hemoglobins A, S, C and F in newborn babies with different hemoglobinopathies using high pressure liquid chromatography, *Hemoglobin*, 6 (1982) 1-13.

196 Gazdak, M. and Szepesi, G., Separation of large polypeptides by high-performance liquid chromatography, *J. Chromatogr.*, 218 (1981) 603-612.

197 Gentleman, S., Lowney, L.I., Cox, B.M. and Goldstein, A., Rapid purification of β-endorphin by high-performance liquid chromatography, *J. Chromatogr.*, 153 (1978) 274-278.

198 Germershausen, J. and Karkas, J.D., Preparative high speed gel permeation
      chromatography of proteins on Toyopearl HW55F, *Biochem. Biophys. Res.
      Commun.*, 99 (1981) 1020-1027.
199 Glad, M., Ohlson, S., Larsson, P.O. and Mosbach, K., High performance liquid
      affinity chromatography (HPLAC) with boronic acid silica [perspectives in
      separation of glycoproteins], *International Symposium on Affinity Chro-
      matography and Related Techniques, Veldhoven, June 22-26, 1981, Abstracts*,
      B-56, Katholieke Universiteit, Nijmegen.
200 Glasel, J.A., Separation of neurohypophyseal proteins by reversed-phase
      high-pressure liquid chromatography, *J. Chromatogr.*, 145 (1978) 469-472.
201 Gooding, K.M., Lu Kai-Chun and Regnier, F.E., High-performance liquid
      chromatography of hemoglobins 1. Determination of hemoglobin A_2, *J.
      Chromatogr.*, 164 (1979) 506-509.
202 Gooding, K.M. and Regnier, F.E., The use of high performance liquid chroma-
      tography in the clinical laboratory [review], *New Dev. Clin. Instrum.*,
      CRC Press, Boca Raton, FL, 1981, 71-101; *C.A.*, 96 (1982) 16 757v.
203 Goodman, M., Keogh, P. and Anderson, H., Racemization studies in peptide
      synthesis using high-pressure liquid chromatography, *Bioorg. Chem.*, 6
      (1977) 239-247.
204 Greber, G. and Schott, H., Separation of nucleosides on polymer gels con-
      taining thymine or cytidine [base-pairing interaction], *Angew. Chem., Int.
      Ed. Engl.*, 9 (1970) 68-69.
205 Grego, B. and Hearn, M.T.W., Pairing ion selectivities in the reversed-
      phase separation of peptides, *Proc. Otago Univ. Med. Sch.*, 58 (1980) 41-43;
      C.A., 93 (1980) 182 155e.
206 Grego, B., Lambrou, F. and Hearn, M.T.W., Retention behaviour of human growth
      hormone tryptic peptides isolated by reversed phase high performance liquid
      chromatography: a comparative study using different chromatographic condi-
      tions and predicted elution positions based on retention coefficients, *J.
      Chromatogr.*, 266 (1983) 89-103.
207 Greibrokk, T., Purification of peptide hormones by reversed phase HPLC
      [review], *Z. Naturforsch.*, (special issue) in press.
208 Greibrokk, T., Currie, B.L., Johansson, K.N.-G., Hansen, J.J., Folkers, K.
      and Bowers, C.Y., Purification of a prolactin inhibiting hormone and the
      revealing of hormone D-GHIH which inhibits the release of growth hormone,
      Biochem. Biophys. Res. Commun., 59 (1974) 704-709.
209 Greibrokk, T., Hansen, J., Knudsen, R., Lam, Yiu-Kuen, Folkers, K. and
      Bowers, C.Y., The isolation of a prolactin inhibiting factor (hormone),
      Biochem. Biophys. Res. Commun., 67 (1975) 338-344.
210 Gribnau, T.C.J., Visser, J. and Nivard, R.J.F. (Editors), *Affinity Chroma-
      tography and Related Techniques* [including rapid separation of biopolymers],
      *Proceedings of the 4th International Symposium, Veldhoven, The Netherlands,
      1981 (Analytical Chemistry Symposia Series*, Vol. 9), Elsevier, Amsterdam,
      1981, pp. 584.
211 Grimaud, E., Lecoq, J.C., Boschetti, E. and Corgier, M., Comparison of gels
      used for molecular sieving of proteins by electron microscopy and pore
      parameters determinations, *J. Chromatogr.*, 166 (1978) 37-45.
212 Gross, S.K. and McCluer, R.H., High-performance liquid chromatographic
      analysis of neutral glycosphingolipids and their per-O-benzoylated deriva-
      tives, *Anal. Biochem.*, 102 (1980) 429-433.
213 Gruber, C.A. and Koets, M.D., Quantitation of hemoglobin A_{1a+b} and hemoglo-
      bin A_{1c} by automated "high-performance" liquid chromatography, *Clin. Chem.
      (Winston-Salem, N.C.)*, 25 (1979) 1970-1971.
214 Gruber, K.A., Whitaker, J.M. and Morris, M., Molecular weight separation
      of proteins and peptides with a new high-pressure liquid chromatography
      column, *Anal. Biochem.*, 97 (1979) 176-183.
215 Gum, E.K. Jr. and Brown, R.D. Jr., Two alternative HPLC separation methods
      for reduced and normal cellooligosaccharides, *Anal. Biochem.*, 82 (1977)
      372-375.

216 Gurkin, M. and Patel, V., Aqueous gel filtration chromatography of enzymes, proteins, oligosaccharides and nucleic acids [review], *Am. Lab. (Fairfield, Conn.)*, 14 (1982) 64, 66, 68-70, 72-73; *C.A.*, 96 (1982) 138 998v.
217 Guy, M.N., Roberson, G.M. and Barnes, L.D., Analysis of angiotensins I, II, III and iodinated derivatives by high-performance liquid chromatography, *Anal. Biochem.*, 112 (1981) 272-277.
218 Hagemeier, E., Bornemann, S., Boss, K.S. and Schlimme, E., High-performance liquid-chromatographic method for separation of dinucleotides, *J. Chromatogr.*, 237 (1982) 174-177.
219 Hagiwara, N., Okazaki, M. and Hara, I., A quantitation reagent specific for choline-containing phospholipids in high-performance liquid chromatography of serum lipoproteins, *Yukagaku*, 31 (1982) 262-267; *C.A.*, 97 (1982) 20 256u.
220 Hagnauer, G.L., Size exclusion chromatography [review], *Anal. Chem.*, 54 (1982) 265R-276R.
221 Haller, W., Chromatography on glass of controlled pore size [separation of viruses], *Nature (London)*, 206 (1965) 693-696.
222 Haller, W., Tympner, K.D. and Hannig, K., Preparation of immunoglobulin concentrates from human serum by chromatography on controlled pore glass, *Anal. Biochem.*, 35 (1970) 23-31.
223 Hanash, S.M. and Shapiro, D.N., Separation of human hemoglobins by ion-exchange high-performance liquid chromatography, *Hemoglobin*, 5 (1981) 165-175; *C.A.*, 94 (1981) 204 702f.
224 Hanash, S.M., Kavadella, M., Amanullah, A., Scheller, K. and Bunnell, K., High performance liquid chromatography of hemoglobins: factors affecting resolution [review], in Hanash, S.M. and Brewer, G.J. (Editors), *Prog. Clin. Biol. Res.*, 60 (1981) (*Adv. Hemoglobin Anal.*), 53-67; *C.A.*, 95 (1981) 164 791w.
225 Hancock, W.S. (Editor), *CRC Handbook of HPLC for the Separation of Amino Acids, Peptides and Proteins*, Vol. I (489 pp.), Vol. II (522 pp.), CRC Press, Boca Raton, FL, 1984.
226 Hancock, W.S., Bishop, C.A., Battersby, J.E., Harding, D.R.K. and Hearn, M.T.W., High-pressure liquid chromatography of peptides and proteins XI. The use of cationic reagents for the analysis of peptides by high-pressure liquid chromatography, *J. Chromatogr.*, 168 (1979) 377-384.
227 Hancock, W.S., Bishop, C.A., Gotto, A.M., Harding, D.R.K., Lamplugh, S.M. and Sparrow, J.T., Separation of the apoprotein components of human very low density lipoproteins by ion-paired, reversed-phase high performance liquid chromatography, *Lipids*, 16 (1981) 250-259.
228 Hancock, W.S., Bishop, C.A. and Hearn, M.T.W., High-pressure liquid chromatography in the analysis of underivatized peptides using a sensitive and rapid procedure [No. I of series "HPLC of peptides and proteins"], *FEBS Lett.*, 72 (1976) 139-142.
229 Hancock, W.S., Bishop, C.A. and Hearn, M.T.W., The application of ion-pair high pressure liquid chromatography to the rapid analysis and isolation of underivatized amino acids, peptides and proteins [review], *Chem. N.Z.*, 43 (1979) 17-24; *C.A.*, 91 (1979) 123 968h.
230 Hancock, W.S., Bishop, C.A., Mayer, L.J., Harding, D.R.K. and Hearn, M.T.W., High-pressure liquid chromatography of peptides and proteins. VI. Rapid analysis of peptides by high-pressure liquid chromatography with hydrophobic ion-pairing amino groups, *J. Chromatogr.*, 161 (1978) 291-298.
231 Hancock, W.S., Bishop, C.A., Prestidge, R.L., Harding, D.R.K. and Hearn, M.T.W., High-pressure liquid chromatography of peptides and proteins. II. The use of phosphoric acid in the analysis of underivatised peptides by reversed-phase high-pressure liquid chromatography, *J. Chromatogr.*, 153 (1978) 391-398.
232 Hancock, W.S., Bishop, C.A., Prestidge, R.L., Harding, D.R.K. and Hearn, M.T.W., Reversed-phase, high-pressure liquid chromatography of peptides and proteins with ion-pairing reagents, *Science (Washington, D.C.)*, 200 (1978) 1168-1170.

233 Hancock, W.S., Bishop, C.A., Prestidge, R.L. and Hearn, M.T.W., The use of high pressure liquid chromatography (HPLC) for peptide mapping of proteins IV, *Anal. Biochem.*, 89 (1978) 203-212.

234 Hancock, W.S., Capra, J.D., Bradley, W.A. and Sparrow, J.T., The use of reversed-phase high-performance liquid chromatography with radial compression for the analysis of peptide and protein mixtures, *J. Chromatogr.*, 206 (1981) 59-70.

235 Hancock, W.S., Pownall, H.J., Gotto, A.M. and Sparrow, J.T., Separation of apolipoproteins A-I and A-II by ion-paired reversed-phase high-performance liquid chromatography, *J. Chromatogr.*, 216 (1981) 285-293.

236 Hancock, W.S. and Sparrow, J.T., Use of mixed-mode, high performance liquid chromatography for the separation of peptide and protein mixtures, *J. Chromatogr.*, 206 (1981) 71-82.

237 Hancock, W.S. and Sparrow, J.T., The separation of proteins by reversed-phase high performance liquid chromatography [review], in Horváth, Cs. (Editor), *High Performance Liquid Chromatography (Advances and Perspectives*, Vol. 3), Academic Press, New York, 1983, pp. 49-85; *C.A.*, 99 (1983) 18 948u.

238 Hansen, J.J., Greibrokk, T., Currie, B.L., Johansson, K.N.-G. and Folkers, K., High-pressure liquid chromatography of peptides, *J. Chromatogr.*, 135 (1977) 155-164.

239 Hara, I., Okazaki, M. and Ohno, Y., Rapid analysis of cholesterol of high density lipoprotein and low density lipoprotein in human serum by high performance liquid chromatography, *J. Biochem. (Tokyo)*, 87 (1980) 1863-1865.

240 Hara, I., Shiraishi, K. and Okazaki, M., High performance liquid chromatography of human serum lipoproteins. Selective detection of triglycerides by enzymatic reaction, *J. Chromatogr.*, 239 (1982) 549-557.

241 Harding, D.R.K., Bishop, C.A., Tarttelin, M.F. and Hancock, W.S., Use of perfluoroalkanoic acids as volatile ion pairing reagents in preparative HPLC. Synthesis, purification and biological testing of the proposed anorexigenic peptide Pyr-His-Gly, *Int. J. Pept. Protein Res.*, 18 (1981) 214-220.

242 Hartman, A. and Persson, B., Gel chromatography of proteins in cheese, *Science Tools*, 27 (1980) 57-59.

243 Hase, S., Ikenaka, T. and Matsushima, Y., A highly sensitive method for analyses of sugar moieties of glycoproteins by fluorescence labeling, *J. Biochem. (Tokyo)*, 90 (1981) 407-414; *C.A.*, 95 (1981) 93 213w.

244 Hase, S., Ikenaka, T. and Matsushima, Y., A highly sensitive method for determination of molecular weights of neutral oligosaccharides by fluorescence labeling, *J. Biochem. (Tokyo)*, 90 (1981) 1275-1279; *C.A.*, 96 (1982) 16 820k.

245 Hase, S., Okawa, K. and Ikenaka, T., Identification of the trimannosyl-chitobiose structure in sugar moieties of japanese quail ovomucoid [HPLC of chemical and exoglycosidase digests of modified major oligosaccharide chain], *J. Biochem. (Tokyo)*, 91 (1982) 735-737; *C.A.*, 96 (1982) 138 282g.

246 Hashimoto, T., Sasaki, H., Aiura, M. and Kato, Y., High-speed aqueous gel-permeation chromatography of proteins [using TSK-Gel (PW) packings], *J. Chromatogr.*, 160 (1978) 301-305.

247 Hashimoto, T., Sasaki, H., Aiura, M. and Kato, Y., High-speed aqueous gel-permeation chromatography [testing of TSK-gels, type PW; dextran and other polymeric substances], *J. Polym. Sci., Polym. Phys. Ed.*, 16 (1978) 1789-1800; *C.A.*, 89 (1978) 215 941f.

248 Havel, E., Tweeten, T.N., Seib, P.A., Wetzel, D.L., Liang, Y.T. and Smith, O.B., Oligosaccharides released during hydration of textured soy as determined by high performance liquid chromatography, *J. Food Sci.*, 42 (1977) 666-668; *C.A.*, 87 (1977) 20 699j.

249 Hawk, G.L., Cameron, J.A. and DuFault, L., Chromatography of biological materials on polyethylene glycol treated controlled-pore glass, *Prep. Biochem.*, 2 (1972) 193-203; *C.A.*, 77 (1972) 58 423w.

250 Hearn, M.T.W., The implications of modern separation techniques for peptides and proteins in endocrinology [Organon Lecture], *Proc. Endocrinol. Soc. Aust.*, 23 (1980) 23.

251 Hearn, M.T.W., High-performance liquid chromatography of amino acids, peptides and proteins. XXVI. The use of reversed phase high-performance liquid chromatography for the structural mapping of polypeptides and proteins [review of analytical and preparative applications], *J. Liq. Chromatogr.*, 3 (1980) 1255-1276.

252 Hearn, M.T.W., Ion-pair chromatography on normal- and reversed-phase systems [review], *Adv. Chromatogr.*, 18 (1980) 59-100.

253 Hearn, M.T.W., High performance liquid chromatography and its application in protein chemistry [review], *Adv. Chromatogr.*, 20 (1982) 1-82.

254 Hearn, M.T.W., Separation of peptides by HPLC [review], in Horváth, Cs. (Editor), *High Performance Liquid Chromatography (Advances and Perspectives*, Vol. 3), Academic Press, New York, 1983, pp. 87-155; *C.A.*, 99 (1983) 35 386b.

255 Hearn, M.T.W., Bishop, C.A., Hancock, W.S., Harding, D.R.K. and Reynolds, G.D., Application of reversed phase high performance liquid chromatography in solid phase peptide synthesis, *J. Liq. Chromatogr.*, 2 (1979) 1-21.

256 Hearn, M.T.W. and Grego, B., Solvent effects in the purification of protein hormones by reversed phase HPLC, *Chromatographia*, 14 (1981) 589-592.

257 Hearn, M.T.W. and Grego, B., High-performance liquid chromatography of amino acids, peptides and proteins. XXVII. Solvophobic considerations for the separation of unprotected peptides on chemically bonded hydrocarbonaceous stationary phases, *J. Chromatogr.*, 203 (1981) 349-363.

258 Hearn, M.T.W. and Grego, B., High-performance liquid chromatography of amino acids, peptides and proteins. XXXVI. Organic solvent modifier effects in the separation of unprotected peptides by reversed-phase liquid chromatography, *J. Chromatogr.*, 218 (1981) 497-507.

259 Hearn, M.T.W. and Grego, B., Selectivity effects of peptidic positional isomers and oligomers separated by reversed phase high performance liquid chromatography, *J. Chromatogr.*, 266 (1983) 75-87.

260 Hearn, M.T.W. and Grego, B., Further studies on the role of organic solvents in reversed phase liquid chromatography of polypeptides: implications for gradient optimisation, *J. Chromatogr.*, 255 (1983) 125-136.

261 Hearn, M.T.W., Grego, B., Bishop, C.A. and Hancock, W.S., High-speed gel permeation chromatography of human thyroglobulin and sheep liver aldehyde dehydrogenase, *J. Liq. Chromatogr.*, 3 (1980) 1549-1560.

262 Hearn, M.T.W., Grego, B. and Bishop, C.A., High performance liquid chromatography of amino acids, peptides and proteins. XXXV. The semipreparative separation of peptides on reversed phase silica packed into radially compressed flexible-walled columns, *J. Liq. Chromatogr.*, 4 (1981) 1725-1744.

263 Hearn, M.T.W., Grego, B. and Chapman, G., Isolation and structural elucidation of tryptic peptides of human growth hormone (hGH) and the 20k dalton hGH variant, *J. Liq. Chromatogr.*, 6 (1983) 215-228; *C.A.*, 98 (1983) 119 840n.

264 Hearn, M.T.W., Grego, B. and Hancock, W.S., High-performance liquid chromatography of amino acids, peptides and proteins. XX. Investigation of the effect of pH and ion-pair formation on the retention of peptides on chemically-bonded hydrocarbonaceous stationary phases, *J. Chromatogr.*, 185 (1979) 429-444.

265 Hearn, M.T.W. and Hancock, W.S., Ion pair partition reversed phase HPLC (A new method for the rapid analysis and isolation of underivatised amino acids, peptides and proteins.) [review], *Trends Biochem. Sci.*, 4 (1979) N58-N62; *C.A.*, 90 (1979) 164 146m.

266 Hearn, M.T.W. and Hancock, W.S., The role of ion-pair reversed phase HPLC in peptide and protein chemistry [review], *Chromatogr. Sci.*, 12 (1979) 243-271; *C.A.*, 92 (1980) 36 964f.

267 Hearn, M.T.W., Hancock, W.S., Hurrell, J.G., Fleming, R.J. and Kemp, B.,
    High performance liquid chromatography of peptides and proteins. XVIII.
    The analysis of insulin-related peptides by reversed-phase high-perfor-
    mance liquid chromatography, *J. Liq. Chromatogr.*, 2 (1979) 919-933.
268 Hearn, M.T.W., Regnier, F.E. and Wehr, C.T. (Editors), *High-Performance
    Liquid Chromatography of Proteins and Peptides (Proceedings of the First
    International Symposium)*, Academic Press, New York, 1983, 288 pp.
269 Hearn, M.T.W., Regnier, F.E. and Wehr, C.T., HPLC of peptides and proteins
    [review], *Inter. Lab.*, 13 (1983) 16-35.
270 Hearn, M.T.W., Stanton, P.G., Simpson, R.J. and Lambrou, F., Analytical and
    semi-preparative separation of several pituitary proteins by high per-
    formance ion-exchange chromatography, *J. Chromatogr.*, 266 (1983) 273-279.
271 Hearn, M.T.W., Su Suew, J. and Grego, B., High performance liquid chroma-
    tography of amino acids, peptides and proteins. Part XXXIII. Pairing ion
    effects in the reversed phase high performance liquid chromatography of
    peptides in the presence of alkylsulfonates, *J. Liq. Chromatogr.*, 4 (1981)
    1547-1567.
272 Hefti, F., High-performance size exclusion chromatography: a buffer for the
    reliable determination of molecular weights of proteins, *Anal. Biochem.*,
    121 (1982) 378-381.
273 Henderson, L.E., Sowder, R. and Oroszlan, S., Protein and peptide purifica-
    tion by reversed-phase high pressure [liquid] chromatography using volatile
    solvents, in Liu, T.Y., Schechter, A.N., Heinrikson, R. and Condliffe, P.G.
    (Editors), *Chemical Synthesis and Sequencing of Peptides and Proteins
    (Developments in Biochemistry*, 1981), Elsevier, Amsterdam, 1981, pp. 251-
    260; *C.A.*, 96 (1982) 213 548d.
274 Henderson, L., Sowder, R. and Oroszlan, S., Reversed phase (RP) HPLC of
    proteins and peptides and its impact on protein microsequencing, in
    Elzinga, M. (Editor), *Methods in Protein Sequence Analysis*, Humana Press,
    Clifton, NJ, 1982, pp. 409-416.
275 Hendrix, D.L., Lee, R.E., Baust, J.G. and James, H., Separation of carbo-
    hydrates and polyols by a radially compressed high-performance liquid
    chromatographic silica column modified with tetraethylenepentamine
    [including oligosaccharides], *J. Chromatogr.*, 210 (1981) 45-53.
276 Henschen-Edman, A. and Lottspeich, F., Aspects on automated and microscale
    sequencing [HPLC of PTH-amino acids], in Birr, Ch. (Editor), *Methods in
    Peptide and Protein Sequence Analysis (Proc. 3rd Int. Conf., 1979)*,
    Elsevier, Amsterdam, 1980, pp. 105-114; *C.A.*, 94 (1981) 43 661z.
277 Hermodson, M. and Mahoney, W.C., Peptide separation by high pressure liquid
    chromatography as a prelude to sequencing of peptides and proteins, in
    Liu, T.Y., Schechter, A.N., Heinrikson, R. and Condliffe, P.G. (Editors),
    *Chemical Synthesis and Sequencing of Peptides and Proteins (Developments
    in Biochemistry*, 1981), Elsevier, Amsterdam, 1981, pp. 119-130; *C.A.*, 96
    (1982) 177 211c.
278 Hiatt, C.W., Shelokov, A., Rosenthal, E.J. and Galimore, J.M., Treatment
    of controlled pore glass with poly(ethylene oxide) to prevent adsorption
    of rabies virus, *J. Chromatogr.*, 56 (1971) 362-364.
279 Hill, D.F., Williams, V.P. and Popjak, G., Characterization of peptides
    in human cerebrospinal fluid [combination of separation methods including
    HPLC], *Peptides (N.Y.)*, 2 (Suppl. 1) (1981) 79-82; *C.A.*, 95 (1981) 146 281a.
280 Hillen, W., Klein, R.D. and Wells, R.D., Preparation of milligram amounts
    of 21 deoxyribonucleic acid restriction fragments, *Biochemistry*, 20 (1981)
    3748-3756.
281 Himmel, M.E. and Squire, P.G., High-performance size exclusion chromato-
    graphy of sea worm chlorocruorin and other large proteins, viruses and
    polysaccharides on a TSK G5000PW preparative column, *J. Chromatogr.*, 210
    (1981) 443-452.
282 Hjertén, S., HPLC of biopolymers on columns of agarose, *Acta Chem. Scand. B*,
    36 (1982) 203-209.

283 Hjertén, S., High performance liquid chromatography on matrices of agarose, *Protides Biol. Fluids*, 30 (1983) 9-17.

284 Hjertén, S. and Yao Kunquan, High-performance liquid chromatography of macromolecules on agarose and its derivatives, *J. Chromatogr.*, 215 (1981) 317-322.

285 Hjertén, S., Yao Kunquan and Patel, V., High performance hydrophobic interaction chromatography of biopolymers, in Gribnau, T.C.J., Visser, J. and Nivard, R.J.F. (Editors), *Affinity Chromatography and Related Techniques (Analytical Chemistry Symposia Series*, Vol. 9), Elsevier, Amsterdam, 1982, pp. 483-489; *C.A.*, 96 (1982) 118 344a.

286 Hobbs, A.A., Grego, B., Smith, M.G. and Hearn, M.T.W., Analysis of rat caseins by high performance liquid chromatography, *J. Liq. Chromatogr.*, 4 (1981) 651-659.

287 Hokse, H., Analysis of cyclodextrins by high-performance liquid chromatography, *J. Chromatogr.*, 189 (1980) 98-100.

288 Hollaway, W.L., Bhown, A.S., Mole, J.E. and Bennett, J.C., Quantitative yields of underivatized peptides by HPLC, *J. High Resolut. Chromatogr. Chromatogr. Commun.*, 1 (1978) 177-178.

289 Hollaway, W.L., Bhown, A.S., Mole, J.E. and Bennett, J.C., HPLC in the structural studies of proteins [including separation of polypeptides], *Chromatogr. Sci.*, 10 (1979) 163-176.

290 Hollaway, W.L., Kemp, M.C., Bhown, A.S., Compans, R.W. and Bennett, J.C., Reverse phase ion-pair chromatography of viral glycopeptides, *J. High Resolut. Chromatogr. Chromatogr. Commun.*, 2 (1979) 149-150.

291 Hollaway, W.L., Prestidge, R.L., Bhown, A.S., Mole, J.E. and Bennett, J.C., Hydrophilic ion-pair reversed-phase high-performance liquid chromatography of peptides and proteins, in Frigerio, A. and McCamish, M. (Editors), *Recent Developments in Chromatography and Electrophoresis*, (Analytical Chemistry Symposia Series, Vol. 3), Elsevier, Amsterdam, 1980, pp. 131-139; *C.A.*, 94 (1981) 12 345v.

292 Honda, S., Matsuda, Y., Takahashi, M., Kakehi, K., Honda, A., Ganno, S. and Ito, M., Fluorimetric analysis of carbohydrates by using aliphatic amines and its application to automated analysis of carbohydrates [mono- and disaccharides; carbohydrates in serum and fucose/galactose ratios of nondialyzable urinary carbohydrates from cancer patients], *J. Pharmacobio-Dyn.*, 3 (1980) S-11, S-31; *C.A.*, 93 (1980) 163 927s.

293 Horváth, Cs., Pellicular ion exchangers [review], in Grushka, E. (Editor), *Bonded Stationary Phases in Chromatography*, Ann Arbor Science Publishers, Ann Arbor, MI, 1974, pp. 59-91; *C.A.*, 82 (1975) 59 192k.

294 Horváth, Cs. (Editor), *High Performance Liquid Chromatography (Advances and Perspectives*, Vol. 2), Academic Press, New York, 1981, pp. 340; *C.A.*, 94 (1981) 149 774d.

295 Horváth, Cs. (Editor), *High-Performance Liquid Chromatography (Advances and Perspectives*, Vol. 3.) [includes nucleic acid fragments, proteins, peptides], Academic Press, New York, 1983, 256 pp.

296 Hostomská-Chytilová, Z., Mikeš, O., Vrátný, P. and Smrž, M., Chromatography of cellodextrins and enzymatic hydrolysates of cellulose on ion exchange derivatives of Spheron, *J. Chromatogr.*, 235 (1982) 229-236.

297 Hostomská, Z. and Mikeš, O., Analytical medium pressure liquid chromatography of cellulolytic enzymes on Spheron ion exchangers, *J. Chromatogr.*, 267 (1983) 355-366.

298 Hostomská, Z. and Mikeš, O., Separation of the cellulolytic system of *Trichoderma viride-reesei* mutant using medium pressure liquid chromatography and characterization of a new exo-cellobiohydrolase, *Int. J. Pept. Protein Res.*, 23 (1984) 402-410.

299 Hruby, V.J., Viswanatha, V. and Young C.S. Yang, Synthesis of S-benzyl-DL-[1-^{13}C]cysteine and its incorporation into oxytocin and [8-arginine]vasopressin and related compounds by total synthesis. Separation of diastereoisomers by partition chromatography and HPLC, *J. Labelled Compds. Radiopharm.*, 17 (1980) 801-812; *C.A.*, 94 (1981) 175 527v.

300 Hughes, G.J., Winterhalter, K.H. and Wilson, K.J., Microsequence analysis:
     I. peptide isolation using high-performance liquid chromatography, *FEBS
     Lett.*, 108 (1979) 81-86.
301 Hughes, G.J., Winterhalter, K.H. and Wilson, K.J., From the isolation of
     proteins to amino acid analysis: the exclusive use of standard HPLC
     equipment, in Lottspeich, F., Henschen, A. and Hupe, K.P. (Editors), *High
     Performance Chromatography in Protein and Peptide Chemistry*, Walter de
     Gruyter, Berlin, New York, 1981, pp. 175-191; *C.A.*, 97 (1982) 52 011b.
302 Hui Koon-Sea, Salschutz, M., Davis, B.A. and Lajtha, A., Separation of
     alkylaminonaphthylenesulfonyl peptides and amino acids by high-performance
     liquid chromatography. Methods for measuring melanotropin inhibiting
     factor breakdown, *J. Chromatogr.*, 192 (1980) 341-350.
303 Huisman, T.H.J., Altay, C., Webber, B., Reese, A.L., Gravely, M.E., Okonjo,
     K. and Wilson, J.B., Quantitation of three types of γ chain of HbF by
     high-pressure liquid chromatography; application of this method to the
     HbF of patients with sickle cell anemia or the S-HPFH condition, *Blood*,
     57 (1981) 75-82; *C.A.*, 94 (1981) 79 518x.
304 Huisman, T.H.J., Gardiner, M.B. and Wilson, J.B., Experiences with the quan-
     titation of human hemoglobin types by high pressure liquid chromatography,
     Prog. Clin. Biol. Res., 60 (1981) 69-82 (*Advances in Hemoglobin Analysis*;
     Alan R. Liss, New York, 1981).
305 Huisman, T.H.J., Webber, B., Okonjo, K., Reese, A.L. and Wilson, J.B., The
     separation of human hemoglobin chains by high pressure liquid chromato-
     graphy, *Prog. Clin. Biol. Res.*, 60 (1981) 23-38 (*Advances in Hemoglobin
     Analysis*; Alan R. Liss, New York, 1981); *C.A.*, 95 (1981) 146 294g.
306 Hunter, C., Sugden, K. and Lloyd-Jones, J.G., HPLC of peptides and peptide
     diastereoisomers on ODS- and cyanopropyl-silica gel column packing
     materials, *J. Liq. Chromatogr.*, 3 (1980) 1335-1352.
307 Hurrell, J.G.R., Fleming, R.J. and Hearn, M.T.W., A high performance liquid
     chromatographic assessment of the isolation of bovine proinsulin and
     asynthetic proinsulin fragment, *J. Liq. Chromatogr.*, 3 (1980) 473-494.
308 Ibarra, B., Franco-Gamboa, E., Ramírez, M.L., Cantú, J.M., Wilson, J.B.,
     Lam, H. and Huisman, T.H.J., HB Chiapas α2 114 Pro → Arg β2: Identifica-
     tion by high pressure liquid chromatography, *Hemoglobin*, 5 (1981) 605-608;
     C.A., 96 (1982) 117 760w.
309 Ikehara, M. and Ohtuska, E., Chemical synthesis of genes [separation and
     purification of oligonucleotides; review in Japanese], *Tanpakushitsu
     Kakusan Koso*, 26 (1981) 531-541; *C.A.*, 95 (1981) 62 506g.
310 Ishiguro, I. and Shinohara, R., Separation of isoenzymes. I. Chromatographic
     methods [review in Japanese], *Rinsho Kensa*, 25 (1981) 183-188; *C.A.*, 94
     (1981) 170 153k.
311 Ishii, M., Highly sensitive method of the determination of blood hemoglobin
     by high performance aqueous gel permeation chromatography - atomic
     absorption spectrometry [in Japanese], *Kyorin Igakkai Zaashi*, 12 (1981)
     247-252; *C.A.*, 96 (1982) 48 592d.
312 Iskandarani, Z. and Pietrzyk, D.J., Liquid chromatographic separation of
     amino acids, peptides and derivatives on a porous polystyrene-divinyl-
     benzene copolymer [RPC], *Anal. Chem.*, 53 (1981) 489-495.
313 Ivie, K.F., High-performance liquid chromatography in the analysis of sugars
     [discussion in Spanish], *Sugar Azucar*, 77 (1982) 80-81, 86-87; *C.A.*, 96
     (1982) 183 126v.
314 Ivory, C.F., Several aspects of the chromatography of macromolecules on
     porous support, *Diss. Abstr. Int. B*, 41 (1980) 277-278; *C.A.*, 93 (1980)
     163 728c.
315 James, L.B., Resolution of γ-glutamyl peptides, *J. Chromatogr.*, 172 (1979)
     481-483.
316 Janák, J., Čoupek, J., Krejčí, M., Mikeš, O. and Turková, J., Sorbents, in
     Deyl, Z., Macek, K. and Janák, J. (Editors), *Liquid Column Chromatography*,
     Elsevier, Amsterdam, 1975, pp. 187-202.

317 Jenik, R.A. and Porter, J.W., High-performance liquid chromatography of proteins by gel permeation chromatography, *Anal. Biochem.*, 111 (1981) 184-188.

318 Johnson, C.S., Moyes, D., Schroeder, W.A., Shelton, J.B., Shelton, J.R. and Beutler, E., Hemoglobin Pasadena, $\alpha_2\beta_2^{75}$(E19)Leu$\rightarrow$Arg. Identification by high performance liquid chromatography of a new unstable variant with increased oxygen affinity, *Biochim. Biophys. Acta*, 623 (1980) 360-367.

319 Johnson, P., Effective peptide fractionation using an amino acid analyzer ion-exchange resin, *J. Chromatogr. Sci.*, 17 (1979) 406-409.

320 Jones, B.N., Lewis, R.V., Paabo, S., Kojima, K., Kimura, S. and Stein, S., Effects of flow rate and eluant composition on the high-performance liquid chromatography of proteins, *J. Liq. Chromatogr.*, 3 (1980) 1373-1383.

321 Jones, D.P., Determination of pyridine dinucleotides in cell extracts by high-performance liquid chromatography, *J. Chromatogr.*, 225 (1981) 446-449.

322 Jost, W., Unger, K. and Schill, G., Reverse-phase ion-pair chromatography of polyvalent ions using oligonucleotides as model substances, *Anal. Biochem.*, 119 (1982) 214-223.

323 Jost, W., Unger, K.K., Lipecky, R. and Gassen, H.G., Application of a weakly basic dimethylamino-modified silica ion exchanger to the separation of oligonucleotides, *J. Chromatogr.*, 185 (1979) 403-412.

324 Jungalwala, F.B., Hayes, L. and McCluer, R.H., Determination of less than a nanomole of cerebrosides by high performance liquid chromatography with gradient elution analysis, *J. Lipid Res.*, 18 (1977) 285-292.

325 Kainuma, K., Nakakuki, T. and Ogawa, T., High-performance liquid chromatography of maltosaccharides, *J. Chromatogr.*, 212 (1981) 126-131.

326 Kalász, H. and Horváth, Cs., Preparative-scale separation of polymyxins with an analytical high-performance liquid chromatography system by using displacement chromatography, *J. Chromatogr.*, 215 (1981) 295-302.

327 Karger, B.L. and Giese, R.W., Reversed phase liquid chromatography and its application to biochemistry [review], *Anal. Chem.*, 50 (1978) 1048A-1073A.

328 Kasche, V., Buchholz, K. and Galunsky, B., Resolution in high-performance liquid affinity chromatography. Dependence on eluite diffusion into the stationary phase [separation of proteases on immobilized soy bean trypsin inhibitor], *J. Chromatogr.*, 216 (1981) 169-174.

329 Kasche, V. and Galunsky, B., Ligand/ligate interactions in heterogenous systems: their influence on operational properties of affinity chromatographic and binding assay adsorbents [for various proteins, especially enzymes], in Gribnau, T.C.J., Visser, J. and Nivard, R.J.F. (Editors), *Affinity Chromatography and Related Techniques*, Elsevier, Amsterdam, 1982, pp. 93-110; *C.A.*, 96 (1982) 100 502r.

330 Kato, T. and Kinoshita, T., Fluorometric detection and determination of carbohydrates by high-performance liquid chromatography using ethanolamine, *Anal. Biochem.*, 106 (1980) 238-243.

331 Kato, Y., Komiya, K., Iwaeda, T., Sasaki, H. and Hashimoto, T., Comparison of resolutions of Toyopearl HW55S and HW55F [resolution of bovine serum albumin and myoglobin in dependence on flow rate], *J. High Resolut. Chromatogr. Chromatogr. Commun.*, 4 (1981) 135.

332 Kato, Y., Komiya, K., Iwaeda, T., Sasaki, H. and Hashimoto, T., Packing of Toyopearl column for gel filtration. I. Influence of packing velocity on column performance, *J. Chromatogr.*, 205 (1981) 185-188.

333 Kato, Y., Komiya, K., Iwaeda, T., Sasaki, H. and Hashimoto, T., Packing of Toyopearl columns for gel filtration. II. Dependence of optimal packing velocity on column size, *J. Chromatogr.*, 206 (1981) 135-138.

334 Kato, Y., Komiya, K., Iwaeda, T., Sasaki, H. and Hashimoto, T., Packing of Toyopearl columns for gel filtration. III. Semi-constant-pressure packing, *J. Chromatogr.*, 208 (1981) 71-73.

335 Kato, Y., Komiya, T., Iwaeda, T., Sasaki, H. and Hashimoto, T., Packing of Toyopearl columns for gel filtration. IV. Gravitational packing and influence of slurry reservoir size, *J. Chromatogr.*, 211 (1981) 383-387.

336 Kato, Y., Komiya, K., Sasaki, H. and Hashimoto, T., Separation range and
    separation efficiency in high-speed gel filtration on TSK-GEL SW columns
    [for globular proteins, dextrans etc.], *J. Chromatogr.*, 190 (1980) 297-303.
337 Kato, Y., Komiya, K., Sasaki, H. and Hashimoto, T., High-speed gel filtra-
    tion of proteins in sodium dodecyl sulphate aqueous solution on TSK-gel
    SW Type, *J. Chromatogr.*, 193 (1980) 29-36.
338 Kato, Y., Komiya, K., Sasaki, H. and Hashimoto, T., Comparison of TSK-GEL
    PW type and SW type in high-speed aqueous gel-permeation chromatography
    [including proteins, dextran], *J. Chromatogr.*, 193 (1980) 311-315.
339 Kato, Y., Komiya, K., Sasaki, H. and Hashimoto, T., High-speed gel filtra-
    tion of proteins in 6 M guanidine hydrochloride on TSK-GEL SW columns,
    J. Chromatogr., 193 (1980) 458-463.
340 Kato, Y., Komiya, K., Sasaki, H. and Hashimoto, T., Separation of human
    serum proteins by high-speed gel filtration on TSK-GEL G3000SWG, *J. High
    Resolut. Chromatogr. Chromatogr. Commun.*, 3 (1980) 145.
341 Kato, Y., Komiya, K., Sawada, Y., Sasaki, H. and Hashimoto, T., Purification
    of enzymes by high-speed gel filtration on TSK-GEL SW columns, *J.
    Chromatogr.*, 190 (1980) 305-310.
342 Kearney, R.D. and McGann, T.C.A., Application of controlled pore glass
    chromatography to milk proteins [separation of reconstituted low-heat
    skins milk powder casein-micelles], in Epton, R. (Editor), *Chromatography
    of Synthetic and Biological Polymers, Vol. 1, (Lect. Chem. Soc. Int. Symp.
    1976)*, Ellis Horwood, Chichester, 1978, pp. 269-274; *C.A.*, 89 (1978)
    127 803g.
343 Kehl, M., Lottspeich, F. and Henschen, A., Analysis of human fibrinopeptides
    by high-performance liquid chromatography, *Hoppe-Seyler's Z. Physiol.
    Chem.*, 362 (1981) 1661-1664; *C.A.*, 96 (1982) 82 121j.
344 Kelmers, A.D., Heatherly, D.E. and Egan, B.Z., Miniature reversed-phase
    chromatography systems for the rapid resolution of transfer RNA's and
    ribosomal RNA's [review], in Moldave, K. (Editor), *RNA Protein Synth.
    1981*, Academic Press, New York, 1981, pp. 29-32; *C.A.*, 96 (1982) 118 292q.
345 Kelmers, A.D., Weeren, H.O., Weiss, J.F., Pearson, R.L., Stulberg, M.P. and
    Novelli, G.D., Reversed-phase chromatography systems for transfer
    ribonucleic acids - preparatory-scale methods [includes also compilation
    of papers on RPC of tRNAs from various sources], in Moldave, K. (Editor),
    RNA Protein Synth. 1981, Academic Press, New York, 1981, pp. 3-28; *C.A.*,
    96 (1982) 139 078g.
346 Kemp, M.C., Hollaway, W.L., Prestidge, R.L., Bennett, J.C. and Compans, R.W.,
    Reversed-phase ion-pair high-performance liquid chromatography of viral
    tryptic glycopeptides, *J. Liq. Chromatogr.*, 4 (1981) 587-598.
347 Kennedy, J.F. and Fox, J.E., Fully automatic ion-exchange chromatographic
    analysis of neutral monosaccharides and oligosaccharides [review], *Methods
    Carbohydr. Chem.*, 8 (1980) 3-12; *C.A.*, 93 (1980) 230 282t.
348 Kennedy, J.F., Fox, J.E. and Skirrow, J.C., Automated computer calculated,
    qualitative and quantitative detailed analyses of starch components and
    related mono- and oligosaccharides [review], *Starch/Stärke*, 32 (1980)
    309-316; *C.A.*, 93 (1980) 188 088g.
349 Kent, S.B.H., Mitchell, A.R., Barany, G. and Merrifield, R.B., Test for
    racemization in model peptide synthesis by direct chromatographic separa-
    tion of diastereomers of the tetrapeptide leucylalanylglycylvaline, *Anal.
    Chem.*, 50 (1978) 155-159.
350 Kikta, E.J. Jr. and Grushka, E., Bonded peptide stationary phases for the
    separation of amino acids and peptides using liquid chromatography, *J.
    Chromatogr.*, 135 (1977) 367-376.
351 Kim, C.J., Hamielec, A.E. and Benedek, A., Characterization of dextrans by
    size exclusion chromatography using DRI/LALLSP [differential refractometer/
    low-angle laser light-scattering photometer] detector system, *J. Liq.
    Chromatogr.*, 5 (1982) 425-441.

352 Kimura, M. and Itokawa, Y., Separation and determination of thiamine-binding proteins in rats by high-performance liquid chromatography, *J. Chromatogr.*, 211 (1981) 290-294.

353 Kimura, S., Lewis, R.V., Gerber, L.D., Brink, L., Rubinstein, M., Stein, S. and Udenfriend, S., Purification to homogeneity of camel pituitary pro-opiocortin, the common precursor of opiod peptides and corticotropin, *Proc. Natl. Acad. Sci. U.S.A.*, 76 (1979) 1756-1759.

354 Kimura, T., Takai, M., Masui, Y., Morikawa, T. and Sakakibara, S., Strategy for the synthesis of large peptides; an application [of HPLC] to the total synthesis of human parathyroid hormone hPTH(1-84), *Biopolymers*, 20 (1981) 1823-1832; *C.A.*, 95 (1981) 204 420h.

355 Kimura, Y., Kitamura, H., Araki, T., Noguchi, K., Baba, M. and Hori, M., Analytical and preparative methods for polymyxin antibiotics using high-performance liquid chromatography with a porous styrene-divinylbenzene copolymer packing, *J. Chromatogr.*, 206 (1981) 563-572.

356 Kiso, Y., Miyazaki, T., Satomi, M., Hiraiwa, H. and Akita, T., An "active ester"-type mixed anhydride method for peptide synthesis. Use of the new reagent, norborn-5-ene-2,3-dicarboximido diphenyl phosphate (NDPP) [RP HPLC-racemization tests], *J. Chem. Soc. Chem. Commun.*, (1980) 1029-1030.

357 Kiso, Y., Satomi, M., Miyazaki, T., Hiraiwa, H. and Akita, T., A new method for racemization test in peptide synthesis: use of reversed-phase high performance liquid chromatography, in Okawa, K. (Editor), *Peptide Chemistry*, Vol. 18, Protein Research Foundation, Osaka, 1981, pp. 71-74; *C.A.*, 95 (1981) 115 988g.

358 Kitamura, T., Matsuda, T., Suenaga, A. and Horikiri, H., Studies on the packing method of Toyopearl. 1. Packing by the gravity method [in Japanese], *Toyo Soda Kenkyu Hokoku*, 24 (1980) 123-131; *C.A.*, 94 (1981) 60 843c.

359 Kitamura, T., Sugio, N., Shimamura, S., Yamamoto, H., Matsuda, T. and Horikiri, H., Studies on the packing method of Toyopearl. 2. Packing by using a peristatic pump [in Japanese], *Toyo Soda Kenkyu Hokoku*, 24 (1980) 133-137; *C.A.*, 94 (1981) 40 769e.

360 Klee, C.B., Oldewurtel, M.D., Williams, J.F. and Lee, J.W., Analysis of Ca^{2+}-binding proteins by high performance liquid chromatography, *Biochem. Int.*, 2 (1981) 485-493; *C.A.*, 95 (1981) 38 292w.

361 Klickstein, L.B. and Wintroub, B.U., Separation of angiotensins and assay of angiotensin-generating enzymes by high-performance liquid chromatography, *Anal. Biochem.*, 120 (1982) 146-150.

362 Knight, M., Cayley, P.J., Silverman, R.H., Wreschner, D.H., Gilbert, C.S., Brown, R.E. and Kerr, I.M., Radioimmune, radiobinding and HPLC analysis of 2-5A and related oligonucleotides from intact cells, *Nature (London)*, 288 (1980) 189-192.

363 Koeppe, R.E. II. and Weiss, L.B., Resolution of linear gramicidins by preparative reversed-phase high-performance liquid chromatography, *J. Chromatogr.*, 208 (1981) 414-418.

364 Kojima, K., Manabe, T., Okuyama, T., Tomono, T., Suzuki, T. and Tokunaga, E., Two-dimensional separation system for analysis of proteins employing isoelectric focusing and high-performance liquid chromatography, *J. Chromatogr.*, 239 (1982) 565-570.

365 Komiya, H., Nishikawa, K., Ogawa, K. and Takemura, S., Methods for sequencing oligoribonucleotides and RNA by high performance liquid chromatography [RNase digests], *J. Biochem. (Tokyo)*, 6 (1979) 1081-1088; *C.A.*, 91 (1979) 206 621r.

366 Komiya, K., Kato, Y., Furukawa, K., Sasaki, H. and Watanabe, H., Introduction and evaluation of TSK-gel IEX-500 series columns for high speed ion-exchange chromatography, *Toyo Soda Kenkyu Hokoku*, 25 (1981) 115-132; *C.A.*, 96 (1982) 65 065b.

367 Konaka, R. and Shoji, J., High performance liquid chromatography of peptide antibiotics [review in Japanese], *Kagaku no Ryoiki, Zokan*, 133 (1981) 151-166; *C.A.*, 95 (1981) 156 629j.

B589

368 Kopaciewicz, W. and Regnier, F.E., Nonideal size-exclusion chromatography of proteins: effects of pH at low ionic strenght, *Anal. Biochem.*, 126 (1982) 8-16.
369 Kopper, R.A. and Singhal, R.P., Stability and separation of aminoacyl-transfer RNAs on chromatography columns [of polystyrene anion exchangers and reversed phase matrix], *Int. J. Biol. Macromol.*, 1 (1979) 65-72.
370 Kratzin, H., Yang Chao-yuh, Krusche, J.U. and Hilschmann, N., Präparative Auftrennung des tryptischen Hydrolysats eines Proteins mit Hilfe der Hochdruck-Flüssigkeitschromatographie. Die Primärstruktur einer monoklonalen L-Kette vom K-Typ, Subgruppe I (Bence-Jones-Protein Wes). (Preparative separation of the tryptic hydrolysate of protein by high-pressure liquid chromatography. The primary structure of a monoclonal L-chain of K-type, subgroup I (Bence-Jones protein Wes)). *Hoppe-Seyler's Z. Physiol. Chem.*, 361 (1980) 1591-1598; *C.A.*, 94 (1981) 63 482v.
371 Kroeff, E.P. and Pietrzyk, D.J., Investigation of the retention and separation of amino acids, peptides and derivatives on porous copolymers by high performance liquid chromatography, *Anal. Chem.*, 50 (1978) 502-511.
372 Kroeff, E.P. and Pietrzyk, D.J., High performance liquid chromatographic study of the retention and separation of short chain peptide diastereomers on a C_8 bonded phase, *Anal. Chem.*, 50 (1978) 1353-1358.
373 Krol, G.J., Banovsky, J.M., Mannan, C.A., Pickering, R.E. and Kho, B.T., Trace analysis of the MIF* analogue pareptide in blood plasma by high-performance liquid chromatography and short-wavelength excitation fluorometry [*melanocyte stimulating hormone release-inhibiting factor], *J. Chromatogr.*, 163 (1979) 383-389.
374 Krol, G.J., Mannan, C.A., Pickering, R.E., Amato, D.V., Kho, B.T. and Sonnenschein, A., Short excitation wavelength fluorimetric detection in high-pressure liquid chromatography of indole peptide, naphthyl, and phenol compounds, *Anal. Chem.*, 49 (1977) 1836-1839.
375 Krstulovic, A.M. and Brown, P.R., *Reversed-Phase High-Performance Liquid Chromatography. Theory, Practice and Biomedical Applications* [including applications in peptide and protein chemistry], Wiley-Interscience, New York, 1982, pp. 296; *C.A.*, 96 (1982) 196 166h.
376 Krumen, K., HPLC in the analysis and separation of pharmaceutically important peptides [review], *J. Liq. Chromatogr.*, 3 (1980) 1243-1254.
377 Krummen, K. and Frei, R.W., The operation of nonapeptides by reversed-phase high-performance liquid chromatography, *J. Chromatogr.*, 132 (1977) 27-36.
378 Krummen, K. and Frei, R.W., Quantitative analysis of nonapeptides in pharmaceutical dosage forms by high-performance liquid chromatography, *J. Chromatogr.*, 132 (1977) 429-436.
379 Krummen, K., Maxl, F. and Nachtmann, F., The use of HPLC in the quality control of oxytocin, *Pharm. Technol. Int.*, 2 (1979) 37-43, 62; *C.A.*, 94 (1981) 7832.
380 Kudirka, P.J., Busby, M.G., Carey, R.N. and Toren, E.C. Jr., Separation of creatine kinase isoenzymes by high-pressure liquid chromatography, *Clin. Chem. (Winston-Salem, N.C.)*, 21 (1975) 450-452.
381 Kudirka, P.J., Schroeder, R.R., Hewitt, T.E. and Toren, E.C. Jr., High-pressure liquid-chromatographic separation of lactate dehydrogenase isoenzymes, *Clin. Chem. (Winston-Salem, N.C.)*, 22 (1976) 471-474.
382 Kullmann, W., Monitoring of protease-catalyzed peptide synthesis by high performance liquid chromatography, *J. Liq. Chromatogr.*, 4 (1981) 1121-1134.
383 Kumanotani, J., Oshima, R., Yamauchi, Y., Takai, N. and Kurosu, Y., Preparative high-performance gel chromatography for acidic and neutral saccharides, *J. Chromatogr.*, 176 (1979) 462-464.
384 Kundu, S.K., Chakravarty, S.K., Roy, S.K. and Roy, A.K., DEAE-silica gel and DEAE-controlled porous glass as ion exchangers for isolation of glycolipids, *J. Chromatogr.*, 170 (1979) 65-72.
385 Kundu, S.K. and Roy, S.K., A rapid and quantitative method for the isolation of gangliosides and neutral glycosphingolipids by DEAE-silica gel chromatography, *J. Lipid Res.*, 19 (1978) 390-395.

386 Kuo, J.C. and Yeung, E.S., Determination of carbohydrates in urine by high-performance liquid chromatography and optical activity detection, *J. Chromatogr.*, 223 (1981) 321-329.

387 Kusunose, E., Kaku, M., Nariyama, M., Kusunose, M., Ichihara, K., Funae, Y. and Kotake, A.N., High-performance liquid chromatography of cytochrome P-450 from rabbit liver, kidney cortex and intestinal mucosa microsomes, *Biochem. Int.*, 3 (1981) 399-406; *C.A.*, 96 (1982) 3150s.

388 Kuwata, S., Yamada, T., Miyazawa, T., Dejima, K. and Watanabe, K., Separation of diastereoisomers of protected peptides and a convenient test for racemization in peptide synthesis, by reversed-phase high-performance liquid chromatography, in Okawa, K. (Editor), *Peptide Chemistry 1980*, Vol. 18, Protein Research Foundation, Osaka, 1981, pp. 65-70; *C.A.*, 95 (1981) 115 987f.

389 Ladisch, M.R., Huebner, A.L. and Tsao, G.T., High-speed liquid chromatography of cellodextrins and other saccharide mixtures using water as the eluent, *J. Chromatogr.*, 147 (1978) 185-193.

390 Ladisch, M.R. and Tsao, G.T., Theory and practice of rapid liquid chromatography [of saccharides] at moderate pressures using water as eluent, *J. Chromatogr.*, 166 (1978) 85-100.

391 Lambert, P.W. and Roos, B.A., Paired-ion reversed-phase high-performance liquid chromatography of human and rat calcitonin, *J. Chromatogr.*, 198 (1980) 293-299.

392 Lam-Thanh, H., Fermandjian, S. and Fromageot, P., Reversed-phase, ion-pair separation of L-methionine and L-methionine dipeptides complexed with palladium(II), *J. Liq. Chromatogr.*, 4 (1981) 681-688.

393 Lam-Thanh, H., Fermandjian, S. and Fromageot, P., High-performance liquid chromatography and magnetic circular dichroism: a study of the "palladium (II)-thioether peptide" complexes, *J. Chromatogr.*, 235 (1982) 139-147.

394 Larsen, B., Fox, B.L., Burke, M.F. and Hruby, V.J., The separation of peptide hormone diastereoisomers by reverse phase high pressure liquid chromatography, *Int. J. Peptide Protein Res.*, 13 (1979) 12-21.

395 Larsen, B., Viswanatha, V., Chang, Sai Y. and Hruby, V.J., Reverse phase high pressure liquid chromatography for the separation of peptide hormone diastereoisomers, *J. Chromatogr. Sci.*, 16 (1978) 207-210.

396 Larsson, P.O., Griffin, T. and Mosbach, K., Some new techniques related to affinity chromatography [review including HPLC], *Colloq.-Inst. Nat. Sante Rech. Med.*, 86 (*Chromatogr. Affinite Interact. Mol.*) (1979) 91-97; *C.A.*, 92 (1980) 106 620s.

397 Lawrence, J.G., Analysis of carbohydrates by high performance liquid chromatography [including disaccharides], *Scan*, 5 (1974) 19-24; *C.A.*, 85 (1976) 71 766v.

398 Lawson, A.M., Lim, C.K. and Richmond, W. (Editors), *Clinical Research Center Symposium, No. 1: Current Developments in the Clinical Applications of HPLC, GC and MS*, Academic Press, London, 1980, 301 pp.; *C.A.*, 94 (1981) 170 638x.

399 Lebl, M., Amino acids and peptides. CLXVII. High-performance liquid chromatography of carba-analogs of oxytocin; a method for the determination of sulfides and sulfoxides in peptides, *Collect. Czech. Chem. Commun.*, 45 (1980) 2927-2937; *C.A.*, 94 (1981) 152 773h.

400 Lebl, M., Amino acids and peptides. CLXXVIII. Correlation between hydrophobicity of substituents in the phenylalanine moiety of oxytocin carba-analogues and reversed-phase-chromatographic k' values, *J. Chromatogr.*, 242 (1982) 342-345.

401 Lebl, M., Carba analogues of neurohypophysial hormones, in Hancock, W.S. (Editor), *Handbook of Use of High Performance Liquid Chromatography for the Analytical and Preparative Separation of Amino Acids, Peptides and Proteins*, CRC Press, Boca Raton, FL, 1984, pp. 169-178.

402 Lebl, M., Amino acids and peptides. CLXXXI. Separation of diastereoisomers of oxytocin analogues, *J. Chromatogr.*, 264 (1983) 459-462.

403 Ledeen, R.W. and Yu, K., Gangliosides: structure, isolation and analysis [HPLC separation techniques are also reviewed], *Methods Enzymol.*, 83 (Complex Carbohydrates, part D) (1982) 139-191.

404 Lee, G. Jia-Long, High-performance liquid-chromatography determination of disaccharides resulting from enzymic degradation of glycosaminoglycans and its application in chemical diagnosis of mucopolysaccharidoses, *Diss. Abstr. Int. B*, 41 (1980) 2167; *C.A.*, 94 (1981) 79 540y.

405 Lee, G. Jia-Long, Evans, J.E. and Tieckelmann, H., Rapid and sensitive determination of enzymatic degradation products of isomeric chondroitin sulfates by high-performance liquid chromatography, *J. Chromatogr.*, 146 (1978) 439-448.

406 Lee, G. Jia-Long, Evans, J.E., Tieckelmann, H., Dulaney, J.T. and Naylor, E.W., Enzymatic studies of urinary isomeric chondroitin sulfates from patients with mucopolysaccharidosis. The application of high performance liquid chromatography, *Clin. Chim. Acta*, 104 (1980) 65-75.

407 Lee, G. Jia-Long, Liu, Der-Wu, Pav, J.W. and Tieckelmann, H., Separation of reduced disaccharides derived from glycosaminoglycans by high-performance liquid chromatography, *J. Chromatogr.*, 212 (1981) 65-73.

408 Lee, G. Jia-Long and Tieckelmann, H., High-performance liquid chromatographic determinations of disaccharides resulting from enzymatic degradation of isomeric chondroitin sulfates, *Anal. Biochem.*, 94 (1979) 231-236.

409 Lee, G. Jia-Long and Tieckelmann, H., High-performance liquid chromatographic separation of unsaturated disaccharides derived from heparan sulfate and heparin, *J. Chromatogr.*, 195 (1980) 402-406.

410 Lee, G. Jia-Long and Tieckelmann, H., The application of high-performance liquid chromatography in enzymatic assays of chondroitin sulfate isomers in normal and human urine, *J. Chromatogr.*, 222 (1981) 23-31.

411 Lee, W.M.F., Westrick, M.A. and Macher, B.A., High-performance liquid chromatography of long-chain neutral glycosphingolipids and gangliosides, *Biochim. Biophys. Acta*, 712 (1982) 498-504.

412 Leemann, H.G., Erni, F. and Schreiber, B., Interdisciplinary application of modern trace analysis to biologically active substances [review including HPLC of cyclosporin A], in Malissa, H., Grasserbauer, M. and Belcker, R. (Editors), *Nature, Aim and Methods of Microchemistry (Proc. 8th Internat. Microchem. Symp.)*, Springer-Verlag, Wien, New York, 1980, pp. 101-120; *C.A.*, 95 (1981) 183 207b.

413 Lenfant, M., Millerioux-DiGiusto, L., Masson, A. and Gasc, J.C., Shortened purification procedure of a spleen-derived immunosuppressive peptide [RP HPLC], *J. Chromatogr.*, 206 (1981) 177-180.

414 Leutzinger, E.E., Miller, P.S. and Ts'o, P.O.P., Application of high-pressure liquid chromatography in the synthesis of oligonucleotides, in Leroy, B., Townsend, L.B. and Tipson, R.S. (Editors), *Nucleic Acid Chemistry, Methods and Techniques*, Part 2, Wiley, 1978, pp. 1037-1043; *C.A.*, 93 (1980) 8430f.

415 Lewis, R.V., Fallon, A., Stein, S., Gibson, K.D. and Udenfriend, S., Supports for reversed-phase high-performance liquid chromatography of large proteins, *Anal. Biochem.*, 104 (1980) 153-159.

416 Lim, C.K., High performance liquid chromatography [a comprehensive review for biological analysis including hormones], *Horm. Blood*, 3 (1979) 63-127; *C.A.*, 92 (1980) 159 886q.

417 Lin, Shen-Nan, Smith, L.A. and Caprioli, R.M., Analysis of dipeptide mixtures by the combination of ion-pair reversed-phase high-performance liquid chromatographic and gas chromatographic-mass spectrometric techniques, *J. Chromatogr.*, 197 (1980) 31-41.

418 Lindeberg, G., Separation of vasopressin analogs by reversed-phase high-performance liquid chromatography, *J. Chromatogr.*, 193 (1980) 427-431.

419 Live, D.H., Analysis of fluorescent labeled peptides by high pressure liquid chromatography, in Goodman, M. and Meienhofer, J. (Editors), *Peptides (Proc. 5th Am. Peptide Symp.)*, Halsted Press and Wiley, New York, 1977, pp. 44-47; *C.A.*, 88 (1978) 185 503z.

420 Loeber, J.G. and Verhoef, J., High-pressure liquid chromatography and
    radioimmunoassay for the specific and quantitative determination of
    endorphins and related peptides, *Methods Enzymol.*, 73 (Immunochemical
    Techniques, Part B) (1981) 261-275; *C.A.*, 95 (1981) 217 173e.
421 Lou, M.F., Morrison, W.H. and Dueber, Ch., A split-stream ion-exchange
    chromatographic method for isolating glycopeptides from biological fluid,
    Biochem. Med., 23 (1980) 47-54; *C.A.*, 93 (1980) 3219j.
422 Lowe, Ch.R., Glad, M., Larsson, P.-O., Ohlson, S., Small, D.A.P., Atkinson,
    T. and Mosbach, K., High-performance liquid affinity chromatography of
    proteins on Cibacron Blue F3G-A bonded silica, *J. Chromatogr.*, 215 (1981)
    303-316.
423 Lu, Kai-Chun, Gooding, K.M. and Regnier, F.E., Rapid analysis of bilirubin
    in neonatal serum. I. The binding of bilirubin to albumin, *Clin. Chem.
    (Winston-Salem, N.C.)*, 25 (1979) 1608-1612; *C.A.*, 91 (1979) 188 992f.
424 Lukas, T.J., Prystowsky, M.B. and Erickson, B.W., Solid-phase peptide
    synthesis under continuous-flow conditions [RP HPLC analysis], *Proc. Natl.
    Acad. Sci. U.S.A.*, 78 (1981) 2791-2795.
425 Lundanes, E. and Greibrokk, T., Reversed-phase chromatography of peptides,
    J. Chromatogr., 149 (1978) 241-254.
426 Lundanes, E. and Greibrokk, T., Retention of dipeptides in reversed phase
    HPLC, in Hancock, W.S. (Editor), *Handbook of Use of HPLC for the Separa-
    tion of Amino Acids, Peptides and Proteins*, CRC Press, Boca Raton, FL,
    1984, pp. 49-52.
427 Mabuchi, H. and Nakahashi, H., Systematic separation of medium-sized
    biologically active peptides by high-performance liquid chromatography,
    J. Chromatogr., 213 (1981) 275-286.
428 Mabuchi, H. and Nakahashi, H., Analysis of small peptides in uremic serum
    by high performance liquid chromatography. *J. Chromatogr.*, 228 (1982)
    292-297.
429 Macek, K., Deyl, Z., Čoupek, J. and Sanitrák, J., Separation of collagen
    types I and III by high-performance column liquid chromatography, *J.
    Chromatogr.*, 222 (1981) 284-290.
430 Macklin, W.B., Pickart, L. and Woodward, D.O., Partial purification of
    microsomal proteolipid(s) from *Neurospora crassa* by high-performance
    liquid chromatography on silica gel, *J. Chromatogr.*, 210 (1981) 174-179.
431 Macrae, R., Applications of high pressure liquid chromatography to food
    analysis [review including beside others also carbohydrates], *J. Food
    Technol.*, 15 (1980) 93-110; *C.A.*, 93 (1980) 6164y.
432 Macrae, R., Recent applications of high-pressure liquid chromatography to
    food analysis [review], *J. Food Technol.*, 16 (1981) 1-11; *C.A.*, 94 (1981)
    172 929s.
433 Macrae, R. (Editor), *HPLC in Food Analysis* (A Volume in the Food Science
    and Technology Series) [in addition to other substances includes also
    carbohydrates and peptides], Academic Press, New York, 1982, 356 pp.
434 Maerki, W., Brown, M. and Rivier, J.E., Bombesin analogs: effect on thermo-
    regulation and glucose metabolism [RP HPLC of synthetic peptides], *Peptides
    (N.Y.)*, 2 (Suppl. 2) (1981) 169-177.
435 Mahoney, W.C. and Hermodson, M.A., Separation of large denatured peptides
    by reverse phase high performance liquid chromatography. Trifluoroacetic
    acid as a peptide solvent, *J. Biol. Chem.*, 255 (1980) 11199-11203.
436 Majors, R.E., Liquid chromatography column technology [general review
    including also oligosaccharides, oligonucleotides, peptides and proteins],
    in Marton, L.J. and Kabra, P.M. (Editors), *Liquid Chromatography in
    Clinical Analysis*, Humana Press, Clifton, NJ, 1981, pp. 51-94; *C.A.*, 95
    (1981) 57 332a.
437 Majors, R.E., Barth, H.G. and Lochmüller, Ch.H., Column liquid chromato-
    graphy [review], *Anal. Chem.*, 54 (1982) 323R-363R.

438 Makuch, B., Arendt, A. and Kolodziejczyk, A.M., Use of high efficiency liquid chromatography to obtain pure diastereoisomers of protected Ac-Phe-Phe-OEt dipeptide, *Pol. J. Chem.*, 55 (1981) 701-704; *C.A.*, 96 (1982) 181 612b.

439 Margolis, J. and Rhoades, P.H., Preparation of high-purity factor VIII by controlled pore glass chromatography, *Lancet*, ii (1981) 446-449; *C.A.*, 96 (1982) 40 782v.

440 Margolis, S.A. and Konash, P.J., Non-peptide impurities in angiotensin I and other commercial peptides, *J. High Resolut. Chromatogr. Chromatogr. Commun.*, 3 (1980) 317-318.

441 Margolis, S.A. and Longenbach, P.J., Analysis of purity of commercial peptides by high-resolution liquid chromatography, in Barker, J.L. and Smith, T.G., Jr. (Editors), *The Role of Peptides in Neuronal Function*, Marcel Dekker, New York, 1980, pp. 49-67; *C.A.*, 94 (1981) 170 366g.

442 Margolis, S.A. and Longenbach, P.J., Separation of structurally similar, biologically active peptides from their impurities, *J. High Resolut. Chromatogr. Chromatogr. Commun.*, 2 (1979) 255-256.

443 Margolis, S.A. and Schaffer, R., Development of a standard reference material for angiotensin I, *National Bureau of Standards Report NBSIR 79-1947*, National Institutes of Health, Bethesda, MD, 1979, pp. 1-31.

444 Markussen, J. and Larsen, U.D., The application of HPLC to the analysis of radioiodinated tracers of glucagon and insulin, in Brandenburg, D. and Wollmer, A. (Editors), *Insulin: Chemistry, Structure and Function of Insulin and Related Hormones (Proc. Int. Insulin Symp., 2nd, 1979)*, Walter de Gruyter, Berlin, 1980, pp. 161-168; *C.A.*, 94 (1981) 60 981w.

445 Martinelli, R.A. and Scheraga, H.A., Assay of bovine fibrinopeptides by high performance liquid chromatography, *Anal. Biochem.*, 96 (1979) 246-249.

446 Marx-Figini, M. and Soubelet, O., Size exclusion chromatography (GPC) of cellulose nitrate using modified silica gel as stationary phase, *Polym. Bull. (Berlin)*, 6 (1982) 501-508; *C.A.*, 96 (1982) 204 488j.

447 Matsumoto, K., Isozyme analysis by high-performance liquid chromatography [theory, use; in Japanese], *Kensa to Gijutsu*, 9 (1981) 359-366; *C.A.*, 94 (1981) 187 510.

448 May, H.E., Tanzer, F.S., Fridland, G.H., Wakelin, C. and Desiderio, D.M., High performance liquid chromatography and proteolytic enzyme characterization of peptides in tooth pulp extract, *J. Liq. Chromatogr.*, 5 (1982) 2135-2154; *C.A.*, 98 (1983) 30 788d.

449 McCluer, R.H. and Evans, J.E., Preparation and analysis of benzoylated cerebrosides, *J. Lipid Res.*, 14 (1973) 611-617; *C.A.*, 80 (1974) 11 577w.

450 McCluer, R.H. and Evans, J.E., Quantitative analysis of brain galactosyl-ceramides by high performance liquid chromatography of their perbenzoyl derivatives, *J. Lipid Res.*, 17 (1976) 412-418; *C.A.*, 85 (1976) 89 528d.

451 McCluer, R.H. and Ullman, M.D., Preparative and analytical high performance liquid chromatography of glycolipids, *ACS Symp. Ser.*, 128 (Cell Surf. Glycolipids) (1980) 1-13; *C.A.*, 93 (1980) 109 940c.

452 McDermott, J.R., Smith, A.I., Biggins, J.A., Al-Noaemi, M.C. and Dwardson, J.A., Characterization and determination of neuropeptides by high-performance liquid chromatography and radioimmunoassay, *J. Chromatogr.*, 222 (1981) 371-379.

453 McFarland, G.D. and Borer, P.N., Separation of oligo-RNA [nucleotides] by reverse-phase HPLC, *Nucleic Acids Res.*, 7 (1979) 1067-1080; *C.A.*, 92 (1980) 37 034g.

454 McLaughlin, L.W., Cramer, F. and Sprinzl, M., Rapid analysis of modified tRNA^{Phe} from yeast by high-performance liquid chromatography: chromatography of oligonucleotides after RNase T$_1$ digestion on Aminopropylsilica and assignment of the fragments based on nucleoside analysis by chromatography on C$_{18}$-silica, *Anal. Biochem.*, 112 (1981) 60-69.

455 Meek, J., Prediction of peptide retention times in high-pressure liquid chromatography on the basis of amino acid composition, *Proc. Natl. Acad. Sci. U.S.A.*, 77 (1980) 1632-1636.

B594

456 Meek, J.L., Separation of neuropeptides by HPLC, in Costa, E. and Trabucchi, M. (Editors), *Neural Peptides and Neuronal Communication* (*Adv. Biochem. Psychopharmacol.*, Vol. 22), Raven Press, New York, 1980, pp. 145-151; *C.A.*, 93 (1980) 145 637t.
457 Meek, J.L. and Bohan, T.P., Use of high pressure liquid chromatography (HPLC) to study enkephalins, in Costa, E. and Trabucchi, M. (Editors), *The Endorphins* (*Adv. Biochem. Psychopharmacol.*, Vol. 18), Raven Press, New York, 1978, pp. 141-147; *C.A.*, 89 (1978) 38 875q.
458 Meek, J.L. and Rossetti, Z.L., Factors affecting retention and resolution of peptides in high-performance liquid chromatography, *J. Chromatogr.*, 211 (1981) 15-28.
459 Melander, W.R., Jacobson, J. and Horváth, Cs., Effect of molecular structure and conformational change of proline-containing dipeptides in reversed-phase chromatography, *J. Chromatogr.*, 234 (1982) 269-276.
460 Melander, W.R., Nahum, A. and Horváth, Cs., Mobile phase effects in reversed-phase chromatography. III. Changes in conformation and retention of oligo(ethylenglycol)derivatives with temperature and eluent composition [as a model for studying chromatographic behavior of biological substances like peptides and oligonucleotides], *J. Chromatogr.*, 185 (1979) 129-152.
461 Mellis, S.J. and Baenziger, J.U., Separation of neutral oligosaccharides by high-performance liquid chromatography, *Anal. Biochem.*, 114 (1981) 276-280.
462 Mencer, H.J. and Grubisic-Gallot, Z., Gel permeation chromatography on chemically modified silica using different solvents [including chromatography of polydextran], *J. Chromatogr.*, 241 (1982) 213-216.
463 Meredith, S.C. and Nathans, G.R., Gel permeation chromatography of asymmetric proteins, *Anal. Biochem.*, 121 (1982) 234-243.
464 Meuser, F., Klingler, R.W. and Niediek, E.A., Trennung von Stärkemolekülen durch Hochdruckflüssigkeitschromatographie. (Separation of starch molecules by high-pressure liquid chromatography), *Getreide, Mehl, Brot*, 33 (1979) 295-299; *C.A.*, 92 (1980) 74 461x.
465 Meyers, C.A., Coy, D.H., Huang, W.Y., Schally, A.V. and Redding, T.W., Highly active position eight analogues of somatostatin and separation of peptide diastereomers by partition chromatography [resolution monitored by RP HPLC], *Biochemistry*, 17 (1978) 2326-2331.
466 Mikeš, O. (Editor), *Laboratory Handbook of Chromatographic and Allied Methods*, Ellis Horwood, (Halsted Press, Wiley), Chichester, 1979, 764 pp.
467 Mikeš, O., Rapid separation of proteins and their high-molecular fragments by means of Spheron ion-exchangers [lecture], *Int. J. Peptide Protein Res.*, 14 (1979) 393-401.
468 Mikeš, O., Rapid chromatography analysis of enzymes and other proteins [review prepared for EURO FOOD CHEM I Congress, Vienna, 1981], *Ernährung/Nutrition*, 5 (1981) 88-98 (English version without references); *C.A.*, 94 (1981) 152 760b.
469 Mikeš, O., Rychlá chromatografická analyza enzymů a jiných bílkovin [Czech translation of the review "Rapid chromatography analysis of enzymes and other proteins" published in *Ernährung/Nutrition*, 5 (1981) 88-98, with all references], *Chem. Listy*, 76 (1982) 59-79; *C.A.*, 96 (1982) 118 297u.
470 Mikeš, O., Rapid chromatographic analysis of enzymes and other proteins [lecture], in Baltes, W., Czedik-Eysenberg, P.B. and Pfannhauser, W. (Editors), *Recent Developments in Food Analysis* (*Proceedings of the First European Conference on Food Chemistry EURO FOOD CHEM I*, Vienna, February 17-20, 1981), Verlag Chemie, Weinheim, 1982, pp. 306-321.
471 Mikeš, O., Sedláčková, J., Rexová-Benková, L'. and Omelková, J., High performance liquid chromatography of pectic enzymes, *J. Chromatogr.*, 207 (1981) 99-114.
472 Mikeš, O., Štrop, P. and Čoupek, J., Ion-exchange derivatives of Spheron. I. Characterization of polymeric supports, *J. Chromatogr.*, 153 (1978) 23-36.

473 Mikeš, O., Štrop, P., Hostomská, Z., Smrž, M., Čoupek, J., Frydrychová, A. and Bareš, M., Ion-exchange derivatives of Spheron. IV. Phosphate derivatives, *J. Chromatogr.*, 261 (1983) 363-379.

474 Mikeš, O., Štrop, P. and Sedláčková, J., Rapid chromatographic separation of technical enzymes on Spheron ion exchangers, *J. Chromatogr.*, 148 (1978) 237-245.

475 Mikeš, O., Štrop, P., Smrž, M. and Čoupek, J., Ion-exchange derivatives of Spheron. III. Carboxylic cation exchangers, *J. Chromatogr.*, 192 (1980) 159-172.

476 Mikeš, O., Štrop, P., Zbrožek, J. and Čoupek, J., Chromatography of biopolymers and their fragments on ion-exchange derivatives of the hydrophilic macroporous synthetic gel Spheron, *J. Chromatogr.*, 119 (1976) 339-354.

477 Mikeš, O., Štrop, P., Zbrožek, J. and Čoupek, J., Ion-exchange derivatives of Spheron. II. Diethylaminoethyl Spheron, *J. Chromatogr.*, 180 (1979) 17-30.

478 Minasian, E., Sharma, R.S., Leach, S.J., Grego, B. and Hearn, M.T.W., A comparative study on the separation of the tryptic peptides of the β-chain of normal and abnormal hemoglobins, *J. Liq. Chromatogr.*, 6 (1983) 199-213; *C.A.*, 98 (1983) 194 367s.

479 Mirabel, B., Séparation des protéines par chromatographie préparative. (Separation of proteins by preparative chromatography), *Actual. Chim.*, (1980) 39-44; *C.A.*, 93 (1980) 148 214p.

480 Mizutani, T., Adsorption of proteins by porous glass and its application to chromatography [in Japanese], in Osawa, K., Tanaka, Y. and Kitami, S. (Editors), *Kyodai Ryushi no Gerupamieishion Kuromatogurafi: Seitai Ryushi no Ryukei Bunri*, Nagoya City University, Nagoya, 1980, pp. 243-260; *C.A.*, 95 (1981) 146 292e.

481 Mizutani, T., Separation of proteins on silicone-coated porous glass, *J. Chromatogr.*, 207 (1981) 276-280.

482 Mizutani, T. and Mizutani, A., Prevention of adsorption of protein on controlled-pore glass with amino acid buffer, *J. Chromatogr.*, 111 (1975) 214-216.

483 Mizutani, T. and Mizutani, A., Use of controlled-pore glass for adsorption chromatography of proteins, *J. Chromatogr.*, 120 (1976) 206-210.

484 Mizutani, T. and Mizutani, A., Adsorption of cationic biological materials on controlled pore glass [basic amino acids, aminosugars, proteins], *Anal. Biochem.*, 83 (1977) 216-221.

485 Mizutani, T. and Mizutani, A., Comparison of elution patterns of proteins chromatographed on controlled-pore glass and carboxymethylcellulose, *J. Chromatogr.*, 168 (1979) 143-150.

486 Mizutani, T. and Narihara, T., Adsorption chromatography of proteins on siliconized porous glass, *J. Chromatogr.*, 239 (1982) 755-760.

487 Mizutani, T. and Narihara, T., Adsorption chromatography of nucleic acids on siliconized porous glass, *Nucleic acids Res.*, *Symp. Ser.*, 11 (1982) 127-130; *C.A.*, 98 (1983) 68 218f.

488 Moehring, J., Boehlen, P., Shoun, J., Mellet, M., Suess, U., Schmidt, M. and Pliska, V., Comparison of radioimmunoassay, chemical assay (HPLC) and bioassay for arginine vasopressin in synthetic standards and posterior pituitary tissue, *Acta Endocrinol. (Copenhagen)*, 99 (1982) 371-378; *C.A.*, 96 (1982) 136 057p.

489 Moench, W. and Dehnen, W., High performance liquid chromatography of peptides [RPC], *J. Chromatogr.*, 140 (1977) 260-262.

490 Moench, W. and Dehnen, W., High performance liquid chromatography of polypeptides and proteins on a reversed-phase support, *J. Chromatogr.*, 147 (1978) 415-418.

491 Molko, D., Derbyshire, R., Gui, A., Roget, A., Teoule, R. and Boucherle, A., Exclusion column for high-performance liquid chromatography of oligonucleotides, *J. Chromatogr.*, 206 (1981) 493-500.

492 Molnár, I. and Horváth, Cs., Separation of amino acids and peptides on nonpolar stationary phases by high-performance liquid chromatography, *J. Chromatogr.*, 142 (1977) 623-640.

493 Molnár, I. and Horváth, Cs., Rapid analysis of peptide mixtures by high performance liquid chromatography with nonpolar stationary phases, in Goodman, M. and Meienhofer, J. (Editors), *Peptides (Proc. 5th Am. Peptide Symp.)*, Wiley, New York, 1977, pp. 48-51; *C.A.*, 88 (1978) 166 205n.

494 Momoi, T., Ando, S. and Nagai, Y., High performance preparative column chromatography of lipids using a new porous silica, Iatrobeads. II. High resolution preparative column chromatographic system for gangliosides using DEAE-Sephadex and a new porous silica, Iatrobeads, *Biochim. Biophys. Acta*, 441 (1976) 488-497.

495 Montenecourt, B.S., Kelleher, T.J., Eveleigh, D.E. and Pettersson, L.G., Biochemical nature of cellulases from mutants of *Trichoderma reesei* [HPLC on DEAE-silica], *Biotechnol. Bioeng. Symp.*, 10 (1980) 15-26; *C.A.*, 94 (1981) 26 631x.

496 Moore, G.J., Reversed phase high-pressure liquid chromatography for the identification and purification of neuropeptides [review], *Life Sci.*, 30 (1982) 995-1002; *C.A.*, 96 (1982) 155 601c.

497 Mopper, K., Dawson, R., Liebezeit, G. and Hansen, H.-P., Borate complex ion exchange chromatography with fluorimetric detection for determination of saccharides [both reducing and nonreducing including oligosaccharides], *Anal. Chem.*, 52 (1980) 2018-2022.

498 Moriya, F., Makino, K., Suzuki, N., Rokushika, S. and Hatano, H., Studies on spin-trapped radicals in γ-irradiated aqueous solutions of glycylglycine and glycyl-L-alanine by high-performance liquid chromatography and ESR spectroscopy, *J. Phys. Chem.*, 84 (1980) 3614-3619; *C.A.*, 94 (1981) 121 927g.

499 Moriya, F., Makino, K., Suzuki, N., Rokushika, S. and Hatano, H., Studies on spin-trapped radicals in γ-irradiated aqueous solutions of L-alanyl-glycine and L-alanyl-L-alanine by high-performance liquid chromatography and ESR spectroscopy, *J. Am. Chem. Soc.*, 104 (1982) 830-836.

500 Morris, H.R., Dell, A., Etienne, T. and Taylor, G.W., Mass spectrometric methods for the structure determination of neuropeptides [after HPLC-separation], proteins and glycoproteins, *Dev. Biochem.*, 10 (Front. Protein. Chem.) (1980) 193-209; *C.A.*, 95 (1981) 20 660e.

501 Murayama, K. and Kinoshita, T., Determination of glutathione by high performance liquid chromatography using N-chlorodansylamide (NCDA), *Anal. Lett.*, 14 (B 15) (1981) 1221-1232; *C.A.*, 96 (1982) 82 147x.

502 Nabeshima, T., Hiramatsu, M., Noma, S., Ukai, M., Amano, M. and Kameyama, T., Determination of methionine-enkephalin, norepinephrine, dopamine, 3,4-dihydroxyphenylacetic acid (DOPAC) and 3-methoxy-4-hydroxyphenylacetic acid (HVA) in brain by high-pressure liquid chromatography with electrochemical detector, *Res. Commun. Chem. Pathol. Pharmacol.*, 35 (1982) 421-442; *C.A.*, 96 (1982) 193 559c.

503 Nachtmann, F., High-performance liquid chromatography of intermediates in the oxytocin synthesis, *J. Chromatogr.*, 176 (1979) 391-397.

504 Nagai, Y. and Iwamori, M., A new approach to the analysis of ganglioside molecular species [including silica gel or Iatrobeads column chromatography], in Svennerholm, L., Dreyfus, H. and Urban, P.-F. (Editors), *Structure and Function of Gangliosides*, Plenum, New York, 1980, pp. 13-21; *C.A.*, 93 (1980) 3211a.

505 Nagai, Y. and Iwamori, M., Brain and thymus gangliosides: their molecular diversity and its biological implications and a dynamic annular model for their function in cell surface membranes [includes Iatrobeads column chromatography], *Mol. Cell. Biochem.*, 29 (1980) 81-90; *C.A.*, 93 (1980) 90 262j.

506 Naider, F., Sipzner, R., Steinfeld, A.S. and Becker, J.M., Separation of protected hydrophobic oligopeptides by normal-phase high-pressure liquid chromatography, *J. Chromatogr.*, 176 (1979) 264-269.

507 Nakamura, H., Zimmerman, C.L. and Pisano, J.J., Analysis of histidine-containing dipeptides, polyamines and related amino acids by high-performance liquid chromatography: application to Guinea pig brain, *Anal. Biochem.*, 93 (1979) 423-429.

508 Narihara, T., Fujita, Y. and Mizutani, T., Fractionation of tRNA on siliconized porous glass coated with trialkylmethylammonium chloride, *J. Chromatogr.*, 236 (1982) 513-518.

509 Natowicz, M. and Baenziger, J.U., A rapid method for chromatographic analysis of oligosaccharides on Bio-Gel P-4, *Anal. Biochem.*, 105 (1980) 159-164.

510 Nguyen Phi Nga, Bradley, J.L. and McGuire, P.M., Resolution of RNA by paired-ion reversed-phase high-performance liquid chromatography, *J. Chromatogr.*, 236 (1982) 508-512.

511 Ng Ying-Kin, N.M.K. and Wolfe, L.S., High-performance liquid chromatographic analysis of oligosaccharides and glycopeptides accumulating in lysozomal storage disorders, *Anal. Biochem.*, 102 (1980) 213-219.

512 Nibbering, N.M.M., Survey, of ionization methods with emphasis on liquid chromatography-mass spectrometry [review including beside others oligopeptides, steroid oligoglycosides, oligonucleotides, oligosaccharides], *J. Chromatogr.*, 251 (1982) 93-104.

513 Nice, E.C., Capp, M. and O'Hare, M.J., Use of hydrophobic interaction methods in the isolation of proteins from endocrine and paraendocrine tissues and cells by high-performance liquid chromatography, *J. Chromatogr.*, 185 (1979) 413-427.

514 Nice, E.C. and O'Hare, M.J., High-pressure liquid chromatography of polypeptide hormones, *J. Endocrinol.*, 79 (1978) 47P-48P; *C.A.*, 90 (1979) 182 491z.

515 Nice, E.C. and O'Hare, M.J., Simultaneous separation of β-lipotropin, adrenocorticotropic hormone, endorphins and enkephalins by high-performance liquid chromatography, *J. Chromatogr.*, 162 (1979) 401-407.

516 Niemann, M.A., Hollaway, W.L. and Mole, J.E., Purification of some biologically significant serum proteins by molecular exclusion high pressure liquid chromatography, *J. High Resolut. Chromatogr. Chromatogr. Commun.*, 2 (1979) 743-745.

517 Nika, H. and Hultin, T., An analyzer and monitor for rapid microscale peptide separations, *Anal. Biochem.*, 98 (1979) 178-183.

518 Nishikawa, A.H., Roy, S.K. and Puchalski, R., High-pressure hydrophobic chromatography of proteins, in Gribnau, T.C.J., Visser, J. and Nivard, R.J.F. (Editors), *Affinity Chromatography and Related Techniques (Analytical Chemistry Symposia Series*, Vol. 9), Elsevier, Amsterdam, 1982, pp. 471-482; *C.A.*, 96 (1982) 158 485d.

519 Nys, P.S., Petyushenko, R.M. and Savitskaya, E.M., Vybor uslovii razdeleniya smesi peptidov pri pomoshchi ionitov (Selection of conditions for the separation of a mixture of peptides using ion exchangers), *Zh. Fiz. Khim.*, 49 (1975) 2330-2334; *C.A.*, 84 (1976) 71 081x.

520 O'Hare, M.J. and Nice, E.C., Hydrophobic high-performance liquid chromatography of hormonal polypeptides and proteins on alkylsilane-bonded silica, *J. Chromatogr.*, 171 (1979) 209-226.

521 Ohlson, S., Glad, M., Larsson, P.O. and Mosbach, K., High performance liquid affinity chromatography (HPLAC) as a tool for the separation of biomolecules, *International Symposium on Affinity Chromatography and Related Techniques, Veldhoven, June 22-26, 1981, Abstracts*, B-57, Katholieke Universiteit, Nijmegen.

522 Ohlson, S., Hansson, L., Larsson, P.-O. and Mosbach, K., High performance affinity chromatography (HPLAC) and its application to the separation of enzymes and antigens, *FEBS Lett.*, 93 (1978) 5-9.

523 Ohno, Y., Okazaki, M. and Hara, I., Fractionation of human serum lipoproteins by high performance liquid chromatography, *J. Biochem. (Tokyo)*, 89 (1981) 1675-1680; *C.A.*, 95 (1981) 38 296a.

524 Okada, S., Kitahata, S., Yoshikawa, S. and Yoshida, M., Determination of maltooligosylsucrose by high-performance liquid chromatography [in Japanese], *Dempun Kagaku*, 25 (1978) 229-233; *C.A.*, 91 (1979) 138 947z.

525 Okazaki, M., Hagiwara, N. and Hara, I., Quantitation method for choline-containing phospholipids in human serum lipoproteins by high performance liquid chromatography [GPC-HPLC of serum lipoproteins], *J. Biochem. (Tokyo)*, 91 (1982) 1381-1389; *C.A.*, 96 (1982) 213 563e.

526 Okazaki, M. and Hara, I., Analysis of cholesterol in high density lipo-protein subfractions by high performance liquid chromatography, *J. Biochem. (Tokyo)*, 88 (1980) 1215-1218; *C.A.*, 93 (1980) 217 377h.

527 Okazaki, M., Hara, I., Tanaka, A., Kodama, T. and Yokoyama, S., Decreased serum HDL_3 cholesterol levels in cirrhosis of the liver [letter to the editor; GPC-HPLC for serum lipoprotein fractionation], *New England J. Med.*, (1981) 1608.

528 Okazaki, M., Ohno, Y. and Hara, I., High-performance aqueous gel permeation chromatography of human serum lipoproteins, *J. Chromatogr.*, 221 (1980) 257-264.

529 Okazaki, M., Ohno, Y. and Hara, I., Rapid method for the quantitation of cholesterol in human serum lipoproteins by high performance liquid chro-matography [GPC-HPLC fractionation of lipoproteins], *J. Biochem. (Tokyo)*, 89 (1981) 879-887; *C.A.*, 94 (1981) 135 262y.

530 Okazaki, M., Shiraishi, K., Ohno, Y. and Hara, I., High-performance aqueous gel permeation chromatography of serum lipoproteins: selective detection of cholesterol by enzymatic reaction, *J. Chromatogr.*, 223 (1981) 285-293.

531 Oldewurtel, M.D., Krinks, M.H., Lee, J.W., Williams, J.F. and Klee, C.B., Isolation and characterization of intracellular Ca^{2+}-binding proteins by high performance liquid chromatography, *Protides Biol. Fluids*, 30 (1983) 713-716; *C.A.*, 98 (1983) 139 959m.

532 Olieman, C., Sedlick, E. and Voskamp, D., *In situ* silylation of an octadecylsilyl-silica stationary phase applied to the analysis of peptides, such as secretin and glucagon, *J. Chromatogr.*, 207 (1981) 421-424.

533 Oroszlan, S., A model study for monitoring Merrifield solid phase peptide synthesis by high-pressure liquid chromatography, *Chromatogr. Sci.*, 10 (1979) 199-224.

534 Orth, P. and Engelhardt, H., Trennung von Zuckern an chemisch modifizierten Kieselgelen (Separation of sugars on chemically modified silica gel) [including oligosaccharides], *Chromatographia*, 15 (1982) 91-96.

535 Oshima, G., Shimabukuro, H. and Nagasawa, K., Separations of amino acid homo-oligopeptides, *J. Chromatogr.*, 152 (1978) 579-584.

536 Oshima, R., Kurosu, Y. and Kumanotani, J., Separation of acid and neutral saccharides by high-performance gel chromatography on cation-exchange resins in the H^+ form with acidic eluents, *J. Chromatogr.*, 179 (1979) 376-380.

537 Oshima, R., Takai, N. and Kumanotani, J., Improved separation of anomers of saccharides by high-performance liquid chromatography on macroreticular anion-exchange resin in the sulphate form, *J. Chromatogr.*, 192 (1980) 452-456.

538 Pearson, J.D., Lin, N.T. and Regnier, F.E., The importance of silica type for reversed-phase protein separations, *Anal. Biochem.*, 124 (1982) 217-230.

539 Pearson, J.D., Mahoney, W.C., Hermodson, M.A. and Regnier, F.E., Reversed-phase supports for the resolution of large denatured protein fragments, *J. Chromatogr.*, 207 (1981) 325-332.

540 Pearson, J.D. and Regnier, F.E., High performance anion exchange chromato-graphy of oligonucleotides, *J. Chromatogr.*, 255 (1983) 137-149.

541 Perrett, D., Application of high-performance liquid chromatography (HPLC) to biochemical analysis [review], *Techniques in Metabolic Research*, B215 (1979) 1-22; *C.A.*, 92 (1980) 36 973h.

542 Persiani, C., Cukor, P. and French, K., Aqueous GPC of water soluble polymers by high pressure liquid chromatography using glyceryl CPG columns [proteins], *J. Chromatogr. Sci.*, 14 (1976) 417-421.

543 Petrides, P.E. and Boehlen, P., The mitogenic activity of insulin: an
    intrinsic property of the molecule [RP HPLC], *Biochem. Biophys. Res.*
    Commun., 95 (1980) 1138-1144.
544 Petrides, P.E., Jones, R.T. and Boehlen, P., Reverse-phase high-performance
    liquid chromatography of proteins: the separation of hemoglobin chain
    variants, *Anal. Biochem.*, 105 (1980) 383-388.
545 Petrides, P.E., Levine, A.E. and Shooter, E.M., Preparative reverse phase
    HPLC: an efficient procedure for the rapid purification of large amounts
    of biologically active proteins, in Rich, D.H. and Gross, E. (Editors),
    Peptides: Synthesis-Structure-Function (Proc. 7th Am. Peptide Symp.),
    Pierce, Rockford, IL, 1981, pp. 781-783; *C.A.*, 96 (1982) 193 556z.
546 Petyushenko, R.M., Savitskaya, E.M. and Nys, P.S., Sorbtsionnye i elektro-
    khimicheskie konstanty peptidov raznogo stroeniya (Sorption and electro-
    chemical constants of peptides of various structure), *Zh. Obsch. Khim.*,
    49 (1975) 422-427; *C.A.*, 83 (1975) 10 806v.
547 Pfannkoch, R., Lu Kai-Chun, Regnier, F.E. and Barth, H.G., Characterization
    of some commercial high performance size-exclusion chromatography columns
    for water-soluble polymers, *J. Chromatogr. Sci.*, 18 (1980) 430-441.
548 Pietrzyk, D.J., Cahil, W.J. and Stodola, J.D., Preparative liquid chromato-
    graphic separation of amino acids and peptides on Amberlite XAD-4, *J.*
    Liq. Chromatogr., 5 (1982) 443-461.
549 Poll, D.J., Knighton, D.R., Harding, D.R.K. and Hancock, W.S., Use of ion-
    paired, reversed-phase thin-layer chromatography for the analysis of
    peptides. A simple procedure for the monitoring of preparative reversed-
    phase high-performance liquid chromatography, *J. Chromatogr.*, 236 (1982)
    244-248.
550 Porthault, M., Introduction to high-performance liquid chromatography; a
    promising method for peptide analysis and preparation, in Rosselin, G.,
    Fromageot, P. and Bonfils, S. (Editors), *Hormone Receptors in Digestion*
    and Nutrition (Proc. 2nd Int. Symp. Horm. Rec. Dig. Fract. Physiol.),
    Elsevier, Amsterdam, 1979, pp. 69-77; *C.A.*, 92 (1980) 90 154c.
551 Pradayrol, L., Significance of high-performance liquid chromatography in
    polypeptide biochemistry [discussion in French], *Gastroenterol. Clin.*
    Biol., 6 (1982) 13-15; *C.A.*, 96 (1982) 118 298p.
552 Prestidge, R.L., Harding, D.R.K., Moore, C.H. and Hancock, W.S., An approach
    to the semisynthesis of acyl carrier protein [preparative HPLC-purification
    of peptides], *Bioorg. Chem.*, 10 (1981) 277-282.
553 Prystowsky, M.B., Lukas, T.J. and Erickson, B.W., Continuous-flow solid-phase
    peptide synthesis using Pam-resins [RP-HPLC used to measure by-products],
    in Gross, E. and Meienhofer, J. (Editors), *Peptides: Structure and*
    Biological Function (Proc. 6th Am. Peptide Symp.), Pierce, Rockford, IL,
    1979, pp. 349-352; *C.A.*, 94 (1981) 157 227t.
554 Rabenstein, D.L. and Saetre, R., Analysis for glutathione in blood by high-
    performance liquid chromatography, *Clin. Chem. (Winston-Salem, N.C.),* 24
    (1978) 1140-1143.
555 Radhakrishnan, A.N., Stein, S., Licht, A., Gruber, K.A. and Udenfriend, S.,
    High-efficiency cation exchange chromatography of polypeptides and
    polyamines in the nanomole range, *J. Chromatogr.*, 132 (1977) 552-555.
556 Rand, P.G. and Nelson, Ch., Improvements in hemoglobin A_1 determination by
    rapid cation-exchange chromatography, *Clin. Chem. (Winston-Salem, N.C.),*
    26 (1980) 1209-1212.
557 Regnier, F.E., Bonded carbohydrate stationary phases for chromatography
    [glycerol-bonded controlled porosity glass packing glycophase for HPLC of
    proteins], *U.S. Pat.*, 3 983 299 (1976); *C.A.*, 86 (1977) 123 391m.
558 Regnier, F.E., High-performance liquid chromatography of proteins and
    peptides [discussion], *Analytika (Johannesburg),* (1981) 17-22 (Publ. in
    Chemsa, 7, 10); *C.A.*, 96 (1982) 139 008j.
559 Regnier, F.E., High-performance ion-exchange chromatography of proteins:
    the current status [review], *Anal. Biochem.*, 126 (1982) 1-7.

560 Regnier, F.E. and Chang Shung-Ho, Rapid analysis of hemoglobin variants by high speed liquid chromatography [using DEAE-glycophase and CM-glycophase], *U.S. Pat.*, 4 108 603 (1977); *C.A.*, 90 (1979) 68 874b.

561 Regnier, F.E. and Chang Shung-Ho, Apparatus for enzyme detection [on-line for HPLC], *U.S. Pat.*, 4 243 753 (1981); *C.A.*, 95 (1981) 2653a.

562 Regnier, F.E. and Gooding, K.M., High performance liquid chromatography of proteins [review], *Anal. Biochem.*, 103 (1980) 1-25.

563 Regnier, F.E. and Gooding, K.M., Clinical analysis of endogenous human biochemicals. Proteins [review], in Kabra, P.M. and Marton, L.J. (Editors), *Liquid Chromatography in Clinical Analysis*, Humana Press, Clifton, NJ, 1981, pp. 323-352; *C.A.*, 95 (1981) 38 231a.

564 Regnier, F.E., Gooding, K.M. and Chang Shung-Ho, High-speed liquid chromatography of proteins [review], in Hercules, D., Hieftje, G., Snyder, L.R. and Evenson, M.A. (Editors), *Contemporary Topics of Analysis in Clinical Chemistry*, Vol. 1., Plenum Press, New York, 1977, pp. 1-48; *C.A.*, 87 (1977) 196 576f.

565 Regnier, F.E. and Noel, R., Glycerolpropylsilane bonded phases in the steric exclusion chromatography of biological macromolecules [enzymes and other proteins, nucleic acids, polysaccharides (dextrans)], *J. Chromatogr. Sci.*, 14 (1976) 316-320.

566 Rexová-Benková, L'., Omelková, J., Mikeš, O. and Sedláčková, J., Medium pressure liquid chromatography of Leozym, a pectic enzyme preparation, on ion exchange derivatives of Spheron, *J. Chromatogr.*, 238 (1982) 183-192.

567 Reynolds, G.D., Harding, D.R.K. and Hancock, W.S., Use of the chromogenic p-(p-(dimethylamino)phenylazo)benzyl (Az) ester in the synthesis of leu-enkephalin [HPLC of the product.], *Int. J. Pept. Protein Res.*, 17 (1981) 231-234.

568 Richey, J., FPLC: A comprehensive separation technique for biopolymers, *Int. Lab.*, 13 (Jan./Feb.) (1983) 50-75.

569 Richmond, W., The separation of clinically important peptides and proteins - a review, *Clin. Res. Cent. Symp. (Harrow, Engl.)*, 1 (*Curr. Dev. Clin. Appl. HPLC, GC MS*) (1980) pp. 85-96; *C.A.*, 94 (1981) 152 730s.

570 Richter, K. and Woelk, H.U., Die Hochdruck-Flüssigkeitschromatographie als Methode zur Untersuchung von Stärkehydrolysaten (High-pressure liquid chromatography as method for the analysis of starch hydrolyzates) [Lecture], *Staerke*, 29 (1977) 273-277; *C.A.*, 87 (1977) 153 670q.

571 Richter, W.O. and Schwandt, P., Simultaneous separation of peptide hormones occuring in the digestive tract [HPLC], *Fresenius' Z. Anal. Chem.*, 311 (1982) 430-431.

572 Riedmann, M. and Wagner, E., Anwendungsbereiche der Flüssigkeits-Chromatographie (HPLC) in der Medizin (Applications of high-pressure-liquid chromatography (HPLC) in medicine [review including some peptides], *Medizintechnik (Stuttgart)*, 100 (1980) 42-46; *C.A.*, 93 (1980) 54 045g.

573 Rittinghaus, K. and Franzen, K.H., HPLC in protein analysis: an alternative to gel-filtration and gel-electrophoresis, *Fresenius' Z. Anal. Chem.*, 301 (1980) 144; *C.A.*, 93 (1980) 3225h.

574 Rivier, J.E., Use of trialkyl ammonium phosphate (TAAP) buffers in reverse phase HPLC for high resolution and high recovery of peptides and proteins, *J. Liq. Chromatogr.*, 1 (1978) 343-366.

575 Rivier, J. and Burgus, R., Application of reverse phase high pressure liquid chromatography to peptides [review], *Biological/Biomedical Applications of Liquid Chromatography (Liquid Chromatography Symp. I), Chromatogr. Sci.*, 10 (1979) 147-161; *C.A.*, 90 (1979) 199 681j.

576 Rivier, J., Desmond, J., Spiess, J., Perrin, M., Vale, W., Eksteen, R. and Karger, B., Peptide and amino acid analysis by RP-HPLC, in Gross, E. and Meienhofer, J. (Editors), *Peptides: Structure and Biological Function (Proc. 6th Am. Peptide Symp.)*, Pierce, Rockford, IL, 1979, pp. 125-128; *C.A.*, 93 (1980) 234 151x.

577 Rivier, J., Spiess, J., Perrin, M. and Vale, W., Application of HPLC in the isolation of unprotected peptides, *Biological/Biomedical Applications of Liquid Chromatography 2*, *Chromatogr. Sci.*, 12 (1979) 223-241; *C.A.*, 92 (1980) 37 020g.

578 Rivier, J., Wolbers, R. and Burgus, R., Application of high pressure liquid chromatography to peptides, in Goodman, M. and Meienhofer, J. (Editors), *Peptides (Proc. 5th Am. Peptide Symp.)*, Wiley, New York, 1977, pp. 52-55; *C.A.*, 89 (1978) 2482v.

579 Rokushika, S., Ohkawa, T. and Hatano, H., High-speed aqueous gel permeation chromatography of proteins, *J. Chromatogr.*, 176 (1979) 456-461.

580 Rossetti, Z.L., Separation of neuropeptides by high performance liquid chromatography (HPLC), *Riv. Farmacol. Ter.*, 12 (1981) 35-39; *C.A.*, 95 (1981) 146 324s.

581 Roumeliotis, P., Kinkel, J. and Unger, K., Assessment and optimization of system parameter in the size exclusion separation of proteins and enzymes on Diol-modified silicas by means of HPLC, *International Symposium on Affinity Chromatography and Related Techniques, Veldhoven, June 22-26, 1981, Abstracts*, B-58, Katholieke Universiteit, Nijmegen.

582 Roumeliotis, P. and Unger, K.K., Preparative separation of proteins and enzymes in the mean molecular-weight range 10 000-100 000 on Lichrosorb Diol® packing by high-performance size exclusion chromatography, *J. Chromatogr.*, 185 (1979) 445-452.

583 Roumeliotis, P. and Unger, K.K., Assessment and optimization of system parameters in size exclusion separation of proteins on diol-modified silica columns, *J. Chromatogr.*, 218 (1981) 535-546.

584 Roumeliotis, P., Unger, K.K., Kinkel, J., Brunner, G., Wieser, R. and Tschank, G., Potential and limitations of high performance size-exclusion chromatography (HPSEC) in the analytical separation and preparative isolation of biopolymers, in Lottspeich, F., Henschen, A. and Hupe, K.P. (Editors), *High Performance Chromatography in Protein and Peptide Chemistry*, Walter de Gruyter, Berlin, New York, 1981, pp. 71-82; *C.A.*, 97 (1982) 51 933s.

585 Rubinstein, M., Preparative high-performance liquid partition chromatography of proteins [review, "normal phase" and "reverse-phase"], *Anal. Biochem.*, 98 (1979) 1-7.

586 Rubinstein, M., Chen-Kiang S., Stein, S. and Udenfriend, S., Characterization of proteins and peptides by high performance liquid chromatography and fluorescence monitoring of their tryptic digest, *Anal. Biochem.*, 95 (1979) 117-121.

587 Rubinstein, M., Rubinstein, S., Familletti, P.C., Gross, M.S., Miller, R.S., Waldman, A.A. and Pestka, S., Human leukocyte interferon purified to homogeneity, *Science (Washington, D.C.)*, 202 (1978) 1289-1290.

588 Rubinstein, M., Rubinstein, S., Familletti, P.C., Miller, R.S., Waldman, A.A. and Pestka, S., Human leukocyte interferon: Production, purification to homogeneity, and clinical characterization, *Proc. Natl. Acad. Sci. U.S.A.*, 76 (1979) 640-644.

589 Rubinstein, M., Stein, S., Gerber, L.D. and Udenfriend, S., Isolation and characterization of the opioid peptides from rat pituitary: β-lipotropin [HPLC], *Proc. Natl. Acad. Sci. U.S.A.*, 74 (1977) 3052-3055.

590 Rubinstein, M., Stein, S. and Udenfriend, S., Characterization of pro-opiocortin [M_r approximately 30,000], a precursor of opioid peptides and corticotropin, *Proc. Natl. Acad. Sci. U.S.A.*, 75 (1978) 669-671.

591 Saint-Blancard, J., Fourcard, J., Limonne, F., Girot, P. and Boschetti, E., New ion exchangers Trisacryl: use and application to human plasma protein fractionation [in French], *Ann. Pharm. Fr.*, 39 (1981) 403-409; *C.A.*, 96 (1982) 74 550h.

592 Saint-Blancard, J., Kirzin, J.M., Riberon, P., Petit, F., Fourcart, J., Girot, P. and Boschetti, E., A simple and rapid procedure for large scale preparation of IgG's and albumin from human plasma by ion exchange and affinity chromatography [HPLC], in Gribnau, T.C.J., Visser, J. and Nivard, R.J.F. (Editors), *Affinity Chromatography and Related Techniques (Analytical Chemistry Symposia Series*, Vol. 9.), Elsevier, Amsterdam, 1982, pp. 305-312.

593 Saito, Y. and Hayano, S., Application of high-performance aqueous gel permeation chromatography to humic substances from marine sediment, *J. Chromatogr.*, 177 (1979) 390-392.
594 Samuelson, O., Chromatography of oligosaccharides and related compounds on ion-exchange resins, *Adv. Chromatogr.*, 16 (1978) 113-150.
595 Šatava, J., Mikeš, O. and Štrop, P., Separation of oligonucleotides on ion-exchange derivatives of Spheron, *J. Chromatogr.*, 180 (1979) 31-37.
596 Schaettle, E., Schnelles Trennverfahren für die Biochemie (Rapid separation methods for biochemistry) [review], *Lab. Praxis*, 2 (1978) 36-38, 40; *C.A.*, 91 (1979) 35 023r.
597 Schauwecker, P., Frei, R.W. and Erni, F., Trace enrichment techniques in reversed-phase high-performance liquid chromatography [peptides in urine], *J. Chromatogr.*, 136 (1977) 63-72.
598 Schiebel, H.M. and Schulten, H.-R., Field desorption mass spectrometry of nucleic acids. VII. Identification of protected deoxyribonucleotides by field desorption mass spectrometry in fractions from high performance liquid chromatography, *Z. Naturforsch. B: Anorg. Chem., Org. Chem.*, 36B (1981) 967-973; *C.A.*, 95 (1981) 187 583u.
599 Schifreen, R.S., Hickingbotham, J.M. and Bowers, G.N. Jr., Accuracy, precision and stability in measurement of hemoglobin A_{1c} by "high-performance" cation-exchange chromatography, *Clin. Chem. (Winston-Salem, N.C.)*, 26 (1980) 466-472; *C.A.*, 92 (1980) 176 748p.
600 Schlabach, T.D. and Abbott, S.R., Serum protein profiles by high-performance liquid chromatography with detection at multiple wavelengths, *Clin. Chem. (Winston-Salem, N.C.)*, 26 (1980) 1504-1508; *C.A.*, 93 (1980) 200 302g.
601 Schlabach, T.D., Alpert, A.J. and Regnier, F.E., Rapid assessment of isoenzymes by high-performance liquid chromatography, *Clin. Chem. (Winston-Salem, N.C.)*, 24 (1978) 1351-1360.
602 Schlabach, T.D., Chang Shung-Ho, Gooding, K.M. and Regnier, F.E., A continuous-flow enzyme detector for liquid chromatography, *J. Chromatogr.*, 134 (1977) 91-106.
603 Schlabach, T.D., Fulton, J.A., Mockridge, P.B. and Toren, E.C. Jr., New developments in analysis of isoenzymes separated by "high-performance" liquid chromatography, *Clin. Chem. (Winston-Salem, N.C.)*, 25 (1979) 1600-1608; *C.A.*, 91 (1979) 153 312g.
604 Schlabach, T.D., Fulton, J.A., Mockridge, P.B. and Toren, E.C. Jr., Determination of serum isoenzyme activity profiles by high performance liquid chromatography [lactate dehydrogenase, creatin kinase, both absorbance and fluorescence detection, diagnosis of myocardial infarction], *Anal. Chem.*, 52 (1980) 729-733; *C.A.*, 92 (1980) 159 393b.
605 Schlabach, T.D., Fulton, J.A., Mockridge, P.B. and Toren, E.C. Jr., Interferences appearing in fluorometrically measured liquid-chromatographic profiles of creatine kinase isoenzymes in serum [lactate dehydrogenase, serum albumin, lipoprotein, prealbumin], *Clin. Chem. (Winston-Salem, N.C.)*, 26 (1980) 707-711; *C.A.*, 93 (1980) 21 320q.
606 Schlabach, T.D. and Regnier, F.E., Techniques for detecting enzymes in high performance liquid chromatography, *J. Chromatogr.*, 158 (1978) 349-364.
607 Schmidt, D.E. Jr., Giese, R.W., Conron, D. and Karger, B.L., High performance liquid chromatography of proteins on a Diol-bonded silica gel stationary phase, *Anal. Chem.*, 52 (1980) 177-182.
608 Schoeneshoefer, M. and Fenner, A., Hydrophilic ion-pair reversed-phase chromatography of biogenic peptides prior to immunoassay, *J. Chromatogr.*, 224 (1981) 472-476.
609 Schoeneshoefer, M., Fenner, A. and Molnar, I., Heterogeneity of corticotropin-immunoreactive compounds in human body fluids, *Clin. Chem. (Winston-Salem, N.C.)*, 27 (1981) 1875-1877; *C.A.*, 96 (1982) 15 348a.
610 Schoeneshoefer, M., Kage, A., Kage, R. and Fenner, A., A convenient technique for the specific isolation of ^{125}I-labelled peptide molecules, *Fresenius' Z. Anal. Chem.*, 311 (1982) 429-430.

611 Schott, H., Preparative isolation of oligonucleotides from chemically
    degradated DNA, *Nucleic Acids Res.* (*Symp. Ser.*, No. 7) (1980) 203-214.
612 Schott, H., Präparative Isolierung von Oligothymidinphosphaten aus partial
    Hydrolysaten chemisch abgebauter DNA mit Hilfe der Template-Chromatographie
    (Preparative isolation of oligothymidine phosphates from partial
    hydrolysates of chemically degraded DNA using template chromatography
    [Watson-Crick base pairing], *J. Chromatogr.*, 237 (1982) 429-438.
613 Schott, H. and Greber, G., Separation of nucleosides on cytidine-containing
    polymer gels in DMSO/CHCl₃ or water [Watson-Crick base pairing], *Angew.
    Chem.*, *Int. Ed. Engl.*, 9 (1970) 465-466.
614 Schott, H. and Greber, G., Synthesis of crosslinked suspension polymers
    containing covalently bound uridine, cytidine and guanosine residues
    [useful for chromatographic separation of nucleic acids; in German],
    Makromol. Chem., 144 (1971) 333-335; *C.A.*, 75 (1971) 37 106p.
615 Schott, H. and Greber, G., Wechselwirkung zwischen Nucleosiden und Poly-
    mergelen mit covalent eingebauten Nucleosidresten (Interaction between
    nucleosides and polymer gels containing covalently linked nucleoside
    groups) [preparation of gels; separation of nucleotide mixtures using the
    principle of Watson-Crick base pairing], *Makromol. Chem.*, 145 (1971) 11-20;
    C.A., 75 (1971) 118 518b.
616 Schott, H. and Greber, G., Säulenchromatographische Trennungen von Nukleosid/
    Nukleotid- und Nukleosid/Dinukleosidphosphatgemischen an Nukleosidgelen
    (Column chromatographic separation of nucleoside/nucleotide and nucleoside/
    dinucleoside phosphate mixtures on nucleoside gels), *Makromol. Chem.*, 149
    (1971) 253-260; *C.A.*, 76 (1972) 46 437v.
617 Schott, H. and Schrade, H., Preparative isolation of purine oligonucleotides
    from chemically degraded DNA, *Nucleic Acids Res.* (*Symp. Ser.*, No. 9),
    (1981) 187-190; *C.A.*, 94 (1981) 116 994x.
618 Schott, H. and Watzlawick, H., Präparative Isolierung von Oligoguanosinphos-
    phaten aus DNA-Partialhydrolysaten mit Hilfe der Template-Chromatographie
    (Preparative isolation of oligoguanosine phosphates from partial hydro-
    lyzates of DNA using template chromatography), *J. Chromatogr.*, 243 (1982)
    57-70.
619 Schroeder, R.R., Kudirka, P.J. and Toren, E.C. Jr., Enzyme selective detector
    systems for high-pressure liquid chromatography [IEC of lactic dehydro-
    genase isoenzymes], *J. Chromatogr.*, 134 (1977) 83-90.
620 Schroeder, W.A., Shelton, J.B. and Shelton, J.R., Separation of hemoglobin
    peptides by high performance liquid chromatography (HPLC), *Hemoglobin*,
    4 (1980) 551-559; *C.A.*, 94 (1981) 1667a.
621 Schroeder, W.A., Shelton, J.B., Shelton, J.R. and Powars, D., Separation of
    peptides by high pressure liquid chromatography for the identification of
    a hemoglobin variant, *J. Chromatogr.*, 174 (1979) 385-392.
622 Schroeder, W.A., Shelton, J.B. and Shelton, J.R., High performance liquid
    chromatography in the identification of human hemoglobin variants [review],
    Prog. Clin. Biol. Res., 60 (Adv. Hemoglobin Anal.) (1981) 1-22; *C.A.*, 95
    (1981) 146 235p.
623 Schwabe, Ch. and McDonald, J.K., Demonstration of a pyroglutamyl residue at
    the N terminus of the β-chain of porcine relaxin, *Biochem. Biophys. Res.
    Commun.*, 74 (1977) 1501-1504.
624 Schwandt, P., Richter, W.O. and Weisweiler, P., Separation of human C-apo-
    lipoproteins by high-performance liquid chromatography, *J. Chromatogr.*,
    225 (1981) 185-188.
625 Schwartz, E.R., Stevens, J. and Schmidt, D.E. Jr., Proteoglycan analysis by
    high-performance liquid chromatography, *Anal. Biochem.*, 112 (1981) 170-175.
626 Scobell, H.D. and Brobst, K.M., Rapid high-resolution separation of oligo-
    saccharides on silver form cation-exchange resins, *J. Chromatogr.*, 212
    (1981) 51-64.

627 Scoble, H.A. and Brown, P.R., Reverse-phase chromatography of nucleic acid fragments [review], in Horváth, Cs. (Editor), *High Performance Liquid Chromatography (Advances and Perspectives*, Vol. 3), Academic Press, New York, 1983, pp. 1-47; *C.A.*, 99 (1983) 35 385a.

628 Sebille, B., Thuaud, N. and Tillement, J.-P., Study of binding of low-molecular-weight ligand to biological macromolecules by high-performance liquid chromatography. Evaluation of binding parameters for two drugs bound to serum albumin, *J. Chromatogr.*, 167 (1978) 159-170.

629 Seidah, N.G., Dennis, M., Corvol, P., Rochemont, J. and Chrétien, M., A rapid high-performance liquid chromatography purification method of iodinated polypeptide hormones, *Anal. Biochem.*, 109 (1980) 185-191.

630 Seliger, H., Bach, T.C., Görtz, H.-H., Happ, E., Holupirek, M., Seemann-Preising, B., Schiebel, H.-M. and Schulten, H.-R., High performance liquid chromatography in combination with field desorption mass spectrometry: separation and identification of building blocks for polynucleotide synthesis, *J. Chromatogr.*, 253 (1982) 65-79; *C.A.*, 98 (1983) 139 925x.

631 Seliger, H., Bach, T.C., Siewert, G., Boidol, W., Töpert, M., Schulten, H.-R. and Schiebel, H.-M., Synthesis of deoxyoligonucleotide linker fragments for genetic engineering purposes using improved preparative and analytical techniques HPLC, *Nucleic Acids Res.*, in preparation.

632 Sellers, J.P. and Clark, H.G., High-pressure liquid chromatography of fibrinopeptides derived from eight mammalian fibrinogens, *Thromb. Res.*, 23 (1981) 91-95; *C.A.*, 95 (1981) 164 877d.

633 Schechter, I., Separation of proteins by high-speed pressure liquid chromatography, *Anal. Biochem.*, 58 (1974) 30-38.

634 Shelton, J.B., Shelton, J.R. and Schroeder, W.A., Further experiments in the separation of globin chains by high-performance liquid chromatography, *J. Liq. Chromatogr.*, 4 (1981) 1381-1392.

635 Shelton, J.B., Shelton, J.R., Schroeder, W.A. and DeSimone, J., Detection of Hb-Papio B, a silent mutation of the baboon β chain, by high performance liquid chromatography. Improved procedures for the separation of globin chain by HPLC, *Hemoglobin*, 6 (1982) 451-464; *C.A.*, 98 (1983) 68 206a.

636 Shimizu, K., Wilson, J.B. and Huisman, T.H.J., The determination of the percentages of G$_\gamma$ and A$_\gamma$ chains in human fetal hemoglobin by HPLC, *Hemoglobin*, 4 (1980) 487-496.

637 Shioiri, T. (Editor), *Peptide Chemistry 1981 (Proceedings of the 19th Symposium on Peptide Chemistry, Nagoya, 1981)*, Peptide Institute, Protein Research Foundation, Minoh, Osaka, 1982, 210 pp.

638 Shoji, J., Kato, T., Terabe, S. and Konaka, R., Resolution of peptide antibiotics, cerexins, octapeptins and tridecaptins, by high-performance liquid chromatography, in Izumaya, N. (Editor), *Peptide Chemistry 1978 (Proc. 16th Symp.)*, Protein Research Foundation, Osaka, 1979, pp. 73-78; *C.A.*, 93 (1980) 155 761t.

639 Shoji, J., Kato, T., Terabe, S. and Konaka, R., Resolution of peptide antibiotics, cerexins and tridecaptins, by high performance liquid chromatography. (Studies on antibiotics from the genus Bacillus. XXVI.), *J. Antibiot.*, 32 (1979) 313-319.

640 Shum Wai-King B. and Crothers, D.M., Simplified methods for large scale enzymic synthesis of oligoribonucleotides [RPC-IEC], *Nucleic Acids Res.*, 5 (1978) 2297-2311; *C.A.*, 89 (1978) 125 612v.

641 Singhal, R.P., Separation and analysis of nucleic acids and their constituents by ion-exclusion and ion-exchange column chromatography [review], *Sep. Purif. Methods*, 3 (1974) 339-398; *C.A.*, 82 (1975) 166 788f.

642 Singhal, R.P., HPLC of transfer RNAs (with special emphasis on the separation of transfer RNAs from mammalian sources), *J. Chromatogr.*, 266 (1983) 359-383.

643 Singhal, R.P., Bajaj, R.K., Buess, Ch.M., Smoll, D.B. and Vakharia, V.N., Reversed-phase boronate chromatography for the separation of O-methylribose nucleosides and aminoacyl-tRNAs, *Anal. Biochem.*, 109 (1980) 1-11.

644 Singhal, R.P., Griffin, G.D. and Novelli, G.D., Separation of transfer ribonucleic acids on polystyrene anion exchangers, *Biochemistry*, 15 (1976) 5083-5087.

645 Sjoedahl, J., High-resolution liquid chromatography of proteins, *Science Tools*, 27 (1980) 54-56.

646 Small, D.A.P., Atkinson, T. and Lowe, Ch.R., High performance liquid affinity chromatography of enzymes on silica-immobilised triazine dyes, *J. Chromatogr.*, 216 (1981) 175-190.

647 Small, D.A.P., Lowe, C.R. and Atkinson, T., Triazine dyes as affinity ligands in high performance liquid affinity chromatography, *International Symposium on Affinity Chromatography and Related Techniques*, Veldhoven, June 22-26, 1981, *Abstracts*, B-43, Katholieke Universiteit, Nijmegen.

648 Smith, J.A. and McWilliams, R.A., High performance liquid chromatography of peptides [review in two parts], *Am. Lab. (Fairfield, Conn.)*, 12 (1980) No. 5, 23, 25, 26, 29; No. 6, 25-30; *C.A.*, 93 (1980) 95 657b.

649 Sokolowski, A., Balgobin, N., Josephson, S., Chattopadhyaya, J.B. and Schill, G., Reversed-phase ion-pair chromatography of oligodeoxyribonucleotides, *Chem. Scr.*, 18 (1981) 189-191; *C.A.*, 96 (1982) 65 071a.

650 Spindel, E., Pettibone, D., Fisher, L., Fernstrom, J. and Wurtman, R., Characterization of neuropeptides by reversed-phase, ion-pair liquid chromatography with post-column detection by radioimmunoassay. Application to thyrotropin-releasing hormone, substance P, and vasopressin, *J. Chromatogr.*, 222 (1981) 381-387.

651 Stein, S., Ultramicro isolation and analysis of peptides and proteins [review, HPLC of opioid peptides with fluorescence detection], *Miles Int. Symp. Ser.*, 12 (1980) 77-85; *C.A.*, 93 (1980) 234 144x.

652 Stein, S., Ultramicroanalysis of peptides and proteins by high-performance liquid chromatography and fluorescence detection [review-discussion], *Peptides (N.Y.)*, 4 (1981) 185-216; *C.A.*, 95 (1981) 183 184s.

653 Stein, S., Rubinstein, M. and Undenfriend, S., Ultramicro analysis of peptides [review on fluorometric detection], *Psychopharmacol. Bull.*, 14 (1978) 29-30; *C.A.*, 90 (1979) 35 676k.

654 Stoklosa, J.T., Ayi, B.K., Shearer, C.M. and DeAngelis, N.J., Separation of minute quantities of impurities in nonapeptides by reverse-phase high-performance liquid chromatography: critical nature of the water/acetonitrile ratio, *Anal. Lett.*, B11 (1978) 889-899; *C.A.*, 90 (1979) 87 873s.

655 Stoming, T.A., Garver, F.A., Gangarosa, M.A., Harrisson, J.M. and Huisman, T.H.J., Separation of the A_γ and G_γ cyanogen bromide peptides of human fetal hemoglobin by high-pressure liquid chromatography, *Anal. Biochem.*, 96 (1979) 113-117.

656 Strickler, M.P., Gemski, M.J. and Doctor, B.P., Purification of commercially prepared bovine trypsin by reverse phase high performance liquid chromatography, *J. Liq. Chromatogr.*, 4 (1981) 1765-1775.

657 Štrop, P. and Čechová, D., Separation of α- and β-trypsin by hydrophobic interaction chromatography, *J. Chromatogr.*, 207 (1981) 55-62.

658 Štrop, P., Mikeš, F. and Chytilová, Z., Hydrophobic interaction chromatography of proteins and peptides on Spheron P-300, *J. Chromatogr.*, 156 (1978) 239-254.

659 Sugden, K., Hunter, C. and Lloyd-Jones, J.G., Separation of the diastereoisomers of pyroglutamylhistidyl-3,3-dimethylprolineamide by ligand-exchange chromatography [HPLC], *J. Chromatogr.*, 204 (1981) 195-200.

660 Su Seuw J., Grego, B., Niven, B. and Hearn, M.T., High performance liquid chromatography of amino acids, peptides and proteins. XXXVII. Analysis of group retention contributions for peptides separated by reverse phase high performance liquid chromatography, *J. Liq. Chromatogr.*, 4 (1981) 1745-1764.

661 Suzuki, A., Handa, S. and Yamakawa, T., Separation of molecular species of higher glycolipids by high performance liquid chromatography of their O-acetyl-N-*p*-nitrobenzoyl derivatives, *J. Biochem. (Tokyo)*, 82 (1977) 1185-1187; *C.A.*, 87 (1977) 196 600j.

662 Suzuki, A., Kundu, S.K. and Marcus, D.M., An improved technique for separation of neutral glycosphingolipids by high-performance liquid chromatography [acetylnitrobenzoyl derivatives], *J. Lipid Res.*, 21 (1980) 473-477.

663 Suzuki, K.T., Direct connection of high-speed liquid chromatograph (equipped with gel permeation column) to atomic absorption spectrophotometer for metalloprotein analysis: metallothinein, *Anal. Biochem.*, 102 (1980) 31-34.

664 Suzuki, K.T., Analysis of metalloproteins, especially metallothioneins, by HPLC-AAS [review on atomic absorption spectrometry detection; in Japanese], *Kagaku no Ryoiki, Zokan*, 133 (1981) 79-92; *C.A.*, 95 (1981) 164 835p.

665 Szepesi, G. and Gazdag, M., Improved high-performance liquid chromatographic method for the analysis of insulins and related compounds, *J. Chromatogr.*, 218 (1981) 597-602.

666 Tabor, C.W. and Tabor, H., An automated ion-exchange assay for glutathione, *Anal. Biochem.*, 78 (1977) 543-553.

667 Takagaki, Y., Gerber, G.E., Nihei, K. and Khorana, H.G., Amino acid sequence of the membranous segment of rabbit liver cytochrome b_5. Methodology for separation of hydrophobic peptides, *J. Biol. Chem.*, 255 (1980) 1536-1541.

668 Takagi, T., High-performance liquid chromatography of protein polypeptides on porous silica gel columns (TSK-GEL SW) in the presence of sodium dodecyl sulfate: comparison with SDS-polyacrylamide gel electrophoresis, *J. Chromatogr.*, 219 (1981) 123-127.

669 Takagi, T., Takeda, K. and Okuno, T., Effect of salt concentration on the elution properties of complexes formed between sodium dodecylsulphate and protein polypeptides in high-performance silica gel chromatography, *J. Chromatogr.*, 208 (1981) 201-208.

670 Takahagi, H., Matsueda, R. and Maruyama, H., High pressure liquid chromatographic monitoring of solid phase peptide synthesis, *Sankyo Kenkyusho Nempo (Ann. Rep. Sankyo Res. Lab.)*, 30 (1978) 57-64; *C.A.*, 91 (1979) 57 485v.

671 Takahashi, N., Isobe, T., Kasai, H., Seta, K. and Okuyama, T., An analytical and preparative method for peptide separation by high performance liquid chromatography on a macroreticular anion-exchange resin, *Anal. Biochem.*, 115 (1981) 181-187.

672 Takaki, T., Low angle laser light scattering photometer. Its application to molecular weight measurement by high pressure silica gel chromatography [proteins, dextrans; in Japanese], in Osawa, K., Tanaka, Y. and Kitami, S. (Editors), *Kyodai Ryushi no Gerupamieishion Kuromatogurafi: Seitai Ryushi no Ryukei Bunri*, Kitami Shobo, Tokyo, 1980, pp. 319-335; *C.A.*, 95 (1981) 146 293f.

673 Tanzer, F.S., Desiderio, D.M. and Yamada, S., HPLC isolation and FD-MS [field desorption mass spectrometry] quantification of picomole amounts of Met-enkephalin in canine tooth pulp, in Rich, D.H. and Gross, E. (Editors), *Peptides: Synthesis-Structure-Function (Proc. 7th Am. Peptide Symp.)*, Pierce, Rockford, IL, 1981, pp. 761-764; *C.A.*, 96 (1982) 193 555y.

674 Tayot, J.L., Tardy, M., Gattel, P., Plan, R. and Roumiantzeff, M., Industrial ion exchange chromatography of proteins on DEAE dextran derivatives of porous silica beds, in Epton, R. (Editor), *Chromatography of Synthetic and Biological Polymers*, Vol. 2, [*Lect. Chem. Soc. Int. Symp. 1976*], Ellis Horwood, Chichester, 1978, pp. 95-110; *C.A.*, 90 (1979) 18 449w.

675 Terabe, S., Konaka, R. and Inouye, K., Separation of some polypeptide hormones by high-performance liquid chromatography, *J. Chromatogr.*, 172 (1979) 163-177.

676 Terabe, S., Konaka, R. and Shoji, J., Separation of polymixins and octapeptins by high-performance liquid chromatography, *J. Chromatogr.*, 173 (1979) 313-320.

677 Tischio, J.P. and Hetyei, N., Isocratic reversed-phase high-performance liquid chromatographic separation of underivatized tyrosine-related peptides of thymopoietin$_{32-36}$ pentapeptide, *J. Chromatogr.*, 236 (1982) 237-243.

678 Titani, K., Sasagwa, T., Resing, K. and Walsh, K., A simple and rapid
    purification of commercial trypsin and chymotrypsin by reverse-phase high-
    performance liquid chromatography, *Anal. Biochem.*, 123 (1982) 408-412.
679 Tjaden, U.R., Krol, J.H., Van Hoeven, R.P., Oomen-Meulemans, E.P.M. and
    Emmelot, P., High-pressure liquid chromatography of glycosphingolipids
    (with special reference to gangliosides), *J. Chromatogr.*, 136 (1977) 233-
    243.
680 Tomono, T., Yoshida, S. and Tokunaga, E., Isolation of plasma proteins by
    high-performance liquid chromatography, *J. Polym. Sci., Polym. Lett. Ed.*,
    17 (1979) 335-341; *C.A.*, 91 (1979) 86 677q.
681 Tonnaer, J.A.D.M., Verhoef, J., Wiegant, V.M. and De Jong, W., Separation
    and quantification of angiotensins and some related peptides by high-
    performance liquid chromatography, *J. Chromatogr.*, 183 (1980) 303-309.
682 Torii, M., Alberto, B.P., Tanaka, S., Tsumuraya, Y., Misaki, A. and Sawai,
    T., Preparative use of the analytical column of a sugar autoanalyser for
    resolution of gluco-oligosaccharides of the same molecular weight, *J.
    Biochem. (Tokyo)*, 85 (1979) 883-886; *C.A.*, 90 (1979) 164 157r.
683 Torii, M. and Sakakibara, K., Column chromatographic separation and quantita-
    tion of α-linked glucose oligosaccharides, *J. Chromatogr.*, 96 (1974)
    255-257.
684 Torii, M., Sakakibara, K., Misaki, A. and Sawai, T., Degradation of alpha-
    linked D-glucooligosaccharides and dextrans by an isomaltodextranase
    preparation from *Arthobacter globiformis* T6, *Biochem. Biophys. Res.
    Commun.*, 70 (1976) 459-464.
685 Torii, M., Tanaka, S. and Sawai, T., An epitopic structure of dextran B-1355
    [IEC of oligosaccharides], *Microbiol. Immunol.* 25 (1981) 969-973; *C.A.*,
    96 (1982) 66 955d.
686 Toth, M., László, E. and Morvai, J., Rapid determination of carbohydrate
    content in beer wort and beer using high pressure liquid chromatography
    [in Hungarian], *Soripar*, 28 (1981) 127-133; *C.A.*, 96 (1982) 141 082x.
687 Traylor, T.D., Koontz, D.A. and Hogan, E.L., High-performance liquid chro-
    matographic resolution of p-nitrobenzyloxyamine derivatives of brain
    gangliosides, *J. Chromatogr.*, 272 (1983) 9-20.
688 Tronquet, C., Guimbard, J.-P. and Paolucci, F., Applications of preparative
    HPLC for the purification of somatostatin and of its chemical precursor,
    in Rosselin, G., Fromageot, P. and Bonfils, S. (Editors), *Hormone Receptors
    in Digestion and Nutrition (Proceedings of the 2nd International Symposium
    on Hormonal Receptors in Digestive Tract Physiology, Montpellier, France,
    1979)*, Elsevier, Amsterdam, 1979, pp. 89-94; *C.A.*, 92 (1980) 106 668p.
689 Tsuji, K. and Robertson, J.H., Improved high-performance liquid chromato-
    graphic method for polypeptide antibiotics and its application to study
    the effects of treatments to reduce microbial levels in bacitracin powder,
    J. Chromatogr., 112 (1975) 663-672.
690 Tsuji, K., Robertson, J.H. and Bach, J.A., Quantitative high-pressure liquid
    chromatographic analysis of bacitracin, a polypeptide antibiotic, *J.
    Chromatogr.*, 99 (1974) 597-608.
691 Turková, J., Specific sorbents for high performance liquid affinity chro-
    matography and large scale isolations of proteinases, in Gribnau, T.C.J.,
    Visser, J. and Nivard, R.J.F. (Editors), *Affinity Chromatography and
    Related Techniques (Analytical Chemistry Symposia Series, Vol. 9)*, Elsevier,
    Amsterdam, 1982, pp. 513-528; *C.A.*, 96 (1982) 139 071z.
692 Turková, J., Bláha, K., Horáček, J., Vajčner, J., Frydrychová, A. and Čoupek,
    J., Hydroxyalkyl methacrylate gels derivatized with epichlorohydrin as
    supports for large-scale and high-performance affinity chromatography,
    J. Chromatogr., 215 (1981) 165-179.
693 Tweeten, T.N. and Euston, C.B., Application of high-performance liquid
    chromatography in the food industry [review], *Food Technol. (Chicago)*, 34
    (1980) 29-32, 34, 36, 37; *C.A.*, 94 (1981) 45 643u.

694 Ui, N., Rapid estimation of the molecular weights of protein polypeptide chains using high-pressure liquid chromatography in 6 M quanidine hydrochloride, *Anal. Biochem.*, 97 (1979) 65-71.

695 Ullman, M.D. and McCluer, R.H., Quantitative analysis of plasma neutral glycosphingolipids by high performance liquid chromatography of their perbenzoyl derivatives, *J. Lipid Res.*, 18 (1977) 371-378.

696 Ullman, M.D. and McCluer, R.H., Quantitative microanalysis of perbenzoylated neutral glycosphingolipids by high-performance liquid chromatography with detection at 230 nm, *J. Lipid Res.*, 19 (1978) 910-913.

697 Ullman, M.D., Pyeritz, R.E., Moser, H.W., Wenger, D.A. and Kolodny, E.H., Application of "high-performance" liquid chromatography to the study of sphingolipidoses, *Clin. Chem. (Winston-Salem, N.C.)*, 26 (1980) 1499-1503; *C.A.*, 93 (1980) 165 577b.

698 Ul'yashin, V.V., Deigin, V.I., Ivanov, V.T. and Ovchinnikov, Yu.A., High-performance size-exclusion liquid chromatography of protected peptides, *J. Chromatogr.*, 215 (1981) 263-277.

699 Unger, K.K. and Roumeliotis, P., Application of high performance liquid chromatography to the separation and isolation of biopolymers, in Gribnau, T.C.J., Visser, J. and Nivard, R.J.F. (Editors), *Affinity Chromatography and Related Techniques (Analytical Chemistry Symposia Series*, Vol. 9), Elsevier, Amsterdam, 1982, pp. 455-470; *C.A.*, 96 (1982) 100 253k.

700 Upreti, R.K. and Holoubek, V., Separation of proteins of nuclear ribonucleoprotein particles by high-performance liquid chromatography, *Anal. Chim. acta*, 131 (1981) 239-245.

701 Vacik, D.N. and Toren, E.C. Jr., Separation and measurement of isoenzymes and other proteins by high-performance liquid chromatography [review], *J. Chromatogr.*, 228 (1982) 1-31.

702 Valent, B.S., Darvill, A.G., McNeil, M., Robertsen, B.K. and Albersheim, P., A general and sensitive chemical method for sequencing the glycosyl residues of complex carbohydrates [RP HPLC of peralkylated oligosaccharides], *Carbohydr. Res.*, 79 (1980) 165-192; *C.A.*, 92 (1980) 215 666g.

703 Vandenberghe, A., Nelles, L. and De Wachter, R., High-pressure liquid chromatography analysis of oligo- and monoribonucleotide mixtures, with special reference to ribosomal RNA constituents, *Anal. Biochem.*, 107 (1980) 369-376.

704 Vandenberghe, A., Van Broeckhoven, C. and De Wachter, R., Separation of oligoribonucleotides by high performance liquid chromatography, *Arch. Int. Physiol. Biochim.*, 87 (1979) 848-849; *C.A.*, 92 (1980) 106 680m.

705 Van der Rest, M., Bennett, H.P.J., Solomon, S. and Glorieux, F.H., Separation of collagen cyanogen bromide-derived paptides by reversed-phase high-performance liquid chromatography, *Biochem. J.*, 191 (1980) 253-256.

706 Van der Rest, M., Cole, W.G. and Glorieux, F.H., Human collagen "fingerprints" produced by clostridiopeptidase A digestion and high-pressure liquid chromatography, *Biochem. J.*, 161 (1977) 527-534.

707 Van der Wal, S. and Huber, J.F.K., Performance of nonrigid ion-exchange packings of small particle size in the separation of cytochrome c and derivatives by high-pressure liquid chromatography, *Anal. Biochem.*, 105 (1980) 219-229.

708 Vaněček, G. and Regnier, F.E., Variables in the high-performance anion exchange chromatography of proteins, *Anal. Biochem.*, 109 (1980) 345-353.

709 Vaněček, G. and Regnier, F.E., Macroporous high-performance anion-exchange supports for proteins, *Anal. Biochem.*, 121 (1982) 156-169.

710 Van Eikeren, P. and McLaughlin, H., Analysis of the lysozyme-catalyzed hydrolysis and transglycosylation of N-acetyl-D-glucosamine oligomers by high-pressure liquid chromatography, *Anal. Biochem.*, 77 (1977) 513-522.

711 Van Haastert, P.J.M., High-performance liquid chromatography of nucleobases, nucleosides and nucleotides. I. Mobile phase composition for the separation of charged solutes by reversed-phase chromatography, *J. Chromatogr.*, 210 (1981) 229-240.

712 Van Haastert, P.J.M., High-performance liquid chromatography of nucleobases, nucleosides and nucleotides. II. Mobile phase composition for the separation of charged solutes by ion-exchange chromatography [S-adenosyl-L-methionine], *J. Chromatogr.*, 210 (1981) 241-254.

713 Viswanatha, V., Larsen, B. and Hruby, V.J., Synthesis of DL-[2-^{13}C]leucine and its use in the preparation of (3-DL-[2-^{13}C]leucine)oxytocin and (8-DL-[2-^{13}C]leucine)oxytocin. Preparative separation of diastereoisomeric peptides by partition chromatography and high pressure liquid chromatography, *Tetrahedron*, 35 (1979) 1575-1580; *C.A.*, 92 (1980) 181 619b.

714 Vivilecchia, R.V., Lightbody, B.G., Thimot, N.Z. and Quinn, H.M., The use of microparticulates in gel permeation chromatography [including an example of human plasma fractionation], *J. Chromatogr. Sci.*, 15 (1977) 424-433.

715 Voelter, W., Bauer, H., Fuchs, S. and Pietrzik, E., Preparative high-performance liquid chromatography of thyrotropin-releasing hormone analogs, *J. Chromatogr.*, 153 (1978) 433-442.

716 Vondruška, M., Šudřich, M. and Mládek, M., Determination of some gel permeation chromatographic parameters of the gel Spheron P-1000 [several proteins], *J. Chromatogr.*, 116 (1976) 457-461.

717 Voskamp, D., Olieman, C. and Beyerman, H.C., The use of trifluoroacetic acid in the reverse-phase liquid chromatography of peptides including secretin, *Recl. Trav. Chim. Pays-Bas*, 99 (1980) 105-108; *C.A.*, 93 (1980) 3223f.

718 Vrátný, P., Čoupek, J., Vozka, S. and Hostomská, Z., Accelerated reversed-phase chromatography of carbohydrate oligomers, *J. Chromatogr.*, 254 (1983) 143-155.

719 Vrátný, P., Mikeš, O., Farkaš, J., Štrop, P., Čopíková, J. and Nejepínská, K., Chromatography of mixtures of oligo- and monosaccharides on DEAE-Spheron, *J. Chromatogr.*, 180 (1979) 39-44.

720 Vrátný, P., Ouhrabková, J. and Čopíková, J., Liquid chromatography of nonreducing oligosaccharides: a new detection principle, *J. Chromatogr.*, 191 (1980) 313-317.

721 Wajcman, H. and Kitzis, A., Glycosylation of hemoglobin and minor hemoglobins: techniques and diagnostic utility [automated MPLC detection in hemolysates of diabetic patients; in French], *Ann. Biol. Clin. (Paris)*, 39 (1981) 227-231; *C.A.*, 96 (1982) 3156y.

722 Wakizaka, A., Kurosaka, K. and Okuhara, E., A rapid separation of dinucleotides and trinucleotides in DNase digest of DNA according to base composition using high pressure liquid chromatography (HPLC) with a weak anion-exchange column, *IRCS Med. Sci., Biochem. Biomed. Technol.*, 6 (1978) 485; *C.A.*, 90 (1979) 50 797u.

723 Warren III, P.H. and Singhal, R.P., Changes in alcohol dehydrogenase activity NADH and ADP levels in rats induced by stress [IEC-HPLC], *Trans. Kans. Acad. Sci.*, 82 (1979) 25-35.

724 Waterfield, M.D. and Scrace, G.T., Peptide separation by liquid chromatography using size exclusion and reverse-phase columns, in Hawk, G.L., Champlin, P.B., Hutton, R.F. and Mol, Ch. (Editors), *Chromatogr. Sci.*, 18 (Biological/Biomedical Applications of Liquid Chromatography III) (1981) 135-158; *C.A.*, 96 (1982) 118 335y.

725 Wehr, T.C., *Nucleic Acid Constituents by High Performance Liquid chromatography* [brochure], Varian, Palo Alto, CA, 1980, 108 pp.

726 Wehr, C.T., Separation of low molecular weight peptides [α-endorphin, methionine- and leucine-enkephalins, (D-ala^2)-met-enkephalin, ranatensin] by reverse phase chromatography, *Liquid Chromatography at Work/84* (Pamphlet issued by Varian Instrument Division, Palo Alto, CA, 94 303.).

727 Wehr, C.T. and Abbott, S.R., High-speed steric exclusion chromatography of biopolymers, *J. Chromatogr.*, 185 (1979) 453-462.

728 Wehr, C.T., Correia, L. and Abbott, S.R., Evaluation of stationary and mobile phases for reversed-phase high performance liquid chromatography of peptides, *J. Chromatogr. Sci.*, 20 (1982) 114-119.

B610

729 Wells, G.B. and Lester, R.L., Rapid separation of acetylated oligosaccharides by reverse-phase high-pressure liquid chromatography, *Anal. Biochem.*, 97 (1979) 184-190.

730 Wells, G.B., Turco, S.J., Hanson, B.A. and Lester, R.L., Separation of dolichylpyrophosphoryloligosaccharides by liquid chromatography, *Anal. Biochem.*, 110 (1981) 397-406.

731 Wells, G.B., Turco, S.J., Hanson, B.A. and Lester, R.L., Resolution of dolichylpyrophosphoryl oligosaccharides by high-pressure liquid chromatography, *Methods Enzymol.*, 83 (Complex Carbohydrates, Part D) (1982) 137-139.

732 White, Ch.A., Corran, P.H. and Kennedy, J.F., Analysis of underivatised D-Gluco-oligosaccharides (d.p. 2-20) by high-pressure liquid chromatography, *Carbohydr. Res.*, 87 (1980) 165-173.

733 Wideman, J., Using HPLC and fluorometric methods in studying endogenous peptides [enkephalins, endorphins], in Way, E.L. (Editor), *Endog. Exog. Opiate Agonists Antagonists (Proc. Int. Narc. Res. Club. Conf. 1979)*, Pergamon, Elmsford, NY, 1980, pp. 341-344; *C.A.*, 94 (1981) 170 378n.

734 Wideman, J., Brink, L. and Stein, S., New automated fluorimetric peptide microassay for carnosine in mouse olfactory bulb [RP HPLC], *Anal. Biochem.*, 86 (1978) 670-678.

735 Williams, B.R.G., Golgher, R.R., Brown, R.E., Gilbert, C.S. and Kerr, I.M., Natural occurence of 2-5A in interferon-treated EMC virus-infected L cells, *Nature (London)*, 282 (1979) 582-586.

736 Wilson, J.B., Lam, H., Pravatmuang, P. and Huisman, T.H.J., Separation of tryptic peptides of normal and abnormal α, β, γ and δ hemoglobin chains by high-performance liquid chromatography, *J. Chromatogr.*, 179 (1979) 271-290.

737 Wilson, K.J., Berchtold, M., Honegger, A. and Hughes, G.J., The use of HPLC in peptide/protein chemistry [review], in Lottspeich, F., Henschen, A. and Hupe, K.P. (Editors), *High-Performance Chromatography in Protein and Peptide Chemistry*, Walter de Gruyter, Berlin, New York, 1981, pp. 159-174; *C.A.*, 97 (1982) 88 030w.

738 Wilson, K.J., Honegger, A. and Hughes, G.J., Comparison of buffers and detection systems for high-pressure liquid chromatography of peptide mixtures, *Biochem. J.*, 199 (1981) 43-51.

739 Wilson, K.J., Honegger, A., Stoetzel, R.P. and Hughes, G.J., The behavior of peptides on reverse-phase supports during high-pressure liquid chromatography, *Biochem. J.*, 199 (1981) 31-41.

740 Wilson, K.J. and Hughes, G.J., High-performance liquid chromatography of peptides and proteins. Pharmaceutical and bio-medical applications, *Chimia*, 35 (1981) 327-333; *C.A.*, 96 (1982) 31 009w.

741 Wilson, K.J., Van Wieringen, E., Klauser, S., Berchtold, M.W. and Hughes, G.J., Comparison of the high-performance liquid chromatography of peptides and proteins on 100- and 300-Å reversed-phase supports, *J. Chromatogr.*, 237 (1982) 407-416.

742 Wong How-Yan S., Roles of high-performance liquid chromatography in nuclear medicine [review including also macromolecules], *Adv. Chromatogr.*, 19 (1980) 1-36; *C.A.*, 94 (1981) 162 806k.

743 Wood, R., Cummings, L. and Jupille, T., Recent developments in ion-exchange chromatography [review including also oligosaccharides, peptides, proteins], *J. Chromatogr. Sci.*, 18 (1980) 551-558.

744 Wooton, M. and Chaudhry, M.A., In vitro digestion of hydroxypropyl derivatives of wheat starch. III. Analysis of oligosaccharide fractions using high pressure liquid chromatography, *Starch/Stärke*, 33 (1981) 200-202; *C.A.*, 95 (1981) 60 022e.

745 Wu, K.M., Sloan, J.W. and Martin, W.R., Development of a combined high-performance liquid chromatographic-fluorometric quantitative assay for enkephalins, *J. Chromatogr.*, 202 (1980) 500-503.

746 Wuensch, E., Peptide factors: definition of purity, in Rosselin, G., Fromageot, P. and Bonfils, S. (Editors), *Hormone Receptors in Digestion and Nutrition* (*Proceedings of the 2nd International Symposium on Hormonal Receptors in Digestive Tract Physiology*, Montpellier, 1979), Elsevier, Amsterdam, 1979, pp. 115-125.

747 Yahara, S., Kishimoto, Y. and Poduslo, J., High performance liquid chromatography of membrane glycolipids. Assessment of cerebrosides on the surface of myelin, *ACS Symp. Ser.*, 128 (1980) 15-33; *C.A.*, 93 (1980) 109 941d.

748 Yamada, S. and Desiderio, D.M., Measurement of endogenous Leu-enkephalin in canine caudate nuclei and hypothalami with high performance liquid chromatography and field desorption mass spectrometry, *Anal. Biochem.*, 127 (1982) 213-221.

749 Yamakawa, T., Handa, S., Yamazaki, T. and Suzuki, A., Analysis of red blood cell glycolipids by high-performance liquid chromatography, *Proc. Int. Congr. Pure Appl. Chem.*, 27 (1980) 351-358; *C.A.*, 93 (1980) 109 942e.

750 Yamashiro, D. and Li Choh Hao, Partition and high-performance liquid chromatography of β-lipotropin and synthetic β-endorphin analogues, *J. Chromatogr.*, 215 (1981) 255-261.

751 Yamazaki, T., Suzuki, A., Handa, S. and Yamakawa, T., Consecutive analysis of sphingolipids on the basis of sugar and ceramide moieties by high performance liquid chromatography, *J. Biochem. (Tokyo)*, 86 (1979) 803-809; *C.A.*, 91 (1979) 188 987h.

752 Yanagawa, H., Analytical method of oligodeoxynucleotides by high-speed liquid chromatography, *Nucleic Acids Res., Spec. Publ.*, 5 (1978) 461-464; *C.A.*, 90 (1979) 83 055d.

753 Yang Chao Yuh, Pauly, E., Kratzin, H. and Hilschmann, N., Chromatographie und Rechromatographie in der Hochdruckflüssigkeitschromatographie von Peptidgemischen. Die vollständige Primärstruktur einer Immunoglobulin L-Kette vom K-Typ, Subgruppe I (Bence-Jones Protein Den) (Chromatography and rechromatography in the high-performance liquid chromatography of peptide mixtures. The complete primary structure of an immunoglobin L-chain of K-type, subgroup I (Bence-Jones protein Den)), *Hoppe-Seyler's Z. Physiol. Chem.*, 362 (1981) 1131-1146; *C.A.*, 95 (1981) 201 816f.

754 Yang, H.S., Studebaker, J.F. and Parravano, C., The study of disulfide bond pairing in proteins and protein fragments with HPLC, *Chromatogr. Sci.*, 10 (1979) 247-260.

755 Yoshida, H. and Imai, H., High performance liquid chromatography of biologically active peptides [review in Japanese], *Tanpakushitsu Kakusan Koso*, 26 (1981) 1135-1141; *C.A.*, 95 (1981) 93 123s.

756 Yu, T.J., Schwartz, H., Giese, R.W., Karger, B.L. and Vorous, P., Analysis of N-acetyl-N,O,S-permethylated peptides by combined liquid chromatography - mass-spectrometry, *J. Chromatogr.*, 218 (1981) 519-533.

757 Yuan, P.M., Pande, H., Clark, B.R. and Shively, J.E., Microsequence analysis of peptides and proteins. I. Preparation of samples by reverse-phase liquid chromatography, *Anal. Biochem.*, 120 (1982) 289-301.

758 Zarytova, V.F. and Lebedev, A.V., Issledovanie mechanizma chimicheskogo sinteza oligonukleotidov. XI. Razdelenie trizameschennykh pirofosfatov - aktivnykh proizvodnykh dinukleotidov - metodom ionoobmennoi khromatografii (Study of the mechanism of the chemical synthesis of oligonucleotides. XI. Separation of trisubstituted pyrophosphates - active dinucleotide derivatives - by ion-exchange chromatography), *Bioorg. Khim.*, 3 (1977) 1211-1218; *C.A.*, 88 (1978) 38 097s.

759 Zeng, Yi-tao, Headlee, M.E., Henson, J., Lam, H., Wilson, J.B. and Huisman, T.H.J., Hemoglobin G-Philadelphia (α68 Asn → Lys) and hemoglobin Matsue-Oki (α75 Asp → Asn) in a black infant: Identification by high pressure chromatography and microsequencing, *Biochim. Biophys. Acta*, 707 (1982) 206-212; *C.A.*, 97 (1982) 179 691a.

760 Zeng Yi-tao, Huang Shu-zheng, Reynolds, A., Lam, H., Webber, B.B., Wilson, J.B. and Huisman, T.H.J., Application of high pressure liquid chromatography and ultra-micro sequencing methodology in the structural analysis of human hemoglobin variants, *Sci. Sin., Ser. B (Engl. Ed.)*, 26 (1983) 836-849; *C.A.*, 99 (1983) 191 035s.

761 Zsadon, B., Otta, K.H., Tüdös, F. and Szejtli, J., Separation of cyclodextrins by high-performance liquid chromatography, *J. Chromatogr.*, 172 (1979) 490-492.

762 Zygmunt, L.C., High pressure liquid chromatographic determination of mono- and disaccharides in presweetened cereals: collaborative study, *J. Assoc. Off. Anal. Chem.*, 65 (1982) 256-264; *C.A.*, 96 (1982) 179 545p.

BIBLIOGRAPHY ADDENDUM

Because the majority of citations 1-762 presented in the above Bibliography does not include periods exceeding the year 1982, this Addendum was prepared, containing more recent citations, namely from the years 1983 to 1986. These Addendum citations (763-1304) correspond to the references given in the Appendix to the Register (Table 14.5).

763 Abraham, D., Daniel, P., Dell, A., Oates, J., Sidebotham, R. and Winchester, B., Structural analysis of the major urinary oligosaccharides in feline α-mannosidosis, *Biochem. J.*, 233 (1986) 899-904.

764 Acharya, A.S., Sussman, L.G. and Manjula, B.N., Application of reductive dihydroxypropylation of amino groups of proteins in primary structural studies: Identification of phenylthiohydantoin derivative of ε-dihydroxypropyl-lysine residues by high-performance liquid chromatography, *J. Chromatogr.*, 297 (1984) 37-48.

765 Aguilar, M.-I., Hodder, A.N. and Hearn, M.T.W., High-performance liquid chromatography of amino acids, peptides and proteins. LXV. Studies on the optimisation of the reversed-phase gradient elution of polypeptides: Evaluation of retention relationships with β-endorphin-related polypeptides, *J. Chromatogr.*, 327 (1985) 115-138.

766 Alexander, P.W., Haddad, P.R. and Trojanowicz, M., Potentiometric flow-injection determination of sugars using a metallic copper electrode, *Anal. Lett.*, 18 (1985) 1953-1978.

767 Allinquant, B., Musenger, C. and Schuller, E., Reversed-phase high-performance liquid chromatography of nucleotides and oligonucleotides, *J. Chromatogr.*, 326 (1985) 281-291.

768 Alpert, A.J., Cation-exchange high-performance liquid chromatography of proteins on poly(aspartic acid)-silica, *J. Chromatogr.*, 266 (1983) 23-37.

769 Alpert, A.J., High-performance hydrophobic-interaction chromatography of proteins on a series of poly(alkyl aspartamide)-silicas, *J. Chromatogr.*, 359 (1986) 85-97.

770 Alpert, C.A., Dörschug, M., Saffen, D., Frank, R., Deutscher, J. and Hengstenberg, W., The bacterial phosphoenolpyruvate-dependent phosphotransferase system. Isolation of active site peptides by reversed-phase high-performance liquid chromatography and determination of their primary structure, *J. Chromatogr.*, 326 (1985) 363-371.

771 Anbari, M., Adachi, K., Ip, C.Y. and Asakura, T., Stability of symmetrical hybrid hemoglobins. Determination by anaerobic high performance liquid chromatography, *J. Biol. Chem.*, 260 (1985) 15522-15525.

772 Anderson, D.J., Anhalt, J.S. and Walters, R.R., High-performance affinity chromatography of divalent concanavalin A on matrices of variable ligand density, *J. Chromatogr.*, 359 (1986) 369-382.

773 Anderson, D.J., Taylor, A.J. and Reischer, R.J., Reversed-phase high-performance liquid chromatography of protected oligodeoxynucleotides, *J. Chromatogr.*, 350 (1985) 313-316.

774 Andersson, T., Carlsson, M., Hagel, L., Pernemalm, P.-Å. and Janson, J.-C., Agarose-based media for high-resolution gel filtration of biopolymers, *J. Chromatogr.*, 326 (1985) 33-44.

775 Ando, S., Kon, K., Isobe, M., Nagai, Y. and Yamakawa, T., Existence of glucosaminyl lactosyl ceramide (amino CTH-I) in human erythrocyte membranes as a possible precursor of blood group-active glycolipids, *J. Biochem. (Tokyo)*, 79 (1976) 625-632.

776 Ando, S. and Yu, R.K., Isolation and characterization of a novel trisialo-ganglioside, G_{T1a}, from human brain, *J. Biol. Chem.*, 18 (1977) 6247-6250.

777 Andrews, A.T., Taylor, M.D. and Owen, A.J., Rapid analysis of bovine milk proteins by fast protein liquid chromatography, *J. Chromatogr.*, 348 (1985) 177-185.

778 Antoniotti, H., Fagot-Revurat, P., Esteve, J.P., Fourmy, D., Pradayrol, L. and Ribet, A., Purification of radioiodinated somatostatin-related peptides by reversed-phase high-performance liquid chromatography, *J. Chromatogr.*, 296 (1984) 181-188.

779 Ashman, K. and Wittmann-Liebold, B., A new isocratic HPLC separation for PTH-amino acids, based on 2-propanol, *FEBS Lett.*, 190 (1985) 129-132; *C.A.*, 103 (1985) 192 549a.

780 Ashworth, D.J., Baird, W.M., Chang, C.J., Ciupek, J.D., Busch, K.L. and Cooks, R.G., Chemical modification of nucleic acids. Methylation of calf thymus DNA investigated by mass spectrometry and liquid chromatography, *Biomed. Mass Spectrom.*, 12 (1985) 309-318; *C.A.*, 103 (1985) 192 654f.

781 Babasaki, K., Takao, T., Shimonishi, Y. and Kurahashi, K., Subtilosin A, a new antibiotic peptide produced by *Bacillus subtilis* 168: Isolation, structural analysis, and biogenesis, *J. Biochem. (Tokyo)*, 98 (1985) 585-693.

782 Banes, A.J., Link, G.W. and Snyder, L.R., Comparison of reversed-phase columns for the separation of tryptic peptides by gradient elution. Correlation of experimental results and model prediction, *J. Chromatogr.*, 326 (1985) 419-431.

783 Bansal, S.K., Love, J.H. and Gurtoo, H.L., Resolution of multiple forms of cytochrome P-450 by high-performance liquid chromatography, *J. Chromatogr.*, 297 (1984) 119-127.

784 Beavis, R.C., Bolbach, G., Ens, W., Main, D.E., Schueler, B. and Standing, K.G., Automated dry fraction collection for microbore high-performance liquid chromatography-mass spectrometry, *J. Chromatogr.*, 359 (1986) 489-497.

785 Becker, C.R., Efcavitch, J.W., Heiner, C.R. and Kaiser, N.F., Use of C_4 column for reversed-phase high-performance liquid chromatographic purification of synthetic oligonucleotides, *J. Chromatogr.*, 326 (1985) 293-299.

786 Belew, M., Peterson, E.A. and Porath, J., A high-capacity hydrophobic adsorbent for human serum albumin, *Anal. Biochem.*, 151 (1985) 438-441.

787 Bennett, H.P.J., Isolation of pituitary peptides by reversed-phase high-performance liquid chromatography. Expansion of the resolving power of reversed-phase columns by manipulating pH and the nature of the ion-pairing reagent, *J. Chromatogr.*, 266 (1983) 501-510.

788 Bennett, H.P.J., Use of ion-exchange Sep-Pak cartridges in the batch fractionation of pituitary peptides, *J. Chromatogr.*, 359 (1986) 383-390.

789 Bennett, H.P.J. and Solomon, S., Use of Pico-Tag methodology in the chemical analysis of peptides with carboxyl-terminal amides, *J. Chromatogr.*, 359 (1986) 221-230.

790 Bermann, C., Legrand, M., Geoffroy, P. and Fritig, B., High-performance liquid chromatography of proteins. Rapid chromatofocusing of plant enzymes, *J. Chromatogr.*, 348 (1985) 167-175.

791 Bhatnagar, G.M., Supakar, P.C. and Bhatnagar, Y.M., A high performance
    liquid chromatography method to obtain rat epidermal filaggrin, *Biochem.*
    Biophys. Res. Commun., 132 (1985) 1196-1203.
792 Bican, P., The application of the high-performance ion-exchange chromato-
    graphy for the analysis of bovine milk proteins, *Experientia*, 41 (1985)
    958-960; *C.A.*, 103 (1985) 103 562j.
793 Bietz, J.A., High performance liquid chromatography: How proteins look in
    cereals [review], *Cereal Chem.*, 62 (1985) 201-212; *C.A.*, 103 (1985) 86 575w.
794 Bietz, J.A. and Cobb, L.A., Improved procedures for rapid wheat varietal
    identification by reversed-phase high-performance liquid chromatography of
    gliadin, *Cereal Chem.*, 62 (1985) 332-339; *C.A.*, 103 (1985) 192 538w.
795 Bischoff, R. and McLaughlin, L.W., Nucleic acid resolution by mixed-mode
    chromatography, *J. Chromatogr.*, 296 (1984) 329-337.
796 Bisse, E., Abraham, A., Stallings, M., Perry, R.E. and Abraham, E.C., High-
    performance liquid chromatographic separation and quantitation of
    glycosylated hemoglobin A_2 as an alternate index of glycemic control,
    J. Chromatogr., 374 (1986) 259-269.
797 Bláha, I., Ježek, J., Zaoral, M., Hamšiková, E., Závadová, H. and Vonka, V.,
    The synthesis of hemagglutinin fragments of Influenza virus, in Zaoral, M.,
    Havlas, Z., Mikeš, O. and Procházka, Ž. (Editors), *Synthetic Immuno-*
    modulators and Vaccines, Czechoslovak Academy of Sciences, Prague, 1986,
    pp. 203-207.
798 Blaszcyk, M., Ross, A.H., Ernst, C.S., Marchisio, M., Atkinson, B.F., Pak,
    K.Y., Steplewski, Z. and Koprowski, H., A fetal glycolipid expressed on
    adenocarcinomas of the colon, *Int. J. Cancer*, 33 (1984) 313-318.
799 Bobleter, O. and Schwald, W., Gel permeation chromatography for the evalua-
    tion of the molecular weight distribution of underived celluloses in
    Cadoxen, *Papier (Darmstadt)*, 39 (1985) 437-441 [in German]; *C.A.*, 104
    (1986) 7334m.
800 Born, J., Hoppe, P., Schwarz, W., Tiedemann, H. and Wittmann-Liebold, B.,
    An embryonic inducing factor: Isolation by high performance liquid chro-
    matography and chemical properties, *Biol. Chem. Hoppe-Seyler*, 366 (1985)
    729-735.
801 Boschetti, E., Egly, J.M. and Monsigny, M., The place of affinity chromato-
    graphy in the production and purification of biomolecules from cultured
    cells, *Trends Anal. Chem.*, 5 (1986) 4-10.
802 Boyle, D.M., Wiehle, R.D., Shahabi, N.A. and Wittliff, J.L., Rapid, high-
    resolution procedure for assessment of estrogen receptor heterogeneity
    in clinical samples, *J. Chromatogr.*, 327 (1985) 369-376.
803 Braco, L., Abad, C., Campos, A. and Fugueruelo, J.E., Time-dependent
    monomerization of gramicidin A, enhanced by phosphatidylcholine in non-
    polar solvents. A high-performance liquid chromatographic and spectro-
    fluorometric study, *J. Chromatogr.*, 353 (1986) 181-192.
804 Brennan, S.O., The separation of globin chains by high pressure cation ex-
    change chromatography, *Hemoglobin*, 9 (1985) 53-63; *C.A.*, 103 (1985) 67 649g.
805 Bretscher, M.S., Heparan sulfate proteoglycans and their polypeptide chains
    from BHK cells, *EMBO J.*, 4 (1985) 1941-1944; *C.A.*, 103 (1985) 210 428s.
806 Brooks, T.L. and Stevens, A., Preparative HPLC purification of IgG and IgM
    monoclonal antibodies, *Am. Lab. (Fairfield, Conn.)*, 17 (1985) 54-64; *Int.*
    Lab., 18 (1985) 72-81; *C.A.*, 103 (1985) 212 876k.
807 Brotherton, J.E. and Widholm, J.M., High-performance liquid chromatography
    of anthranilate synthase using gel filtration and a post-column reactor,
    J. Chromatogr., 350 (1985) 332-335.
808 Brückner, H. and Przybylski, M., Isolation and structural characterization
    of polypeptide antibiotics of the peptaibol class by high-performance
    liquid chromatography with field desorption and fast atom bombardment
    mass spectrometry, *J. Chromatogr.*, 296 (1984) 263-275.
809 Brunner, H. and Mann, H., Combination of conventional and high-performance
    liquid chromatographic techniques for the isolation of so-called "uraemic
    toxins", *J. Chromatogr.*, 297 (1984) 405-416.

810 Budĕšínský, M., Ježek, J., Krchňák, V., Lebl, M. and Zaoral, M., Synthetic fragments of bacterial cell walls, physicochemical and biological properties, in K. Bláha and P. Maloň (Editors), *Peptides 1982, Proc. 17th Eur. Pept. Symp., Praha, Aug. 29 - Sept. 3, 1982*, Walter de Gruyter, Berlin, New York, 1983, pp. 305-310.

811 Burke, D.J., Duncan, J.K., Dunn, L.C., Cummings, L., Siebert, C.J. and Ott, G.S., Rapid protein profiling with a novel anion-exchange material, *J. Chromatogr.*, 353 (1986) 425-437.

812 Burke, D.J., Duncan, J.K., Siebert, C. and Ott, G.S., Rapid cation-exchange chromatography of hemoglobins and other proteins, *J. Chromatogr.*, 359 (1986) 533-540.

813 Cachia, P.J., Van Eyk, J., Chong, P.C.S., Taneja, A. and Hodges, R.S., Separation of basic peptides by cation-exchange high-performance liquid chromatography, *J. Chromatogr.*, 266 (1983) 651-659.

814 Cachia, P.J., Van Eyk, J., McCubbin, E.D., Kay, C.M. and Hodges, R.S., Ion-exchange high-performance liquid chromatographic purification of bovine cardiac and rabbit skeletal muscle troponin subunits, *J. Chromatogr.*, 343 (1985) 315-328.

815 Calam, D.H. and Davidson, J., Isolation of influenza viral protein by size-exclusion and ion-exchange high-performance liquid chromatography: the influence of conditions on separation, *J. Chromatogr.*, 296 (1984) 285-292.

816 Calam, D.H., Davidson, J. and Ford, A.W., Investigations of the allergens of cocksfoot grass (*Dactylis glomerata*) pollen, *J. Chromatogr.*, 266 (1983) 293-300.

817 Calam, D.H., Davidson, J. and Harris, R., High-performance liquid chromatographic investigations on some enzymes of papaya latex, *J. Chromatogr.*, 326 (1985) 103-111.

818 Candiano, G., Ghiggeri, G.M., Delfino, G. and Queirolo, C., Fractionation of human serum albumin isoforms with chromatofocusing, *J. Chromatogr.*, 375 (1986) 405-410.

819 Carlquist, M. and Rökaeus, Å., Isolation of a proform of porcine secretin by ion-exchange and reversed-phase high-performance liquid chromatography, *J. Chromatogr.*, 296 (1984) 143-151.

820 Čech, M., Jelinková, M. and Čoupek, J., High pressure chromatography of tobacco mosaic virus on Spheron gels, *J. Chromatogr.*, 135 (1977) 435-440.

821 Chaiken, I.M., Kanmera, T., Sequeira, R.P. and Swaisgood, H.E., High-performance liquid chromatography and studies of neurophysin-neurohypophysial hormone pathways, *J. Chromatogr.*, 336 (1984) 63-71.

822 Chang, T.-M., Erway, B. and Chey, W.Y., Rapid, small-scale preparation of gastrointestinal hormones by high-performance liquid chromatography on a C_{18} column, *J. Chromatogr.*, 326 (1985) 121-127.

823 Chase, H.A., Prediction of the performance of preparative affinity chromatography, *J. Chromatogr.*, 297 (1984) 179-202.

824 Cheetham, N.W.H. and Teng, G., Some applications of reversed-phase high-performance liquid chromatography to oligosaccharide separations, *J. Chromatogr.*, 336 (1984) 161-172.

825 Chicz, R.M., Shi, Z. and Regnier, F.E., Preparation and evaluation of inorganic anion-exchange sorbents not based on silica, *J. Chromatogr.*, 359 (1986) 121-130.

826 Chien, J.-L. and Hogan, E.L., Novel pentahexosyl ganglioside of the globo series purified from chicken muscle, *J. Biol. Chem.*, 258 (1983) 10727-10730.

827 Chikuma, T., Ishii, Y. and Kato, T., Highly sensitive assay for PZ-peptidase activity by high-performance liquid chromatography, *J. Chromatogr.*, 348 (1985) 205-212.

828 Clezardin, P., Bourgo, G. and McGregor, J.L., Tandem purification of IgM monoclonal antibodies from mouse ascites fluids by anion-exchange and gel fast protein liquid chromatography, *J. Chromatogr.*, 354 (1986) 425-433.

829 Clezardin, P., McGregor, J.L., Manach, M., Robert, F., Dechavanne, M. and Clementson, K.J., Isolation of thrombospondin released from thrombin-stimulated human platelets by fast protein liquid chromatography on an anion-exchange Mono-Q column, *J. Chromatogr.*, 296 (1984) 249-256.

B616

830 Cohen, S.A., Benedek, K.P., Dong, S., Tapuhi, Y. and Karger, B.L., Multiple peak formation in reversed-phase liquid chromatography of papain, *Anal. Chem.*, 56 (1984) 217-221.
831 Coles, E., Reinhold, V.N. and Carr, S.A., Fluorescent labeling of carbo-hydrates and analysis by liquid chromatography. Comparison of derivatives using mannosidosis oligosaccharides, *Carbohydr. Res.*, 139 (1985) 1-11; *C.A.*, 103 (1985) 101 350q.
832 Colin, H., Lowy, G. and Cazes, J., Design and performance of a preparative-scale HPLC, *Am. Biotechnol. Lab.*, 3 (1985) 36-44; *C.A.*, 104 (1986) 2950z.
833 Collinge, A., Use of high-performance liquid chromatography for analysis of foods [review with 52 references], *Tijdschr. Belg. Ver. Laboratoriumtechnol.*, 12 (1985) 91-96, 105-118; *C.A.*, 103 (1985) 69 844r.
834 Colpan, M. and Riesner, D., High-performance liquid chromatography of high-molecular-weight nucleic acids on the macroporous ion exchanger, Nucleogen, *J. Chromatogr.*, 296 (1984) 339-353.
835 Conlon, J.M. and Göke, B., Metabolism of substance P in human plasma and in the rat circulation, *J. Chromatogr.*, 296 (1984) 241-247.
836 Cook, N.D. and Peters, T.J., A sensitive high-performance liquid chromato-graphy assay for γ-glutamyl hydrolase, *Biochem. Soc. Trans.*, 13 (1985) 1226-1227; *C.A.*, 103 (1985) 156 195y.
837 Cooper, E.H., Johns, E.A., Itoh, Y. and Webb, J.R., Chromatographic study of the interrelationships of immunoglobulin A and α_1-microglobulin in myelomatosis, *J. Chromatogr.*, 327 (1985) 179-188.
838 Cooper, E.H., Turner, R., Webb, J.R., Lindblom, H. and Fagerstam, L., Fast protein liquid chromatography scale-up procedures for the preparation of low-molecular-weight proteins from urine, *J. Chromatogr.*, 327 (1985) 269-277.
839 Cooper, J.D.H. and Turnell, D.C., Automated preparation of biological samples prior to high pressure liquid chromatography: Part II - The combined use of dialysis and trace enrichment for analyzing biological material, *J. Autom. Chem.*, 7 (1985) 181-184; *C.A.*, 104 (1986) 17 336c.
840 Co-Sarno, M.E., Tapang, M. Asuncion and Luckhurst, D.G., Determination of polymer and purification of albumin by high-performance liquid chromato-graphy, *J. Chromatogr.*, 266 (1983) 105-113.
841 Crabb, J.W. and Heilmeyer, L.M.G., Jr., Micropreparative protein purifica-tion by reversed-phase high-performance liquid chromatography, *J. Chro-matogr.*, 296 (1984) 129-141.
842 Crooks, P.A., Krechniak, J.W., Olson, J.W. and Gillespie, M.N., High-per-formance liquid chromatographic analysis of pulmonary metabolites of Leu- and Met-enkephalins in isolated perfused rat lung, *J. Pharm. Sci.*, 74 (1985) 1010-1012.
843 Crowley, S.C. and Walters, R.R., Determination of immunoglobulins in blood serum by high-performance affinity chromatography, *J. Chromatogr.*, 266 (1983) 157-162.
844 Cunico, R.L., Simpson, R., Correia, L. and Wehr, C.T., High-sensitivity phenylthiohydantoin amino acid analysis using conventional and microbore chromatography, *J. Chromatogr.*, 336 (1984) 105-113.
845 Cunico, R.L. and Schlabach, T., Comparison of ninhydrin and *o*-phthalaldehyde post-column detection techniques for high-performance liquid chromatography of free amino acids, *J. Chromatogr.*, 266 (1983) 461-470.
846 Davis, T.P. and Culling-Berglund, A., High-performance liquid chromatographic analysis of *in vitro* central neuropeptide processing, *J. Chromatogr.*, 327 (1985) 279-292.
847 Dawkins, J.V., Lloyd, L.L. and Warner, F.P., Chromatographic characteristics of polymer-based high-performance liquid chromatography packings, *J. Chromatogr.*, 352 (1986) 157-167.
848 Debowski, J., Grassini-Strazza, G. and Sybilska, D., α- and β-Cyclodextrins as selective agents for the separation of isomers by reversed-phase high-performance thin-layer and column liquid chromatography, *J. Chromatogr.*, 349 (1985) 131-136.

849 Deibler, G.E., Boyd, L.F., Martenson, R.E. and Kies, M.W., Isolation of tryptic peptides of myelin basic protein by reversed-phase high-performance liquid chromatography, *J. Chromatogr.*, 326 (1985) 433-442.

850 De Jong, G.J., Lammers, N., Spruit, F.J., Frei, R.W. and Brinkman, U.A.T., Analytical implications of the half-life of the chemiluminiscence signal in the peroxyoxalate detection system for liquid chromatography, *J. Chromatogr.*, 353 (1986) 249-257.

851 Delucas, L.J. and Muccio, D.D., Purification of bovine rhodopsin by high-performance size-exclusion chromatography, *J. Chromatogr.*, 296 (1984) 121-128.

852 Dennis, M., Lazure, C., Seidah, N.G. and Chrétien, M., Characterization of β-endorphin immunoreactive peptides in rat pituitary and brain by coupled gel and reversed-phase high-performance liquid chromatography, *J. Chromatogr.*, 266 (1983) 163-172.

853 Desiderio, D.M., Kai, M., Tanzer, F.S., Trimble, J. and Wakelyn, C., Measurement of enkephalin peptides in canine brain regions, teeth and cerebrospinal fluid with high-performance liquid chromatography and mass spectrometry, *J. Chromatogr.*, 297 (1984) 245-260.

854 Di Bussolo, J.M. and Gant, J.R., Vizualization of protein retention and migration in reversed-phase liquid chromatography, *J. Chromatogr.*, 327 (1985) 67-76.

855 Djordjevic, S.P., Batley, M. and Redmond, J.W., Preparative gel chromatography of acidic oligosaccharides using a volatile buffer, *J. Chromatogr.*, 354 (1986) 507-510.

856 Dölling, R., Winter, R., Schmidt, M., Furkert, J., Otto, H.-A., Liebermann, H. and Bienert, M., The synthesis of viral peptide sequences using a biphasic solvent system, in Zaoral, M., Havlas, Z., Mikeš, O. and Procházka, Ž., *Synthetic Immunomodulators and Vaccines*, Czechoslovak Academy of Sciences, Prague, 1986, pp. 218-227.

857 Dong, M.V. and Gant, J.R., High-speed liquid chromatographic analysis of amino acids by post-column sodium hypochlorite-phthalaldehyde reaction, *J. Chromatogr.*, 327 (1985) 17-25.

858 Doris, P.A., Analysis of plasma angiotensins by reversed phase HPLC and radioimmunoassay, *J. Liq. Chromatogr.*, 8 (1985) 2017-2034.

859 Dornburg, R., Földi, P. and Hofschneider, P.H., Increase of cloning efficiencies by using high-performance liquid chromatography-purified vectors and linkers, *J. Chromatogr.*, 296 (1984) 379-385.

860 Drager, R.R. and Regnier, F.E., Application of the stoichiometric displacement model of retention to anion-exchange chromatography of nucleic acids, *J. Chromatogr.*, 359 (1986) 147-155.

861 Drouen, A.C.J.H., Billiet, H.A.H. and De Galan, L., Practical procedure for the optimization of reversed-phase separations with quaternary mobile phase mixtures, *J. Chromatogr.*, 352 (1986) 127-139.

862 Drumheller, A.L., Bachelard, H., St-Pierre, S. and Jolicoeur, F.B., Determination of picomole levels of neurotensin, bombesin, and related peptide fragments using gradient elution high performance liquid chromatography coupled with electrochemical detection, *J. Liq. Chromatogr.*, 8 (1985) 1829-1843.

863 Drysdale, A.C., McEwan, J., Electricwala, A. and Sherwood, R., Investigation of a peptide fraction in the plasma of pain patients by high-performance liquid chromatography, *J. Chromatogr.*, 375 (1986) 376-379.

864 Dunaway, H.E., Hutchens, T.W. and Besch, P.K., Isolation of unliganded steroid receptor proteins by high-performance size-exclusion chromatography. An investigation of steroid-dependent structural alterations, *J. Chromatogr.*, 327 (1985) 221-235.

865 Eberendu, A.R.N., Venables, B.J. and Daugherty, K.E., Lipid separations. A review of chromatographic techniques for the analysis of glycolipids and phospholipids, *LC, Liq. Chromatogr. HPLC Mag.*, 3 (1985) 424-432; *C.A.*, 103 (1985) 100 985p.

866 Eggert, F.M. and Jones, M., Measurement of neutral sugars in glycoproteins as dansyl derivatives by automated high-performance liquid chromatography, *J. Chromatogr.*, 333 (1985) 123-131.

867 Ehrat, M., Cecchini, D.J. and Giese, R.W., Neonucleoproteins. Preparation and high-performance liquid chromatographic characterization of succinyl-lysozyme-diaminooctyl-polycytidylic acid and related polycytidylic acid conjugates, *J. Chromatogr.*, 326 (1985) 311-320.

868 Ellis, B. and Jolly, J., Purification of a uridine-specific acid nuclease from chicken liver by fast protein chromatography, *J. Chromatogr.*, 326 (1985) 157-161.

869 El Rassi, Z. and Horváth, Cs., High-performance displacement chromatography of nucleic acid fragments in a tandem enzyme reactor-liquid chromatograph system, *J. Chromatogr.*, 266 (1983) 319-340.

870 El Rassi, Z. and Horváth, Cs., High-performance liquid chromatography of tRNAs on novel stationary phases, *J. Chromatogr.*, 326 (1985) 79-90.

871 El Rassi, Z. and Horváth, Cs., Metal chelate-interaction chromatography of proteins with iminodiacetic acid-bonded stationary phases on silica support, *J. Chromatogr.*, 359 (1986) 241-253.

872 El Rassi, Z. and Horváth, Cs., Tandem columns and mixed-bed columns in high-performance liquid chromatography of proteins, *J. Chromatogr.*, 359 (1986) 255-264.

873 Elton, T.S. and Reeves, R., The effects of oxidation on the reverse-phase high-performance liquid chromatography characteristics of the high mobility groups 1 and 2 proteins, *Anal. Biochem.*, 149 (1985) 316-321.

874 Eriksson, S., Glad, G., Pernemalm, P.-Å. and Westman, E., Separation of DNA restriction fragments by ion-pair chromatography, *J. Chromatogr.*, 359 (1986) 265-274.

875 Fägerstam, L.G., Lizana, J., Axiö-Fredriksson, U.-B. and Wahlström, L., Fast chromatofocusing of human serum proteins with special reference to α_1-antitrypsin and Gc-globulin, *J. Chromatogr.*, 266 (1983) 523-532.

876 Farkaš, J., Ledvina, M., Brokeš, J., Ježek, J., Zajíček, J. and Zaoral, M., The synthesis of O-(2-acetamido-2-deoxy-β-D-glucopyranosyl)-(1→4)-N-acetylnormuramoyl-L-α-aminobutyryl-D-isoglutamine, *Carbohydr. Res.*, (1987) in press.

877 Fausnaugh, J.L. and Regnier, F.E., Solute and mobile phase contributions to retention in hydrophobic interaction chromatography of proteins, *J. Chromatogr.*, 359 (1986) 131-146.

878 Felix, A.M., Heimer, E.P., Lambros, T.J., Swistok, J., Tarnowski, S.J. and Wang, C.-T., Analysis of different forms of recombinant human leukocyte interferons and synthetic fragments by high-performance liquid chromatography, *J. Chromatogr.*, 327 (1985) 359-368.

879 Fell, A.F., Clark, B.J. and Scott, H.P., Analysis and characterization of aromatic amino acids, metabolites and peptides by rapid-scanning photo-diode array detection in high-performance liquid chromatography, *J. Chromatogr.*, 297 (1984) 203-214.

880 Findlay, J.B.C., Uses of high-performance liquid chromatographic methods in microsequencing [review with 9 refs.], *Biochem. Soc. Trans.*, 13 (1985) 1071-1073; *C.A.*, 103 (1985) 119 037c.

881 Fleming, L.H. and Reynolds, N.C., Jr., Separation and detection of closely related endorphins by liquid chromatography-electrochemistry, *J. Chromatogr.*, 375 (1986) 65-73.

882 Floyd, T.R., Crowther, J.B. and Hartwick, R.A., HPLC of nucleic acids: Trends in the separation of synthetic single-stranded oligonucleotides and double-stranded DNA fragments [a review with 101 references], *LC, Liq. Chromatogr. HPLC Mag.*, 3 (1985) 508-520; *C.A.*, 103 (1985) 215 653w.

883 Foriers, A., Lauwereys, M. and De Neve, R., Complete high-performance liquid chromatographic separation of phenylthiohydantoin- and 4-N,N-dimethylamino-azobenzene 4'-thiohydantoin-amino acids on an Ultrasphere ODS column with the same buffer system, *J. Chromatogr.*, 297 (1984) 75-82.

884 Frank, B.H., Beckage, M.J. and Willey, K.A., High-performance liquid chromatographic preparation of single-site carrier-free pancreatic polypeptide hormone radiotracers, *J. Chromatogr.*, 266 (1983) 239-248.
885 Frantíková, V., Borvák, J., Kluh, I. and Morávek, L., Amino acid sequence of the N-terminal region of human hemopexin, *FEBS Lett.*, 178 (1984) 213-216.
886 Funae, Y., Seo, R. and Imaoka, S., Two-step purification of cytochrome P-450 from rat liver macrosomes using high-performance liquid chromatography, *J. Chromatogr.*, 374 (1986) 271-278.
887 Gabriel, M.K. and McGuinness, T., Ion-pair reverse-phase high-performance liquid chromatography. Application to the study of chicken liver NAD⁺ kinase, *Anal. Biochem.*, 149 (1985) 339-343; *C.A.*, 103 (1985) 118 756z.
888 Gaertner, H. and Puigserver, A., Separation of some peptides and related isopeptides by high-performance liquid chromatography, *J. Chromatogr.*, 350 (1985) 279-284.
889 Gallagher, J.T., Morris, A. and Dexter, T.M., Identification of two binding sites for wheat-germ agglutinin on polylactosamine-type oligosaccharides, *Biochem. J.*, 231 (1985) 115-122.
890 Gallo, P., Olsson, O. and Sidén, Å., Methods for chromatofocusing of cerebrospinal fluid and serum immunoglobulin G, *J. Chromatogr.*, 327 (1985) 293-299.
891 Gallo, P., Olsson, O. and Sidén, Å., Small-column chromatofocusing of cerebrospinal fluid and serum immunoglobulin G, *J. Chromatogr.*, 375 (1986) 277-283.
892 Garcia, S. and Liautard, J.P., Separation of macromolecular RNAs by reversed-phase high-performance liquid chromatography, *J. Chromatogr.*, 296 (1984) 355-362.
893 Gazzotti, G., Sonnino, S. and Ghidoni, R., Normal-phase high-performance liquid chromatographic separation of non-derivatized ganglioside mixtures, *J. Chromatogr.*, 348 (1985) 371-378.
894 Gazzotti, G., Sonnino, S., Ghidoni, R., Kirschner, G. and Tettamanti, G., Analytical and preparative high-performance liquid chromatography of gangliosides, *J. Neurosci. Res.*, 12 (1984) 179-192.
895 Gazzotti, G., Sonnino, S., Ghidoni, R., Orlando, P. and Tettamanti, G., Preparation of the tritiated molecular forms of gangliosides with homogeneous long chain base composition, *Glycoconjugate J.*, 1 (1984) 111-121.
896 Geng, X. and Regnier, F.E., Retention model for proteins in reversed-phase liquid chromatography, *J. Chromatogr.*, 296 (1984) 15-30.
897 Glajch, J.L., Quarry, M.A., Vasta, J.F. and Snyder, L.R., Separation of peptide mixtures by reversed-phase gradient elution. Use of flow rate changes for controlling band spacing and improving resolution, *Anal. Chem.*, 58 (1986) 280-285.
898 Glöckner, G. and Van den Berg, J.H.M., Precipitation and adsorption phenomena in polymer chromatography, *J. Chromatogr.*, 352 (1986) 511-522.
899 Goheen, S.C. and Chow, T.M., Reversed-phase high-performance liquid chromatography of red cell membranes, *J. Chromatogr.*, 359 (1986) 297-305.
900 Goheen, S.C. and Matson, R.S., Purification of human serum gamma globulins by hydrophobic interaction high-performance liquid chromatography, *J. Chromatogr.*, 326 (1985) 235-241.
901 Gonwa, T.A., Westrick, M.A. and Macher, B.A., Inhibition of mitogen- and antigen-induced lymphocyte activation by human leukemia cell gangliosides, *Cancer Res.*, 44 (1984) 3467-3470.
902 Gooding, D.L., Schmuck, M.N., Nowlan, M.P. and Gooding, K.M., Optimization of preparative hydrophobic interaction chromatographic purification methods, *J. Chromatogr.*, 359 (1986) 331-337.
903 Gooding, K.M. and Schmuck, M.N., Purification of trypsin and other basic proteins by high-performance cation-exchange chromatography, *J. Chromatogr.*, 266 (1983) 633-642.
904 Gooding, K.M. and Schmuck, M.N., Ion selectivity in the high-performance cation-exchange chromatography of proteins, *J. Chromatogr.*, 296 (1984) 321-328.

905 Gooding, K.M. and Schmuck, M.N., Comparison of weak and strong high-performance anion-exchange chromatography, *J. Chromatogr.*, 327 (1985) 139-146.

906 Gooding, D.L., Schmuck, M.N. and Gooding, K.M., Analysis of proteins with new, mildly hydrophobic high-performance liquid chromatography packing materials, *J. Chromatogr.*, 296 (1984) 107-114.

907 Goworek, J., Nooitgedacht, F., Rijkhof, M. and Poppe, H., Determination of silanol groups on packing for high-performance liquid chromatography by means of isotopic exchange, *J. Chromatogr.*, 352 (1986) 399-406.

908 Graeve, L., Goemann, W., Földi, P. and Kruppa, J., Fractionation of biologically active messenger RNAs by gel filtration, *Biochem. Biophys. Res. Commun.*, 107 (1982) 1559-1565.

909 Grandier-Vazeille, X. and Tetaert, D., Methodology for purification of large hydrophobic peptides by high-performance liquid chromatography, *J. Chromatogr.*, 296 (1984) 301-308.

910 Green, E.D., Baenziger, J.U. and Boime, I., Cell-free sulfation of human and bovine pituitary hormones. Comparison of the sulfated oligosaccharides of lutropin, follitropin, and thyrotropin, *J. Biol. Chem.*, 260 (1985) 15631-15638.

911 Green, E.D., van Halbeek, H., Boime, I. and Baenziger, J.U., Structural elucidation of the disulfated oligosaccharide from bovine lutropin, *J. Biol. Chem.*, 260 (1985) 15623-15630.

912 Green, J.D., Detection of femtomole quantities of proteases by high-performance liquid chromatography, *Anal. Biochem.*, 152 (1986) 83-88.

913 Grego, B., Baldwin, G.S., Knessel, J.A., Simpson, R.J., Morgan, F.J. and Hearn, M.T.W., High-performance liquid chromatography of amino acids, peptides and proteins. LVIII. Application of reversed-phase high-performance liquid chromatography to the separation of tyrosine-specific phosphorylated polypeptides related to human growth hormone, *J. Chromatogr.*, 297 (1984) 21-29.

914 Grego, B. and Hearn, M.T.W., High-performance liquid chromatography of amino acids, peptides and proteins. LXIII. Reversed-phase high performance liquid chromatographic characterisation of several polypeptide and protein hormones, *J. Chromatogr.*, 336 (1984) 25-40.

915 Grego, B., Lambrou, F. and Hearn, M.T.W., High-performance liquid chromatography of amino acids, peptides and proteins. XLVIII. Retention behaviour of tryptic peptides of human growth hormone isolated by reversed-phase high-performance liquid chromatography: A comparative study using different chromatographic conditions and predicted elution behaviour based on retention coefficients, *J. Chromatogr.*, 266 (1983) 89-103.

916 Grego, B., Nice, E.C. and Simpson, R.J., Use of scanning diode array detector with reversed-phase microbore columns for the real-time spectral analysis of aromatic amino acids in peptides and proteins at the submicrogram level. Applications to peptide and protein microsequencing, *J. Chromatogr.*, 352 (1986) 359-368.

917 Grushka, E. and Cohen, A.S., Chromatographic behaviour of ternary nucleotide-metal cation-amino acid complexes, *J. Chromatogr.*, 353 (1986) 389-398.

918 Guerini, D. and Krebs, J., Separation of various calmodulins, calmodulin tryptic fragments, and different homologous Ca^{2+}-binding proteins by reversed-phase, hydrophobic interaction, and ion-exchange high-performance liquid chromatography technique, *Anal. Biochem.*, 150 (1985) 178-187.

919 Guiochon, G. and Martin, M., Theoretical investigation of the optimum particle size for the resolution of proteins by size-exclusion chromatography, *J. Chromatogr.*, 326 (1985) 3-32.

920 Gunzer, G. and Hennrich, N., Purification of α_1-proteinase inhibitor by triazine dye affinity chromatography, ion-exchange chromatography and gel filtration on fractogel TSK, *J. Chromatogr.*, 296 (1984) 221-229.

921 Guo, D., Mant, C.T., Taneja, A.K. and Hodges, R.S., Prediction of peptide retention times in reversed-phase high-performance liquid chromatography. II. Correlation of observed and predicted peptide retention times and factors influencing the retention times of peptides, *J. Chromatogr.*, 359 (1986) 519-532.

922 Guo, D., Mant, C.T., Taneja, A.K., Parker, J.M.R. and Hodges, R.S., Prediction of peptide retention times in reversed-phase high-performance liquid chromatography. I. Determination of retention coefficients of amino acid residues of model synthetic peptides, *J. Chromatogr.*, 359 (1986) 499-518.

923 Gurley, L.R., D'Anna, J.A., Blumenfeld, M., Valdez, J.G., Sebring, R.J., Donahue, P.R., Prentice, D.A. and Spall, W.D., Preparation of histone variants and high-mobility group proteins by reversed-phase high-performance liquid chromatography, *J. Chromatogr.*, 297 (1984) 147-165.

924 Gurley, L.R., Prentice, D.A., Valdez, J.G. and Spall, W.D., High-performance liquid chromatography of chromatin histones, *J. Chromatogr.*, 266 (1983) 609-627.

925 Gurley, L.R., Valdez, J.G., Prentice, D.A. and Spall, W.D., Histone fractionation by high-performance liquid chromatography, *Anal. Biochem.*, 129 (1983) 132-144.

926 Haff, L.A., Fägerstam, L.G. and Barry, A.R., Use of electrophoretic titration curves for predicting optimal chromatographic conditions for fast ion-exchange chromatography of proteins, *J. Chromatogr.*, 266 (1983) 409-425.

927 Hakam, A., McLick, J. and Kun, E., Separation of poly(ADP-ribose) by high-performance liquid chromatography, *J. Chromatogr.*, 296 (1984) 369-377.

928 Hakam, A., McLick, J. and Kun, E., Simultaneous determination of mono- and poly(ADP-ribose) *in vivo* by tritium labelling and direct high-performance liquid chromatographic separation, *J. Chromatogr.*, 359 (1986) 275-284.

929 Hakomori, S.-i., Nudelman, E., Kannagi, R. and Levery, S.B., The common structure in fucosyllactoaminolipids accumulating in human adenocarcinomas, and its possible absence in normal tissue, *Biochem. Biophys. Res. Commun.*, 109 (1982) 36-44.

930 Halls, T.D.J., Raju, M.S., Wenkert, E., Zuber, M., Lefrancier, P. and Lederer, E., The anomeric configuration of the immunostimulant N-acetyl-muramoyl-dipeptide and some of its derivatives, *Carbohydr. Res.*, 81 (1980) 173-176.

931 Hamacher, K. and Sahm, H., Characterization of enzymatic degradation products of carboxymethyl cellulose by gel chromatography, *Carbohydr. Polym.*, 5 (1985) 319-327; *C.A.*, 104 (1986) 7343p.

932 Hamšíková, E., Závadová, H., Zaoral, M., Ježek, J., Bláha, K. and Vonka, V., Immunogenicity of a synthetic peptide corresponding to a portion of the heavy chain of H3N2 influenza virus haemogglutinin, *J. Gen. Virol.*, (1987) submitted for publication.

933 Hansson, K.-A., Purification of NADP by high-performance anion-exchange chromatography, *J. Chromatogr.*, 266 (1983) 395-399.

934 Harada, H., Kamei, M., Yui, S. and Koyama, F., Lectin affinity high-performance liquid chromatography columns for the resolution of nucleotide sugars, *J. Chromatogr.*, 355 (1986) 291-295.

935 Hataka, K., Motoyoshi, K., Ishizaka, Y., Saito, M., Takaku, F. and Miura, Y., Purification of human colony-stimulating factor by high-performance liquid chromatography, *J. Chromatogr.*, 344 (1985) 339-344.

936 Hatano, H. (Guest Editor), proceedings of the *International Symposium on High-Performance Liquid Chromatography, Kyoto, Jan. 28-30, 1985; and Annual Meeting on High-Performance Liquid Chromatography of the Japanese Research Group on Automatic Liquid Chromatography, Kyoto, Jan. 31 - Feb. 1, 1985; J. Chromatogr.*, 332 (1985) 1-300.

937 Hearn, M.T.W. (Guest Editor), proceedings of the *International Symposium on High-Performance Liquid Chromatography in the Biological Sciences, Melbourne, Feb. 20-22, 1984; J. Chromatogr.*, 336 (1984) 1-466.

938 Hearn, M.T.W. and Aguilar, M.I., High-performance liquid chromatography of amino acids, peptides and proteins. LXVII. Evaluation of bandwidth relationships of peptides related to human β-endorphin, separated by gradient-elution reversed-phase high-performance liquid chromatography, *J. Chromatogr.*, 352 (1986) 35-66.

B622

939 Hearn, M.T.W. and Aguilar, M.I., High-performance liquid chromatography of amino acids, peptides and proteins. LXVIII. Evaluation of retention and bandwidth relationships of peptides related to luteinising hormone-releasing hormone and growth hormone-releasing factor, separated by gradient elution reversed-phase high-performance liquid chromatography, *J. Chromatogr.*, 359 (1986) 31-54.

940 Hearn, M.T.W. and Grego, B., High-performance liquid chromatography of amino acids, peptides and proteins. XLVI. Selectivity effects of peptidic positional isomers and oligomers separated by reversed-phase high-performance liquid chromatography, *J. Chromatogr.*, 266 (1983) 75-87.

941 Hearn, M.T.W. and Grego, B., High-performance liquid chromatography of amino acids, peptides and proteins. LV. Studies on the origin of band broadening of polypeptides and proteins separated by reversed-phase high-performance liquid chromatography, *J. Chromatogr.*, 296 (1984) 61-82.

942 Hearn, M.T.W. and Grego, B., High-performance liquid chromatography of amino acids, peptides and proteins. LVI. Detergent-mediated reversed-phase high-performance liquid chromatography of polypeptides and proteins, *J. Chromatogr.*, 296 (1984) 309-319.

943 Hearn, M.T.W., Hodder, A.N. and Aguilar, M.-I., High-performance liquid chromatography of amino acids, peptides and proteins. LXVI. Investigations on the effects of chromatographic dwell in the reversed-phase high-performance liquid chromatographic separation of proteins, *J. Chromatogr.*, 327 (1985) 47-66.

944 Hearn, M.T.W., Regnier, F.E. and Wehr, C.T. (Editors), *High Performance Liquid Chromatography of Proteins and Peptides (Proceedings of the First International Symposium, Washington, DC, 1981)*, Academic Press, New York, 1983, 288 pp.

945 Hearn, M.T.W., Regnier, F.E. and Wehr, C.T. (Guest Editors), proceedings of the *Second International Symposium on High-Performance Liquid Chromatography of Proteins, Peptides and Polynucleotides, Baltimore, MD, Dec. 6-8, 1982; J. Chromatogr.*, 266 (1983) 1-666.

946 Hecker, R., Colpan, M. and Riesner, D., High-performance liquid chromatography of DNA restriction fragments, *J. Chromatogr.*, 326 (1985) 251-261.

947 Helfpenny, A.P. and Brown, P.R., Simultaneous high-performance liquid chromatographic assay of the activities of erythrocytic hypoxanthine-guanine phosphoribosyl transferase and purine nucleoside phosphorylase, *J. Chromatogr.*, 349 (1985) 275-282.

948 Henschen, A., Hupe, K.-P., Lottspeich, F. and Voelter, W., *High-Performance Liquid Chromatography in Biochemistry*, Verlag Chemie, Weinheim, 1985, 638 pp.; *C.A.*, 104 (1986) B 48 306w.

949 Henson, G.W., The HPLC of immunoglobulins, *Immunol. Methods*, 3 (1985) 111-124; *C.A.*, 104 (1986) 18 363w.

950 Hermann, P., Jannasch, R. and Lebl, M., Application of high-performance liquid chromatography and capillary isotachophoresis to the purification and characterization of products of peptide synthesis, *J. Chromatogr.*, 351 (1986) 283-293.

951 Herring, S.W. and Enns, R.K., Rapid purification of leukocyte interferons by high-performance liquid chromatography, *J. Chromatogr.*, 266 (1983) 249-256.

952 Herring, S.W., Shitanishi, K.T., Moody, K.E. and Enns, R.K., Isolation of human factor VIII:C by preparative high-performance size-exclusion chromatography, *J. Chromatogr.*, 326 (1985) 217-224.

953 Heubner, A., Manz, B., Grill, H.-J. and Pollow, K., High-performance and ion-exchange chromatography and chromatofocusing of the human uterine progesterone receptor: Its application to the identification of 21-[3H]dehydro Org 2058-labelled receptor, *J. Chromatogr.*, 297 (1984) 301-311.

954 Heukeshoven, J. and Dernick, R., Characterization of a solvent system for separation of water-insoluble poliovirus proteins by reversed-phase high-performance liquid chromatography, *J. Chromatogr.*, 326 (1985) 91-101.

955 Hew, C.L., Joshi, S. and Wang, N.-C., Analysis of fish antifreeze poly-
    peptides by reversed-phase high-performance liquid chromatography, *J.
    Chromatogr.*, 296 (1984) 213-219.
956 Hirani, S., Lambris, J.D. and Müller-Eberhard, H.J., Structural analysis of
    the aspargine-linked oligosaccharides of human complement component C3,
    Biochem. J., 233 (1986) 613-616.
957 Hjertén, S. and Eriksson, K.-O., High-performance molecular sieve chromato-
    graphy of proteins on agarose columns: the relation between concentration
    and porosity of the gel, *Anal. Biochem.*, 137 (1984) 313-317.
958 Hjertén, S., Liu, Z.-Q. and Yang, D., Some studies on the resolving power
    of agarose-based high-performance liquid chromatographic media for the
    separation of macromolecules, *J. Chromatogr.*, 296 (1984) 115-120.
959 Hjertén, S., Yao, K., Eriksson, K.-O. and Johansson, B., Gradient and
    isocratic high-performance hydrophobic interaction chromatography of
    proteins on agarose columns, *J. Chromatogr.*, 359 (1986) 99-109.
960 Hjertén, S., Yao, K., Liu, Z.-Q., Yang, D. and Wu, B.-L., Simple method to
    prepare non-charged, amphiphilic agarose derivatives for instance for
    hydrophobic interaction chromatography, *J. Chromatogr.*, 354 (1986) 203-210.
961 Hjertén, S. and Zhu, M.-D., Micropreparative version of high-performance
    electrophoresis. The electrophoretic counterpart of narrow-bore high-
    performance liquid chromatography, *J. Chromatogr.*, 327 (1985) 157-164.
962 Holmes, E.H. and Hakomori, S.-i., Isolation and characterization of a new
    fucoganglioside accumulated in precancerous rat liver and in rat hepatome
    induced by N-2-acetylaminofluorene, *J. Biol. Chem.*, 257 (1982) 7698-7703.
963 Holmquist, L. and Carlson, L.A., Subfractionation and characterization of
    native and incubation enlarged human plasma high density lipoprotein
    particles by high performance gel filtration, *Lipids*, 20 (1985) 378-388;
    C.A., 103 (1985) 84 345x.
964 Holmskov-Nielsen, U., Erb, K. and Jensenius, J.Chr., Semi-automatic analysis
    of proteins and protein complexes by automated enzyme immuno assay after
    separation by high-performance gel-permeation chromatography. Size distribu-
    tion of C3-IgG complexes, *J. Chromatogr.*, 297 (1984) 225-233.
965 Homandberg, G.A., Evans, D.B., Kramer, J. and Erickson, J.W., Interaction
    between fluorescence-labeled fibronectin fragments studied by gel high-
    performance liquid chromatography, *J. Chromatogr.*, 327 (1985) 343-349.
966 Honda, K., Miyaguchi, K. and Imai, K., Evaluation of fluorescent compounds
    for peroxyoxalate chemiluminescence detection, *Anal. Chim. Acta*, 177 (1985)
    111-120.
967 Hounsell, E.F., Jones, N.J. and Stoll, M.S., The application of high-per-
    formance liquid chromatography to the purification of oligosaccharides
    containing neutral and acetamido sugars, *Biochem. Soc. Trans.*, 13 (1985)
    1061-1064; *C.A.*, 103 (1985) 156 641r.
968 Huber, J.F.K., Welte, S. and Reich, G., Prediction of partition coefficients
    in ternary liquid-liquid systems as a function of the phase system com-
    position by factor analysis for chromatography, *Anal. Chim. Acta*, 177
    (1985) 1-14.
969 Huebner, F.R. and Bietz, J.A., Detection of quality differences among
    wheats by high-performance liquid chromatography, *J. Chromatogr.*, 327 (1985)
    333-342.
970 Husseini, H.S. and Balzer, H.O., Use of TSK-SW columns for the high-perfor-
    mance liquid chromatographic analysis of proteins, isolated from sym-
    pathetic nerves and fractionated by Fractogel TSK-HW chromatography.
    Purification of L-DOPA decarboxylase, *J. Chromatogr.*, 297 (1984) 375-383.
971 Hutchens, T.W., Dunaway, H.E. and Besch, P.K., High-performance chromato-
    focusing of steroid receptor proteins in the presence and absence of
    steroid. Investigation of steroid-dependent alterations in surface charge
    heterogenicity, *J. Chromatogr.*, 327 (1985) 247-259.
972 Hutchens, T.W., Gibbons, W.E. and Besch, P.K., High-performance chromato-
    focusing and size-exclusion chromatography. Separation of human uterine
    estrogen-binding proteins, *J. Chromatogr.*, 297 (1984) 283-299.

973 Hutchens, T.W., Li, C.M. and Besch, P.K., Development of focusing buffer systems for generation of wide-range pH gradients during high-performance chromatofocusing, *J. Chromatogr.*, 359 (1986) 157-168.
974 Hutchens, T.W., Li, C.M. and Besch, P.K., Performance evaluation of a focusing buffer developed for chromatofocusing on high-performance anion-exchange columns, *J. Chromatogr.*, 359 (1986) 169-179.
975 Hutchens, T.W., Wiehle, R.D., Shahabi, N.A. and Wittliff, J.L., Rapid analysis of estrogen receptor heterogeneity by chromatofocusing with high-performance liquid chromatography, *J. Chromatogr.*, 266 (1983) 115-128.
976 Hyder, S.M., Baldi, A., Crespi, M. and Wittliff, J.L., Rapid purification of topoisomerase I from human breast cancer cells by high-performance liquid chromatography, *J. Chromatogr.*, 359 (1986) 433-447.
977 Hyder, S.M., Wiehle, R.D., Brandt, D.W. and Wittliff, J.L., High-performance hydrophobic-interaction chromatography of steroid hormone receptors, *J. Chromatogr.*, 327 (1985) 237-246.
978 Iadarola, P., Bonferoni, C., Ferri, G., Stoppini, M. and Zapponi, M.C., Reversed-phase high-performance liquid chromatographic separation and partial structural characterization of chloroplast glyceraldehyde-3-phosphate dehydrogenase subunits, *J. Chromatogr.*, 359 (1986) 423-432.
979 Ichimura, T., Amano, Y., Isobe, T. and Okuyama, T., (High-performance liquid chromatography of proteins and peptides on an octadecyl-bonded glass support.), *Bunseki Kagaku*, 34 (1985) 653-658; *C.A.*, 104 (1986) 48 102c.
980 Imaoka, S. and Funae, Y., Ion-exchange high-performance liquid chromatography of membrane-bound protein cytochrome P-450, *J. Chromatogr.*, 375 (1986) 83-90.
981 Ingraham, R.H., Lau, S.Y.M., Taneja, A.K. and Hodges, R.S., Denaturation and the effects of temperature on hydrophobic-interaction and reversed-phase high-performance liquid chromatography of proteins, Bio-Gel TSK-phenyl-5-PW column, *J. Chromatogr.*, 327 (1985) 77-92.
982 Inouye, K., Nakamura, K., Mitoma, Y., Matsumoto, M. and Igarashi, T., Application of new ion-exchanger TSK-GEL DEAE-5PW to the purification of Cu,Zn-superoxide dismutase of bovine erythrocytes, *J. Chromatogr.*, 327 (1985) 301-311.
983 Ishii, I., Takahashi, N., Kato, S., Akamatsu, N. and Kawazoe, Y., High-performance liquid chromatographic analysis of changes of asparagine-linked oligosaccharides in regenerating rat liver, *J. Chromatogr.*, 345 (1985) 134-139.
984 Ito, N., Noguchi, K., Kazama, M., Shimura, K. and Kasai, K.-I., Separation of human Glu-plasminogen and plasmin by high-performance affinity chromatography on Asahipak GS-Gel coupled with *p*-aminobenzamidine, *J. Chromatogr.*, 348 (1985) 199-204.
985 Ito, N., Naguchi, K., Shimura, K. and Kasai, K.-I., High-performance affinity chromatography of trypsins on Asahipak GS-Gel coupled with *p*-aminobenz-amidine, *J. Chromatogr.*, 333 (1985) 107-114.
986 Iwamamoto, T., Yoshiura, M., Iriyama, K., Tomizawa, N., Kurihara, S., Lee, T. and Suzuki, N., Identification of glutathione in reversed-phase liquid chromatogram of a lens, *Jikeikai Med. J.*, 32 (1985) 245-249; *C.A.*, 103 (1985) 210 170b.
987 Iwamori, M. and Nagai, Y., Isolation and characterization of a novel ganglioside, monosialosyl pentahexaosyl ceramide from human brain, *J. Biochem. (Tokyo)*, 84 (1978) 1601-1608.
988 Iwamori, M. and Nagai, Y., Monosialogangliosides of rabbit skeletal muscle. Characterization of N-acetylneuraminosyl lacto-N-noroctaosyl ceramide, *J. Biochem. (Tokyo)*, 89 (1981) 1253-1264.
989 Jackson, P.S. and Gurley, L.R., Analysis of nucleoproteins by direct injection of dissolved nuclei or chromosomes into a high-performance liquid chromatographic system, *J. Chromatogr.*, 326 (1985) 199-216.
990 James, S. and Bennett, M.P.J., Use of reversed-phase and ion-exchange batch extraction in the purification of bovine pituitary peptides, *J. Chromatogr.*, 326 (1985) 329-338.

991 Jandera, P., Method for characterization of selectivity in reversed-phase liquid chromatography. I. Derivation of the method and verification of the assumptions, *J. Chromatogr.*, 352 (1986) 91-110.

992 Jandera, P., Method for characterization of selectivity in reversed-phase liquid chromatography. II. Possibilities for the prediction of retention data, *J. Chromatogr.*, 352 (1986) 111-126.

993 Janson, J.C., Large-scale chromatography [a review with 9 refs.], *GBF Monogr. Ser.*, 7 (1984) 13-20; *C.A.*, 103 (1985) 210 029n.

994 Jehl, F., Gallion, C., Jaeger, A., Flesch, F. and Minck, R., Determination of α-amantin and β-amantin in human biological fluids by high-performance liquid chromatography, *Anal. Biochem.*, 149 (1985) 35-42.

995 Jeppsson, J.-O., Källman, I., Lindgren, G. and Fägerstam, L.G., Hb-Linköping (β36 Pro→Thr): A new hemoglobin mutant characterized by reversed-phase high-performance liquid chromatography, *J. Chromatogr.*, 297 (1984) 31-36.

996 Jeppsson, J.-O., Lilja, H. and Johansson, M., Isolation and characterization of two minor fractions of α_1-antitrypsin by high-performance liquid chromatographic chromatofocusing, *J. Chromatogr.*, 327 (1985) 173-177.

997 Ježek, J., Glykopeptidy bakteriálních stěn (Glycopeptides of bacterial cell walls) [a review in Czech, containing 234 citations], *Chem. Listy*, 80 (1986) 337-365.

998 Ježek, J., Straka, R., Zaoral, M., Ryba, M., Rotta, J. and Krchňák, V., Synthesis of peptide and glycopeptide fragments of bacterial cell walls, *Collect. Czech. Chem. Commun.*, (1987) in press.

999 Jinno, K., Retention prediction of phenylthiohydantoin amino acid derivatives in reversed-phase liquid chromatography, *Chromatographia*, 20 (1985) 743-746.

1000 Jonák, J., Petersen, E., Meloun, B. and Rychlik, I., Histidine residues in elongation factor EF-Tu from *Escherichia coli* protected by aminoacyl-tRNA against photo-oxidation, *Eur. J. Biochem.*, 144 (1984) 295-303.

1001 Jonák, J., Pokorná, K., Meloun, B. and Karas, K., Structural homology between elongation factors EF-Tu from *Bacillus stearothermopilus* and *Escherichia coli* in the binding site for aminoacyl-tRNA, *Eur. J. Biochem.*, 154 (1986) 355-362.

1002 Jones, B.N. and Gilligan, J.P., *o*-Phthaldialdehyde precolumn derivatization and reversed-phase high-performance liquid chromatography of polypeptide hydrolysates and physiological fluids, *J. Chromatogr.*, 266 (1983) 471-482.

1003 Jones, B.L. and Lookhart, G.L., High performance liquid chromatographic separation of peptides for sequencing studies, *Cereal Chem.*, 62 (1985) 89-96; *C.A.*, 103 (1985) 84 341t.

1004 Josić, Dj., Hofmann, W., Wieland, B., Nuck, R. and Reutter, W., Anion-exchange high-performance liquid chromatography of membrane proteins from liver and Morris hepatomas, *J. Chromatogr.*, 359 (1986) 315-322.

1005 Josić, D., Mattia, E., Ashwell, G. and van Renswoude, J., Quantitative determination of intracellular, ferritin-associated radioactive iron by high-performance liquid chromatography and immunoprecipitation, *Anal. Biochem.*, 152 (1986) 42-47.

1006 Josić, D., Schütt, W., van Renswoude, J. and Reutter, W., High-performance liquid chromatographic methods for antibodies, glycosidases and membrane proteins, *J. Chromatogr.*, 353 (1986) 13-18.

1007 Judd, R.C. and Caldwell, H.D., Identification and isolation of surface-exposed portions of the major outer membrane protein of *Chlamydia trachomatis* by two-dimensional peptide mapping and high-performance liquid chromatography, *J. Liq. Chromatogr.*, 8 (1985) 1559-1571.

1008 Kai, M., Miura, T., Ishida, J. and Ohkura, Y., High-performance liquid chromatographic method for monitoring leupeptin in mouse serum and muscle by precolumn fluorescence derivatization with benzoin, *J. Chromatogr.*, 345 (1985) 259-265.

1009 Kalhorn, T.F., Thummel, K.E., Nelson, S.D. and Slattery, J.T., Analysis of oxidized and reduced pyridine dinucleotides in rat liver by high-performance liquid chromatography, *Anal. Biochem.*, 151 (1985) 343-347.

1010 Kaliszan, R., Osmialowski, K., Tomellini, S.A., Hsu, S.-H., Fazio, S.D. and
Hartwick, R.A., Quantitative retention relationships as a function of
mobile and C$_{18}$ stationary phase composition for non-cogeneric solutes,
J. Chromatogr., 352 (1986) 141-155.
1011 Kanedo, N., Noro, Y. and Nagatsu, T., Highly sensitive assay for
acetylcholinesterase activity by high-performance liquid chromatography
with electrochemical detection, *J. Chromatogr.*, 344 (1985) 93-100.
1012 Kaplan, P. and Abreu, S.L., A high-yield chromatographic method for the
purification of rat fibroblast interferon to a high specific activity,
J. Interferon Res., 5 (1985) 415-422; *C.A.*, 104 (1986) 4381p.
1013 Kårsnäs, P., Moreno-Lopez, J. and Kristiansen, T., Bovine viral Diarrhea
virus. Purification of surface proteins in detergent-containing buffers
by fast protein liquid chromatography, *J. Chromatogr.*, 266 (1983) 643-
649.
1014 Kato, Y., Kitamura, T. and Hashimoto, T., High-performance hydrophobic
interaction chromatography of proteins, *J. Chromatogr.*, 266 (1983) 49-54.
1015 Kato, Y., Kitamura, T. and Hashimoto, T., Resin-based support for reversed-
phase chromatography of proteins, *J. Chromatogr.*, 333 (1985) 93-106.
1016 Kato, Y., Kitamura, T. and Hashimoto, T., Preparative high-performance
hydrophobic interaction chromatography of proteins on TSK Gel phenyl-5PW,
J. Chromatogr., 333 (1985) 202-210.
1017 Kato, Y., Matsuda, T. and Hashimoto, T., New gel permeation column for the
separation of water-soluble polymers, *J. Chromatogr.*, 332 (1985) 39-46.
1018 Kato, Y., Nakamura, K. and Hashimoto, T., New ion exchanger for the separation of proteins and nucleic acids, *J. Chromatogr.*, 266 (1983) 385-394.
1019 Kato, Y., Nakamura, K. and Hashimoto, T., High-performance metal chelate
affinity chromatography of proteins, *J. Chromatogr.*, 354 (1986) 511-517.
1020 Kato, S. and Sada, E., Purification of fermentation products by affinity
chromatography [a review with 9 refs.], *Kakko Kogaku Kaishi*, 63 (1985)
473-475 (in Japanese); *C.A.*, 103 (1985) 213 239s.
1021 Kato, Y., Sasaki, M., Hashimoto, T., Murotsu, T., Fukushige, S. and
Matsubara, K., Operational variables in high-performance gel filtration
of DNA fragments and RNAs, *J. Chromatogr.*, 266 (1983) 341-349.
1022 Katz, E.D., Ogan, K. and Scott, R.P.W., Distribution of a solute between
two phases. The basic theory and its application to the prediction of
chromatographic retention, *J. Chromatogr.*, 352 (1986) 67-90.
1023 Kawaguchi, S., Marui, Y., Arisue, K., Koda, K., Hayashi, C. and Miyai, K.,
Automatic analysis of serum lactate dehydrogenase isoenzymes by high-
performance ion-exchange chromatography, *J. Chromatogr.*, 374 (1986) 45-50.
1024 Kawasaki, T., Takahashi, S. and Ikeda, K., Hydroxyapatite high-performance
liquid chromatography: Column performance for proteins, *Eur. J. Biochem.*,
152 (1985) 361-371.
1025 Kedersha, N.L., Tkacz, J.S. and Berg, R.A., Characterization of the oligosaccharides of prolyl hydroxylase, a microsomal glycoprotein, *Biochemistry*,
24 (1985) 5952-5960.
1026 Kennedy, L.A., Kopaciewicz, W. and Regnier, F.E., Multimodal liquid chromatography columns for the separation of proteins in either the anion-
exchange or hydrophobic-interaction mode, *J. Chromatogr.*, 359 (1986) 73-84.
1027 Kerlavage, A.R., Weitzmann, C.J., Hasan, T. and Cooperman, B.S., Reversed-
phase high-performance liquid chromatography of the separation of a
complex protein mixture, *J. Chromatogr.*, 266 (1983) 225-237.
1028 Kinkel, J.N., Anspach, B., Unger, K.K., Wieser, R. and Brunner, G.,
Separation of plasma membrane proteins of cultured human fibroblasts by
affinity chromatography on bonded microparticulate silicas, *J. Chromatogr.*,
297 (1984) 167-177.
1029 Kirkland, J.J. and Yau, W.W., Thermal field-flow fractionation of water-
soluble macromolecules, *J. Chromatogr.*, 353 (1986) 95-107.

1030 Knighton, D.R., Harding, D.R., Napier, J.R. and Hancock, W.S., Purification of synthetic lipid associating peptides and the monitoring of the deformylation of N^{in}-formyltryptophan by reversed-phase high-performance liquid chromatography, *J. Chromatogr.*, 347 (1985) 237-248.

1031 Knox, J.H. (Guest Editor), proceedings of the *9th International Symposium on Column Chromatography, Edinburgh, July 1-5, 1985*, Part I: *J. Chromatogr.*, 352 (1986) 1-525; Part II: *J. Chromatogr.*, 353 (1986) 1-468.

1032 Knox, J.H., Kaur, B. and Millward, G.R., Structure and performance of porous graphitic carbon in liquid chromatography, *J. Chromatogr.*, 352 (1986) 3-25.

1033 Köck, A. and Luger, T.A., Purification of human interleukin 1 by high-performance liquid chromatography, *J. Chromatogr.*, 296 (1984) 293-300.

1034 Köck, A. and Luger, T.A., High-performance liquid chromatographic separation of distinct epidermal cell-derived cytokines, *J. Chromatogr.*, 326 (1985) 129-136.

1035 Kodama, C., Ototani, N., Isemura, M., Aikawa, J. and Yosizawa, Z., Liquid-chromatographic determination of urinary glycosaminoglycane for differential diagnosis of genetic mucopolysaccharidoses, *Clin. Chem. (Winston-Salem, N.C.)*, 32 (1986) 30-34.

1036 Köhler, J., Chase, D.B., Farlee, R.D., Vega, A.J. and Kirkland, J.J., Comprehensive characterization of some silica-based stationary phases for high-performance liquid chromatography, *J. Chromatogr.*, 352 (1986) 275-305.

1037 Kolbe, H.V.J., Lu, R.C. and Wohlrab, H., Reversed-phase high-performance liquid chromatographic separation and quantitation of phenylthiohydantoin derivatives of 25 amino acids, including those of cysteic acid, 4-hydroxy-proline, methionine sulfone, S-carboxymethylcysteine and S-methylcysteine, *J. Chromatogr.*, 327 (1985) 1-7.

1038 Kopaciewicz, W., Rounds, M.A., Fausnaugh, J. and Regnier, F.E., Retention model for high-performance ion-exchange chromatography, *J. Chromatogr.*, 266 (1983) 3-21.

1039 Kostenko, V.G., Methods of identifying organic compounds using high-performance liquid chromatography with ultraviolet detection, *J. Chromatogr.*, 355 (1986) 296-301.

1040 Kratzin, H.D., Kruse, T., Maywald, F., Thinnes, F.P., Götz, H., Egert, G., Pauly, E., Friedrich, J., Yang, C.-Y., Wernet, P. and Hilschmann, N., Primary structure of human class II histocompability antigen. Reversed-phase high-performance liquid chromatography for integral membrane proteins, *J. Chromatogr.*, 297 (1984) 1-11.

1041 Krchňák, V., Ježek, J. and Zaoral, M., Preparation of N-acetylmuramyl-α-aminobutyryl-D-isoglutaminyl-lysyl-lysyl-lysine amide, *Collect. Czech. Chem. Commun.*, 48 (1983) 2079-2081.

1042 Krstulović, A.M., Friedman, M.J., Colin, H., Guiochon, G., Gaspar, M. and Pajer, K.A., Analytical methodology for assays of serum tryptophan metabolites in control subjects and newly abstinent alcoholics: Preliminary investigation by liquid chromatography with amperometric detection, *J. Chromatogr.*, 297 (1984) 271-281.

1043 Krysl, S. and Smolkova-Keulemannova, E., Cyklodextriny a jejich použití v chromatografických metodách (Cyclodextrins and their use in chromatographic methods) [a review with 119 refs.], *Chem. Listy*, 79 (1985) 919-942 (in Czech); *C.A.*, 104 (1986) 28 154t.

1044 Kuffer, A.D. and Herington, A.C., Partial purification of a specific inhibitor of the insulinlike growth factors by reversed-phase high-performance liquid chromatography, *J. Chromatogr.*, 336 (1984) 87-92.

1045 Kumagaye, K.Y., Takai, M., Chino, N., Kimura, T. and Sakakibara, S., Comparison of reversed-phase and cation-exchange high-performance liquid chromatography for separating closely related peptides. Separation of Asp^{76}-human parathyroid hormone (1-84) from Asn^{76}-human parathyroid hormone (1-84), *J. Chromatogr.*, 327 (1985) 327-332.

1046 Kundu, S.K., DEAE-Silica gel and DEAE-controlled porous glass as ion exchangers for the isolation of glycolipids, *Methods Enzymol.*, 72 (Part D, Lipids) (1981) 174-204.

1047 Kundu, S.R. and Scott, D.D., Rapid separation of gangliosides by high-performance liquid chromatography, *J. Chromatogr.*, 232 (1982) 19-27.

1048 Kundu, S.R. and Suzuki, A., Simple micro-method for the isolation of gangliosides by reversed-phase chromatography, *J. Chromatogr.*, 224 (1981) 249-256.

1049 Kunitani, M., Hirtzer, P., Johnson, D., Halenbeck, R., Boosman, A. and Koths, K., Reversed-phase chromatography of interleukin-2 muteins, *J. Chromatogr.*, 359 (1986) 391-402.

1050 Kuwata, K., Era, S., Inouye, H., Sogami, M. and Sasaki, H., Ion-exchange high-performance liquid chromatographic studies on sulphydryl-catalysed structural alterations of bovine mercaptalbumin, *J. Chromatogr.*, 332 (1985) 29-37.

1051 Ladron de Guevara, O., Estrada, G., Antonio, S., Alvarado, X., Guereca, L., Zamudio, F. and Bolivar, F., Identification and isolation of human insulin A and B chains by high-performance liquid chromatography, *J. Chromatogr.*, 349 (1985) 91-98.

1052 Lahm, H.-W. and Stein, S., Characterization of recombinant human inter-leukin-2 with micromethods, *J. Chromatogr.*, 326 (1985) 357-361.

1053 Mabert, D.T. and Lerner, A.B., Optimization of a melantropin-receptor binding assay by reversed-phase high-performance liquid chromatography, *J. Chromatogr.*, 266 (1983) 567-576.

1054 Lambotte, P., Van Snick, J. and Boon, T., Partial purification of a membrane glycoprotein antigen by high-pressure size-exclusion chromatography without loss of antigenicity, *J. Chromatogr.*, 297 (1984) 139-145.

1055 Lange, R.D., Andrews, R.B., Trent, D.J., Reyniers, J.P., Draganac, P.S. and Farkas, W.R., Preparation of purified erythropoietin by high performance liquid chromatography, *Blood Cells*, 10 (1984) 305-314; *C.A.*, 103 (1985) 206 714q.

1056 Lazure, C., Rochemont, J., Seidah, N.G. and Chrétien, M., Novel approach to rapid and sensitive localization of protein disulfide bridges by high-performance liquid chromatography and electrochemical detection, *J. Chromatogr.*, 326 (1985) 339-348.

1057 Lebl, M. and Gut, V., Calculation of the rate constant of a reversible reaction from the chromatographic peaks for muramyldipeptide anomers, *J. Chromatogr.*, 260 (1983) 478-482.

1058 Lederer, E., Muramylpeptides and their use in vaccines [opening lecture], in Zaoral, M., Havlas, Z., Mikeš, O. and Procházka, Ž. (Editors), *Synthetic Immunomodulators and Vaccines*, Czechoslovak Academy of Sciences, Prague, 1986, pp. 3-39.

1059 Lefrancier, P. and Lederer, E., Chemistry of synthetic immunomodulant muramyl peptides, *Fortschr. Chem. Org. Naturst. (Prog. Chem. Organ. Natur. Products)*, 40 (1981) 1-47.

1060 Lenzi, S., Sampietro, T., Giampietro, O., Miccoli, R., Masoni, A. and Navalesi, R., An optimized method to measure glycosyl-proteins: HPLC assay of 5-hydroxymethyl-2-furfuraldehyde (5-HMF), *Protides Biol. Fluids*, 33 (1985) 567-570; *C.A.*, 104 (1986) 2962e.

1061 Lindberg, W., Ohman, J., Wold, S. and Martens, H., Simultaneous determina-tion of five different food proteins by high-performance liquid chromato-graphy and partial least-squares multivariate calibration, *Anal. Chim. Acta*, 174 (1985) 41-51.

1062 Lindblom, H., Rapid chromatographic method for the isolation of glucose-6-phosphate dehydrogenase from yeast enzyme concentrate, *J. Chromatogr.*, 266 (1983) 265-271.

1063 Lindblom, H., Söderberg, L., Cooper, E.H. and Turner, R., Urinary protein isolation by high-performance ion-exchange chromatography, *J. Chromatogr.*, 266 (1983) 187-196.

1064 Lindgren, G., Lundström, B., Källman, I. and Hansson, K.-A., Physical characteristics and properties of new chromatographic packing materials for the separation of peptides and proteins, *J. Chromatogr.*, 296 (1984) 83-95.

1065 L'Italien, J.J., Microscale elucidation of an N-linked glycosylation site by comparative high-performance liquid chromatographic peptide mapping, *J. Chromatogr.*, 359 (1986) 213-220.

1066 Lonardi, S., Ottofaro, F. and Mosconi, L., (Automation development in HPLC. Apparatus for automatic shifting of the mobile phase), *Boll. Chim. Farm.*, 124 (1985) 109-114; *C.A.*, 103 (1985) 226 709v.

1067 Lonsdorfer, M., Clements, N.C., Jr. and Wittliff, J.L., Use of high-performance liquid chromatography in the evaluation of the synthesis and binding of fluorescein-linked steroids to estrogen receptors, *J. Chromatogr.*, 266 (1983) 129-139.

1068 Lork, K.D., Unger, K.K. and Kinkel, J.N., Role of the functional group in n-octylmethylsilanes in the synthesis of C_8 reversed-phase silica packings for high-performance liquid chromatography, *J. Chromatogr.*, 352 (1986) 199-211.

1069 Lottspeich, F., Microscale isocratic separation of phenylthiohydantoin amino acid derivatives, *J. Chromatogr.*, 326 (1985) 321-327.

1070 Lottspeich, F., High-performance liquid chromatography in microsequencing, *Trends Anal. Chem.*, 4 (1985) 244-246.

1071 Low, G.K.C., Haddad, P.R. and Duffield, A.M., Proposed model for peak splitting in reversed-phase ion-pair high-performance liquid chromatography with computer prediction of eluted peak profiles, *J. Chromatogr.*, 336 (1984) 15-24.

1072 Low, T.L.K., McClure, J.E., Naylor, P.H., Spangelo, B.L. and Goldstein, A.L., Isolation of thymosin α_1 from thymosin fraction 5 of different species by high-performance liquid chromatography, *J. Chromatogr.*, 266 (1983) 533-544.

1073 Lozier, J., Takahashi, N. and Putnam, F.W., Purification of cyanogen bromide fragments from β-2-glycoprotein I by high-performance liquid chromatography, *J. Chromatogr.*, 266 (1983) 545-554.

1074 Lu, X.M., Benedek, K. and Karger, B.L., Conformational effect in the high-performance liquid chromatography of proteins. Further studies of the reversed-phase chromatographic behavior of ribonuclease A, *J. Chromatogr.*, 359 (1986) 19-29.

1075 Lundahl, P., Greijer, E., Lindblom, H. and Fägerstam, L.G., Fractionation of human red cell membrane proteins by ion-exchange chromatography in detergent on Mono Q, with special reference to the glucose transporter, *J. Chromatogr.*, 297 (1984) 129-137.

1076 Magnani, J.L., Nilsson, B., Brockhaus, M., Zopf, D., Steplewski, Z., Koprowski, H. and Ginsburg, V., Monoclonal antibody-defined antigen associated with gastrointestinal cancer in a ganglioside containing sialylated lacto-N-fucopentaose II, *J. Biol. Chem.*, 257 (1982) 14365-14369.

1077 Mahieu, J.-P., Sebille, B., Craescu, C.T., Rhoda, M.-D. and Beuzard, Y., Determination of the dissociation constant of oligomeric proteins by size-exclusion high-performance liquid chromatography. Application to human haemoglobin, *J. Chromatogr.*, 327 (1985) 313-325.

1078 Mahieu, J.P., Sebille, B., Gosselet, M., Garel, M.C. and Beuzard, Y., Study of protein reactivity with thiol reagents by anion-exchange high-performance liquid chromatography, *J. Chromatogr.*, 359 (1986) 461-474.

1079 Majors, R.E., Practical aspects of preparative liquid chromatography, *LC, Liq. Chromatogr. HPLC Mag.*, 3 (1985) 862-866; *C.A.*, 103 (1985) 217 405r.

1080 Makino, K., Wada, H., Ozaki, H., Takeuchi, T., Hatano, H., Fukui, T., Noguchi, K. and Yanagihara, Y., Separation of single-stranded oligonucleotides on polyvinyl alcohol gel column, *Chromatographia*, 20 (1985) 713-716.

1081 Manalan, A.S., Newton, D.L. and Klee, C.B., Purification and peptide
     mapping of calmodulin and its chemically modified derivatives by reversed-
     phase high-performance liquid chromatography, *J. Chromatogr.*, 326 (1985)
     387-397.
1082 Månsson, J.-E., Rosengren, B. and Svennerholm, L., Separation of gan-
     gliosides by anion-exchange chromatography on Mono Q, *J. Chromatogr.*,
     322 (1985) 465-472.
1083 Mant, C.T. and Hodges, R.S., General method for the separation of cyanogen
     bromide digest of proteins by high-performance liquid chromatography.
     Rabbit skeletal troponin I, *J. Chromatogr.*, 326 (1985) 349-356.
1084 Mant, C.T. and Hodges, R.S., Separation of peptides by strong cation-
     exchange high-performance liquid chromatography, *J. Chromatogr.*, 327
     (1985) 147-155.
1085 Marceau, F., Drumheller, A., Gendreau, M., Lussier, A. and St.-Pierre, S.,
     Rapid assay of human plasma carboxypeptidase N by high-performance liquid
     chromatographic separation of hippuryl-lysine and its products, *J.
     Chromatogr.*, 266 (1983) 173-177.
1086 Marriage, H.J. and Harvey, D.J., Resolution of mouse hepatic cytochrome
     P-450 isozymes by chromatofocusing, *J. Chromatogr.*, 354 (1986) 383-392.
1087 Marshall, R.J., Turner, R., Yu, H. and Cooper, E.H., Cluster analysis of
     chromatographic profiles of urine proteins, *J. Chromatogr.*, 297 (1984)
     235-244.
1088 Matson, R.S. and Goheen, S.C., Use of high-performance size exclusion
     chromatography to determine the extent of detergent solubilization of
     human erythrocyte ghosts, *J. Chromatogr.*, 359 (1986) 285-295.
1089 Maurich, V., Pitotti, A., Vio, L. and Mamolo, M.G., Ion-pair liquid chro-
     matographic assay of angiotensin-converting enzyme activity, *J. Pharm.
     Biomed. Anal.*, 3 (1985) 425-432.
1090 May, H.E., Cotter, L. and Lee, C., HPLC Separation of geometric isomers
     of carbamyl peptides, *J. Liq. Chromatogr.*, 8 (1985) 1397-1412; *C.A.*, 103
     (1985) 156 632p.
1091 McCown, S.M., Southern, D. and Morrison, B.E., Solvent properties and their
     effects on gradient elution high-performance liquid chromatography. III.
     Experimental findings for water and acetonitrile, *J. Chromatogr.*, 352
     (1986) 493-509.
1092 McCown, S.M., Southern, D., Morrison, B.E. and Garteiz, D., Solvent
     properties and their effects on gradient elution high-performance liquid
     chromatography. I. Bulk and molecular properties of water and acetonitrile,
     J. Chromatogr., 352 (1986) 465-482.
1093 McCown, S.M., Southern, D., Morrison, B.E. and Garteiz, D., Solvent
     properties and their effects on gradient elution high-performance liquid
     chromatography. II. Temperature gradient in high-performance liquid
     chromatographic columns, *J. Chromatogr.*, 352 (1986) 483-492.
1094 McDermott, J.R. and Kidd, A.M., Ion-exchange gel-filtration and reversed-
     phase high-performance liquid chromatography in the isolation of
     neurotensin-degrading enzymes from rat brain, *J. Chromatogr.*, 296 (1984)
     231-239.
1095 McGregor, J.L., Clezardin, P., Manach, M., Gronlund, S. and Dechavanne, M.,
     Tandem separation of labelled human blood platelet membrane glycoproteins
     by anion-exchange and gel fast protein liquid chromatography, *J.
     Chromatogr.*, 326 (1985) 179-190.
1096 McGuffin, V.L., and Zare, R.N., Laser fluorescence detection in microcolumn
     liquid chromatography: Application to derivatized carboxylic acids, *Appl.
     Spectrosc.*, 39 (1985) 847-853; *C.A.*, 104 (1986) 45 124a.
1097 Meek, J.L., Derivatizing reagents for high-performance liquid chromatography
     detection of peptides at the picomole level, *J. Chromatogr.*, 266 (1983)
     401-408.
1098 Mehrens, H.A. and Reimerdes, E.H., (Separation of proteins and peptides by
     HPLC), *GIT-Suppl.*, (1985) 65-68; *C.A.*, 103 (1985) 210 161z.

1099 Meloun, B., Baudyš, M., Kostka, V., Hausdorf, G., Frömmel, C. and Höhne,
W.E., Complete primary structure of thermitase from *Thermoactinomyces
vulgaris* and its structural features related to the subtilisin-type
proteinases, *FEBS Lett.*, 183 (1985) 195-200.

1100 Meloun, B., Baudyš, M., Pavlik, M., Kostka, V., Hausdorf, G. and Höhne,
W.E., Thermitase from *Thermoactinomyces vulgaris*: Amino acid sequence of
the large N-terminal cyanogen bromide peptide, *Collect. Czech. Chem.
Commun.*, 50 (1985) 885-896.

1101 Meloun, B., Jonáková, V. and Henschen, A., Acidic acrosin inhibitors from
bull seminal plasma, structural differences, *Biol. Chem. Hoppe-Seyler*,
366 (1985) 1155-1160.

1102 Meloun, B., Pohl, J., Baudyš, M. and Kostka, V., Homology of bovine spleen
cathepsin B with other SH-proteinases, *Biol. Chem. Hoppe-Seyler*, 367
(Supplement, Abstracts, 17th FEBS Meeting, 1986, Berlin) (1986) 365,
Abstr. FRI, 06.03.50.

1103 Ménez, J.F., Berthou, F., Meskar, A., Picart, D., Le Bras, R. and Bardou,
L.G., Glycosylated haemoglobin. High-performance liquid chromatographic
determination of 5-(hydroxymethyl)-2-furfuraldehyde after haemoglobin
hydrolysis, *J. Chromatogr.*, 297 (1984) 339-350.

1104 Miedziak, I., Jozefaciuk, G. and Waksmundzki, A., (Preparation of carriers
for affinity chromatography on the basis of silica gel I. Modification
of silica gel surface by carbonization-silanization.), *Chem. Anal. (Warsaw)*,
30 (1985) 275-279 (in Polish); *C.A.*, 103 (1985) 226 728a.

1105 Mikeš, O., Hostomská, Z., Štrop, P., Smrž, M., Slováková, S., Vrátný, P.,
Rexová, L'., Kolář, J. and Čoupek, J., Ion exchange derivatives of
Spheron. VI. Strongly basic derivatives, *J. Chromatogr.*, 440 (1988) 287-
304.

1106 Mikeš, O. and Rexová, L'., High performance liquid chromatography (HPLC)
of pectic enzymes, *Methods Enzymol.*, 161B (1988) 385-399.

1107 Mikeš, O., Štrop, P., Hostomská, Z., Smrž, M., Slováková, S. and Čoupek, J.,
Ion-exchange derivatives of Spheron. V. Sulphate and sulpho derivatives,
J. Chromatogr., 301 (1984) 93-105.

1108 Miller, N.T. and Karger, B.L., High-performance hydrophobic-interaction
chromatography on ether-bonded phases. Chromatographic characteristics and
gradient optimization, *J. Chromatogr.*, 326 (1985) 45-61.

1109 Mimura, T. and Romano, J.C., Muramic acid measurements for bacterial
investigations in marine environments by high-pressure liquid chromatography, *Appl. Environ. Microbiol.*, 50 (1985) 229-237; *C.A.*, 104 (1986)
10 243m.

1110 Minganti, C., Ganesh, K.N., Sproat, B.S. and Gait, M.J., Comparison of
controlled pore glass and kieselguhr-polyphase synthesis of 23-residue
oligodeoxyribonucleotides in milligram amounts, *Anal. Biochem.*, 147
(1985) 63-74.

1111 Mino, Y., Tsutsui, S. and Ota, N., Separation of acetylated inulin by
reversed-phase high performance liquid chromatography, *Chem. Pharm. Bull.*,
33 (1985) 3503-3506.

1112 Miyaguchi, K., Honda, K., Toyo'oka, T. and Imai, K., Application of a
microbore high-performance liquid chromatography-chemiluminescence detection system to the N-terminal amino acid analysis of bradykinin, *J.
Chromatogr.*, 352 (1986) 255-260.

1113 Morávek, L., Kühnemund, O., Havlíček, J., Kopecký, P. and Pavlik, M., Type
1M protein of *Staphylococcus pyogenes* N-terminal sequence and peptic
fragments, *FEBS Lett.*, 208 (1986) 435-438.

1114 Mortensen, S.B. and Kilian, M., A rapid method for the detection and
quantitation of IgA protease activity by macrobore gel-permeation chromatography, *J. Chromatogr.*, 296 (1984) 257-262.

1115 Mosca, A. and Carpinelli, A., Automated determination of glycated
hemoglobins with a new high-performance liquid chromatography analyzer,
Clin. Chem. (Winston-Salem, N.C.), 32 (1986) 202-203.

1116 Muccio, D.D. and DeLucas, L.J., Isolation of detergent-solubilized monomers of bacteriorhodopsin by size-exclusion high-performance liquid chromatography, *J. Chromatogr.*, 326 (1985) 243-250.

1117 Muller, D., Ndoume-Nze, M. and Jozefonvicz, J., High-pressure size-exclusion chromatography of anticoagulant materials, *J. Chromatogr.*, 297 (1984) 351-358.

1118 Muller, D., Yu, X.J., Fischer, A.M., Bros, A. and Jozefonvicz, J., High-performance affinity chromatography of human thrombin on modified polystyrene resins, *J. Chromatogr.*, 359 (1986) 351-357.

1119 Müller, W., Fractionation of DNA restriction fragments with ion-exchangers for high-performance liquid chromatography, *Eur. J. Biochem.*, 155 (1986) 203-212.

1120 Müller, W., New phase supports for liquid-liquid partition chromatography of biopolymers in aqueous poly(ethyleneglycol)-dextran systems. Synthesis and application for the fractionation of DNA restriction fragments, *Eur. J. Biochem.*, 155 (1986) 213-222.

1121 Murata, K. and Yokoyama, Y., Enzymatic analysis with chondrosulfatases of constituent disaccharides of sulfated chondroitin sulfate and dermatan sulfate isomers by high-performance liquid chromatography, *Anal. Biochem.*, 149 (1985) 261-268.

1122 Murata, K. and Yokoyama, Y., Analysis of hyaluronic acid and chondroitin by high-performance liquid chromatography of the constituent disaccharide units, *J. Chromatogr.*, 374 (1986) 37-44.

1123 Murphy, R., Furness, J.B. and Costa, M., Measurement and chromatographic characterization of vasoactive intestinal peptide from guinea-pig enteric nerves, *J. Chromatogr.*, 336 (1984) 41-50.

1124 Muto, N. and Tan, L., Purification of oestrogen synthetase by high-performance liquid chromatography. Two membrane-bound enzymes from the human placenta, *J. Chromatogr.*, 326 (1985) 137-146.

1125 Nakabayashi, H., Iwamori, M. and Nagai, Y., Analysis and quantitation of gangliosides as *p*-bromophenacyl derivatives by high-performance liquid chromatography, *J. Biochem.*, 96 (1984) 977-984.

1126 Nakagawa, T., Shibukawa, A., Kaihara, A., Itamochi, T. and Tanaka, H., Liquid chromatography with crown ether-containing mobile phases. VI. Molecular recognition of amino acids and peptides, *J. Chromatogr.*, 353 (1986) 399-408.

1127 Nakamura, K., Nagashima, M., Sekine, M., Igarashi, M., Ariga, T., Atsumi, T., Miyatake, T., Suzuki, A. and Yamakawa, T., Gangliosides of hog skeletal muscle, *Biochim. Biophys. Acta*, 752 (1983) 291-300.

1128 Nice, E.C., Grego, B. and Simpson, R.J., Application of short microbore HPLC "guard" columns for the preparation of samples for protein microsequencing, *Biochem. Int.*, 11 (1985) 187-195; *C.A.*, 103 (1985) 137 916t.

1129 Nice, E.C., Lloyd, C.J. and Burgess, A.W., The role of short microbore high-performance liquid chromatography columns for protein separation and trace enrichment, *J. Chromatogr.*, 296 (1984) 153-170.

1130 Nika, H., Reaction detector system for the simultaneous monitoring of primary amino groups and sulfhydryl groups in peptides eluted by high-performance liquid chromatography, *J. Chromatogr.*, 297 (1984) 261-270.

1131 Noyes, C.M., Optimization of complex separations in high-performance liquid chromatography. Application to phenylthiohydantoin amino acids, *J. Chromatogr.*, 266 (1983) 451-460.

1132 Nyberg, F., LeGreves, P., Moberg, U., Pernow, C. and Winter, A., Reversed phase high performance liquid chromatography in studies of the substance P turnover, *Protides Biol. Fluids*, 33 (1985) 571-574; *C.A.*, 104 (1986) 920r.

1133 Nyberg, F., Pernow, C., Moberg, U. and Eriksson, R.B., High-performance liquid chromatography and diode-array detection for the identification of peptides containing aromatic amino acids in studies of endorphin-degrading activity in human cerebrospinal fluid, *J. Chromatogr.*, 359 (1986) 541-551.

1134 Okazaki, M., Kinoshita, M., Naito, C. and Hara, I., High-performance liquid chromatography of apolipoproteins in serum high-density lipoproteins, *J. Chromatogr.*, 336 (1984) 151-159.

1135 O'Keefe, J.H., Sharry, L.F. and Jones, A.J., Macroscale high-performance liquid chromatographic separation and instrumental identification of components of diethylaminoethyl murine epidermal growth factor, *J. Chromatogr.*, 336 (1984) 73-85.

1136 Ortel, T.L., Takahashi, N. and Putnam, F.W., Separation of limited tryptic fragments of human ceruloplasmin by gel-permeation high-performance liquid chromatography, *J. Chromatogr.*, 266 (1983) 257-263.

1137 Ou, C.-N., Buffone, G.J., Reimer, G.L. and Alpert, A.J., High-performance liquid chromatography of human hemoglobins on a new cation exchanger, *J. Chromatogr.*, 266 (1983) 197-205.

1138 Ovchinnikov, Yu.A., Abdulaev, N.G. and Bogachuk, A.S., (Use of covalent chromatography for the study of membrane proteins structure.), *Biol. Membr.*, 2 (1985) 962-975; *C.A.*, 104 (1986) 2967k.

1139 Ozaki, H., Wada, H., Takeuchi, T., Makino, K., Fukui, M. and Kato, Y., Behaviour of single-stranded oligodeoxyribonucleotides on a DEAE-5PW anion-exchange column, *J. Chromatogr.*, 332 (1985) 243-253.

1140 Pabst, R. and Dose, K., Rapid delipidation of the membrane protein bacterio-rhodopsin by high-performance liquid chromatography, *J. Chromatogr.*, 350 (1985) 427-434.

1141 Pan, Y.-C.E., Wideman, J., Blacher, R., Chang, M. and Stein, S., Use of high-performance liquid chromatography for preparing samples for micro-sequencing, *J. Chromatogr.*, 297 (1984) 13-19.

1142 Parente, E.S. and Wetlaufer, D.B., Relationship between isocratic and gradient retention times in the high-performance ion-exchange chromatography of proteins. Theory and experiment, *J. Chromatogr.*, 355 (1986) 29-40.

1143 Patience, R.L. and Rees, L.H., Comparison of reversed-phase and anion-exchange high-performance liquid chromatography for the analysis of human growth hormones, *J. Chromatogr.*, 352 (1986) 241-253.

1144 Pavlik, M., Jehnička, J. and Kostka, V., A modification of the on-line principle of analysis of PTH amino acids during sequential degradation of proteins and peptides using optical sensor controlled injection, *J. Chromatogr.*, (1987) submitted for publication.

1145 Pavlik, M. and Kostka, V., Modification of serine residues during sequential degradation of proteins and peptide by the phenylisothiocyanate method, *Anal. Biochem.*, 151 (1985) 520-525.

1146 Pavlu, B., Johansson, U., Nyhlén, C. and Wichman, A., Rapid purification of monoclonal antibodies by high-performance liquid chromatography, *J. Chromatogr.*, 359 (1986) 449-460.

1147 Pearson, J.D., High-performance liquid chromatography column length designed for submicrogram scale protein isolation, *Anal. Biochem.*, 152 (1986) 189-198.

1148 Pearson, S.D., Dixon, J.D., Nothwehr, S.F. and Kurosky, A., Isolation of high-specific-activity subunits of cholera toxin by reversed-phase high-performance liquid chromatography, *J. Chromatogr.*, 359 (1986) 413-421.

1149 Phelan, M.A. and Cohen, K.A., Gradient optimization principles in reversed-phase high-performance liquid chromatography and the separation of influenza virus components, *J. Chromatogr.*, 266 (1983) 55-66.

1150 Phillips, L.R., Nishimura, O. and Fraser, B.A., Synthesis and fast-atom-bombardment-mass spectrometry of N-acetylmuramoyl-L-alanyl-D-isoglutamine (MDP), *Carbohydr. Res.*, 132 (1984) 275-286.

1151 Phillips, T.M., High performance immunoaffinity chromatography [a review with 38 refs.], *LC, Liq. Chromatogr. HPLC Mag.*, 3 (1985) 962-972; *C.A.*, 104 (1986) 4342b.

1152 Phillips, T.M., Queen, W.D., More, N.S. and Thompson, A.M., Protein A-coated glass beads. Universal support medium for high-performance immunoaffinity chromatography, *J. Chromatogr.*, 327 (1985) 213-219.

1153 Phillips, T.M., More, N.S., Queen, W.D. and Thompson, A.M., Isolation and quantitation of serum IgE levels by high-performance immunoaffinity chromatography, *J. Chromatogr.*, 327 (1985) 205-211.

1154 Pick, J., Vajda, J., Anh-Tuan, N., Leisztner, L. and Hollan, S.R., Class fractionation of acidic glycolipids and further separation of gangliosides by OPTLC [over-pressured thin-layer chromatography], *J. Liq. Chromatogr.*, 7 (1984) 2777-2791.

1155 Power, S.D., Lochrie, M.A. and Poyton, R.O., Reversed-phase high-performance liquid chromatographic purification of subunits of yeast cytochrome *c* oxidase, *J. Chromatogr.*, 266 (1983) 585-598.

1156 Rajakylä, E., Use of reversed-phase chromatography in carbohydrate analysis, *J. Chromatogr.*, 353 (1986) 1-12.

1157 Ray, K.P. and Wallis, M., Use of high-performance liquid chromatography in the purification of human somatomedin C., *Biochem. Soc. Trans.*, 13 (1985) 1233-1234; *C.A.*, 103 (1985) 116 430q.

1158 Regnier, F.E., Hearn, M.T.W., Unger, K.K., Wehr, C.T. and Janson, J.-Ch. (Guest Editors), proceedings of the *4th International Symposium on High-Performance Liquid Chromatography of Proteins, Peptides and Polynucleotides, Baltimore, MD, December 10-12, 1984*, Part I: *J. Chromatogr.*, 326 (1985) 1-445; Part II: *J. Chromatogr.*, 327 (1985) 1-383.

1159 Reimerdes, E.H. and Rothkitt, K.D., Analysis of lactulose and epilactose in the presence of lactose and other carbohydrates in milk products by ion exchange chromatography, *Z. Lebensm.-Unters. Forsch.*, 181 (1985) 408-411; *C.A.*, 104 (1986) 18 703g.

1160 Renauer, D., Oesch, F., Kinkel, J., Unger, K.K. and Wieser, R.J., Fractionation of membrane proteins on immobilized lectins by high-performance liquid affinity chromatography, *Anal. Biochem.*, 151 (1985) 424-427.

1161 Ricard, C.S. and Sturman, L.S., Isolation of the subunits of the coronavirus envelope glycoprotein E2 by hydroxyapatite high-performance liquid chromatography, *J. Chromatogr.*, 326 (1985) 191-197.

1162 Rice, K.G., Kim, Y.S., Grant, A.X., Merchant, Z.M. and Linhart, R.J., High-performance liquid chromatographic separation of heparin-derived oligosaccharides, *Anal. Biochem.*, 150 (1985) 325-331.

1163 Rideout, J.M., Smith, G.D., Lim, C.K. and Peters, T.J., Comparative study of several phase systems of the high-performance liquid chromatographic separation of monoiodoinsulins, *Biochem. Soc. Trans.*, 13 (1985) 1225-1226; *C.A.*, 103 (1985) 116 429w.

1164 Riffe, A., Delpech, M., Levy-Favatier, F., Boissel, J.-P. and Kruh, J., Analysis and preparation of chromosomal high-mobility group proteins by ion-exchange high-performance liquid chromatography, *J. Chromatogr.*, 344 (1985) 332-338.

1165 Robinet, D., Sarmini, H., Lesure, J. and Funes, A., Separation of haemoglobins using a monodisperse cation exchanger, *J. Chromatogr.*, 297 (1984) 333-337.

1166 Rogers, M.E., Adlard, M.W., Saunders, G. and Holt, G., Derivatization techniques for high-performance liquid chromatographic analysis of β-lactams, *J. Chromatogr.*, 297 (1984) 385-391.

1167 Rogers, M.E., Adlard, M.W., Saunders, G. and Holt, G., High-performance liquid affinity chromatography of β-lactamase, *J. Chromatogr.*, 326 (1985) 163-172.

1168 Ross, S.E. and Carson, S.D., Rapid chromatographic purification of apolipoproteins A-I and A-II from human plasma, *Anal. Biochem.*, 149 (1985) 166-168.

1169 Roughley, P.J. and Mort, J.S., Resolution of cartilage proteoglycan and its proteolytic degradation products by high-performance liquid chromatography using a gel filtration system, *Anal. Biochem.*, 149 (1985) 136-141.

1170 Roumeliotis, P., Kurganov, A.A. and Davankov, V.A., Effect of the hydrophobic spacer in bonded [Cu(L-hydroxyprolyl)-alkyl]$^+$ silicas on retention and enantioselectivity of α-amino acids in high-performance liquid chromatography, *J. Chromatogr.*, 266 (1983) 439-450.

1171 Roy, S.K., McGregor, W.C. and Orichowskyj, S.T., Automated high-performance immunosorbent assay for recombinant leukocyte A interferon, *J. Chromatogr.*, 327 (1985) 189-192.

1172 Rusconi, L. and Montecucchi, P.C., Reversed-phase high-performance liquid chromatography of natural and synthetic sauvagines, *J. Chromatogr.*, 346 (1985) 390-395.

1173 Rusconi, L., Perseo, G., Franzoi, L. and Montecucchi, P.C., Reversed-phase high-performance liquid chromatographic characterization of a novel proline-rich tryptophyllin, *J. Chromatogr.*, 349 (1985) 117-130.

1174 Russo, T., Salvatore, F. and Cimino, F., Determination of pseudouridine in tRNA and in acid-soluble tissue extracts by high-performance liquid chromatography, *J. Chromatogr.*, 296 (1984) 387-393.

1175 Saeed, A.M.K. and Greenberg, R.N., Preparative purification of *Escherichia coli* heat-stable enterotoxin, *Anal. Biochem.*, 151 (1985) 431-437.

1176 Saito, H., Expectation for chemical engineering in liquid chromatography purification, *Kagaku Kogaku*, 49 (1985) 902-903; *C.A.*, 103 (1985) 217 401m.

1177 Saleh, A.K. and Moussa, M.A.A., An automated liquid-chromatographic system for convenient determination of glycated hemoglobin A_{1c}, *Clin. Chem. (Winston-Salem, N.C.)*, 31 (1985) 1872-1876.

1178 Sasagawa, T. and Hashimoto, T., (Very high efficiency chromatography of peptides) [a review with 117 refs.], *Tanpakushitsu Kakusan Koso*, 30 (1985) 1320-1335; *C.A.*, 103 (1985) 210 036n.

1179 Sato, N., Hyder, S.M., Chang, L., Thais, A. and Wittliff, J.L., Interaction of estrogen receptor isoforms with immobilized monoclonal antibodies, *J. Chromatogr.*, 359 (1986) 475-487.

1180 Sauter, A. and Frick, W., Determination of neuropeptides in discrete regions of the rat brain by high-performance liquid chromatography with electrochemical detection, *J. Chromatogr.*, 297 (1984) 215-223.

1181 Scanlon, D., Haralambidis, J., Southwell, C., Turton, J. and Tregear, G., Purification of synthetic oligodeoxyribonucleotides by ion-exchange high-performance liquid chromatography, *J. Chromatogr.*, 336 (1984) 189-198.

1182 Schlabach, T.D., Dual-detector methods for selective identification of prolyl residues and amide-blocked N-terminal groups in chromatographically separated peptides, *J. Chromatogr.*, 266 (1983) 427-437.

1183 Schmerr, M.J.F., Goodwin, K.R., Lehmkuhl, H.D. and Cutlip, R.C., Preparation of sheep and cattle immunoglobulins with antibody activity by high-performance liquid chromatography, *J. Chromatogr.*, 326 (1985) 225-233.

1184 Schmuck, M.N., Gooding, D.L. and Gooding, K.M., Comparison of porous silica packing materials for preparative ion-exchange chromatography, *J. Chromatogr.*, 359 (1986) 323-330.

1185 Scholze, H., Determination of phenylthiocarbamyl amino acids by reversed-phase high-performance liquid chromatography, *J. Chromatogr.*, 350 (1985) 453-460.

1186 Schott, H. and Eckstein, H., High-performance liquid chromatographic separations of isomeric pyrimidine oligodeoxynucleotides, *J. Chromatogr.*, 296 (1984) 363-368.

1187 Schultz, E. and Nissinen, E., Determination of tyrosine hydroxylase activity by high-performance liquid chromatography with on-line radio-chemical detection, *J. Chromatogr.*, 375 (1986) 141-146.

1188 Schütte, H., Hummel, W. and Kula, M.-R., Improved enzyme screening by automated fast protein liquid chromatography, *Anal. Biochem.*, 151 (1985) 547-553.

1189 Schwandt, P., Richter, W.O., Heinemann, V. and Weisweiler, P., Characterization of human apolipoprotein A-I by reversed-phase high-performance liquid chromatography, *J. Chromatogr.*, 345 (1985) 145-149.

1190 Seidah, N.G., Pélaprat, D., Rochemont, J., Lambelin, P., Dennis, M., Chan, J.S.D., Hamelin, J., Lazure, C. and Chrétien, M., Enzymatic maturation of pro-opiomelanocortin by anterior pituitary granules. Methodological approach leading to definite characterization of cleavage sites by means of high-performance liquid chromatography and microsequencing, *J. Chromatogr.*, 266 (1983) 212-224.

1191 Seltzer, J.L., Eschbach, M.L. and Eisen, A.Z., Purification of gelatin-specific neutral protease from human skin by conventional and high-performance liquid chromatography, *J. Chromatogr.*, 326 (1985) 147-155.

1192 Shimohigashi, Y., Lee, R. and Chen, H.-C., Gel high-performance liquid chromatographic studies on the elution behavior of chemically deglyco-sylated human chorionic gonadotropin and its subunits, *J. Chromatogr.*, 266 (1983) 555-565.

1193 Shimura, K., Kasai, K.-I. and Ishii, S.-I., High-performance affinity chromatography of anhydrochymotrypsin on a hydrophilic vinyl-polymer gel coupled with tryptophan, *J. Chromatogr.*, 350 (1985) 265-272.

1194 Shimura, K., Kazama, M. and Kasai, K.-I., High-performance affinity chromatography of plasmin and plasminogen on a hydrophilic vinyl-polymer gel coupled with *p*-aminobenzamidine, *J. Chromatogr.*, 292 (1984) 369-382.

1195 Siddiqui, B., Buehler, J., DeGregorio, M.W. and Macher, B.A., Differential expression of ganglioside G_{D3} by human leukocytes and leukemia cells, *Cancer Res.*, 44 (1984) 5262-5265.

1196 Simmaco, M., Barra, D. and Bossa, F., Separation of phenylthiohydantoin-amino acids by high-performance liquid chromatography and some applications in dansyl Edman sequence analysis, *J. Chromatogr.*, 349 (1985) 99-103.

1197 Simmons, W.H. and Meisenberg, G., Separation of Dns-amino acid amides by high-performance liquid chromatography, *J. Chromatogr.*, 266 (1983) 483-489.

1198 Simonian, M.H. and Capp, M.W., Purification of poly(A)-messenger ribonucleic acid by reversed-phase high-performance liquid chromatography, *J. Chromatogr.*, 266 (1983) 351-358.

1199 Singh, A.V. and Morris, H.R., An HPLC I push-pull perfusion technique for investigating peptide metabolism, *Biochem. Biophys. Res. Commun.*, 130 (1985) 37-42.

1200 Singhal, R.P., High-performance liquid chromatography of transfer RNAs. Separation of transfer RNAs from mammalian sources, *J. Chromatogr.*, 266 (1983) 359-383.

1201 Sipila, H.T. and Saarni, H.K., A new coaxial-type tube radioactivity detector and its use in the synthesis of 2-[^{18}F]fluorodeoxyglucose, *Turun Yliopiston Julk.*, Sar. D, 17 (1984) 156-159; *C.A.*, 104 (1986) 10 563r.

1202 Sledzinska, B. and Kowalczyk, J.S., Evaluation of the possibility of increasing output of preparative liquid chromatographic system by intensification of sampling, *Chem. Anal. (Warsaw)*, 30 (1985) 3-9; *C.A.*, 103 (1985) 226 508d.

1203 Small, D.A.P., Atkinson, T. and Lowe, C.R., Preparative high-performance liquid affinity chromatography, *J. Chromatogr.*, 266 (1983) 151-156.

1204 Šmíd, F., Bradová, V., Mikeš, O. and Sedláčková, J., Rapid ion-exchange separation of human brain gangliosides, *J. Chromatogr.*, 377 (1986) 69-78.

1205 Smith, J.A. and O'Hara, M.J., Ion-exchange high-performance liquid chromatography of mouse epidermal growth factor and its congeners. Mobile phase optimization with ion-pairing additives, *J. Chromatogr.*, 345 (1985) 168-172.

1206 Sogami, N., Era, S., Nagaoka, S., Kuwata, K., Kida, K., Shigemi, J., Miura, K., Suzuki, E., Muto, Y., Tomita, E., Hayano, S., Sawada, S., Noguchi, K. and Miyata, S., High-performance liquid chromatographic studies on non-mercapt ⇄ mercapt conversion of human serum albumin II, *J. Chromatogr.*, 332 (1985) 19-27.

1207 Soiefer, A.I., Miller, M.S., Sabri, M.I. and Spencer, P.S., The separation and identification of enolase isozymes of brain and sciatic nerve by high-pressure liquid anion-exchange chromatography, *Brain Res.*, 342 (1985) 196-199; *C.A.*, 103 (1985) 156 242m.

1208 Sonnino, S., Ghidoni, R., Gazzotti, G., Kirschner, G., Galli, G. and
     Tettamanti, G., High performance liquid chromatography preparation of
     the molecular species of GM1 and GD1a gangliosides with monogeneous long
     chain base composition, *J. Lipid Res.*, 25 (1984) 620-629.
1209 Spatola, A.F. and Benowitz, D.E., Improved detection limits in the analysis
     of tyrosine-containing polypeptide hormones by using electrochemical
     detection, *J. Chromatogr.*, 327 (1985) 165-171.
1210 Speedie, M.K., Wong, D.L. and Ciaranello, R.D., Characterization of the
     subunit structure of dopamine β-hydroxylase by anion-exchange high-
     performance liquid chromatography, *J. Chromatogr.*, 327 (1985) 351-357.
1211 Stadalius, M.A., Gold, H.S. and Snyder, L.R., Optimization model for the
     gradient elution separation of peptide mixtures by reversed-phase high-
     performance liquid chromatography. Verification of retention relation-
     ships, *J. Chromatogr.*, 296 (1984) 31-60.
1212 Stadalius, M.A., Gold, H.S. and Snyder, L.R., Optimization model for the
     gradient elution separation of peptide mixtures by reversed-phase high-
     performance liquid chromatography. Verification of band width relation-
     ships for acetonitrile-water mobile phases, *J. Chromatogr.*, 327 (1985)
     27-45.
1213 Stadalius, M.A., Quarry, M.A. and Snyder, R.L., Optimization model for the
     gradient elution separation of peptide mixtures by reversed-phase high-
     performance liquid chromatography. Application to method development and
     the choice of column configuration, *J. Chromatogr.*, 327 (1985) 93-113.
1214 Stanton, P.G., Grego, B. and Hearn, M.T.W., High-performance liquid
     chromatography of amino acids, peptides and proteins. LVII. Analysis of
     radioiodinated thyrotropin polypeptides by reversed-phase high-perfor-
     mance liquid chromatography, *J. Chromatogr.*, 296 (1984) 189-197.
1215 Stanton, P.G., Simpson, R.J., Lanbrou, F. and Hearn, M.T.W., High-perfor-
     mance liquid chromatography of amino acids, peptides and proteins. XLVII.
     Analytical and semi-preparative separation of several pituitary proteins
     by high-performance ion-exchange chromatography, *J. Chromatogr.*, 266
     (1983) 273-279.
1216 Stec, W.J., Zon, G. and Uznański, B., Reversed-phase high-performance liquid
     chromatographic separation of diastereomeric phosphorothioate analogues
     of oligodeoxyribonucleotides and other backbone-modified congeners of
     DNA, *J. Chromatogr.*, 326 (1985) 263-280.
1217 Stenman, U.-H., Laatikainen, T., Salminen, K., Muhtala, M.-L. and
     Leppäluoto, J., Rapid extraction and separation of plasma β-endorphin by
     cation-exchange high-performance liquid chromatography, *J. Chromatogr.*,
     297 (1984) 399-403.
1218 Stenman, U.-H., Pesonen, K., Ylinen, K., Huhtala, M.-L. and Teramo, K.,
     Rapid chromatographic quantitation of glycosylated haemoglobins, *J.
     Chromatogr.*, 297 (1984) 327-332.
1219 Stocchi, V., Cucchiarini, L., Piccoli, G. and Magnani, M., Complete high-
     performance liquid chromatographic separation of 4-N,N-dimethylaminoazo-
     benzene-4'-thiohydantoin and 4-dimethylaminoazobenzene-4'-sulphonyl
     chloride amino acids utilizing the same reversed-phase column at room
     temperature, *J. Chromatogr.*, 349 (1985) 77-82.
1220 Stone, K.L. and Williams, K.R., High-performance liquid chromatographic
     peptide mapping and amino acid analysis in the subnanomole range, *J.
     Chromatogr.*, 359 (1986) 203-212.
1221 Stout, R.W. and DeStefano, J.J., A new, stabilized, hydrophilic silica
     packing for the high-performance gel chromatography of macromolecules,
     J. Chromatogr., 326 (1985) 63-78.
1222 Stout, R.W., Sivakoff, S.I., Ricker, R.D. and Snyder, L.R., Separation of
     proteins by gradient elution from ion-exchange columns. Optimizing
     experimental conditions, *J. Chromatogr.*, 253 (1986) 439-463.
1223 Strahler, J.R., Rosenblum, B.B., Hanash, S. and Butkunas, R., Separation
     of transferrin types in human plasma by anion-exchange high-performance
     liquid chromatography, *J. Chromatogr.*, 266 (1983) 281-291.

1224 Strang, R.H.C., Purification of egg white lysozyme by ion-exchange chro-
     matography, *Biochem. Educ.*, 12 (1984) 57-59; *C.A.*, 103 (1985) 140 975s.
1225 Strasberg, P.M., Warren, I., Skomorowski, M.A. and Lowden, J.A., HPLC
     Analysis of neutral glycolipids: An aid in the diagnosis of lysosomal
     storage disease, *Clin. Chim. Acta*, 132 (1983) 29-41.
1226 Sugii, A., Harada, K. and Ogawa, N., Liquid chromatographic characterization
     of porous vinylpyridine polymer as a weak anion-exchange column packing,
     J. Chromatogr., 354 (1986) 211-217.
1227 Sugimoto, N., Ozutsumi, K. and Matsuda, M., Purification by high-performance
     chromatography of *Clostridium perfringens* type A enterotoxin prepared from
     high toxin procedures selected by 4 toxin-antitoxin halo, *Eur. J.
     Epidemiol.*, 1 (1985) 131-138; *C.A.*, 103 (1985) 117 602j.
1228 Sullivan, R.C., Shing, Y.W., D'Amore, P.A. and Klagsbrun, M., Use of size-
     exclusion and ion-exchange high-performance liquid chromatography for the
     isolation of biologically active growth factors, *J. Chromatogr.*, 266
     (1983) 301-311.
1229 Suomela, H., Himberg, J.-J. and Kuronen, T., High-performance liquid chro-
     matography in the quality control of immunoglobulin preparations during
     production and storage, *J. Chromatogr.*, 297 (1984) 369-373.
1230 Suzuki, Y., Hirabayashi, Y., Suzuki, T. and Matsumoto, M., Occurrence of
     O-glycosidically peptide-linked oligosaccharides of poly-N-acetyl-
     lactosamine type (erythroglycan II) in the I-antigenically active Sendai
     virus receptor sialoglycoprotein GP-2, *J. Biochem. (Tokyo)*, 98 (1985)
     1653-1659.
1231 Svoboda, M., Lambert, M., Moroder, L. and Christophe, J., One-step isocratic
     high-performance liquid chromatographic purification of radioiodinated
     and radioiodinated-photoactivable derivatives of cholecystokinin, *J.
     Chromatogr.*, 296 (1984) 199-211.
1232 Svoboda, V. and Vočková, J., Optimization of eluent composition and column
     length in liquid chromatography. Separation of nucleotides by ion-pair
     chromatography, *J. Chromatogr.*, 353 (1986) 139-152.
1233 Swaisgood, H.E. and Chaiken, I.M., Quantitative affinity high-performance
     liquid chromatography of neuroendocrine polypeptides using porous and
     non-porous glass derivatives, *J. Chromatogr.*, 327 (1985) 193-204.
1234 Takahashi, N., Ishioka, N., Takahashi, Y. and Putnam, F.W., Automated
     tandem high-performance liquid chromatographic system for separation of
     extremely complex peptide mixtures, *J. Chromatogr.*, 326 (1985) 407-418.
1235 Takahashi, N., Takahashi, Y., Heiny, M.E. and Putnam, F.W., Purification
     of hemopexin and its domain fragments by affinity chromatography and
     high-performance liquid chromatography, *J. Chromatogr.*, 326 (1985) 373-
     385.
1236 Takahashi, N., Takahashi, Y., Ishioka, N., Blumberg, B.S. and Putnam, F.W.,
     Application of an automated tandem high-performance liquid chromatographic
     system to peptide mapping of genetic variants of human serum albumin,
     J. Chromatogr., 359 (1986) 181-191.
1237 Takahashi, N., Takahashi, Y. and Putnam, F.W., Two-dimensional high-perfor-
     mance liquid chromatography and chemical modification in the strategy
     of sequence analysis. Complete amino acid sequence of the lambda light
     chain human immunoglobulin D, *J. Chromatogr.*, 266 (1983) 511-522.
1238 Takeuchi, T., Saito, T. and Ishii, D., Microcolumn high-performance size-
     exclusion chromatographic separation of proteins, *J. Chromatogr.*, 351
     (1986) 295-301.
1239 Tan, L., Simple, efficient ternary solvent system for the separation of
     luteinizing hormone-releasing hormone and enkephalins by reversed-phase
     high-performance liquid chromatography, *J. Chromatogr.*, 266 (1983) 67-74.
1240 Tan, L. and Muto, N., HPLC of NADPH cytochrome P-450 reductase, *LC, Liq.
     Chromatogr. HPLC Mag.*, 3 (1985) 522-526; *C.A.*, 103 (1985) 67 179d.
1241 Tandy, N.E., Dilley, R.A. and Regnier, F.E., High-performance liquid chro-
     matographic purification of the hydrophobic ω subunit of the chloroplast
     energy coupling complex, *J. Chromatogr.*, 266 (1983) 599-607.

1242 Tarvers, R.C., Calcium-dependent changes in properties of human prothrombin: A study using high-performance size-exclusion chromatography and gel-permeation chromatography, *Arch. Biochem. Biophys.*, 241 (1985) 639-648.

1243 Tempst, P., Woo, D.D.-L., Teplow, D.B., Aebersold, R., Hood, L.E. and Kent, S.B.H., Microscale structure analysis of a high-molecular-weight, hydrophobic membrane glycoprotein fraction with platelet-derived growth factor-dependent kinase activity, *J. Chromatogr.*, 359 (1986) 403-412.

1244 Thody, A.J., Fisher, C., Kendal-Taylor, P., Jones, M.T., Price, J. and Abraham, R.R., The measurement and characterization by high-pressure liquid chromatography of immunoreactive α-melanocyte-stimulating hormone in human plasma, *Acta Endocrinol.*, 110 (1985) 313-318; *C.A.*, 104 (1986) 937b.

1245 Tomono, T., Ikeda, H. and Tokunaga, E., High-performance ion-exchange chromatography of plasma proteins, *J. Chromatogr.*, 266 (1983) 39-47.

1246 Toren, E.C., Jr., Vacik, D.N. and Mockridge, P.B., Cation-exchange, high-performance liquid chromatographic determination of hemoglobin A$_{1c}$, *J. Chromatogr.*, 266 (1983) 207-212.

1247 Traylor, T.D., High-performance liquid chromatographic resolution of p-nitrobenzoyloxyamine derivative of brain gangliosides, *J. Chromatogr.*, 272 (1983) 9-20.

1248 Turková, J., *Affinity Chromatography (Journal of Chromatography Library, Vol. 12)*, Elsevier, Amsterdam, 1978, 405 pp.

1249 Turková, J., Chaiken, I.M. and Hearn, M.T.W. (Guest Editors), proceedings of the *6th International Symposium on Bioaffinity Chromatography and Related Techniques*; *J. Chromatogr.*, 376 (1986) 1-452.

1250 Tweeten, K.A. and Tweeten, T.N., Reversed-phase chromatography of proteins on resin-based wide-pore packings, *J. Chromatogr.*, 359 (1986) 111-119.

1251 Tyler, G.A. and Rosenblatt, M., Semi-preparative high-performance liquid chromatographic purification of a 28-amino acid synthetic parathyroid hormone antagonist, *J. Chromatogr.*, 266 (1983) 313-318.

1252 Uemura, T., Aoi, Y., Horiguchi, Y., Wadano, A., Goshima, N. and Sakaguchi, G., Purification of *Clostridium perfringens* enterotoxins by high-performance liquid chromatography, *FEMS Microbiol. Lett.*, 29 (1985) 293-297; *C.A.*, 103 (1985) 208 203w.

1253 Ueno, A., Hong, Y.-M., Arakaki, N. and Takeda, Y., Insulin-stimulating peptide from a tryptic digest of bovine serum albumin: Purification and characterization, *J. Biochem. (Tokyo)*, 98 (1985) 269-278.

1254 Ueno, H., Pospischil, M.A., Kluger, R. and Manning, J.M., Methyl acetyl phosphate: A novel acetylating agent. Its site-specific modification of human hemoglobin A, *J. Chromatogr.*, 359 (1986) 193-201.

1255 Unger, K.K. and Chatziathanassiou, M., Size exclusion chromatography of proteins on improved bonded silica column, *Protides Biol. Fluids*, 33 (1985) 563-565; *C.A.*, 104 (1986) 2961d.

1256 Unger, K.K., Jilge, G., Kinkel, J.N. and Hearn, M.T.W., Evaluation of advanced silica packings for the separation of biopolymers by high-performance liquid chromatography. II. Performance of non-porous mono-disperse 1.5-μm silica beads in the separation of proteins by reversed-phase gradient elution high-performance liquid chromatography, *J. Chromatogr.*, 359 (1986) 61-72.

1257 Unger, K.K., Kinkel, J.N., Anspach, B. and Giesche, H., Evaluation of advanced silica packings for the separation of biopolymers by high-performance liquid chromatography. I. Design and properties of parent silicas, *J. Chromatogr.*, 296 (1984) 3-14.

1258 Unger, K., Regnier, F.E., Hearn, M.T.W., Wehr, C.T. and Janson, J.-Ch. (Guest Editors), proceedings of the *3rd International Symposium on High-Performance Liquid Chromatography of Proteins, Peptides and Polynucleotides, Monte Carlo, November 14-16, 1983*; Part I: *J. Chromatogr.*, 296 (1984) 1-406; Part II: *J. Chromatogr.*, 297 (1984) 1-420.

1259 Urdal, D.L., Mochizuki, D., Conlon, P.J., March, C.J., Remerowski, M.L., Eisenman, J., Ramthun, C. and Gillis, S., Lymphokine purification by reversed-phase high-performance liquid chromatography, *J. Chromatogr.*, 296 (1984) 171-179.

1260 Van der Wal, S., Comparison of some high-performance liquid chromatographic fluorescence detector flow cells, *J. Chromatogr.*, 352 (1986) 351-358.

1261 Van der Zee, R. and Welling, G.W., Detection of Sendai virus protein by reversed-phase high-performance liquid chromatography combined with immuno-chromatography, *J. Chromatogr.*, 327 (1985) 377-380.

1262 Van der Zee, R., Welling-Wester, S. and Welling, G.W., Purification of detergent-extracted Sendai virus proteins by reversed-phase high-performance liquid chromatography, *J. Chromatogr.*, 266 (1983) 577-584.

1263 Vardanic, A., Ion-exchange columns in high-performance liquid chromatography of proteins: a simple loading technique that improves resolution, *J. Chromatogr.*, 350 (1985) 299-303.

1264 Vensel, W.H., Fujita, V.S., Tarr, G.E., Margoliash, E. and Kayser, H., Two-dimensional peptide mapping by reversed-phase column chromatography applied to the sequence determination of cytochrome *c* from the wild type and a mutant of the butterfly, *Pieris brassicae, J. Chromatogr.*, 266 (1983) 491-500.

1265 Verillon, F. and Glandian, R., A new robotic unit for HPLC sample preparation and injection, *Am. Lab. (Fairfield, Conn.)*, 17 (1985) 142-147; *C.A.*, 103 (1985) 226 713s.

1266 Vespalec, R., Cigankova, M. and Viska, J., Effect of hydrothermal treatment in the presence of salts on the chromatographic properties of silica gel, *J. Chromatogr.*, 354 (1986) 129-143.

1267 Viljoen, G.J., Mills, M.J., Neitz, A.W.H., Potgieter, D.J.J. and Vermeulen, N.M.J., Determination of anti-protease homogeneity, *J. Chromatogr.*, 297 (1984) 359-367.

1268 Vingad, C.E., Iqbal, M., Griffin, M. and Smith, F.J., Separation of hordein proteins from European barley by high-performance liquid chromatography: Its application to the identification of barley-cultivars, *Chromatographia*, 21 (1986) 49-54.

1269 Vrátný, P., Mikeš, O., Štrop, P., Čoupek, J., Rexová-Benková, L'. and Chadimová, D., High-performance anion-exchange chromatography of organic acids [including oligouronic acids], *J. Chromatogr.*, 257 (1983) 23-35.

1270 Wagner, R.M., Exopeptidase-high-performance liquid chromatography peptide mapping of small peptides, *J. Chromatogr.*, 326 (1985) 399-405.

1271 Walton, K.M. and Schnaar, R.L., Ganglioside glycosyltransferase assay using ion-exchange chromatography, *Anal. Biochem.*, 152 (1986) 154-159.

1272 Watanabe, K. and Tomono, Y., One-step fractionation of neutral and acidic glycosphingolipids by high-performance liquid chromatography, *Anal. Biochem.*, 139 (1984) 367-372.

1273 Wehr, C.T., Hearn, M.T., Janson, J.-Ch., Regnier, F.E. and Unger, K.K. (Guest Editors), proceedings of the *5th International Symposium on High-Performance Liquid Chromatography of Proteins, Peptides and Polynucleotides, Toronto, November 4-6, 1985; J. Chromatogr.*, 359 (1986) 1-556.

1274 Weisweiler, P., Friedl, C. and Schwandt, P., Fast protein chromatofocusing of human very-low-density lipoproteins, *Biochim. Biophys. Acta*, 875 (1986) 48-51.

1275 Welinder, B.S., Linde, S. and Hansen, B., Reversed-phase high-performance liquid chromatography of insulin and insulin derivatives. A comparative study, *J. Chromatogr.*, 348 (1985) 347-361.

1276 Welling, G.W., Groen, G., Slopsema, K. and Welling-Wester, S., Combined size-exclusion and reversed-phase high-performance liquid chromatography of a detergent extract of Sendai virus, *J. Chromatogr.*, 326 (1985) 173-178.

1277 Welling, G.W., Groen, G. and Welling-Wester, S., Isolation of Sendai virus F protein by anion-exchange high-performance liquid chromatography in the presence of Triton X-100, *J. Chromatogr.*, 266 (1983) 629-632.

1278 Welling, G.W., Nijmeijer, J.R.J., Van der Zee, R., Groen, G., Wilterdink, J.B. and Welling-Wester, S., Isolation of detergent-extracted Sendai virus proteins by gel-filtration, ion-exchange and reversed-phase high-performance liquid chromatography and the effect on immunological activity, *J. Chromatogr.*, 297 (1984) 101-109.

1279 Welling, G.W., Slopsema, K. and Welling-Wester, S., Size-exclusion high-performance liquid chromatography of Sendai virus membrane proteins in different detergents. A comparison of different columns, *J. Chromatogr.*, 359 (1986) 307-314.

1280 Wells, R.D., High-performance liquid chromatography of DNA [review], *J. Chromatogr.*, 336 (1984) 3-14.

1281 Wenclawiak, B., Kleiböhmer, W. and Krebs, B., High-performance liquid chromatographic investigations on the time-dependent reaction of cis-Pt(NH$_3$)$_2$Cl$_2$, Pt(en)Cl$_2$ and Pt(pn)Cl$_2$ with RNA fragments, *J. Chromatogr.*, 296 (1984) 395-401.

1282 Westrick, M.A., Lee, W.M.F. and Macher, B.A., Isolation and characterization of gangliosides from chronic myelogenous leukemia cells, *Cancer Res.*, 43 (1983) 5890-5894.

1283 Wetlaufer, D.B. and Koenigbauer, M.R., Surfactant-mediated protein hydrophobic-interaction chromatography, *J. Chromatogr.*, 359 (1986) 55-60.

1284 Wettenhall, R.E.H. and Quinn, M.J., Separation of tryptic phosphopeptides of ribosomal origin by reversed-phase high-performance liquid chromatography, *J. Chromatogr.*, 336 (1984) 51-61.

1285 Wiehle, R.D. and Wittliff, J.L., Isoforms of estrogen receptors by high-performance ion-exchange chromatography, *J. Chromatogr.*, 297 (1984) 313-326.

1286 Williams, R.F., Aivaliotis, M.J., Barnes, L.D. and Robinson, A.K., High-performance liquid chromatographic application of the Hummel and Dreyer method for the determination of colchicine-tubulin binding parameters, *J. Chromatogr.*, 266 (1983) 141-150.

1287 Williams, R.R., Wehr, C.T., Rogers, T.J. and Bennett, R.W., High-performance liquid chromatography of staphylococcal enterotoxin B, *J. Chromatogr.*, 266 (1983) 179-186.

1288 Winkler, G., Heinz, F.X., Guirakhoo, F. and Kunz, C., Separation of Flavivirus membrane and capsid proteins by multistep high-performance liquid chromatography optimized by immunological monitoring, *J. Chromatogr.*, 326 (1985) 113-119.

1289 Winkler, G., Heinz, F.X. and Kunz, C., Exclusive use of high-performance liquid chromatographic techniques for the isolation, 4-dimethylamino-azobenzene-4'-sulphonyl chloride amino acid analysis and 4-N,N-dimethyl-aminoazobenzene-4'-isothiocyanate-phenyl isothiocyanate sequencing of viral membrane protein, *J. Chromatogr.*, 297 (1984) 63-73.

1290 Witting, L.A., Gisch, D.J., Ludwig, R. and Eksteen, R., Bonded-phase selection in the high-performance liquid chromatography of proteins, *J. Chromatogr.*, 296 (1984) 97-105.

1291 Wolfe, R.A., Casey, J., Familletti, P.C. and Stein, S., Isolation of proteins from crude mixtures with silica and silica-based adsorbents, *J. Chromatogr.*, 296 (1984) 277-284.

1292 Wong, W.W., Jack, R.M., Smith, J.A., Kennedy, C.A. and Fearon, D.T., Rapid purification of the human C3b/C4b receptor (CR1) by monoclonal antibody affinity chromatography, *J. Immunol. Methods*, 82 (1985) 303-313; *C.A.*, 103 (1985) 176 692m.

1293 Wu, A.H.B. and Gornet, T.G., Measurement of creatine kinase MM sub-type by anion-exchange liquid chromatography, *Clin. Chem. (Winston-Salem, N.C.)*, 31 (1985) 1841-1845.

1294 Wu, S.-L., Benedek, K. and Karger, B.L., Thermal behavior of proteins in high-performance hydrophobic-interaction chromatography. On-line spectroscopic and chromatographic characterization, *J. Chromatogr.*, 359 (1986) 3-17.

1295 Wunderwald, P., Schrenk, W.J., Port, H. and Kresze, G.-B., Removal of
     endoproteinases from biological fluids by "Sandwich affinity chromato-
     graphy" with α_2-macroglobulin bound to zinc chelate-Sepharose, *J. Appl.
     Biochem.*, 5 (1983) 31-42.
1296 Xia, X., Cui, Z. and Gu, T., (Liver cancer and glycosphingolipids. II.
     Isolation and purification of the hepatome-associated gangliosides),
     Shengwu Huaxue Zazhi, 1 (1985) 13-17; *C.A.*, 104 (1986) 17 227t.
1297 Yao, Z., (High performance affinity chromatography) [a review with 15 refs.],
     Shengwu Huaxue Yu Shengwu Wuli Jinzhan, 64 (1985) 60-62; *C.A.*, 103 (1985)
     210 024g.
1298 Yasukawa, K., Kasai, M., Yanagihara, Y. and Noguchi, K., High-performance
     liquid chromatographic analysis of peptides on an Asahipak GS-320 column
     packed with hydrophilic polymer gel, *J. Chromatogr.*, 332 (1985) 287-295.
1299 Yip, K.F. and Albarella, J.P., Use of high-performance liquid chromato-
     graphy in the preparation of flavin adenine dinucleotide analyte
     conjugates, *J. Chromatogr.*, 326 (1985) 301-310.
1300 Young, W.W., Jr., Laine, R.A. and Hakomori, S., An improved method for the
     covalent attachment of glycolipids to solid supports and macromolecules,
     J. Lipid Res., 20 (1979) 275-278.
1301 Yu, P.H., Bailey, B.A. and Durden, D.A., High-performance liquid chromato-
     graphy of aldehydes and acids formed in monoamine oxidase-catalyzed
     reactions, *Anal. Biochem.*, 152 (1986) 160-166.
1302 Zaoral, M., Havlas, Z., Mikeš, O. and Procházka, Ž. (Editors), *Synthetic
     Immunomodulators and Vaccines (Proc. Int. Symp., October 14-18, 1985,
     Třeboň)*, Czechoslovak Academy of Sciences, Prague, 1986, 353 pp.
1303 Zucali, J.R. and Sulkowski, E., Purification of human urinary erythropoietin
     on controlled-pore glass and silicic acid, *Exp. Hematol. (Copenhagen)*, 13
     (1985) 833-837; *C.A.*, 103 (1985) 190 063p.
1304 Zukauskas, D., Improved separation of biomolecules using a new ion exchange
     support, *Am. Biotechnol. Lab.*, 3 (1985) 22-25; *C.A.*, 104 (1986) 2949f.

Chapter 16

PROSPECTS FOR HIGH-PRESSURE AND MEDIUM-PRESSURE COLUMN LIQUID CHROMATOGRAPHY OF
BIOPOLYMERS AND BIOOLIGOMERS

The aim of this short chapter is to present a concise discussion of the future
of modern liquid chromatography with respect to the classes of substances for
which separations have been described in preceding chapters. It is clear that
any prognostic considerations should be based on precise analyses of the history
and the present-day state of a particular process, and extrapolation of the trends
found. Therefore, as a first step an attempt will be made to review briefly
the main statistical features of the evolution of chromatographic methods in the
last decade. In addition, opinions concerning future trends in LC of some
prominent workers in this field will be considered.

16.1 SURVEY OF NUMBERS OF PAPERS PUBLISHED ON CHROMATOGRAPHIC AND OTHER SEPARA-
TION METHODS IN THE LAST DECADE

The basis for such a statistical analysis is as complete as possible a
bibliography of chromatographic methods. Two sources of data were used when
compiling this chapter: (1) The "Bibliography Sections" included as supplements
in some issues of the *Journal of Chromatography* or (later) published as independent volumes and (2) the "Fundamental Reviews" in *Analytical Chemistry*,
published biennially, which contain specialized reviews on liquid chromatography.
Because the format of the "Bibliography Sections" of *Journal of Chromatography*
has changed slightly over the years, Tables 16.1 and 16.2 were constructed, as
a key to the supplements and to help the reader find quickly the necessary
bibliographic data for possible independent studies. From all the data in the
sources compiled in Tables 16.1 and 16.2, a survey of the total number of papers
describing research on and applications of separation methods during 1980-86 was
made and is summarized in Table 16.3.
Table 16.3 reveals that there are separation methods for which the amount of
research and applications has decreased such a way that their independent registration is no longer necessary; a typical example is paper chromatography.
In contrast, the number of papers on liquid column chromatography (LCC) have
not only exceeded by far the number on gas chromatography (the main chromato-

TABLE 16.1

KEY TO THE COMPLETE BIBLIOGRAPHY ON CHROMATOGRAPHIC (AND ELECTROPHORETIC) SEPARATION METHODS FOR THE PERIOD 1980-1982 BASED ON THE BIBLIOGRAPHY SECTIONS OF *JOURNAL OF CHROMATOGRAPHY*

B represents pagination in the Bibliography Supplements (to the normal volumes of *J. Chromatogr.*); GC = gas chromatography); LCC = liquid column chromatography; PC = paper chromatography [only up to Vol. 250 (1982)]; TLC = thin-layer chromatography; ET = electrophoretic techniques; Indexes = subject index and index of chromatographed substances.

Year	Volume	Separation mode					Indexes
		GC	LCC	PC	TLC	ET	
1980	189	B1-B15	B15-B62	B62-B66	B66-B85	B85-B109	
	194	B111-B125	B125-B167	B167-B171	B171-B189	B189-B203	
	197	B205-B225	B225-B255	B255-B259	B259-B278	B278-B300	
	200	B301-B323	B324-B367	B367-B373	B373-B392	B392-B410	B415-B474
1981	207	B1-B14	B14-B70	B70-B73	B74-B95	B96-B111	
	211	B113-B143	B143-B182	B182-B186	B186-B208	B208-B236	
	214	B237-B253	B254-B309	B309-B314	B314-B337	B337-B352	
	219	B353-B375	B375-B407	B408-B411	B412-B432	B432-B446	B449-B499
1982	237	B1-B14	B15-B47	B47-B52	B52-B74	B74-B89	
	242	B91-B110	B110-B155	B155-B160	B160-B181	B181-B195	
	245	B197-B228	B228-B272	B272-B275	B276-B296	B297-B309	
	250	B311-B340	B341-B398	B398-B401	B401-B420	B421-B427	B441-B498

TABLE 16.2

KEY TO THE COMPLETE BIBLIOGRAPHY ON CHROMATOGRAPHIC (AND ELECTROPHORETIC) SEPARATION METHODS FOR THE PERIOD 1983-1986 BASED ON THE BIBLIOGRAPHY SECTIONS OF *JOURNAL OF CHROMATOGRAPHY*

Since 1983 (Vol. 258), the Bibliography Sections have started with LCC, PC means "planar chromatography" and includes paper chromatography + thin-layer chromatography; paper chromatography is no longer treated independently. Since 1984 (Vol. 304) the Bibliography Sections have been issued in the form of independent volumes and not as supplements to normal volumes. Other symbols are identical with those in Table 16.1.

Year	Volume and No.	Separation mode				Indexes
		LCC	GC	PC	ET	
1983	258	B1-B62	B63-B80	B81-B100	B101-B114	
	261	B115-B168	B169-B198	B199-B217	B218-B241	
	265	B243-B289	B290-B328	B329-B351	B352-B371	
	269	B373-B437	B438-B471	B472-B491	B492-B511	B515-B577
1984	304					
	No. 1	B1-B39	B40-B72	B73-B90	B91-B108	
	No. 2	B109-B168	B169-B196	B197-B212	B213-B228	
	No. 3	B229-B279	B280-B314	B315-B340	B341-B366	
	No. 4	B367-B450	B451-B481	B482-B500	B501-B525	
	No. 5	B527-B577	B578-B618	B619-B631	B632-B643	B647-B715
1985	335					
	No. 1	B1-B86	B87-B115	B117-B146	B147-B173	
	No. 2	B175-B222	B223-B244	B245-B262	B263-B280	
	No. 3	B281-B335	B337-B355	B357-B373	B375-B396	
	No. 4	B397-B452	B453-B478	B479-B495	B497-B514	
	No. 5	B515-B568	B569-B598	B599-B619	B621-B648	
	No. 6	B649-B689	B691-B718	B719-B736	B737-B758	B761-B839

(Continued on p. B646)

TABLE 16.2 (continued)

Year	Volume and No.	Separation mode				Indexes
		LCC	GC	PC	ET	
1986	372					
	No. 1	B1-B59	B61-B82	B83-B98	B99-B123	
	No. 2	B125-B172	B173-B202	B203-B220	B221-B240	
	No. 3	B241-B293	B294-B321	B322-B337	B338-B356	
	No. 4	B357-B404	B405-B428	B429-B443	B444-B462	
	No. 5	B463-B530	B531-B582	B583-B600	B601-B631	
	No. 6	B633-B681	B682-B705	B706-B726	B727-B747	B751-B826

TABLE 16.3

TOTAL NUMBER OF PAPERS PUBLISHED IN YEARS 1980-1986 DESCRIBING RESEARCH ON AND
APPLICATIONS OF THE MAIN SEPARATION METHODS

This Table is based on the list of papers published regularly in the Bibliography
Sections of *Journal of Chromatography* (cf., Tables 16.1 and 16.2).

Year	Volumes	LCC*	GC*	PC*	ET*	Total
1980	189-200	2735	1100	1367 (177)**	1336	6538
1981	207-219	3168	1340	1529 (155)**	1180	7217
1982	237-250	3139	1516	1431 (141)**	928	7014
1983	258-269	4069	2068	1189	1248	8574
1984	304	4781	2749	1295	1527	10 352
1985	335	5529	2282	1655	2135	11 601
1986	372	5163	2740	1347	2133	11 383
Total		28 584	13 795	9813 (473)**	10 487	62 679

*LCC = Liquid column chromatography; GC = gas chromatography; PC = planar chromatography (i.e., thin-layer chromatography + paper chromatography); ET = electrophoretic techniques.
**In parentheses are given the numbers of bibliographic data on paper chromatography alone, which are included in the above data on planar chromatography. Since 1983 the Bibliography Sections of the *Journal of Chromatography* have not included separate data on paper chromatography.

graphic technique in the recent past)* but the number of LCC papers nearly
doubled in the period 1980-86. The ratio of LCC to the total number of papers on
separation methods increased from 41.8% in 1980 to 45.4% in 1986, whereas the
corresponding proportion of papers on planar chromatographic methods decreased
from 20.9% in 1980 to 11.8% in 1986. The ratio of papers on gas chromatography
to the total number of papers increased from 16.8% in 1980 to 24.1% in 1986,
whereas that of electrophoretic methods decreased from 20.4% in 1980 to 18.7%
in 1986. It can be seen from Table 16.3, that these numbers are not accidental
swings, but systematic changes (trends), and therefore we may conclude that at

*This is the reason why since 1983 LCC has been placed ahead of GC in the Bibliography Sections; see Tables 16.1 and 16.2.

TABLE 16.4

TOTAL NUMBER OF PAPERS ON THE LCC SEPARATION OF BIOPOLYMERS, BIOOLIGOMERS AND SOME SELECTED TOPICS PUBLISHED IN THE PERIOD 1980-1986

These data were calculated from the list of papers presented in the Bibliography Sections of *Journal of Chromatography* (cf., Tables 16.1-16.3).

Topic	1980	1981	1982	1983	1984	1985	1986	Total
Amino acids and their derivatives	84	87	119	115	139	153	142	839
Peptides and peptidic (proteinaceous) hormones	68	102	72	127	131	208	171	879
General techniques for elucidation of structure of proteins	39	47	15	24	27	65	70	287
Proteins (total)	257	324	199	310	355	567	606	2618
Enzymes (total)	330	377	206	236	212	345	449	2155
Nucleoproteins	5	4	1	3	2	-	2	17
Ribonucleic acids	24	16	8	12	19	25	27	131
Deoxyribonucleic acids	10	15	9	10	12	21	10	87
Structural studies of nucleic acids	12	12	5	4	8	9	9	59
Carbohydrates (total)	121	141	118	155	157	229	205	1126
Lipids and their constituents	33	32	40	60	58	105	98	426
Lipoproteins and their constituents	6	11	12	15	10	33	36	123
Cells and cellular particles	9	11	14	18	21	13	25	111
Total	998	1179	818	1089	1151	1773	1850	8858
Quantitative analysis	4	-	10	2	5	6	5	32
Preparative-scale chromatography	12	18	14	8	17	14	17	100
Automation, computerization	17	4	6	15	19	38	31	130

present liquid column chromatography is the main and most developed separation method, followed by gas chromatography, with about half its importance. The significance of planar and electrophoretic techniques remains smaller (of course, only from the statistical point of view).

In the following part we shall consider only liquid column chromatography. In Table 16.4 some specialized data concerning topics of interest have been selected from the general number of papers given in the LCC column of Table 16.3. It is evident at first sight that interest in some topics is increasing rapidly, whereas in other areas topics stagnation can be observed. In the Table 16.4, the ratio of papers on the separation of peptides to the total number of registered papers (in the year in question) was 6.8% in 1980 and 9.1% in 1986. Similarly, the corresponding numbers for proteins are 25.8% in 1980 and 32.2% in 1986, for lipids 3.3% in 1980 and 5.2% in 1986 and for cells (or cellular particles) 0.9% in 1980 and 1.3% in 1986. A great increase can be observed with lipoproteins and constituents, from 0.6% in 1980 to 1.9% in 1986. Some branches have kept their levels nearly unchanged, e.g., amino acids and derivatives with 8.4% and 7.5% in 1980 and 1986, respectively, and carbohydrates with 12.1% and 10.9% in 1980 and 1986, respectively. Surprisingly, the relative number of papers on enzymes decreased from 33.1% in 1980 to 23.9% in 1986 and that on nucleic acids (i.e., RNA + DNA + structural studies) from 4.6% in 1980 to 2.4% in 1986. For nucleoproteins the decrease can be seen even in the absolute numbers, so that the relative data need not be calculated. In absolute numbers, preparative chromatography has attracted permanent but not increasing interest, whereas the increase in interest in automation (computerization) is obvious.

An important series of reviews on column liquid chromatography (focused predominantly on HPLC) was published in *Analytical Chemistry* [Walton (1980); Majors et al. (1982, 1984); Barth et al. (1986)], summarizing the most important papers on LC published in English, French, German and Russian in 2-year periods. From these reviews some data were selected and compiled in Table 16.5. The number of books published in the period 1980-86 increased from 8 to 52 per 2 years, i.e. more than 5-fold, and the number of reviews increased more than 10-fold. It is hardly possible to interpret the given data on chromatographic modes, derivatization and preparative chromatography unambigously, because they do not represent the total real values, but are products of a preliminary selection (by the cited authors) as to the originality and contributions to fundamental developments. A surprising feature is the decrease in the number of cited papers on RPC + IPC. However, the number of papers on SEC remains high and has required an independent presentation in two last reviews. Another trait deserving attention is the constant increase in the number of papers devoted to the application of LC to physico-chemical measurements, which is in agreement with the ideas of Horváth (cf., Chapter 3 in Part A).

TABLE 16.5

SELECTED BIBLIOGRAPHIC DATA FROM REVIEWS ON LIQUID COLUMN CHROMATOGRAPHY

Sources: Walton (1980); Majors et al. (1982, 1984); Barth et al. (1986).

Topic	1980	1982	1984	1986
Total number of references in the cited reviews	544	956	1344	1144
Books	8	21	20	52
Reviews	15	116	171	n.a.**
Modes* of LC:				
IEC	(158)	93	36	71
LSC	(16)	55	45	12
NPC	26	19	8	12
RPC + IP	133	94	87	49
SEC	n.a.	n.a.	229	195
General theory + optimization	n.a.	35	21	25
Derivatization:				
Pre-column	n.a.	59	73	38
Post-column	n.a.	68	83	41
Total	53	127	156	79
Preparative LC	7	40	53	15
Application of LC to physico-chemical measurements	n.a.	26	35	54

*Chromatography modes: IEC = ion-exchange; LSC = liquid-solid; NPC = normal-phase; RPC = reversed-phase; IP = ion-pair; SEC = size-exclusion.
**n.a. = Data not available.

One of the most important items in a discussion of the development of LC are the columns, supports and instrumentation. There are summarized in Table 16.6; these data are also based on reviews cited in *Analytical Chemistry*. In the period 1980-1986, the total number of papers nearly doubled. Remarkable developments in microcolumns can be seen in recent years. However, the majority of papers devoted to instrumentation deal with detectors. Absorbance detectors are the most often used systems for monitoring column effluents, and have been developed to a high degree of precision, and also refractive index detectors, for which the number of papers published during 1980-86 has increased by one order of magnitude. However, other types of detectors are also important, e.g., fluorescence, electrochemical and mass spectrometric detectors. On the basis of Table 16.6 it is clear that in spite of the existence of many detectors of various types, some other detection principles have been considered, as can be seen from the entry "miscellaneous detectors", because the number of papers published is not only very high, but represents a sharply rising trend.

TABLE 16.6

SELECTED BIBLIOGRAPHY DATA ON INSTRUMENTATION OF LIQUID COLUMN CHROMATOGRAPHY

Sources: Walton (1980); Majors et al. (1982, 1984); Barth et al. (1986).

Topics	1980	1982	1984	1986
Instrumentation (total)	n.a.*	323	490	506
Instrumentation (general)	n.a.	26	48	40
Columns and packing materials	n.a.	44	71	29
Microcolumns	n.a.	18	19	59
Detectors:				
General	n.a.	18	36	32
Absorbance + refractive index	8	20	31	71
Fluorescence	5	19	16	13
Electrochemical	18	45	95	59
LC-mass spectrometry	9	41	46	37
Miscellaneous	21	110	147	225
Total	61	253	371	437

*n.a. = Data nor available.

An interesting comparison between the data on applications of chromatographic techniques in the period 1970-85 was presented by Macek (1987), who studied developments of HPLC for drug level monitoring in the context of analytical methods in pharmaceutical, biomedical and forensic science. This comparison is illustrated in Table 16.7. A great increase in the number of papers on the HPLC of (i) drugs, (ii) biogenic amines, (iii) antibiotics and (iv) environmental analysis in the period 1970-85 can be clearly seen.

After this brief statistical analysis of the summarized data on papers on LCC, we may now come closer to a prognostic discussion. However, one important note must be added to the above discussion: it is valid only for the analysis of published papers on LC and is not directly transferrable to the production and sale of liquid chromatographs and their components. Many liquid chromatographs and columns are bought for repeated analyses only as a part of the overall equipment for analytical and other service laboratories, which do not carry out original work (or which publish only very seldom). Therefore, the bibliographic data and the number of papers derived from Tables 16.1-16.6 do not give decisive indications for a commercial orientation. Another type of consid-

TABLE 16.7

APPLICATIONS OF CHROMATOGRAPHIC TECHNIQUES ON SELECTED TOPICS IN THE PERIOD 1970-1985 AND THE NUMBER OF PAPERS IN 1985 CLASSIFIED ACCORDING TO THE CHROMATOGRAPHIC MODES

Reprinted from Macek (1987).

Class of compounds	Total papers		Increase (%)	Papers in 1985		
	1970	1985		LC*	GC*	PC*
1. Drugs	179	1559	771	892	288	379
2. Proteins and enzymes	412	1213	194	1184	8	21
3. Environmental analyses (including pesticides)	119	672	464	205	348	119
4. Inorganic compounds	185	511	176	344	94	73
5. Amino acids and peptides	112	501	347	378	53	70
6. Lipids	161	443	175	149	89	205
7. Organic acids	122	405	231	225	110	70
8. Carbohydrates	116	400	244	285	48	67
9. Amines	57	359	529	233	62	64
10. Steroids	150	351	134	188	48	115
11. Hydrocarbons	70	348	397	120	184	44
12. Nucleic acids and their constituents	227	297	30	244	13	40
13. Vitamins	82	234	185	182	27	25
14. Antibiotics	37	212	472	143	8	61
Total	3732	10 569	183	6632	2282	1655

*LC = Liquid column chromatography; GC = gas chromatography; PC = planar chromatography (i.e., paper + thin-layer chromatography).

eration is necessary for this purpose, based on an analysis of supply and demand; such data are not available to the present author*.

Nevertheless, several pieces of knowledge can be derived from the data presented in this section, which may be interesting not only for scientists but also from the business point of view.

The author feels that, within these general considerations on the development of liquid chromatography, acknowledgement should be made to commercial firms producing liquid chromatographs, their components and chromatographic packings. Without their great efforts in developing and perfecting equipment and materials,

---

*Some commercially interesting information can be found in the introductory parts of the reviews by Majors et al. (1982, 1984), where other sources of information on this subject are also cited.

it would not have been possible to reach the present advanced state of liquid chromatographic separations. One reason is that expert research workers in commercial firms not only collaborate closely with university laboratories and other research institutes and hence are in permanent contact with the present "state of the art", but also they themselves are very often first-class experienced scientists who publish original results in international journals and have contributed in this way to the development of liquid chromatography. This is an important feature that should be duly appreciated in research laboratories of leading firms. It is also reflected in the high quality of various technical brochures and application notes issued by firms, from which much useful and often seriously presented information can be acquired.

16.2 PROSPECTS FOR LIQUID CHROMATOGRAPHY

An attempt will now be made to discuss the prognosis of liquid chromatography for the near future. First of all, the opinions of some prominent scientists working in this field (published on various occassions) will be considered.

Majors (1980) discussed the trends in HPLC columns and technology, which nowadays are already reality. He emphasized the importance of microparticulate column packings (5-10 μm) and the increasing spread of RPC pre-packed columns. In contrast, interest in pellicular materials was transferred from analytical to guard columns. The prospects for packings of diameters > 20 μm were shifted to preparative LC. In the section "Future Developments" Majors discussed the application of 5-μm packings in shorter columns in comparison with longer columns and wrote, "Often, one can achieve the same separation in less time on 10 cm x 3-4 mm I.D. columns than on the longer 10-μm columns. When operating at the same flow-rate, the 15-cm columns use half the solvent and give twice the separation speed, but give twice the pressure drop". The applicability of 3-μm and smaller particles in HPLC was critically discussed, and also the application of microbore columns (d_c = 0.5-1 mm) and micropacked capillary (50 μm $\leq d_c \leq$ 200 μm) and open-tubular (10 μm $\leq d_c \leq$ 60 μm) columns. Methods for improving the stability of silica-based packings and the problems in preparative LC and multi-dimensional chromatography were reviewed briefly from the point of view of further developments, in addition to other topics.

Regnier and Gooding (1981) predicted that at least 80% of all column fractionations of proteins in research laboratories will be carried out with HPLC supports because of their greater resolving power and shorter separation times. Higher resolution and the introduction of new separation modes (not being used today) may be expected. The simply automated HPLC systems and the absence of lengthy sample preparations are the major reasons why HPLC has the potential to gain wide acceptance in the field of protein separations. Regnier and Gooding also mentioned the prospects for multiple detectors, which may facilitate the isolation of some other proteins (in addition to the required one) during the column chromatography of natural biological samples, because these proteins are also resolved and normally go to waste. Multiple detectors could be used for simultaneous assays of several classes of compounds.

Unger and Roumeliotis (1982), in an up-to-date review on applications of HPLC in the separation and isolation of biopolymers, discussed some problems concerning future trends. They emphasized that with 5-μm particles in pre-packed columns, 20 000 theoretical plates can be achieved on a 25-cm column under isocratic conditions. Peak symmetry (which may create difficulties) can be improved by adding suitable modifiers to the eluent. The linear sample capacity (as a measure of column loadability) ranges from 0.1 to 1 mg of solute per gram of packing. For analytical purposes 25 cm x 4 mm I.D. columns are most often used, but 10 mm I.D. columns can be employed for both analytical and semipreparative purposes; large-bore microparticulate columns (22 or 30 mm I.D.) for preparations are also commercially available, but they are extremely expensive owing to the large content of expensive packings. To improve the selectivity when analysing complex mixtures the column switching method was introduced, using columns containing packings of different selectivity. Fluorescence detection for monitoring of biopolymer separations was discussed. In conclusion, Unger and Roumeliotis emphasized three recommendations for the future: (i) attempts should be made to cast more light on the retention mechanism of biopolymers on HPLC phases in order to be able to predict and evaluate separation schemes; (ii) more thought than at present must be given to the role of the surface structure of the packing and the way it affects retention, selectivity and performance; and (iii) activities must be concerned with the still wide-open field of detection (to evaluate the potential offered by post-column reactors).

From the prognostic point of view, a very interesting treatise on technical aspects of biochemical HPLC was published by Sjödahl et al. (1985). After a historical introduction and a chapter on design aspects of biochemically optimized HPLC instrumentation (where the question of chemical inertness of the construction material is emphasized) a chapter on "Trends in HPLC" follows, containing some important ideas. They are focused on the future development of VHPLC (very-high-performance liquid chromatography). Two major trends are discussed: (i) a "micro-HPLC" concept and (ii) multi-channel detection by linear photodiode-array (PDA) devices.

The micro-concept may facilitate "very-high-speed HPLC" using 3-μm spherical particles in relatively short columns. An example of the present state of "very-high-speed gel filtration" was given, in which a mixture of thyroglobulin, γ-globulin, ovalbumin, myoglobin and vitamin B_{12} was separated on a 7.5 cm x 7.5 mm I.D. LKB 2135-075 TSK GWSP (10 μm) column in 3 min using 0.1 M sodium phosphate (pH 6.8) for elution at a flow-rate of 1 ml/min and detection at 280 nm. The authors stated that run times of less then 1 min should be possible with smaller diameter packing materials. Suitable packing materials for the separation of biomacromolecules using other chromatographic modes (e.g. RPC) should be developed, but "3-μm RPC" places greater demands on instrumentation. In addition, the problem of microbore columns was discussed with respect to the advantages of the low solvent consumption and also the necessity for minimization of injection volumes, internal volumes and all dead volumes was considered.

The photodiode array (PDA) integrated system, containing a "subsequent mono-chromator", facilitates the rapid capture of complete UV spectra from sample molecules passing through the flow cell. In this way the combination of the optical unit with a data processing unit enables complete information of the UV absorbance of separated components to be obtained, without interfering with the chromatographic process. This type of chromatography with UV detection can now be three-dimensional (absorbance, time and wavelength). This technique, already adopted within analytical chemistry and commercially available, will probably become essential also for biochemists. The handling and ordering of such data using computer assistance was discussed. Simultaneous multi-channel optical detection is one of the trends in HPLC, with the promise of new quality in monitoring column effluents in biochemical separations.

The last contribution to prognosis was published by Regnier (1987) in a review on the chromatography of complex protein mixtures, in the section "Future Opportunities and Prospects". Further developments in this field will depend on the needs of both the life sciences research community and the biotechnology industry. There is a trend towards continuing development of microvolume columns and suitable gradient elution equipment. Higher recoveries and less sample dilution can be obtained in this way. Higher selectivity in separation processes will be required in order to facilitate the discrimination of mutant proteins - the products of developing biotechnology. There is a need to develop more efficient preparative separations of proteins in larger systems. The simplest way, i.e., to make the analytical mini-scale system much larger, would be possible only for the isolation of very expensive products because this approach has obvious economic limitations. The necessity to prepare a generation of true preparative packing materials will probably emerge in the next 5 years. Regnier characterized such packings as follows: (i) mechanical stability to more than 100 bar; (ii) loading capacity 1.5-2 times higher than in analytical materials; (iii) 20% of the cost of current HPLC sorbents; (iv) dynamic loading capacity equal to 70% of that obtained in the static mode; and (v) good chemical stability from pH 2 to 11. Displacement chromatography will be one of the important techniques followed (especially for the preparation of smaller peptides). Bioaffinity materials have great possibilities for specific isolations of biomaterials. Non-porous sorbents will lead to extremely rapid separations on the micro-scale (with run times of less than 30 s).

The presentations at the last International Symposium on Column Liquid Chromatography (Amsterdam, June 28th-July 3rd, 1987) indicated that the main trends in the application of LCC in biochemistry and for the separation of biopolymers and biooligomers are as follows: (1) development of highly effective sorbents (synthetic polymers, particle size 1-3 µm); (2) miniaturization of the equipment (microbore and capillary columns); (3) detectors and detection methods (derivatization, photodiode array combined with mass spectrometry, enzymic detectors, laser spectrometry focused on microcolumns); (4) sample preparation (combined with bioaffinity chromatography, multi-columns, solid-phase extraction, pre- and post-column reactions, automation, robotics); (5) preparative chromatography (both supports and equipment); (6) protein structural changes during LC; and (7) separation of optical isomers.

It can be seen from the above survey that the opinions of the various authors cited agreed in many respects. They are also mostly in agreement with the opinion of the present author. We can now consider what should be added to these prognoses.

The requirements to the chemical stability of future packing materials (especially for large-scale separations) can be ensured by a change from silica-based materials to synthetic organic packings. This is the fundamental means of escaping from the serious problems of the gradual dissolution of the support in alkaline media. A reduction in the particle diameter leads to higher speeds of the chromatographic separation process, but with the disadvantage of higher pressure. This approach may be useful for micro-scale and analytical purposes only. Shortening of the elution time from minutes to several tens of seconds has little practical importance if accompanied by the need for much more expensive equipment and higher constraints on the precision of the experimental micro-technique. Let us consider the importance with regard to the total time requirements that extreme shortening of the chromatographic run time has if the preliminary preparation of the biological material for the experiment requires a period of time that may be several orders of magnitude higher than these run times. Considering non-porous sorbents, it must be realized that they have extremely low sorption capacities and can be applied only in special analyses in repeated series; there is a risk that they will follow the fate of pelliculars. The present author believes that the less expensive medium-pressure techniques and packings will often satisfy the requirements of many research workers. The application of medium-pressure techniques is indicated especially for large-scale separations, because the use of high-pressure techniques in very large-scale separations seem to be economically impossible for most separated substances. In addition, not only the equipment but also the present packings for medium-pressure applications are substantially cheaper. The same reasons lead the author to predict the success of applications of concentrated cross-linked agarose and its derivatives for the preparative separation of biopolymers.

There is one fundamental question that must be answered when the prognostics of HPLC are discussed: how long will the lifetime of the method be? We could follow the course of the development and applications of other chromatographic methods, e.g., paper chromatography, which was found to be very useful and helped considerably in set down the prerequisites for the subsequent development of biochemistry and also helped in other fields of chemistry. Many thousands of papers have been published using this convenient separation method (cf., Hais and Macek, 1963) but it is now used only very seldom. It is obvious that LC cannot be compared with paper chromatography in many respects, but we must realize one general rule: particular chromatographic methods have only a limited lifetime, whereas the general chromatographic separation principle (see the introductory part of Chapter 2, Vol. A) seems to be nearly "eternal". From this philosophical point of view the above "fundamental question" relates only to the present well known form of HPLC (which was described in detail in both Volumes A and B of this book) and not to the general chromatographic separation principle. By observing carefully the data in Table 16.3, we may come to the conclusion, illustrated in Fig. 16.1, that the number of papers published per year on both LCC and GC seems to have reached a maximum, when considered in relation to all papers published on separation methods[*]. The curve at the maximum is flat, but appears to be significant. This course of both curves leads the present author to the conclusion that gas chromatography reached its maximum interest for research workers roughly in the mid-1980s, and the LCC will have its zenith before the end of the 1980s (i.e., after roughly two decades of intensive development of HPLC). In no way does this conclusion mean that there will be a

---

[*]It must be realized that the number of papers considered in a bibliography does not represent absolute data and depends on the selection policy of the author who prepared the bibliography; the number of papers that are not included in bibliographies is greater every year. Therefore, Fig. 16.1 must be judged with this reservation.

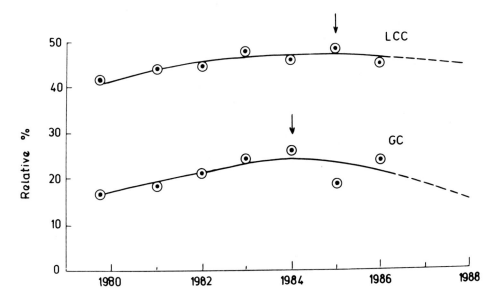

Fig. 16.1. Course of the percentage of the number of papers published on liquid column chromatography (LCC) and gas chromatography (GC) based on the total number of papers on separation methods published in the respective year (according to Table 16.3). The arrows indicate maxima. The dashed lines represent an attempt at extrapolation.

decline in the usage of either method. On the contrary, it means that they have been developed to such perfection that workers undertaking fundamental research will find this area to be "exhausted" in many respects and will focus their interest on other separation processes (which may be chromatographic in their general principle, but based on other approaches). However, the precise equipment and packing materials developed for HPLC will find numerous uses both in standard serial methods and research efforts in chemistry, biochemistry, pharmacology, biomedical applications and other life sciences, biotechnology, the food industry, agricultural studies and in many branches for both analytical and preparative purposes. Therefore, even if in the next decade HPLC will not be developed so intensively in itself as it has been in the present decade, we need not be afraid of its successful future.

REFERENCES

Barth, H.G., Barber, W.E., Lochmüller, Ch.H., Majors, R.E. and Regnier, F.E., *Anal. Chem.*, 58 (1986) 211R-250R.
Hais, I.M. and Macek, K. (Editors), *Paper Chromatography*, Academic Press, New York and Prague, 3rd ed., 1963, 955 pp.

Macek, K., in Tagliaro, F. (Editor), *Developments in Analytical Methods in Pharmaceutical, Biomedical and Forensic Sciences (Proceedings of International Conference, Verona 1986)*, Plenum, New York, 1987, pp. 1-17.

Majors, R.E., *J. Chromatogr. Sci.*, 18 (1980) 488-511; *C.A.*, 94 (1981) 24 494f.

Majors, R.E., Barth, H.G. and Lochmüller, C.H., *Anal. Chem.*, 54 (1982) 323R-363R.

Majors, R.E., Barth, H.G. and Lochmüller, C.H., *Anal. Chem.*, 56 (1984) 300R-349R.

Regnier, F.E., *J. Chromatogr.*, 418 (1987) 115-143.

Regnier, F.E. and Gooding, K.M., in Laurence, J.M., and Kabra, P.K. (Editors), *Liquid Chromatography in Clinical Analysis*, Humana Press, Clifton, NJ, 1981, pp. 323-353.

Sjödahl, J., Dilley, K.J., Eriksson, R., Pellerin, J., Parvez, S.H. and Arlinger, L., *Prog. HPLC*, 1 (1985) 179-217.

Unger, K.K. and Roumeliotis, P., in Gribnau, T.C.J., Visser, J. and Nivard, R.J.F. (Editors), *Affinity Chromatography and Related Techniques (Analytical Chemistry Symposia Series*, Vol. 9), Elsevier, Amsterdam, 1982, pp. 455-470; *C.A.*, 96 (1982) 100 253k.

Walton, H.F., *Anal. Chem.*, 52 (1980) 15R-27R.

SUBJECT INDEX

This combined subject index comprizes entries both from Part A (designated A) and Part B (designated B). The register of chromatographed substances (Chapter 14) and the full title bibliography (Chapter 15) are not included in this subject index.

—, primary amine groups in the amino-
propyl derivative of glass A231
—, remaining pendant double bonds
A212
—, scanning electron microscopy A230
—, swelling A212
—, thermal vacuum depolymerization
A212
—, chromatographic methods B425
Analysis of nucleotides by IEC
B205
Analysis time A47, A318
Angiotensin I A325; B172, B398
Angiotensin I and II B 150
Angiotensin-converting enzyme B398
Angiotensin-generating enzymes B157
Angiotensins B84, B127, B144, B156-
B158
Aniline-Sepharose A82
Anilino-thiazolinone derivative of the
N-terminal amino acid B389
Anilinothiazolinones (ATZ), auto-
matic conversion into PTH-amino
acids B392
Animal viruses, see "Viruses, animal"
Anion exchanger, medium basic A221
Anion exchanger AX-10A B226
Anion exchangers, moderately basic,
preparation of A177
Anion-exchange silica A187
Anorexigenic peptide B125
Anthrone-sulphuric acid reaction B312
Antibiotics B161, B164, B361, B651
Antibodies B337
—, separation of A117
Antibodies that reacted with predeter-
mined sites on proteins B354
Antibody-antigen complex
—, derivatization of epitopes B348
—, X-ray crystallography B348
Antigen A111
Antigen associated with gastrointestinal
cancer - a ganglioside B334
Antigen fragments, modern approach to
preparation B352
Antigen of cRNA animal virus, group
specific B145
Antigenic determinants
—, conformational and sequentional
B347
—, experimental techniques for
searching B348
—, theoretical prediction B347
Antigenic fragments of proteins B347
Antigenic protein determinants, prediction of B354
Antigenic proteins
—, atomic mobility in sites B347
—, segmental mobility in B347

Antigenic sites
—, mapping of B354
—, searching for B347
Antigenic structure of proteins,
reappraisal of B354
Antigens B337
—, foetal, tumour-associated B128
—, immunologically active segments
of B347
—, segments of B347
—, synthetic, chromatography of B348
Antigens against influenza virus,
chronological review of examinations
B350
Anti-HSA-silica A114, A215
Anti-human serum albumin A197
Anti-Ig/LiChrospher A215
α_1-Antitrypsin B24, B45
Antiviral vaccines, discussion B354
AP/Silochrom A215
Apolipoproteins in human HD lipo-
proteins B45
Apomyoglobin A58
Apparatus for packing of chromato-
graphic columns A311, A312
Application of HPLC
—, biochemical, biomedical, pharma-
ceutical; books on A5
—, books on A5
—, food analysis and foodstuffs
industry A5
Application of LCC in biochemistry,
main trends in B656
Applications of chromatographic
techniques, number of published
papers B652
Applications of cyclodextrins A119,
A120
Aprotinine B122
Aquachrom A153, A158
Aquapore-OH A153, A158
Aquapore-RP A171
Ar+ laser optics B277
Argentation chromatography A119
[Arg]vasopressin B133, B152, B153
[Arg8] vasopressin B149
—, diastereoisomers B120
Arg-vasotocin B152
Aromatic halogen compounds, position
in the triangle diagram A315
Aromatic hydrocarbons, position in
the triangle diagram A315
Artificial intelligence A286
Aryl sulphatase isoenzymes B74
Asahipak GS columns B241, B242
Aspartame sweetener B164
Aspartylglucosaminuria B356
Aspergillus viridi-mutants B365
Associated globulins, separation of
B51

Ion-pair (liquid) chromatography, see
  also "Paired-ion chromatography"
Ion-pair reagents
—, alkylsulphonates A102
—, list of A102-A104
—, quadrupole ion-pairs A102
—, zwitterionic pairing agents A102
Ion-pair reversed-phase chromatography
  A54, A100, A306
—, comprehensive treatise A110
—, counter ions A101
—, hydrophilic ion-pairing A110
—, hydrophobic ion-pairing A110
—, ion-pairs A101
—, laboratory practise of A108
—, mechanism of A102
—, phosphoric acid counter ions A101
—, practical guidelines for A109
—, quadrupole ion-pairs A102
—, resolution of RNAs A110
—, secondary equilibria A110
—, selection of suitable conditions
  A108, A109
—, separation of oligodeoxyribonucleotides A110
—, sulphonate counter ions A101
—, tetrabutylammonium counter ions
  A101
—, theory of A104-A107
—, various counter ions A102
Ion-pairing reagents for peptide RP-
  HPLC
—, aliphatic carboxylic acids B124
—, alkylsulphonates B124
—, d-camphor-10-sulphonate B124
—, n-hexylsulphonate B124
—, perfluoroalkanoic acids B124, B125
Ionic strength A69
Ionization control of chromatographic
  separation A94
Ionized compounds, detectors for A279,
  A280
IP-LC (ion-pair liquid chromatography)
  A100
IP-RPC (ion-pair reversed-phase chromatography) A54, A100
IP-RPC, theory, see "Theory of IP-RPC"
IP-RPC mechanism
—, dynamic ion exchangers A104
—, ion interaction hypothesis A104
—, ion-exchange hypothesis A102, A104
—, ion-pair hypothesis A102
IP-RPC of polyvalent ions
—, capacity factor A107
—, capacity of the stationary phase
  A107
—, equilibrium constant for A107
Irreversible sorption of proteins
—, Carbowax desactivation A130

—, minimizing of A130
—, poly(ethylene glycol) desactiva-
  tion A130
—, poly(ethylene oxide) desactivation
  A130
—, siliconizing of CPG A130
Isoamylases from saliva and pancreatic
  juice B77
Isocratic elution A47
Isoelectric points, pI A54
—, of proteins A64, A73, A154; B6
Isoenzyme detectors A264, A265, A290-
  A296; B65
Isoenzymes B80
—, detectors for A290-A292
—, diagnostic importance B66
—, specific detection methods B65
Isoenzymes in biomedical applications
  B80
Isoglutamine or isoasparagine deriva-
  tives B340
Isomaltitol B281
Isomaltodextrins B261
Isomaltose oligosaccharides B280
Isomeric dipeptides B88
Isopentenyladenine or "wye" (formerly
  "Y") base B184
2-p-Isothiocyanophenyl-3-phenylindenone
  B397, B414
Isotocin B152
ITH (diphenylindenonyl thiohydantoin)
  B414
ITH amino acids, UV detection limit
  B397
ITH-Ile B397
ITH-Leu B397

J

Jasco HP-01 A142

K

Kangaroo and horse meat identification
  in processed meats B164
Karo corn syrup, acetylated oligo-
  saccharides B260
Karo light syrup oligosaccharides
  B254
Karo syrup maltooligosaccharides B292
Katal B65
Kel F A194, A205, A206; B192, B198
—, ion exchangers A194
Keratin, biodegradation of B40
Ketones, position in the triangle
  diagram A315

ERRATA TO PART A

Page A7, 7th line from the top, "of" should read "or".

Page A14, 13th line from the top, "Lederer F." should read "Lederer E.".

Page A66, 2nd line from the bottom, "1984" should read "1984a".

Page A70, 4th line from the bottom in Table 3.5, "Te$^+$" should read "Tl$^+$".

Page A74, 2nd line from the bottom, "1984" should read "1984a".

Page A78, 14th line from the bottom, "developing" should read "developed".

Page A97, 2nd line legend to Fig. 3.16, "25 m" should read "25 cm".

Page A98, 3rd line from the bottom, "are not no longer" should read "are no
          longer".

Page A99, 4th line from the bottom, "work" should read "works".

Page A117, 7th line from the top, "ar" should read "at".

Page A125, 16th line from the top, "Schöneschöfer" should read "Schöneshöfer".
          25th line from the top, "Pullamnn" should read "Pullmann".

Page A147, 5th line from the bottom, "of Grignard" should read "or Grignard".

Page A152, last line, "wing" should read "ring".

Page A153, 12th line from the top, "(1977)." should read "(1977),".

Page A200, 12th line from the top, "of" should read "or".

Page A204, 15th line from the top, "Y-5981" should read "Y-5918".

Page A218, 8th line from the top in Table 4.11, "20% cross-linked" should read
          "20%, cross-linked".

Page A239, 3rd line from the bottom, "IN = integrator" should be inserted in the
          legend.

Page A242, 4th line from the top, "Maintenace" should read "Maintenance".

Page A246, 21st line from the top, "of" should read "or".

Page A307, 17th line from the bottom, "constant velocity" should read
          "constant-velocity".

Page A309, 6th line from the top, "may" should read "many".

Page A321, first formula, "Fluorescentamino" should read "Fluorescent amino".
          6th line from the top, "Fluorescamine has" should read "Fluorescamine
          reaction has".

Page A333, 6th line from the bottom, "low-through" should read "flow-through".

Page A353, 18th line from the bottom, "A276" should read "A279".

JOURNAL OF CHROMATOGRAPHY LIBRARY

A Series of Books Devoted to Chromatographic and Electrophoretic Techniques and their Applications

Although complementary to the *Journal of Chromatography*, each volume in the Library Series is an important and independent contribution in the field of chromatography and electrophoresis. The Library contains no material reprinted from the journal itself.

Other volumes in this series